BIOLOGICAL AND MEDICAL PHYSICS
BIOMEDICAL ENGINEERING

BIOLOGICAL AND MEDICAL PHYSICS BIOMEDICAL ENGINEERING

The fields of biological and medical physics and biomedical engineering are broad. multidisciplinary and dynamic. They lie at the crossroads of forntier research in physics, biology, chemistry, and medicine. The Biological & Medical Physics/Biomedical Engineering Series is intended to be comprehensive, covering a broad range of topics important to the study of the physical, chemical and biological sciences. Its goal is to provide scientists and engineers with textbooks, monographs, and reference works to address the growing need for information.

S.H. Chung, O.S. Andersen, and
V. Krishnamurthy (Eds.)

Biological Membrane Ion Channels

Dynamics, Structure, and Applications

Editors:

Shin-Ho Chung
Res. School of Biological Sciences
Australian National University,
Canberra, Australia
Canberra ACT 0200
E-mail: shin-ho.chung@anu.edu.au

Olaf S. Andersen
Weill Medical College Dept. Physiology
Cornell University, New York, NY, USA
1300 York Avenue
New York 10021
E-mail: sparre@med.cornell.edu

Vikram Krishnamurthy
Dept. of Elect. & Comp. Engi.
MacLeod Bulding
University of British Columbia, Vancouver,
BC, Canada
E-mail: vikramk@ece.ubc.ca

Library of Congress Control Number: 2006926448

ISBN-10: 0-387-33323-1
ISBN-13: 978-0387-33323-6

Printed on acid-free paper.

9 8 7 6 5 4 3 2 1

springer.com

Preface

Ion channels are water-filled, biological "sub-nanotubes" formed by large protein molecules. They constitute a class of membrane proteins that serve as conduits for rapid, regulated ion movement across cellular membranes. Ion channels thereby provide the molecular substrate for rapid, electrical signaling in excitable tissues. In addition to playing this important role, ion channels regulate the release of hormones and neurotransmitters and control cell and body electrolyte and volume homeostasis. They are also involved in the transduction of external stimuli to sensory signals. Proper ion channel function is a prerequisite for normal cell, organ and body function—and disorders in ion channel function, channelopathies, underlie many human diseases, such as, cardiac arrythmias, cystic fibrosis, some cases of diabetes mellitus and epilepsy, myotonias and myasthenias. The list is growing. Not surprisingly, ion channels, which long were considered to be rather specialized entities studied by electrophysiologists, are attracting increasing interest.

In most, maybe all, ion channels, ion movement occurs as an electrodiffusive barrier crossing by which selected ions move through a water-filled pore. As the free energy profile the permeating ions have to traverse is relative flat, the throughput is high, of the order of 10^7 ions per second. It thus becomes possible to observe the function of single ion channels in real time using electrophysiological recording methods. Indeed, the first single-molecule measurements were single-channel measurements made almost 40 years ago on ion channels incorporated into planar lipid bilayers (Bean, R.C., W.C. Shepherd, M. Chan, and J. Eichner. Discrete conductance fluctuations in lipid bilayer protein membranes. *J. Gen. Physiol.* 53:741–757, 1969)—and the first single-channel recordings in biological membranes were made 30 years ago (Neher, E., and B. Sakmann. Single-channel currents recorded from membrane of denervated frog muscle fibers. *Nature* 260:779–802, 1976).

Electrophysiological methods improved, the power of molecular and structural biology was unleashed, and ion channels are no longer "black boxes" but molecular entities. Mutations in the DNA sequences encoding channel subunits cause well-defined changes in channel function, which range from mutations that compromise the delivery of the channels to their proper destination, over mutations that cause dysregulation of channel function, to mutations that alter the rate of ion movement. The mechanistic interpretation of these studies is guided by the availability of atomic-resolution structures of a growing number of channels, as well as by increasingly sophisticated computational studies ranging from *ab initio* calculations, over molecular and Brownian dynamics simulations, to continuum descriptions. Taken together these different approaches provide for unprecedented insights into molecular function.

Preface

The current interest in ion channels, however, arises not only from their biological importance; their high turnover numbers make ion channels well suited to serve as switches in sensors. Ion channels also are targets for a growing number of drugs. In many cases ion channels are the desired target(s), but serious side effects may arise from unintended (and unexpected) drug-induced changes in channel function. It is important to develop methods that allow for efficient screening for unintended side effects.

Though the basic functions of ion channels are well understood, at least in comparison with other classes of membrane proteins, ion channels continue to pose a wide range of problems for which the principles and practices of biophysics, nanotechnology design, statistical signal processing can provide elegant and efficient solutions. Indeed, the cross fertilization of ideas in these disparate disciplines will eventually enable us to relate the atomic structure of an ion channel to its experimentally measurable properties through the fundamental processes operating in electrolyte solutions or the basic laws of physics.

The aim of the present book is to provide an introduction to ion channels as molecular entities. It is aimed at researchers and graduate students in the life sciences, biophysics, engineering and computational physics who are interested in acquiring an understanding of the key research results in ion channels. Given the breadth of the field, we do not aim for a comprehensive coverage but focus on the physical description of channel function, the power of computational approaches toward obtaining mechanistic insight into this important class of molecules, and the possibility of the future developments in ion channel research. Thus, this volume is intended to extract from the vast literature in ion channels the central ideas and essential methods regarding the dynamics, structure and application of ion channels.

The chapters in this book are organized as follows. P. Jordan in the first chapter gives a lucid account of the major advances made in the ion channel research over the past 50 years. In the following 11 chapters, some of the current issues in the main classes of ion channels are reviewed. These are: the gramicidin channel (O. Andersen, R. Koeppe II, and B. Roux), voltage-gated ion channels (F. Bezanilla), voltage-gated potassium channel (S. Korn and J. Trapani), BK_{Ca} channels (D. Cox), voltage-gated sodium channels (D. Hanck and H. Fozzard), calcium channels (B. Corry and L. Hool), ClC channels (M. Pusch), ligand-gated channels (J. Lynch and P. Barry), mechanosensitive channels (B. Martinac), TRP channels (T. Voets, Owsianik, and Nilius) and ion channels in epithelian cells (L. Palmer). These are followed by four chapters dealing with theoretical and computational approaches to studying the permeation of ions across biological ion channels. These chapters highlight the strengths and weaknesses of the main tools of physics that are employed in this endeavor, together with examples of how they are applied. The theoretical approaches that are covered here are the Poisson-Nernst-Planck theory (R. Coalson and M. Kurnikova), semi-microscopic Monte Carlo method (P. Jordan), stochastic dynamics (S. Chung and V. Krishnamurthy) and molecular dynamics (A. Grottesi, S. Haider, and M. Sansom). The final three chapters deal with new emerging technology in microfabricated patch-clamp electrodes (F. Sigworth and K. Klemics), an ion channel based biosensor device (F. Separovic and B. Cornell) and hidden Markov model signal processing techniques for extracting small signals from channel currents (V. Krishnamurthy and S. Chung).

The chapters appearing in this book thus comprehensively summarize our current understanding of biological ion channels and the state-of-the-art experimental and computational methodologies used in this field. We hope that the chapters contained in this volume will assist in advancing the boundaries of our understanding of the workings of ion channels and enhance multi-disciplinary research in ion channels.

Shin-Ho Chung
Olaf Andersen
Vikram Krishnamurthy

Contents

List of Contributors

Olaf S. Andersen

Department of Physiology and Biophysics
Weill Medical College of Cornell University
1300 York Avenue, Rm C-501B
New York, NY 10021-4896
E-mail: sparre@med.cornell.edu

Peter H. Barry

Department of Physiology and Pharmacology
School of Medical Sciences
The University of New South Wales
Sydeny, New South Wales 2052
Australia
E-mail: P. Barry @unsw.edu.au

Francisco Bezanilla

Institute for Molecular Pediatric Science
CIS Building
929 E 57th Street
University of Chicago
Chicago, IL 60637
E-mail: fbezanilla@uchicago.edu

Shin-Ho Chung

Research School of Biological Sciences
Australian National University
Canberra, Act
Australia
E-mail: shin-ho.chung@anu.edu.au

Rob D. Coalson

Department of Chemistry
University of Pittsburgh
Pittsburgh, PA 15213
E-mail: rob@mercury.chem.pitt.edu

Bruce A. Cornell

Ambri Ltd.
126 Greville Street
Chatswood, NSW 2067
Australia
E-mail: BruceC@ambri.com.au

Ben Corry

School of Biomedical, Biomoleculer and Chemical Sciences
The University of Western Australia
Crawley WA 6009
Australia
E-mail: ben@theochem.uwa.edu.au

List of Contributors

Daniel H. Cox

Department of Neuroscience
New England Medical Center, MCRI
Tufts University School of Medicine
750 Washington Street
Box 7868
Boston, MA 02111
E-mail: dan.cox@tufts.edu

Harry A. Fozzard

Department of Medicine
University of Chicago
5841 S Maryland Avenue
Chicago, IL 60637
E-mail: foz@hearts.bsd.uchicago.edu

Alessandro Grottesi

CASPUR Consorzio Interuniversitario per le Applicazioni
del Supercalcolo per Universita' e Ricerca
Via dei Tizii, 6b.
00185 Roma - Italy
E-mail: alegrot@caspur.it

Shozeb Haider

BioMolecular Structure Group
The School of Pharmacy
University of London
Bloomsbury, London WC1N 1AX
England
E-mail: shozeb.haider@pharmacy.ac.uk

Dorothy Hanck

Department of Medicine
University of Chicago
5841 S. Maryland M/C 6094
Chicago, IL 60637
E-mail: dhanck@uchicago.edu

Livia C. Hool

School of Biomedical Biomolecula and Chemical Sciences
The University of Western Australia
Crawley, WA 6009
Australia
E-mail: lhool@cyllene.uwa.edu.au

Peter C. Jordan

Department of Chemistry, MS-015
Brandeis University
POB 549110
Waltham, MA 02454-9110
E-mail: jordan@brandeis.edu

Kathryn G. Klemic

Department of Cellular and Molecular Physiology
Yale University
New Haven, CT 06520
E-mail: kathryn.klemic@yale.edu

Roger E. Koeppe II

G11 Phoenix House
University of Arkansas

Fayetteville, AR 72701
E-mail: rk2@uark.edu

Stephen Korn

NINDS
Neuroscience Center
6001 Executive Blvd.
National Institute of Health
Bethesda, MD 20892
Email: korns@ninds.nih.gov

Vikram Krishnamurthy

Department of Electrical and
Computer Engineering
University of British Columbia
Vancouver V6T 1Z4
E-mail: vikramk@ece.ubc.ca

Maria Kurnikova

Department of Chemistry
Carnegie Mellon University
4400 Fifth Avenue
Pittsburgh, PA 15213
E-mail: kurnikova@cmu.edu

Joe Lynch

School of Biomedical Sciences
University of Queensland
Brisbane, Queensland 4072
Australia
E-mail: j.lynch@uq.edu.ac

Boris Martinac

School of Biomedical Sciences
University of Queensland
Brisbane, Queensland 4072
Australia
E-mail: b.martinac@uq.edu.au

Bernd Nilius

Laboratorium voor Fysiologie
Campus Gasthuisberg, O&N 1
Herestraat 40, Bus 802
KU Leuven
B-300 Leuven, Belgium
E-mail: bernd.nilius@med.kuleuven.be

Grzegorz Owsianik

Laboratorium voor Fysiologie
Campus Gasthuisberg, O&N 1
Herestraat 40, Bus 802
KU Leuven
B-300 Leuven, Belgium
E-mail: grzegorz.owsianik@med.kuleuven.be

Lawrence G. Palmer

Dept. of Physiology and
Biophysics
Weill Medical College of
Cornell U.
1300 York Ave.
New York, NY 10021
E-mail: lgpalm@med.cornell.edu

List of Contributors

Michael Pusch

Istituto di Biofisica
CNR
Via de Marini 6
I-16149 Genova
Italy
E-mail: pusch@ge.ibf.cnr.it

Benoit Roux

Institute for Molecular Pediatric Science
CIS Building
929 E 57th Street
University of Chicago
Chicago, IL 60637
E-mail: roux@uchicago.edu

Mark S.P. Sansom

Department of Biochemistry
University of Oxford
South Parks Road
Oxford OX1 3QU
England
E-mail: mark.sansom@biop.ox.ac.uk

Frances Separovic

School of Chemistry
University of Melbourne
Melbourne, Victoria 3010
Australia
E-mail: fs@unimelb.edu.au

Fred J. Sigworth

Department of Cellular and Molecular Physiology
Yale University
New Haven, CT 06520
E-mail: fred.sigworth@yale.edu

Josef G. Trapani

The Volllum Institute and Oregon Hearing Research Center
Oregon Health & Science University
Portland, OR 97239
E-mail: trapanij@ohsu.edu

Thomas Voets

Laboratorium voor Fysiologie
Onderwijs en Navorsing Gasthuisberg
KU Leuven, Herestraat 49 bus 802
B-300 Leuven, Belgium
E-mail: thomas.voets@med.kuleuven.be

Part I
Introduction

1 Ion Channels, from Fantasy to Fact in Fifty Years[1]

Peter C. Jordan

1.1 Introduction

Biologists have long recognized that the transport of ions and of neutral species across cell membranes is central to physiological function. Cells rely on their biomembranes, which separate the cytoplasm from the extracellular medium, to maintain the two electrolytes at very different composition. Specialized molecules, essentially biological nanodevices, have evolved to selectively control the movement of all the major physiological species. As should be clear, there have to be at least two distinct modes of transport. To maintain the disequilibrium, there must be molecular assemblies that drive ions and other permeable species against their electrochemical potential gradients. Such devices require energy input, typically coupling a vectorial pump with a chemical reaction, the dephosphorylation of ATP (adenosine triphosphate). These enzymes (*biochemical catalysts*) control highly concerted, and relatively slow, process, with turnovers of $\sim 100\ s^{-1}$.

Another class of enzymes, the focus of this chapter, controls the transmembrane flux of ions and other permeant species down their electrochemical potential gradients. Two types of molecules are immediate candidates for this purpose. Nature could have designed specialized carrier molecules, which first bind ions or other lipophobic species at the water–membrane interface and then diffuse across the membrane. Alternatively, transport could be carried out by channel-forming molecules, whose water-filled interiors form electrical shunts that provide essentially barrierless pathways for the transport of charged and polar species. Both types exist, and have similar design features: lipophilic exteriors, stabilizing their interaction with membranes, and polar interiors, stabilizing their interaction with charged and polar species. Selective ion channels, which can support fluxes as high as $\sim 10^8\ s^{-1}$, control biological electrical signaling. All such assemblies exhibit three crucial properties: they are highly permeable; they are highly selective; their opening and closing is exquisitely controlled. Understanding their behavior hinges on determining structure and relating it to function. A different class of selective channels exhibits physiologically equally important, but diametrically opposed behavior; aquaporins forbid

[1] Portions reprinted, with permission, from P.C. Jordan, "Fifty Years of Progress in Ion Channel Research," *IEEE Transactions on Nanobioscience*, **4**:3–9 [2005], IEEE.

ion passage of any kind, but allow the passage of water and, not infrequently, other small neutral species.

Hodgkin and Huxley's 1952 (Hodgkin and Huxley, 1952b) analysis of electrical activity in squid giant axon established that both Na^+ and K^+ contributed to the ionic current, and that the fluxes were opposed. It suggested channels as the ionic pathways, but it took 20 more years until Hladky and Haydon (Hladky and Haydon, 1972) definitively demonstrated the existence of ion channels, and then only in studies of the small antibiotic, gramicidin. When incorporated in lipid bilayer membranes bathed by electrolytes, it exhibited what has come to be the characteristic electrical signature of an ion channel: quantized bursts of current of variable duration arising in response to application of a transmembrane electric field.

From the time when ion channels were but a reasonable hypothesis, ever more reliable pictures have evolved. The major steps in transforming this idea from a plausible description of the biological assemblies responsible for controlling passive ion transport across membranes to established fact, thus linking structure to function, have involved great technical strides in electrophysiology, biochemistry, molecular biology, structure determination, computer power, theoretical chemistry, and bioinformatics. Three Nobel prizes, in 1963, 1991, and 2003, specifically cite studies devoted to ion channels. This overview describes important results of the past half century, beginning with one Nobel bookend, the Hodgkin–Huxley model (Hodgkin and Huxley, 1952b), that postulated independent sodium and potassium pathways, and ending with the revolution spawned by another, celebrating Agre's (Denker et al., 1988; Smith and Agre, 1991) discovery of aquaporins and MacKinnon's (Doyle et al., 1998) atomic level determination of a potassium channel's structure. Important electrophysiological, biochemical, molecular biological, structural, and theoretical tools are discussed in the context of the transition from studying systems containing many channels of varying degrees of complexity to investigating single-channel behavior. Examples are taken from a range of channel families illustrating different aspects of channel behavior: the model peptide gramicidin, the nicotinic receptor family, the voltage-regulated cation channel family, chloride channels, and aquaporins. Particular emphasis is placed on recent developments and some questions of current interest are posed.

1.2 Classical Biophysics

When electrically stimulated, a polarization wave propagates along the length of an axon. As squid giant axon is over a meter long, Hodgkin et al. (1952), using 1950s technology, could perfect the "voltage clamp" method in which an electrode is inserted in the axon interior, permitting local perturbation of the membrane potential from its resting value of −60 mV (a Nernst potential reflecting the fact that the axon interior concentrates potassium to which the membrane is selectively permeable), and the transmembrane current measured. They studied how varying both membrane potential and extracellular electrolyte composition altered electrical response, distinguished and separately measured the currents carried by Na^+ and K^+ (Hodgkin and Huxley, 1952a) and determined their voltage-dependent kinetics. The ionic pathways were separate; each flux was activated and deactivated in response to voltage; each was electrodiffusive, controlled by its own resting potential. Properly parameterized, their kinetic model for axon behavior (Hodgkin and Huxley, 1952b) accounted for all salient features of the action potential. After a local partial depolarization, there is a large inward flow of Na^+ and the interior potential rises sharply, inverting membrane polarity; the sodium pathway then shuts down and the potassium pathway opens, restoring the resting potential, which causes it to shut down. The initial large depolarization provides the stimulus that propagates the polarization wave further along the axon. Further studies of the potassium pathway provided more detailed insight, indicating it was multiply occupied in its conducting state (Hodgkin and Keynes, 1955).

Hodgkin and Huxley demonstrated the existence of independent pathways for K^+ and Na^+ flow. They showed that the ionic fluxes were Nernst-like, due to electrodiffusive potential differences. Further studies showed membranes were selectively permeable to other ions as well, suggesting individual pathways for other physiologically important ions, i.e., Ca^{2+}, H^+, Cl^-, and HCO_3^-. By varying the composition of the external electrolyte, a pathway's relative permeability to different ions could be established, thus determining a selectivity sequence. For the axon's potassium pathway it is

$$K^+ \geq Rb^+ > Cs^+ \gg Na^+ \geq Li^+. \tag{1.1}$$

If electrodiffusion governs permeation, such relationships must reflect the underlying thermodynamics. What is the free energy change in removing an ion from electrolyte and inserting it into the transmembrane pathway? Are there general molecular level principles governing selectivity? Electrostatic design obviously permits discrimination based on ionic polarity. Eisenman (1962) noted that for the five alkali cations, only 11 of 120 possible selectivity sequences were commonly found. Why might this be the case? Could electrostatic influences dominate here as well? His answer developed from the study of glass electrodes and the thermodynamics of ion exchange. For an electrode that selects B^+ over A^+, the relative free energy for binding to glass must be more favorable than the relative free energy for hydration, i.e.,

$$G_B(\text{glass}) - G_A(\text{glass}) < G_B(\text{aqueous}) - G_A(\text{aqueous}). \tag{1.2}$$

Hydration free energy differences for isovalent ions are known. He modeled ion–electrode interaction as purely electrostatic, governed by the contact distance between the bound cation and the anionic site in the glass. For (univalent) alkali cations, the interaction energy, E, is

$$E = z_A e^2 N_{Av}/(4\pi\varepsilon\varepsilon_0[r_A + r_C]), \tag{1.3}$$

where z_A is the anionic valence, e the electronic charge, N_{Av} Avogadro's number, ε the dielectric constant, ε_0 the vacuum permittivity, r_C the cation radius, and r_A the radius of the anionic binding site. For large anions, ionic hydration (the right-hand side of (1.2)) governs the equilibrium; the electrode (or peptide) would select for Cs^+. As the anion becomes smaller, the left-hand side of (1.2) becomes ever more important, ultimately leading to selection for the smallest cation, Li^+. As the radius of the anionic site decreases, this simple electrostatic theory generates the 11 observed selectivity sequences. Nothing about the Eisenman argument is limited to glass electrodes. Similar qualitative considerations apply to binding sites along transmembrane ionic pathways and Eisenman's sequence III or IV (Eisenman, 1962) describes the selectivity of the potassium pathway (1.1). Recent work suggests that slightly modified, these electrostatic considerations could quantitatively account for selectivity in the potassium channel (Noskov et al., 2004b).

As transmembrane ionic pathways are narrow water-filled shunts surrounded by protein and lipid, electrostatics is central to understanding the influence of these surroundings. Parsegian (1969) was the first to provide quantitative estimates of how the associated permittivity differences could affect ionic transport. The energy barrier associated with ionic motion from aqueous electrolyte ($\varepsilon \sim 80$) directly into a membrane ($\varepsilon \sim 2$) is prohibitive. It is much reduced, but not eliminated, for water-filled transmembrane conduits. Even if water in the path is dielectrically equivalent to bulk water, an ion induces charges along the water–lipid interface, which impede its translocation. Were trapped water electrically inequivalent to ambient water, its ability to reorganize and shield an ion from the low ε surroundings would be reduced, thus increasing the barrier. In addition, a charging energy would be associated with ionic transfer to lower ε surroundings, also increasing the barrier. These dielectrically induced energy barriers, impeding electrodiffusion, are reduced due to interaction with Eisenman-like binding sites along the path.

Classical electrodiffusion views ion flow as movement down an electrochemical potential gradient, modulated by travel over a sequence of wells and barriers, reflecting a series of ion binding sites. This is naturally treated by biochemical kinetics, which, at its simplest, invokes a two-step translocation process: ion transfer from water to the binding well from one side of the membrane, followed by dissociation to the other side. This is mechanistically expressed as

$$\mathrm{I_L} + \mathrm{Ch} \leftrightarrow \mathrm{I} \cdot \mathrm{Ch} \tag{1.4a}$$

$$\mathrm{I} \cdot \mathrm{Ch} \leftrightarrow \mathrm{I_R} + \mathrm{Ch}, \tag{1.4b}$$

where I_L and I_R represent ions to the left and right of the membrane, Ch is the channel and $I \cdot Ch$ the ion at the binding site. The rate of entry from the left is $R_f^L = k_f[I_L][Ch]$ and the rate of back reaction is $R_b^L = k_b\,[I \cdot Ch]$; k_f and k_b are rate constants and square brackets signify species concentrations. Similar expressions describe processes occurring on the right. Analysis of electrophysiological data can establish the rate constants. Microscopic interpretation involves using these parameters to deduce the associated energetics. The classic work of Arrhenius (1887) showed that rate constants take the form

$$k = A\exp(-E_A/RT), \tag{1.5}$$

where E_A is an "activation energy," T the absolute temperature, and R the gas constant. The topological picture identifies E_A as the energy required to surmount an activation barrier along the electrodiffusive pathway. The biochemical problem requires deconvoluting A and E_A. In ordinary chemical kinetics this is done by determining how k varies with T. It is more difficult biochemically: wide temperature ranges cannot be accessed; proteins denature; membranes undergo phase transitions. From quantum statistics Eyring (1935) developed absolute reaction rate theory where, given an energy profile, rate constants may be computed. His expression takes the form

$$k = \nu\exp(-\Delta G^{\ddagger}/RT) = \nu\exp(\Delta S^{\ddagger}/R)\exp(-\Delta H^{\ddagger}/RT), \tag{1.6}$$

where $\Delta G^{\ddagger}$ is the standard free energy change in forming the "activated complex" from reactants and $\Delta H^{\ddagger}$ and $\Delta S^{\ddagger}$ are the corresponding standard enthalpy and entropy changes. The activated complex is a saddle point on a multidimensional potential surface describing interacting species and ν is the frequency with which molecules in this activated state proceed to products. While the correspondence between (1.5) and (1.6) is superficially seductive, in the absence of temperature variation studies extracting activation enthalpies or free energies from (1.6) is fraught with difficulty since neither the structure reorganizational term, $\Delta S^{\ddagger}$, nor the frequency factor, ν, are easy to estimate (Jordan, 1999).

1.3 Pharmacology and Single Channels

Hodgkin and Huxley's work (Hodgkin and Huxley, 1952a,b; Hodgkin et al., 1952) was done on whole cell preparations. Distinguishing potassium and sodium currents and eliminating perturbations from other membrane components required both clever technique and a system relatively rich in sodium and potassium pathways. More general study entailed suppressing the competing currents or developing techniques for pathway isolation.

Neurotoxins, which selectively block the pathways, pharmacologically separate sodium and potassium contributions to the action potential. Tetrodotoxin (TTX), which makes fugu such a risky delicacy, eliminates the sodium current (Narahashi et al., 1964; Nakamura et al., 1965), thus isolating the potassium pathway. The quaternary amine, tetraethyl ammonium (TEA), has a complementary effect, blocking the potassium pathway (Hagiwara and Saito, 1959; Armstrong and Binstock, 1965). The discovery that each ionic pathway could be individually suppressed showed that sodium and potassium crossed the membrane by separate localized paths (Hille, 1970). The possibility that they shared a common pathway (Mullins, 1959, 1968), with selectivity arising from small changes in pathway structure, was effectively eliminated as the dominant influence (however, there is evidence that allotropic mechanisms can occasionally affect selectivity (Callahan and Korn, 1994; Immke et al., 1999)). It further provided strong inferential evidence for the notion that these were, in fact, transmembrane channels. Most importantly, it implied these channels were molecular receptors, with toxins acting at specific sites. Viewed this way, channels are enzymes that facilitate ion flux and toxins are reversible inhibitors, described by the physical chemist's Langmuir adsorption isotherm or the biochemist's Michaelis–Menten expression. The 1972 Hladky–Haydon (Hladky and Haydon, 1972) studies on gramicidin provided final and compelling evidence for the existence of an ion channel, albeit a simple one.

The acceptance of the channel paradigm radically altered the conceptual framework, focusing interest on identifying and characterizing specific structural features, and developing putative models, "cartoons," which often bore an uncanny resemblance to the actual structures found years later. Studies by Hille (1971, 1973), with a set of organic cations, estimated the dimensions of the sodium and potassium channel pores, showing that in neither case could occlusion account for selectivity. A chemically attractive hypothesis, similar to Eisenman's selectivity theory (Eisenman, 1962), came immediately to mind, that the orifices were surrounded by rings of carbonyl oxygen atoms, regions sufficiently negatively charged to compensate for the free energy of dehydration (Bezanilla and Armstrong, 1972; Hille, 1973). Thirty years later, this idea was confirmed by MacKinnon's X-ray structure (Doyle et al., 1998) of a potassium channel. While no sodium channel structure has yet been determined, subsequent work has demonstrated that here charged residues, not carbonyl groups, control selectivity (Heinemann et al., 1992; Yang et al., 1993).

Sodium channel kinetics is complex. Whole cell studies showed that, in addition to open and closed states, there is a functionally distinct "inactivated" state. When the channel is open, Na^+ streams down the electrochemical potential gradient entering the cell, toward a locally negative region. But the channel does not long remain conducting. It shuts down, and requires potential reversal to slowly undo inactivation. These kinetics are consistent with reversible block of the pore's inner mouth by a positively charged group, which Armstrong intuited acted like a "ball and chain" (Armstrong et al., 1973), a large tethered group that swings into the pore's inner mouth when it becomes negatively charged. Work of the last 30 years has confirmed this hypothesis: large organic cations, injected into the cell, immobilize recovery from inactivation (Yeh and Armstrong, 1978); excision of the "ball and chain" domain, eliminates inactivation (Hoshi et al., 1990; Zagotta et al., 1990). Channel opening and closing modify structure somewhere along a permeation pathway; in inactivation a bulky charged group occludes a channel entrance.

Inspired pharmacology clearly provides insight at the molecular level. But, once gramicidin was proved to be an ion channel (Hladky and Haydon, 1972), the race was on to isolate individual channel proteins, reconstitute them in bilayer membranes and make single-channel measurements free from electrical interference by other channels. Two groups reported success in 1976 (Miller and Racker, 1976; Schein et al., 1976), totally changing the experimental landscape. Even though whole cell measurements made accurately enough can in principle provide even more information than single-channel studies, the complexity of a membrane mosaic makes teasing this out unfeasible. Thus, if a single-channel conductance is sufficiently large, the pharmacological techniques developed to study whole cell preparations are immediately transferable, permitting study with greater accuracy and in more detail.

The discovery of ClC chloride channels provides striking evidence of the value of single-channel studies. Unitary cation conductance measurements showed that there is a unique ion pathway associated with each channel protein. Chloride channel current records were strikingly different. They could not be rationalized in terms of a single ionic pathway (Miller, 1982). The data implied each protein had two identical pathways opening in separate fast processes and an additional slow step activating the whole dimer. The system behaved like a double-barreled shotgun. This remarkable cartoon was confirmed 20 years later, by X-ray structure determination (Dutzler et al., 2002).

1.4 Patch Clamp, Sequencing, and Mutagenesis

Isolation of single-channel proteins by biochemical separation methods is difficult and laborious. Electrophysiology on whole cell preparations is severely limited in the systems that can be studied. In 1976, Neher and Sakmann, in work honored by the 1991 Nobel Prize, reported a way to observe single-channel currents from tiny patches of living cellular material (Neher and Sakmann, 1976), a technique that was refined to permit fusing cell membranes with the tip of a micropipette (Hamill et al., 1981). The contents of the pipette bathing the membrane surface could be adjusted at will; the patching protocol could expose either membrane surface to this electrolyte. Thus single-channel recording became simple, accurate and reliable, and there was no limit to the cells that could be studied.

Patch clamp recording methods gave unprecedented freedom in assessing how channel function was altered by different stimuli. But mechanistic understanding was still limited to cartoon models. Channel proteins are linear arrays comprising up to a few thousand amino acid residues. They fold and form channels, but what groups give rise to ionic pathways, which to the assemblies that respond to gating stimuli, and which are responsible for channel selectivity? New tools were needed to relate structure and function. These were soon forthcoming. Numa and coworkers, applying recombinant DNA techniques to the acetylcholine receptor (AChR) channel of the electric eel, developed efficient methods to determine its primary amino acid sequence (Noda et al., 1982). This assembly, formed from five similar subunits (Raftery et al., 1980) plays a crucial role in neuromuscular transmission. Located at synaptic junctions, it governs nerve–muscle communication. The subunit sequences exhibited considerable homology, i.e., substantial residue similarities

(Noda et al., 1983). Coupling these results with rules for identifying secondary structural elements (Chou and Fasman, 1978) and regions of hydrophilicity (Hopp and Woods, 1981), provided the first solid hypothesis for channel protein architecture (Noda et al., 1983).

The AChR sequences from both calf and electric eel are substantially homologous, suggesting that the similar regions are functionally important. Using cloning techniques Numa's group was able to create a hybrid AChR, mixing subunits from the two organisms (Takai et al., 1985). In a collaboration with Sakmann, its electrical properties were measured; the hybrid and its parents had the same unitary conductance strongly suggesting that channel conductance was controlled by a string of 22 amino acids, highly homologous in each of the five subunits (Sakmann et al., 1985). By selectively mutating specific residues in the protein they confirmed this speculation (Imoto et al., 1988). The era of site-directed mutagenesis had arrived.

These tools form the everyday arsenal of modern electrophysiology. The patch clamp permits near total control of the environment in studying ion channels. Site-directed mutation permits near complete freedom in protein design. Their applications showed that the cation channels—sodium, potassium, and calcium—form a superfamily. Na^+ and Ca^{2+} channel proteins are single stranded, linking four similar peptides; K^+ channel proteins are homologous, with sequences similar to individual Na^+ or Ca^{2+} peptide subunits (Jan and Jan, 1990). Clever thermodynamics connected these observations, showing K^+ channels to be tetramers (MacKinnon, 1991). But what makes them unique? Primary sequence comparisons identify "signature" domains that control selectivity; these regions are residue stretches common to, e.g., all K^+ channels, etc. In K^+ channels, the conserved feature is a five-peptide sequence (Yellen et al., 1991). In Ca^{2+} channels the signature is a set of four negatively charged residues, one from each subunit. This filter is remarkable, favoring passage of Ca^{2+} over Na^+, even though the latter is $\sim$100 times more prevalent physiologically. Na^+ channels have residues with a net -1 charge in place of the four negative signature residues of Ca^{2+} channels. Mutational signature interconversion makes a Na^+ channel selective for Ca^{2+}, and vice versa (Heinemann et al., 1992; Yang et al., 1993).

Selectivity gives channels chemical individuality. Gating, which can be coupled with permeation in anion channels (Pusch et al., 1995), provides functional control. A branch of the cation superfamily is voltage regulated, with the sensor a highly charged domain. Unlike selectivity, which is basically understood, molecular details of gating are still controversial. Like the tale of the blind men and the elephant, different experiments suggest different structural interpretations. In one picture the sensor snuggles up to countercharged regions of the protein (Bezanilla, 2002; Gandhi and Isacoff, 2002; Horn, 2002). Another suggests the sensor is on the outer side of the assembly, attracting water to stabilize its charges, and that gating involves large sensor movement (Jiang et al., 2003b). While the jury is still out, recent work favors smaller sensor motions (Chanda et al., 2005; Posson et al., 2005; Revell Phillips et al., 2005).

Patch clamp, sequencing, and mutagenesis provided much functional information. However, the structures inferred were still cartoons. The 15-residue, channel-forming peptide gramicidin is small, amenable to NMR structure determination (Arseniev et al., 1985). Long before X-ray structures of channel proteins were available, designed chemical mutation of this cation selective channel provided detailed insights connecting structural modification to changes in channel behavior (Koeppe and Anderson, 1996). Slight sequence variations can have major consequences, making the channel voltage sensitive (Durkin et al., 1993) or totally altering its fold (Salom et al., 1995).

1.5 Structure

With the exception of gramicidin, atomic resolution structures were unavailable until 1998. Channel portraits were either cartoons or silhouettes. The histories of gramicidin and AChR are illustrative. As gramicidin has only 15 residues, conformational analysis combined with secondary structure deduction tools (Chou and Fasman, 1978; Hopp and Woods, 1981) might possibly suffice for reliable structure determination. The original hypothesis, a head-to-head, left-handed β-helix, both sufficiently long to be membrane spanning and sufficiently narrow to be valence selective, was very close to the mark (Urry, 1971). A stable channel arises when carbonyl and amino groups six residues apart form hydrogen bonds to one another. However, the screw is right-handed as shown by Arseniev's 1985 pioneering NMR study (Arseniev et al., 1985). Cation selectivity arises from interaction with channel lining carbonyls, Eisenman's selectivity sequence II (Eisenman, 1962). Since gramicidin is readily deformable its structure quite sensitive to its surroundings. Comparison of the channel's structure in sodium dodecyl sulfate (SDS) micelles with that in oriented dimyristoyl phosphatidylcholine (DMPC) bilayers suggests that lipids may play important roles in channel stabilization. Channel pitch and the orientation of one of the Trp groups anchoring the channel to the water–membrane interface (surprisingly, the one furthest from the interface) are both environmentally dependent (Ketchem et al., 1997; Townsley et al., 2001).

AChR, responsible for communication at the synaptic junction, is a pentameric channel formed by self-assembly of four different, but homologous, peptides with stoichiometry $\alpha_2\beta\gamma\delta$. Its structure is important not just for itself but because it is a representative of a ligand-gated superfamily of neurotransmitters that includes the glycine, glutamate, and $GABA_A$ receptors. Oriented two-dimensional preparations have been studied for 20 years by electron microscopy (EM). From the outset they portrayed gross channel architecture (Brisson and Unwin, 1985; Mitra et al., 1989), but were inadequate to reliably determine the molecular architecture of the interface between the protein and the water-filled channel constriction. Even exceptionally high (4 Å) resolution data (Miyazawa et al., 2003) might still have been inadequate for unambiguous structural inference. However in 2001, Sixma's group obtained a high-resolution X-ray structure of a related polypeptide; the acetyl choline binding protein

(AChBP) (Brejc et al., 2001); its sequence is highly homologous to the N-terminal domain of the α-subunit of AChR and provides a template facilitating interpretation of cryo-EM data, leading to a definitive picture of AChR. The importance of the AChBP structure as a surrogate for higher resolution is clearly demonstrated in studies of water channels, the aquaporins. There are significant, resolution related, differences between cryo-EM results, at 3.8 Å (Murata et al., 2000), and X-ray determinations, at 2.2 Å (Sui et al., 2001). Cryo-EM's advantage is the small size of its samples; the corollary is a limit to the number of images that can be acquired. Without guidance from AChBP, it is questionable whether cryo-EM would have solved the AChR problem.

Analysis of the AChR structure poses an interesting interpretive problem. It has been imaged in both open (Unwin, 1995) and closed states (Miyazawa et al., 2003; Unwin, 2005); in each instance the pore radius is sufficiently large to permit ion passage, with constriction diameters of ~7 and ~3.5 Å respectively. Why won't small cations like Na^{+}, K^{+}, and Ca^{2+} flow through the narrower constriction? Certainly not because it is occluded. The interior of the pentameric channel is formed of ~20 rings of predominantly nonpolar residues, which present a hydrophobic surface that would repel water from the channel's interior making it inhospitable to ions. However, two of these rings are negatively charged; these could attract cations, which could permit ionic entry into the pore and possibly cause binding and block. Theoretical analysis provides a clear pathway to discriminate between these options.

In 1998, MacKinnon solved two daunting problems: isolating and purifying channel proteins "in bulk" and crystallizing them. X-ray methods, with their superior signal to noise ratios, were then applied, determining the structure of an ion-occupied bacterial potassium channel, KcsA (Doyle et al., 1998). It completely substantiated 40 years of cartooning. The selectivity filter, formed of residues identified a decade earlier (Yellen et al., 1991), is a narrow cylinder near the channel's external mouth, ~15 Å long, just adequate to accommodate two potassium ions easily; further in is a water-filled pool, with a third ion, all substantiating the prediction of multiple occupancy (Hodgkin and Keynes, 1955). Ions in the filter are stabilized by interaction with carbonyl oxygens of the filter residues. This lining, in some ways reminiscent of gramicidin, is only possible because the two glycines of the filter have more folding options than any other residues. However, dipolar orientation of the carbonyls in gramicidin and KcsA differ substantially: in gramicidin they parallel the channel axis; in KcsA they are perpendicular to it. A pair of membrane spanning helices from each KcsA subunit surrounds the central water pool; the N to C orientation of short pore helices surrounding the filter stabilizes the central cation. Both the filter fold and the aqueous cavity were utterly unexpected.

KcsA was crystallized in a nonconducting state, its interior end too narrow to permit ion passage. Another bacterial potassium channel, MthK, was trapped with its interior mouth open (Jiang et al., 2002a). Even this coarse grained picture (only the C_{α} coordinates could be resolved) provided further insight into K-channels, again consistent with electrophysiological canon; it suggested strongly that a glycine residue was the gating hinge (Jiang et al., 2002b). These channels are proton regulated and calcium activated respectively. A snapshot of a third bacterial potassium channel, KirBac1.1, shines a different spotlight on how structural details influence channel behavior (Kuo et al., 2003). The filter region replicates that of KcsA. But this channel is inwardly rectifying; in its open state, permeating cations flow preferentially into the cell. Mutational analysis of eukaryotic inward rectifiers implicated three residues as crucial for inward rectification (Lu and MacKinnon, 1994; Yang et al., 1995). As expected from electrophysiological inference, homology analysis shows two of these form rings of negative charge in an intracellular C-terminal domain where they can strongly attract polyvalent cations to impede outward potassium flow.

The three bacterial potassium channels share a common feature; they have two transmembrane helices, not the six characteristic of voltage-gated assemblies. In 2003, MacKinnon crystallized a thermophilic voltage-gated bacterial K-channel, KvAP (Jiang et al., 2003a). Its two interior helices, which surround the filter, differ little from those just discussed. As expected, the paddle-like voltage sensor was sited external to and quite independent of the filter assembly. However, its orientation with respect to the interior domain was significantly at odds with years of experimental inference. Instead of the sensor's four basic residues facing the filter, they abutted the lipid, a structure that generated immense interest and a firestorm of controversy (Gandhi et al., 2003; Laine et al., 2003). Why the differences? A voltage sensor, by its very nature is balanced on a hair trigger; crystallization may have severely reoriented it. Alternatively, eukaryotic and prokaryotic organelles (cells with and without nuclei) might differ fundamentally. Electron paramagnetic resonance (EPR) studies provided structural evidence to bridge the gap (Cuello et al., 2004); the sensor's charged groups interface with the lipid, but are oriented in a fashion that shields the charge. Recent structural work from MacKinnon's lab characterized the first eukaryotic voltage-gated assembly (Long et al., 2005a,b). This channel from the Shaker family, which forms the basis of most electrophysiological voltage-gating studies, was crystallized with lipid present, quite different (and much milder) conditions than those needed to stabilize KvAP. Its sensor paddle also floats freely but the charge group orientations no longer affront years of electrophysiological study. Two of the four basic groups nestle up to the filter assembly and the other two abut the lipid; while there are still some differences, this picture goes a long way toward reconciling structural, spectroscopic, and biochemical studies.

A bacterial ClC chloride assembly suggests there can be important differences between prokaryotes and eukaryotes. The crystal structure of one such ClC (Dutzler et al., 2002) confirmed Miller's cartoon with its two identical parallel pathways (Miller, 1982). Another structure indicates how chloride ions get into the path, but not how they exit (Dutzler et al., 2003) and suggests that the fast gate reflects a small conformational change involving a single residue, a strictly conserved glutamate of the selectivity filter. The bacterial protein exhibits signature regions characteristic of chloride selectivity, but its conductance is too small for single-channel analysis. In fact, even though its protein is significantly homologous with that of eukaryotes, it is not a channel but a pump with exceptionally high turnover. For every two chloride ions that pass in one direction, a proton goes the other way (Accardi and Miller,

2004). The turnover is a thousand times faster than a typical pump and 100 times slower than a channel. Might it be a missing link, which could provide insight into subtle differences that interconvert pumps and channels? Recent results show that some eukaryotic members of the ClC family are also antiporters (Picollo and Pusch, 2005; Scheel et al., 2005).

Ion channels are selective shunts promoting electrical activity. Aquaporins are just as selective, but toward a different end. Instead of catalyzing ion passage they rigorously forbid it, while allowing high fluxes of water and other small neutral, polar species. The puzzle is how a narrow water channel can totally discriminate against proton flow. Experience with gramicidin suggests that water readily forms hydrogen-bonded chains in narrow pores, alignments conducive to proton transfer along a proton wire via a Grotthus mechanism (Pomès and Roux, 1996, 2002). What makes aquaporins different? Their structures are built on a few recurring motifs (Fu et al., 2000; Sui et al., 2001; Harries et al., 2004). There are two strictly conserved sequences (asparagine–proline–alanine, NPA) near channel midpoints. About 10 Å distant is a constriction (the selectivity filter, SF), with a conserved arginine. Regions ~10 Å to either side of the NPA domain form the narrow pore, beyond which are wide vestibules. One face of the pore lining is hydrophobic and the other is formed of oriented carbonyls, the oxygens of which point in opposite directions to either side of the NPA domains. This structural feature might specifically impact proton flow since the dipolar inversion would promote water inversion to either side of the NPA, which could break a water wire. A number of these channels' electrical features might account for their general ability to reject ions: the narrow region contains positively charged residues, oriented helix dipoles, and the conserved arginine and asparagines, all of which could deter cation passage; negatively charged inner vestibule groups and the constriction's carbonyl oxygens could deter anion entry into the pore. Theory provides ways to analyze these options in detail.

1.6 Spectroscopy

Crystallography provides unparalleled structural detail. However, the pictures are static and channel stabilization sometimes requires aggressive biochemical intervention, which may massively perturb a native structure (Zhou et al., 2001; Jiang et al., 2003a). Electro-optical methods, that correlate conductance and spectral properties, go a long way toward filling this gap (Mannuzzu et al., 1996; Cha and Bezanilla, 1997). While these approaches provide a less detailed picture than crystallography, the structural modifications needed to render proteins spectroscopically active are much milder. Coupling spectroscopy with electrophysiology is an ideal way to investigate dynamic behavior, and provide insight into details of voltage sensor motion.

In the voltage-gated cation channel superfamily (Jan and Jan, 1990), potassium channels are tetrameric (MacKinnon, 1991) while the single-stranded sodium and calcium channel proteins are built from four internally homologous repeats, each strongly resembling a potassium channel monomer. Their hydrophobicity patterns indicated that each repeat (and the K-channel monomer) contains six α-helical regions separated by linkers. One of these, denoted S4 and universally present, has an amino acid composition $([R/K]XX)_n$, i.e., a run of up to eight arginines (or lysines) each separated by two apolar residues. Electrophysiological work indicates that the gating process controlling channel opening requires moving ~12–16 charges through the transmembrane electric field (Schoppa et al., 1992; Aggarwal and MacKinnon, 1996); the discovery of this highly charged domain immediately suggested identifying it as the channels' voltage sensor. Ingenious mutational experiments demonstrated that the arginines did indeed move as the transmembrane voltage changed, as expected for a voltage sensor. The mutation R1448C in the channel $Na_V1.4$ was studied. This arginine, located in repeat IV, was expected to be on the channel's extracellular side. The cysteine, with its exposed SH group, readily reacts with methane thiosulfonate (MTS) moieties (Akabas et al., 1992; Karlin and Akabas, 1998). When designed with permanent charges, which cannot cross the membrane, they can be used to rigorously establish whether processes occur *cis* or *trans*. The R1448C mutant reacts with extracellular MTS reagents, but not intracellular ones. Even more importantly reaction is voltage sensitive. The Cys is only accessible if the channel is depolarized but nonreactive if the channel is hyperpolarized, just as expected, since the positively charged S4 segment would move outward in depolarization (Yang and Horn, 1995). Similar studies showed complementary movement of other S4 basic groups, those near the intracellular side (Bezanilla, 2000). What they fail to do is establish how much motion takes place and whether it occurs in stages or a single step.

Spectroscopy, coupled with site-directed cysteine mutagenesis, is a powerful tool for quantitatively establishing channel geometry and monitoring the detailed consequences of electrical perturbation. The cysteine's SH group is a "hook" on which to hang an almost endless array of substituents. Fluorescent dyes can be attached at targeted locations within the protein and directly monitored, whether buried in the transmembrane domain or located at a water–peptide interface, while simultaneously recording currents (Mannuzzu et al., 1996; Cha and Bezanilla, 1997). Not unexpectedly, electro-optical studies of S4 are totally consistent with MTS accessibility measurements. Correlating optical and electrical measurement yields detailed kinetic information suggesting that voltage sensor motion takes place in stages (Baker et al., 1998). Electro-optical study links voltage and spectral changes, providing dynamic insights not possible via crystallography.

The kinetics of resonant energy transfer is highly sensitive to donor (D) and acceptor (A) distances, falling off as R^{-6}. By engineering complementary fluorescent or luminescent moieties into the same peptide it is possible to measure D–A distances (Cha et al., 1999; Glauner et al., 1999). The voltage sensor in potassium channels provides a striking case. A mutation in the potassium channel sequence is expressed at equivalent sites in each of the four monomeric strands. By properly adjusting donor and acceptor concentrations, it is possible to synthesize a mutated channel with predominantly D_3A stoichiometry. If the four spectroscopically active moieties are roughly at the corners of a square, there are two possible D–A distances, differing roughly by a factor of $\sqrt{2}$, a prediction borne out experimentally (Cha et al., 1999). Electro-optical studies mutating a series of residues in the S3–S4 linker determines the voltage dependence of a set of inter-residue distances and provides a dynamic monitor for correlating sensor motion with changes in voltage. While the behavior of the voltage sensor is still a matter of controversy (Jiang et al., 2003b; Long et al., 2005a), the most recent evidence indicates that gating occurs without large sensor displacements (Chanda et al., 2005; Posson et al., 2005; Revell Phillips et al., 2005).

A complementary approach to dynamically establishing the influence of voltage on structure relies on EPR spectroscopy. Here the label is the paramagnetic nitroxide moiety, linked to a peptide at a Cys-mutated site (Hubbell et al., 1996). EPR analysis of spin-labeled mutants provide a somewhat different perspective, establishing structural properties and mobilities of the labeled sites (Perozo et al., 1998, 2002). When applied to KvAP (Cuello et al., 2004), it clearly showed that the structure of this channel's voltage sensor is partially consistent with "traditional" and "paddle" models, and that both pictures required modification, an observation confirmed by crystallographic analysis of a eukaryotic potassium channel (Long et al., 2005a).

1.7 Theory

As experimental tools developed, so did theoretical ways to relate structure and function. Early studies, based on cartoon structures, treated channel–water–membrane ensembles as problems in electrostatics. They focused on discriminating among these pictures and gave qualitative insights (Parsegian, 1969; Levitt, 1978; Jordan, 1983). The aqueous shunt was presumed electrically equivalent to bulk water. Studies of narrow, selective channels were limited to treating reaction field effects that reflected presumed system geometries and the permittivity differences between functionally distinct domains: pore, protein, electrolyte, and membrane. Läuger presented the chemical kinetic, Eyring-like view, conductance as passage over a series of structurally induced barriers (Laüger, 1973), while Levitt considered conductance as an electrodiffusional process (Levitt, 1986) in a field created by the potential energy surface. Such treatments, when judiciously employed, can provide semiquantitative physical insight (Jordan, 1987; Jordan et al., 1989; Cai and Jordan, 1990), but they remain most useful as correlational tools.

Once a reasonable structure for gramicidin was available, its behavior became the focus for applying two powerful molecular level methods, Brownian dynamics (BD) (Cooper et al., 1985) and molecular dynamics (MD) (Mackay et al., 1984), and developing a more general electrodiffusional approach, Poisson–Nernst–Planck (PNP) theory (Chen et al., 1992; Eisenberg, 1999). PNP views ions as diffuse charge clouds, an adequate model for wide channels but problematical for narrow, selective ones (Corry et al., 2000, 2003; Edwards et al., 2002). When properly modified to incorporate ion discreteness, it can be applied to narrow channels (Mamonov et al., 2003), but its great strength, physical simplicity, is lost. For many applications it is a powerful, flexible tool, generating current–voltage (I–V) profiles for direct comparison with data. By contrast, BD treats ions as discrete entities; it tracks their motion through a pore, also generating I–V profiles. The aqueous pore is viewed as a viscous, dielectric continuum and stochastic ionic motion occurs in the potential field of the protein, frictionally retarded by pore water. Both PNP and BD impose severe dielectric assumptions, similar to the earlier electrodiffusional approaches (Levitt, 1986), ones which must be used cautiously (Schutz and Warshel, 2001). Their special strength is their ability to efficiently correlate the electrophysiological effects of alterations in protein structure and charge distribution with experimental I–V data, thus providing insight into likely structural possibilities. MD is less constrained. It describes atomic level motion, governed by empirical force fields, but even now limits to computational power preclude direct determination of I–V profiles. Wilson's pioneering study (Mackay et al., 1984) focused on ion–water–peptide correlations in gramicidin, concluding that water, in these confined surroundings, formed an oriented, hydrogen-bonded chain even in the ion-free channel.

With a single exception (Long et al., 2005a) crystallography has provided structures of bacterial ion channels, not the systems generally studied electrophysiologically. While the prokaryotic and eukaryotic assemblies have important sequence features in common (signatures for secondary structure, selectivity filters, etc.), their overall homology may be as low as 15%. Irrespective of the approach employed, two general strategies inform structure-based theoretical study of channel conduction. Both are speculative. The behavior of the bacterial assembly is analyzed, even though prokaryotic systems may not even be channels (Accardi and Miller, 2004), and correlated with observed behavior in eukaryotes, essentially arguing by analogy. Alternatively, bioinformatic alignment techniques (Thompson et al., 1994) provide ways to go from known bacterial structures to plausible model structures for the systems investigated in the wet lab (Corry et al., 2004). As these hypothetical eukaryotic structures, no matter how reasonable, aren't verifiable, there remains an irreducible fortuitous component to agreement (or disagreement) between computed permeation behavior and that observed experimentally.

While highly idealized, BD can provide significant insight. Studies on potassium-like channels provide a detailed view of the permeation process (Chung et al., 2002a). Open and closed state structures demonstrate that K-channels' inner mouths are very flexible (Doyle et al., 1998; Jiang et al., 2002a). BD shows that small changes in the size of the inner mouth easily accounts for the observed 100-fold spread in K-channel conductances (Chung et al., 2002b). Even though the bacterial ClC chloride assembly isn't a channel (Accardi and Miller, 2004), its pore may well still be a template for true channels and a entry for generating likely model structures for members of the eukaryotic ClC channel family. With these as input, BD studies account for observed conductance behavior (Corry et al., 2004), providing evidence for the essential validity of the models.

Although attempts have been made using a microscopic–mesoscopic approach (Burykin et al., 2002), MD hasn't yet directly reproduced I–V profiles; however, there is an indirect pathway—computing a potential of mean force (PMF), in essence the permeation free energy for ion transfer from bulk electrolyte to the channel interior. This has provided surprising insights, predicting a K-channel ion binding site (Bernèche and Roux, 2001) before experimental confirmation (Zhou et al., 2001) and generating a PMF that, when coupled with a BD treatment of field-driven diffusion, reproduces conductance measurements (Bernèche and Roux, 2003).

The origin of gating in AChR poses a challenge. Its closed pore is still quite wide, ~3.1 Å radius (Miyazawa et al., 2003). What is the exclusionary mechanism? Very likely a hydrophobic one (Beckstein et al., 2001; Beckstein and Sansom, 2004). Water naturally tends to be expelled from the greasy interior of a narrow, nonpolar pore and ion entry is facile only if the ion is fully hydrated (Beckstein and Sansom, 2004). The critical radii, 3.5 Å and 6.5 Å respectively, closely mimic the radii of closed (Unwin, 1995) and open forms of AChR (Miyazawa et al., 2003).

Atomic level modeling of permeation through ClC channels, which is coupled with the opening of the fast gate, is a knotty problem. The X-ray structures indicate that movement of the selectivity filter glutamate is needed for permeation (Dutzler et al., 2002) and strongly imply that this only occurs after it is protonated (Dutzler et al., 2003). Atomic level computations, based on modeling the bacterial pore, demonstrates this to be a plausible mechanism for ion entry from the external electrolyte (Miloshevsky and Jordan, 2003; Bostick and Berkowitz, 2004; Cohen and Schulten, 2004; Faraldo-Gomez and Roux, 2004). Modeling reproduces the observed ion binding sites and suggests the possibility that occupancy of an additional external site plays an important role in proton-assisted, fast gating (Bostick and Berkowitz, 2004). While artificially opened channels allow ionic transit (Cohen and Schulten, 2004; Corry et al., 2004), none of the analyses provides a natural mechanism for ion transit from the central binding site into the cytoplasm; the block created by the filter's serine and tyrosine residues remains impassable.

Aquaporins ability to reject most ions has been studied extensively and is readily explained as reflecting channel electrostatics. However, how a water-filled tube absolutely forbids proton passage is more controversial. If proton interaction with the channel mimicked that of other cations, electrostatics would dominate (de Groot et al., 2003; Chakrabarti et al., 2004; Ilan et al., 2004; Miloshevsky and Jordan, 2004b). However, if a water wire formed the protonic charge could be delocalized, which would reduce the electrostatic penalty for permeation. Conceivably it is disruption of a pore spanning water wire that inhibits proton passage (Tajkhorshid et al., 2002; Jensen et al., 2003); this could be induced by the reversal in carbonyl orientation to either side of the NPR motif, coupled with the influence of the NPR itself. An alternate explanation implicates two contributing factors: electrostatics and the partial dehydration of a proton upon pore entry (Burykin and Warshel, 2003, 2004). The role of the conserved arginine and asparagines is less contentious. Both are crucial determinants in excluding cations from the pore (Miloshevsky and Jordan, 2004b).

The signature property of the potassium channel filter is its selectivity. The larger ion, potassium, permeates while the smaller one, sodium, does not. Why is this? If the filter were sufficiently rigid, Born model arguments suggest that its structure would favor binding potassium over sodium (Doyle et al., 1998). All analyses suggest that some filter flexibility is still consistent with this basic physical picture (Allen et al., 2000; Bernèche and Roux, 2001; Luzhkov and Åqvist, 2001; Burykin et al., 2003; Garofoli and Jordan, 2003). However, recent MD studies suggest the filter is so flexible that it can cradle potassium and sodium equally well (Noskov et al., 2004a), in which case an alternate explanation is needed. The answer may well be found in an extension of Eisenman's model (Eisenman, 1962), where selectivity reflects the strength of the local electric field sensed by an ion at its binding site.

But there remain problems. Gramicidin, which is structurally superbly characterized (Ketchem et al., 1997) and for which the electrophysiological data set is essentially unlimited (Koeppe and Anderson, 1996), is the critical testing ground. Here the picture is mixed. PMF computations based on standard force fields yield energy barriers implying conductances $\sim 10^7$-fold too small (Allen et al., 2003). What has gone wrong? Possibly methodological artifacts, which can be corrected for (Allen et al., 2004), or possibly the force fields (Dorman and Jordan, 2004; Miloshevsky and Jordan, 2004a). The question remains open.

Given the present limitations of high-level theory, studies of simplified abstractions are still valuable. A physical explanation of selectivity in sodium and calcium channels derives from the observations that the cation channels form a superfamily (Jan and Jan, 1990) and that exchange of signature sequences interconverts sodium and calcium channel selectivity behavior (Heinemann et al., 1992; Yang et al., 1993). The Ca channel filter has a net charge of –4 (Glu–Glu–Glu–Glu) while in the Na filter the charge is only –1 (Asp–Glu–Lys–Ala). If the residues face into an aqueous domain similar in size to that of a potassium channel's inner pore, there will be huge differences in the local electric fields. Charge compensation in a Ca channel requires residence by two Ca^{2+} or four Na^{+}; selection for calcium arises because sodium occupancy is inhibited due to crowding and additional electrostatic repulsion (Nonner et al., 2000). Extensions of the approach rationalize the calcium channel's preference for Ca^{2+} over Ba^{2+} and the effect of ion size on monovalent cation residency (Boda et al., 2001); suitably modified it also accounts for selectivity properties of the sodium channel filter (Boda et al., 2002).

1.8 What's Next?

Predicting scientific advances is a fool's errand, but irresistible. A bacterial sodium channel has been crystallized and, with good fortune, its structure will soon appear and put speculation to rest. The next few years should resolve the mechanism of voltage gating. The functional basis for differences between prokaryotic and eukaryotic ClCs should be clarified, providing insight into the structural features distinguishing channels from pumps. New theoretical tools, applicable to millisecond processes, will provide insight into gating, the slow conformational changes that control channel opening; attempts have already been made (Gullingsrud and Schulten, 2003). Channels will be engineered in novel ways, creating practical nanodevices, which has already been done with gramicidin; it is only a matter of time until similar modifications are grafted onto biological channels. Detailed structural knowledge will lead to hosts of pharmaceutical products, specifically targeted at biological ion channels. Theory will be more prominent, providing quantitative physical explanations of the mechanisms of permeation, selectivity, and gating.

Acknowledgments

I thank Gennady Miloshevsky and Michael Partensky for helpful comments and I am indebted to Shin-Ho Chung for his encouragement and his critique. This work was supported by a grant from the National Institutes of Health, GM-28643.

References

Accardi, A., and C. Miller. 2004. Secondary active transport mediated by a prokaryotic homologue of ClC Cl-channels. *Nature* 427:803–807.

Aggarwal, S.K., and R. MacKinnon. 1996. Contribution of the S4 segment to gating charge in the Shaker K^+ channel. *Neuron* 16:1169–1177.

Akabas, M.H., D.A. Stauffer, M. Xu, and A. Karlin. 1992. Acetylcholine receptor channel structure probed in cysteine-substitution mutants. *Science* 258:307–310.

Allen, T.W., O.S. Andersen, and B. Roux. 2004. Energetics of ion conduction through the gramicidin channel. *Proc. Natl. Acad. Sci. USA* 101:117–122.

Allen, T.W., T. Bastug, S. Kuyucak, and S.H. Chung. 2003. Gramicidin a channel as a test ground for molecular dynamics force fields. *Biophys. J.* 84:2159–2168.

Allen, T.W., A. Bilznyuk, A.P. Rendell, S. Kuyucak, and S.H. Chung. 2000. The potassium channel: Structure, selectivity and diffusion. *J. Chem. Phys.* 112:8191–8204.

Armstrong, C.M., F. Bezanilla, and E. Rojas. 1973. Destruction of sodium conductance inactivation in squid axons perfused with pronase. *J. Gen. Physiol.* 62:375–391.

Armstrong, C.M., and L. Binstock. 1965. Anomalous rectification in the squid giant axon injected with tetraethylammonium chloride. *J. Gen. Physiol.* 48:859–872.

Arrhenius, S. 1887. Einfluss der Neutralsalze auf der Reactionsgeschwindigkeit der Verseifung von Äthylacetat. *Zeitschrift für Physikalisches Chemie* 1:110–133.

Arseniev, A.S., I.L. Barsukov, V.F. Bystrov, A.L. Lomize, and Y.A. Ovchinnikov. 1985. 1H-NMR study of gramicidin A transmembrane ion channel. Head-to-head right-handed, single-stranded helices. *FEBS Lett.* 186:168–174.

Baker, O.S., H.P. Larsson, L.M. Mannuzzu, and E.Y. Isacoff. 1998. Three transmembrane conformations and sequence-dependent displacement of the S4 domain in Shaker K^+ channel gating. *Neuron* 20:1283–1294.

Beckstein, O., P.C. Biggin, and M.S.P. Sansom. 2001. A hydrophobic gating mechanism for nanopores. *J. Phys. Chem. B* 105:12902–12905.

Beckstein, O., and M.S.P. Sansom. 2004. The influence of geometry, surface character, and flexibility on the permeation of ions and water through biological pores. *Phys. Biol.* 1:43–52.

Bernèche, S., and B. Roux. 2001. Energetics of ion conduction through the K^+ channel. *Nature* 414:73–77.

Bernèche, S., and B. Roux. 2003. A microscopic view of ion conduction through the K^+ channel. *PNAS* 100:8644–8648.

Bezanilla, F. 2000. The voltage sensor in voltage-dependent ion channels. *Physiol. Rev.* 80:555–592.

Bezanilla, F. 2002. Voltage sensor movements. *J. Gen. Physiol.* 120:465–473.

Bezanilla, F., and C.M. Armstrong. 1972. Negative conductance caused by entry of sodium and cesium ions into the potassium channels of squid axon. *J. Gen. Physiol.* 60:588–608.

Boda, D., D.D. Busath, B. Eisenberg, D. Henderson, and W. Nonner. 2002. Monte Carlo simulations of ion selectivity in a biological Na channel: Charge-space competition. *Phys. Chem. Chem. Phys.* 4:5154–5160.

Boda, D., D. Henderson, and D.D. Busath. 2001. Monte Carlo study of the effect of ion and channel size on the selectivity of a model calcium channel. *J. Phys. Chem. B* 105:11574–11577.

Bostick, D.L., and M.L. Berkowitz. 2004. Exterior site occupancy infers chloride-induced proton gating in a prokaryotic homolog of the ClC chloride channel. *Biophys. J.* 87:1686–1696.

Brejc, K., W.J. van Dijk, R.V. Klaassen, M. Schuurmans, J. van Der Oost, A.B. Smit, and T.K. Sixma. 2001. Crystal structure of an ACh-binding protein reveals the ligand-binding domain of nicotinic receptors. *Nature* 411:269–276.

Brisson, A., and P.N. Unwin. 1985. Quaternary structure of the acetylcholine receptor. *Nature* 315:474–477.

Burykin, A., M. Kato, and A. Warshel. 2003. Exploring the origin of the ion selectivity of the KcsA potassium channel. *Proteins* 52:412–426.

Burykin, A., C.N. Schutz, J. Villa, and A. Warshel. 2002. Simulations of ion current in realistic models of ion channels: The KcsA potassium channel. *Proteins* 47:265–280.

Burykin, A., and A. Warshel. 2003. What really prevents proton transport through aquaporin? Charge self-energy versus proton wire proposals. *Biophys. J.* 85:3696–3706.

Burykin, A., and A. Warshel. 2004. On the origin of the electrostatic barrier for proton transport in aquaporin. *FEBS Lett.* 570:41–46.

Cai, M., and P.C. Jordan. 1990. How does vestibule surface charge affect ion conduction and toxin binding in a sodium channel? *Biophys. J.* 57:883–891.

Callahan, M.J., and S.J. Korn. 1994. Permeation of Na^+ through a delayed rectifier K^+ channel in chick dorsal root ganglion neurons. *J. Gen. Physiol.* 104:747–771.

Cha, A., and F. Bezanilla. 1997. Characterizing voltage-dependent conformational changes in the Shaker K^+ channel with fluorescence. *Neuron* 19:1127–1140.

Cha, A., G.E. Snyder, P.R. Selvin, and F. Bezanilla. 1999. Atomic scale movement of the voltage-sensing region in a potassium channel measured via spectroscopy. *Nature* 402:809–813.

Chakrabarti, N., E. Tajkhorshid, B. Roux, and R. Pomes. 2004. Molecular basis of proton blockage in aquaporins. *Structure (Camb)* 12:65–74.

Chanda, B., O.K. Asamoah, R. Blunck, B. Roux, and F. Bezanilla. 2005. Gating charge displacement in voltage-gated ion channels involves limited transmembrane movement. *Nature* 436:852–856.

Chen, D.P., V. Barcilon, and R.S. Eisenberg. 1992. Constant fields and constant gradients in open ionic channels. *Biophys. J.* 61:1372–1393.

Chou, P.Y., and G.D. Fasman. 1978. Empirical predictions of protein conformation. *Annu. Rev. Biochem.* 47:251–276.

Chung, S.H., T.W. Allen, and S. Kuyucak. 2002a. Conducting-state properties of the KcsA potassium channel from molecular and Brownian dynamics simulations. *Biophys. J.* 82:628–645.

Chung, S.H., T.W. Allen, and S. Kuyucak. 2002b. Modeling diverse range of potassium channels with Brownian dynamics. *Biophys. J.* 83:263–277.

Cohen, J., and K. Schulten. 2004. Mechanism of anionic conduction across ClC. *Biophys. J.* 86:836–845.

Cooper, K., E. Jakobsson, and P. Wolynes. 1985. The theory of ion transport through membrane channels. *Prog. Biophys. Mol. Biol.* 46:51–96.

Corry, B., S. Kuyucak, and S.H. Chung. 2000. Tests of continuum theories as models of ion channels. II. Poisson–Nernst–Planck theory versus Brownian dynamics. *Biophys. J.* 78:2364–2381.

Corry, B., S. Kuyucak, and S.H. Chung. 2003. Dielectric self-energy in Poisson–Boltzmann and Poisson–Nernst–Planck models of ion channels. *Biophys. J.* 84:3594–3606.

Corry, B., M. O'Mara, and S.-H. Chung. 2004. Conduction mechanisms of chloride ions in ClC-type channels. *Biophys. J.* 86:846–860.

Cuello, L.G., D.M. Cortes, and E. Perozo. 2004. Molecular architecture of the KvAP voltage-dependent K^+ channel in a lipid bilayer. *Science* 306:491–495.

de Groot, B.L., T. Frigato, V. Helms, and H. Grubmüller. 2003. The mechanism of proton exclusion in the aquaporin-1 water channel. *J. Mol. Biol.* 333: 279–293.

Denker, B., B. Smith, F. Kuhajda, and P. Agre. 1988. Identification, purification, and partial characterization of a novel Mr 28,000 integral membrane protein from erythrocytes and renal tubules. *J. Biol. Chem.* 263:15634–15642.

Dorman, V.L., and P.C. Jordan. 2004. Ionic permeation free energy in gramicidin: A semimicroscopic perspective. *Biophys. J.* 86:3529–3541.

Doyle, D.A., J. Morais-Cabral, R.A. Pfuetzner, A. Kuo, J.M. Gulbis, S.L. Cohen, B.T. Chait, and R. MacKinnon. 1998. The structure of the potassium channel: Molecular basis of K^+ conduction and selectivity. *Science* 280:69–77.

Durkin, J.T., L.L. Providence, R.E. Koeppe 2nd, and O.S. Andersen. 1993. Energetics of heterodimer formation among gramicidin analogues with an NH2-terminal addition or deletion. Consequences of missing a residue at the join in the channel. *J. Mol. Biol.* 231:1102–1121.

Dutzler, R., E.B. Campbell, M. Cadene, B.T. Chait, and R. MacKinnon. 2002. X-ray structure of a ClC chloride channel at 3.0 A reveals the molecular basis of anion selectivity. *Nature* 415:287–294.

Dutzler, R., E.B. Campbell, and R. MacKinnon. 2003. Gating the selectivity filter in ClC chloride channels. *Science* 300:108–112.

Edwards, S., B. Corry, S. Kuyucak, and S.H. Chung. 2002. Continuum electrostatics fails to describe ion permeation in the gramicidin channel. *Biophys. J.* 83:1348–1360.

Eisenberg, R.S. 1999. From structure to function in open ionic channels. *J. Membr. Biol.* 171:1–24.

Eisenman, G. 1962. Cation selective glass electrodes and their mode of operation. *Biophys. J.* 2(2, Pt 2):259–323.

Eyring, H. 1935. The activated complex in chemical reactions. *J. Chem. Phys.* 1:107–115.

Faraldo-Gomez, J.D., and B. Roux. 2004. Electrostatics of ion stabilization in a ClC chloride channel homologue from *Escherichia coli*. *J. Mol. Biol.* 339:981–1000.

Fu, D., A. Libson, L.J. Miercke, C. Weitzman, P. Nollert, J. Krucinski, and R.M. Stroud. 2000. Structure of a glycerol-conducting channel and the basis for its selectivity. *Science* 290:481–486.

Gandhi, C.S., E. Clarck, E. Loots, A. Pralle, and E.Y. Isacoff. 2003. The orientation and molecular movement of a K^+ channel voltage-sensing domain. *Neuron* 40:515–525.

Gandhi, C.S., and E.Y. Isacoff. 2002. Molecular models of voltage sensing. *J. Gen. Physiol.* 120:455–463.

Garofoli, S., and P.C. Jordan. 2003. Modeling permeation energetics in the KcsA potassium channel. *Biophys. J.* 84:2814–2830.

Glauner, K.S., L.M. Mannuzzu, C.S. Gandhi, and E.Y. Isacoff. 1999. Spectroscopic mapping of voltage sensor movement in the Shaker potassium channel. *Nature* 402:813–817.

Gullingsrud, J., and K. Schulten. 2003. Gating of MscL studied by steered molecular dynamics. *Biophys. J.* 85:2087–2099.

Hagiwara, S., and N. Saito. 1959. Voltage–current relations in nerve cell membrane of *Onchidium verruculatum*. *J. Physiol.* 148:161–179.

Hamill, O.P., A. Marty, E. Neher, B. Sakmann, and F.J. Sigworth. 1981. Improved patch–clamp techniques for high-resolution current recording from cells and cell-free membrane patches. *Pflügers Arch* 391:85–100.

Harries, W.E.C., D. Akhavan, L.J.W. Miercke, S. Khademi, and R.M. Stroud. 2004. The channel architecture of aquaporin 0 at a 2.2-A resolution. *PNAS* 101:14045–14050.

Heinemann, S.H., H. Terlau, W. Stuhmer, K. Imoto, and S. Numa. 1992. Calcium channel characteristics conferred on the sodium channel by single mutations. *Nature* 356:441–443.

Hille, B. 1970. Ionic channels in nerve membranes. *Prog. Biophys. Mol. Biol.* 21:1–32.

Hille, B. 1971. The permeability of the sodium channel to organic cations in myelinated nerve. *J. Gen. Physiol.* 58:599–619.

Hille, B. 1973. Potassium channels in myelinated nerve. Selective permeability to small cations. *J. Gen. Physiol.* 61:669–686.

Hladky, S.B., and D.A. Haydon. 1972. Ion transfer across lipid membranes in the presence of gramicidin A. I. Studies of the unit conductance channel. *Biochim. Biophys. Acta* 274:294–312.

Hodgkin, A.L., and A.F. Huxley. 1952a. Currents carried by sodium and potassium ions through the membrane of the giant axon of Loligo. *J. Physiol.* 116:449–472.

Hodgkin, A.L., and A.F. Huxley. 1952b. A quantitative description of membrane current and its application to conduction and excitation in nerve. *J. Physiol.* 117:500–544.

Hodgkin, A.L., A.F. Huxley, and B. Katz. 1952. Measurement of current–voltage relations in the membrane of the giant axon of *Loligo*. *J. Physiol.* 116:424–448.

Hodgkin, A.L., and R.D. Keynes. 1955. The potassium permeability of a giant nerve fibre. *J. Physiol.* 128:61–88.

Hopp, T.P., and K.R. Woods. 1981. Prediction of protein antigenic determinants from amino acid sequences. *Proc. Natl. Acad. Sci. USA* 78:3824–3828.

Horn, R. 2002. Coupled movements in voltage-gated ion channels. *J. Gen. Physiol.* 120:449–453.

Hoshi, T., W.N. Zagotta, and R.W. Aldrich. 1990. Biophysical and molecular mechanisms of Shaker potassium channel inactivation. *Science* 250:533–538.

Hubbell, W.L., H.S. McHaourab, C. Altenbach, and M.A. Lietzow. 1996. Watching proteins move using site-directed spin labeling. *Structure* 4:779–783.

Ilan, B., E. Tajkhorshid, K. Schulten, and G.A. Voth. 2004. The mechanism of proton exclusion in aquaporin channels. *Proteins* 55:223–228.

Immke, D., M. Wood, L. Kiss, and S.J. Korn. 1999. Potassium-dependent changes in the conformation of the Kv2.1 potassium channel pore. *J. Gen. Physiol.* 113:819–836.

Imoto, K., C. Busch, B. Sakmann, M. Mishina, T. Konno, J. Nakai, H. Bujo, Y. Mori, K. Fukuda, and S. Numa. 1988. Rings of negatively charged amino acids determine the acetylcholine receptor channel conductance. *Nature* 335:645–648.

Jan, L.Y., and Y.N. Jan. 1990. A superfamily of ion channels. *Nature* 345:672.

Jensen, M.O., E. Tajkhorshid, and K. Schulten. 2003. Electrostatic tuning of permeation and selectivity in aquaporin water channels. *Biophys. J.* 85:2884–2899.

Jiang, Y., A. Lee, J. Chen, M. Cadene, B.T. Chait, and R. MacKinnon. 2002a. Crystal structure and mechanism of a calcium-gated potassium channel. *Nature* 417:515–522.

Jiang, Y., A. Lee, J. Chen, M. Cadene, B.T. Chait, and R. MacKinnon. 2002b. The open pore conformation of potassium channels. *Nature*, 417:523–526.

Jiang, Y., A. Lee, J. Chen, V. Ruta, M. Cadene, B.T. Chait, and R. MacKinnon. 2003a. X-ray structure of a voltage-dependent K^+ channel. *Nature* 423:33–41.

Jiang, Y., V. Ruta, J. Chen, A. Lee, and R. MacKinnon. 2003b. The principle of gating charge movement in a voltage-dependent K^+ channel. *Nature* 423:42–48.

Jordan, P.C. 1983. Electrostatic modeling of ion pores. II. Effects attributable to the membrane dipole potential. *Biophys. J.* 41:189–195.

Jordan, P.C. 1987. How pore mouth charge distributions alter the permeability of transmembrane ionic channels. *Biophys. J.* 51:297–311.

Jordan, P.C. 1999. Ion permeation and chemical kinetics. *J. Gen. Physiol.* 114:601–603.

Jordan, P.C., R.J. Bacquet, J.A. McCammon, and P. Tran. 1989. How electrolyte shielding influences the electrical potential in transmembrane ion channels. *Biophys. J.* 55:1041–1052.

Karlin, A., and M.H. Akabas. 1998. Substituted-cysteine accessibility method. *Methods Enzymol.* 293:123–145.

Ketchem, R., B. Roux, and T. Cross. 1997. High-resolution polypeptide structure in a lamellar phase lipid environment from solid state NMR derived orientational constraints. *Structure* 5:1655–1669.

Koeppe, R.E., 2nd, and O.S. Anderson. 1996. Engineering the gramicidin channel. *Annu. Rev. Biophys. Biomol. Struct.* 25:231–258.

Kuo, A., J.M. Gulbis, J.F. Antcliff, T. Rahman, E.D. Lowe, J. Zimmer, J. Cuthbertson, F.M. Ashcroft, T. Ezaki, and D.A. Doyle. 2003. Crystal structure of the potassium channel KirBac1.1 in the closed state. *Science* 300:1922–1926.

Laine, M., M.C. Lin, J.P. Bannister, W.R. Silverman, A.F. Mock, B. Roux, and D.M. Papazian. 2003. Atomic proximity between S4 segment and pore domain in Shaker potassium channels. *Neuron* 39:467–481.

Laüger, P. 1973. Ion transport through pores: A rate-theory analysis. *Biochim. Biophys. Acta* 311:423–441.

Levitt, D.G. 1978. Electrostatic calculations for an ion channel. I. Energy and potential profiles and interactions between ions. *Biophys. J.* 22:209–219.

Levitt, D.G. 1986. Interpretation of biological ion channel flux data – Reaction-rate versus continuum theory. *Annu. Rev. Biophys. Biophys. Chem.* 15:29–57.

Long, S.B., E.B. Campbell, and R. MacKinnon. 2005a. Crystal structure of a mammalian voltage-dependent Shaker family K^+ channel. *Science* 309:897–903.

Long, S.B., E.B. Campbell, and R. MacKinnon. 2005b. Voltage sensor of Kv1.2: Structural basis of electromechanical coupling. *Science* 309:903–908.

Lu, Z., and R. MacKinnon. 1994. Electrostatic tuning of Mg^{2+} affinity in an inward-rectifier K^+ channel. *Nature* 371:243–246.

Luzhkov, V.B., and J. Åqvist. 2001. K(+)/Na(+) selectivity of the KcsA potassium channel from microscopic free energy perturbation calculations. *Biochim. Biophys. Acta* 1548:194–202.

Mackay, D.H.J., P.H. Berens, K.R. Wilson, and A.T. Hagler. 1984. Structure and dynamics of ion transport through gramicidin A. *Biophys. J.* 46:229–248.

MacKinnon, R. 1991. Determination of the subunit stoichiometry of a voltage-activated potassium channel. *Nature* 350:232–235.

Mamonov, A.B., R.D. Coalson, A. Nitzan, and M.G. Kurnikova. 2003. The role of the dielectric barrier in narrow biological channels: A novel composite approach to modeling single-channel currents. *Biophys. J.* 84:3646–3661.

Mannuzzu, L.M., M.M. Moronne, and E.Y. Isacoff. 1996. Direct physical measure of conformational rearrangement underlying potassium channel gating. *Science* 271:213–216.

Miller, C. 1982. Open-state substructure of single chloride channels from Torpedo electroplax. *Philos. Trans. R. Soc. Lond. B Biol. Sci.* 299:401–411.

Miller, C., and E. Racker. 1976. Ca^{++}-induced fusion of fragmented sarcoplasmic reticulum with artificial planar bilayers. *J. Membr. Biol.* 30:283–300.

Miloshevsky, G.V., and P.C. Jordan. 2003. Theoretical study of the passage of chloride ions through a bacterial ClC chloride channel. *J. Gen. Physiol.* 122:32A.

Miloshevsky, G.V., and P.C. Jordan. 2004a. Permeation in ion channels: The interplay of structure and theory. *Trends Neurosci.* 27:308–314.

Miloshevsky, G.V., and P.C. Jordan. 2004b. Water and ion permeation in bAQP1 and GlpF channels: A kinetic Monte Carlo study. *Biophys. J.* 87:3690–3702.

Mitra, A.K., M.P. McCarthy, and R.M. Stroud. 1989. Three-dimensional structure of the nicotinic acetylcholine receptor and location of the major associated 43-kD cytoskeletal protein, determined at 22 A by low dose electron microscopy and x-ray diffraction to 12.5 A. *J. Cell Biol.* 109:755–774.

Miyazawa, A., Y. Fujiyoshi, and N. Unwin. 2003. Structure and gating mechanism of the acetylcholine receptor pore. *Nature* 424:949–955.

Mullins, L. 1959. The penetration of some cations into muscle. *J. Gen. Physiol.* 42:817–829.

Mullins, L.J. 1968. A single channel or a dual channel mechanism for nerve excitation. *J. Gen. Physiol.* 52:550–556.

Murata, K., K. Mitsuoka, T. Hirai, T. Walz, P. Agre, J.B. Heymann, A. Engel, and Y. Fujiyoshi. 2000. Structural determinants of water permeation through aquaporin-1. *Nature* 407:599–605.

Nakamura, Y., S. Nakajima, and H. Grundfest. 1965. The action of tetrodotoxin on electrogenic components of squid giant axons. *J. Gen. Physiol.* 48:985–996.

Narahashi, T., J.W. Moore, and W.R. Scott. 1964. Tetrodotoxin blockage of sodium conductance increase in lobster giant axons. *J. Gen. Physiol.* 47:965–974.

Neher, E., and B. Sakmann. 1976. Single-channel currents recorded from membrane of denervated frog muscle fibres. *Nature* 260:799–802.

Noda, M., H. Takahashi, T. Tanabe, M. Toyosato, Y. Furutani, T. Hirose, M. Asai, S. Inayama, T. Miyata, and S. Numa. 1982. Primary structure of alpha-subunit precursor of *Torpedo californica* acetylcholine receptor deduced from cDNA sequence. *Nature* 299:793–797.

Noda, M., H. Takahashi, T. Tanabe, M. Toyosato, S. Kikyotani, Y. Furutani, T. Hirose, H. Takashima, S. Inayama, T. Miyata, and S. Numa. 1983. Structural homology of *Torpedo californica* acetylcholine receptor subunits. *Nature* 302:528–532.

Nonner, W., L. Catacuzzeno, and B. Eisenberg. 2000. Binding and selectivity in L-type calcium channels: A mean spherical approximation. *Biophys. J.* 79:1976–1992.

Noskov, S.Y., S. Berneche, and B. Roux. 2004a. Control of ion selectivity in potassium channels by electrostatic and dynamic properties of carbonyl ligands. *Nature* 431:830–834.

Noskov, S.Y., S. Berneche, and B. Roux. 2004b. The microscopic origin of ion selectivity in potassium channels. *Biophys. J.* 86:351a–352a.

Parsegian, A. 1969. Energy of an ion crossing a low dielectric membrane: Solutions to four relevant electrostatic problems. *Nature* 221:844–846.

Perozo, E., D.M. Cortes, and L.G. Cuello. 1998. Three-dimensional architecture and gating mechanism of a K^+ channel studied by EPR spectroscopy. *Nat. Struct. Biol.* 5:459–469.

Perozo, E., L.G. Cuello, D.M. Cortes, Y.S. Liu, and P. Sompornpisut. 2002. EPR approaches to ion channel structure and function. *Novartis Found Symp.* 245:146–158; discussion 158–164, 165–168.

Picollo, A., and M. Pusch. 2005. Chloride/proton antiporter activity of mammalian CLC proteins ClC-4 and ClC-5. *Nature* 436:420–423.

Pomès, R., and B. Roux. 1996. Structure and dynamics of a proton wire: A theoretical study of H^+ translocation along the single-file water chain in the gramicidin a channel. *Biophys. J.* 71:19–39.

Pomès, R., and B. Roux. 2002. Molecular mechanism of H^+ conduction in the single-file water chain of the gramicidin channel. *Biophys. J.* 82:2304–2316.

Posson, D.J., P. Ge, C. Miller, F. Bezanilla, and P.R. Selvin. 2005. Small vertical movement of a K^+ channel voltage sensor measured with luminescence energy transfer. *Nature* 436:848–851.

Pusch, M., U. Ludewig, A. Rehfeldt, and T.J. Jentsch. 1995. Gating of the voltage-dependent chloride channel CLC-0 by the permeant anion. *Nature* 373:527–531.

Raftery, M.A., M.W. Hunkapiller, C.D. Strader, and L.E. Hood. 1980. Acetylcholine receptor: Complex of homologous subunits. *Science* 208:1454–1456.

Revell Phillips, L., M. Milescu, Y. Li-Smerin, J.A. Mindell, J.I. Kim, and K.J. Swartz. 2005. Voltage-sensor activation with a tarantula toxin as cargo. 436:857–860.

Sakmann, B., C. Methfessel, M. Mishina, T. Takahashi, T. Takai, M. Kurasaki, K. Fukuda, and S. Numa. 1985. Role of acetylcholine receptor subunits in gating of the channel. *Nature* 318:538–543.

Salom, D., M.C. Bano, L. Braco, and C. Abad. 1995. HPLC demonstration that an all Trp–Phe replacement in gramicidin A results in a conformational rearrangement from beta-helical monomer to double-stranded dimer in model membranes. *Biochem. Biophys. Res. Commun.* 209:466–473.

Scheel, O., A.A. Zdebik, S. Lourdel, and T.J. Jentsch. 2005. Voltage-dependent electrogenic chloride/proton exchange by endosomal CLC proteins. *Nature* 436:424–427.

Schein, S.J., M. Colombini, and A. Finkelstein. 1976. Reconstitution in planar lipid bilayers of a voltage-dependent anion-selective channel obtained from paramecium mitochondria. *J. Membr. Biol.* 30:99–120.

Schoppa, N.E., K. McCormack, M.A. Tanouye, and F.J. Sigworth. 1992. The size of gating charge in wild-type and mutant Shaker potassium channels. *Science* 255:1712–1715.

Schutz, C.N., and A. Warshel. 2001. What are the dielectric "constants" of proteins and how to validate electrostatic models? *Proteins* 44:400–417.

Smith, B., and P. Agre. 1991. Erythrocyte Mr 28,000 transmembrane protein exists as a multisubunit oligomer similar to channel proteins. *J. Biol. Chem.* 266:6407–6415.

Sui, H., B.G. Han, J.K. Lee, P. Walian, and B.K. Jap. 2001. Structural basis of water-specific transport through the AQP1 water channel. *Nature* 414:872–878.

Tajkhorshid, T., P. Nollert, R.M. Stroud, and K. Schulten. 2002. Global orientational tuning controls selectivity of the AQP water channel family. *Science* 296:525–530.

Takai, T., M. Noda, M. Mishina, S. Shimizu, Y. Furutani, T. Kayano, T. Ikeda, T. Kubo, H. Takahashi, T. Takahashi et al. 1985. Cloning, sequencing and expression of cDNA for a novel subunit of acetylcholine receptor from calf muscle. *Nature* 315:761–764.

Thompson, J.D., D.G. Higgins, and T.J. Gibson. 1994. CLUSTAL W: Improving the sensitivity of progressive multiple sequence alignment through sequence weighting, position-specific gap penalties and weight matrix choice. *Nucleic Acids Res.* 22:4673–4680.

Townsley, L.E., W.A. Tucker, S. Sham, and J.F. Hinton. 2001. Structures of gramicidins A, B, and C incorporated into sodium dodecyl sulfate micelles. *Biochemistry* 40:11676–11686.

Unwin, N. 1995. Acetylcholine receptor channel imaged in the open state. *Nature* 373:37–43.

Unwin, N. 2005. Refined structure of the nicotinic acetylcholine receptor at 4 A resolution. *J. Mol. Biol.* 346:967–989.

Urry, D.W. 1971. The gramicidin A transmembrane channel: A proposed π(L,D) helix. *Proc. Natl. Acad. Sci. USA* 68:672–676.

Yang, J., P.T. Ellinor, W.A. Sather, J.F. Zhang, and R.W. Tsien. 1993. Molecular determinants of Ca^{2+} selectivity and ion permeation in L-type Ca^{2+} channels. *Nature* 366:158–161.

Yang, J., Y.N. Jan, and L.Y. Jan. 1995. Control of rectification and permeation by residues in two distinct domains in an inward rectifier K^+ channel. *Neuron* 14:1047–1054.

Yang, N., and R. Horn. 1995. Evidence for voltage-dependent S4 movement in sodium channels. *Neuron* 15:213–218.

Yeh, J.Z., and C.M. Armstrong. 1978. Immobilisation of gating charge by a substance that simulates inactivation. *Nature* 273:387–389.

Yellen, G., M.E. Jurman, T. Abramson, and R. MacKinnon. 1991. Mutations affecting internal TEA blockade identify the probable pore-forming region of a K^+ channel. *Science* 251:939–942.

Zagotta, W.N., T. Hoshi, and R.W. Aldrich. 1990. Restoration of inactivation in mutants of Shaker potassium channels by a peptide derived from ShB. *Science* 250:568–571.

Zhou, Y., J.H. Morais-Cabral, A. Kaufman, and R. MacKinnon. 2001. Chemistry of ion coordination and hydration revealed by a K^+ channel-Fab complex at 2.0 A resolution. *Nature* 414:43–48.

Part II

Specific Channel Types

2 Gramicidin Channels: Versatile Tools

Olaf S. Andersen, Roger E. Koeppe II, and Benoît Roux

2.1 Overview

Gramicidin channels are miniproteins in which two tryptophan-rich subunits associate by means of transbilayer dimerization to form the conducting channels. That is, in contrast to other ion channels, gramicidin channels do not open and close; they appear and disappear. Each subunit in the bilayer-spanning channel is tied to the bilayer/solution interface through hydrogen bonds that involve the indole NH groups as donors and water or the phospholipid backbone as acceptors. The channel's permeability characteristics are well-defined: gramicidin channels are selective for monovalent cations, with no measurable permeability to anions or polyvalent cations; ions and water move through a pore whose wall is formed by the peptide backbone; and the single-channel conductance and cation selectivity vary when the amino acid sequence is varied, even though the permeating ions make no contact with the amino acid side chains. Given the plethora of available experimental information—for not only the wild-type channels but also for channels formed by amino acid-substituted gramicidin analogues—gramicidin channels continue to provide important insights into the microphysics of ion permeation through bilayer-spanning channels. For similar reasons, gramicidin channels constitute a system of choice for evaluating computational strategies for obtaining mechanistic insights into ion permeation through the more complex channels formed by integral membrane proteins.

2.2 Introduction

Gramicidin channels are formed by the linear gramicidins, a family of peptide antibiotics produced by the soil bacillus *B. brevis* (Dubos, 1939; Hotchkiss, 1944), or by semisynthesis or total synthesis of gramicidin analogues (Bamberg et al., 1978; Morrow et al., 1979; Heitz et al., 1982; Greathouse et al., 1999). The linear gramicidins were the first antibiotics used in clinical practice (Herrell and Heilman, 1941). They exert their antibacterial activity by increasing the cation permeability of target bacterial plasma membranes (Harold and Baarda, 1967) through the formation of bilayer-spanning channels (Hladky and Haydon, 1970).

Compared to channels formed by many other antibiotics, the gramicidin channels are exceptionally well-behaved. Once formed, the bilayer-spanning channels have a single conducting state, which taken together with the channels' cation selectivity (Myers and Haydon, 1972) and well-understood single-stranded (SS) $\beta^{6.3}$-helical structure (Urry, 1971, 1972; Arseniev et al., 1986; Ketchem et al., 1997; Townsley et al., 2001; Allen et al., 2003) make gramicidin channels important tools for understanding the physicochemical basis ion permeation through bilayer-spanning channels. Not surprisingly, the gramicidin channels have served as the prototypical channels in the development of many physical approaches toward understanding channel (or membrane protein) structure and function (Busath, 1993).

Gramicidin channels are arguably the best understood ion channels. Atomic (or near-atomic) resolution structures have been reported for an increasing number of complex ion channels formed by integral membrane proteins (Weiss et al., 1991; Cowan et al., 1992; Chang et al., 1998; Doyle et al., 1998; Bass et al., 2002; Jiang et al., 2002, 2003; Kuo et al., 2003; Long et al., 2005; Unwin, 2005). Yet the gramicidin channels embody a unique combination of features that sets them apart from other channels: first, the advantages construed by the wealth of information about the channel's ion permeability taken together with an atomic-resolution structure; second, the ion permeability can be modulated by defined chemical modifications whose influence on structure as well as function can be determined experimentally; finally, the wild-type and amino-substituted analogue channels are large enough to be nontrivial, yet small enough to be amenable to detailed computational studies.

In this chapter we first review the channel structure, dynamics, and function. We show that even though the linear gramicidins are conformationally polymorphic in organic solvents, they fold into one predominant conformer (the channel structure) in lipid bilayers and bilayer-like environments—yet the conformational preference may change in cases of extreme bilayer-channel hydrophobic mismatch. We then discuss the channels' permeability properties and show how it is possible to develop discrete-state models to describe the kinetics of ion movement through the channels. Finally, we briefly show how molecular dynamics (MD) simulations have advanced to the level of achieving semiquantitative agreement between observed and predicted ion permeabilities.

2.3 Structure

The linear gramicidins, exemplified by [Val^1]gramicidin A (gA), have an alternating (L-D)-amino acid sequence (Sarges and Witkop, 1965):

Formyl-L-Val^1-(D)-Gly-L-Ala^3-D-Leu-L-Ala^5-
-D-Val- L-Val^7-D-Val-L-Trp^9-D-Leu-L-Trp^{11}-
-D-Leu-L-Trp^{13}-L-Trp^{15}-ethanolamide.

(Only the L residues are numbered.) In *B. brevis*, the linear gramicidins are synthesized by nonribosomal peptide synthesis (Lipmann, 1980). The synthesis is catalyzed by a complex of four nonribosomal peptide synthases (Kessler et al., 2004), which encompass a total of 16 modules that catalyze the successive activation and condensation (peptide bond formation) of amino acids to produce the specific sequence. The D-residues in the sequence are produced by epimerization of the corresponding L-residues. (For an overview of nonribosomal peptide synthesis, see Sieber and Marahiel, 2005.) The first module, which activates the amino-terminal L-Val, contains a putative formylation domain (Kessler et al., 2004). In the growing chain, the last amino acid is Gly; the peptide is released from the synthase template in a two-step NADH/NADPH-dependent reduction of this Gly to yield the carboxyterminal ethanolamide (Schracke et al., 2005). Due to relaxed substrate specificity in the first and eleventh modules, the naturally occurring gramicidin (gramicidin "D" after R. Dubos) is a mixture of [Val^1]- and [Ile^1]gramicidins A (with Trp^{11}), B (with Phe^{11}), and C (Tyr^{11}). [Val^1]gA constitutes about 70% of the natural mixture (as deduced from HPLC tracings in Koeppe et al., 1985).

For modern electrophysiological studies, the gramicidins are synthesized on peptide synthesizers using standard, solid-state peptide chemical methods (Greathouse et al., 1999). The synthesized products are purified, to a purity of better than 99%, using a two-stage reversed-phase HPLC procedure (Weiss and Koeppe, 1985). This is particularly important for single-channel experiments (Greathouse et al., 1999), which usually rely on the ability to identify the unique electrophysiological "fingerprints" of channels formed by different sequence-substituted gA analogues.

The alternating (L-D) gramicidin sequence allows the molecule to fold as a β-helix with the side chains projecting from the exterior surface of a cylindrical tube formed by the peptide backbone (Urry, 1971; Ramachandran and Chandrasekaran, 1972). Many such folding patterns are possible, and gA is conformationally polymorphic (Veatch et al., 1974). At least seven different helical conformers have been described in organic solvents (Veatch and Blout, 1974), where gA forms a variety of double-stranded (DS), intertwined dimers—the so-called ππ-helices (Bystrov and Arseniev, 1988; Langs, 1988; Abdul-Manan and Hinton, 1994; Burkhart et al., 1998).

In aqueous solutions, gA usually forms poorly defined aggregates (Kemp and Wenner, 1976). At very low aqueous concentrations, in the low picomolar range that is used in electrophysiological studies where the gramicidins usually are added to the aqueous solutions at the two sides of the bilayer, gA seems to fold into a stable monomer structure (Jagannadham and Nagaraj, 2005).

2.3.1 One Conducting Channel Conformation

In spite of the conformational polymorphism observed in organic solvents, gramicidin channels adopt a unique structure in lipid bilayer membranes or in bilayer-like environments (such as micelles, with a nonpolar/polar interface). The channel's electrophysiological "fingerprints" (single-channel current traces, Fig. 2.1; and current transition amplitude and lifetime distributions, Fig. 2.2) show that a single predominant conducting channel structure is formed by both the wild-type gA and the [D-Ala^2,Ser^3]gA analogue—and most other gA analogues.

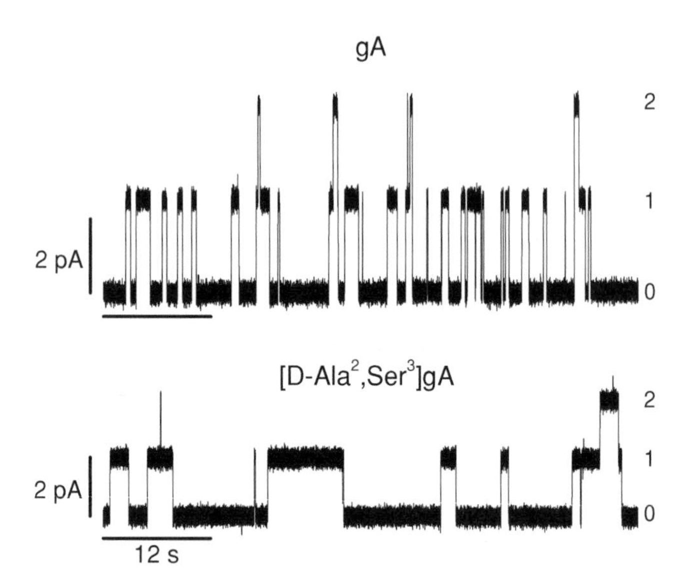

Fig. 2.1 Gramicidin single-channel current traces. Top: single-channel current trace obtained with gA. Bottom: single-channel current trace obtained with the gA analogue [D-Ala^2,Ser^3]gA. Both gA and [D-Ala^2,Ser^3]gA form one predominant type of ion-conducting channels, which are "visible" as discrete, well-defined transitions changes in the current through the membrane. The numbers to the right of the traces denote the number of conducting channels at the different current levels. The conducting states are "stable," meaning that the current noise of the conducting state is indistinguishable from that of the baseline. In the gA experiment, the root-mean-square (RMS) of the current in levels 0 through 2 varies between 0.148 and 0.150 pA; in the [D-Ala^2,Ser^3]gA experiment, the RMS of the current in levels 0 through 2 varies between 0.133 and 0.136 pA. The experimental methods were done as described in Greathouse et al. (1999). Experimental conditions: diphytanoylphosphatidylcholine/*n*-decane bilayers; electrolyte solution, 1.0 M NaCl, pH 7; applied potential 200 mV; current signal filtered at 500 Hz.

Both the single-channel current transition amplitudes and channel lifetimes vary when the amino acid sequence is varied; different gA analogues usually form channels with distinct electrophysiological characteristics—even though the amino acid side chains do not come in direct contact with the permeating ions. For example, the Gly^2→D-Ala^2 substitution by itself causes a 10–15% increase in the single-channel current transition amplitude and a fourfold increase in the average channel lifetime (τ) (Mattice et al., 1995). The Ala^3→Ser^3 substitution behaves similarly to other polar amino acid substitutions in the formyl-NH-terminal half of the sequence (Morrow et al., 1979; Russell et al., 1986; Durkin et al., 1990; Koeppe et al., 1990), to cause reductions in both the current transition amplitude and lifetime.

On occasion, one may observe transitions between two different current levels in a "stable" bilayer-spanning gA channel (Busath and Szabo, 1981). Some reports also have suggested that up to 25–50% of channels formed by purified gA may be variants (affectionately called "minis") whose current transition amplitudes fall above or (usually) below the main peak (Hladky and Haydon, 1972; Busath and Szabo, 1981;

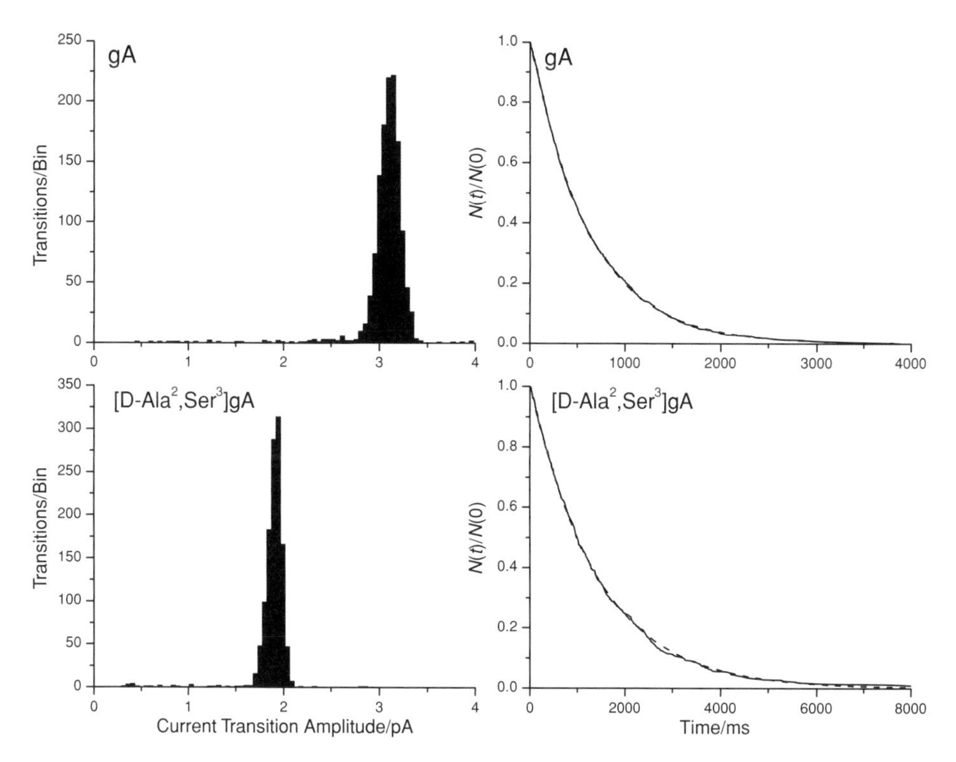

Fig. 2.2 Characterization of gramicidin channel function. Top: results for gA channels. Bottom: results for [D-Ala^2,Ser^3]gA channels. Left: current transition amplitude histograms. The histograms are assembled from the current transition amplitudes, the absolute value of the difference in current level just before and just after a current transition (Andersen, 1983a), and only the transition amplitudes are plotted. The histogram thus differs from a current level (or all-points) histogram. The single predominant peaks around 3.1 pA (for gA channels) and 1.9 pA (for [D-Ala^2,Ser^3]gA channels) comprise 1246 out of 1294 transitions (for gA) and 1175 out of 1202 transitions (for [D-Ala^2,Ser^3]gA), meaning that we can account for 96–97% of the observed current transitions. Right: lifetime distributions as normalized survivor plots; note the twofold difference in the scale in the abscissa. The interrupted curves are fits of $N(t)/N(0) = \exp\{-t/\tau\}$ to the results, where $N(t)$ is the number of channels with lifetime longer than time t, and τ is the average lifetime (610 ms for gA channels and 1410 ms for [D-Ala^2,Ser^3]gA channels). Experimental conditions as in Fig. 2.1.

Urry et al., 1984). Though the structural basis for these conductance variants remains enigmatic, they somehow result from subtle experimental "artifacts" (Busath et al., 1987). They will not be considered further here.

In addition to establishing that there is a single predominant channel conformation, electrophysiological and other functional studies provided important structural constraints that were instrumental in establishing the subunit fold. Based upon the channel's permeability to alkali metal cations, H^+ and water, and its impermeability to urea (Hladky and Haydon, 1972; Myers and Haydon, 1972; Finkelstein, 1974), it was concluded that the peptide backbone surrounds an aqueous pore with luminal diameter ~4 Å. Even among the β-helices, different folding patterns with ~4.4 and

~6.3 residues per turn are possible. The functional results allowed for refinement of the originally proposed β-helical structure (Urry, 1971), eventually leading to a $\beta^{6.3}$-helical structure with ~6.3 residues per helical turn (Urry, 1972). Together with results from gA-dependent conductance relaxations (Bamberg and Läuger, 1973), single-channel experiments on modified gramicidins (Bamberg et al., 1977; Veatch and Stryer, 1977; Cifu et al., 1992) and nuclear magnetic resonance (NMR) spectroscopy on labeled gramicidins (Weinstein et al., 1979, 1980), the constraints imposed by the physiological results established gA channels to be antiparallel SS $\beta^{6.3}$-helical dimers. More details concerning the early structural and functional evidence for this structure are summarized in Andersen and Koeppe (1992) and Andersen et al. (1999).

The SS $\beta^{6.3}$-helical channel structure is remarkably resilient to a large variety of single amino acid substitutions in the gA sequence (Mazet et al., 1984; Russell et al., 1986; Durkin et al., 1990; Becker et al., 1991; Fonseca et al., 1992; Koeppe et al., 1994a; Mattice et al., 1995; Koeppe and Andersen, 1996; Killian et al., 1999), but see also Durkin et al. (1992) and Andersen et al. (1996). This structural resilience makes gA channels valuable as prototypical channels.

2.3.2 Atomic Resolution Structure

The atomic resolution structure was determined by independent solution and solid-state NMR experiments (Arseniev et al., 1986; Ketchem et al., 1997; Townsley et al., 2001). Though the solution NMR studies were done on gA incorporated into sodium dodecylsulfate micelles (SDS) and the solid-state NMR experiments were done on gA incorporated into oriented dimyristoylphosphatidylcholine (DMPC) bilayers, the results have converged to agreement upon one channel structure (Fig. 2.3). The channel is an antiparallel formyl-NH-terminal-to-formyl-NH-terminal dimer (Cross et al., 1999) formed by right-handed (RH), SS $\beta^{6.3}$-helical subunits, which are joined by six intermolecular hydrogen bonds (Fig. 2.4). Minor differences between the reported Protein Data Bank (PDB) structures, based on solid-state and solution NMR (PDB:1MAG and PDB:1JNO, respectively), can be reconciled through molecular dynamics analysis of the structures (Allen et al., 2003) and by considerations of degenerate Trp side-chain conformations that satisfy the solid-state NMR data (Koeppe et al., 1994b; Hu and Cross, 1995).

The intermolecular hydrogen bonds at the subunit interface that stabilize the bilayer-spanning dimer in RH $\beta^{6.3}$-helices are formed between the L-residues in the two subunits (Fig. 2.4); they are topologically equivalent to those of antiparallel β-sheets. By contrast, the intramolecular hydrogen bonds that define the subunit fold are topologically equivalent to those of parallel β-sheets (Urry, 1971). Apart from the helix being right-handed, rather than left-handed, the structure is remarkably similar to the one proposed by D.W. Urry more than 30 years ago (Urry, 1971, 1972).

An atomic resolution structure is available only for the gA dimer, not for the nonconducting monomer. Small-angle x-ray scattering experiments on *t*-BOC-gA (He et al., 1994), with the formyl group replaced by a *tert*-butyloxycarbonyl moiety,

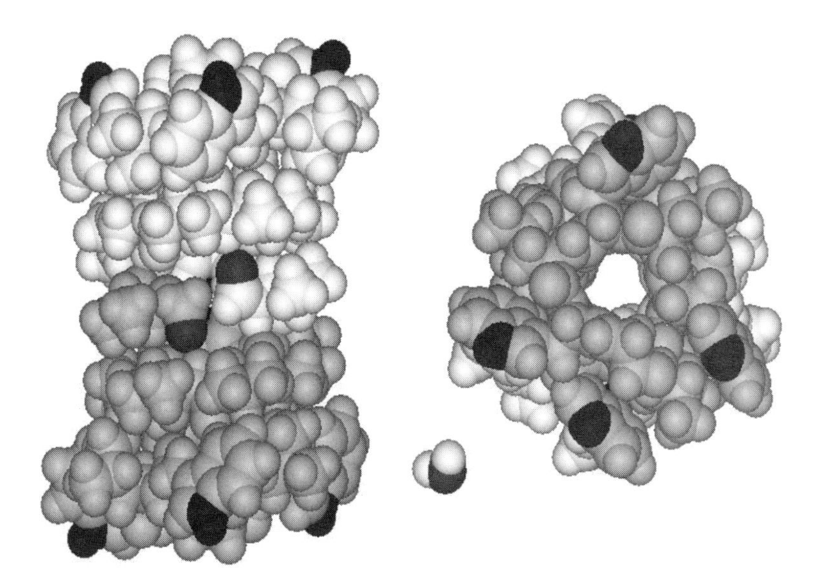

Fig. 2.3 Side and end views of the bilayer-spanning gramicidin A channel. The energy-minimized structure represents a composite that is consistent with the several NMR-determined structures (Arseniev et al., 1986; Ketchem et al., 1997; Townsley et al., 2001). The two subunits are shaded differently. The formyl oxygens and indole NH's are black. The four Trp indole rings cluster near each membrane/solution interface. (A water molecule is shown for comparison.)

which destabilizes the ion-conducting dimer by some five orders of magnitude relative to gA, show that bilayer-incorporated monomers also fold into $\beta^{6.3}$-helices. These monomers are imbedded in the bilayer with their (average) axis orientation parallel to the bilayer normal.

Even though the gA channel weighs in at only ~4 kDa, the results in Figs. 2.1–2.3 show that gA channels have the structural and functional definitions expected for more complex membrane proteins. These combined features continue to put gA channels in a class of their own. In fact, the current transitions in Fig. 2.1 show less "excess noise" than the current transitions observed in channels formed by integral membrane proteins, e.g., KcsA channels (Nimigean and Miller, 2002) or Ca^{2+}-activated potassium channels (Park et al., 2003). With gramicidin channels, one therefore can with some certainty relate the measured function to a specific, albeit dynamic (Allen et al., 2003), structure. Gramicidin channels should be considered to be miniproteins!

2.3.3 Importance of the Tryptophans for Channel Structure

The four Trp residues in each subunit are important determinants of the channel fold because the indole NH groups seek to form hydrogen bonds to polar residues at the bilayer/solution interface (O'Connell et al., 1990), which would favor SS over DS conformers (Durkin et al., 1992). The propensity for Trp to avoid the bilayer center

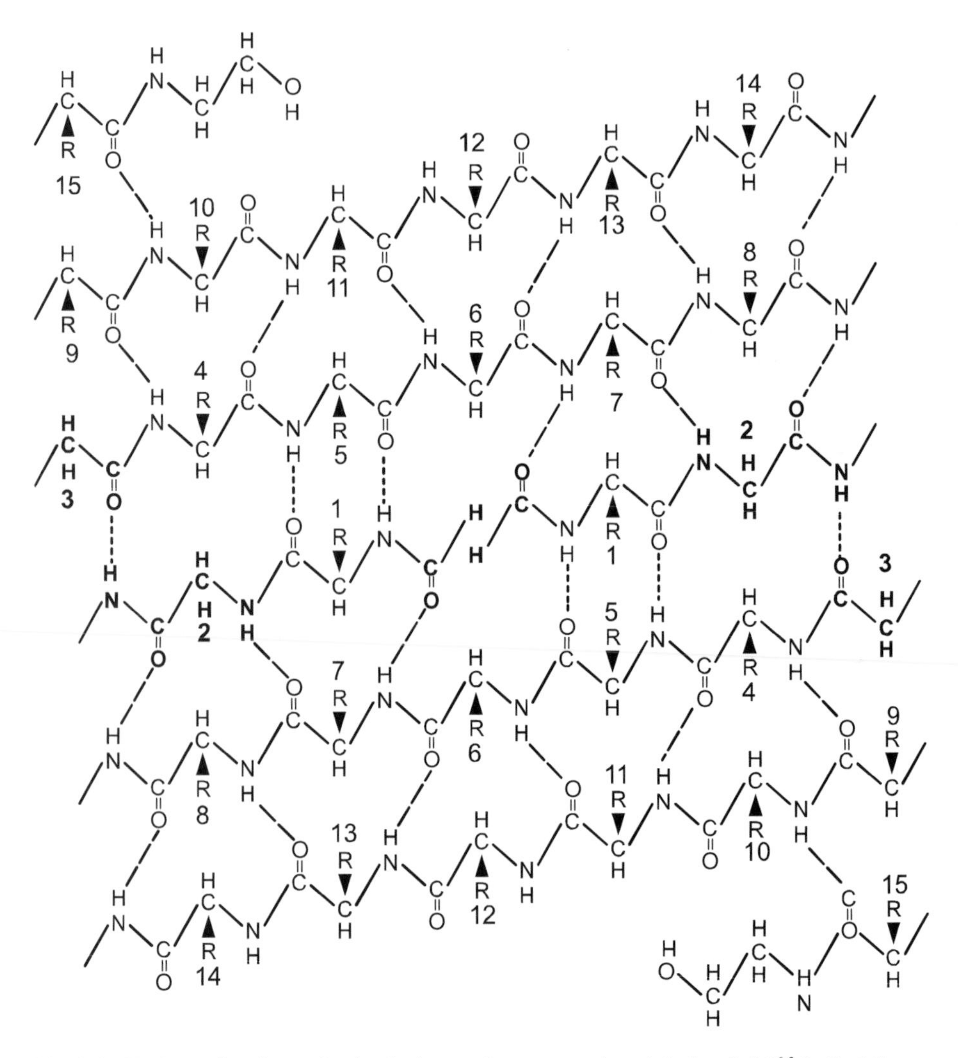

Fig. 2.4 Hydrogen bond organization in the membrane-spanning, right-handed $\beta^{6.3}$-helical dimer. The intramolecular hydrogen bonds that stabilize each $\beta^{6.3}$-helical subunit are denoted by – –. The six intermolecular hydrogen bonds that stabilize the subunit interface are denoted by - - -. The residues in each subunit are numbered; residues 2 and 3, which were modified in [D-Ala^{2},Ser^{3}]gA are in **bold**.

is further highlighted by the very short lifetimes of channels formed by gramicidins that have N-formyl-Trp[1] instead of N-formyl-Val[1] (Mazet et al., 1984). Indeed, the conformational plethora observed in organic solutes shows that the intrinsic gA folding preference is to form intertwined DS structures in which the Trp residues are distributed along the dimer surface (Fig. 2.5).

In a lipid bilayer or bilayer-like environment, nevertheless, the folding preference becomes dramatically altered, driven by the energetic necessity to rearrange

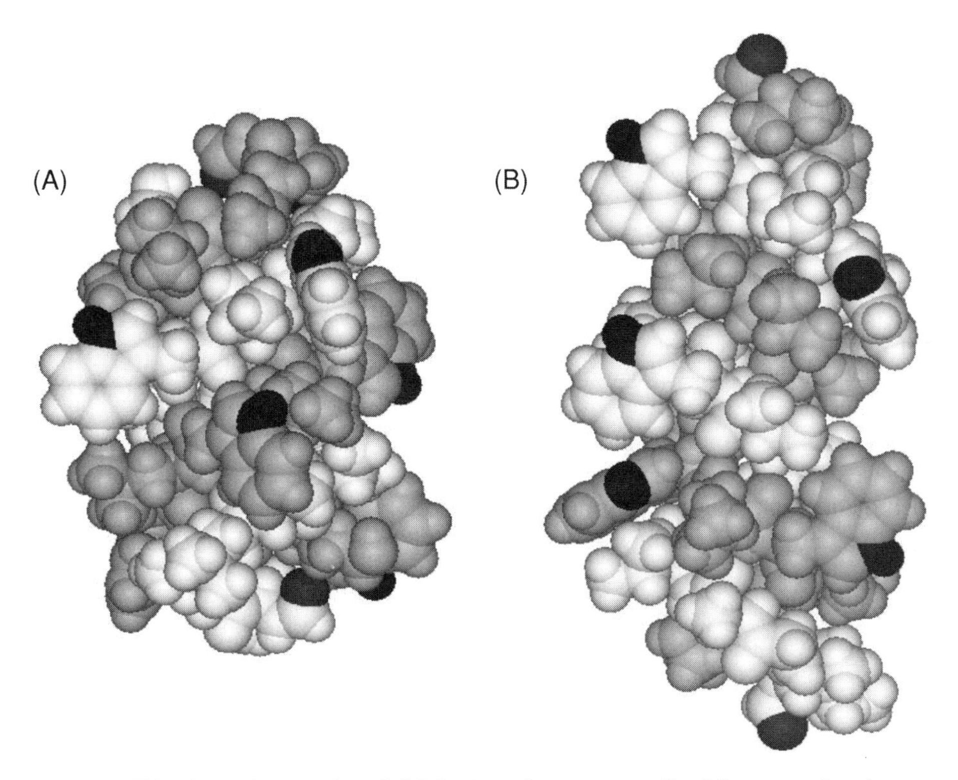

Fig. 2.5 Side views of two antiparallel DS gA conformers crystallized from organic solvents. (A) The $\overrightarrow{\pi}\overleftarrow{\pi}^{7.2}$ structure, with 7.2 residues per helical turn, crystallized from CsCl/methanol or from acetic acid, PDB:1AV2 (Burkhart et al., 1998). (B) The $\overrightarrow{\pi}\overleftarrow{\pi}^{5.6}$structure, with 5.6 residues per helical turn, crystallized from ethanol, PDB:1ALZ (Langs, 1988). As in the $\beta^{6.3}$-helical structure in Fig. 2.3, the two subunits are shaded differently, and the formyl oxygens and indole NH's are black. In contrast to Fig. 2.3, the Trp indole rings in the DS structures do *not* cluster near the ends (the membrane/solution interfaces), but are distributed quite evenly along the dimer.

the Trp indole rings (cf. Figs. 2.3 and 2.5). When the DS conformers encounter lipid bilayer membranes, they unfold/refold into the Trp-anchored SS structure in Fig. 2.3 (Andersen et al., 1999). In electrophysiological single-channel experiments, the conducting channels form by the transbilayer dimerization of two nonconducting ($\beta^{6.3}$-helical) subunits (O'Connell et al., 1990)—as opposed to a direct interconversion of nonconducting DS dimers to conducting SS dimers. In this $\beta^{6.3}$-helical conformation, the seemingly very hydrophobic (Segrest and Feldmann, 1974) gA molecules cross lipid bilayers poorly (O'Connell et al., 1990)—providing further evidence for the importance of the hydrogen bond-stabilized anchoring to the bilayer/solution interface.

The preference to fold into SS channels in lipid bilayers, and in bilayer-like environments, arises from the energetic penalty associated with burying the Trp indole rings in the bilayer hydrophobic core, as observed initially with [Trp^{1}]gA (Mazet et al., 1984). Furthermore, gA analogues with multiple Trp→Phe substitutions form

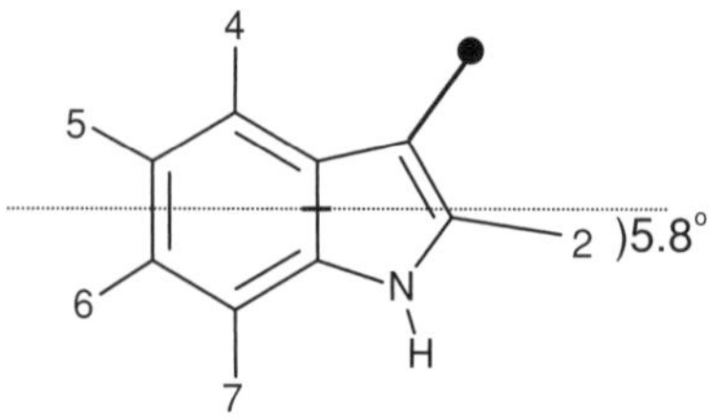

Fig. 2.6 Trp indole ring geometry with the positions of deuteration numbered; after Pulay et al. (2005). The indole ring geometry was determined from ab initio calculations and experimental ^{2}H-NMR spectra (Koeppe et al., 2003). A critical geometric feature is the 5.8° angle of the C2–^{2}H bond with respect to the normal to the ring bridge.

a variety of DS structures in lipid bilayers (Salom et al., 1995, 1998; Rawat et al., 2004). These DS conformers generally are nonconducting, although some specialized combinations of Trp→Phe substitutions together with changes in backbone length and side-chain stereochemistry can lead to the formation of DS conducting channels (Durkin et al., 1992).

The interfacial localization of Trp, and Tyr, in gA channels seems to be a general characteristic of membrane proteins (Schiffer et al., 1992; Cowan and Rosenbusch, 1994; Killian and von Heijne, 2000). The organization of Trp and Tyr residues in integral membrane proteins presumably is due to the same energetic principles that have been deduced from the detailed analysis of gA channels.

The characterization of the orientations and dynamics of the Trp indole rings has been facilitated by solid-state ^{2}H-NMR spectra from gramicidins with labeled tryptophans, which have been incorporated into oriented multilayers (Cross, 1994; Davis and Auger, 1999). In addition to providing information about the dynamics of the gramicidin channels, these studies also have led to an improved understanding of the indole ring geometry—specifically the C2–H bond angle (Koeppe et al., 2003). Ab initio calculations, together with analysis of experimentally observed quadrupolar splittings from ^{2}H-NMR spectra, have converged to reveal that the indole C2–H bond makes an angle of 5.8° with respect to the normal to the bridge between the 5- and 6-membered rings (Fig. 2.6) (Koeppe et al., 2003).

Further calculations have established all of the tensor elements of the electric field gradient for each carbon–deuterium bond in the ring of deuterated 3-methylindole (Pulay et al., 2005). The off-bond tensor elements permit one to calculate an asymmetry parameter $\eta = (|V_{yy}| - |V_{xx}|)/|V_{zz}|$ for each position on the indole ring. These asymmetry parameters in turn allow for improved descriptions of the average orientation and dynamics of each of the four Trp indole rings that anchor gA channels in their transmembrane orientation (Fig. 2.3), which in turn allows for detailed descriptions of not only the average orientations but also the dynamics of backbone and side chains (Pulay et al., 2005). These developments could be implemented because of the unique advantages conferred by gA, and they have implications for understanding the average orientation, dynamics, and functional role(s) of Trp residues in integral membrane proteins.

At the NMR time scale, the bilayer-spanning gramicidin channels have their pore axes remarkably parallel to the bilayer normal (Nicholson et al., 1987; Cornell et al., 1988; Killian et al., 1992). From solid-state NMR spectra, the average wobble is zero and the principal order parameter is 0.93 (Separovic et al., 1993). The outer pair of Trp residues (#13 and #15) have principal order parameters identical to the backbone order parameter (Pulay et al., 2005). The inner pair of Trp residues (#9 and #11) wobble only slightly more than the backbone (Pulay et al., 2005). Additionally, there may be rapid transitions of Trp^9 between different "conventional" rotamer positions (Allen et al., 2003). The four Val side chains likewise occupy canonical rotamer positions, with two rigid and two hopping (Lee et al., 1995). Two rotamers have been modeled for Trp^9 in lipid bilayers (Koeppe et al., 1994b; Ketchem et al., 1997), one of which is similar to the dominant rotamer in SDS micelles (Arseniev et al., 1986; Townsley et al., 2001). As shown by MD analysis of the Ketchem et al. and the Townsley et al. structures (Allen et al., 2003), both Trp^9 rotamers are compatible with a variety of solid-state NMR measurements. A weighted average, with rapid interconversion between the rotamers, is compatible with not only MD simulations but with fluorescence results that suggest that two of the Trps quench each other (Scarlata, 1988; Mukherjee and Chattopadhyay, 1994).

2.3.4 Importance of the Bilayer-Channel Hydrophobic Match

The shift in conformational preference between the polymorphic behavior in organic solvents and the unique structure in lipid bilayers is but one example of environment-dependent folding. Because the conducting channels form by the transbilayer dimerization of two nonconducting subunits (Fig. 2.7), the preference to fold into SS $\beta^{6.3}$-helical dimers, for example, depends on the hydrophobic "match" between bilayer thickness and channel length (Greathouse et al., 1994; Mobashery et al., 1997; Galbraith and Wallace, 1998).

Specifically, if the length of the channel's hydrophobic exterior (l) is much greater (Greathouse et al., 1994) or much less (Mobashery et al., 1997; Galbraith and Wallace, 1998) than the average hydrophobic thickness of the unperturbed host bilayer (d_0), then gA may fold into structures other than the standard SS $\beta^{6.3}$-helical dimers.

This bilayer thickness-dependent folding preference arises because: first, the energetic cost associated with exposing hydrophobic residues to water (Kauzmann, 1957; Tanford, 1980; Engelman et al., 1986; Dill, 1990; Sharp et al., 1991; White and Wimley, 1999) is substantial, varying between 2.5 and 7.5 kcal mol^{-1} nm^{-2}) (Dill et al., 2005); and second, lipid bilayers are not just thin sheets of liquid hydrocarbon, stabilized by the polar head groups, but liquid crystals with well-defined material properties. Fully hydrated, liquid-crystalline phospholipid bilayers are elastic bodies with volumetric compressibility moduli that are of the order of $\sim 10^9$ N m^{-2} (Liu and Kay, 1977; Tosh and Collings, 1986)—one to two orders of magnitude less than the moduli for globular proteins in water (Gekko and Noguchi, 1979). Because lipid

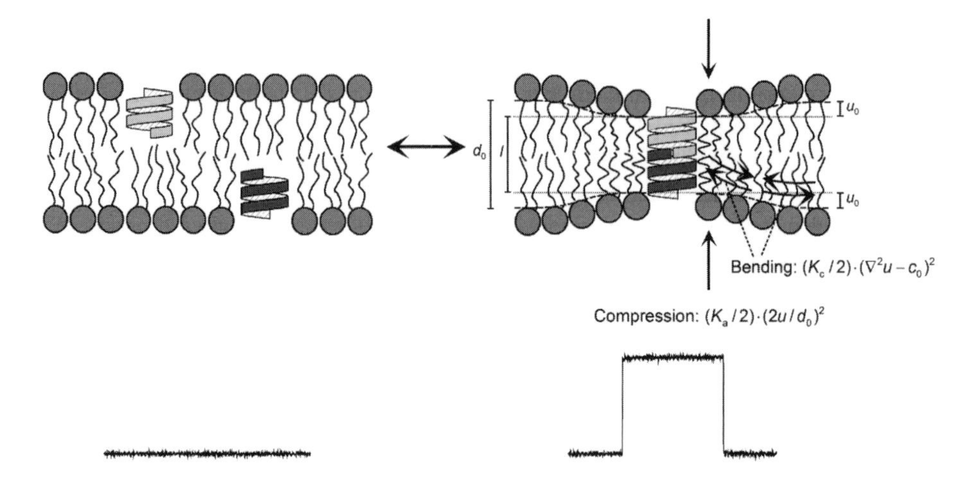

Fig. 2.7 gA channels form by the transbilayer dimerization of two RH, SS $\beta^{6.3}$-helical subunits. Top: schematic representation of the kinetics of channel appearance/disappearance (association/dissociation). Bottom: the associated current signals. When the channel's hydrophobic length (l) differs from the average thickness of the unperturbed bilayer hydrophobic core (d_0), channel formation will be associated with a bilayer deformation, which can be decomposed into the compression (with energy density $(K_a/2)\cdot(2u/d_0)^2$) and bending (with energy density $(K_c/2)\cdot(\nabla^2 u - c_0)^2$) of the two bilayer leaflets (Huang, 1986), where K_a and K_c denote the bilayer compression and bending moduli and $2u$ the local bilayer deformation.

bilayers are more flexible than the imbedded proteins (see below), the host lipid bilayer will deform whenever the hydrophobic length of a bilayer-spanning protein differs from the average thickness of the unperturbed bilayer (Owicki et al., 1978; Mouritsen and Bloom, 1984), cf. Fig. 2.7.

In principle, both the bilayer and the protein will deform in response to a hydrophobic mismatch, $d_0 \neq l$. In practice, the proteins can be approximated as being rigid bodies because lipid bilayers are very soft materials. The area expansion moduli (K_a) for liquid-crystalline phospholipid bilayers are $\sim$250 mN m^{-1} (Rawicz et al., 2000), and the bilayer hydrophobic thickness is $\sim$3 nm (Lewis and Engelman, 1983; Simon and McIntosh, 1986; Rawicz et al., 2000) meaning that the moduli for bilayer thickness compressibility, which is given by K_a/d_0 is $\sim 10^{-8}$ N m^{-2}, cf. Evans and Hochmuth (1978). That is, lipid bilayers are 100- to 1000-fold softer than the imbedded proteins. Because lipid bilayers are so soft, the bilayer/solution interface undergoes substantial thermal fluctuations (Wiener and White, 1992), which involve the local movement of individual phospholipid molecules, as well as more global bilayer undulations and peristaltic motions (Wiener and White, 1992; Lindahl and Edholm, 2000). The bilayer/solution interface thus is "fuzzy"; but the *average* bilayer thickness, which is relevant here, is a well-defined parameter.

The bilayer thickness-dependent folding arises because a bilayer deformation has an associated energetic cost, the bilayer deformation energy (ΔG^0_{def}) (Huang, 1986), which contributes to the overall free energy change (ΔG^0_{tot}) associated with the gA monomer↔dimer equilibrium (Fig. 2.7):

$$\mathsf{M} + \mathsf{M} \underset{k_{-1}}{\overset{k_1}{\rightleftharpoons}} \mathsf{D}$$

Scheme I

where M and D denote the nonconducting monomers (one in each bilayer leaflet) and conducting dimers, respectively. The dimerization constant (K_D) is given by

$$K_{\mathrm{D}} = \frac{k_1}{k_{-1}} = \frac{[\mathsf{D}]}{[\mathsf{M}]^2} = \exp\left\{\frac{-\Delta G^0_{\mathrm{tot}}}{k_{\mathrm{B}}T}\right\} = \exp\left\{\frac{-\left(\Delta G^0_{\mathrm{prot}} + \Delta\Delta G^{\mathrm{M}\rightarrow\mathrm{D}}_{\mathrm{def}}\right)}{k_{\mathrm{B}}T}\right\}, \quad (2.1)$$

where $\Delta G^0_{\mathrm{prot}}$ denotes the free energy contributions that arise from the channel (protein) per se, k_{B} Boltzmann's constant, T the temperature in Kelvin, and [M] and [D] the mole-fractions of gA monomers and dimers in the bilayer. The $\Delta\Delta G^{\mathrm{M}\rightarrow\mathrm{D}}_{\mathrm{def}}$ contribution to $\Delta G^0_{\mathrm{tot}}$ arises from the cost of the bilayer compression and monolayer bending that occur when the bilayer-spanning dimer forms (Huang, 1986; Nielsen and Andersen, 2000), cf. Fig. 2.7. For any conformation of a bilayer-spanning protein the standard free energy of the protein-induced bilayer deformation ($\Delta G^0_{\mathrm{def}}$) is given by

$$\Delta G^0_{\mathrm{def}} = \int_{r_0}^{\infty} \left\{K_{\mathrm{a}} \cdot (2u/d_0)^2 + K_{\mathrm{c}} \cdot (\nabla^2 u - c_0)^2\right\} \cdot \pi \cdot dr - \int_{r_0}^{\infty} K_{\mathrm{c}} \cdot (\nabla^2 u - c_0)^2 \cdot \pi \cdot dr, \quad (2.2)$$

where r_0 is the channel radius, K_{a} and K_{c} the bilayer compression and bending moduli, $2u$ the local channel-bilayer mismatch ($u = (d - l)/2$, where d is the local bilayer thickness, is the local monolayer deformation) and c_0 the intrinsic (or spontaneous) monolayer curvature (Gruner, 1985). Eq. 2.2 can be expressed as (Nielsen et al., 1998; Nielsen and Andersen, 2000; Lundbæk et al., 2004):

$$\Delta G^0_{\mathrm{def}} = H_{\mathrm{B}} \cdot (d_0 - l)^2 + H_{\mathrm{X}} \cdot (d_0 - l) \cdot c_0 + H_{\mathrm{C}} \cdot c_0^2, \quad (2.3)$$

where the coefficients H_{B}, H_{X}, and H_{C} are functions of K_{a}, K_{c}, d_0 and r_0 (Nielsen et al., 1998). The thickness-dependent conformational preference thus arises because $\Delta G^0_{\mathrm{def}}$ varies as a function of $(d_0 - l)^2$. As $|d_0 - l|$ increases, the energetic penalty for inserting a bilayer-spanning SS dimer increases, eventually becoming so large that other conformers, including various DS structures, become favored (Greathouse et al., 1994; Mobashery et al., 1997; Galbraith and Wallace, 1998).

Given this context, the RH, SS $\beta^{6.3}$-helical (dimer) conformation is remarkably robust: it is preserved in bilayers with acyl chain lengths varying between C_{10}

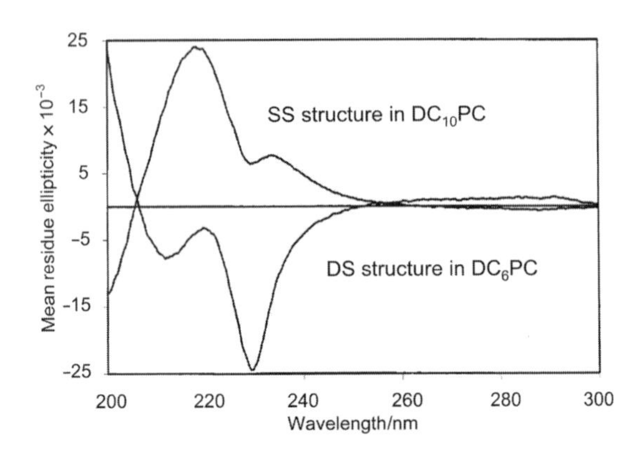

Fig. 2.8 CD spectroscopic "signatures" for the channel conformation of gA in Di-C_{10}-PC bilayers, and for the nonchannel (intertwined, DS) conformation in Di-C_6-PC micelles. Adapted from Greathouse et al. (1994).

and $C_{20:1}$ (Cornell et al., 1989; Greathouse et al., 1994; Mobashery et al., 1997; Galbraith and Wallace, 1998), and even in sodium dodecyl sulfate (SDS) micelles (Arseniev et al., 1986; Townsley et al., 2001; Allen et al., 2003). Fig. 2.8 shows the characteristic circular dichroism (CD) signatures for gA incorporated into di-C_{10}-phosphatidylcholine ($DC_{10}PC$), where it is in the RH, SS $\beta^{6.3}$-helical channel conformation, and into DC_6PC, where it is in a DS conformation.

2.3.5 Structural Equivalence of gA Mutants

Not only is the channel structure remarkably stable and well-defined, but the basic fold and peptide backbone organization also do not vary when the amino acid sequence is varied (as long as the alternating L-D-sequence, and the aromatic–aliphatic organization in the carboxy-ethanolamide half of the sequence, are maintained).

To test whether a mutant gramicidin, e.g. [D-Ala^2,Ser^3]gA, forms channels that are structurally equivalent to the native gA channels, one can exploit the following features (Durkin et al., 1990): first, gA channels are symmetric, antiparallel dimers formed by the transmembrane association of nonconducting, $\beta^{6.3}$-helical subunits residing in each leaflet of the bilayer (Fig. 2.7); and second, gA analogues usually form only a single channel type. Then, if two different gA analogues (A and B) form symmetric, homodimeric (AA and BB) channels that have the same structure, meaning the same peptide backbone fold, one should be able to observe the formation of asymmetric, heterodimeric (AB and BA) channels (Veatch and Stryer, 1977; Durkin et al., 1990, 1993). If the AA and BB channels differ in their current transition amplitudes, the heterodimeric (AB or BA) channels would be expected to have current transition amplitudes in between those of the homodimeric AA and BB channels. In general, if the potential of mean force for ion movement through the heterodimeric channels is asymmetric, which almost invariably is the case, the current in the A→B direction will differ from the current in the B→A direction. (In the limit where the

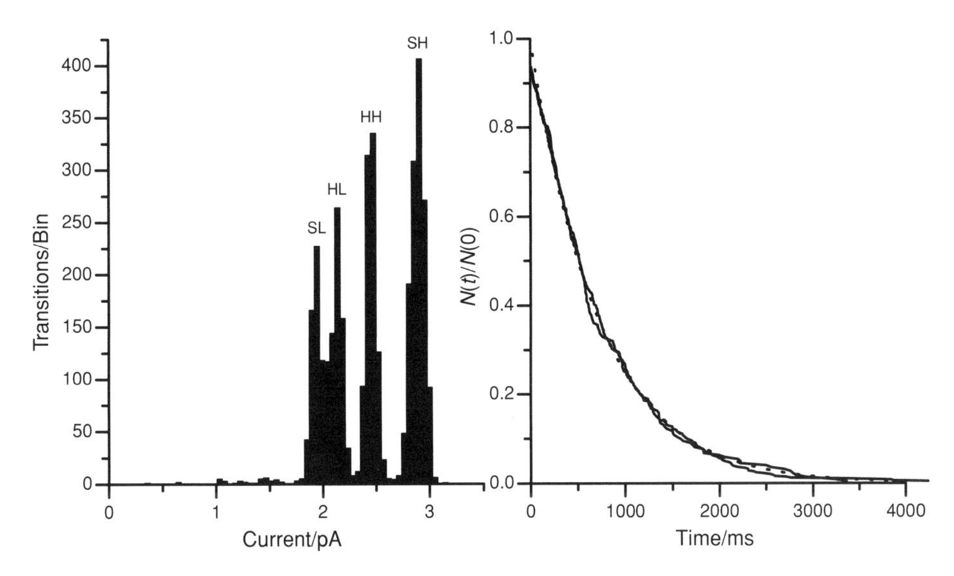

Fig. 2.9 Structural equivalence of gA and [D-Ala^2,Ser^3]gA channels. Left: current transition amplitude histogram obtained when [D-Ala^2,Ser^3]gA and gA both were added to both sides of a lipid bilayer. The four peaks represent (labeled from left to right): SL (symmetric low conductance) denotes [D-Ala^2,Ser^3]gA homodimers (at ~1.9 pA); HL (hybrid low conductance) denotes [D-Ala^2,Ser^3]gA/gA heterodimers (at ~2.2 pA); HH (hybrid high conductance) denotes gA/[D-Ala^2,Ser^3]gA heterodimers (at ~2.5 pA); and SH (symmetric high conductance) denotes gA homodimers (at ~2.9 pA). Right: lifetime distributions for the heterodimeric [D-Ala^2,Ser^3]gA/gA and gA/[D-Ala^2,Ser^3]gA channels. The interrupted curves are fits of $N(t)/N(0) = \exp\{-t/\tau\}$ to the results, where $N(t)$ is the number of channels with lifetime longer than time t, and τ is the average lifetime (770 ms for [D-Ala^2,Ser^3]gA/gA channels and 750 ms for gA/[D-Ala^2,Ser^3]gA channels). Experimental conditions as in Fig. 2.1.

applied voltage (V) $\rightarrow 0$, the current, or more precisely the conductance, will be the same in the two directions.)

Figure 2.9 shows results from an experiment with the wild-type gA and the [D-Ala^2,Ser^3]gA analogue.

In the current transition amplitude histogram (Fig. 2.9A), there are four peaks. In addition to the symmetric gA/gA and [D-Ala^2,Ser^3]gA/[D-Ala^2,Ser^3]gA peaks (at ~2.9 and ~1.9 pA, respectively), which can be identified by comparison to the results with just a single gramicidin (Figs. 2.1 and 2.2), there are two new peaks (at ~2.2 and ~2.5 pA, respectively). These new peaks represent the [D-Ala^2,Ser^3]gA/gA and gA/[D-Ala^2,Ser^3]gA heterodimeric channels, respectively. The identities of the heterodimers were determined by adding [D-Ala^2,Ser^3]gA to only side of a bilayer and gA to only the other side, which defines the orientation of the heterodimeric channels because the gA and gA analogues cross lipid bilayers very poorly (O'Connell et al., 1990; Fonseca et al., 1992).

Formally, heterodimer (hybrid channel) formation can be described as the interconversion between two symmetric homodimeric channel types (AA and BB) and the corresponding asymmetric heterodimeric channel types (AB and BA), cf. Durkin et al. (1990):

$$\mathrm{AA} + \mathrm{BB} \rightleftarrows \mathrm{AB} + \mathrm{BA}.$$

Scheme II

The equilibrium constant (K) for heterodimer formation is given by (Durkin et al., 1993)

$$K = \frac{[\mathsf{AB}] \cdot [\mathsf{BA}]}{[\mathsf{AA}] \cdot [\mathsf{BB}]}, \tag{2.4}$$

and the standard free energy for heterodimer formation ($\Delta\Delta G^0$) is defined as

$$\begin{aligned}\Delta\Delta G^0 &= -k_\mathrm{B}T \cdot \ln\{K\} = -\frac{k_\mathrm{B}T}{2} \cdot \ln\left\{\frac{K_\mathrm{AB} \cdot K_\mathrm{BA}}{K_\mathrm{AA} \cdot K_\mathrm{BB}}\right\} \\ &= \frac{(\Delta G^0_\mathrm{AB} + \Delta G^0_\mathrm{BA}) - (\Delta G^0_\mathrm{AA} + \Delta G^0_\mathrm{BB})}{2},\end{aligned} \tag{2.5}$$

where K_AB and ΔG^0_AB denote the association constant and standard free energy of formation for AB (the reaction $\mathsf{A} + \mathsf{B} \rightleftarrows \mathsf{AB}$), and so on. (The factor 1/2 arises because we wish to measure $\Delta\Delta G^0$ per mole of heterodimer, or subunit interface, cf. Durkin et al., 1990, 1993.)

The time-averaged "concentration" (channels/bilayer area) of AB is given by:

$$[\mathsf{AB}] = f_\mathsf{AB} \cdot \tau_\mathsf{AB}, \tag{2.6}$$

where $f_\mathsf{AB} (= k_\mathsf{AB} \cdot [\mathsf{A}] \cdot [\mathsf{B}])$ and $\tau_\mathsf{AB} (= 1/k_{-\mathsf{AB}})$ denote the channel appearance rate and lifetime, and k_AB and $k_{-\mathsf{AB}}$ are the association and dissociation rate constants. $\Delta\Delta G^0$ thus can be expressed in terms of experimental observables as

$$\Delta\Delta G^0 = -\frac{k_\mathrm{B}T}{2} \cdot \ln\left\{\frac{(f_\mathsf{AB} \cdot \tau_\mathsf{AB}) \cdot (f_\mathsf{BA} \cdot \tau_\mathsf{BA})}{(f_\mathsf{AA} \cdot \tau_\mathsf{AA}) \cdot (f_\mathsf{BB} \cdot \tau_\mathsf{BB})}\right\}. \tag{2.7}$$

The activation energy for heterodimer formation relative to the symmetric channels ($\Delta\Delta G^\ddagger_\mathrm{f}$) can be defined (and determined) by a reasoning that parallels the above:

$$\Delta\Delta G^\ddagger_\mathrm{f} = -\frac{k_\mathrm{B}T}{2} \cdot \ln\left\{\frac{f_\mathsf{AB} \cdot f_\mathsf{BA}}{f_\mathsf{AA} \cdot f_\mathsf{BB}}\right\}. \tag{2.8}$$

If $\Delta\Delta G^\ddagger_\mathrm{f} = 0$, there are no subunit-specific barriers to the formation of heterodimers, as compared to homodimers (Durkin et al., 1990, 1993), meaning that the different subunits have the same fold. In this case, the distribution between the AA, BB, AB, and BA channels will be given by (Durkin et al., 1993):

$$f_{\mathrm{AB}} \cdot f_{\mathrm{BA}} = f_{\mathrm{AA}} \cdot f_{\mathrm{BB}}. \tag{2.9}$$

If also $\Delta\Delta G^0 = 0$, there are no subunit-specific interactions between the different subunits in the bilayer-spanning dimers, meaning that AA and BB are structurally equivalent (Durkin et al., 1990, 1993).

The relative heterodimer appearance rates in Fig. 2.9 conform to the predictions of Eq. 2.9, as $f_{\mathrm{AB}} \cdot f_{\mathrm{BA}}/(f_{\mathrm{AA}} \cdot f_{\mathrm{BB}}) = 0.96$ (or $\Delta\Delta G_{\mathrm{f}}^{\ddagger} \approx 0$ kcal mol^{-1}). We therefore conclude that the Ala→Ser (and Gly→D-Ala) substitutions are well tolerated within the $\beta^{6.3}$-helical fold. Similar results have been obtained with many other gA mutants (Mazet et al., 1984; Durkin et al., 1990; Becker et al., 1991; Fonseca et al., 1992; Durkin et al., 1993; Jude et al., 1999), and the approach has been verified by solution NMR (Mattice et al., 1995; Sham et al., 2003).

Even though $\Delta\Delta G_{\mathrm{f}}^{\ddagger} \approx 0$ kcal mol^{-1}, $\Delta\Delta G^0$ may be different from 0 kcal mol^{-1}. If $\Delta\Delta G^0$ is positive, there may be a strain at the subunit interface (Durkin et al., 1990, 1993); to relieve this strain the heterodimeric channels may switch between different conductance states (Durkin et al., 1993), and may even exhibit voltage-dependent transitions between closed and open channel states (Oiki et al., 1994, 1995).

In the case of gA and [D-Ala^2,Ser^3]gA, $\Delta\Delta G^0$ is positive ($\sim$0.25 kcal mol^{-1}) because the heterodimeric channel lifetimes (Fig. 2.9) are *less* than would be predicted from the lifetimes of the homodimeric channels (Fig. 2.2) in the absence of subunit-specific interactions: $\tau_{\mathrm{AB}} \cdot \tau_{\mathrm{BA}}/(\tau_{\mathrm{AA}} \cdot \tau_{\mathrm{BB}}) = 1$. The modest (relative) destabilization of the heterodimeric channels can be understood by examination of the gA channel structure. Because residue 3 in subunit A is in close proximity to residue 3 in subunit B (Fig. 2.10), the homodimeric [D-Ala^2,Ser^3]gA channels may be stabilized by hydrogen bond formation between the subunits—a bond that cannot be formed in the heterodimeric channels. The well-defined gA channel structure thus allows for quantitative studies on the interaction between amino acid residues at the channel/bilayer interface.

2.4 Channel Function

gA channels are water-filled pores that span lipid bilayers to catalyze selective ion movement across the membrane. The small pore radius restricts ions and water to move in single file through the pore (Levitt et al., 1978; Rosenberg and Finkelstein, 1978); but the rate of ion movement through the channels is large enough to allow for single-channel measurements to be done over a wide range of permeant ion concentrations and potentials (Hladky and Haydon, 1972; Neher et al., 1978; Andersen, 1983a; Cole et al., 2002). Because gA channels are seemingly ideally selective for monovalent cations (Myers and Haydon, 1972), single-channel measurements

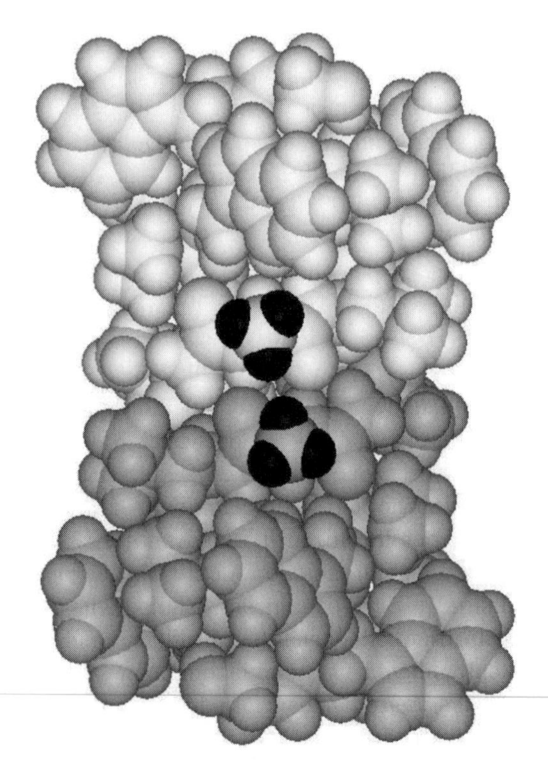

Fig. 2.10 Side view of the bilayer-spanning gramicidin A channel. The energy-minimized structure is the same as in Fig. 2.3, but turned 180° along the channel axis. The black atoms are the Ala^3 side-chain hydrogens on the respective subunits.

provide direct information about the net cation flux through the channel, meaning that it is possible to elucidate the kinetics of ion movement through the channel from current–concentration–voltage studies (see below). It is in this context important that: first, the $\beta^{6.3}$-helical channel conformation does not vary as function of the permeant ion type or concentration (Wallace et al., 1981; Katsaras et al., 1992; Tian and Cross, 1999); and second, the current through gA channels usually has little "excess" noise (cf. Fig. 2.1, where the SD of the current noise does not vary when a channel appears/disappears). These features should ensure that one indeed can relate function to the molecular structure and dynamics—and to use gA channels to critically evaluate computational strategies for understanding ion channel and, more generally, membrane protein function.

Before discussing the channels' permeability properties in detail, it is instructive to compare the measured conductances to the predictions of the simplest model of a water-filled pore, e.g., Hille (2001):

$$g_{\mathrm{pred}} \approx \lambda^{\mathrm{o}} \cdot C \cdot \frac{\pi \cdot r_{\mathrm{p}}^{2}}{l_{\mathrm{p}}}, \tag{2.10a}$$

where g_{pred} is the predicted single-channel conductance, λ^o the limiting equivalent conductivity, C the permeant ion concentration in the bulk aqueous phase, r_p the pore radius, and l_p the pore length. In this estimate, we neglect the access resistance to the channel (Hille, 1968; Läuger, 1976; Andersen, 1983c), which becomes important at low permeant ion concentrations (Andersen, 1983c).

For Na^+ permeation through gA channels ($\lambda^o_{Na} = 50.1$ S cm^2 mol^{-1} (Robinson and Stokes, 1959), 1.0 M NaCl, $r_p = 2$ Å and $l_p = 25$ Å), $g_{pred} \approx 250$ pS according to Eq. 2.10. Eq. 2.10 provides an overestimate, however; as noted by Ferry (1936), the relevant radius is not the geometric pore radius, but the difference between the pore and ion radius (r_i), the so-called capture radius (r_o), because the ion centers are constrained to move within a relatively narrow column of radius $r_p - r_i$. Incorporating this geometric constraint, the "corrected" g_{pred} (g^{corr}_{pred}) becomes

$$g^{corr}_{pred} \approx \lambda^o \cdot C \cdot \frac{\pi \cdot (r_p - r_i)^2}{l_p}, \qquad (2.10b)$$

and g^{corr}_{pred} (in 1.0 M NaCl) becomes ~70 pS ($r_{Na} \approx 0.95$ Å (Hille, 1975)). Because g^{corr}_{pred} is only fivefold higher than the measured conductance (Fig. 2.2), the channel cannot impose a major barrier for ion movement across the bilayer—a surprising result, given the significant electrostatic barrier associated with traversing the low dielectric constant bilayer hydrophobic core (Parsegian, 1969; Levitt, 1978). Even more surprising, in 1.0 M CsCl ($r_{Cs} \approx 1.63$ Å), $g^{corr}_{pred} \approx 13$ pS, which is almost fourfold less than the measured conductance of 50 pS. The channel has a *higher* conductance than the equivalent column of water!

That is, though gA channels are water-filled pores, they are not just water-filled pores. Even in the absence of a kinetic analysis of ion movement, one may conclude that favorable, short-range ion-channel interactions effectively compensate for the electrostatic barrier for ion movement through the low-dielectric bilayer core—such that the overall energetic barriers for ion movement are small (only a few $k_B T$).

2.4.1 Cation Binding in Gramicidin Channels

The low-dielectric bilayer hydrophobic core imposes a large electrostatic barrier for the transmembrane movement of alkali metal cations (Neumcke and Läuger, 1969; Parsegian, 1969).[1] Ion channels, and other membrane proteins, lower this barrier by

[1] The magnitude of this barrier can be estimated from the Born approximation for the free energy of transfer (ΔG^0_{trans}) for transferring an ion of valence z and radius r_i from a bulk phase of dielectric constant ε_1 to another bulk phase of dielectric constant ε_2 (Finkelstein and Cass, 1968; Parsegian, 1969):

$$\Delta G^0_{trans} \approx \frac{(ze_0)^2}{8 \cdot \pi \cdot r_i \cdot \varepsilon_0} \cdot \left(\frac{1}{\varepsilon_2} - \frac{1}{\varepsilon_1}\right),$$

where e_0 is the elementary charge and ε_0 is the permittivity of free space. The Born expression overestimates, cf. (Bockris and Reddy, 1970, Ch. 2), and provides only a rough estimate of the energies

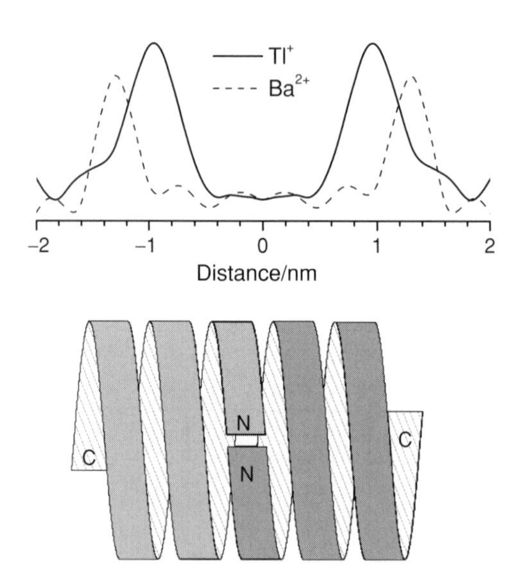

Fig. 2.11 Position of ion binding sites in gramicidin channels. Top: electron density profiles for Tl^+, with maxima at 9.6 ± 0.3 Å from the channel center (the maximum for K^+ is at the same position), and Ba^{2+}, with maxima at 13.0 ± 0.2 Å from the channel center. Redrawn after Olah et al. (1991). Bottom: schematic representation of the RH, SS $\beta^{6.3}$-helical dimer drawn on the same scale as the top panel.

providing a local environment with a higher dielectric constant (Parsegian, 1969; Levitt, 1978; Jordan, 1981), but the electrostatic barrier is still $\sim 10\ k_B T$ (Jordan, 1981)—far too high to be compatible with the measured conductances. Efficient ion movement through gA channels depend on ion solvation by the pore-lining residues and the single file of pore water (Andersen and Procopio, 1980; Mackay et al., 1984; Allen et al., 2004a). Indeed, the free energies of transfer of small monovalent cations from water to formamide or dimethylformamide are negative, whereas they are positive for small monovalent anions (Cox et al., 1974). Not surprisingly, therefore, monovalent alkali metal cations bind with surprisingly high affinity to the channel and, even though there are no residues that form specific binding sites, the ions tend to be localized to delimited regions, or "binding sites." Fig. 2.11 shows the electron density profiles for Tl^+ and Ba^{2+} based on x-ray scattering experiments on gA incorporated into oriented DMPC multilayers (Olah et al., 1991).

For both Ba^{2+} and Tl^+, there are two major binding sites, with no significant ion occupancy outside these two sites. As would be expected from the Born expression (Footnote 2), the Ba^{2+} sites are located furthest away from the channel center, at the channel entrance; the Tl^+ sites are 3 Å deeper within the pore—within the

involved, $\sim 70\ k_B T$ for moving a monovalent ion of radius 1 Å from water ($\varepsilon = 80$) to a bulk hydrocarbon ($\varepsilon \approx 2$).

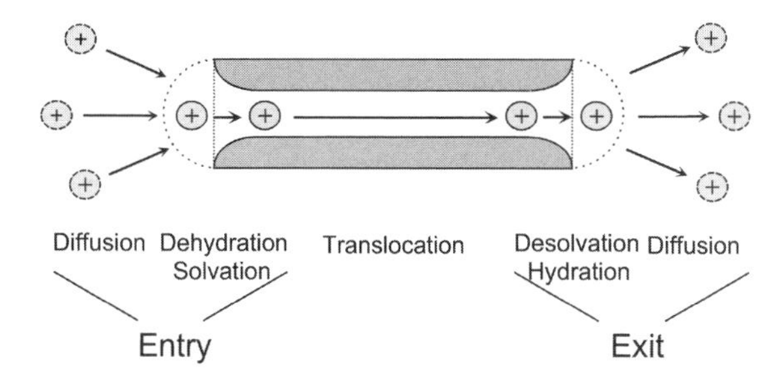

Fig. 2.12 The kinetic steps involved in ion movement through gA channels.

single-filing region. (The mono- and divalent cation binding sites also have been characterized by solution NMR on micelle-incorporated gA (Urry et al., 1982, 1989; Jing et al., 1995; Jing and Urry, 1995) solid-state NMR on gA in oriented DMPC bilayers (Tian and Cross, 1999). The results obtained with these different methods are in reasonable agreement.)

2.4.2 Discrete-State Kinetics of Channel-Catalyzed Ion Movement

Based on the preceding arguments, and the results in Fig. 2.11, ion movement through a gramicidin channel can be decomposed into the following steps (Fig. 2.12): ion entry—diffusion to channel entrance and association with the pore (dehydration and resolvation by polar groups in the pore wall); translocation through the channel interior; and ion exit—dissociation from the pore (desolvation/rehydration) and diffusion out into the other bulk solution.

This decomposition of the ion transfer into a series of discrete steps is an approximation because each step represents an electrodiffusive barrier crossing; yet it summarizes the essential features of channel-mediated ion permeation. Features that also can be deduced by inspection of the potential of mean force (PMF) for ion movement through the channels (see Section 2.5). Figure 2.12 thus serves as a convenient reference for the analysis of experiments on channels formed by gA and amino acid-substituted gA analogues.

Figure 2.11 shows the existence of two major ion binding "sites," with little occupancy outside of these sites, but not whether they are occupied simultaneously. It thus becomes necessary to consider a hierarchy of kinetic models when modeling ion movement through gramicidin channels. In the simplest case, the ion movement can be described in terms of transitions among three different states: OO, in which there is no ion in the pore; IO, in which there is an ion in the left "binding site"; and OI, in which there is an ion in the right "binding site." Moreover, the transit time between these states is short compared to the residence times. In this case, the kinetic model for ion permeation becomes a three-barrier-two-site-one-ion (3B2S1I) model (Scheme III).

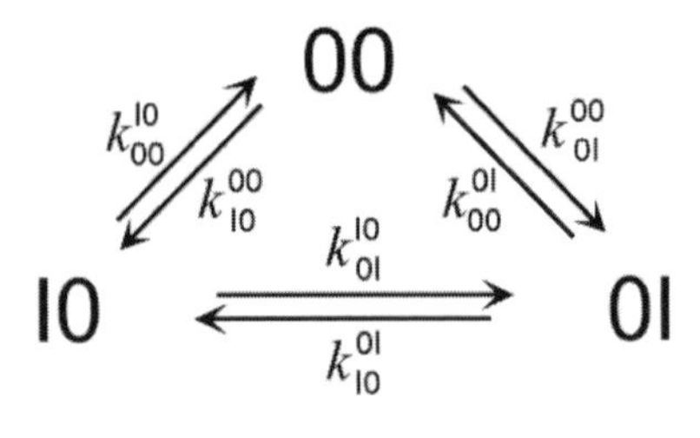

Scheme III

The (voltage-dependent) rate constants for the interconversions between the different states are denoted as $k_{\mathrm{ff}}^{\mathrm{ii}}$ (where the super- and subscripts denote the initial and final states, respectively), and the kinetic equations associated with Scheme III become:

$$\begin{aligned}\frac{dW(00)}{dt} &= -\left(k_{10}^{00}\cdot[\mathrm{I}]_{\mathrm{l}}+k_{01}^{00}\cdot[\mathrm{I}]_{\mathrm{r}}\right)\cdot W(00)+k_{00}^{10}\cdot W(10)+k_{00}^{01}\cdot W(01)\\ \frac{dW(10)}{dt} &= -\left(k_{00}^{10}+k_{01}^{10}\right)\cdot W(10)+k_{10}^{00}\cdot[\mathrm{I}]_{\mathrm{l}}\cdot W(00)+k_{10}^{01}\cdot W(01)\\ \frac{dW(01)}{dt} &= -\left(k_{00}^{01}+k_{10}^{01}\right)\cdot W(01)+k_{01}^{00}\cdot[\mathrm{I}]_{\mathrm{r}}\cdot W(00)+k_{01}^{10}\cdot W(10)\end{aligned} \tag{2.11}$$

subject to the conservation:

$$W(00)+W(10)+W(01)=1, \tag{2.12}$$

where $W(00)$, etc., denotes the probability of being in the state 00, etc. In the steady state, which is of interest here,

$$\frac{W(00)}{dt}=\frac{W(10)}{dt}=\frac{W(01)}{dt}=0 \tag{2.13}$$

and Eqs. 2.11–2.13 can be solved algebraically or by the graph-theoretical method of King and Altman (1956). The flux (j) through the channel is given by

$$\begin{aligned}j &= k_{01}^{10}\cdot W(10)-k_{10}^{01}\cdot W(01)\\ &= \frac{k_{10}^{00}\cdot k_{01}^{10}\cdot k_{00}^{01}\cdot[\mathrm{I}]_{\mathrm{l}}-k_{01}^{00}\cdot k_{10}^{01}\cdot k_{00}^{10}\cdot[\mathrm{I}]_{\mathrm{r}}}{\mathrm{DD}},\end{aligned} \tag{2.14}$$

where the flux from left to right is positive and DD is given by, cf. Andersen (1989)

$$\begin{aligned}\mathrm{DD} &= k_{01}^{10}\cdot k_{00}^{01}+k_{10}^{01}\cdot k_{00}^{10}+k_{00}^{10}\cdot k_{00}^{01}\\ &\quad +k_{10}^{00}\cdot[\mathrm{I}]_{\mathrm{l}}\cdot\left(k_{01}^{10}+k_{10}^{01}+k_{00}^{01}\right)\\ &\quad +k_{01}^{00}\cdot[\mathrm{I}]_{\mathrm{r}}\cdot\left(k_{01}^{10}+k_{10}^{01}+k_{00}^{10}\right)\end{aligned} \tag{2.15}$$

with the rate constants subject to the constraint imposed by detailed balance, e.g., Amdur and Hammes (1966), which for monovalent cation movement becomes:

$$\frac{k_{\mathrm{I0}}^{00} \cdot k_{\mathrm{0I}}^{\mathrm{I0}} \cdot k_{00}^{\mathrm{0I}}}{k_{\mathrm{0I}}^{00} \cdot k_{\mathrm{I0}}^{\mathrm{0I}} \cdot k_{00}^{\mathrm{I0}}} = \exp\left\{\frac{e_0 \cdot \Delta V}{k_{\mathrm{B}}T}\right\}, \tag{2.16}$$

where ΔV is the potential difference applied across the membrane ($\Delta V = V_{\mathrm{l}} - V_{\mathrm{r}}$).

Generally, the voltage dependence of the rate constants can be expressed as:

$$k_{\mathrm{ff}}^{\mathrm{ii}} = \kappa_{\mathrm{ff}}^{\mathrm{ii}} \cdot f_{\mathrm{ff}}^{\mathrm{ii}}(\Delta V), \tag{2.17}$$

where $\kappa_{\mathrm{ff}}^{\mathrm{ii}}$ denotes the 0 mV value for the rate constant in question, which can be determined from the PMF following Kramers (1940), see also Andersen (1989) and Roux et al. (2004), and $f_{\mathrm{ff}}^{\mathrm{ii}}(\Delta V)$ its voltage dependence. For an ion channel with equal permeant ion concentrations in the two aqueous phases, $[\mathrm{I}]_{\mathrm{l}} = [\mathrm{I}]_{\mathrm{r}} = [\mathrm{I}]$, there is no net flux at $\Delta V = 0$, and the detailed balance constraint, Eq. 2.16, can be expressed as

$$\frac{\kappa_{\mathrm{I0}}^{00} \cdot \kappa_{\mathrm{0I}}^{\mathrm{I0}} \cdot \kappa_{00}^{\mathrm{0I}}}{\kappa_{\mathrm{0I}}^{00} \cdot \kappa_{\mathrm{I0}}^{\mathrm{0I}} \cdot \kappa_{00}^{\mathrm{I0}}} = 1 \tag{2.18a}$$

and

$$\frac{f_{\mathrm{I0}}^{00}(\Delta V) \cdot f_{\mathrm{0I}}^{\mathrm{I0}}(\Delta V) \cdot f_{00}^{\mathrm{0I}}(\Delta V)}{f_{\mathrm{0I}}^{00}(\Delta V) \cdot f_{\mathrm{I0}}^{\mathrm{0I}}(\Delta V) \cdot f_{00}^{\mathrm{I0}}(\Delta V)} = \exp\left\{\frac{e_0 \cdot \Delta V}{k_{\mathrm{B}}T}\right\}. \tag{2.18b}$$

The single-channel current (i) is given by $e_0 \cdot j$,

$$i = e_0 \cdot \frac{k_{\mathrm{I0}}^{00} \cdot k_{\mathrm{0I}}^{\mathrm{I0}} \cdot k_{00}^{\mathrm{0I}} \cdot [\mathrm{I}]_{\mathrm{l}} - k_{\mathrm{0I}}^{00} \cdot k_{\mathrm{I0}}^{\mathrm{0I}} \cdot k_{00}^{\mathrm{I0}} \cdot [\mathrm{I}]_{\mathrm{r}}}{\mathrm{DD}}, \tag{2.19}$$

and the single-channel conductance (g) is given by $i/(\Delta V - E)$, where E is the equilibrium (or Nernst) potential for the ion, $E = (k_{\mathrm{B}}T/e_0)\cdot\ln\{[\mathrm{I}]_{\mathrm{l}}/[\mathrm{I}]_{\mathrm{r}}\}$ (where we disregard activity coefficient corrections).

In the limit $\Delta V \rightarrow 0$, and symmetric permeant concentrations, the single-channel conductance–concentration (g–C) relation becomes

$$g = \frac{g_{\mathrm{max}} \cdot [\mathrm{I}]}{K_{\mathrm{I}} + [\mathrm{I}]}. \tag{2.20}$$

For a symmetric channel ($\kappa_{\mathrm{I0}}^{00} = \kappa_{0\mathrm{I}}^{00}$, etc.), the expressions for $g_{\max}$ and K_{I} become

$$g_{\max} = \frac{e_0^2}{k_B T} \cdot \frac{k_{0\mathrm{I}}^{\mathrm{I0}} \cdot k_{00}^{0\mathrm{I}}}{2 \cdot \left(2 \cdot k_{0\mathrm{I}}^{\mathrm{I0}} + k_{00}^{0\mathrm{I}}\right)} \tag{2.21a}$$

and

$$K_{\mathrm{I}} = \frac{k_{00}^{\mathrm{I0}}}{2 \cdot k_{\mathrm{I0}}^{00}}. \tag{2.21b}$$

Though Eq. 2.20 provides a satisfactory fit of the g–C relation for Na^+ through gA channels, with $K_{\mathrm{I}} = 0.20$ M and $g_{\max} = 15.8$ pS (Andersen et al., 1995), Scheme III does not provide a satisfactory description of the kinetics of Na^+ movement over an extended voltage range. More complex permeation models are called for.

Indeed, the two cation "binding sites" (Fig. 2.11) can be occupied simultaneously (Schagina et al., 1978), such that it becomes necessary to extend Scheme III to include also a kinetic state in which both sites are occupied (II). The minimal kinetic model for ion permeation thus becomes a three-barrier-two-site-two-ion (3B2S2I) model (Scheme IV):

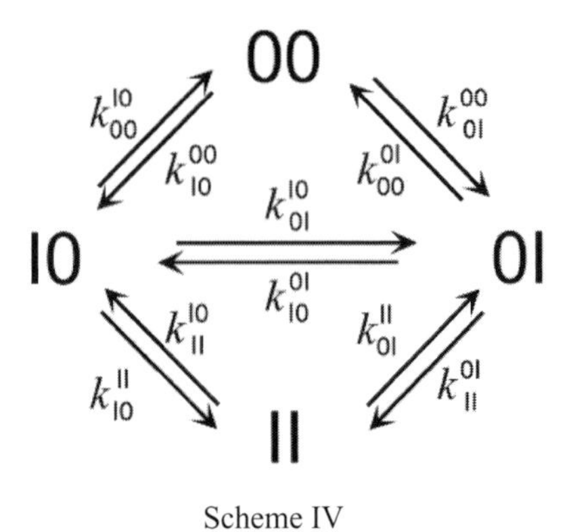

Scheme IV

The kinetic equations associated with Scheme IV can be solved algebraically, as above, but it usually is simpler to solve the equations associated with complex kinetic schemes using the King and Altman (1956) graph-theoretical approach, cf. Hille and Schwarz (1978). For a symmetric channel in the limit $V \to 0$, the g–C relation becomes (Heckmann, 1965; Urban et al., 1978; Finkelstein and Andersen, 1981):

$$g = \frac{e_0^2}{k_B T} \cdot \frac{[\mathrm{I}]}{K_{\mathrm{I}} + [\mathrm{I}] + [\mathrm{I}]^2 / K_{\mathrm{II}}} \cdot \frac{\kappa_{0\mathrm{I}}^{\mathrm{I0}} \cdot \left(\kappa_{00}^{0\mathrm{I}} + \kappa_{\mathrm{II}}^{0\mathrm{I}} \cdot [\mathrm{I}]\right)}{2 \cdot \left(2 \cdot \kappa_{0\mathrm{I}}^{\mathrm{I0}} + \kappa_{00}^{0\mathrm{I}} + \kappa_{\mathrm{II}}^{0\mathrm{I}} \cdot [\mathrm{I}]\right)}, \tag{2.22}$$

where

$$K_{\mathsf{I}} = \frac{\kappa_{00}^{\mathsf{I0}}}{2 \cdot \kappa_{\mathsf{I0}}^{00}} \quad \text{and} \quad K_{\mathsf{II}} = \frac{2 \cdot \kappa_{0\mathsf{I}}^{\mathsf{II}}}{\kappa_{\mathsf{II}}^{0\mathsf{I}}} \tag{2.23}$$

(the factors of 2 arise because an ion can enter an empty channel, 00, and leave a double-occupied channel, II, from either side). The second term in Eq. 2.22 denotes the probability of finding the channel in a single-occupied state, $W(\mathsf{I0}) + W(0\mathsf{I})$, which reaches a maximum when $[\mathsf{I}] = \sqrt{K_{\mathsf{I}} \cdot K_{\mathsf{II}}}$; the third term keeps track of transitions in single-occupied channels, it reaches the limiting value $\kappa_{0\mathsf{I}}^{\mathsf{I0}}/4$ as $[\mathsf{I}] \to \infty$.

As evident by inspection of Eqs. 2.20–2.21 and Eqs. 2.22–2.23, g–C relations do not impart sufficient information (independent parameters) to allow for the determination of all the underlying rate constants in Schemes III and IV. For both schemes, however, it is possible to determine all the rate constants (and their associated voltage-dependent terms) from an analysis of the current–concentration–voltage (i–C–V) relations. In the case of Scheme IV, it even is possible to determine the 0 mV values for the rate constants by measuring the g–C relation and the correlation factor (f), which is defined as the ratio between the channel's equilibrium (or net) permeability coefficient (p_{I}) and the tracer permeability coefficient (p_{I}^*), cf. Heckmann (1972). The equilibrium and tracer permeability coefficients are given by (Finkelstein and Andersen, 1981; Andersen, 1989):

$$p_{\mathsf{I}} = \frac{k_{\mathrm{B}}T}{e_0^2} \cdot \frac{g}{[\mathsf{I}]} = \frac{1}{K_{\mathsf{I}} + [\mathsf{I}] + [\mathsf{I}]^2/K_{\mathsf{II}}} \cdot \frac{\kappa_{0\mathsf{I}}^{\mathsf{I0}} \cdot \left(\kappa_{00}^{0\mathsf{I}} + \kappa_{\mathsf{II}}^{0\mathsf{I}} \cdot [\mathsf{I}]\right)}{2 \cdot \left(2 \cdot \kappa_{0\mathsf{I}}^{\mathsf{I0}} + \kappa_{00}^{0\mathsf{I}} + \kappa_{\mathsf{II}}^{0\mathsf{I}} \cdot [\mathsf{I}]\right)} \tag{2.24}$$

and

$$p_{\mathsf{I}}^* = \frac{1}{K_{\mathsf{I}} + [\mathsf{I}] + [\mathsf{I}]^2/K_{\mathsf{II}}} \cdot \frac{\kappa_{0\mathsf{I}}^{\mathsf{I0}} \cdot \left(2 \cdot \kappa_{00}^{0\mathsf{I}} + \kappa_{\mathsf{II}}^{0\mathsf{I}} \cdot [\mathsf{I}]\right)}{2 \cdot \left(2 \cdot (2 \cdot \kappa_{0\mathsf{I}}^{\mathsf{I0}} + \kappa_{00}^{0\mathsf{I}}) + \kappa_{\mathsf{II}}^{0\mathsf{I}} \cdot [\mathsf{I}]\right)}, \tag{2.25}$$

such that the correlation factor becomes:

$$f = \frac{p_{\mathsf{I}}^*}{p_{\mathsf{I}}} = \frac{2 \cdot \kappa_{00}^{0\mathsf{I}} + \kappa_{\mathsf{II}}^{0\mathsf{I}} \cdot [\mathsf{I}]}{\kappa_{00}^{0\mathsf{I}} + \kappa_{\mathsf{II}}^{0\mathsf{I}} \cdot [\mathsf{I}]} \cdot \frac{2 \cdot \kappa_{0\mathsf{I}}^{\mathsf{I0}} + \kappa_{00}^{0\mathsf{I}} + \kappa_{\mathsf{II}}^{0\mathsf{I}} \cdot [\mathsf{I}]}{2 \cdot (2 \cdot \kappa_{0\mathsf{I}}^{\mathsf{I0}} + \kappa_{00}^{0\mathsf{I}}) + \kappa_{\mathsf{II}}^{0\mathsf{I}} \cdot [\mathsf{I}]}. \tag{2.26}$$

In single-occupied channels (Scheme III), $\kappa_{\mathsf{II}}^{0\mathsf{I}} = 0$ and $f \equiv 1$; that is, measuring the tracer permeability coefficient does not provide additional information. In double-occupied channels, $0.5 < f < 1$, and f approaches 1 as $[\mathsf{I}] \to 0$ or $[\mathsf{I}] \to \infty$. In this case there should be sufficient information to determine all the underlying rate constants—unless $\kappa_{0\mathsf{I}}^{\mathsf{I0}} < \kappa_{00}^{0\mathsf{I}}$ in which case $f \approx 1$.

Even Scheme IV needs to be enhanced, however, because the diffusional entry step (Fig. 2.12) constitutes a significant barrier to ion movement through gA channels (Andersen, 1983c). Diffusion limitation (DL) becomes important because incoming ions must "hit" the pore entrance rather precisely—meaning that the channel is "hidden" behind a diffusion resistance. The existence of this diffusion limitation will complicate the mechanistic interpretation of structure–function studies because one may not be able to discern the "true" consequences of a sequence substitution. In addition to this complication, a potential difference applied across the bilayer (and channel) will change the interfacial ion concentrations. This interfacial polarization (IP) will in its own right have an impact on ion movement through the channel (Andersen, 1983b), which becomes increasingly important as the ionic strength (or permeant ion concentration) is reduced. When the 3B2S2I kinetic model is enhanced to incorporate both diffusion limitation and interfacial limitation, the resulting 3B2S2I(DL,IP) model provides an acceptable, discrete-state, kinetic description of ion permeation through gA channels (see Section 2.4.3).

2.4.3 Ion Permeation through gA Channels

The amino acid side chains do not contact the permeating ions, but the gA channels' permeability properties are modulated by amino acid substitutions throughout the sequence (Bamberg et al., 1976; Morrow et al., 1979; Heitz et al., 1982; Mazet et al., 1984; Russell et al., 1986; Koeppe et al., 1990): nonpolar→polar substitutions in the formyl-NH-terminal half of the sequence tend to reduce the conductance; polar→nonpolar substitutions in the carboxy-ethanolamide half tend to reduce the conductance. The primary mechanism by which amino acid substitutions alter the ion permeability appears to be electrostatic interactions between the permeating ions and the side chain dipoles (Mazet et al., 1984; Koeppe et al., 1990; Andersen et al., 1998; Busath et al., 1998).

The kinetics of ion movement through gA channels have been studied using single-channel measurements in glycerolmonoleate (Neher et al., 1978; Urban et al., 1980; Busath et al., 1998; Cole et al., 2002) or DPhPC (Becker et al., 1992) bilayers. Figures 2.13 and 2.14 show results obtained in DPhPC/*n*-decane with Na^+ as the permeant ion, as well as the fit of the 3B2S2I(DL,IP) model to the results.

The data span a large range of voltages and concentrations. One needs such a large data set to evaluate discrete-state kinetic models—as well as more detailed, physical descriptions of ion movement through a channel—because the $i - V$ relations are fairly linear. That is, the individual data points are highly correlated, such that the information content per point is limited. It is particularly important to have results at high potentials (at both low permeant ion concentrations, where the ion entry step is rate limiting, and high permeant concentrations, where ion exit becomes limiting).

When $[Na^+] \leq 0.1$ M or so, the i–V relations tend to level off toward a voltage-independent limit as V increases (Andersen, 1983c) because the rate of ion movement becomes determined by the voltage-independent diffusion step (Fig. 2.12). Yet, the unavoidable interfacial polarization causes the interfacial cation concentration to increase at the positive channel entrance (Andersen, 1983b), such that a bona fide voltage-independent limit is attained only in the limit where the ionic

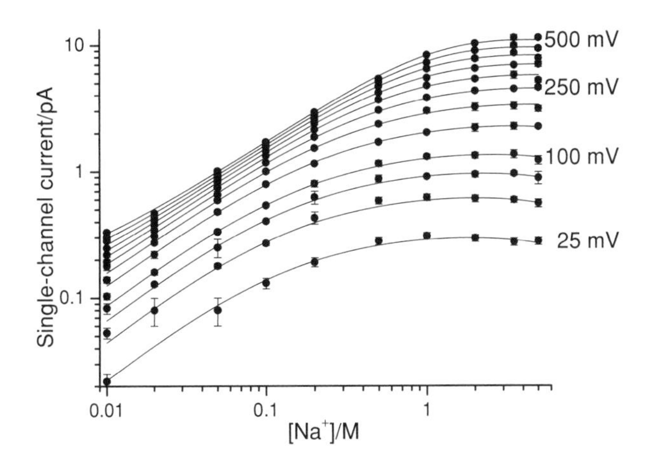

Figure 2.13 Current–voltage–concentration results for Na^+ moving through gA channels in DPhPC bilayers and fit of the 3B2S2I(DL,IP) model to the results (the rate constants are listed in Table 2.1). 25°C. O.S. Andersen and M.D. Becker, unpublished observations; from Andersen et al. (2005) with permission.

strength (adjusted using inert salt, which does not permeate through or block the channels) is much higher than the permeant cation concentration (Andersen, 1983b). This limiting current allows for the determination of the diffusion-limited ion access permeability (p_0), or rate constant (κ_0) (Hille, 1968; Hall, 1975; Läuger, 1976):

$$p_0 = 2\pi \cdot D_{aq} \cdot r_0 \tag{2.27a}$$

$$\kappa_0 = 2\pi \cdot D_{aq} \cdot r_0 \cdot N_A, \tag{2.27b}$$

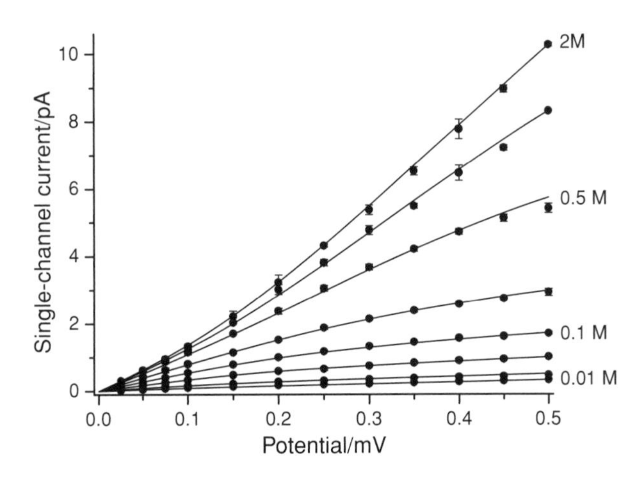

Figure 2.14 Current–voltage results for $[Na^+]$ between 0.01 and 2.0 M and fits of the 3B2S2I(DL,IP) model to the results. Results from Fig. 2.13; from Andersen et al. (2005) with permission.

where D_{aq} is the ion's diffusion coefficient and r_0 the channel's capture radius for the ion, which is equal to the difference between the effective pore and ion radii, cf. Eq. 2.10b and Ferry (1936), and N_A Avogadro's number. p_0 is $\sim 1.8 \times 10^{-13}$ cm^3·s^{-1} for Na^+ and 3×10^{-13} cm^3·s^{-1} for the higher alkali metal cations, corresponding to κ_0 being $\sim 10^8$ M^{-1}·s^{-1} for Na^+ and $\sim 2 \times 10^8$ M^{-1}·s^{-1} for the higher alkali metal cations (Andersen, 1983c). These values are comparable to diffusion-limited association rate constants in enzyme-catalyzed reactions (Eigen, 1974; Fersht, 1985). The diffusion-limited ion access permeability imposes an upper limit on the single-channel conductance (Hille, 1968; Läuger, 1976):

$$g_{\text{lim}} = \frac{e_0^2 \cdot p_0 \cdot [\text{I}]}{2 \cdot k_B T}, \tag{2.28}$$

which in the case of 0.1 M Na^+ is $\sim$34 pS—only about than sevenfold higher than the measured conductance (5.2 pS, Andersen, 1983a).

Assuming that the ion's diffusion coefficient adjacent to the pore entrance is equal to the bulk diffusion coefficient (which is questionable, see König et al., 1994), the channel's capture radius for the alkali metal cations is $\sim$0.3 Å (Andersen, 1983c). The small value of the capture radius provides for an estimate of an ion's thermal velocity (v_0), because the ions' overall collision rate with the pore entrance is $v_0 \cdot 2\pi \cdot r_0^2 \cdot C_0$, where C_0 is the ion concentration at the pore entrance. By contrast, the diffusion-limited rate of ion/channel encounters is $2\pi \cdot r_0 \cdot D \cdot (C_b - C_0)$, where C_b is the bulk ion concentration. When $r_0 > 1$ Å, as is the case for diffusion-limited reactions in bulk solutions, the diffusion-limited step is rate limiting. When $r_0 < 1$ Å, the relative resistance imposed by the collision step becomes increasingly important as r_0 decreases, which makes it possible to estimate v_0 to be $\sim 10^4$ cm s^{-1} (Andersen and Feldberg, 1996), as predicted by Einstein (1907).

When $[Na^+] > 1$ M, the channel usually is occupied by at least one Na^+, the slope of i–V relations increases as V increases (Fig. 2.13) because the rate of ion movement in this case becomes determined by voltage-dependent transitions within the pore (translocation and exit, Fig. 2.12) determine the overall rate of ion movement. That is, by examining the i–V–C relations over a large range of ion concentrations and applied potentials it is possible to "isolate" and explore the various steps in ion movement through the pore—and to determine the underlying rate constants. Table 2.1 summarizes results obtained with gA, as well as two Trp→Phe-substituted gA analogues [Phe^9]gA and [Phe^{15}]gA. These analogues form channels, structurally equivalent to gA channels (Becker et al., 1991; Jordan et al., 2005), with conductances of 6 pS and 11 pS (in 1.0 M NaCl and 200 mV applied potential), respectively, as compared to 15 pS for gA channels (Becker et al., 1991).

Though it is possible to determine all the rate constants in the kinetic scheme, it is important to keep in mind what has been determined—rate constants. Attempts to convert a set of rate constants to a so-called energy profile are fraught with problems (Andersen and Koeppe, 1992; Andersen, 1999), and should be considered unnecessary obfuscation.

Table 2.1 Kinetics of Na^+ permeation through gA channels and Trp→Phe substituted gA channels

Trp→Phe substitution rate constant	None	Position 9	Position 15
κ_0 (M^{-1} $s^{-1}\times 10^7$)	9	9	9
κ_{I0}^{00} (M^{-1} $s^{-1}\times 10^7$)	41	51	7
κ_{00}^{I0} ($s^{-1}\times 10^7$)	11	26	4.6
κ_{0I}^{I0} ($s^{-1}\times 10^6$)	7	2.3	4.7
κ_{II}^{0I} (M^{-1} $s^{-1}\times 10^7$)	0.7	0.02	0.005
κ_{0I}^{II} ($s^{-1}\times 10^7$)	2.8	0.3	0.02

DPhPC/*n*-decane bilayers; 25°C.

The rate constants, apart from κ_0, are defined in Scheme II and Eq. 2.5. The standard deviations, determined from Monte Carlo-based error analysis (Alper and Gelb, 1990), are less than 20%, usually less then 10%. From Becker et al. (1992).

The predicted Na^+ affinities, and relative affinities for the first and second ion that binds, are listed in Table 2.2. Note that even though the *g–C* relation (determined in the limit $V\rightarrow 0$) could be fitted quite well by Eq. 2.20, the deduced ion-channel dissociation constant (0.20 M) is higher than the dissociation deduced from the complete kinetic analysis. Also, the ratio $\kappa_{0I}^{I0}/\kappa_{00}^{I0}$ is less than 1, meaning that the correlation factor is close to 1.0. It is not possible to deduce all the rate constants in the kinetic scheme from the *g–C* relation and concentration dependence of the correlation factor.

To appreciate the channels' affinity for Na^+, it is useful to compare the Na^+ mole-fraction (n_{Na}) in the pore (n_{Na}^{pore}), which has ~6 H_2O molecules in the single-filing stretch (Levitt et al., 1978; Rosenberg and Finkelstein, 1978; Allen et al., 2004a), to n_{Na} in the bulk solution (n_{Na}^{bulk}). At 0.14 M, $n_{Na}^{bulk} \approx 1/400$, whereas $n_{Na}^{pore} \approx 1/10$ (corresponding to a 5 M solution) or ~40-fold higher than in the bulk solution. Na^+ therefore is preferentially solvated by the pore lining residues, as compared to the solvation in bulk water! A similar conclusion can be drawn for other ion-conducting channels (Andersen and Koeppe, 1992).

Table 2.2 Derived parameters for Na^+ permeation through gA and Trp→Phe substituted gA channels

Trp→Phe substitution parameter	None	Position 9	Position 15
K_I (M)	0.14	0.25	0.32
K_{II} (M)	8	38	13
K_{II}/K_I	58	150	39
$\kappa_{II}^{0I}/\kappa_{I0}^{00}$	0.02	0.004	0.007
$\kappa_{0I}^{II}/\kappa_{00}^{I0}$	0.25	0.01	0.007

The entries are calculated based on the parameters in Table 2.1, using a Monte Carlo-based error analysis, e.g. (Alper and Gelb, 1990).

2.4.4 Gramicidin Channels as Enzymes

Gramicidin channels, like other ion channels (and membrane proteins) catalyze ion movement across a lipid bilayer by providing a reaction path that obviates the ion's passage through the lipid bilayer hydrophobic core per se. Gramicidin channels therefore are enzymes, albeit members of a special class of enzymes in which no covalent bonds are made or broken during the catalytic cycle. It now is possible to estimate the channels' catalytic rate enhancement, the rate of channel-mediated ion movement (k_{cat}) relative to the rate noncatalyzed movement through the bilayer, $k_{\text{cat}}/k_{\text{non}}$ (Wolfenden and Snider, 2001).

To a first approximation, $k_{\text{cat}}/k_{\text{non}}$ can be equated with $K_{\text{p/w}}/K_{\text{m/w}}$, where $K_{\text{p/w}}$ and $K_{\text{m/w}}$ denote the ion partition coefficients into the pore and into the bilayer hydrophobic core, respectively, in the limit where the channel's ion occupancy is $\ll 1$. In this limit, one can determine $K_{\text{p/w}}$ from the preceding kinetic analysis: $K_{\text{p/w}} \approx n_{\text{Na}}^{\text{pore}}/n_{\text{Na}}^{\text{bulk}} \approx 10^2$. $K_{\text{m/w}}$ can be estimated to be $\sim 10^{-14}$ based on the conductance ($G_0 \approx 10^{-9}$ S cm^{-2}) of unmodified bilayers in 1.0 M NaCl (Hanai et al., 1965) using the relation (Hodgkin and Katz, 1949; Andersen, 1989):

$$G_0 = N_{\text{A}} \cdot \frac{(ze)^2}{k_{\text{B}}T} \cdot \frac{D_{\text{m}}}{d_0} \cdot K_{\text{m/w}} \cdot C, \tag{2.29}$$

where D_{m} is the ion's diffusion coefficient in the bilayer hydrophobic core ($\sim 10^{-5}$ cm$^2 \cdot$s^{-1}, cf. Schatzberg, 1965). We thus find that the catalytic rate enhancement is in the neighborhood of 10^{15} to 10^{16}, which is comparable to the rate enhancement observed for conventional enzymes (Wolfenden and Snider, 2001).

2.4.5 Ion–Ion Interactions May Be Water-Mediated

Tables 2.1 and 2.2 show that $K_{\text{II}}/K_{\text{I}} \gg 1$, indicative of repulsive ion–ion interactions; but the ratio differs among the channels, suggesting that ion–ion interactions within the doubly-occupied channels are not due solely to electrostatic interactions. This conclusion is supported by examining the ratios of association and dissociation rate constants for the first and the second ion entering (or leaving), the lower two lines in Table 2.2. Both ratios are decreased—with the major decrease being in the ratio of the association rate constants. This surprising result presumably means that the water in the pore (being relatively incompressible) plays an important role in mediating ion–ion interactions, see also Roux et al. (1995). It further suggests that the pore water needs to be considered in Brownian dynamics (BD) simulations of channel-catalyzed ion movement.

2.4.6 Importance of the Trp Residues for Ion Permeation

The four Trp residues at the pore entrance are important for both channel folding and channel function. They are oriented with their dipole moments directed away from the channel center, the NH's toward the aqueous solution (Arseniev et al., 1986; Ketchem et al., 1997; Townsley et al., 2001), which will tend to lower the central electrostatic barrier below that estimated using a structure-less dielectric model (Jordan, 1984; Sancho and Martinez, 1991; Andersen et al., 1998).

Consistent with this idea, gA analogues with one or more Trp→Phe replacements form channels with decreased ion permeabilities (Becker et al., 1991), see Section 2.4.3. When all four Trp residues are replaced by Phe, the Cs^+ conductance is reduced sixfold (Heitz et al., 1982; Fonseca et al., 1992). The basis for the reduced conductance appears to be a greatly reduced rate constant for ion translocation through the channel, κ_{0I}^{I0}, as deduced by Becker et al. (1992) and Caywood et al. (2004), although the rate constant for ion entry also is reduced (Becker et al., 1992; Fonseca et al., 1992). The latter could arise because the amphipathic indoles may be able to move (a little) out beyond the hydrophobic membrane core, which could be important as an incoming ion sheds most of its hydration shell to become solvated by the peptide backbone. The backbone deformation that is needed to optimize ion–oxygen contacts (Noskov et al., 2004) will involve also side chain motions (Urry, 1973).

The changes in Na^+ permeation through single Trp→Phe substituted gramicidins have been examined in detail (Becker et al., 1992); see Tables 2.1 and 2.2. The kinetic analysis provides information about rate constants for ion translocation, but no information about the absolute barrier heights (or well depths) of the free energy profile for ion movement through the pore (Andersen, 1989, 1999). Such information can be extracted only through an appropriate physical theory, such as a hierarchical implementation of MD and BD simulations (Roux et al., 2004). Nevertheless, it is possible to estimate the *changes* in barrier heights and well depths as $-k_B T \cdot \ln\{\kappa^{\text{Analogue}}/\kappa^{\text{Control}}\}$ for each of the transitions along the reaction coordinate (Andersen, 1989). We conclude that single Trp→Phe substitutions, which have little effect on the channel structure whether determined by heterodimer formation (Besker et al., 1991) or solution NMR (Jordan et al., 2005), at the channel pore entrances (one in each subunit) may alter the barrier profile (decrease well depths, increase barrier heights) by several $k_B T$.

Trp→Phe substitutions increase the height of the central barrier. Because the sequence substitutions do not alter the structure of the subunit interface (Becker et al., 1991), the conductance changes most likely result from favorable electrostatic interactions between the indole dipole and the permeant ion (Andersen et al., 1998; Busath et al., 1998). Surprisingly, the sign of the Trp→Phe substitution-induced changes in the entry/exit barrier depends on the position where the substitution is made (Table 2.1), indicating that the side chain dynamics indeed are important for the rate of the ions' hydration/solvation.

2.4.7 Importance of the Lipid Bilayer

Changes in ion permeability as a result of amino acid substitutions are not surprising, even though the side chains do not contact the permeating ions. But only a detailed kinetic analysis would be able to reveal that the changes in the ion entry/exit kinetics

Table 2.3 gA single-channel conductances in bilayers formed by phospholipids having different acyl chains

Phospholipid	Conductance (pS)
diphytanoyl-PC	15.0 ± 0.5
1-palmitoyl-2-oleoyl-PC*	11.3 ± 0.5
dioleoyl-PC*	13.0 ± 0.5
dilinoleoyl-PC*	14.9 ± 0.4

25°C, 1.0 M NaCl, 200 mV.
*Results from Girshman et al. (1997).

depend on residue position. What *is* surprising is that even changes in the bilayer lipid composition—the acyl chain composition—alter the channels' ion permeability (Table 2.3).

Even larger changes than those indicated in Table 2.3 can be observed when the polar head groups of the bilayer-forming lipids are varied. Moreover, the conductance is higher in bilayers formed by ether, as opposed to ester phospholipids (Providence et al., 1995), presumably because the interfacial dipole potential is less positive in bilayers formed by ether phospholipids (Gawrisch et al., 1992). It is important not to consider the lipid bilayer just to be some "inert" host for the channels of interest.

2.5 Molecular Dynamics Analysis of Ion Permeation

To obtain better insights into the molecular basis for channel (or membrane protein) function, it is necessary to establish direct links between the atomic structure of a channel (or protein) and its observed function. When establishing these links, it becomes important to consider explicitly that channel proteins (as well as the water in the pore) are composed of discrete atoms that constantly undergo thermal fluctuations, from the rapid (picosecond) vibrations, through slower (multi-nanosecond) global reorientations and side-chain isomerizations, to long timescale (microsecond to second) conformational changes (Karplus and McCammon, 1981). The energetic consequences of these fluctuations are illustrated in Fig. 2.15, which shows fluctuations in potential energy for a K^+ in a gA channel imbedded in a dimyristoylphosphatidylcholine bilayer.

The fluctuations in energy are large because the pore is narrow, meaning that even small variations in the ion–carbonyl distance become important (Allen et al., 2004b). It is important to consider them when constructing permeation models.

Though MD simulations, in principle, should provide the desired tool for incorporating molecular reality and the consequences of molecular flexibility into descriptions of channel-mediated ion movement (Edwards et al., 2002), one cannot reach the necessary level of insight through "brute force" MD simulations (Roux, 2002; Roux et al., 2004), because the measured ion flux typically corresponds to ion transit times of $\sim$100 ns—much longer than typical MD trajectories (Roux, 2002).

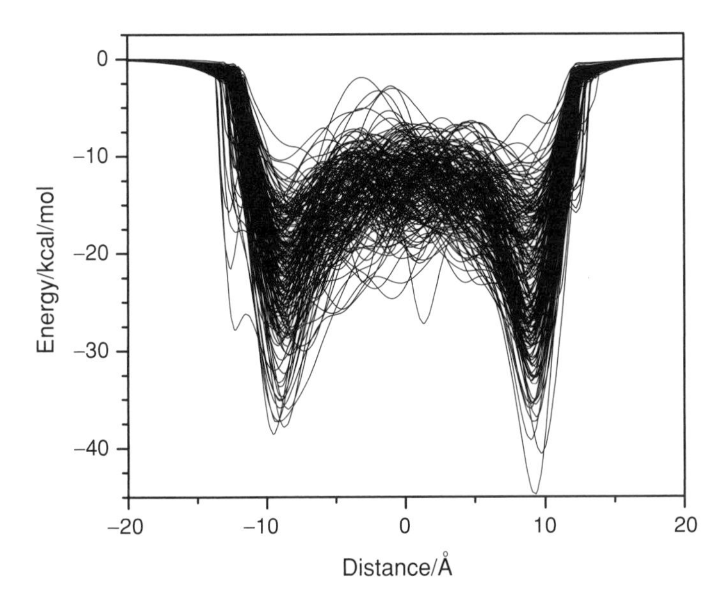

Figure 2.15 Small, Angstrom-scale thermal fluctuations in the gA structure produce large variations in the energetics of ion permeation, when used in models based on any one rigid structure. The figure shows Poisson solutions for 194 frames from a 4 ns MD trajectory initiated with the PDB:1JNO gA structure, and using a 1.4 Å water probe to define the high-dielectric constant region in the pore. Each sample was oriented such that the channel axis coincides, as closely as possible, with the fixed z-axis of the system. Distance is measured from the channel center. The variations in the potential profile are as much as 39 kcal mol^{-1}. After Allen et al. (2004b).

Indeed, when using MD simulations to predict the PMF for ion movement through gA channels, the barrier heights are too high by many kcal mol^{-1} (Roux and Karplus, 1993; Allen et al., 2003), meaning that the predicted conductances are several orders of magnitude less than those observed experimentally. This is of concern because the small and relatively well-behaved gA channels should be particularly amenable to in-depth theoretical analysis and computer simulations.

There is reason for optimism, nevertheless. Fig. 2.16 shows two recently determined PMFs (free energy profiles) for K^+ permeation through gA channels (Allen et al., 2004a).

The main structural features of the PMFs, i.e., two cation binding sites near the channel's end separated by a central barrier, are qualitatively consistent with Fig. 2.11 which has been deduced from experiments (Urry et al., 1989; Olah et al., 1991). From a quantitative point of view, however, the "uncorrected" PMF in Fig. 2.16 continues to display a central barrier that is markedly too high, a problem similar to what has been observed in previous studies and which must be addressed and resolved.

After correcting for two potentially serious artifacts, the resulting "corrected" PMF in Fig. 2.16 leads to predicted experimental observables that are in semiquantitative agreement with experimental results. The first artifact is introduced by the periodic boundary conditions of the finite simulation system, which cause a spurious

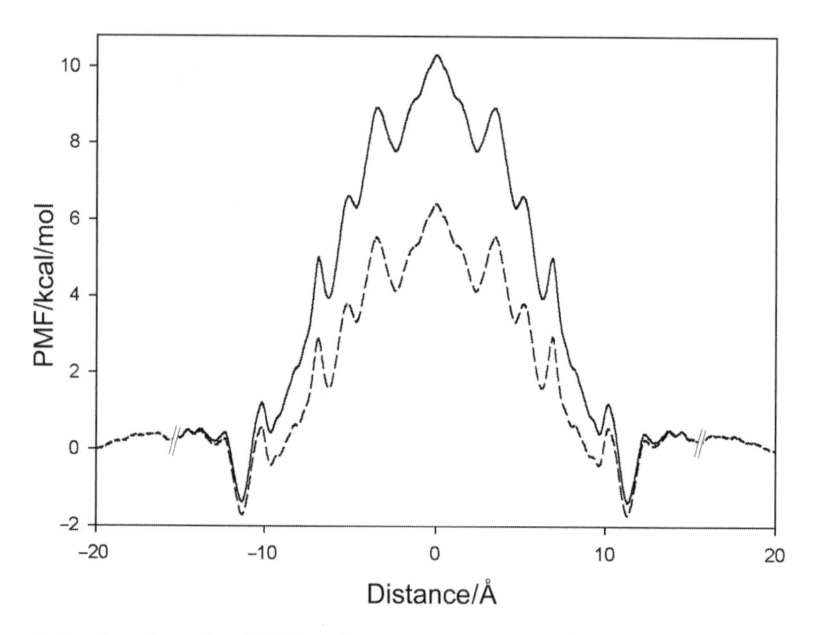

Figure 2.16 One-dimensional PMF, or free energy profile, for K^+ along the gA axis. The upper solid curve denotes the initial result of MD simulations. The free energy profile is not meaningful beyond the cuts at $z = \pm 15$ Å because no absolute reference of such a one-dimensional PM can be defined for large z (Roux et al., 2004). The lower dashed curve denotes the PMF corrected for size, and membrane dielectric constant (see text). After Allen et al. (2004a); modified from Andersen et al. (2005) with permission.

destabilization of the permeating ion in the channel relative to the bulk, which disappears if the system becomes exceedingly large (Hünenberger and McCammon, 1999). The second artifact arises because the hydrocarbon chains of the lipids are not polarizable in MD force fields, such that this region is treated as corresponding to a dielectric constant of 1, whereas it should be ~2 (Huang and Levitt, 1977; Simon and McIntosh, 1986; Åqvist and Warshel, 1989), i.e., the dielectric constant of bulk hydrocarbons (Smyth, 1955), or even higher (Mamanov et al., 2003).

It is possible to correct for these artifactual contributions to the energy profile by a continuum electrostatic approximation using the configurations of the MD trajectories to average over protein and single-file water configurations (Allen et al., 2004a). It thus turns out that the periodic boundary conditions cause the ion of interest to interact with phantom charges in the infinite array of channels in the simulation system, which causes a spurious energy barrier on the order of 2 kcal mol^{-1} when the ion is in the middle of the channel. Moreover, a change of dielectric constant for the bilayer core from 1 to 2 stabilizes the ion in the channel by about 2 kcal mol^{-1} (Allen et al., 2006). The barrier in the corrected PMF in Fig. 2.16 therefore is about 4 kcal mol^{-1} lower than in the original PMF, which in a single ion occupied channel permeation model (Scheme III) corresponds to a 1000-fold increase in the predicted ion permeability. Indeed, the maximal conductance for K^+ (again assuming single ion occupancy) is predicted to be ~0.8 pS—"only" 30-fold less than the measured value of ~25 pS in 1.0–2.0 M KCl and diphytanoylphosphatidylcholine bilayers (Andersen, 1983a; Bingham et al., 2003).

Some problems persist because the location and depth of the predicted binding sites differ somewhat from the experimental values (Urry et al., 1989; Olah et al., 1991): the binding sites are too far from the channel center and too shallow to be compatible with the observed ion affinities; but the predicted single-ion dissociation constant ($K_I \approx 0.34$ M) differs only by a factor five from the one predicted from the kinetic analysis of i–V–C results in gA channels ($K_I \approx 0.07$ M, O.S. Andersen and M.D. Becker, unpublished results). Both problems are likely to result from the use of nonpolarizable force fields to evaluate the ion–peptide backbone interactions, which therefore will tend to be underestimated. Another problem, which has not been fully resolved, is that the water models used in MD have been developed to describe bulk solution properties, which may differ from those of the single-file column of water in the channel pore. Because the PMF effectively arises as the sum of contributions from the permeating ion's interaction with the channel peptide, the single-filing water in the pore, and the bilayer hydrophobic core (Allen et al., 2004a), both the position and depth of the energy minima are likely to be particularly sensitive to the choice of force fields.

Given the progress that is taking place in terms of developing force fields that include induced polarization (Lamoureux and Roux, 2003; Lamoureux et al., 2003; Anisimov et al., 2004), there is every reason to be optimistic about the future. Indeed, even the present generation of force fields predicts changes in the central barrier (which should reflect primarily long-range electrostatic interactions) that are in near-quantitative agreement with experimental results (Allen et al., 2006), which suggests that one with some confidence can begin to use MD to understand the basis for amino acid substitution-induced *changes* in channel function.

2.6 Conclusion

The gramicidin channels possess remarkably well-defined structural and functional features that allow for detailed insights into the molecular basis for channel function. The channel's permeability properties can be modulated by changes in the amino acid sequence and the channels' bilayer environment. Single amino acid substitutions do not induce major changes in channel structure. It is difficult, however, to exclude that "minor" changes in side chain orientation or dynamics could be important for the observed changes in ion permeability. It is in this context important that the modest channel size allows for detailed computational analysis, which promises to provide atomic-level insights into the microphysics underlying ion permeation. This combination of features remains unprecedented, and suggests that the gA channels will become increasingly important as test beds for developing theoretical models of channel-catalyzed ion permeation and to critically evaluate (and improve) atomistic simulations of ion permeability.

Acknowledgments

This work was supported in part by the NIH under grants GM21342 (OSA), RR15569 (REK), GM62342 (BR), and GM70971 (OSA, REK and BR). We thank T. W. Allen, M. D. Becker, A. E. Daily, D. V. Greathouse, E. A. Hobart, H. Ingolfsson, T. Olson, S. Shobana, and S. E. Tape for stimulating discussions, technical assistance, and comments on the manuscript.

References

Abdul-Manan, N., and J.F. Hinton. 1994. Conformational states of gramicidin A along the pathway to the formation of channels in model membranes determined by 2D NMR and circular dichroism spectroscopy. *Biochemistry* 33:6773–6783.

Allen, T.W., O.S. Andersen, and B. Roux. 2003. The structure of gramicidin A in a lipid bilayer environment determined using molecular dynamics simulations and solid-state NMR data. *J. Am. Chem. Soc.* 125:9868–9878.

Allen, T.W., O.S. Andersen, and B. Roux. 2004a. Energetics of ion conduction through the gramicidin channel. *Proc. Natl. Acad. Sci. USA* 101:117–122.

Allen, T.W., O.S. Andersen, and B. Roux. 2004b. On the importance of atomic fluctuations, protein flexibility and solvent in ion permeation. *J. Gen. Physiol.* 124:679–690.

Allen, T.W., O.S. Andersen, and B. Roux. 2006. Ion permeation through a narrow channel: Using gramicidin to ascertain all-atom molecular dynamics potential of mean force methodology and biomolecular force fields. *Biophys. J.* 90:3447–3468.

Allen, T.W., T. Bastug, S. Kuyucak, and S.-H. Chung. 2003. Gramicidin A channel as a test ground for molecular dynamics force fields. *Biophys. J.* 84:2159.

Alper, J.S., and R.I. Gelb. 1990. Standard errors and confidence intervals in nonlinear regression: Comparison of Monte Carlo and parametric statistics. *J. Phys. Chem.* 94:4747–4751.

Amdur, I., and G.G. Hammes. 1966. Chemical Kinetics: Principles and Selected Topics. McGraw-Hill, New York.

Andersen, O.S. 1983a. Ion movement through gramicidin A channels. Single-channel measurements at very high potentials. *Biophys. J.* 41:119–133.

Andersen, O.S. 1983b. Ion movement through gramicidin A channels. Interfacial polarization effects on single-channel current measurements. *Biophys. J.* 41:135–146.

Andersen, O.S. 1983c. Ion movement through gramicidin A channels. Studies on the diffusion-controlled association step. *Biophys. J.* 41:147–165.

Andersen, O.S. 1989. Kinetics of ion movement mediated by carriers and channels. *Methods Enzymol.* 171:62–112.

Andersen, O.S. 1999. Graphic representation of the results of kinetic analyses. *J. Gen. Physiol.* 114:589–590.

Andersen, O.S., H.-J. Apell, E. Bamberg, D.D. Busath, R.E. Koeppe II, F.J. Sigworth, G. Szabo, D.W. Urry, and A. Woolley. 1999. Gramicidin channel controversy—the structure in a lipid environment. *Nat. Struct. Biol.* 6:609.

Andersen, O.S., and S.W. Feldberg. 1996. The heterogeneous collision velocity for hydrated ions in aqueous solutions is $\sim 10^4$ cm/s. *J. Phys. Chem.* 100:4622–4629.

Andersen, O.S., D.V. Greathouse, L.L. Providence, M.D. Becker, and R.E. Koeppe II. 1998. Importance of tryptophan dipoles for protein function: 5-fluorination of tryptophans in gramicidin A channels. *J. Am. Chem. Soc.* 120:5142–5146.

Andersen, O.S., and R.E. Koeppe II. 1992. Molecular determinants of channel function. *Physiol. Rev.* 72:S89–S158.

Andersen, O.S., R.E. Koeppe II, and B. Roux. 2005. Gramicidin channels. *IEEE Trans. Nanobioscience* 4:10–20.

Andersen, O.S., J.A. Lundbæk, and J. Girshman. 1995. Channel function and channel-lipid bilayer interactions. *In*: Dynamical Phenomena in Living Systems. E. Mosekilde and O.G. Mouritsen, editors. Springer, New York, pp. 131–151.

Andersen, O.S., and J. Procopio. 1980. Ion movement through gramicidin A channels. On the importance of the aqueous diffusion resistance and ion–water interactions. *Acta Physiol. Scand.; Suppl.* 481:27–35.

Andersen, O.S., G. Saberwal, D.V. Greathouse, and R.E. Koeppe II. 1996. Gramicidin channels—a solvable membrane "protein" folding problem. *Ind. J. Biochem. Biophys.* 33:331–342.

Anisimov, V.M., I.V. Vorobyov, G. Lamoureux, S. Noskov, B. Roux, and A.D. MacKerell Jr. 2004. CHARMM all-atom polarizable force field parameter development for nucleic acids. *Biophys. J.* 86:415a.

Åqvist, J., and A. Warshel. 1989. Energetics of ion permeation through membrane channels. Solvation of Na^+ by gramicidin A. *Biophys. J.* 56:171–182.

Arseniev, A.S., A.L. Lomize, I.L. Barsukov, and V.F. Bystrov. 1986. Gramicidin A transmembrane ion-channel. Three-dimensional structure reconstruction based on NMR spectroscopy and energy refinement. *Biol. Membr.* 3:1077–1104.

Bamberg, E., H.J. Apell, and H. Alpes. 1977. Structure of the gramicidin A channel: Discrimination between the $\pi_{L,D}$ and the β helix by electrical measurements with lipid bilayer membranes. *Proc. Natl. Acad. Sci. USA* 74:2402–2406.

Bamberg, E., H.J. Apell, H. Alpes, E. Gross, J.L. Morell, J.F. Harbaugh, K. Janko, and P. Läuger. 1978. Ion channels formed by chemical analogs of gramicidin A. *Fed. Proc.* 37:2633–2638.

Bamberg, E., and P. Läuger. 1973. Channel formation kinetics of gramicidin A in lipid bilayer membranes. *J. Membr. Biol.* 11:177–194.

Bamberg, E., K. Noda, E. Gross, and P. Läuger. 1976. Single-channel parameters of gramicidin A, B, and C. *Biochim. Biophys. Acta* 419:223–228.

Bass, R.B., P. Strop, M. Barclay, and D.C. Rees. 2002. Crystal structure of *Escherichia coli* MscS, a voltage-modulated and mechanosensitive channel. *Science* 298:1582–1587.

Becker, M.D., D.V. Greathouse, R.E. Koeppe II, and O.S. Andersen. 1991. Amino acid sequence modulation of gramicidin channel function. Effects of tryptophan-to-phenylalanine substitutions on the single-channel conductance and duration. *Biochemistry* 30:8830–8839.

Becker, M.D., R.E. Koeppe II, and O.S. Andersen. 1992. Amino acid substitutions and ion channel function: Model-dependent conclusions. *Biophys. J.* 62:25–27.

Bingham, N.C., N.E. Smith, T.A. Cross, and D.D. Busath. 2003. Molecular dynamics simulations of Trp side-chain conformational flexibility in the gramicidin A channel. *Biopolymers* 71:593–600.

Bockris, J.O'.M., and A.K.N. Reddy. 1970. Modern Electrochemistry, Vol. 1. Plenum, New York.

Burkhart, B.M., N. Li, D.A. Langs, W.A. Pangborn, and W.L. Duax. 1998. The conducting form of gramicidin A is a right-handed double-stranded double helix. *Proc. Natl. Acad. Sci. USA* 95:12950–12955.

Busath, D.D. 1993. The use of physical methods in determining gramicidin channel structure and function. *Annu. Rev. Physiol.* 55:473–501.

Busath, D.D., O.S. Andersen, and R.E. Koeppe II. 1987. On the conductance heterogeneity in membrane channels formed by gramicidin A. A cooperative study. *Biophys. J.* 51:79–88.

Busath, D.D., and G. Szabo. 1981. Gramicidin forms multi-state rectifying channels. *Nature* 294:371–373.

Busath, D.D., C.D. Thulin, R.W. Hendershot, L.R. Phillips, P. Maughan, C.D. Cole, N.C. Bingham, S. Morrison, L.C. Baird, R.J. Hendershot, M. Cotten, and T.A. Cross. 1998. Noncontact dipole effects on channel permeation. I. Experiments with (5F-indole)Trp^{13} gramicidin A channels. *Biophys. J.* 75:2830–2844.

Bystrov, V.F., and A.S. Arseniev. 1988. Diversity of the gramicidin A spatial structure: Two-dimensional proton NMR study in solution. *Tetrahedron* 44:925–940.

Caywood, D., J. Durrant, P. Morrison, and D.D. Busath. 2004. The Trp potential deduced from gramicidin A/gramicidin M channels. *Biophys. J.* 86:55a.

Chang, G., R.H. Spencer, A.T. Lee, M.T. Barclay, and D.C. Rees. 1998. Structure of the MscL homolog from *Mycobacterium tuberculosis*: A gated mechanosensitive ion channel. *Science* 282:2220–2226.

Cifu, A.S., R.E. Koeppe II, and O.S. Andersen. 1992. On the supramolecular structure of gramicidin channels. The elementary conducting unit is a dimer. *Biophys. J.* 61:189–203.

Cole, C.D., A.S. Frost, N. Thompson, M. Cotten, T.A. Cross, and D.D. Busath. 2002. Noncontact dipole effects on channel permeation. VI. 5F- and 6F-Trp gramicidin channel currents. *Biophys. J.* 83:1974–1986.

Cornell, B.A., F. Separovic, A.J. Baldassi, and R. Smith. 1988. Conformation and orientation of gramicidin A in oriented phospholipid bilayers measured by solid state carbon-13 NMR. *Biophys. J.* 53:67–76.

Cornell, B.A., F. Separovic, D.E. Thomas, A.R. Atkins, and R. Smith. 1989. Effect of acyl chain length on the structure and motion of gramicidin A in lipid bilayers. *Biochim. Biophys. Acta* 985:229–232.

Cowan, S.W., and J.P. Rosenbusch. 1994. Folding pattern diversity of integral membrane proteins. *Science* 264:914–916.
Cowan, S.W., T. Schirmer, G. Rummel, M. Steiert, R. Ghosh, R.A. Pauptit, J.N. Jansonius, and J.P. Rosenbusch. 1992. Crystal structures explain functional properties of two *E. coli* porins. *Nature* 358:727–733.
Cox, B.G., G.R. Hedwig, A.J. Parker, and D.W. Watts. 1974. Solvation of ions. XIX Thermodynamic properties for transfer of single ions between protic and dipolar aprotic solvents. *Aust. J. Chem.* 27:477–501.
Cross, T.A. 1994. Structural biology of peptides and proteins in synthetic membrane environments by solid-state NMR spectroscopy. *Annu. Rep. NMR Spetrosc.* 29:123–167.
Cross, T.A., A. Arseniev, B.A. Cornell, J.H. Davis, J.A. Killian, R.E. Koeppe II, L.K. Nicholson, F. Separovic, and B.A. Wallace. 1999. Gramicidin channel controversy-revisited. *Nat. Struct. Biol.* 6:610–611; discussion 611–612.
Davis, J.H., and M. Auger. 1999. Static and magic angle spinning NMR of membrane peptides and proteins. *Progr. Nucl. Mag. Res. Spectr.* 35:1–84.
Dill, K.A. 1990. Dominant forces in protein folding. *Biochemistry* 29:7133–7155.
Dill, K.A., T.M. Truskett, V. Vlachy, and B. Hribar-Lee. 2005. Modeling water, the hydrophobic effect, and ion solvation. *Annu. Rev. Biophys. Biomol. Struct.* 34:173–199.
Doyle, D.A., J. Morais Cabral, R.A. Pfuetzner, A. Kuo, J.M. Gulbis, S.L. Cohen, B.T. Chait, and R. MacKinnon. 1998. The structure of the potassium channel: Molecular basis of K^+ conduction and selectivity. *Science* 280:69–77.
Dubos, R.J. 1939. Studies on a bactericidal agent extracted from a soil bacillus I. Preparation of the agent. Its activity*in vitro*. *J. Exp. Med.* 70:1–10.
Durkin, J.T., R.E. Koeppe II, and O.S. Andersen. 1990. Energetics of gramicidin hybrid channel formation as a test for structural equivalence. Side-chain substitutions in the native sequence. *J. Mol. Biol.* 211:221–234.
Durkin, J.T., L.L. Providence, R.E. Koeppe II, and O.S. Andersen. 1992. Formation of non-β-helical gramicidin channels between sequence-substituted gramicidin analogues. *Biophys. J.* 62:145–159.
Durkin, J.T., L.L. Providence, R.E. Koeppe II, and O.S. Andersen. 1993. Energetics of heterodimer formation among gramicidin analogues with an NH_2-terminal addition or deletion. Consequences of a missing residue at the join in channel. *J. Mol. Biol.* 231:1102–1121.
Edwards, S., B. Corry, S. Kuyucak, and S.-H. Chung. 2002. Continuum electrostatics fails to describe ion permeation in the gramicidin channel. *Biophys. J.* 83:1348.
Eigen, M. 1974. Diffusion control in biochemical reactions. *In*: Quantum Statistical Mechanics in the Natural Sciences. S.L. Mintz and S.M. Widmayer, editors. Ed. Plenum Press, New York, pp. 37–61.
Einstein, A. 1907. Theoretische Betrachtungen über der Brownsche Bewegungen. *Zeit. f. Elektrochemie* 13:41–42.

Engelman, D.M., T.A. Steitz, and A. Goldman. 1986. Identifying nonpolar transbilayer helices in amino acid sequences of membrane proteins. *Annu. Rev. Biophys. Biophys. Chem.* 15:321–353.

Evans, E.A., and R.M. Hochmuth. 1978. Mechanochemical properties of membranes. *Curr. Top. Membr. Transp.* 10:1–64.

Ferry, J.D. 1936. Statistical evaluation of sieve constants in ultrafiltration. *J. Gen Physiol.* 20:95–104.

Fersht, A. 1985. Enzyme Structure and Mechanism, 2nd Ed. W.H. Freeman and Co., New York.

Finkelstein, A. 1974. Aqueous pores created in thin lipid membranes by the antibiotics nystatin, amphotericin B and gramicidin A. Implications for pores in plasma membranes. *In*: Drugs and Transport Processes. B.A. Callingham, editor. MacMillan, London, pp. 241–250.

Finkelstein, A., and O.S. Andersen. 1981. The gramicidin A channel: A review of its permeability characteristics with special reference to the single-file aspect of transport. *J. Membr. Biol.* 59:155–171.

Finkelstein, A., and A. Cass. 1968. Permeability and electrical properties of thin lipid membranes. *J. Gen. Physiol.* 52:145s–172s.

Fonseca, V., P. Daumas, L. Ranjalahy-Rasoloarijao, F. Heitz, R. Lazaro, Y. Trudelle, and O.S. Andersen. 1992. Gramicidin channels that have no tryptophan residues. *Biochemistry* 31:5340–5350.

Galbraith, T.P., and B.A. Wallace. 1998. Phospholipid chain length alters the equilibrium between pore and channel forms of gramicidin. *Faraday Discuss.* 111:159–164; discussion 225–246.

Gawrisch, K., D. Ruston, J. Zimmerberg, V.A. Parsegian, R.P. Rand, and N. Fuller. 1992. Membrane dipole potentials, hydration forces, and the ordering of water at membrane surfaces. *Biophys. J.* 61:1213–1223.

Gekko, K., and H. Noguchi. 1979. Compressibility of globular proteins in water at 25 °C. *J. Phys. Chem.* 83:2706–2714.

Girshman, J., J.V. Greathouse, R.E. Koeppe, II, and O.S. Andersen. 1997. Gramicidin channels in phospholipid bilayers having unsaturated acyl chains. *Biophys. J.* 73:1310–1319.

Greathouse, D.V., J.F. Hinton, K.S. Kim, and R.E. Koeppe II. 1994. Gramicidin A/short-chain phospholipid dispersions: Chain length dependence of gramicidin conformation and lipid organization. *Biochemistry* 33:4291–4299.

Greathouse, D.V., R.E. Koeppe II, L.L. Providence, S. Shobana, and O.S. Andersen. 1999. Design and characterization of gramicidin channels. *Methods Enzymol.* 294:525–550.

Gruner, S.M. 1985. Intrinsic curvature hypothesis for biomembrane lipid composition: A role for nonbilayer lipids. *Proc. Natl. Acad. Sci. USA* 82:3665–3669.

Hall, J.E. 1975. Access resistence of a small circular hole. *J. Gen. Physiol.* 66:531–532.

Hanai, T., D.A. Haydon, and J. Taylor. 1965. The variation of capacitance and conductance of bimolecular lipid membranes with area. *J. Theor. Biol.* 9:433–443.

Harold, F.M., and J.R. Baarda. 1967. Gramicidin, valinomycin, and cation permeability of *Streptococcus faecalis*. *J. Bacteriol.* 94:53–60.

He, K., S.J. Ludtke, Y. Wu, H.W. Huang, O.S. Andersen, D. Greathouse, and R.E. Koeppe II. 1994. Closed state of gramicidin channel detected by X-ray in-plane scattering. *Biophys. Chem.* 49:83–89.

Heckmann, K. 1965. Zur Theorie der "single file" diffusion. I. *Z. Phys. Chem. N.F.* 44:184–203.

Heckmann, K. 1972. Single file diffusion. *Biomembranes* 3:127–153.

Heitz, F., G. Spach, and Y. Trudelle. 1982. Single channels of 9,11,13,15-destryptophyl-phenylalanyl-gramicidin A. *Biophys. J.* 40:87–89.

Herrell, W.E., and D. Heilman. 1941. Experimental and clinical studies on gramicidin. *J. Clin. Invest.* 20:583–591.

Hille, B. 1968. Pharmacological modifications of the sodium channels of frog nerve. *J. Gen. Physiol.* 51:199–219.

Hille, B. 1975. Ionic selectivity of Na and K channels in nerve membranes. *In*: Membranes. Lipid Bilayers and Biological Membranes: Dynamic Properties. G. Eisenman, editor. Marcel Dekker, Inc., New York, pp. 255–323.

Hille, B. 2001. Ionic Channels of Excitable Membranes, 3rd Ed. Sinauer, Sunderland, MA.

Hille, B., and W. Schwarz. 1978. Potassium channels as multi-ion single-file pores. *J. Gen. Physiol.* 72:159–162.

Hladky, S.B., and D.A. Haydon. 1970. Discreteness of conductance change in bimolecular lipid membranes in the presence of certain antibiotics. *Nature* 225:451–453.

Hladky, S.B., and D.A. Haydon. 1972. Ion transfer across lipid membranes in the presence of gramicidin A. I. Studies of the unit conductance channel. *Biochim. Biophys. Acta* 274:294–312.

Hodgkin, A.L., and B. Katz. 1949. The effect of sodium ions on the electrical activity of the giant axon of the squid. *J. Physiol.* 108:37–77.

Hotchkiss, R.D. 1944. Gramicidin, tyrocidine, and tyrothricin. *Adv. Enzymol.* 4:153–199.

Hu, W., and T.A. Cross. 1995. Tryptophan hydrogen bonding and electric dipole moments: Functional roles in the gramicidin channel and implications for membrane proteins. *Biochemistry* 34:14147–14155.

Huang, H.W. 1986. Deformation free energy of bilayer membrane and its effect on gramicidin channel lifetime. *Biophys. J.* 50:1061–1070.

Huang, W., and D.G. Levitt. 1977. Theoretical calculation of the dielectric constant of a bilayer membrane. *Biophys. J.* 17:111–128.

Hünenberger, P.H., and J.A. McCammon. 1999. Ewald artifacts in computer simulations of ionic solvation and ion–ion interaction: A continuum electrostatics study. *J. Chem. Phys.* 110:1856.

Jagannadham, M.V., and R. Nagaraj. 2005. Conformation of gramicidin a in water: Inference from analysis of hydrogen/deuterium exchange behavior by matrix assisted laser desorption ionization mass spectrometry. *Biopolymers* 80:708–713.

Jiang, Y., A. Lee, J. Chen, M. Cadene, B.T. Chait, and R. MacKinnon. 2002. Crystal structure and mechanism of a calcium-gated potassium channel. *Nature* 417:515–522.

Jiang, Y., A. Lee, J. Chen, V. Ruta, M. Cadene, B.T. Chait, and R. MacKinnon. 2003. X-ray structure of a voltage-dependent K^+channel. *Nature* 423:33–41.

Jing, N., K.U. Prasad, and D.W. Urry. 1995. The determination of binding constants of micellar-packaged gramicidin A by ^{13}C-and ^{23}Na-NMR. *Biochim. Biophys. Acta* 1238:1–11.

Jing, N., and D.W. Urry. 1995. Ion pair binding of Ca^{2+} and Cl^- ions in micellar-packaged gramicidin A. *Biochim. Biophys. Acta* 1238:12–21.

Jordan, P.C. 1981. Energy barriers for passage of ions through channels. Exact solution of two electrostatic problems. *Biophys. Chem.* 13:203–212.

Jordan, P.C. 1984. The total electrostatic potential in a gramicidin channel. *J. Membr. Biol.* 78:91–102.

Jordan, J.B., P.L. Easton, and J.F. Hinton. 2005. Effects of phenylalanine substitutions in gramicidin A on the kinetics of channel formation in vesicles and channel structure in SDS micelles. *Biophys. J.* 8:224–234.

Jude, A.R., D.V. Greathouse, R.E. Koeppe II, L.L. Providence, and O.S. Andersen. 1999. Modulation of gramicidin channel structure and function by the aliphatic "spacer" residues 10, 12, and 14 between the tryptophans. *Biochemistry* 38:1030–1039.

Karplus, M., and J.A. McCammon. 1981. The internal dynamics of globular proteins. *CRC Crit. Rev. Biochem.* 9:293–349.

Katsaras, J., R.S. Prosser, R.H. Stinson, and J.H. Davis. 1992. Constant helical pitch of the gramicidin channel in phospholipid bilayers. *Biophys. J.* 61:827– 830.

Kauzmann, W. 1957. Some factors in the interpretation of protein denaturation. *Adv. Protein Chem.* 14:1–63.

Kemp, G., and C. Wenner. 1976. Solution, interfacial, and membrane properties of gramicidin A. *Arch. Biochem. Biophys.* 176:547–555.

Kessler, N., H. Schuhmann, S. Morneweg, U. Linne, and M.A. Marahiel. 2004. The linear pentadecapeptide gramicidin is assembled by four multimodular nonribosomal peptide synthetases that comprise 16 modules with 56 catalytic domains. *J. Biol. Chem.* 279:7413–7419.

Ketchem, R.R., B. Roux, and T.A. Cross. 1997. High-resolution polypeptide structure in a lamellar phase lipid environment from solid state NMR derived orientational constraints. *Structure* 5:1655–1669.

Killian, J.A., S. Morein, P.C. van der Wel, M.R. de Planque, D.V. Greathouse, and R.E. Koeppe 2nd. 1999. Peptide influences on lipids. *Novartis Found. Symp.* 225:170–183; discussion 183–187.

Killian, J.A., M.J. Taylor, and R.E. Koeppe II. 1992. Orientation of the valine-1 side chain of the gramicidin transmembrane channel and implications for channel functioning. A^2 H NMR study. *Biochemistry* 31:11283–11290.

Killian, J.A., and G. von Heijne. 2000. How proteins adapt to a membrane-water interface. *TIBS* 25:429–434.

King, E.L., and C. Altman. 1956. A schematic method of deriving the rate laws for enzyme-catalyzed reactions. *J. Phys. Chem.* 60:1375–1378.

Koeppe, R.E., II, and O.S. Andersen. 1996. Engineering the gramicidin channel. *Annu. Rev. Biophys. Biomol. Struct.* 25:231–258.

Koeppe, R.E., II, D.V. Greathouse, A. Jude, G. Saberwal, L.L. Providence, and O.S. Andersen. 1994a. Helix sense of gramicidin channels as a "nonlocal" function of the primary sequence. *J. Biol. Chem.* 269:12567–12576.

Koeppe, R.E. II, J.A. Killian, and D.V. Greathouse. 1994b. Orientations of the tryptophan 9 and 11 side chains of the gramicidin channel based on deuterium nuclear magnetic resonance spectroscopy. *Biophys. J.* 66:14–24.

Koeppe, R.E., II, J.-L. Mazet, and O.S. Andersen. 1990. Distinction between dipolar and inductive effects in modulating the conductance of gramicidin channels. *Biochemistry* 29:512–520.

Koeppe, R.E., II, J.A. Paczkowski, and W.L. Whaley. 1985. Gramicidin K, a new linear channel-forming gramicidin from *Bacillus brevis*. *Biochemistry* 24:2822–2827.

Koeppe, R.E., II, H. Sun, P.C. van der Wel, E.M. Scherer, P. Pulay, and D.V. Greathouse. 2003. Combined experimental/theoretical refinement of indole ring geometry using deuterium magnetic resonance and ab initio calculations. *J. Am. Chem. Soc.* 125:12268–12276.

König, S., E. Sackmann, D. Richter, R. Zorn, C. Carlile, and T.M. Bayerl. 1994. Molecular dynamics of water in oriented DPPC multilayers studied by quasielastic neutron scattering and deuterium–nuclear magnetic resonance relaxation. *J. Chem. Phys.* 100:3307–3316.

Kramers, H.A. 1940. Brownian motion in a field of force and the diffusion model of chemical reactions. *Physica* 7:284–304.

Kuo, A., J.M. Gulbis, J.F. Antcliff, T. Rahman, E.D. Lowe, J. Zimmer, J. Cuthbertson, F.M. Ashcroft, T. Ezaki, and D.A. Doyle. 2003. Crystal structure of the potassium channel KirBac1.1 in the closed state . *Science* 300:1922–1926.

Lamoureux, G., A.D. MacKerell Jr., and B. Roux. 2003. A simple polarizable water model based on classical Drude oscillators. *J. Chem. Phys.* 119:5185–5197.

Lamoureux, G., and B. Roux. 2003. Modeling induced polarizability with classical Drude oscillators: Theory and molecular dynamics simulation algorithm. *J. Chem. Phys.* 119:3025–3039.

Langs, D.A. 1988. Three-dimensional structure at 0.86 Å of the uncomplexed form of the transmembrane ion channel peptide gramicidin A. *Science* 241:188–191.

Läuger, P. 1976. Diffusion-limited ion flow through pores. *Biochim. Biophys. Acta* 455:493–509.

Lee, K.C., S. Huo, and T.A. Cross. 1995. Lipid–peptide interface: Valine conformation and dynamics in the gramicidin channel. *Biochemistry* 34:857–867.

Levitt, D.G. 1978. Electrostatic calculations for an ion channel. I. Energy and potential profiles and interactions between ions. *Biophys. J.* 22:209–219.

Levitt, D.G., S.R. Elias, and J.M. Hautman. 1978. Number of water molecules coupled to the transport of sodium, potassium and hydrogen ions via gramicidin, nonactin or valinomycin. *Biochim. Biophys. Acta* 512:436–451.

Lewis, B.A., and D.M. Engelman. 1983. Lipid bilayer thickness varies linearly with acyl chain length in fluid phosphatidylcholine vesicles. *J. Mol. Biol.* 166:211–217.

Lindahl, E., and O. Edholm. 2000. Mesoscopic undulations and thickness fluctuations in lipid bilayers from molecular dynamics simulations. *Biophys. J.* 79:426.

Lipmann, F. 1980. Bacterial production of antibiotic polypeptides by thiol-linked synthesis on protein templates. *Adv. Microb. Physiol.* 21:227–266.

Liu, N., and R.L. Kay. 1977. Redetermination of the pressure dependence of the lipid bilayer phase transition. *Biochemistry* 16:3484–3486.

Long, S.B., E.B. Campbell, and R. Mackinnon. 2005. Crystal structure of a mammalian voltage-dependent Shaker family K^+ channel. *Science* 309:897–903.

Lundbæk, J.A., P.H.A.J. Birn, R. Søgaard, C. Nielsen, J. Girshman, M.J. Bruno, S.E. Tape, J. Egebjerg, D.V. Greathouse, G.L. Mattice, R.E. Koeppe II, and O.S. Andersen. 2004. Regulation of sodium channel function by bilayer elasticity: The importance of hydrophobic coupling: Effects of micelle-forming amphiphiles and cholesterol. *J. Gen. Physiol.* 123:599–621.

Mackay, D.H.J., P.H. Berens, K.R. Wilson, and A.T. Hagler. 1984. Structure and dynamics of ion transport through gramicidin A. *Biophys. J.* 46:229–248.

Mamanov, A.B., R.D. Coalson, A. Nitzan, and M.G. Kurnikova. 2003. The role of the dielectric barrier in narrow biological channels: A novel composite approach to modeling single-channel currents. *Biophys. J.* 84:3646–3661.

Mattice, G.L., R.E. Koeppe II, L.L. Providence, and O.S. Andersen. 1995. Stabilizing effect of D-alanine^2 in gramicidin channels. *Biochemistry* 34:6827–6837.

Mazet, J.L., O.S. Andersen, and R.E. Koeppe II. 1984. Single-channel studies on linear gramicidins with altered amino acid sequences. A comparison of phenylalanine, tryptophan, and tyrosine substitutions at positions 1 and 11. *Biophys. J.* 45:263–276.

Mobashery, N., C. Nielsen, and O.S. Andersen. 1997. The conformational preference of gramicidin channels is a function of lipid bilayer thickness. *FEBS Lett.* 412:15–20.

Morrow, J.S., W.R. Veatch, and L. Stryer. 1979. Transmembrane channel activity of gramicidin A analogs: Effects of modification and deletion of the amino-terminal residue. *J. Mol. Biol.* 132:733–738.

Mouritsen, O.G., and M. Bloom. 1984. Mattress model of lipid–protein interactions in membranes. *Biophys. J.* 46:141–153.

Mukherjee, S., and A. Chattopadhyay. 1994. Motionally restricted tryptophan environments at the peptide–lipid interface of gramicidin channels. *Biochemistry* 33:5089–5097.

Myers, V.B., and D.A. Haydon. 1972. Ion transfer across lipid membranes in the presence of gramicidin A. II. Ion selectivity. *Biochim. Biophys. Acta* 274:313–322.

Neher, E., J. Sandblom, and G. Eisenman. 1978. Ionic selectivity, saturation, and block in gramicidin A channels. II. Saturation behavior of single channel conductances and evidence for the existence of multiple binding sites in the channel. *J. Membr. Biol.* 40:97–116.

Neumcke, B., and P. Läuger. 1969. Nonlinear electrical effects in lipid bilayer membranes II. Integration of the generalized Nernst–Planck equations. *Biophys. J.* 9:1160–1170.

Nicholson, L.K., F. Moll, T.E. Mixon, P.V. LoGrasso, J.C. Lay, and T.A. Cross. 1987. Solid-state 15N NMR of oriented lipid bilayer bound gramicidin A'. *Biochemistry* 26:6621–6626.

Nielsen, C., M. Goulian, and O.S. Andersen. 1998. Energetics of inclusion-induced bilayer deformations. *Biophys. J.* 74:1966–1983.

Nielsen, C., and O.S. Andersen. 2000. Inclusion-induced bilayer deformations: Effects of monolayer equilibrium curvature. *Biophys. J.* 79:2583–2604.

Nimigean, C.M., and C. Miller. 2002. Na^+ block and permeation in a K^+ channel of known structure. *J. Gen. Physiol.* 120:323.

Noskov, S.Y., S. Bernèche, and B. Roux. 2004. Control of ion selectivity in potassium channels by electrostatic and dynamic properties of carbonyl ligands. *Nature* 431:830–834.

O'Connell, A.M., R.E. Koeppe II, and O.S. Andersen. 1990. Kinetics of gramicidin channel formation in lipid bilayers: Transmembrane monomer association. *Science* 250:1256–1259.

Oiki, S., R.E. Koeppe II, and O.S. Andersen. 1994. Asymmetric gramicidin channels. Heterodimeric channels with a single F^6 Val^1 residue. *Biophys. J.* 66:1823–1832.

Oiki, S., R.E. Koeppe II, and O.S. Andersen. 1995. Voltage-dependent gating of an asymmetric gramicidin channel. *Proc. Natl. Acad. Sci. USA* 92:2121–2125.

Olah, G.A., H.W. Huang, W.H. Liu, and Y.L. Wu. 1991. Location of ion-binding sites in the gramicidin channel by X-ray diffraction. *J. Mol. Biol.* 218:847–858.

Owicki, J.C., M.W. Springgate, and H.M. McConnell. 1978. Theoretical study of protein–lipid and protein–protein interactions in bilayer membranes. *Proc. Natl. Acad. Sci. USA* 75:1616–1619.

Park, J.B., H.J. Kim, P.D. Ryu, and E. Moczydlowski. 2003. Effect of phosphatidylserine on unitary conductance and Ba^{2+} block of the BK Ca^{2+}-activated K^+ channel: Re-examination of the surface charge hypothesis. *J. Gen. Physiol.* 121:375–398.

Parsegian, A. 1969. Energy of an ion crossing a low dielectric membrane: Solutions to four relevant electrostatic problems. *Nature* 221:844–846.

Providence, L.L., O.S. Andersen, D.V. Greathouse, R.E. Koeppe II, and R. Bittman. 1995. Gramicidin channel function does not depend on phospholipid chirality. *Biochemistry* 34:16404–16411.

Pulay, P., E.M. Scherer, P.C. van der Wel, and R.E. Koeppe II. 2005. Importance of tensor asymmetry for the analysis of 2H NMR spectra from deuterated aromatic rings. *J. Am. Chem. Soc.* 127:17488–17493.

Ramachandran, G.N., and R. Chandrasekaran. 1972. Studies on dipeptide conformation and on peptides with sequences of alternating L and D residues with special reference to antibiotic and ion transport peptides. *Progr. Pept. Res.* 2:195–215.

Rawat, S.S., D.A. Kelkar, and A. Chattopadhyay. 2004. Monitoring gramicidin conformations in membranes: A fluorescence approach. *Biophys. J.* 87:831–843.

Rawicz, W., K.C. Olbrich, T. McIntosh, D. Needham, and E. Evans. 2000. Effect of chain length and unsaturation on elasticity of lipid bilayers. *Biophys. J.* 79:328–339.

Robinson, R.A., and R.H. Stokes. 1959. Electrolyte Solutions, 2nd Ed. Butterworth, London.

Rosenberg, P.A., and A. Finkelstein. 1978. Interaction of ions and water in gramicidin A channels: Streaming potentials across lipid bilayer membranes. *J. Gen. Physiol.* 72:327–340.

Roux, B. 2002. Computational studies of the gramicidin channel. *Acc. Chem. Res.* 35:366–375.

Roux, B., T.W. Allen, S. Bernèche, and W. Im. 2004. Theoretical and computational models of biological ion channels. *Q. Rev. Biophys.* 37:15–103.

Roux, B., and M. Karplus. 1993. Ion transport in the gramicidin channel: Free energy of the solvated right-handed dimer in a model membrane. *J. Am. Chem. Soc.* 115:3250–3262.

Roux, B., B. Prod'hom, and M. Karplus. 1995. Ion transport in the gramicidin channel: Molecular dynamics study of single and double occupancy. *Biophys. J.* 68:876–892.

Russell, E.W.B., L.B. Weiss, F.I. Navetta, R.E. Koeppe II, and O.S. Andersen. 1986. Single-channel studies on linear gramicidins with altered amino acid side chains. Effects of altering the polarity of the side chain at position 1 in gramicidin A. *Biophys. J.* 49:673–686.

Salom, D., M.C. Baño, L. Braco, and C. Abad. 1995. HPLC demonstration that an all Trp→Phe replacement in gramicidin A results in a conformational rearrangement from beta-helical monomer to double-stranded dimer in model membranes. *Biochem. Biophys. Res. Commun.* 209:466–473.

Salom, D., E. Perez-Paya, J. Pascal, and C. Abad. 1998. Environment- and sequence-dependent modulation of the double-stranded to single-stranded conformational transition of gramicidin A in membranes. *Biochemistry* 37:14279–14291.

Sancho, M., and G. Martinez. 1991. Electrostatic modeling of dipole–ion interactions in gramicidin like channels. *Biophys. J.* 60:81–88.

Sarges, R., and B. Witkop. 1965. Gramicidin A. V. The structure of valine- and isoleucine-gramicidin A. *J. Am. Chem. Soc.* 87:2011–2019.

Scarlata, S.F. 1988. The effects of viscosity on gramicidin tryptophan rotational motion. *Biophys. J.* 54:1149–1157.

Schagina, L.V., A.E. Grinfeldt, and A.A. Lev. 1978. Interaction of cation fluxes in gramicidin A channels in lipid bilayer membranes. *Nature* 273:243–245.

Schatzberg, P. 1965. Diffusion of water through hydrocarbon liquids. *J. Polym. Sci. C* 10:87–92.

Schiffer, M., C.-H. Chang, and F.J. Stevens. 1992. The functions of tryptophan residues in membrane proteins. *Protein Engng.* 5:213–214.

Schracke, N., U. Linne, C. Mahlert, and M.A. Marahiel. 2005. Synthesis of linear gramicidin requires the cooperation of two independent reductases. *Biochemistry* 44:8507–8513.

Segrest, J.P., and R.J. Feldmann. 1974. Membrane proteins: Amino acid sequence and membrane penetration. *J. Mol. Biol.* 87:853–858.

Separovic, F., R. Pax, and B. Cornell. 1993. NMR order parameter analysis of a peptide plane aligned in a lyotropic liquid crysta. *Mol. Phys.* 78:357–369.

Sham, S.S., S. Shobana, L.E. Townsley, J.B. Jordan, J.Q. Fernandez, O.S. Andersen, D.V. Greathouse, and J.F. Hinton. 2003. The structure, cation binding, transport, and conductance of Gly15-gramicidin A incorporated into SDS micelles and PC/PG vesicles. *Biochemistry* 42:1401–1409.

Sharp, K.A., A. Nicholls, R.F. Fine, and B. Honig. 1991. Reconciling the magnitude of the microscopic and macroscopic hydrophobic effects. *Science* 252:106–109.

Sieber, S.A., and M.A. Marahiel. 2005. Molecular mechanisms underlying nonribosomal peptide synthesis: Approaches to new antibiotics. *Chem. Rev.* 105:715–738.

Simon, S.A., and T.J. McIntosh. 1986. Depth of water penetration into lipid bilayers. *Methods Enzymol.* 127:511–521.

Smyth, C.P. 1955. Dielectric Behavior and Structure. Mcgraw-Hill, New York.

Tanford, C. 1980. The Hydrophobic Effect: Formation of Micelles and Biological Membranes, 2nd Ed. Wiley, New York.

Tian, F., and T.A. Cross. 1999. Cation transport: An example of structural based selectivity. *J. Mol. Biol.* 285:1993–2003.

Tosh, R.E., and P.J. Collings. 1986. High pressure volumetric measurements in dipalmitoylphosphatidylcholine bilayers. *Biochim. Biophys. Acta* 859:10–14.

Townsley, L.E., W.A. Tucker, S. Sham, and J.F. Hinton. 2001. Structures of gramicidins A, B, and C incorporated into sodium dodecyl sulfate micelles. *Biochemistry* 40:11676–11686.

Unwin, N. 2005. Refined structure of the nicotinic acetylcholine receptor at 4 Å resolution. *J. Mol. Biol.* 346:967–989.

Urban, B.W., S.B. Hladky, and D.A. Haydon. 1978. The kinetics of ion movements in the gramicidin channel. *Fed. Proc.* 37:2628–2632.

Urban, B.W., S.B. Hladky, and D.A. Haydon. 1980. Ion movements in gramicidin pores. An example of single-file transport. *Biochim. Biophys. Acta* 602:331–354.

Urry, D.W. 1971. The gramicidin A transmembrane channel: A proposed $\pi_{(L,D)}$ helix. *Proc. Natl. Acad. Sci. USA* 68:672–676.

Urry, D.W. 1972. Protein conformation in biomembranes: Optical rotation and absorption of membrane suspensions. *Biochim. Biophys. Acta* 265:115–168.

Urry, D.W. 1973. Polypeptide conformation and biological function: β-helices ($\pi_{L,D}$-helices) as permselective transmembrane channels. *Jerusalem Symp. Quant. Chem. Biochem.* 5:723–736.

Urry, D.W., S. Alonso-Romanowski, C.M. Venkatachalam, R.J. Bradley, and R.D. Harris. 1984. Temperature dependence of single channel currents and the peptide libration mechanism for ion transport through the gramicidin A transmembrane channel. *J. Membr. Biol.* 81:205–217.

Urry, D.W., T.L. Trapane, C.M. Venkatachalam, and R.B. McMichens. 1989. Ion interactions at membranous polypeptide sites using nuclear magnetic resonance: Determining rate and binding constants and site locations. *Methods Enzymol.* 171:286–342.

Urry, D.W., T.L. Trapane, J.T. Walker, and K.U. Prasad. 1982. On the relative lipid membrane permeability of Na^+ and Ca^{2+}. A physical basis for the messenger role of Ca^{2+}. *J. Biol. Chem.* 257:6659–6661.

Veatch, W., and L. Stryer. 1977. The dimeric nature of the gramicidin A transmembrane channel: Conductance and fluorescence energy transfer studies of hybrid channels. *J. Mol. Biol.* 113:89–102.

Veatch, W.R., and E.R. Blout. 1974. The aggregation of gramicidin A in solution. *Biochemistry* 13:5257–5264.

Veatch, W.R., E.T. Fossel, and E.R. Blout. 1974. The conformation of gramicidin A. *Biochemistry* 13:5249–5256.

Wallace, B.A., W.R. Veatch, and E.R. Blout. 1981. Conformation of gramicidin A in phospholipid vesicles: Circular dichroism studies of effects of ion binding, chemical modification, and lipid structure. *Biochemistry* 20:5754–5760.

Weinstein, S., B.A. Wallace, E.R. Blout, J.S. Morrow, and W. Veatch. 1979. Conformation of gramicidin A channel in phospholipid vesicles: A carbon-13 and fluorine-19 nuclear magnetic resonance study. *Proc. Natl. Acad. Sci. USA* 76:4230–4234.

Weinstein, S., B.A. Wallace, J.S. Morrow, and W.R. Veatch. 1980. Conformation of the gramicidin A transmembrane channel: A 13C nuclear magnetic resonance study of 13C-enriched gramicidin in phosphatidylcholine vesicles. *J. Mol. Biol.* 143:1–19.

Weiss, L.B., and R.E. Koeppe II. 1985. Semisynthesis of linear gramicidins using diphenyl phosphorazidate (DPPA). *Int. J. Pept. Protein Res.* 26:305–310.

Weiss, M.S., A. Kreusch, E. Schiltz, U. Nestel, W. Welte, J. Weckesser, and G.E. Schulz. 1991. The structure of porin from *Rhodobacter capsulatus* at 1.8 Å resolution. *FEBS Lett.* 280:379–382.

White, S.H., and W.C. Wimley. 1999. Membrane protein folding and stability: Physical principles. *Annu. Rev. Biophys. Biomol. Struct.* 28:319–365.

Wiener, M.C., and S.H. White. 1992. Structure of a fluid dioleoylphosphatidylcholine bilayer determined by joint refinement of x-ray and neutron diffraction data. III. Complete structure. *Biophys. J.* 61:437–447.

Wolfenden, R., and M.J. Snider. 2001. The depth of chemical time and the power of enzymes as catalysts. *Acc. Chem. Res.* 12:938–945.

3 Voltage-Gated Ion Channels

Francisco Bezanilla

3.1 Introduction

The bit of information in nerves is the action potential, a fast electrical transient in the transmembrane voltage that propagates along the nerve fiber. In the resting state, the membrane potential of the nerve fiber is about −60 mV (negative inside with respect to the extracellular solution). When the action potential is initiated, the membrane potential becomes less negative and even reverses sign (overshoot) within a millisecond and then goes back to the resting value in about 2 ms, frequently after becoming even more negative than the resting potential. In a landmark series of papers, Hodgkin and Huxley studied the ionic events underlying the action potential and were able to describe the conductances and currents quantitatively with their classical equations (Hodgkin and Huxley, 1952). The generation of the rising phase of the action potential was explained by a conductance to Na^+ ions that increases as the membrane potential is made more positive. This is because, as the driving force for the permeating ions (Na^+) was in the inward direction, more Na^+ ions come into the nerve and make the membrane more positive initiating a positive feedback that depolarizes the membrane even more. This positive feedback gets interrupted by the delayed opening of another voltage-dependent conductance that is K-selective. The driving force for K^+ ions is in the opposite direction of Na^+ ions, thus K^+ outward flow repolarizes the membrane to its initial value. The identification and characterization of the voltage-dependent Na^+ and K^+ conductances was one of the major contributions of Hodgkin and Huxley. In their final paper of the series, they even proposed that the conductance was the result of increased permeability in discrete areas under the control of charges or dipoles that respond to the membrane electric field. This was an insightful prediction of ion channels and gating currents.

Many years of electrical characterization, effects of toxins on the conductances, molecular biological techniques, and improvement of recording techniques led to the identification of separate conducting entities for Na^+ and K^+ conductances. These conductances were finally traced to single-membrane proteins, called ion channels, that can gate open and closed an ion conducting pathway in response to changes in membrane potential.

3.2 Voltage-Dependent Ion Channels Are Membrane Proteins

The first voltage-dependent ion channel that was isolated and purified was extracted from the eel electroplax where there is a large concentration of Na^+ channels (Agnew et al., 1978). Several years later, the sequence of the eel Na^+ channel was deduced from its mRNA (Noda et al., 1984). The first K^+ channel sequence was deduced from the *Shaker* mutant of *Drosophila melanogaster* (Tempel et al., 1987). These initial sequences were the basis to subsequent cloning of a large number of Na^+, K^+, and Ca^{2+} channels in many different species. Hydropathy plots were helpful in deciding which parts of the sequence were transmembrane or intra- or extracellular. A basic pattern emerged from all these sequences: the functional channels are made up of four subunits (K^+ channels) or one protein with four homologous domains (Na^+ and Ca^{2+} channels). Each one of the domains or subunits has six transmembrane segments and a pore loop (see Fig. 3.1). The fifth and sixth transmembrane segments (S5 and S6) and the pore loop were found to be responsible for ion conduction. The

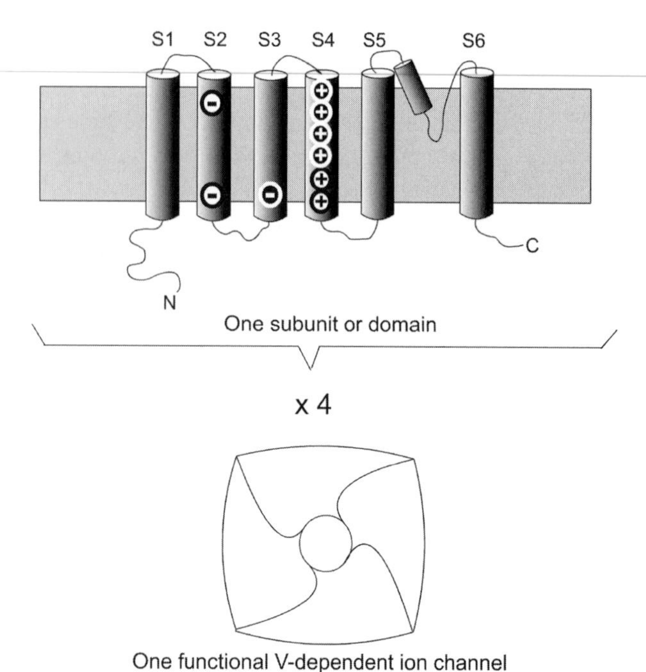

Fig. 3.1 The general architecture of voltage-gated channels. Top part shows the basic subunit (or domain in the case of Na^+ and Ca^{2+} channels). The gray background represents the lipid bilayers. The cylinders are transmembrane segments. The region between S5 and S6 forms the pore. Segments 1 through 4 are called the voltage sensor part of the channel. The + or – signs in white indicate charges that have been implicated in voltage sensing. In the bottom, a schematic view of the channel from the outside showing the assembled four subunits or domains.

fourth transmembrane segment (S4) contains several basic residues, arginines or lysines and was initially postulated to be the voltage sensor (Noda et al., 1984). In addition, S2 and S3 contain acidic residues such as aspartate and glutamate. Most of the channels have additional subunits that modify the basic function but they are not necessary for voltage sensing and ion conduction.

3.3 The Parts of the Voltage-Dependent Channel

We think of voltage-dependent channels as made of three basic parts: the *voltage sensor*, the *pore* or *conducting pathway*, and the *gate*. These three parts can be roughly mapped in the putative secondary structure (Fig. 3.1). The pore and the gate are in the S5-loop-S6 region and the voltage sensor in the S1–S4 region. As the conduction is dependent on the voltage across the membrane, a useful analogy is a field effect transistor (FET; see Fig. 3.1). If we take a typical voltage-dependent K channel, its voltage sensor corresponds to the gate of a p-channel FET, the conducting pathway of the ion channel corresponds to the p-channel (Fig. 3.2) and the gate of the channel is the space charge in the p-channel. As we will see below, this analogy is useful to discuss the parts of the channel but it cannot be pushed very far because, although the functions are similar, the actual mechanisms are quite different.

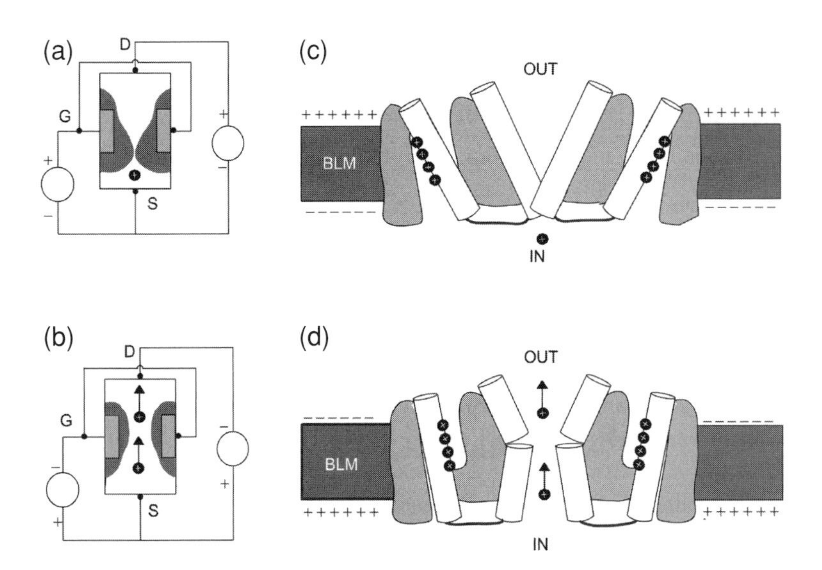

Fig. 3.2 A comparison between a field effect transistor and a voltage-gated ion channel. The FET transistor is represented as a p-channel device to make a closer analogy to a cation selective voltage-gated channel. (a) and (c) are the closed states; (b) and (d) are the open states. Notice that, in contrast with the FET, the gate in the voltage-gated channel indicates the actual point of flow interruption. In the FET, D is the drain, S is the source, G is the gate. For details see text.

3.3.1 The Conducting Pathway

Living cells, and in particular nerve fibers, are surrounded by a thin membrane made of a bimolecular layer of lipids. The permeability of ions through the lipid bilayer is extremely low because it takes a large amount of energy to put a charged ion species inside the low-dielectric constant lipid bilayer (Parsegian, 1969). The conducting pathway of ion channels lowers that energy barrier by providing a favorable local environment and thus allowing large flows under an appropriate driving force. Details of the ion conduction pore structure, conductance, and selectivity are covered in other chapters in this book (see Chapters 4, 5 and 16). What is important to emphasize here is that the ion flow is proportional to the driving force for the selected ion. The driving force corresponds to the difference between the voltage applied, V and the voltage at which there is no flow, or reversal potential E. If the channel is perfectly selective to one type of ion, say K^+, then E is the Nernst potential, otherwise E is predicted by the Goldman–Hodgkin and Katz equation that considers concentrations and relative permeabilities. Knowing the conductance of the conducting pathway γ, we can compute the current flow i through the open conducting pore as,

$$i = \gamma(V - E). \tag{3.1}$$

The i–V curve of an open channel may be nonlinear because in general, γ is voltage dependent.

3.3.2 The Gate

The ion conduction through the pore may be interrupted by closing a gate (see Fig. 3.2). Thermal fluctuations will close and open the gate randomly and the current will have interruptions. In voltage-dependent channels, the probability that the gate is open, P_o, depends on the membrane potential. In the majority of voltage-dependent Na^+, K^+, and Ca^{2+} channels from nerve and muscle the P_o increases with membrane depolarization (i.e., decrease in the resting potential). There are a few cases, such as Kat1 channel, where P_o increases on hyperpolarization.

The operation of the gate can be seen by recording the current flowing through a single ion channel. This is possible with the patch clamp technique (Hamill et al., 1981) that records currents from a very small patch of membrane with a small glass pipette and a low-noise system that can resolve currents in the order 1 pA. An example of the operation of one K^+ channel is shown in a simulation in Fig. 3.3. As the internal concentration of K^+ is more than 10 times higher in the cell as compared to the extracellular space, the reversal potential E for K^+ channels is around −80 mV. Starting with a negative membrane potential (−100 mV), the channel is closed most of the time. A depolarizing voltage pulse to −30 mV increases the open probability and the channel spends some time in the open state (Fig 3.3a). As we are dealing with one molecule, thermal fluctuations will generate different responses for each repetition of the same pulse (four of such are shown in the figure). A larger depolarization

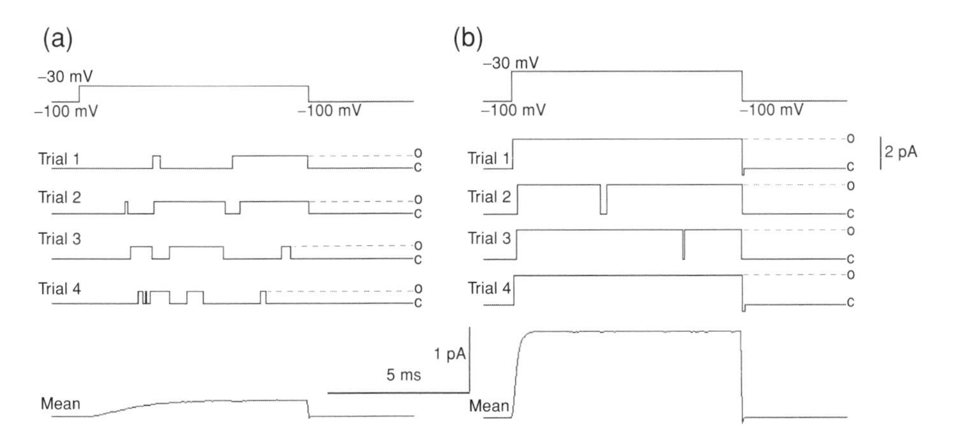

Fig. 3.3 Time course of single channel and macroscopic ionic currents. The applied voltage is in the top trace and the current recorded through one channel is shown for four different trials. The mean current is the result of thousands of trials. (a) Small depolarization to −30 mV, open the channel infrequently. (b) A large depolarization (to +30 mV) opens the channel most of the time. c is the closed state and o is the open state.

(+30 mV) increases the P_o even more by increasing the open times and decreasing the closed times, as seen in Fig. 3.3b. Notice also that the time elapsed between the start of the pulse and the first opening (first latency) is decreased for the larger depolarization. Apart from increasing the open times, the magnitude of the current through the pore was increased by the larger depolarization. This is because the V applied is now further away from E, increasing the driving force for ion movement. Thus, this increase in current is not a result of increasing P_o but is just a passive property of the open pore. An average of several thousands of repetitions gives us the macroscopic ionic currents. Provided the channels do not interact, the average of thousands of repetitions is the same as having thousands of channels operating simultaneously. The bottom trace (labeled mean) in Fig. 3.3a and 3.3b shows the macroscopic currents for −30 and +30 mV, respectively. Notice that the turn-on kinetics is faster for a more positive potential and that the current magnitude is also increased. The kinetics change is the result of an increased P_o while the magnitude increase is the result of both increased P_o and driving force. We can now write the expression for the macroscopic current as,

$$I = P_o(V, t)N\gamma(V - E), \tag{3.2}$$

where N is the channel density and $P_o(V, t)$ is the voltage and time-dependent open probability.

3.3.3 The Voltage Sensor

How does P_o become voltage dependent? It is clear that to detect changes in membrane potential a voltage sensor is needed. The electric field in the bilayer could be detected by electric or magnetic charges or dipoles that change their position according to changes in the field. As there is no evidence of magnetic charges, electric charges or dipoles remain as the prime candidates. We will see below that the actual charges involved in voltage sensing have been identified and a schematic representation of their relocation is shown in Fig. 3.2b. In the resting (hyperpolarized) condition, the membrane is negative inside and the positive charges are located in contact with the interior of the cell. Upon depolarization, the positive charges are driven toward the outside. This movement in the electric field has two consequences: it is coupled to the gate resulting in pore opening (Fig. 3.2b) and the charge translocation produces another membrane current that is transient in nature, called *gating current*. It is called gating current because it ultimately gates the channel open and close and it is transient because the charge locations are bound to limiting positions as they are tethered to the protein.

3.4 Gating Charge and the Voltage Sensor

An understanding of the voltage sensor requires a characterization of the gating charge movement and a correlation of that movement to structural changes in the protein. In this section, we will address two functional questions. The first question is what are the kinetics and steady state properties of the gating charge movement and how does this charge movement relate to channel activation. The second is how many elementary charges move in one channel to fully activate the conductance and how does this movement occur in one channel.

3.4.1 The Gating Currents and the Channel Open Probability

The movement of charge or dipole reorientation is the basic mechanism of the voltage sensor and was predicted by Hodgkin and Huxley (1952). Gating currents are transient and they only occur in the potential range where the sensor responds to the electric field, therefore they behave like a nonlinear capacitance. In addition, as gating currents are small, to record them it is necessary to decrease or eliminate the ionic currents through the pore and eliminate the normal capacitive current required to charge or discharge the membrane. This is normally accomplished by applying a pulse in the voltage range that activates the current and then subtract another pulse or pulses in the voltage range that does not activate the currents to eliminate the linear components (Armstrong and Bezanilla, 1973; Keynes and Rojas, 1974). Using these subtraction techniques, the kinetics of Na^+ gating currents were studied in detail in squid giant axon and other preparations where a high-channel density was found. The combination of gating currents, macroscopic ionic currents, and single-channel recordings were used to propose detailed kinetic models of channel operation (see Vandenberg and Bezanilla, 1991). When voltage-dependent channels were cloned and expressed in oocytes or cell lines, it was possible to achieve large channel densities on the surface membrane and study those channels in virtual absence of currents from other channels. In comparison to the currents in natural tissues such as the squid giant axon, the expression systems gating currents were much larger and made the recording easier and cleaner. In addition, the study of the pore region gave us the possibility of mutating the channel protein to eliminate ionic conduction but maintaining the operation of the gating currents (Perozo et al., 1993). We can illustrate the basic features of gating currents and their relation to ionic currents in recordings from Shaker K^+ channels with fast inactivation removed (Shaker-IR) as shown in Fig. 3.4a.

In this figure, two separate experiments are shown. The top traces are the time course of the ionic currents recorded during pulses that range from -120 to 0 mV, starting and returning to -90 mV. The bottom traces are gating currents recorded for the same set of pulses from Shaker-IR with a mutation that changes a tryptophan into a phenylalanine in position 434 (W434F) that renders the pore nonconducting (Perozo et al., 1993). Several features that are characteristic of most voltage-dependent channels can be observed. First, the ionic currents do not show significant activation for potential more negative than -40 mV while the gating currents are visible for all the pulses applied, implying that there is charge displacement in a region of potentials where most of the channels are closed. Second, the time course of activation of the ionic current is similar to the time course of decay of the gating current. Third, the time course of the return of the charge (gating current "tail") changes its kinetics drastically when returning from a pulse more positive than -40 mV, which is precisely the potential at which ionic currents become clearly visible. The gating tails are superimposable for potentials more positive than -20 mV, showing that most of the charge has moved at -20 mV. The total charge moved at each potential may be computed as the time integral of the gating current for each pulse. As we will see below, it is possible to estimate the total charge moved per channel molecule, therefore the voltage dependence of the charge moved can be plotted as shown in Fig. 3.4b. Knowing the number of channels present (see below), using Eq. 3.2, it is possible to estimate the voltage dependence of P_o, as shown in the same figure. Fig. 3.4b establishes the relationship between the charge movement and channel opening and clearly shows that the opening of the channel is not superimposable with charge movement as expected from a simple two state model. A striking feature of these plots is that the $Q(V)$ relation is displaced to the left of the $P_o(V)$ curve so that there is quite a large charge movement in a region where the P_o is essentially zero. This is an expected feature of a channel that requires several processes to occur to go from closed to open, such as the classical Hodgkin and Huxley model where four independent particles are needed to be simultaneously in the active position for the channel to be open.

Current recordings of the type shown in Fig. 3.4a can be used to formulate kinetic models of channel gating. These models are normally written as a collection of closed and open states interconnected with rate constants that, in general, are a function of the membrane potential. Initially, kinetic models were developed from the macroscopic ionic currents only (Hodgkin and Huxley, 1952). The addition of single-channel recordings and gating current recordings imposes several constraints

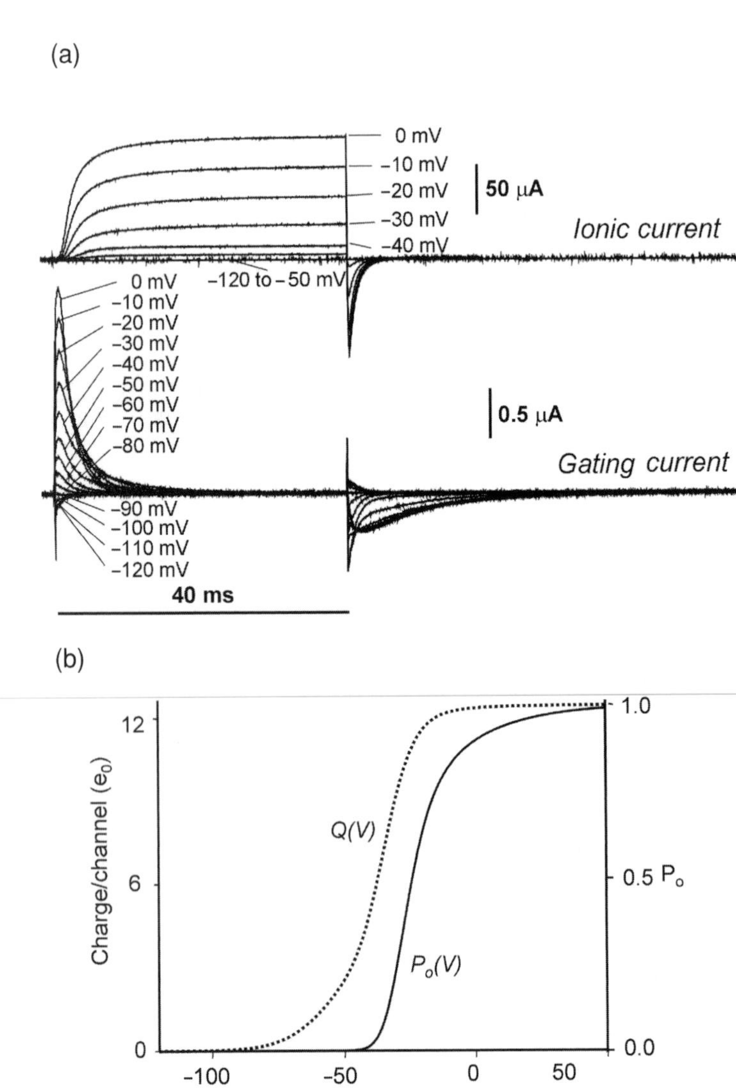

Fig. 3.4 Ionic and gating currents in Shaker-IR K^+ channel. (a) Top traces, time course of ionic currents for pulses to the indicated potentials starting and returning to −90 mV. Bottom traces, time course of the gating currents for the pulses indicated. Notice the difference in the amplitude calibration for ionic as compared to gating currents. (b) The voltage dependence of the open probability, P_o, and the charge moved per channel, $Q(V)$. For details see text.

in the possible models and in the parameters fitted, thus obtaining a more robust representation of the kinetic parameters that characterize the channel.

The common assumption in kinetic models is that the movement of the charge or dipole in the field has a finite number of low-energy positions and energy barriers in between them. According to kinetic theory, the transition rates across the energy barrier are exponentially related to the negative of the free energy amplitude of the barrier. This free energy contains nonelectrical terms and an electrical term such that the membrane voltage can either increase or decrease the total energy barrier resulting in changes in the forward and backward rates of crossing the barriers. There are multiple examples in the literature with varied levels of complexity that describe well the kinetic and steady state properties of several types of voltage-dependent channels (see for example, Vandenberg and Bezanilla, 1991; Bezanilla et al., 1994; Zagotta et al., 1994; Schoppa and Sigworth, 1998).

A more general approach to modeling is based on a representation of a landscape of energy using the charge moved as the reaction coordinate. In this type of modeling, the above-mentioned discrete kinetic models are also represented when the landscape of energy has surges that exceed $4kT$ (Sigg et al., 1999).

When gating currents are recorded with high-bandwidth new components are observed (Sigg et al., 2003). Figure 3.5a shows a typical gating current trace recorded with a bandwidth of 10 kHz. Notice that before the initial plateau and decay of the

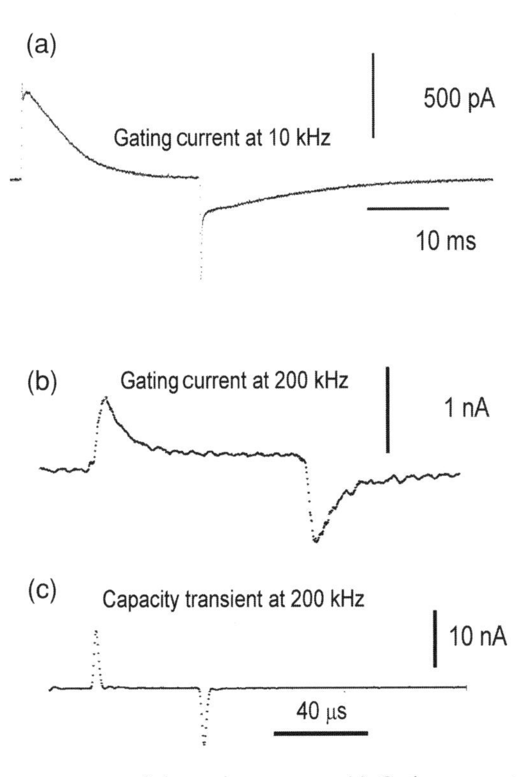

Fig. 3.5 The early component of the gating current. (a) Gating current recorded at 10 kHz bandwidth. (b) Gating current recorded at 200 kHz bandwidth. Notice the differences in the amplitude and time scales. (c) The time course of the charging of the membrane capacitance for the experiment in part (b). Recordings done in collaboration with Dr. Enrico Stefani.

gating current there is a brief surge of current. When the bandwidth is increased to 200 kHz, the first surge of current is the predominant amplitude as shown in the fast time scale recording of Fig. 3.5b. The large spike of current is followed by a long plateau of current that corresponds to the plateau observed at 10 kHz in Fig. 3.5a. However, the area spanned by the large spike is a very small fraction of the total charge moved during the entire time course of the gating current. The interpretation of this early component is best understood using the representation of the gating charge moving in a landscape of energy that undergoes a change in tilt when the membrane potential is changed such that the charge advances in its initial energy well before making the jump across the first energy barrier. Using this approach, Sigg et al. (2003) computed the viscosity encountered by the gating charge in its well of energy.

3.4.1.1 What Have We Learned with Kinetic Modeling?

In fact a great deal. It is now clear that voltage-dependent channels have multiple closed states and, in some cases several open states. In general, the opening of the gate requires all four subunits to be activated, although there are cases where intermediate states have been observed (see Chapman and VanDongen, 2005). Each subunit undergoes several transitions before reaching the active state and in the case of K^+ channels they do not seem to interact until the final step that opens the channel (Horn et al., 2000; Mannuzzu and Isacoff, 2000). The situation is different in the case of the muscle Na^+ channel where site-directed fluorescence studies show that interdomain interactions are manifested prior to channel opening (Chanda et al., 2004). Kinetic modeling has given us a picture of the channel in terms of channel physical states with transitions between them that are regulated by voltage. Kinetic modeling is a critical step in developing a physical model of channel operation because all the predicted features of channel function should be reproduced by the structure of the protein and its voltage-induced conformational changes.

3.4.2 Gating Charge per Channel

When the gating charge moves within the electric field we detect a current in the external circuit. The time integral of that current represents the charge moved times the fraction of the field it traverses, therefore our measurement of gating charge does not represent the exact number of charges displaced because it includes the arrangement of the electric field. We must keep this in mind, when we represent the reaction coordinate of the activation of the channel in the variable q. A channel evolves from $q = 0$ to $q = z_T$ traversing many closed and/or open states. Activation of the channel corresponds to the opening of the pore and, analogous to the chemical potential, we define the *activation potential* as

$$W_a = -kT \ln P_o. \tag{3.3}$$

Then, the activation charge displacement q_a corresponds to the negative gradient of the activation potential,

$$q_{\mathrm{a}} = -\frac{dW_{\mathrm{a}}}{dV} = kT\frac{d\ln P_{\mathrm{o}}}{dV}. \tag{3.4}$$

The equilibrium probabilities in each physical state of the channel can be explicitly written using the Boltzmann distribution knowing the potential of mean force F_i for each state i. Then, by assigning open or closed (or intermediate states) conductances to every state, we can write an expression for P_{o} that includes the voltage dependence of F_i. The final result of the derivation (Sigg and Bezanilla, 1997) gives a relation between $q(V)$, q_a, z_T and the charge moving between open states q_l,

$$q(V) = z_{\mathrm{T}} - q_{\mathrm{a}} - q_{\mathrm{l}}. \tag{3.5}$$

Note that $q(V)$ is the Q–V curve shown in Fig. 3.5. This result is general and includes cases with any number of open and closed states connected in any arbitrary manner. If there is no charge movement between open states ($q_l = 0$), then the Q–V curve superimposes on q_a. In addition, it is possible to estimate z_T, the total charge per channel, by taking the limiting value of q_a that makes $q(V)$ go to zero. In the typical case of a channel that closes at negative potentials, we obtain

$$z_{\mathrm{T}} = \lim_{V\to-\infty} kT\frac{d\ln P_{\mathrm{o}}}{dV}, \tag{3.6}$$

a result that was first obtained by (Almers, 1978) for the special case of a sequential series of closed states ending in an open state. This method has been applied to several types of voltage-dependent channels and the charge per channel obtained ranges between 9 and 14 e_0 (Hirschberg et al., 1996; Noceti et al., 1996; Seoh et al., 1996).

Another way to estimate the charge per channel is to measure the maximum charge from the Q–V curve and divide by the number of channels present, Q/N method. The number of channels can be estimated by noise analysis (Schoppa et al., 1992) or by toxin binding (Aggarwal and MacKinnon, 1996). The value of charge per channel estimated by the Q/N method was 12 to 13 e_0 for the Shaker K^+ channel, a value that was not different from the value obtained by the limiting slope method (Seoh et al., 1996). As the limiting slope measures only the charge involved in opening the channel, the agreement between the two methods imply that in case of the Shaker channel there is no peripheral charge. The large value of 12 to 13 e_0 per channel explains the very steep voltage dependence of the superfamily of voltage-gated ion channels. At very negative V, q_a has a linear dependence on V, so that P_{o} is exponential in V, such that it increases by e in only 2 mV:

$$P_{\mathrm{o}} \propto \exp(z_{\mathrm{T}}V/kT). \tag{3.7}$$

This very steep voltage dependence explains why the P_o–V curve of Fig. 3.4b shows no visible P_o at potentials more negative than −40 mV: in fact there is a finite value of P_o of less than 10^{-5} at potentials as negative as −100 mV.

Voltage-dependent channels that have several open states with charge moving between them ($q_l \neq 0$), show plots of P_o–V that may not be used to compute z_T. One example is the maxi K^+ channel, activated by voltage and Ca^{++}, that shows a P_o–V curve with a slope that decreases as the potential is made more negative. This result is consistent with the multiple-state allosteric model proposed for this channel (Horrigan et al., 1999).

3.4.2.1 Gating Current of One Channel

The previous paragraph shows that we can estimate the total charge that moves in one channel but does not give any ideas of how that charge movement occurs at the single-channel level. Two limiting cases can be proposed. In one case, the time course of the gating current is just a scaled down version of the macroscopic gating current shown in Fig. 3.4. The other case assumes that the charge movement occurs in large elementary jumps and that the macroscopic gating current is the sum of those charge shots. Figure 3.6 is a simulation of the gating shots expected from a channel that is made up of four identical subunits, each having two states: resting and active. The resting position is favored at negative potentials while the active position is favored at positive potentials and there is a large energy barrier that separates the resting from the active position. At negative potential, the charge will cross the energy barrier rarely, due to thermal motion. As the membrane potential is made more positive the energy landscape tilts and the barrier decreases. Then the probability of crossing the barrier increases and discrete jumps occur. The jump of the charge generates a very fast current transient (shown in the figure as vertical bars) whose duration is limited in practice by the frequency response of the recording system. The simulation shows one trial starting at −100 mV and pulsing to 0 mV. Notice that immediately after the pulse, each of the four subunits (I_{g1} through I_{g4}) responds at different times and that there are spontaneous reverse currents in I_{g1} and I_{g2}. The ionic channel current only appears when all four subunits have made the transition as shown in I_{single}. When several trials are averaged, the gating shots produce the macroscopic gating current (Average I_g) and the average of the single-channel currents produce the Average I_{ionic}. In principle, if the charge is moving in discrete packets, it should be possible to detect those elementary events as it has been possible to detect the elementary events of pore conduction. The problem is that, as those events are very fast and extremely small, they are not detectable above background noise with present techniques. However, if those events do exist, they should produce detectable fluctuations in the gating currents recorded from a relatively small number of channels. In the simulation shown in Fig. 3.6, the average Ig shows excess noise during the decay of the gating current. This noise is the result of the contribution of the elementary shots shown for only one trial in the same figure. To detect this excess noise and do the fluctuation analysis, the same voltage pulse is

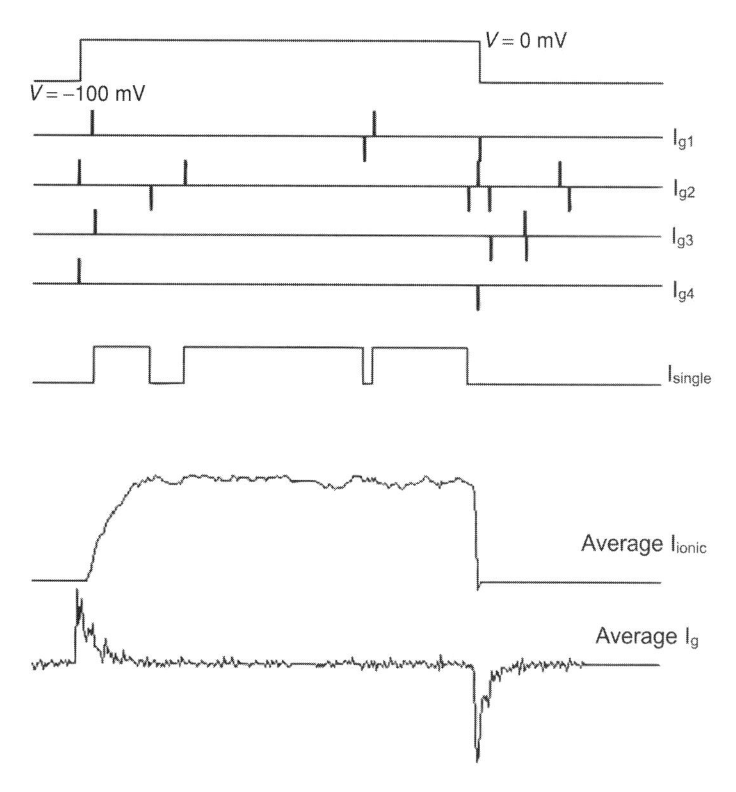

Fig. 3.6 Time course of the gating shots, single-channel current and macroscopic ionic and gating currents. The applied voltage is in the top trace. I_{g1} through I_{g4} represent the current recording of the gating shots for each one of the four subunits. I_{single} represents the time course of the ionic current for one trial as a result of the movements of the four sensors. Average I_{ionic} is the average of the single-channel currents for 80 trials. Average I_g is the average of all the subunits gating shots for a total of 80 trials.

repeated many times and a large number of gating current traces are recorded. From these traces, an ensemble mean value and an ensemble variance are obtained that allows the estimation of the elementary event (Fig. 3.7)

Gating current noise analysis was first done in Na^+ channels expressed in oocytes (Conti and Stuhmer, 1989) where indeed it was possible to detect fluctuations that were used to estimate an elementary charge of 2.2 e_0. A similar analysis was done in Shaker-IR K^+ channel (Sigg et al., 1994) and the elementary charge was found to be 2.4 e_0. This value corresponds to the maximum shot size and if it were per subunit it would account for only 9.6 e_0 or 3.6 e_0 less than the 13 e_0 measured for the whole channel. This means that there is a fraction of the total charge that produces less noise and then it may correspond to smaller shots or even a continuous process. Figure 3.7 shows the mean and variance for a large pulse (to +30 mV) and a smaller depolarization (to −40 mV). For the small depolarization (Fig. 3.6b), gating current noise increases by the time that more than half of the charge has moved. This

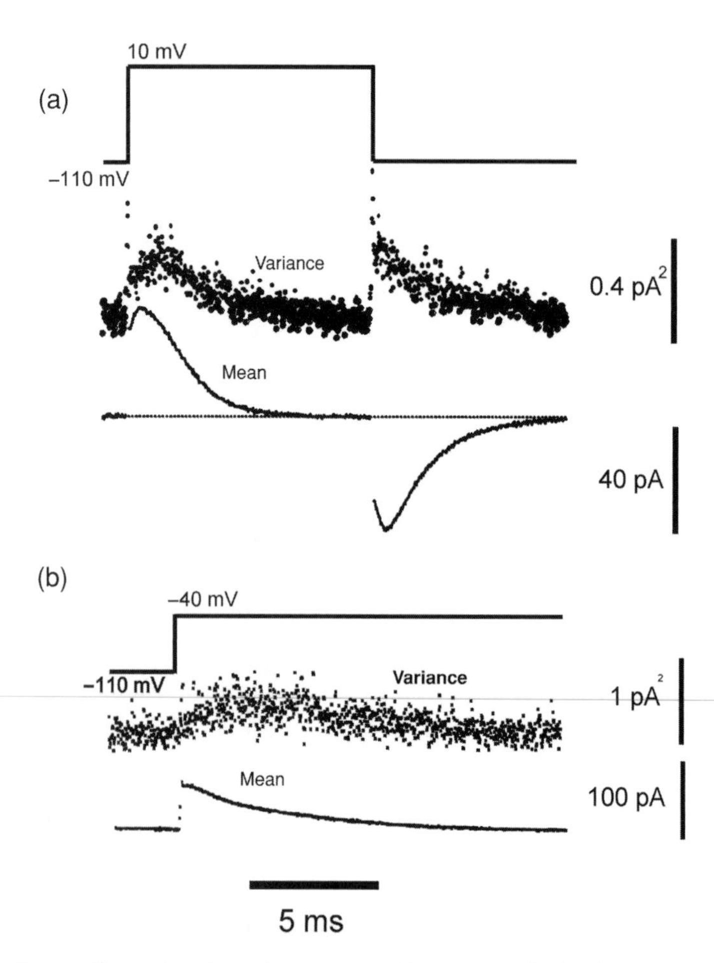

Fig. 3.7 Current fluctuations in gating currents. (a) Top trace is the time course of a pulse to +10 mV, middle trace is the variance and bottom trace is the mean computed from several hundred traces. (b) Top trace is the time course of a smaller pulse to −40 mV, middle trace is the variance and lower trace is the mean computed from hundreds of traces. Modified from Sigg et al. (1994).

indicates that at early times the elementary event is smaller than at longer times. The gating current decreases with two exponential components, which are more separated in time at small depolarizations. Therefore, it is possible to attribute the small shots to the fast component and the large shots to the slow component. This would indicate that the early transitions of the gating current are made up of several steps each carrying a small elementary charge. The total movement of charge in the early transitions would have to account for the 3.6 e_0 needed to make the total of 13 e_0 per channel. We conclude from these experiments that the gating current of a single channel is made up of small shots at early times followed big shots of currents that in the average gating current show two exponential decays.

3.5 Structural Basis of the Gating Charges

In this section, we will address the relation between the function of the voltage sensor and its structural basis. We will first ask where are the charges or dipoles in the structure of the channel and then how those charges or dipoles move in response to changes in membrane potential.

3.5.1 The Structures Responsible for the Gating Charge Movement

There are many ways one could envision how the charge movement is produced by the channel protein. The putative α-helical transmembrane segments have an intrinsic dipole moment that upon tilting in the field would produce an equivalent charge movement. Also, induced dipoles of amino acids side chains could accomplish the same. However, 13 e_0 per channel is very large and charged amino acids become the most likely possibility. Since the first channel was cloned, it was recognized that the S4 segment with its basic residues would be the prime candidate for the voltage sensor (Noda et al., 1984). By introducing mutations that neutralize the charges in S4, several studies found that there were clear changes in the voltage dependence of the conductance. However, shifts in the voltage dependence of the P_o–V curve or even apparent changes in slope in the voltage range of detectable conductance do not prove that charge neutralization is decreasing the gating charge. The proof requires the measurement of the charge per channel (z_T) for each one of the neutralizations. If after neutralization of a charged residue the charge per channel decreases, one may assume such charge is part of the gating current. Using the methods to measure total charge per channel discussed above in Shaker, two groups found that the four most extracellular positive charges in S4 (Aggarwal and MacKinnon, 1996; Seoh et al., 1996) and that one negative charge in the S2 segment were part of the gating charge (Seoh et al., 1996) (see white symbols in Fig. 3.1). It is interesting to note that in several instances a neutralization of one particular residue decreased the total gating charge by more than 4 e_0. This indicates that somehow the charges interact with the electric field where they are located such that the elimination of one charge can affect the field seen by the remaining charges. If most of the gating charge is carried by the S4 segment (4 e_0 per subunit), it gives a total of 16 e_0. Therefore, to account for the 13 e_0 for the total channel obtained from charge/channel measurements, they all must move at least 81% of the membrane electric field ($16 \times 0.81 = 13$).

3.5.2 Movement of the Charges in the Field

Knowing the residues that make up the gating charge is a big advance because it makes it possible to test their positions as a function of the voltage and thus infer the possible conformational changes.

As we will see below, the literature contains a large number of papers (for reviews, see Yellen, 1998; Bezanilla, 2000) with many types of biophysical experiments testing the accessibility, movement and intramolecular distance changes. With all these data, several models of the voltage sensor movement were proposed that could account for all the experimental observations but in total absence of solid information on the three-dimensional structure of the channel. In the next section, we will discuss many of the biophysical measurements of the voltage sensor and the models that have emerged from them and the recently solved crystal structures of KvAP (Jiang et al., 2003a) and Kv1.2 (Long et al., 2005a).

3.6 Structural Basis of the Voltage Sensor

A three-dimensional structure of the channel, even in only one conformation, would be invaluable as a guideline in locating the charges inside the protein and in proposing the other conformations that account for the charge movement consistent with all the biophysical measurements. We will see below that the first solved crystal structure of a voltage-dependent channel available is not in a native conformation but just recently, a second crystal structure from a mammalian voltage-gated K channel (Kv1.2) appears to be closer to its expected native conformation. In any case, in both cases the crystals are in only one conformation (open inactivated). Therefore, we are still relying on data that can only be used to propose models because the information on the three-dimensional structure of the channel is still uncertain and incomplete.

3.6.1 Crystal Structures of Voltage-Dependent Channels

The long awaited first crystal structure of a voltage-dependent channel, KvAP from *archea Aeropirum pernix*, was published by the MacKinnon group (Jiang et al., 2003a). The structure was a surprise because it showed the transmembrane segments in unexpected positions with respect to the inferred bilayer. For example, the N-terminal was buried in the bilayer whereas it has been known to be intracellular; the S1–S2 linker is also buried although it has been previously shown to be extracellular. The S4 segment, along with the second part of S3 (S3B) formed the paddle structure that was intracellular and lying parallel to the bilayer, a location that would be interpreted as the closed position of the voltage sensor. However, the pore gates in the same crystal structure clearly corresponded to an open state. Thus, the crystal structure is in a conformation that was never observed functionally, raising the question whether that crystal structure of KvAP is indeed representative of the native conformation of the channel in the bilayer. The authors functionally tested the structure by incorporating the channel in bilayers and recording the currents through the channel after Fab fragments were added to the inside or to the outside of the channel. The Fab fragment did not attach from the inside but only from the outside, showing that the position of the S3–S4 shown in the crystal structure (obtained in detergent) did not represent a native conformation of the channel in the bilayer (Jiang et al., 2003b).

Along with the crystal structure of the full KvAP channel, Jiang et al. (2003a) solved the crystal structure of the S1 through S4 region of KvAP (the isolated voltage sensor). This crystal showed similarities but was not identical to the S1–S4 region of the full KvAP crystal structure (Cohen et al., 2003). This second crystal was docked to the pore region of the full crystal to obtain two new structures that were proposed in the open and closed conformations as the *paddle model* of channel activation. It is important to note here that the structures shown in the second paper (Jiang et al., 2003b) do not correspond to the original KvAP structure. In fact, the proposed structures of the model depart so much from the crystal structure that we should treat them just like any other model of activation proposed before.

A recent paper by Long et al. (2005a) reporting the crystal structure of the Kv1.2 channel in the open-slow-inactivated state shows an arrangement of the transmembrane segments that is dramatically different from KvAP but at the same time much closer to the inferred structure from previous biophysical results.

3.6.2 Models of Sensor Movement

A compact way of reviewing the large body of biophysical information on structural changes of the voltage sensor is to describe the models that are currently proposed to explain the operation of the voltage sensor.

Figure 3.8 shows schematically three classes of model that have been proposed to explain the charge movement in voltage-dependent channels. In all cases, only two of the four sensors are shown and the coupling from the sensor to the gate is not shown explicitly. In addition, the charge that moves resides completely in the first four charges of the S4 segment. The common feature of all three models is that the charge is translocated from the inside to the outside upon depolarization. However, there are important differences as of how those charges relocate in the protein structure.

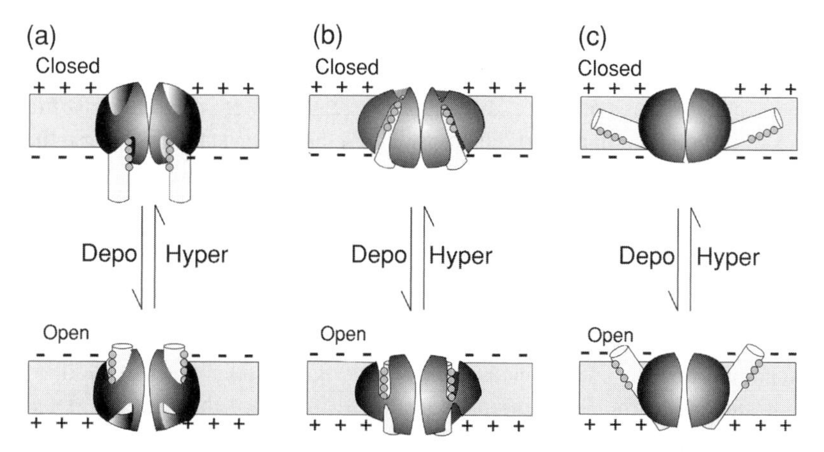

Fig. 3.8 Three models of the voltage sensor. (a) Helical screw model, (b) Transporter model, (c) Paddle model. The charged residues are shown as gray circles. For details see text.

Figure 3.8a shows the conventional *helical screw* model (Caterall, 1986; Durell and Guy, 1992). Although there are variations on this model, the general idea is that upon depolarization the S4 segment rotates along its axis and at the same time translates as a unit perpendicular to the membrane, thus changing the exposure of the charges from the intracellular to the extracellular solution, effectively translocating 4 e_0 per subunit (see Fig. 3.8a). In the original version of the model the change of exposure required a large 16 Å translocation of the S4 segment and the positively charged arginines were making salt bridges with aspartate or glutamates residues that had to be broken to initiate the movement. In more recent versions (Gandhi and Isacoff, 2002; Ahern and Horn, 2004; Durell et al., 2004) the charges are in water crevices in both the closed and open position, decreasing the amount of translation required of the S4 segment.

Figure 3.8b shows the *transporter* model (Bezanilla, 2002; Starace and Bezanilla, 2004; Chanda et al., 2005). In the closed position, the charges are in a water crevice connected to the intracellular solution and in the open position they are in another water crevice connected to the extracellular solution. The translocation of the charges is achieved by a tilt and rotation of the S4 segment with little or no translation. In this case the field is concentrated in a very small region that changes from around the first charge in the closed state to the fourth charge in the open state.

Fig. 3.8c shows the *paddle* model introduced by the MacKinnon group (Jiang et al., 2003b) where the S4 segment is located in the periphery of the channel and the charges are embedded in the bilayer. The S4 segment makes a large translation such that the most extracellular charge goes from exposed to the extracellular medium in the open state to completely buried in the bilayer in the closed state.

The helical screw and transporter models are similar but they differ dramatically from the paddle model in that the gating charges in the paddle model are embedded in the bilayer while in the helical screw and transporter models the charges are surrounded by water, anions or making salt bridges. In contrast with the transporter model, the helical screw model has in common with the paddle model the large translation of the S4 segment.

In the following sections, we will review biophysical experiments that were designed to test the topology of the channel and the extent of the conformational changes of the gating charge. In the absence of a native crystal structure in the open and closed conformations, these experiments are the data that we can use to support or reject the available models of voltage sensor operation.

3.6.3 The Topology of the Channel and Gating Charge Location

Alanine and tryptophan scanning have been used to infer the relative positions of the transmembrane segments (Monks et al., 1999; Li-Smerin et al., 2000) and the results indicate that the S1 segment is in the periphery of the channel. This result is at odds with the recent report by Cuello et al. (2004) where using electron paramagnetic resonance (EPR) scanning, they found that in KvAP the S1 segment is not exposed to the bilayer but rather surrounded by the rest of the protein. This difference could be an inherent limitation in the ala or trp scanning techniques that test the function of the channel or it could be a genuine difference between eukaryotic and prokaryotic channels. Recent results seem to confirm this last possibility. Using LRET measurements, Richardson et al. (2005, 2006) show that in both the prokaryotic voltage-gated channels NaChBac and KvAP the S1 segment is indeed close to the pore in agreement with the results of Cuello et al. (2004) and the recent structure of Long et al. (2005a) shows that in the eukaryotic Kv1.2 voltage-gated channel the S1 is closer to the periphery. With regard to the S4 segment, the results of Cuello et al. (2004) show that the S4 segment seems to be partially exposed to the bilayer, in agreement with the paddle model. However, in Cuello et al. (2004) the two innermost charged residues are protected by the protein and the two outermost charges are in the interface, contrary to the location proposed in the paddle model. The crystal structure of Kv1.2 confirms the results of Cuello et al. (2004) because the two inner charges are protected while the two outermost charges are in the interface. It is interesting to notice that the crystal of Kv1.2 show the two outermost arginines pointing into the bilayer but their alpha carbons are at 13 Å from the center of the bilayer, which corresponds to the polar part of the lipid bilayer. Recent molecular dynamics simulations of the Kv1.2 channel in a lipid bilayer shows that the first two arginines are completely hydrated (Roux, personal communication).

Laine et al. (2003) found that the extracellular part of the S4 segment gets within a few Ångströms of the pore region in the open state of Shaker-IR channel. This result is consistent with the helical screw and the transporter model but is inconsistent with the paddle model presented in (Jiang et al., 2003b) because in their model the extracellular portion of S4 is well separated from the rest of the channel protein giving a distance of about 98 Å between segments across the pore region. In experiments using resonance energy transfer with lanthanides (Cha et al., 1999a,b; Richardson et al., 2005) or with organic fluorophores (Glauner et al., 1999) that distance was found to be around 50 Å also in agreement with measurements done with tethered TEA derivatives (Blaustein et al., 2000). These results indicate that the S4 segment is not extended into the bilayer but it is against the bulk of the channel protein. If the paddle model were modified, so that the S4 segments would be almost perpendicular to the plane of the bilayer in the open state and almost parallel to the bilayer in the closed state but still flush against the rest of the protein it would be consistent with the distance constraints just mentioned. In another report, the MacKinnon group made such modification at least for the open-inactivated state (Jiang et al., 2004). The recent structure of Kv1.2 (Long et al., 2005a) shows that the S4 segment is indeed almost perpendicular to the plane of the membrane. However, even with these modifications the paddle model locates the charges in the bilayer, a proposal that is inconsistent with the EPR results (Cuello et al., 2004) and several other biophysical experiments (Fernandez et al., 1982; Yang and Horn, 1995; Larsson et al., 1996; Yang et al., 1996; Yusaf et al., 1996; Starace et al., 1997; Baker et al., 1998; Islas and Sigworth, 2001; Starace and Bezanilla, 2001; Asamoah et al., 2003, 2004; Starace and Bezanilla, 2004). Locating the charges in the bilayer has been a subject of intense debate because the energy required in moving a charge into the low dielectric constant bilayer is extremely high (Parsegian, 1969). If positively charged gating charges are neutralized by making salt bridges with acidic residues, the energy decreases (Parsegian, 1969) but there would not be any gating charge movement. In a recent molecular dynamics simulation (Freites et al., 2005) the authors conclude that an isolated S4-like segment can be stabilized in a bilayer by making salt bridges between the arginines and the phosphates of the phospholipids producing a constricted 10 Å hydrophobic region. It is not clear from that structure what would be the charge translocated. Most importantly, the structure presented by Freites et al. (2005) is not a good model of the S4 in voltage-gated channels because it has been shown that the arginines are shielded (Cuello et al., 2004; Long et al., 2005a).

Evidence obtained by charge measurements in the squid axon sodium channel, have shown that the gating charges do not move in the bilayer (Fernandez et al., 1982). In these experiments addition of chloroform increased the kinetics of translocation of the hydrophobic ion dipicrylamine while it did not change the kinetics of the sodium gating currents. The conclusion was that the gating charge, unlike hydrophobic ions, does not move in the bilayer. Ahern and Horn (2004a) have explored this subject in more detail. As in the paddle model the S4 segment is in the bilayer, they reasoned that on addition of more charges in the S4 segment, the net gating charge should increase. Their results show that the charge addition at several positions did not increase the gating charge, indicating that only the charges in aqueous crevices are responsible for gating and can sense the changing electric field. Both these experimental results are hard to reconcile with the paddle model where the voltage sensor is immersed in the hydrophobic core of the lipid bilayer.

The S1–S2 loop has been clearly located in the extracellular region by several criteria. One is that a glycosylation site in Shaker occurs in this loop (Santacruz-Toloza et al., 1994). In addition, fluorescence signals from fluorophores attached in this loop are consistent with this region being extracellular (Asamoah et al., 2004), as well as the recent EPR scanning of KvAP (Cuello et al., 2004). These results are in agreement with the helical screw and transporter models but are again inconsistent with the location proposed in the paddle model. In the KvAP crystal structure the S1–S2 linker is buried in the bilayer with an S2 segment almost parallel and also buried in the bilayer. The recent crystal structure of Kv1.2, although not well resolved in this region, confirms that the S1–S2 loop is extracellular.

3.6.4 Voltage-Induced Exposure Changes of the S4 Segment and Its Gating Charges

Testing the exposure of the gating charges in the intra- or extracellular medium would give us an idea whether their voltage-induced movement takes them out of the protein core or from the lipid bilayer. There have been three different approaches to test exposure. The first method, cysteine scanning mutagenesis, consists of replacing the residue in question by a cysteine and then test whether a cysteine reagent can react from the extra- or intracellular solution and whether that reaction is affected by voltage (Yang and Horn, 1995; Larsson et al., 1996; Yang et al., 1996; Yusaf et al., 1996; Baker et al., 1998). The second method, histidine scanning mutagenesis, consists of titrating with protons the charge in the residues. In this case, as the pKa of the arginine is very high, histidine was used as a replacement allowing the titration in a pH range tolerated by the cell expressing the channel (Starace et al., 1997; Starace and Bezanilla, 2001, 2004). The third method consists of replacing the residue in question with a cysteine, followed by tagging it with a biotin group and then testing whether avidin can react from the inside or outside depending on the membrane potential (Jiang et al., 2003b). The outcome of all these experiments is that indeed the charged residues in S4 undergo changes in exposure when the membrane potential is changed. However, it is important to look at each of these procedures with their limitations and compare their results because that may reveal many details of the actual movement of the charge.

3.6.4.1 Cysteine Scanning Mutagenesis

The experiments by Yang and Horn (1995) were the first to test the accessibility of charged groups to the extra- and intracellular solutions and its dependence on membrane potential. The idea of these experiments is to test whether a cysteine reacting moiety can attach to an engineered cysteine in the channel depending where the moiety is and what the electric field is. This method works provided the attachment induces a detectable change in the channel currents. In addition, the attachment of the moiety will depend on the local pH and state of ionization of the cysteine residue. Yang and Horn (1995) showed that the reaction rate of MTSET to the cysteine-replaced most extracellular charge of the S4 segment of the fourth domain of a sodium channel depended on membrane potential. To conclude that a particular site changes its exposure, the reaction rate must differ by more than an order of magnitude between the two conditions. This is important because if one just measures whether the reaction occurred or not, one may be sampling a rarely occurring conformational state. Their result showed for the first time that the accessibility of this group changed with voltage or alternatively that the ionization state was changed by voltage. This work was expanded to the deeper charges in the S4 segment of domain IV of the Na^+ channel (Yang et al., 1996). The conclusion was that upon depolarization the most extracellular charge was exposed and that at hyperpolarized potential it became buried. In addition, the two following charges could be accessed from the inside at hyperpolarized potential and from outside at depolarized potential. The next two deeper charges were always accessible from the inside regardless of the membrane potential. These results strongly suggest that the three outermost charges change exposure with voltage and are consistent with the idea of a conformational change that translocates charges from inside to outside upon depolarization giving a physical basis to the gating currents.

These studies were also done in the Shaker-IR K^+ channel (Larsson et al., 1996; Yusaf et al., 1996; Baker et al., 1998) and showed the same trend: the outermost charges were exposed to the outside upon depolarization and the innermost to the inside on hyperpolarization. While some residues were inaccessible from one side to the cysteine-reacting compound on changing the voltage, there were a few that reacted when the reactive agent was added to the other side.

The conclusion from the cysteine scanning experiments is that the residues bearing the gating charges change their exposure with the voltage-dependent state of the channel. Another very important conclusion is that the charges are exposed to the solutions and not to the lipid bilayer since the SH group in the cysteine has to be ionized to react with the cysteine modifying reagent. This would be energetically unfavorable in the low dielectric medium of the bilayer. It is important to note that the reactivity of thiol groups on some of the sites tested was comparable to their reactivity in free solution.

3.6.4.2 Histidine Scanning Mutagenesis

Testing the titration of the charged residues is a direct way to address exposure of the charges to the solutions. In this approach, each arginine or lysine is exchanged to a histidine residue that can be titrated in a pH range that is tolerated by the expression system. The titration of the histidine with a proton can be easily detected as a change in the gating current, provided the ionic current is blocked. For this reason, most of these experiments were carried out in the nonconducting Shaker-IR K^+ channel that bears the W434F mutation.

The logic of this procedure can be understood by taking some limiting cases. (i) If the histidine is not exposed to the solutions and/or if it does not move in the field, it would not be possible to titrate it from either side or under any membrane potential; therefore, no change in the gating currents are observed. (ii) The histidine can be exposed to the inside or to the outside depending on the membrane potential and thus on the conformation of the voltage sensor. In this case, if a pH gradient is established, every translocation of the voltage sensor would also translocate a proton, thus producing a proton current. This proton current would be maximum at potentials where the sensor is making most excursions, which occurs around the midpoint of the Q–V curve. At extreme potentials, the gating current would be affected but no steady proton current would be observed. Therefore, the I–V curve of such proton current would be bell-shaped. (iii) One particular conformation of the sensor locates the histidine as a bridge between the internal and external solutions forming a proton selective channel. In this case a steady current would be observed that would have an almost linear I–V curve in the range of potentials where that conformation is visited and would be zero otherwise. The analysis of histidine scanning was done in Shaker-IR K^+ channels and six charges were tested starting from the most extracellularly located (Starace et al., 1997; Starace and Bezanilla, 2001, 2004). Fig. 3.9a shows the results where the accessibility of each of the mutant is shown. The first four charges are accessible from both sides depending on the membrane potential, while the next two are not titratable.

(a)

Mutant	Accessibility	Proton Current
R362H	both sides	pore
R365H	both sides	transporter
R368H	both sides	transporter
R371H	both sides	transporter/pore
K374H	not titratable	none
R377H	not titratable	none

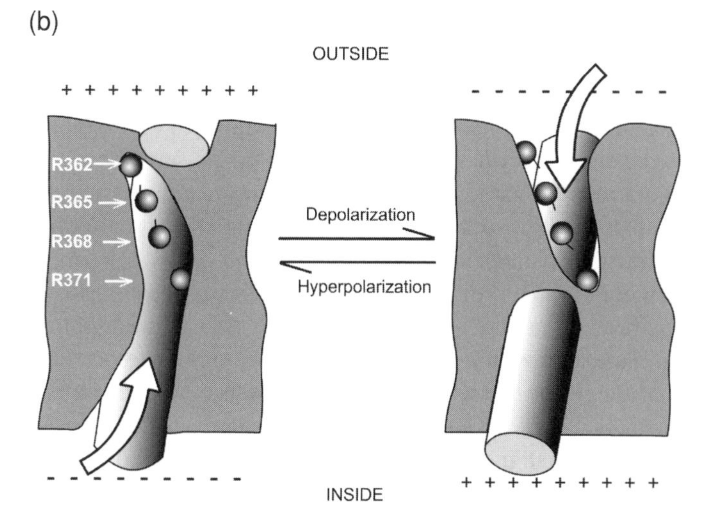

Fig. 3.9 Results of histidine scanning mutagenesis. (a) Each mutant is listed with the type of current observed and their accessibility to the internal and external solutions. In the case of K374H and R377H, the histidine may not be accessible and/or they do not move in the field. (b) Interpretation of the histidine scanning experiments with the transporter model based on a molecular model built with KvAP and KcsA crystal structures. A change in tilt of the S4 segment exposes the first four charges to the outside in the depolarized condition and to the inside in the hyperpolarized condition. In the hyperpolarized condition, there is a very narrow region bridged by histidine in position 362 and in the depolarized condition a bridge is formed by histidine in position 371. For details see text.

These results are consistent with what we know about the role of each charge. Only the first four charges, which are responsible for most of the gating charge, seem to translocate, while the next two either do not move in the field or are never accessible from the solution. The results also show that some of the charges that appeared buried to the cysteine scanning mutagenesis method are accessible to protons indicating that they are pointing into deep crevices that are too narrow for the cysteine-modifying reagent but large enough for protons to reach.

Figure 3.9b shows how the results could be explained in terms of an actual structure of the S4 segment with its associated membrane segments and bilayer, based on a transporter-like molecular model (Chanda et al., 2005) built with the crystal structures of KvAP and KcSA. The alpha-helical S4 segment is represented by a cylinder showing the first four charged arginines, as labeled. The next two arginines are on the back side of the S4 helix and are not visible. The hydrophobic region, that includes other segments and the bilayer, is simply shown as a gray area. In the closed state the first four charges are in contact with the internal solution by way of a water crevice (up arrow); however, the most extracellular charge (R362) is in the boundary between the intracellularly connected water crevice and the extracellular solution. If this arginine were replaced by a histidine it would form a proton pore by making a bridge between both solutions. Upon depolarization the S4 segment changes its tilt and all four charges become exposed to a water crevice (down arrow) connected to the extracellular solution. Now the fourth charge is in the boundary between the extra- and intracellular solutions so that a histidine in this position would form a proton pore. Notice that R365 (and also R368) change exposure in such a way that if each one were replaced by a histidine in the presence of a proton gradient, every transition would be able to shuttle one proton from one side to the other. It was found that R371H is a transporter but it can also form a proton pore at large depolarized potentials (Starace and Bezanilla, 2001).

The consequence of these results is that in the two extreme positions the electric field is concentrated (Islas and Sigworth, 2001) in a very narrow region of the protein: near 362 in the closed position and near 371 in the open position. This means that there is no need for a large movement of the voltage sensor to transport a large amount of charge. Asamoah et al. (2003) have shown that indeed the field is concentrated in this region by using a cysteine reactive electrochromic fluorophore. By attaching this fluorophore in different regions of the channel and comparing the fluorescence signal induced by voltage changes with the same electrochromic group in the bilayer, the field in the S4 region was found to be at least three times more intense. By attaching different lengths adducts to cysteine replacing position 362, Ahern and Horn (2005) have also estimated that the field at that position is concentrated. When residue 362 was replaced by alanine (Tombola et al., 2005) it was found that a current may be recorded at negative potentials (the ω current) a result that is consistent with the results of the 362H histidine pore and confirming the proposal that this region becomes extremely narrow at negative potentials.

The results of histidine and cysteine scanning experiments suggest that the charged arginine residues are stabilized by water and possible anions residing in the water filled crevices. It has been shown by Papazian and collaborators (Papazian et al., 1995; Tiwari-Woodruff, 1997) that the acidic residues in S2 and S3 segments play a stabilizing role in the structure of Shaker-IR channel but they could also lower the energy of the ariginines in the crevices and possibly contribute to the gating charge by focusing the field in that region.

3.6.4.3 Biotin and Streptavidin Scanning

Biotin and its cysteine modifying derivative can react with cysteines engineered in the channel protein. In addition, the biotin group binds avidin, a large soluble protein molecule that is not expected to cross the membrane. Jiang et al. (2003b) studied several positions by mutating residues to cysteine in S3b and S4 segments of KvAP and attached to it a biotin molecule. After incorporating these channels into bilayers, they tested whether the currents were affected by avidin in the external or internal solutions. They found that, depending on the site of biotin attachment, the currents were decreased by externally or internally applied avidin. There were two sites (residues 121 and 122), lying in between the second and third charges of KvAP, where they saw inhibition by avidin from both sides. They indicated that in the conventional model (helical screw or transporter models) it would be extremely difficult to relocate the linker of the biotin along with the charges moving within the protein core. Therefore, they proposed that those sites must move a large distance within the bilayer to reach for the avidin present in the solution and gave them the basic restrictions on the extent that the paddle must move. Although this result is consistent with the paddle model, it is not inconsistent with the transporter model because those residues may in fact be facing the bilayer (Chanda et al., 2005) and thus allowing the biotin to reach to either side of the bilayer to attach to the avidin. It should be noted that no reaction rate was measured in those experiments (Jiang et al., 2003b) nor in the most recent paper (Ruta et al., 2005). Therefore, the accessibilities measured with avidin may well be the result of conformations that were rarely visited because the biotin–avidin reaction is essentially irreversible.

3.6.5 Fluorescence Spectroscopy Reveals Conformational Changes of the Voltage Sensor

Site-directed fluorescence is a powerful technique to follow conformational changes in a protein. In the case of voltage-dependent channels, the idea is to label specific sites of the channel protein with a fluorophore and measure changes in fluorescence induced by changes in the field. The salient feature of this technique is that the changes measured reflect local changes in or near the site where the fluorophore is located in the protein as opposed to electrical measurements that reflect overall conformations of the protein. Two main types of measurements have been done with site-directed fluorescence. One is the detection of fluorescence intensity changes of one fluorophore in the labeled site and the other is the estimation of distance and distance changes between two fluorophores using fluorescence resonance energy transfer (FRET).

3.6.5.1 Site-Directed Fluorescence Changes

In these experiments, the site of interest is mutated to a cysteine and then is reacted to a fluorophore that has a cysteine reactive group and the time course of fluorescence is monitored during pulses applied to open or close the channel (Mannuzzu et al., 1996; Cha and Bezanilla, 1997). To detect a fluorescence change, the conformational change must change the environment around the fluorophore so that the intensity changes because of spectral shifts, quenching or dye reorientation. In most cases the changes in fluorescence have been traced to changes in the quenching environment around the fluorophore (Cha and Bezanilla, 1997, 1998) and in a few cases it is produced by spectral shifts of the fluorophore as the hydrophobicity of the environment changes with the change in conformation (Asamoah et al., 2004). In some cases, the presence of quenching groups in the protein is crucial in obtaining a signal. For example, in the bacterial channel NaChBac the signals are very small or in some sites not detectable (Blunck et al., 2004) although, following the classical pattern, there are four charges in the S4 segment and the total gating charge was recently measured to be about 14 e_0 (Kuzmenkin et al., 2004). This result has been traced to the lack of quenching groups in the structure of this channel (Blunck et al., 2004). To obtain signals from sites near the S4 segment upon changes of membrane potential, a requirement seems to be the presence of glutamate residues in the nearby region that act as quenching groups (Blunck et al., 2004).

Fluorescence changes in site-directed fluorescent labeling have provided information of local conformational changes around the S4 segment, the S3–S4 linker, S1–S2 linker and the pore region as a result of changes in membrane potential (Mannuzzu et al., 1996; Cha and Bezanilla, 1997; Loots and Isacoff, 1998). In the absence of three-dimensional structure, the interpretation of these fluorescence changes is not straightforward because the exact location of the fluorophore is unknown. However, several important qualitative results have been obtained. For example, it has been found that the kinetics of the fluorescence changes in S4 are slower than around the S1–S2 linker, suggesting that there might be earlier conformational changes that precede the main conformational change normally attributed to the S4 segment (Cha and Bezanilla, 1997). Also, the time course of fluorescence changes near the pore region are much slower than channel activation and their kinetics can be traced to another gating process called slow inactivation in Shaker K^+ channel (Cha and Bezanilla, 1997; Loots and Isacoff, 1998).

Probably one of the most informative results have been obtained in the sodium channel because, as this technique detects local changes, it has been possible to distinguish specific functions for each one of the four domains of the Na^+ channel. Fast inactivation is another gate that operates by blocking the channel pore (Hodgkin and Huxley, 1952; Armstrong and Bezanilla, 1977; Hoshi et al., 1990). By labeling the S4 segment of each domain of the Na^+ channel, it was found that the gating charge immobilization produced by inactivation only occurred in domains III and IV, thus locating the regions of the channel that interact with the inactivating particle (Cha et al., 1999a). The kinetics of the fluorescence of sites in S4 was found to be very fast in domains I–III but slower in domain IV, indicating that domain IV followed the other three domains (Chanda and Bezanilla, 2002). Finally, by comparing the effect of a perturbation in one domain to the fluorescence signal in another domain, it was found that all four domains of the Na^+ channel move in cooperative fashion. This result is important in explaining why sodium channels are faster than potassium channels, an absolute requirement in eliciting an action potential (Chanda et al., 2004). For more details on the voltage-dependent sodium channel, see the chapter by Hank in this volume.

3.6.5.2 Resonance Energy Transfer

By labeling the protein with a donor and an acceptor fluorophore it is possible to estimate the distance between them using Förster theory of dipole–dipole interaction (Cantor and Schimmel, 1980). Depending on the fluorophore pair used, those distances can be from a few to about 100 Å, thus enabling the measurements of intramolecular distances and changes in distances.

This technique was used to estimate distances and distance changes between subunits in Shaker-IR channel by Cha et al. (1999b) and Glauner et al. (1999). Cha et al. (1999b) used a variant of FRET, called LRET that uses a lanthanide (terbium) as a donor and has the advantage that the estimation of the distances is more accurate mainly because the orientation factor is bound between tight limits giving a maximum uncertainty of $\pm 10\%$, but frequently is even better because the acceptor is not immobilized (Selvin, 2002). The LRET technique was tested and validated by Cha et al. (1999b) in the Shaker K^+ channel where the measurement of distances between residues in the pore region gave an agreement within 1 Å when compared to an equivalent residue in the KcsA crystal structure. In addition, each measurement using LRET gave two distances that corresponded to the separation between adjacent and opposite subunits in the channel and those measured distances were related by the $\sqrt{2}$, as expected from the tetrameric structure, giving an internal calibration and consistency check of the technique.

Both studies (Cha et al., 1999b; Glauner et al., 1999) showed that the distances between S4 segments were around 50 Å and that it did not change very much from the closed to the open states. The maximum distance change measured by Cha et al. (1999b) was about 3 Å while Glauner et al. (1999) measured about 5 Å. These measurements were all done between subunits therefore they do not completely rule out a translation across the bilayer, as proposed in the helical screw and paddle models.

In Shaker K^+ channel the linker between S3 and S4 is made of about 30 residues. At least six of the residues close to the extracellular part of S4 seem to be in alpha-helical conformation (Gonzalez et al., 2001) suggesting that S4 is extended extracellularly as an alpha helix. This would explain why Cha et al. (1999b) measured changes in distance in the S3–S4 linker as a result of membrane potential changes. One of those changes is a rotation in the linker and the other is a decrease in the tilt of the S4 with its extension upon depolarization, providing a possible mechanism for charge translocation (see Fig. 3.9).

The question of how much translation the S4 undergoes across the bilayer with depolarization has been approached with two other variants of FRET. In the first series of experiments (Starace et al., 2002) green fluorescent protein (eGFP) was inserted after the S6 of Shaker K^+ channel and was used as the donor to an acceptor attached in the extracellular regions of the channel. In this case, the acceptor was a sulforhodamine with an MTS reactive group that can react with an engineered cysteine in the channel but it also can be cleaved off with a reducing agent. This allows the measurement of donor fluorescence (intracellularly located) in presence and absence of acceptor (in the extracellular regions of the channel) to compute the energy transfer and the distance. These measurements were done in multiple sites of the S1–S2 loop, S3–S4 loop, and S4 segment under depolarized and hyperpolarized conditions giving changes in distance that did not exceed 2 Å. In fact, some sites increased while others decreased their distance to the intracellular donor upon depolarization, indicating that there is little translation of the S4 across the membrane.

The second method used the hydrophobic negative ion dipycrilamine (dpa) as an acceptor that distributes in the edges of the bilayer according to the membrane potential (Fernandez et al., 1982; Chanda et al., 2005). The donor was rhodamine attached to specific sites in the S4 segment. The experiment predicts a clear distinction for the outcome depending whether the S4 does or does not make a large translation across the membrane. If the S4 segment undergoes a large translation across the bilayer such that the donor crosses its midpoint, then a *transient* fluorescence decrease is expected because dipycrilamine and the fluorophore start and end in opposite sides of the membrane but there is a period where both donor and acceptor reside simultaneously in both sides of the membrane increasing transfer and consequently decreasing donor fluorescence. On the other hand, if there is no crossing of the bilayer midline by the donor, the fluorescence will increase if the donor is above the midline or decrease if it is below, but no transient should be observed. The results of four sites in S4 are consistent with no crossing of the midline strongly suggesting that the S4 does not make a large translation upon depolarization (Chanda et al., 2005).

A recent detailed experiment using LRET confirms the lack of large translation of the S4 (Posson et al., 2005). In this case, the donor is Tb (in chelate form) attached in several sites of the S4 segment and the S3–S4 linker while the acceptor is in a toxin that blocks the pore of Shaker from the extracellular side. Results show that in all S4 sites measured the distance did not decrease more than 1 Å upon depolarization. When this change is projected as a translation of the S4 segment it gives an upper limit of only 2 Å.

It is interesting to note that in the LRET experiments using an acceptor in the toxin and a terbium chelate in the extracellular part of the S3 segment showed that in fact the distance *increased* by about 2 Å upon depolarization, which is also incompatible with the paddle model that postulates a simultaneous translation of S3 and S4 segments. Another confirmation that the extracellular part of S3 does not get buried in the closed state was provided by a recent paper by Gonzalez et al. (2005) who tested accessibility of these residues and found no state dependence.

Finally, another recent paper using a toxin that binds to the S3–S4 linker of the Shaker K channel also show that there is limited translation of the S4 segment (Phillips et al., 2005).

A recent molecular dynamic calculation of the voltage sensor of a K^+ channel under the influence of an electric field also supports a conformational change that involves minimum translation (Treptow et al., 2004).

In summary, FRET experiments are inconsistent with a large transmembrane displacement of the S4 segment in response to a voltage pulse. As many other experiments, presented above and discussed in recent reviews (Bezanilla, 2002; Cohen et al., 2003; Ahern and Horn, 2004b; Lee et al., 2003; Swartz, 2004) do not support the idea that the charged residues are in the bilayer, the two main features of the paddle model become inconsistent with the available data.

3.7 Coupling of the Sensor to the Gate

In contrast to an FET where gating of the channel is produced by a change in space charge, the gate in voltage-gated channels seem to be a mechanical obstruction to flow (Yellen, 1998). The crystal structure of the bacterial channel KcSA reveals a closed state of the channel and Perozo et al. (1999) found that when it opens, the S6 makes a scissor-like action allowing ions to go through. The crystal structure of MthK, another prokaryotic channel, shows an open pore where a glycine in the S6 segment is shown to break the S6 in two segments (Jiang et al., 2002a). This led MacKinnon to propose that the gate opens when the S6 segment is broken and the intracellular part is pulled apart (Jiang et al., 2002b). In the crystal structure of the voltage-dependent prokaryotic channel KvAP the pore is in the open conformation by a break in the S6 segment in a glycine residue (Jiang et al., 2003a). In the case of the eukaryotic Shaker K^+ channel there is a PVP sequence in the S6 segment that has been proposed to be the actual gate (Webster et al., 2004) and the crystal structure of Kv1.2 seems to indicate that the PVP motif is important. In addition to the main gate formed by the bundle crossing of the S6 segments, there is now very good evidence that the selectivity filter can also stop conduction, thus introducing another gate in series (Bezanilla and Perozo, 2003; Cordero et al., 2006; Blunck et al., submitted). However, there is no evidence or a physical mechanism to couple this filter gate to the movement of the sensor.

The question of how and what kind of physical movement of the sensor, mainly the S4 segment, couples the opening of the pore gate is far from resolved. Most of the proposals, including the paddle model suggest that the change in position of the S4 couples via the intracellular S4–S5 linker to change the position of the S5 segment that in turns allow the opening of S6. The crystal structure of Kv1.2 suggests the same mechanism of opening whereby the S4 segment pulls on S5 to allow channel opening (Long et al., 2005b). In the transporter model (Chanda et al., 2005) the mechanism is more explicit because a closed and an open structures are proposed. In this case the change in tilt of the S4 segment carries the S5 segment away from S6 allowing the break at the glycine residue and thus opening the pore.

3.8 Concluding Remarks

The main component of the sensor of voltage-gated channels is the S4 segment with its basic residues moving in the field. The movement of the sensor in response to changes in the electric field produces the gating currents. The study of gating

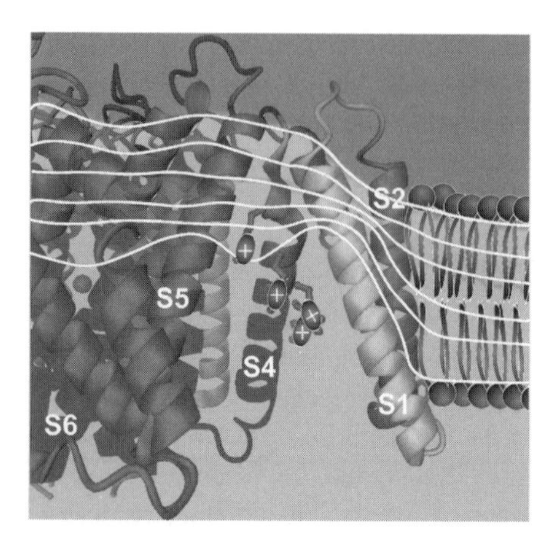

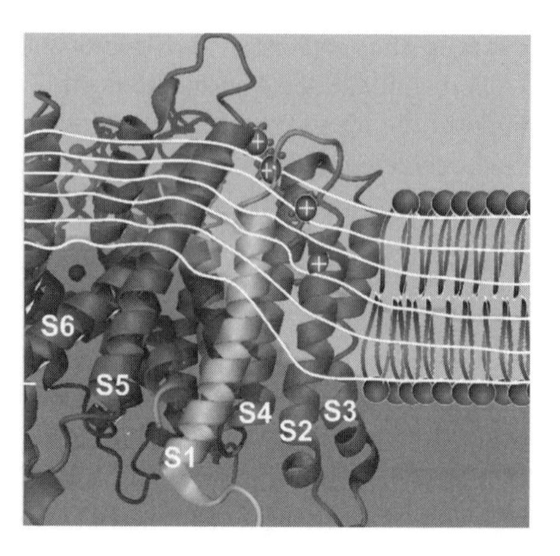

Fig. 3.10 The closed and open conformations and the voltage profile of the Shaker K^+ channel transporter model (Chanda et al., 2005). The isopotential lines divide equally the total voltage applied between the inside and the outside. (a) Closed state is obtained at hyperpolarized potentials (< -100 mV, clear inside). (b) Open state is obtained at depolarized potentials (>50 mV, clear outside). Notice that going from a hyperpolarized to a depolarized potential, the S4 segment has not translated across the membrane but it has undergone a change in tilt that moves the position of S5 which in turn opens the pore by breaking S6. The arginines side chains relocate according to the direction of the field.

currents, single-channel currents, and macroscopic currents has generated detailed kinetic models of channel operation. The sensor couples to the gate possibly via the S4–S5 linker. The channel fully opens after all sensors have moved and there is little cooperativity in the early movements of the sensor in the K^+ channel but strong positive cooperativity in the Na^+ channel. The results of a large variety of biophysical experiments have helped in delineating the conformational changes of the sensor. These data have shown that the S4 segment does not undergo a translational motion across the membrane. Rather, the charges are in water crevices and they only move a small distance because the field is focused in a narrow region within the protein core. In the absence of a crystal structure representative of the channel in its native form in the closed state, we believe that the data support the transporter model.

The essence of the transporter model is the shaping of the electric field that allows the hydration of the arginines in the closed and active conformations of the sensor. As the field is concentrated in a small region of the protein, the movement of the charged arginines is much less than if they were to move in the hydrophobic part of the membrane. Fig. 3.10 shows a detail of the Shaker K channel model in the closed and active conformations as proposed by Chanda et al. (2005). In the closed state there is a large water crevice that penetrates the protein from the intracellular side, thus concentrating all the voltage drop in a narrow region close to the extracellular side (Fig. 3.10a). In this conformation, most of the arginines are in this water crevice but the most extracellular arginine is still within the field. Upon depolarization, the most extracellular arginine senses the change in field and tends to move out thus changing the tilt of the S4 segment as whole. This makes the intracellular crevice decrease locating the other arginines in the field, which will also help moving the S4 segment. Eventually, the internal crevice disappears and a small crevice appears in the extracellular side where the arginines will swing into it (Fig. 3.10b). The whole process involves a solid body motional tilt of the S4, most likely in several steps, and rearrangements of the arginine side chains giving a net transport of charge from the inside to the outside. There is almost no translation of the S4 across the plane of the membrane but there is a significant change in tilt that rearranges many other parts of the protein. In the resting state the extracellular portion of the S3 segment acts as a dielectric cover over the internal crevice and upon depolarization it moves away thus exposing the extracellular cavity that receives the guanidinium group of the arginines. Chanda et al. (2005) used the two proposed molecular structures (closed and open) to compute the net transfer of charge across the membrane solving the Poisson–Nernst–Planck equations (Roux, 1997). The result was a total of 13 e_0, in excellent agreement with experimental data and the solution showed that the most of the charge was contributed by the first four most extracellular arginines of the channel (residues 362, 365, 368, and 371). The explicit molecular models of the closed and open states presented in Chanda et al. (2005) are based on a multitude of biophysical data and the available structural data. At present, these structures and the associated mechanism of the conformational change induced by membrane potential changes should be viewed as a representation of a conceptual model. Details of the positions of each of the relevant residues and their side chains and their trajectories during activation will be required to have a complete description of the voltage sensor operation.

3.9 Outlook

A detailed understanding of voltage-gated channels means that we can represent the landscape of energy of the physical states of the channel at atomic resolution together with the structural changes evoked by the electric field. To achieve this goal, it will not only require a static, high-resolution three-dimensional structure, but also a detailed description of the kinetics of the voltage-induced conformational changes. The latter are expected to be obtained with spectroscopic and computational techniques. Fluorescence and EPR spectroscopies have started to unravel some of the details of conformational changes during gating but the study of detailed kinetics is still developing. For example, in the same way that single-channel current recordings were critical in understanding the operation of the pore and the gate, single molecule fluorescence is expected to show the local conformational changes during voltage sensor operation and channel gating. These techniques are advancing very rapidly and some results have already been published as just the conformational change (Blunck and Bezanilla, 2002; Sonnleitner et al., 2002) or even correlated to current recordings (Blunck et al., 2003; Borisenko et al., 2003).

Acknowledgments

Many thanks to Drs. Rikard Blunck and Baron Chanda for reading the manuscript and for their helpful and insightful comments. This work was supported in part by the U.S.P.H.S., N.I.H. under Grant GM30376.

References

Aggarwal, S.K., and R. MacKinnon. 1996. Contribution of the S4 segment to gating charge in the *Shaker* K^+ channel. *Neuron* 16:1169–1177.

Agnew, W.S., S.R. Levinson, J.S. Brabson, and M.A. Raftery. 1978. Purification of the tetrodotoxin-binding component associated with the voltage-sensitive sodium channel from *Electrophorus electricus* electroplax membranes. *Proc. Natl. Acad. Sci. USA* 75:2602–2610.

Ahern, C.A., and R. Horn. 2004a. Specificity of charge-carrying residues in the voltage sensor of potassium channels. *J. Gen. Physiol.* 123:205–216.

Ahern, C.A., and R. Horn. 2004b. Stirring up controversy with a voltage sensor paddle. *Trends Neurosci.* 27(6):303–307.

Ahern, C.A., and R. Horn. 2005. Focused electric field across the voltage sensor of potassium channels. *Neuron* 48:25–29.

Almers, W. 1978. Gating currents and charge movements in excitable membranes. *Rev. Physiol. Biochem. Pharmacol.* 82:96–190.

Armstrong, C.M., and F. Bezanilla. 1973. Currents related to movement of the gating particles of the sodium channels. *Nature* 242:459–461.

Armstrong, C.M., and F. Bezanilla. 1977. Inactivation of the sodium channel. II. Gating current experiments. *J. Gen. Physiol.* 70:567–590.

Asamoah, O.K, B. Chanda, and F. Bezanilla. 2004. A spectroscopic survey of gating-induced environmental changes in the Shaker potassium channel. *Biophys. J.* 86:432a.

Asamoah, O.K., J.P. Wuskell, L.M. Loew, and F. Bezanilla. 2003. A fluorometric approach to local electric field measurements in a voltage-gated ion channel. *Neuron* 37:85–97.

Baker, O.S., H.P. Larsson, L.M. Mannuzzu, and E.Y. Isacoff. 1998. Three transmembrane conformation and sequence-dependent displacement of the S4 domain in Shaker K^+ channel gating. *Neuron* 20:1283–1294.

Bezanilla, F. 2000. The voltage sensor in voltage-dependent channels. *Phys. Rev.* 80:555–592.

Bezanilla, F. 2002. Perspective: Voltage sensor movements. *J. Gen. Physiol.* 120:465–473.

Bezanilla, F., and E. Perozo. 2003. The voltage sensor and the gate in ion channels. *In*: Advances in Protein Chemistry, Vol. 63. D. Rees, editor. Elsevier Science, New York.

Bezanilla, F., E. Perozo, and E. Stefani. 1994. Gating of Shaker K^+ channels. II. The components of gating currents and a model of channel activation. *Biophys. J.* 66:1011–1021.

Blaustein, R.O., P.A. Cole, C. Williams, and C. Miller. 2000. Tethered blockers as molecular 'tape measures' for a voltage-gated K^+ channel. *Nat. Struct. Biol.* 7:309–311.

Blunck R., and F. Bezanilla. 2002. Fluorescence recordings of a low number of voltage gated K^+ channels. *Biophys. J.* 82:267a.

Blunck R., J. Cordero, L. Cuello, E. Perozo, and F. Bezanilla. 2006. Detection of the opening of the bundle crossing in KcsA with fluorescence lifetime spectroscopy reveals the existence of two gates for ion conduction (submitted).

Blunck, R., D.M. Starace, A.M. Correa, and F. Bezanilla. 2004. Detecting rearrangements of *Shaker* and NaChBac in real-time with fluorescence spectroscopy in patch-clamped mammalian cells. *Biophys. J.* 86:3966–3980.

Blunck, R., J.L. Vazquez-Ibar, Y.S. Liu, E. Perozo, and F. Bezanilla. 2003. Fluorescence measurements of KcsA channels in artificial bilayers. *Biophys. J.* 84(2, Pt 2):124a–125a.

Borisenko, V., T. Lougheed, J. Hesse, E. Fureder-Kitzmuller, N. Fertig, J.C. Behrends, G.A. Woolley, G.J. Schutz. 2003. Simultaneous optical and electrical recording of single gramicidin channels. *Biophys. J.* 84(1):612–622.

Cantor, C.R., and P.R. Schimmel. 1980. Biophysical Chemistry. Part II. Techniques for the Study of Biological Structure and Function. W.H. Freeman and Co., New York.

Caterall, W.A. 1986. Molecular properties of voltage-sensitive sodium channels. *Annu. Rev. Biochem.* 55:953–985.

Cha, A., and F. Bezanilla. 1997. Characterizing voltage-dependent conformational changes in the *Shaker* K^+ channel with fluorescence. *Neuron* 19:1127–1140.

Cha, A., and F. Bezanilla. 1998. Structural implications of fluorescence quenching in the *Shaker* K^+ channel. *J. Gen. Physiol.* 112:391–408.

Cha, A., P.C. Ruben, A.L. George, E. Fujimoto, and F. Bezanilla. 1999a. Voltage sensors in domains III and IV, but not I and II, are immobilized by Na^+ channel fast inactivation. *Neuron* 22: 73–87.

Cha, A., G. Snyder, P.R. Selvin, and F. Bezanilla. 1999b. Atomic scale movement of the voltage-sensing region in a potassium channel measured via spectroscopy. *Nature* 402:809–813.

Chanda, B., O.K. Asamoah, and F. Bezanilla. 2004. Coupling interactions between voltage sensors of the sodium channel as revealed by site-specific measurements. *J. Gen. Physiol.* 123:217–230.

Chanda, B., O.K. Asamoah, R. Blunck, B. Roux, and F. Bezanilla. 2005. Gating charge displacement in voltage-gated channels involves limited transmembrane movement. *Nature* 436:852–856.

Chanda, B., and F. Bezanilla. 2002. Tracking voltage-dependent conformational changes in skeletal muscle sodium channel during activation. *J. Gen. Physiol.* 120:629–645.

Chapman, M.L., and A.M.J. VanDongen. 2005. K channel subconductance levels result from heteromeric pore conformations. *J. Gen. Physiol.* 126:87–103.

Cohen, B.E., M. Grabe, and L.Y. Jan. 2003. Answers and questions from KvAP structure. *Neuron* 39:395–400.

Conti, F., and W. Stuhmer. 1989. Quantal charge redistribution accompanying the structural transitions of sodium channels. *Eur. Biophys. J.* 17:53–59.

Cordero-Marales, J.F., L.G. Cuello, Y. Zhao, V. Jogini, D.M. Cortes, B. Roux, and E. Perozo. 2006. Molecular determinants of gating at the potassium-channel selectivity filter. *Nature Struct. Mol. Biol.* 13:311–318.

Cuello, L.G., M. Cortes, and E. Perozo. 2004. Molecular architecture of the KvAP voltage dependent K^+ channel in a lipid bilayer. *Science* 306:491–495.

Durell, S.R., and H.R. Guy. 1992. Atomic scale structure and functional models of voltage-gated potassium channels. *Biophys. J.* 62:238–250.

Durell, S.R., I.H. Shrivastava, and H.R. Guy. 2004. Models of the structure and voltage-gating mechanism of the Shaker K^+ channel. *Biophys. J.* 87:2116–2130.

Fernandez, J.M., F. Bezanilla, and R.E. Taylor. 1982. Effect of chloroform on the movement of charges within the nerve membrane. *Nature* 297:150–152.

Freites, J.A., D.J. Tobias, G. von Heijne, and S.H. White. 2005. Interface connections of a transmembrane voltage sensor. *PNAS* 102:15059–15064.

Gandhi, C.S., and E.Y. Isacoff. 2002. Molecular models of voltage sensing. *J. Gen. Physiol.* 120:455–463.

Glauner, K.S., L.M. Mannuzzu, C.S. Gandhi, and E.Y. Isacoff. 1999. Spectroscopic mapping of voltage sensor movements in the Shaker potassium channel. *Nature* 402:813–817.

Gonzalez, C., F.J. Morera, E. Rosenmann, and R. Latorre. 2005. S3b amino acid residues do not shuttle across the bilayer in voltage-gated Shaker K^+ channels. *Proc. Natl. Acad. Sci.* 102:5020–5025.

Gonzalez, C., E. Rosenman, F. Bezanilla, O. Alvarez, and R. Latorre. 2001. Periodic perturbations in Shaker K^+ channel gating kinetics by deletions in the S3–S4 linker. *Proc. Natl. Acad. Sci.* 98:9617–9623.

Hamill, O.P., A. Marty, E. Neher, B. Sackmann, and F.J. Sigworth. 1981. Improved patch-clamp techniques for high-resolution current recording from cells and cell-free membrane patches. *Pflugers Arch.* 391:85–100.

Hirschberg, B., A. Rovner, M. Lieberman, and J. Patlak. 1996. Transfer of twelve charges is needed to open skeletal muscle Na^+channels. *J. Gen. Physiol.* 106:1053–1068.

Hodgkin, A.L., and A.F. Huxley. 1952. A quantitative description of membrane current and its application to conduction and excitation in nerve. *J. Physiol.* 117:500–544.

Horn, R., S. Ding, and H.J. Gruber. 2000. Immobilizing the moving parts of voltage-gated ion channels. *J. Gen. Physiol.* 116:461–476.

Horrigan, F.T., J. Cui, and R.W. Aldrich. 1999. Allosteric voltage gating of potassium channels I: mSlo ionic currents in absence of Ca^{2+}. *J. Gen. Physiol.* 114:277–304.

Hoshi, T., W.N. Zagotta, and R.W. Aldrich. 1990. Biophysical and molecular mechanisms of Shaker potassium channel inactivation. *Science* 250:533–538.

Islas, L.D., and F.J. Sigworth. 2001. Electrostatics and the gating pore of *Shaker* potassium channels. *J. Gen. Physiol.* 117:69–89.

Jiang, Y., A. Lee, J. Chen, M. Cadene, B.T. Chait, and R. MacKinnon. 2002a. Crystal structure and mechanism of a calcium-gated potassium channel. *Nature* 417:515–522.

Jiang, Y., A. Lee, J. Chen, M. Cadene, B.T. Chait, and R. MacKinnon. 2002b. The open pore conformation of potassium channels. *Nature* 417:523–526.

Jiang, Y., A. Lee, J. Chen, V. Ruta, M. Cadene, B.T. Chait, and R. MacKinnon. 2003a. X-ray structure of a voltage-dependent K(+) channel. *Nature* 423:33–41.

Jiang, Y., V. Ruta, J. Chen, A. Lee, and R. MacKinnon. 2003b. The principle of gating charge movement in a voltage-dependent K^+channel. *Nature* 423:42–48.

Jiang, Q.-X., D.-N. Wang, and R. MacKinnon. 2004. Electron microscopic analysis of KvAP voltage dependent K^+ channel in an open conformation. *Nature* 430:806–810.

Keynes, R.D., and E. Rojas. 1974. Kinetics and steady-state properties of the charged system controlling sodium conductance in the squid giant axon. *J. Physiol (Lond.)* 239:393–434.

Kuzmenkin, A., F. Bezanilla, and A.M. Correa. 2004. Gating of the bacterial sodium channel NaChBac: Voltage dependent charge movement and gating currents. *J. Gen. Physiol.* 124:349–356.

Laine, M., M.C. Lin, J.P. Bannister, W.R. Silverman, A.F. Mock, B. Roux, and D.M. Papazian. 2003. Atomic proximity between S4 segment and pore domain in Shaker potassium channels. *Neuron* 39:467–481.

Larsson, H.P., O.S. Baker, D.S. Dhillon, and E.Y. Isacoff. 1996. Transmembrane movement of the *Shaker* K^+ channel S4. *Neuron* 16:387–397.

Lee, H.C., J.M. Wang, and K.J. Swartz. 2003. Interaction between extracellular Hanatoxin and the resting conformation of the voltage sensor paddle in KV channels. *Neuron* 40(3):527–536.

Li-Smerin, Y., D.H. Hackos, and K.J. Swartz. 2000. Alpha-helical structural elements within the voltage-sensing region domains of a K^+ channel. *J. Gen. Physiol.* 115:33–50.

Long, S.B., E.B. Campbell, and R. MacKinnon. 2005a. Crystal structure of a mammalian voltage-dependent Shaker family K+ channel. *Science* 309:897–903.

Long, S.B., E.B. Campbell, and R. MacKinnon. 2005b. Structural basis of electromechanical coupling. *Science* 309:903–908.

Loots, E., and E.Y. Isacoff. 1998. Protein rearrangements underlying slow inactivation of the Shaker K^+ channel. *J. Gen. Physiol.* 112:377–389.

Mannuzzu, L.M., and E.Y. Isacoff. 2000. Independence and cooperativity in rearrangements of a potassium channel voltage sensor revealed by single subunit fluorescence. *J. Gen. Physiol.* 115:257–268.

Mannuzzu, L.M., M.M. Moronne, and E.Y. Isacoff. 1996. Direct physical measure of conformational rearrangement underlying potassium channel gating. *Science* 271:213–216.

Monks, S.A., D.J. Needleman, and C. Miller. 1999. Helical structure and packing orientation of the S2 segment in the Shaker K^+ channel. *J. Gen. Physiol.* 113:415–423.

Noceti, F., P. Baldelli, X. Wei, N. Qin, L. Toro, L. Birnbaumer, and E. Stefani. 1996. Effective gating charges per channel in voltage-dependent K^+ and Ca^{2+} channel. *J. Gen. Physiol.* 108:143–155.

Noda, M., S. Shimizu, T. Tanabe, T. Takai, T. Kayano, T. Ikeda, H. Takahashi, H. Nakayama, Y. Kanaoka, and N. Minamino. 1984. Primary structure of *Electrophorus electricus* sodium channel deduced from cDNA sequence. *Nature* 312:121–127.

Papazian, D.M., X.M. Shao, S.-A. Seoh, A.F. Mock, Y. Huang, and D.H. Weinstock. 1995. Electrostatic interactions of S4 voltage sensor in Shaker K^+ channel. *Neuron* 14:1293–1301.

Parsegian, A. 1969. Energy of an ion crossing a low dielectric membrane: Solutions to four relevant electrostatic problems. *Nature* 221:844–846.

Perozo, E., M. Cortes, and L.G. Cuello. 1999. Structural rearrangements underlying K^+-channel activation gating. *Science* 285:73–78.

Perozo, E., R. MacKinnon, F. Bezanilla, and E. Stefani. 1993. Gating currents from a non-conducting mutant reveal open-closed conformations in Shaker K^+ channels. *Neuron* 11:353–358.

Phillips, L.R., M. Milescu, Y. Li-Smerin, J.A. Midell, J.I. Kim, and K.J. Swartz. 2005. Voltage sensor activation with a tarantula toxin as cargo. *Nature* 436:857–860.

Posson, D.J., P. Ge, C. Miller, F. Bezanilla, and P.R. Selvin. 2005. Small vertical movement of a K+ channel voltage sensor measured with luminescence energy transfer. *Nature* 436:848–851.

Richardson, J., P. Ge, P.R. Selvin, F. Bezanilla, and D.M. Papazian. 2006. Orientation of the voltage sensor relative to the pore differs in prokaryotic and eukaryotic voltage-dependent potassium channels [abstract]. *Biophys. J.*

Richardson, J., D.M. Starace, F. Bezanilla, and A.M. Correa. 2005. Scanning NaChBac topology using LRET [abstract]. *Biophys. J.*

Roux, B. 1997. Influence of the membrane potential on the free energy of an intrinsic protein. *Biophys. J.* 73:2980–2989.

Ruta, V., J. Chen, and R. MacKinnon. 2005. Calibrated measurement of gating-charge arginine displacement in the KvAP voltage-dependent K^+ channel. *Cell* 123:463–475.

Santacruz-Toloza, L., Y. Huang, S.A. John, and D.M. Papazian. 1994. Glycosylation of Shaker potassium channel protein in insect cell culture and in Xenopus oocytes. *Biochemistry* 33:5607–5613.

Schoppa, N.E., K. McCormack, M.A. Tanouye, and F.J. Sigworth. 1992. The size of gating charge in wild-type and mutant Shaker potassium channels. *Science* 255:1712–1715.

Schoppa, N.E., and F.J. Sigworth. 1998. Activation of Shaker potassium channels. III. An activation gating model for wild-type and V2 mutant channel. *J. Gen. Physiol.* 111:313–342.

Selvin, P.R. 2002. Principles and biophysical applications of luminescent lanthanide probes. *Annu. Rev. Biophys. Biomol. Struct.* 31:275–302.

Seoh, S.-A., D. Sigg, D.M. Papazian, and F. Bezanilla. 1996. Voltage-sensing residues in the S2 and S4 segments of the Shaker K^+ channel. *Neuron* 16:1159–1167.

Sigg, D., and F. Bezanilla. 1997. Total charge movement per channel: The relation between gating displacement and the voltage sensitivity of activation. *J. Gen. Physiol.* 109:27–39.

Sigg, D., F. Bezanilla, and E. Stefani. 2003. Fast gating in the Shaker K^+ channel and the energy landscape of activation. *PNAS* 100:7611–7615.

Sigg, D., H. Qian, and F. Bezanilla. 1999. Kramers' diffusion theory applied to gating kinetics of voltage dependent channels. *Biophys. J.* 76:782–803.

Sigg, D., E. Stefani, and F. Bezanilla. 1994. Gating current noise produced by elementary transition in Shaker potassium channels. *Science* 264:578–582.

Sonnleitner, A., L.M. Mannuzzu, S. Terakawa, and E.Y. Isacoff. 2002. Structural rearrangements in single ion channels detected optically in living cells. *Proc. Natl. Acad. Sci. USA* 99(20):12759–12764.

Starace, D.M., and F. Bezanilla. 2001. Histidine scanning mutagenesis of basic residues of the S4 segment of the Shaker K^+ channel. *J. Gen. Physiol.* 117:469–490.

Starace, D.M., and F. Bezanilla. 2004. A proton pore in a potassium channel voltage sensor reveals a focused electric field. *Nature* 427:548–552.

Starace, D.M., P.R. Selvin, and F. Bezanilla. 2002. Resonance energy transfer measurements on transmembrane motion of Shaker K^+ channel voltage sensing region. *Biophys. J.* 82:174a.

Starace, D.M., E. Stefani, and F. Bezanilla. 1997. Voltage-dependent proton transport by the voltage sensor of the Shaker K^+ channel. *Neuron* 19:1319–1327.

Swartz, K.J. 2004. Towards a structural view of gating in potassium channels. *Nat. Rev. Neurosci.* 5:905–916.

Tempel, T.M., D.M. Papazian, T.L. Schwarz, Y.N. Jan, and L.Y. Jan. 1987. Sequence of a probable potassium channel component encoded at Shaker locus in Drosophila. *Science* 237:770–775.

Tiwari-Woodruff, S.K., C.T. Schulteis, A.F. Mock, and D.M. Papazian. 1997. Electrostatic interactions between transmembrane segments mediate folding of *Shaker* K^+ channel subunits. *Biophys. J.* 72:1489–1500.

Tombola, F., M.M. Pathak, and E.Y. Isacoff. 2005. Voltage-sensing arginines in a potassium channel permeate and occlude cation-selective pores. *Neuron* 45:379–388.

Treptow, W., B. Maigret, C. Chipot, and M. Tarek. 2004. Coupled motions between the pore and voltage-sensor domains: A model for Shaker B, a voltage-gated potassium channel. *Biophys. J.* 87:2365–2379.

Vandenberg, C.A., and F. Bezanilla. 1991. A sodium channel model of gating based on single channel, macroscopic ionic and gating currents in the squid giant axon. *Biophys. J.* 60:1511–1533.

Webster, S.M., D. Del Camino, J.P. Dekker, and G. Yellen. 2004. Intracellular gate opening in Shaker K^+ channels defined by high-affinity metal bridges. *Nature* 428:864–868.

Yang, N., A.L. George, and R. Horn. 1996. Molecular basis of charge movement in voltage-gated sodium channels. *Neuron* 16:113–122.

Yang, N., and R. Horn. 1995. Evidence for voltage-dependent S4 movement in sodium channels. *Neuron* 15:213–218.

Yellen, G. 1998. The moving parts of voltage-gated ion channels. *Quart. Rev. Biophys.* 31:239–295.

Yusaf, S.P., D. Wray, and A. Sivaprasadarao. 1996. Measurement of the movement of the S4 segment during activation of a voltage-gated potassium channel. *Pflugers Arc. Eur. J. Physiol.* 433:91–97.

Zagotta, W.N., T. Hoshi, J. Dittman, and R. Aldrich. 1994. *Shaker* potassium channel gating III: Evaluation of kinetic models for activation. *J. Gen. Physiol.* 103:321–362.

4 Voltage-Gated Potassium Channels

Stephen J. Korn and Josef G. Trapani

Part I. Overview

Potassium (K^+) channels are largely responsible for shaping the electrical behavior of cell membranes. K^+ channel currents set the resting membrane potential, control action potential duration, control the rate of action potential firing, control the spread of excitation and Ca^{2+} influx, and provide active opposition to excitation. To support these varied functions, there are a large number of K^+ channel types, with a great deal of phenotypic diversity, whose properties can be modified by many different accessory proteins and biochemical modulators.

As with other ion channels, there are two components to K^+ channel operation. First, channels provide a pathway through the cell membrane that selectively allows a particular ion species (in this case, K^+) to flow with a high flux rate. Second, channels have a gating mechanism in the conduction pathway to control current flow in response to an external stimulus. To accommodate their widespread involvement in cellular physiology, K^+ channels respond to a large variety of stimuli, including changes in membrane potential, an array of intracellular biochemical ligands, temperature, and mechanical stretch. Additional phenotypic variation results from a wide range of single-channel conductances, differences in stimulus threshold, and variation in kinetics of three basic gating events. Channel opening (activation) is caused by changes in membrane potential or changes in the concentration of specific ligands, and can occur at different rates, and over different voltage or concentration ranges. Channel closing (deactivation), which occurs upon removal of the activating stimulus, can also proceed at different rates. Although the molecular events that underlie activation and deactivation are known in significant detail, the mechanisms that account for differences in activation and deactivation rates are not well understood. Finally, channels can undergo a process called inactivation, whereby the channel stops conducting even in the presence of the activating stimulus. There are at least two mechanisms of inactivation (Choi et al., 1991; Hoshi et al., 1991). One type of inactivation, which is fast (it occurs and is complete during the first several milliseconds of a strong depolarization), is called N-type because it is structurally associated with the amino (N) terminus of the channel (see Fig. 4.2). Only a few K^+ channels have an N-type inactivation mechanism. Another, slower inactivation mechanism occurs in almost all channels, and is associated with events near the selectivity filter and outer vestibule of the channel. This latter inactivation mechanism, first described in the *Shaker* K^+ channel, was originally called C-type inactivation, because the structural components involved were located more toward the carboxyl (C) terminus than those involved in N-type inactivation (Choi et al., 1991). However, slow inactivation has significantly different characteristics in different channels, which makes it unclear whether slow inactivation involves an identical molecular mechanism in all channels. Whereas fast inactivation is designed to rapidly turn off K^+ channel conduction following activation, slow inactivation modifies the influence of a channel depending on the frequency or duration of an excitatory event (Fig. 4.1).

4.1 Basics of K^+ Channel Structure

The first crystal structure of a K^+ channel, KcsA, was solved in 1998 (Doyle et al., 1998). This simple K^+ channel, which consists of only two transmembrane domains, appears to provide an accurate representation of the conduction pathway of all K^+ channels. All mechanistic investigations of the K^+ channel conduction pathway are now designed and interpreted in light of structural information obtained from X-ray diffraction studies. Recently, additional, larger K^+ channel structures have been resolved by X-ray crystallography (cf. Jiang et al., 2002, 2003a,b; Kuo et al., 2003; Long et al., 2005a). Although two of these crystallized channels were voltage-gated, six transmembrane domain channels (Jiang et al., 2003a,b; Long et al., 2005a), the physical relationship between the voltage-sensing domains and conduction pathway suggested by these crystal structures has been controversial. An extensive discussion of the voltage-sensing mechanism is presented in Chapter 3.

Figure 4.2 illustrates the basic structural components of K^+ channels. Each subunit of voltage-gated potassium channels has six transmembrane domains, labeled S1 through S6 (Fig. 4.2A). The S4 domain contains a repeating series of positively charged amino acid residues, mostly arginines, and is the primary domain responsible for sensing changes in voltage (Aggarwal et al., 1996; Seoh et al., 1996). Upon depolarization, the S4 domain moves, which causes a gate in the pore to open and permit ion flow. Until recently, a wide variety of evidence suggested that membrane depolarization drove the positively charged S4 domain through the membrane in an outward direction (cf. Ding and Horn, 2002, 2003; Ahern and Horn, 2004). More recent evidence, however, suggests that the actual translocation of peptide through the membrane is very minimal, perhaps only one or two angstroms (Chanda et al., 2005). The S5 and S6 domains, connected by a P (pore) loop, form the conduction pathway. The P-loop itself forms the narrow region of the conduction pathway to create the channel's selectivity filter, where selection for the passage of K^+ over the other dominant monovalent ion, Na^+, occurs. In six transmembrane domain channels, both the amino (N) and carboxy (C) termini are in the cytoplasm. A Ca^{2+}-dependent K^+ channel, BK, has a seventh transmembrane domain, S0, which puts the N-terminus on the extracellular side of the membrane. Inward rectifier K^+ channels have just two transmembrane domains, analogous to S5 and S6, with the intervening P-loop, and N and C termini in the cytoplasm.

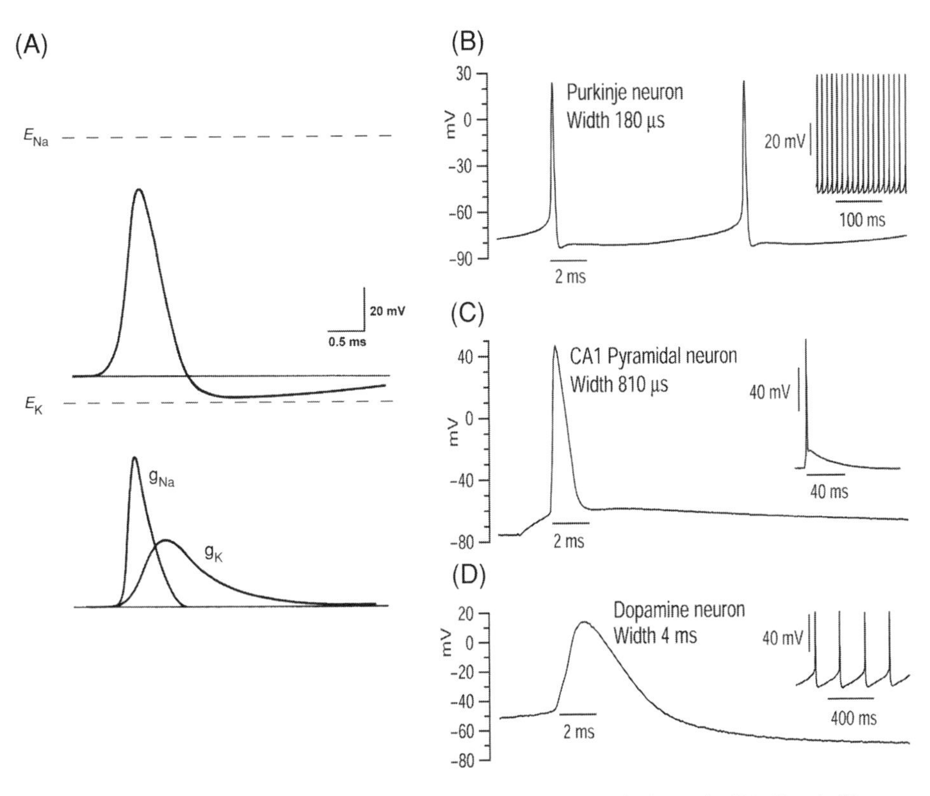

Fig. 4.1 Control of action potential duration and frequency by K^+ channels. (A) Classic illustration of the conductances that generate the squid axon action potential, adapted from Hodgkin and Huxley (1952). There are only two requirements for an action potential. The depolarizing upstroke requires the opening of voltage-gated channels (Na^+ or Ca^{2+}) to produce rapid inward currents. Repolarization requires activation of voltage-gated channels (K^+) that produce outward currents to hyperpolarize the membrane. In this simplest of action potentials, tetrodotoxin-sensitive Na^+ channels (g_{Na}) open rapidly and inactivate within approximately 0.5 ms. Delayed rectifier K^+ channels (g_K) open with a slight delay and close somewhat slowly as the membrane hyperpolarizes. Within a given class of action potentials (i.e., those carried by Na^+ or Ca^{2+}), the duration of the action potential can be essentially attributed to the time course and magnitude of the K^+ conductance. There are many different delayed rectifier channels, which, among other characteristics, open at different membrane potentials and have different rates of opening and closing. A more rapid generation of a larger g_K, via some combination of faster opening kinetics, more negative activation voltage and larger current (due to either more channels opening or larger single-channel conductance), will speed repolarization and shorten the action potential duration. Conversely, repolarization will be slower, and action potential duration longer, if g_K is smaller or more delayed. Note also that activation of gK, combined with inactivation of gNa, drives the membrane more negative to the resting membrane potential. The more quickly K^+ channels close, the sooner the membrane will be ready for another action potential, and vice versa. (B–D) Illustration of different shapes of action potentials in three types of mammalian central neurons (recorded at 35°C). These action potentials have markedly different durations, and fire with different rhythms, due to different complements of voltage-gated channels. Neurons in panels B and D fire spontaneously, but at very different rates. Neurons in panel C fire only single action potentials. (B) Action potentials from a dissociated mouse cerebellar Purkinje neuron. This neuron fired spontaneously at 87 Hz (see inset)

A complete, functional potassium channel consists of four subunits that come together to form a symmetric tetramer. [Tandem pore (2P) channels are an exception to this. These channels have either four or eight transmembrane domains, a second P-loop region, and form complete channels from dimers (Buckingham et al., 2005).] Figure 4.2B illustrates two of four subunits of the KcsA K^+ channel (Doyle et al., 1998). Each subunit consists of an outer (S5) and inner (S6) helix. The P-loop region connects the two helices, and forms the selectivity filter and outer vestibule of the channel. The KcsA pore contains four potassium ions, two in the selectivity filter, one in an inner, water filled cavity and one in the outer vestibule (Zhou et al., 2001b). The two K^+ ions in the selectivity filter are separated by a water molecule, and can occupy two of four locations, positions 1 and 3 (open circles) or positions 2 and 4 (shaded circles). The inner helices (S6 domain) form a bundle crossing at the cytoplasmic entrance to the conduction pathway in KcsA, but are probably arranged in a slightly different orientation in Kv channels (del Camino et al., 2000). This bundle crossing is the general location of the cytoplasmic activation gate, whose opening is linked to the S4 domain movement in voltage-gated channels.

Figure 4.2C illustrates the N- and C-terminal domains in more detail (again, just two subunits are shown). The N-terminal domain of voltage-gated channels contains a tetramerization domain (T1 domain; ribbon), which comes together just underneath the conduction pathway, in a fourfold arrangement resembling a "hanging gondola" (Kobertz et al., 2000). Several, but not all, voltage-gated channels also have a globular, positively charged structure, the N inactivation ball, at the end of the N-terminus. When channels are opened by depolarization, one of the four inactivation balls rapidly enters the conduction pathway and blocks it, leading to N-type inactivation (I_N; Fig. 4.1D). The C-terminal domain, illustrated as a shaded oblong structure, varies considerably in size in different K^+ channels, associates with both the cytoplasmic end of the conduction pathway and the N-terminal domain, and is an important modulator of K^+ channel function.

Figure 4.2D illustrates in cartoon form the four basic gating states of a K^+ channel. Panel 1 illustrates a channel with a closed cytoplasmic gate (note that four potassium ions are illustrated to be in the pore here, but the number of K^+ ions in this

Fig. 4.1 (*continued*) in the absence of current injection, as is typical of Purkinje neurons. Kv3 channels, which open and close very quickly, contribute the dominant delayed rectifier current. The rapid opening and closing kinetics of Kv3 channels allow for very high-frequency firing of action potentials. Note that the action potential is so brief in panel B that Na^+ channels undergo little or no inactivation. Recording by Andrew Swensen. (C) Action potentials in rat CA1 pyramidal neuron recorded from a hippocampal brain slice (stimulated by 2 ms injection of current to reach threshold). Kv1, Kv4 and possibly some Kv2 or Kv3 channels all contribute to action potential repolarization. Recording by Alexia E. Metz. (D) Action potentials in mouse midbrain dopamine neuron, recorded from a brain slice. This cell fired with a typical rhythmic pacemaking activity at ~4 Hz. The K^+ channels responsible for repolarization have not yet been identified. Recording by Michelino Puopolo. Fig. B–D was kindly provided by Bruce Bean and Marco Martina. Details can be found in Martina et al. (1998), Mitterdorfer and Bean (2002), Martina et al. (2003), Puopolo et al. (2005, and references within).

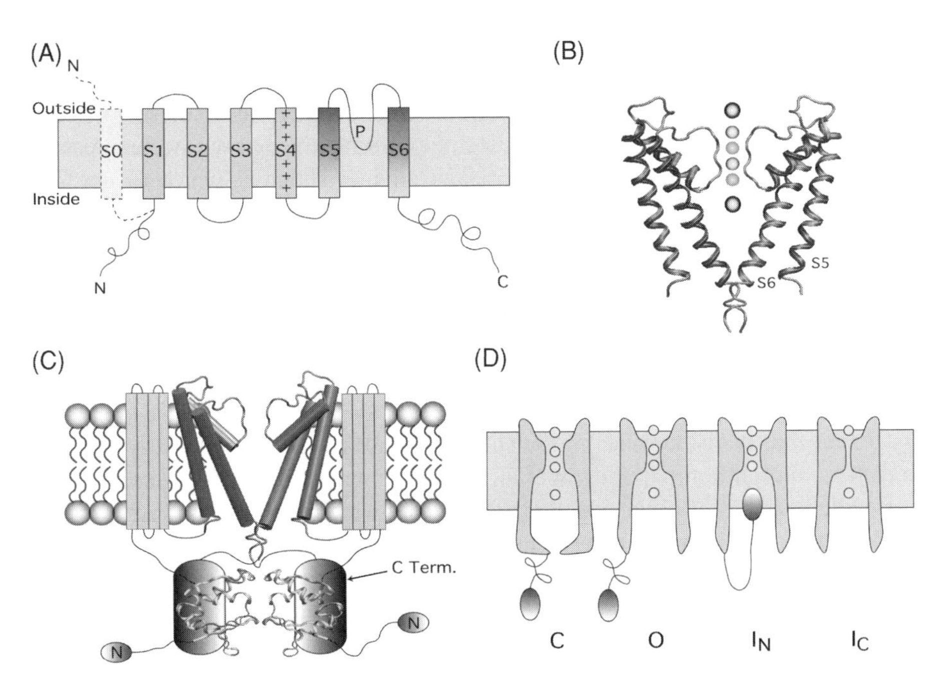

Fig. 4.2 Basics of K^+ channel structure and states. (A) Cartoon of the fundamental K^+ channel subunit. (B) Channels are formed from four subunits arranged symmetrically around a central core. In this panel, two opposing subunits are shown; the front and rear subunits were omitted for illustration clarity. The selectivity filter is formed between the S5–S6 pairs of transmembrane domains. The locations of K^+ binding sites in the conduction pathway denoted by spheres. (C) Schematic of two subunits of a Kv channel, showing N-terminal domain (ribbon) and C-terminal domain (oblong blocks) underneath the pore. In some channels, the distal portion of the N-terminal domain contains a charged, yet hydrophobic peptide sequence (the inactivation particle N). Upon depolarization, the charged peptide is driven into the pore with a delay, causing N-type inactivation (see panel D). (D) Illustration of four channel states, closed (C), open (O), N-type inactivated (I_N), and C-type inactivated (I_C).

and other channel states is currently unknown, and may in fact differ in different K^+ channels). Upon depolarization, the cytoplasmic gate opens (panel 2). In channels with an N inactivation ball, open channels quickly become blocked by the inactivation ball (panel 3), creating a transient current. Almost all K^+ channels, whether they have an N inactivation ball or not, undergo a second inactivation mechanism, which apparently involves a constriction at or near the selectivity filter (panel D).

4.2 Functional Classification

There are over 150 K^+ channel species, which can be broadly characterized into several main types: voltage-gated (Kv), Ca^{2+}-dependent (K_{Ca}), inward rectifier (Kir), and tandem pore (2P). Fully functional channels, formed by primary (α) subunits, are often associated with accessory (β) subunits. The β subunits also come in different flavors, some of which are apparently restricted to the cytoplasm (e.g., β1, β2, KChIPs) and some of which are small membrane spanning subunits (MinK, MiRP). β subunits, each of which can interact with a host of different α subunits, modify α subunit characteristics such as ion selectivity, gating kinetics, surface expression, and pharmacology. In addition, there are also silent α subunits, which do not form functional channels by themselves. They can, however, combine with functional α subunits to form heterotetramers with somewhat altered properties (Kramer et al., 1998; Shepard and Rae, 1999; Ottschytsch et al., 2002; Kerschensteiner et al., 2003). Relatively little is known about the role of these silent α subunits in cell physiology.

4.2.1 Voltage-Gated Channels

There are four major families of Kv channels, Kv1.x to Kv4.x, currently with 2–8 members in each family (Gutman et al., 2003). The first cloned potassium channel was called *Shaker*, after the mutant fruit fly from which it was identified (Papazian et al., 1987). This channel is equivalent in the Kv nomenclature to Kv1.0, and due to its extensive study early on, retained its name and became the prototypical Kv channel.

Kv channels are generally activated by depolarization, and primarily serve to control the duration, shape and firing frequency of action potentials. Thus, rapidly activating, slowly inactivating Kv channels keep action potentials brief (~2 ms), whereas slowly activating Kv channels allow for more prolonged action potentials, (for example, cardiac myocytes have action potentials >200 ms in duration, partially as a result of a very slowly activating Kv channel). Whereas there is certainly overlap of function, different voltage-gated channels are specialized for certain functions. For example, in contrast to most other K^+ channels, current through the Kv2.1 channel increases as external K^+ concentration ($[K^+]$) is elevated (Wood and Korn, 2000). This unusual feature apparently predisposes the Kv2.1 channel to be of special importance during high-frequency firing (Du et al., 2000; Malin and Nerbonne, 2002), when external $[K^+]$ tends to rise and other Kv channel currents diminish due to the consequent reduction in electrochemical driving force. The HERG potassium channel, which is also activated by depolarization, has the unusual property of inactivating faster than it activates with strong depolarizations (such as produced by an action potential). Upon membrane repolarization, channels exit from the inactivated state more quickly than they close. Consequently, current through these channels remains low until after repolarization begins. During repolarization, as current through other open K^+ channels diminishes, current through HERG channels increases, thus adding to the speed of repolarization. The Kv3.x channels are generally localized where information is encoded by the high-frequency firing of action potentials (Rudy et al., 1999). These channels activate rapidly, which hastens repolarization and limits Na^+ channel inactivation. They also deactivate extremely rapidly, which thus shuts off hyperpolarizing current. These two properties keep action potentials brief and the membrane poised to immediately fire a subsequent action potential. Finally, some members of the Kv1, Kv3, and Kv4 channel families also have a rapid inactivation mechanism. Rapidly inactivating channels have several functions, which include regulation of the frequency of rhythmic action potential firing (Connor and Stevens, 1971a,b) and integration of electrical activity in neuronal dendrites (see Part III).

4.2.2 Ca^{2+}-Dependent Channels

In excitable cells, K_{Ca} channels come in two varieties: large conductance (maxi-K or BK) and small conductance (SK). BK channels are also activated by depolarization, but channel activation is greatly facilitated by elevation of intracellular Ca^{2+}, which can result either from depolarization-evoked entry of Ca^{2+} through Ca^{2+}-permeable channels or from release of Ca^{2+} from intracellular stores. As a result of this Ca^{2+} dependence, BK channels produce a long, post-action potential hyperpolarization, which keeps the membrane potential away from firing threshold and thus delays the membrane's ability to undergo another action potential. In contrast, SK channels are activated solely by elevation of intracellular Ca^{2+} (Bond et al., 1999). The strict dependence on intracellular Ca^{2+} means that their influence increases as intracellular Ca^{2+} builds up, such as during high-frequency firing of action potentials. Thus, they may serve primarily to terminate long bursts of action potentials (Swensen and Bean, 2003).

4.2.3 Inward Rectifier Channels

As a generalization, Kir channels help maintain the negative resting potential. They are open at negative membrane potentials, and consequently oppose excitation by small excitatory stimuli. Upon significant depolarization, they are blocked by intracellular small molecules (Mg^{2+} or polyamines) so as not to be a continuing factor during the excitation event. Kir channels can be modulated by G-proteins as well as other intracellular biochemicals, which demonstrates that modulation of their ability to suppress cell excitation is an important functional feature of these channels. One highly complex Kir channel, KATP, is sensitive not only to the intracellular molecules and modulators that influence other Kir channels but also to intracellular ATP. Elevation of intracellular ATP inhibits channel activity, which links the influence of these channels to changes in cell activity and metabolism (cf. Tarasov et al., 2004).

4.2.4 2P Channels

Tandem pore (2P) K^+ channels have two P-loop regions and come in two varieties: one member isolated from yeast (TOK1) has eight transmembrane domains, whereas all other known members have four (Goldstein et al., 2001; Kim, 2003; Buckingham et al., 2005). The role of 2P channels, discovered fairly recently in the overall history of ion channel research, is still unclear. Initially, they were thought to be constitutively open (Goldstein et al., 1996) and thus came to be called, ingloriously, "leak" channels. For decades, the basis of the resting membrane conductance to K^+ remained a mystery. These channels appear to represent the solution to the mystery. In addition, recent work has shown that opening of 2P channels can be regulated by fatty acids, temperature, intracellular pH and membrane stretch (Lopes et al., 2000; Kim, 2003; Kang et al., 2005). Consequently, these channels may not only contribute to the steady resting conductance, but may also turn out to be actively involved in cellular function (cf. Richter et al., 2004).

4.3 Summary

There are more than 150 known potassium channel genes (Goldstein et al., 2001; Gutman et al., 2003), and there can be many K^+ channel species in any given cell membrane. The general role of each channel phenotype in membrane electrophysiology is reasonably well understood. Consequently, much of the effort toward understanding function is now focused on (1) the myriad biochemical mechanisms by which each channel function is modulated and (2) how each of the many channels in a particular cell type specifically influences cell function. In the last 15 years, the maturation of molecular biological approaches has led to an extensive understanding of the structural basis of ion channel function. The structures underlying the voltage-sensor and the fast inactivation gate were found (Aggarwal and MacKinnon, 1996; Seoh et al., 1996; Aldrich, 2001), the structure underlying the selectivity filter was found (Heginbotham et al., 1994; Doyle et al., 1998; Zhou et al., 2001b), and the general mechanism of conduction through multi-ion pores has been reasonably well described. The more recent ability to obtain structural information from X-ray crystallography, and the use of NMR spectroscopy with knowledge of these structures, has once again accelerated the pace of discovery. Knowledge of the structure itself has constrained and suggested mechanism. With knowledge of the fundamental pore structure, molecular studies of gating mechanisms and permeation properties can now be designed and interpreted more astutely. Recently, description of the structures of voltage-gated and Ca^{2+}-dependent channels has led to a more sophisticated understanding of gating. Indeed, perhaps the most invigorating recent publication in the field of ion channel biophysics was the crystallization of a voltage-sensitive K^+ channel (Jiang et al., 2003a,b). The proposed structure of the voltage sensor was at odds with years of biophysical measurements, and the push to reconcile the two led to the rapid design of critical experiments that greatly accelerated our understanding of gating mechanisms (Ahern and Horn, 2004). A more recent crystal structure of another voltage-gated channel (Long et al., 2005a) presents a structural picture that more closely matches results from biophysical experiments, yet incompatibilities between the structural and biophysical data remain. Resolution of these incompatibilities will surely refine our understanding of the ion channel mechanisms. Finally, it is clear that over the last 25 years, most biophysical measurements have been made either on naked α subunits, in the absence of potential modulators, or on α subunits that are complexed with unknown stoichiometry to unknown modulators. The advance of molecular and genetic technology has led to the discovery of an enormous number of these modulators, and with continuing sophistication in molecular, structural, biophysical, and theoretical approaches, a new surge in mechanistic discovery is imminent.

Part II. K^+ Channel Operation

4.4 Control of Single-Channel Conductance

The primary function of an ion channel is to permit the rapid flux of ions across a cell membrane. The potassium channel family displays a large range of single-channel conductances, ranging from 2 to 240 pS. This range certainly reflects the diversity of functions required of K^+ channels. How single-channel conductance is varied so greatly among structurally similar channels is one of the great fundamental questions remaining about ion channel function. Indeed, it is not known whether there is a general structural basis for control of conductance or whether conductance is limited at different locations in different channels.

4.4.1 Role of the Selectivity Filter

One school of thought has been that the selectivity filter would be rate limiting for conductance. This possibility is derived from the fact that the filter is the narrowest, and presumably the most electrically resistive, region of the pore. Furthermore, to enable high selectivity for one ion over a very similar ion, one might infer that the more permeant ion also has a relatively high affinity for the selectivity filter. There is some evidence that subtle variations in the selectivity filter might influence single-channel conductance. For example, mutations can be made to residues near or within the selectivity filter that change single-channel conductance without destroying ion selectivity (Zheng and Sigworth, 1997; Lu et al., 2001; So et al., 2001). However, there are two compelling arguments against a primary role of the selectivity filter in determining single-channel conductance. First, the structure of the K^+ channel selectivity filter is highly conserved among members of the K^+ channel family that have widely varying single-channel conductances. This structural similarity makes the selectivity filter an unlikely location for significant functional variation. Second, it has been suggested, based on the structural data from the KcsA channel, that the selectivity filter is always fully occupied by two K^+ ions, and is designed for very high throughput (Morais-Cabral et al., 2001; Zhou et al., 2001b; Zhou and MacKinnon, 2003, 2004). Interestingly, the conductance of the KcsA channel increases linearly up to very high $[K^+]$, which suggests that the filter presents almost no barrier to conduction at physiological $[K^+]$. Theoretical simulations, based on the structural data, suggest that ions in the selectivity filter of conducting channels are in a stable equilibrium, with little energy needed to push ions through the filter (Bernèche and Roux, 2001; Chung et al., 2002). Whether this is true for all potassium channels remains to be determined. For example, biophysical studies suggest that the selectivity filter in the Kv2.1 channel is not fully occupied in conducting channels (Immke and Korn, 2000).

The interplay between experimental studies and sophisticated simulation models has already produced, and will continue to produce, a plethora of insight into ion conductance through the selectivity filter. For example, in some channels, conduction of K^+ and the very similar Rb^+ can be markedly different. The ability to (1) determine differences in occupancy of the selectivity filter by K^+ and Rb^+ (cf. Zhou and MacKinnon, 2003), (2) manipulate the molecular structure of the selectivity filter, and (3) make rigorously consistent models that account for both the structural and biophysical data, will undoubtedly yield a greater understanding of how the selectivity filter accommodates high flux while maintaining high selectivity, as well as potentially explaining differences in conduction among different K^+ channels (cf. Zhou and MacKinnon, 2004).

4.4.2 Role of the Inner Vestibule

Recently, experimental and theoretical studies have suggested that ion entry into the inner vestibule may be a key determinant of conductance magnitude. The BK channel, so named because of its "big" conductance, has a ring of glutamates near the inner entryway of the channel (Brelidze et al., 2003). Removal of these glutamates significantly reduces single-channel conductance (Brelidze et al., 2003). Conversely, insertion of glutamates into KcsA in this same region significantly increases conductance, although not to the level observed for BK (Nimigean et al., 2003). These studies suggest that the ring of negative charges lowers the energy barrier for ion entry into the pore and thus facilitates conduction. Interestingly, studies of internal quaternary ammonium (QA) block suggest that BK channels have a larger inner vestibule than other K^+ channels (Li and Aldrich, 2004). This is consistent with theoretical studies, which suggest that enlargement of the inner vestibule can dramatically increase single-channel conductance (Chung et al., 2002).

4.4.3 Role of the Outer Vestibule

In some channels, K^+ exit from the selectivity filter into the outer vestibule may also be rate limiting for conductance. An early study demonstrated that a chimeric channel, in which the outer vestibule region of the Kv2.1 channel was replaced by the equivalent region of the Kv3.1 channel, displayed the conductance characteristics of the channel that contributed the outer vestibule region (Hartmann et al., 1991). More recently, it was shown that a positively charged lysine in the outer vestibule of Kv2.1 reduces single-channel conductance by interfering with K^+ exit from the selectivity filter (Consiglio et al., 2003; Trapani and Korn, 2003). Moreover, physiologically relevant conformational changes in the outer vestibule, which reorients this lysine, alters single-channel conductance (Trapani et al., 2006). Thus, it may be that there are multiple locations used by different channels to limit single-channel conductance.

4.5 Activation Gates

Permeation through ion channels is gated, and clearly, understanding the control of current flow through a channel requires a comprehensive understanding of gating mechanisms. At a minimum, there needs to be one gate that opens and closes on command to start and stop flux through the pore. However, recent data suggest that there may be a second activation gate.

4.5.1 The Cytoplasmic Gate

Upon depolarization of the membrane, the voltage sensor moves, the gate inside the pore opens and current flows. In a series of papers published in the early 1970s, Armstrong and Hille presented evidence which suggested that the activation gate in the squid K^+ channel was located toward the intracellular end of the channel entrance (Armstrong and Hille, 1972). Briefly, these studies showed that quaternary ammonium (QA) ions in the intracellular solution gained access to their blocking site after the channel opened, and that they appeared to be trapped within the channel upon channel closing (Armstrong, 1971; Armstrong and Hille, 1972). Subsequent molecular (Choi et al., 1993) and structural (Zhou et al., 2001a) studies demonstrated that internal QA ions bind at the internal entrance to the selectivity filter, deep within a large aqueous central cavity. Furthermore, cysteine accessibility studies demonstrated that even small cations (e.g., Ag^+) could not pass through the activation gate and enter the central cavity region when channels were closed (del Camino and Yellen, 2001; but see Soler-Llavina et al., 2003 for a quantitative description of how closed a closed channel is). These studies confirmed Armstrong's essential conclusion that a voltage-sensitive gate formed a trap door at the intracellular entrance to the pore.

The location of the cytoplasmic gate was refined by cysteine scanning studies (Liu et al., 1997). Single-cysteine substitutions were made along the S6 domain of the *Shaker* channel and a very simple question asked: which cysteines could be modified by applied thiol-reactive chemicals when the channels were closed, and which cysteines required the channel to be open? Using this approach, Liu et al. (1997) demonstrated that several residues located toward the selectivity filter were relatively inaccessible when channels were closed but became accessible upon channel opening. In contrast, a cysteine located five residues closer to the cytoplasmic entrance of the channel was exposed similarly in both closed and open channels. These results placed the cytoplasmic gate at a defined region at the cytoplasmic entrance to the pore. The gate prevented entry of both positively and negatively charged cysteine modifying reagents, which indicated that gate closing was based on steric, not electrostatic, occlusion (del Camino and Yellen, 2001). Subsequently, a completely different approach, which evaluated mutation-induced disruption of gating, also placed the intracellular gate at this same region of the pore (Hackos et al., 2002).

4.5.2 Coupling of the Voltage-Sensor to the Cytoplasmic Gate

It remains to be conclusively determined how the voltage-sensor is coupled to the cytoplasmic gate, and how movement of the voltage-sensor causes the gate to open and close. Some evidence, however, is beginning to emerge. Lu et al. (2001, 2002) made chimeras between *Shaker*, a voltage-gated channel and KcsA, a non-voltage-gated channel. Remarkably, concatenating the voltage-sensing domains from *Shaker* to the two transmembrane domain channel, KcsA, provided the ability to gate KcsA by voltage. Lu et al. (2002) found two regions near the cytoplasmic end of the channel that were essential for coupling of the voltage-sensor to the cytoplasmic gate: the S4–S5 linker region and a critical sequence near the carboxy-terminal end of S6. This latter region corresponds to the location of the cytoplasmic gate defined by studies described above. Studies of this interaction in the HERG potassium channel and the HCN channel (a related, hyperpolarization-activated channel) suggested that movement of the voltage-sensor disrupts a specific, perhaps electrostatic, interaction between the S4–S5 linker and the C terminus near the carboxy-terminal end of S6 (Tristani-Firouzi et al., 2002; Decher et al., 2004). A recently crystallized voltage-sensitive channel provided new structural information regarding the physical association of the S4 domain with the pore (Long et al., 2005b). From these structural data, the authors presented a model as to how movement of the voltage sensor might perform mechanical work on the pore, via the S4–S5 linker, to open a gate.

4.5.3 The Selectivity Filter may be an Activation Gate

Most potassium channels undergo a slow inactivation process, whereby current is curtailed despite a stimulus for channels to be open. Extensive experimental investigation has demonstrated that slow inactivation involves a cut off of ion flux at the selectivity filter (see below). It is thus widely accepted that the selectivity filter can act as an inactivation gate. Although inactivation may be coupled to activation, these two processes can be independently manipulated, and almost certainly represent two independent mechanisms. Whereas there is no direct evidence for a role of the selectivity filter in channel activation in voltage-gated channels, a great deal of circumstantial evidence suggests that it may indeed be an additional activation gate.

4.5.3.1 Voltage-Independent Gating Transitions

Most K^+ channels display rapid close–open transitions, even when fully activated. Bao et al. (1999) studied the kinetics of these transitions in a Kv channel (*Shaker*) that had its voltage-sensor disabled. This channel, which was constitutively activated, displayed rapid close–open transitions with essentially identical kinetics to that observed in the open, wild-type, voltage-sensitive channel. These data were consistent with the presence of a gate whose operation was unaffected when the pore was functionally uncoupled from the voltage sensor. Although one could argue that this gate was the cytoplasmic gate, it seems unlikely that a single structural gate would flicker identically whether coupled to a crippled S4 domain or a functional S4 domain in a fully activated channel.

4.5.3.2 Mutagenesis Near the Selectivity Filter

Liu and Joho (1998) demonstrated that a mutation in the S6 transmembrane domain of the Kv2.1 voltage-gated channel, near the selectivity filter, altered the voltage-independent, rapid gating transitions. In addition, mutation of different amino acids at this location altered single-channel conductance selectively for different ions. These authors postulated that this mutation, located a significant distance from the cytoplasmic gate, influenced the operation of a second gate, which they considered to be the selectivity filter. Their two-gate hypothesis was strengthened by the subsequent demonstration that a different mutation, thought to be within the conduction pathway, influenced the voltage-dependent gating mechanism but not the voltage-independent gating transitions observed with the first mutation (Espinosa et al., 2001).

Experiments that utilized mutagenesis within the selectivity filter of both inward rectifier and Kv channels also supported a model of the selectivity filter as an activation gate. Inward rectifier potassium channels are not gated by voltage but rather, by a channel block mechanism (Lu, 2004). Thus, outward current is prevented upon depolarization due to block by intracellular Mg^{2+} or polyamines. In the absence of blocker, however, single-channel records demonstrate clear open–close transitions. Mutations in the selectivity filter of the Kir2.1 inward rectifier altered the mean open time of the channel, as well as transitions between conductance states (Lu et al., 2001; So et al., 2001). Similarly, in *Shaker*, mutation of the residue that forms the innermost part of the selectivity filter increased channel open time even when channels were fully activated by depolarization (Zheng and Sigworth, 1997). Moreover, heteromeric channels, which contained different numbers of mutations at this selectivity filter location, displayed different conductance sublevels and, at each conductance level, different ion selectivity (Zheng and Sigworth, 1998). These results suggested that the transitions among conductance sublevels in fully activated channels were associated with the operation of the selectivity filter. However, other than the vague postulate that conformational changes account for this flicker, a mechanistic description of how these gating events might occur at the selectivity filter has remained elusive (but see Bernèche and Roux, 2005, for an interesting hypothesis).

4.5.3.3 Accessibility Studies

Proks et al. (2003) asked whether intracellular Ba^{2+} block of the ATP-sensitive inward rectifier, Kir6.2, was altered by closing of the activation gate. ATP binding stabilizes Kir6.2 channels in the closed conformation via an allosteric mechanism. Thus, in the presence of ATP, transitions of Kir6.2 into the closed state are more frequent and prolonged. Ba^{2+} block occurs at the inner end of the selectivity filter (Neyton and Miller, 1988; Jiang and MacKinnon, 2000). Proks et al. (2003) predicted that the rate and apparent affinity of internal Ba^{2+} block would be reduced in the presence of ATP if the gate were closer to the cytoplasmic entrance of the pore than the Ba^{2+} blocking site. In contrast to this prediction, closing of the gate with ATP slightly increased the rate of Ba^{2+} block, and produced little or no change in apparent Ba^{2+} affinity (Proks et al., 2003). These results put the ATP-activated gate above the Ba^{2+} block site, and thus at the selectivity filter.

As described above, differences in cysteine accessibility in open and closed channels can be used to define the location of a gate. This approach was also used to search for the location of the gate in SK channels and cyclic nucleotide-gated (CNG) channels. These two channel types contain six transmembrane domains, and are structurally similar to Kv channels. However, neither is activated by a change in membrane potential. SK channels are K^+-selective channels activated by intracellular Ca^{2+}, CNG channels are nonselective cation channels activated by intracellular cyclic nucleotides. Bruening-Wright et al. (2002) demonstrated that the cysteine-modifying reagent, MTSEA, had equal access to a cysteine in the inner vestibule of open and closed SK channels. This residue is located between the presumed location of the cytoplasmic gate and the selectivity filter. Similarly, Flynn and Zagotta (2001) demonstrated that Ag^+ could access a cysteine at an approximately equivalent location equally well in closed and open CNG channels. Based on the structural information that there isn't a constriction in the pore between the location of a possible cytoplasmic gate and the selectivity filter, these data suggested that the selectivity filter was the activation gate in these two channels. Interestingly, the larger cysteine-modifying reagent, MTSET, had differential access to this inner region of the pore in open and closed channels (Flynn and Zagotta, 2001; Bruening-Wright et al., 2002). The potential significance of this result will be discussed below.

4.5.4 Coupling Between the Cytoplasmic Gate and Selectivity Filter Gate

If there are two activation gates, are they energetically (or mechanistically) coupled? Several studies suggest that they are. First, the subconductance state experiments by Zheng and Sigworth (1997), described above, suggested that the two activation mechanisms were energetically coupled. Second, NMR studies of KcsA indicated that during gating, both the cytoplasmic gate region and the region adjacent to the selectivity filter change conformations (Perozo et al., 1999). Third, an extensive mutant cycle analysis performed on the *Shaker* pore led to two findings: that residues near the cytoplasmic gate and near the selectivity filter are coupled to activation and that residues in these two regions are energetically coupled to each other (Yifrach and MacKinnon, 2002).

The accessibility studies described above present an intriguing possibility about coupling between the gates. In both SK and CNG channels, even though relatively small cysteine-modifying reagents entered the pore equally in closed and open channels, the larger MTSET had differential access to the inner pore in closed and open channels (Flynn and Zagotta, 2001; Bruening-Wright et al., 2002). These results demonstrated that, during activation, the region typically associated with the cytoplasmic gate still undergoes a conformational change, but that it doesn't form a barrier to ion conduction in the closed state. Thus, the conformational changes associated with gating in channels that use either the cytoplasmic gate or the selectivity filter gate may not be all that different. Indeed, it may be that in voltage-gated channels, the cytoplasmic gate fully closes and the rapid conformational changes at the selectivity filter, perhaps controlled, perhaps not, underlie the rapid, close–open transitions. In contrast, channels that are gated by intracellular ligands (e.g., SK, CNG, KATP) may have an open cytoplasmic end of the pore at all membrane potentials. However, ligand-induced conformational changes at the cytoplasmic end of the pore (the location of the voltage-sensitive gate in Kv channels) may trigger gating at the selectivity filter. Thus, the fundamental mechanism of gating may be conserved among all of these different types of channels, and differences in gating mechanisms may then be seen to reflect variations on the basic theme.

4.5.5 Why Have Two Gates in One Channel?

The magnitude of current flow through a single channel is determined by two factors: single-channel conductance and mean open time. With a single gate, current magnitude in a fully activated channel will be determined by the single-channel conductance alone. Adding a second gate provides yet additional diversity to channel operation. Once the inner gate is open, the mean open time will be determined by fast open–close transitions that apparently occur at the selectivity filter. Whereas the cytoplasmic gate opening appears to respond primarily to voltage, gating at the selectivity filter may provide a locus for voltage-independent modulation of gating.

4.6 Functions of the Outer Vestibule

The outer vestibule is relatively small compared to the rest of the potassium channel (Doyle et al., 1998). Nonetheless, it is the site of many important channel functions. In addition to its role in regulating conductance in at least one channel (described above), it is the location for block by many naturally occurring toxins and the prototypical small molecule open channel blocker, tetraethylammonium (TEA), and is involved in the process of slow inactivation. Recent evidence suggests, however, that we may understand less about these latter two events than we think.

4.6.1 The TEA Binding Site

TEA is a cationic K^+ channel pore blocker that acts at an internal binding site when applied from the cytoplasm and at a separate, external binding site when applied from the extracellular solution. It has been used for decades as a tool to investigate K^+ channel pharmacology, permeation, gating, structure, and dynamics of protein movement. TEA inhibits most potassium channels. This lack of selectivity in the outer vestibule has been advantageous, in that much has been learned about the outer vestibule as a result of the variation in TEA potency, which ranges from low mM to >100 mM, among K^+ channels. Mutagenesis studies in the early 1990s demonstrated that the amino acid residue at *Shaker* position 449, believed to be just external to the selectivity filter, had a dramatic impact on TEA potency (MacKinnon and Yellen, 1990; Heginbotham and MacKinnon, 1992). Indeed, channels that are not blocked by TEA typically (but not always) have a positively charged residue at the equivalent position and neutralization of this residue can restore TEA block in an otherwise TEA-insensitive channel. The greatest TEA potency occurs in channels with an aromatic residue at this position (Heginbotham and MacKinnon, 1992). Studies of heterotetramers demonstrated that all four position 449 residues contributed approximately equally to TEA potency (Heginbotham and MacKinnon, 1992; Kavanaugh et al., 1992). These studies led to a model in which TEA was coordinated by the side chains at position 449. Indeed, the position equivalent to 449 in *Shaker* became known as "the TEA binding site."

Molecular dynamics studies that modeled TEA binding using the KcsA channel structural information came to much the same conclusion (Crouzy et al., 2001). In these models, the energy minimum for TEA in the outer vestibule was located very near the equivalent of position 449 residues when aromatic residues (native to KcsA) were placed at this site. With threonine (native to Shaker) at this site, the TEA molecule was less dehydrated and located in a somewhat more external position.

4.6.2 Slow Inactivation

Fluorescence measurements demonstrated that a residue in the outermost region of the outer vestibule moves during inactivation (Loots and Isacoff, 1998), and mutation of this same externally located residue alters inactivation kinetics (Jerng and Gilly, 2002; Kehl et al., 2002). The volume of the inner vestibule, and thus inner vestibule conformation, also appears to change during slow inactivation (Jiang et al., 2003a), which suggests that inactivation involves a widespread structural reorientation. Nonetheless, there is general agreement that closing of the inactivation gate involves a constriction of the pore just external to, and possibly at, the selectivity filter. This picture is derived in large part from four observations, primarily made using the *Shaker* potassium channel as a model system. First, during slow inactivation in *Shaker* (which displays the prototypical slow inactivation mechanism called "C-type" inactivation), the residues at position 449 move relative to their surroundings (Yellen et al., 1994). If cysteines are introduced into this position, the affinity of Cd^{2+} for these cysteines increases ~45,000-fold during inactivation (Yellen et al., 1994). This observation led to a model which postulated that, as the constriction progressed during inactivation, the cysteine side chains at position 449 on each subunit moved closer together, so that an increased number of side chains contributed to the coordination of a single Cd^{2+} (Yellen et al., 1994). Second, and perhaps most persuasive, cysteines at position 448, one residue deeper into the pore, crosslink during inactivation (Liu et al., 1996). Thus, the cysteine side chains at position 448 must move closer together as channels inactivate. Third, occupancy of the selectivity filter by K^+ slows classical C-type inactivation (Baukrowitz and Yellen, 1996; Kiss and Korn, 1998), which suggests that the constriction could not occur with a K^+ in the selectivity filter. Finally, the ability to conduct the smaller Na^+ ion increases during the inactivation process (Starkus et al., 1997; Kiss et al., 1999; Wang et al., 2000), which is again consistent with the diameter of the selectivity filter decreasing during inactivation.

4.6.3 Model of TEA Binding and Slow Inactivation

External TEA slows C-type inactivation, and presumably the underlying constriction (Grissmer and Cahalan, 1989; Choi et al., 1991). The synthesis of results obtained from experiments that examined the role of position 449 residues on TEA potency and C-type inactivation led to a comprehensive and compelling model of TEA binding, and the mechanism underlying its influence on slow inactivation (Yellen, 1998). In this model, TEA was coordinated by the position 449 residues. During C-type inactivation, the position 449 residues moved closer together as a result of the constriction. However, the constriction could not occur if a TEA molecule was coordinated by these residues, and only resumed when TEA came off of its binding site. Thus, TEA slowed inactivation by a "foot-in-the-door" mechanism. Once the inactivation door was shut, external TEA could no longer bind to the channel.

4.6.4 Problems with the Model

Recent evidence, however, is incompatible with the accepted models of TEA binding and slow inactivation. Compared with results obtained in the *Shaker* channel, TEA potency in Kv2.1 is relatively little affected by mutations at the position equivalent to 449 (position 380 in Kv2.1) (Andalib et al., 2004). Introduction of cysteines into position 380 and 449 of Kv2.1 and *Shaker*, respectively, facilitated studies of the role of these residues in TEA binding. Covalent modification of these cysteines by the positively charged sulfhydryl reagent, MTSET, was unaffected by the presence of TEA in the outer vestibule (Andalib et al., 2004). These results clearly demonstrated that TEA binding does not involve direct coordination by the position 380/449 residues. Additional cysteine protection experiments suggested that TEA may occupy a location at a somewhat more external location in the outer vestibule (Andalib et al., 2004).

These cysteine protection experiments were also at odds with the proposed mechanism of slow inactivation. In fully inactivated *Shaker* channels, where the position 449 cysteines were presumed to have moved centrally within the conduction pathway, the presence of TEA in the outer vestibule did not prevent cysteine modification by MTSET (Andalib et al., 2004). However, TEA did inhibit modification of these cysteines by larger MTS reagents in fully inactivated channels, which demonstrated that TEA does bind to inactivated channels. Thus, these results argue against a model whereby the position 449 side chains move to a central location within the conduction pathway during inactivation. Although these data do not rule out the possibility that TEA slows inactivation by a foot-in-the-door mechanism, they do indicate that the position 449 residues are not the door.

The results described above indicate that TEA does not bind where and how it had been thought to bind, and that the mechanism of C-type inactivation is not entirely understood. Indeed, there are many other apparently conflicting findings regarding slow inactivation. For example, whereas classical C-type inactivation, as observed in *Shaker* and the Kv1.3 potassium channel, is slowed by external TEA, slow inactivation in the Kv2.1 channel is virtually unaffected by external TEA. And whereas elevation of external K^+ slows classical C-type inactivation (López-Barneo et al., 1993), it accelerates slow inactivation in Kv2.1 (Immke et al., 1999). Interestingly, the silent α subunit, Kv9.3, has a dramatic influence on slow inactivation in Kv2.1, apparently by subtly changing channel structure near the cytoplasmic activation gate (Kerschensteiner et al., 2003). How this effect relates to slow inactivation at the selectivity filter is unclear. However, slow inactivation appears to proceed at different rates, and via different conformational routes, in closed and open channels (Klemic et al., 1998, 2001). The Kv9.3 effect near the cytoplasmic gate appears to shift the balance between these two routes (Kerschensteiner et al., 2003). These data suggest that, similar to activation, the slow inactivation mechanism at the selectivity filter may be energetically coupled to the cytoplasmic end of the pore. [An alternative possibility is that, despite its similarities, slow inactivation in Kv2 channels is mechanistically unrelated to C-type inactivation in Kv1 channels. Indeed, it has been postulated that Kv4 channels, which display a slow inactivation process that shares many similarities with Kv2, utilize a slow inactivation mechanism distinct from that of Kv1 channels (see Jerng et al., 2004a).] In summary, whereas it seems likely that slow inactivation involves some sort of constriction at or near the selectivity filter, the precise set of structural events that constitute inactivation, and whether inactivation results from just one or multiple mechanisms, remain unknown.

4.6.5 Occupancy of the Outer Vestibule by Cations

In order for K^+ to enter the narrow selectivity filter, it must be almost completely dehydrated (similarly, K^+ exiting the selectivity filter must rehydrate). Crystallization of KcsA in the presence of potassium revealed an outer vestibule K^+ binding site that appears to be the location of K^+ dehydration/rehydration (Zhou et al., 2001b). In fact, it appears that this site may consist of two outer vestibule positions, a more external position where a more hydrated K^+ resides, and a position closer to the selectivity filter where a partially dehydrated K^+ resides. Biophysical data demonstrated that this site has important functional properties in at least one channel. In Kv2.1, a lysine (which is positively charged) located in the external portion of the outer vestibule interferes with the access of K^+ to this site (Consiglio et al., 2003). This interaction reduces single-channel conductance (Trapani and Korn, 2003), reduces K^+ permeability (Consiglio et al., 2003), and reduces the sensitivity of Kv2.1 currents to changes in external K^+ concentration (Andalib et al., 2002). Moreover, changes in outer vestibule conformation that accompany changes in external [K^+] reorient this lysine, which consequently changes the magnitude of its influence on channel conductance (Wood and Korn, 2000; Trapani et al., 2006). Interestingly, this binding site appears to be selective for K^+ and Rb^+; the interactions of Na^+ and Cs^+ with the channel are unaffected by either the presence or movement of the outer vestibule lysine (unpublished data).

The outer vestibule lysines also interfere with TEA binding, but do not interfere with access of the positively charged MTSET to cysteines deep within the outer vestibule. This suggests that the lysines don't simply electrostatically repel TEA but interfere specifically with the interaction of TEA with its binding site. However, TEA and external K^+ do not compete at concentrations that produce a measurable interaction of each with the outer vestibule (unpublished data). Whereas the simple explanation is that K^+ and TEA do not compete for the same cation binding site, this leaves one with the uncomfortable conclusion that there are two completely isolated cation binding sites in the outer vestibule. [Interestingly, external K^+ and TEA do not compete with each other in either Shaker or Kv2.1, but they do compete in Kv1.5 channels with the residue at position 480 (equivalent to Shaker 449) mutated to allow for high-affinity TEA binding.] Additional complexity is added with the observation that an intermediately sized cation, tetramethylammonium (TMA), apparently cannot occupy the outer vestibule in a functionally meaningful way (Andalib et al., 2004).

So, where and how are these cations binding? The results described above suggest that TEA binding does not involve direct coordination by position 380/449 residues in the outer vestibule. In addition, there are no other candidate residues to perform this role. Together, these observations suggest that TEA is more indirectly stabilized in the aqueous outer vestibule. This possibility is supported by molecular dynamics simulations, which suggest that TEA tumbles while at its "binding site" (Crouzy et al., 2001). However, this indirect stabilization mechanism must be selective for TEA; TMA is not similarly stabilized. Similarly, K^+ must be stabilized in the outer vestibule by a mechanism that apparently doesn't influence Cs^+ and Na^+. What is the nature of a cation stabilization site that is selective for one monovalent ion over another? Are TEA and K^+ stabilized differently? A better understanding of the outer vestibule would undoubtedly help toward the development of small molecules that reversibly enhance or block K^+ current through specific K^+ channels via actions in the external mouth of the pore.

4.7 Functions of the N-Terminal Domain

4.7.1 Fast Inactivation

The N-terminal domain (N-terminus) is most notably involved in the process of fast, or N-type, inactivation. The mechanism of N-type inactivation is well understood (Aldrich, 2001; Zhou et al., 2001a), and represents a puzzle that was solved by a series of remarkably insightful investigations. Whereas many investigators contributed to filling in the pieces over the last 55 years (see Aldrich, 2001), four seminal studies provided the bulk of the information. Fast inactivation was initially postulated as an unknown mechanism by which the squid giant axon Na^+ conductance turned off during the action potential (Hodgkin and Katz, 1949). To explain their mathematical model of the Na^+ conductance underlying the squid axon action potential, Hodgkin and Huxley (1952) postulated the existence of "particles," three of which moved to initiate activation and one of which moved to produce inactivation. By combining a variety of physiological evidence with remarkable interpretation, Armstrong and Bezanilla (1977) proposed the "ball and chain" model of inactivation, whereby a charged inactivation ball (or particle) was tethered to the cytoplasmic end of the Na^+ channel by a polypeptide chain. Upon depolarization, the positively charged inactivation ball entered into the conduction pathway and blocked the pore (see Fig. 4.2). In the early 1990s, Aldrich and coworkers demonstrated the enormous power of applying molecular techniques to the study of cloned ion channels when they identified the inactivation ball and chain on the rapidly inactivating, Shaker potassium channel (Hoshi et al., 1990; Zagotta et al., 1990). In a series of elegant studies, these investigators demonstrated that, indeed, fast inactivation was mediated by a charged yet hydrophobic polypeptide "ball" located on the distal portion of the N-terminal domain of the channel. Finally, combined structural and biophysical studies illustrated the mechanism by which the inactivation "ball" enters and docks within the pore (Gulbis et al., 2000; Zhou et al., 2001a). Interestingly, fast, N-type inactivation can be conferred in one of two ways. Some channels, like Shaker, have the inactivation ball tethered directly to the N-terminus of the α subunit. Many K^+ channels, however, lack an intrinsic N-type inactivation mechanism and obtain one by associating with Kvβ subunits (Rettig et al., 1994; Zhou et al., 2001a).

4.7.2 Other Functions

In six transmembrane domain K^+ channels, the N-terminus also contains a molecular structure involved in channel assembly, called the T1 or tetramerization domain (cf. Shen et al., 1993; Long et al., 2005a). This domain appears to prevent the formation of heteromers among subunits derived from different K^+ channel families (Li et al., 1992; Shen and Pfaffinger, 1995). In addition, although it is not required for channel assembly (Kobertz and Miller, 1999), it does appear to facilitate channel formation (Zerangue et al., 2000).

Subtle mutations presumed to be near the interface between the T1 domain and the rest of the channel alter both activation voltage and deactivation rate (Cushman et al., 2000; Minor et al., 2000). The buried location of residues that influence gating suggests the possibility that, rather than being a modulatory domain, the T1 domain may be structurally close to the gating apparatus and mutations in the T1 domain alter gating for structural reasons. Some indirect evidence also supports the possibility that interactions between the N-terminus and the gating apparatus may be physiologically relevant. First, formation of heterotetramers between a silent α

subunit, Kv2.3, and functional Kv2.1 subunits, alter Kv2.1 channel gating apparently via a change in N-terminus properties (Chiara et al., 1999). Second, β subunits, which can alter channel gating properties, bind to the N-terminus (Gulbis et al., 2000; Long et al., 2005a). Nonetheless, other than its role in fast inactivation, the role of the N-terminus in K^+ channel physiology remains conjecture. For example, removal of the entire N-terminus has little effect on K^+ channel function (Kobertz and Miller, 1999). Moreover, mutagenesis and structural studies suggest that the N and C termini directly interact (Schulteis et al., 1996; Ju et al., 2003; Kuo et al., 2003; Sokolova et al., 2003), and it appears that binding of the β subunit to the N-terminus alters the interaction of C-terminal domains with the channel (Sokolova et al., 2003). Thus, experimental manipulations of the N-terminus may influence gating indirectly, via changes in the interaction of C-terminal domains with the channel.

4.8 Modulation at the C-Terminal Domain

Until recently, relatively little attention had been paid to the C-terminal domain (C-terminus) of the K^+ channel. Part of the reason for this probably relates to a lack of understanding of the structural relationship of the C-terminus to the rest of the channel. The recent determination of crystal structures for a Ca^{2+}-dependent K^+ channel (Jiang et al., 2002) and inward rectifier channels (Nishida and MacKinnon, 2002; Kuo et al., 2003) demonstrated, however, that the C-terminal domain was ordered, and formed a symmetrical tetrameric structure adjacent to the cytoplasmic end of the pore. An electron microscopic, low-resolution 3D structure of *Shaker* with and without a large portion of the C-terminal domain suggested that a compact C-terminal domain is juxtaposed to, and surrounding, the T1 domain of the N-terminus (Sokolova et al., 2003). This structural data has provided great insight into previous and subsequent functional data, and has opened the door to detailed mechanistic analysis of C-terminal function.

The length of the C-terminus varies greatly among K^+ channels, ranging from <100 amino acids to ~ 450 amino acids in Kv channels. The C-terminus of the Ca^{2+}-dependent, BK channel, which is critical to its function, is ~ 800 amino acids long. This variation by itself suggested that this channel domain must have some functional significance. Mutagenesis experiments on several K^+ channels have identified the C-terminus as the location of functionally relevant phosphorylation sites (Holmes et al., 1996; Murakoshi et al., 1997; Zilberberg et al., 2000; Sergeant et al., 2004). These studies suggested that, at the very least, the C-terminus is a conduit for biochemical modulation of K^+ channel behavior. However, many studies, including the few described below, suggest an even more central role of the C-terminus in K^+ channel function.

Specific point mutations in the C-terminus in *Shaker*, close to the S6 domain, influence both gating and permeation properties of the channel (Ding and Horn, 2002, 2003). These dual influences suggested that this region is near the cytoplasmic entrance to the pore, and can affect the operation of the cytoplasmic gate.

Mutation-induced shifts in voltage dependence further suggested that this region is relevant to the coupling between the voltage sensor and the cytoplasmic gate (Ding and Horn, 2003). This interpretation was supported by subsequent work on the BK family of K^+ channels (see below; Ca^{2+}-dependent K^+ channel structure and function are described in detail in Chapter 6).

Several elegant studies of the mechanism of Ca^{2+} dependence of the BK channel have provided great insight into the role of the C-terminus in regulation of channel function. BK channels respond to Ca^{2+} at concentrations that range over five orders of magnitude (Xia et al., 2002). To achieve this range of sensitivity, BK channels have at least two different Ca^{2+} regulatory sites in the C-terminus (Xia et al., 2002) and possibly one located elsewhere (Piskorowski and Aldrich, 2002). Remarkably, the C-terminus appears to be a completely modular channel domain. The BK channel homolog, mSlo3, responds not to intracellular Ca^{2+} but rather, to changes in intracellular pH (Schreiber et al., 1998). Swapping the C-terminal domains between BK and mSlo3 completely exchanged ligand sensitivity (Xia et al., 2004). Thus, the C-terminus of mSlo3 conferred pH sensitivity (and lack of Ca^{2+} sensitivity) on BK and vice versa. These results suggest that the C-terminal domains from different channels, which respond to different modulatory ligands, are completely exchangeable. Interestingly, both channels contain an identical sequence that links the C-terminus to the S6 domain. This linker region corresponds to the region in voltage-gated K^+ channels, studied by Ding and Horn (2003), that influences the coupling between the voltage sensor and the cytoplasmic gate. Taken together, these results suggested that changes in C-terminus conformation may influence gating via a direct action on the cytoplasmic gate. Based on structural data, Jiang et al. (2002) suggested a mechanism for this effect, whereby a Ca^{2+}-induced change in conformation of the C-terminus was mechanically coupled to opening of the cytoplasmic gate.

Tests of this hypothesis continued with a set of experiments by Niu et al. (2004), who demonstrated that changes in the length of this linker altered the response of BK channels to Ca^{2+}. In the presence of Ca^{2+}, shortening the linker increased, and lengthening the linker decreased, the probability of channel opening in response to voltage. Niu et al. (2004) proposed a model whereby the C-terminus was connected to the cytoplasmic gate by a spring mechanism, and that changing linker (spring) length, and therefore tension, changed the probability of channel opening at all voltages.

In summary, current models suggest that the C-terminus is a modular channel domain, which exerts mechanical force on the S6 domain in the region of the cytoplasmic gate. In some channels, alteration of this force is directly responsible for opening the gate, whereas in others, it changes the probability of opening in response to another, primary stimulus (e.g., voltage). This model also appears to hold for channels that use the selectivity filter as a gate (described above). For example, in CNG channels, interaction of cyclic nucleotides with the C-terminus produces a conformational change in the region of the cytoplasmic gate, which is associated with opening of the selectivity filter gate (Flynn and Zagotta, 2001; Johnson and Zagotta, 2001). Conversely, interaction of ATP with the C-terminus of the KATP channel closes the selectivity filter gate (Proks et al., 2003). Thus, the great variety of C-terminal structures may largely reflect the need for channels to respond to different ligands and/or modulators. Within this context, the C-terminus may play a generally common role, and act by a fundamentally common mechanism, in many channels.

4.9 The MinK/MiRP Family of Accessory Subunits

The functional properties of voltage-gated K^+ channels are altered by a wide variety of accessory proteins, broadly called β subunits. Co-assembly with β subunits influences trafficking, cell surface expression, and physiological properties of α subunits. The Kvβ class of subunits are cytoplasmic proteins that bind to the N-terminal domains in a fourfold symmetric configuration (Zhou et al., 2001a), and will be discussed in Part III. Another class of β subunits, the MinK/MiRP peptides, are integral membrane proteins with one membrane spanning region (Fig. 4.3A). In a variety of cells, complexes composed of K^+ channel α subunits and MinK/MiRP peptides create a specific endogenous K^+ current. For example, the cardiac I_{Ks} current is derived from the KvLQT channel (the α subunit) and MinK peptide (Fig. 4.3A–C; Barhanin et al., 1996; Sanguinetti et al., 1996). Similarly, a skeletal muscle K^+ channel is derived from co-assembled Kv3.4 subunits and MiRP2 peptides (Fig. 4.3D–G; Abbott et al., 2001). Moreover, a variety of serious, often fatal diseases are associated with point mutations in one or another member of the MinK/MiRP peptide family (cf. Goldstein et al., 2004). For example, several variants of cardiac long Q-T syndrome, a cardiac arrhythmia that often leads to a fatal ventricular fibrillation, are associated with mutations to MinK/MiRP peptides (Abbott et al., 1999, 2001; Abbott and Goldstein, 2002; Goldstein et al., 2004). Mutations in one family member, MiRP1, are responsible for a common form of drug-induced arrhythmia, which often leads to sudden death (Abbott et al., 1999, Sesti et al., 2000).

Whereas their importance has led to extensive investigation, the mechanisms by which they modify α subunits remain obscure. The complexity of this interaction can be illustrated by just a few examples. Co-assembly of the MiRP2 peptide with the Kv2.1 and Kv3.1 potassium channels reduces current density and slows activation rate, but produces *no change* in voltage dependence of current activation (McCrossan et al., 2003). In contrast, co-assembly of MiRP2 with the Kv3.4 potassium channel shifts the voltage dependence of current activation in the *negative direction* by almost 40 mV (Abbott et al., 2001). Yet another contrasting effect is observed with the co-assembly of MiRP1 with the Kv4.2 channel. As with most other MiRP—α subunit interactions, co-assembly of MiRP1 with Kv4.2 slows activation (Zhang et al., 2001). However, MiRP1 also shifts the voltage dependence of Kv4.2 activation in the *positive direction* by about 30 mV (Zhang et al., 2001). Finally, many MinK/MiRP interactions with α subunits result in a substantial change in inactivation rate. As discussed above, activation and inactivation mechanisms are

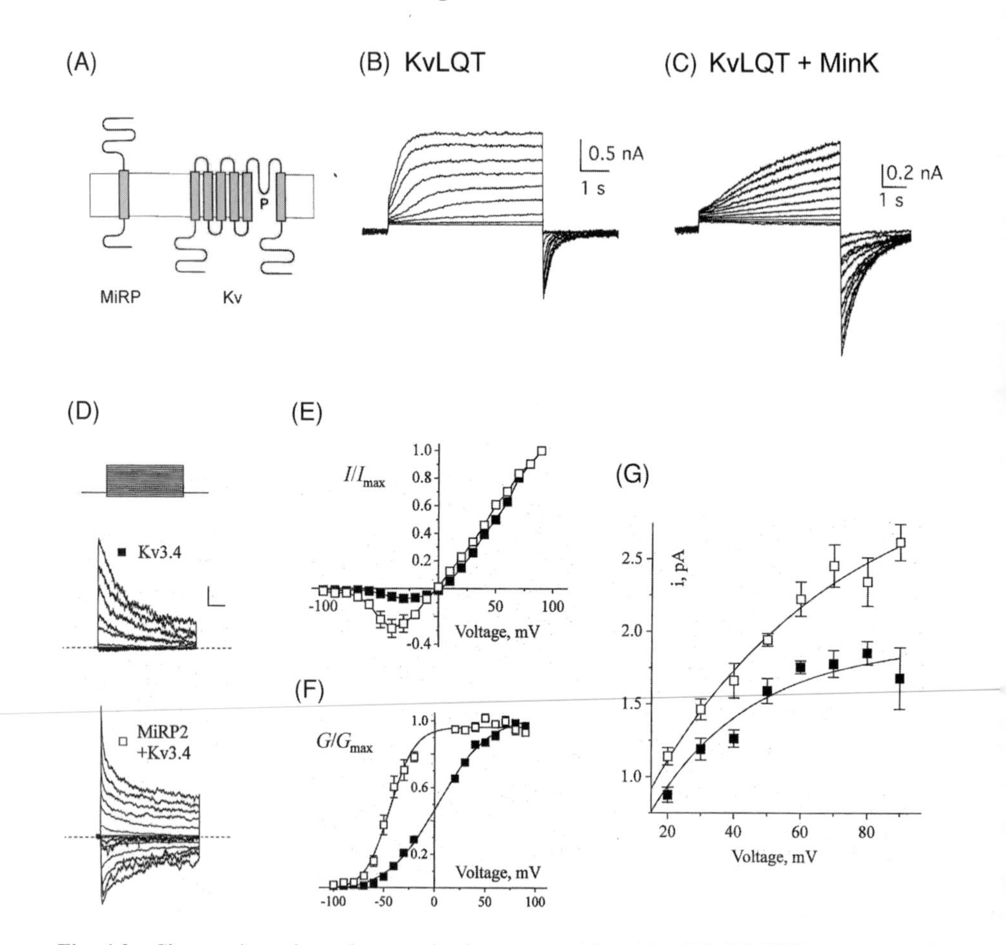

Fig. 4.3 Changes in a channel properties by co-assembly with a MinK/MiRP accessory subunit. (A) Schematic representation of a single transmembrane domain MinK/MiRP subunit and a six transmembrane domain α subunit. (B–C) K^+ currents through channels composed solely of KvLQT1 channel subunits (B) or channels composed of co-assembled KvLQT1 + human MinK subunits (C). Note the dramatic (18-fold) slowing of activation in the co-assembled channel. Co-assembly with MinK also increases single-channel conductance ~4-fold, shifts the voltage dependence of activation ~25 mV in the positive direction and slows channel closing rate by ~3-fold (Sesti and Goldstein, 1998). (D–G) Currents through channels composed solely of Kv3.4 and through channels composed of co-assembled Kv3.4 plus MiRP2. Currents were recorded with 100 mM internal and external K^+. Co-assembly with MiRP2 produces a dramatic change in slope of the activation curve (panel F), which results in the generation of large currents at very negative potentials (panels D, E). The inward current in the co-assembled channels results from the negative shift in the activation curve. Co-assembly with MiRP2 also produces a significant increase in single-channel conductance (panel G). These two effects would not only enhance the subthreshold current, but would tend to dramatically dampen excitation by producing a substantial K^+ conductance at the resting membrane potential. Currents in panels B–C were recorded in oocytes, currents in panel D were recorded in CHO cells. Figure kindly provided by Steve A. N. Goldstein. Adapted from Sesti and Goldstein, 1998 (panels B–C) and Abbott et al., 2001 (panels A, D–G).

structurally distinct. In sum, these small peptides have multiple effects, and the same peptide can have different effects on different channels.

Some headway has been made into understanding the structural interaction between the MinK/MiRP peptides and K^+ channel α subunits. It appears that two MinK subunits co-assemble with the tetrameric KvLQT1 channel (Chen et al., 2003a), although others have come to different conclusions (Tzounopoulos et al., 1995; Wang et al., 1998). It also appears that MinK is exposed to both the internal and external conduction pathway (Wang et al., 1996; Sesti and Goldstein, 1998; Tai and Goldstein, 1998; Chen et al., 2003b). These observations raise two questions. First, how do two MinK peptides fit within the conduction pathway of a symmetric channel composed of four α subunits, without disrupting the precisely tuned selectivity mechanism? Second, how does a small peptide that is exposed to the conduction pathway produce dramatic and widely differing effects on both activation and inactivation in different channels? Indeed, the complex and contrasting effects of these small peptides on α subunit function is currently perplexing. However, once the structural interaction between the different subunits is understood, the complex array of functional effects of these small peptides will undoubtedly provide new insight into the gating mechanisms, and modulation of gating mechanisms, of K^+ channels.

Part III. Specific Properties of Voltage-Gated Channels

4.10 Diversity of Function

The complexity of the brain is derived from the integration of an enormous number of intracellular and intercellular events, occurring both simultaneously and sequentially on timescales of milliseconds to seconds. Individual neurons can have one or more specific functions within a circuit, depending on environmental and/or circumstantial events. To accomplish their different functions, a large number of ion channel species is required within a cell, and different combinations of ion channel species are required across different cell types. Some neurons are merely conduits for faithful transmission of incoming information, and the type of information being transmitted will dictate the need for specific channel types. Other neurons must integrate a variety of inputs, inhibitory and excitatory, of large and small magnitude, occurring at different locations on the cell at different times. This integration results in a constant decision-making process by the neuron as to whether and how to respond to the everchanging set of inputs. This within-neuron integration not only requires a large cohort of channels, but also requires that they can be modulated by a diverse array of second messengers and protein partners. The interaction of channels to form a functional result is extraordinarily complex. Indeed, it would be foolishly simplistic to assign a particular role to a particular channel in shaping neuronal excitation. Nonetheless, phenotypic differences among channels can generally be associated with one or more specific functional roles.

The heart represents a system in which the precise contribution of a multitude of channels to function is well understood (one can find many good, comprehensive descriptions in textbooks and reviews). In contrast, we are only beginning to understand how different K^+ channels coordinate and regulate neuronal behavior. Historically, identification of individual channel roles was examined with pharmacological manipulations. Although this approach, which continues today, has generated significant insight into how different channels shape intra- and intercellular function, an obvious shortcoming is the lack of specificity of available pharmacological agents. Several recent advances, most of which revolve around molecular technology, seem to be the final requirement for a comprehensive description and precise understanding of the role of channel species in neuronal electrophysiology. For example, the ability to generate knockout mice allows the study of circuitry in a system devoid of a single-channel species or modulator. Conversely, knock-in technology allows the expression of exogenously-provided channel species into a neuron. Even more promising is the burgeoning ability to direct expression (or lack of expression) to particular parts of a circuit, in vivo, using cell-specific promoters. This type of approach will lead to a comprehensive understanding of the role of the particular channel in cellular behavior. Moreover, it will ultimately reveal whether the influence of a channel mutation in a particular circuit (or cell) was necessary and sufficient to produce an entire disease, or one or more symptoms of a disease. The application of this technology to circuitry, especially in vivo, is just beginning to occur. In the following section, several examples are provided to illustrate how channel phenotype influences the electrophysiological behavior of a neuron at a cellular level. It must be stressed that these studies are also in an early stage, and even in these examples, our understanding of cellular behavior is incomplete.

4.11 Kv1 Channels

Kv1 channels are widely distributed, and have different functions in different neuronal circuits. Mutations to the Kv1.1 channel are most notably associated with episodic ataxia, which in turn, is associated with many other symptoms (Browne et al., 1994; Klein et al., 2004). To date, how a single mutation to a single channel leads to a diverse set of disease symptoms, or whether, in fact, all of the symptoms are associated with a single Kv1.1 mutation, are unknown.

Kv1.1, Kv1.2, Kv1.5, and Kv1.6, and/or heteromers derived from these channel subunits, can all be present in an individual neuron, and can all regulate subthreshold excitability (a measure of the likelihood that a cell will fire an action potential in response to a small, subthreshold stimulus). Indeed, when studied in isolation, multiple channels may make indistinguishable contributions to membrane currents. Upon closer scrutiny, however, one channel type may contribute to very different aspects of membrane excitability. Specialized functions can be based on cell type, membrane localization, stimulus characteristic, environment or physiological circumstance. Just a few examples will be given to provide a flavor for the association of phenotype with function. Below are two examples of Kv1.1 function, not associated with ataxia, and an example of Kv1.3 function in neuronal electrophysiology.

Auditory neurons of the medial nucleus of the trapezoid body (MNTB) are involved in localizing sound from binaural intensity differences. To accomplish this task, neurons must respond with high-fidelity to high-frequency stimulation. For example, 20 stimuli generated at a very high frequency (hundreds of Hz) should result in 20 action potentials fired with exquisite temporal accuracy relative to the stimuli. MNTB neurons contain Kv1.1, Kv1.2, and Kv1.6 channels, most likely in heteromeric configurations (Brew et al., 2003). Both homomers and heteromers containing Kv1.1 activate at very negative membrane potentials (and thus generate a subthreshold current). Subthreshold currents have several functions. In MNTB neurons, they rapidly terminate a depolarization associated with a synaptic event, and thus limit the number of action potentials (hopefully to one) produced by the individual synaptic stimulus (Brew et al., 2003). In other neurons, subthreshold K^+ currents effectively dampen excitability by opposing the firing of action potentials by small stimuli (cf. McKay et al., 2005 and references within). Knock-out of Kv1.1 reduces subthreshold K^+ current activated at very negative voltages, perhaps by two mechanisms: by reduction of current and also by a positive shift in activation voltage for the current carried by other channel subunits that no longer form with Kv1.1 (cf. Brew et al., 2003). The reduction or loss of subthreshold current, in turn, leads to the firing of multiple action potentials in response to individual stimuli. Thus, binaural encoding of auditory information, which requires high-fidelity pairing of action potentials to stimuli, is made possible by the Kv1.1 channel. However, this Kv1.1 function is not by itself sufficient to ensure temporal fidelity at high frequencies. High-fidelity firing in auditory neurons also requires Kv3 channel function, which will be discussed below.

A second example involves Kv1.1 with a completely different personality. In MNTB cells, Kv1.1 produces a noninactivating outward current. However, Kv1.1 channels can couple to a β subunit, Kvβ1.1, to produce a rapidly inactivating current (Rettig et al., 1994; several Kv1 channels interact with one of three Kvβ subunits). This fast inactivation results from a standard ball and chain mechanism associated with N-type inactivation. In this case, however, the Kvβ subunit provides the inactivation ball to an α subunit that doesn't have one (cf. Zhou et al., 2001a). When coupled, Kv1.1/Kvβ1.1 channels activate at quite negative voltages (the Kv1.1 supplied property) and inactivate rapidly (the Kvβ1.1 property). Rapidly inactivating K^+ currents have a variety of roles, some of which will be discussed further below. But why create a new rapidly inactivating channel when several Kv1 and Kv4 channels have an intrinsic fast inactivation mechanism?

One electrophysiological event that involves rapidly inactivating K^+ channels composed of Kv1.1/Kvβ1.1 is spike broadening. If a rapidly inactivating K^+ channel is involved in repolarization of the action potential, then with high-frequency stimulation, successive action potentials will become longer in duration as K^+ channels inactivate (cf. Giese et al., 1998). This would occur regardless of the channel species utilized to produce the rapidly activating, rapidly inactivating current. However, the precise nature of the electrophysiological event will differ in important ways depending on the channel species involved. For example, the fast inactivation process resulting from Kv1.1/Kvβ1.1 coupling will respond differently to high-frequency firing than will a channel with an intrinsic N-type inactivation mechanism. As action potentials broaden, Ca^{2+} influx during the action potential will increase. As a result, Ca^{2+}-dependent K^+ currents, which mediate slow afterhyperpolarizations, will be increased in magnitude and perhaps prolonged. This will reduce the likelihood of firing and/or reduce firing frequency. In addition, increasing Ca^{2+} influx will produce larger Ca^{2+}-dependent biochemical responses, and also, perhaps, increase the spread of intracellular Ca^{2+}. This will delocalize Ca^{2+}-dependent events. If K^+ channel inactivation increases further, perhaps via increased firing frequency, the magnitude of Ca^{2+} influx would continue to increase. However, when the rapidly inactivating K^+ current is derived from the Kv1.1/Kvβ1.1 complex, elevation of intracellular Ca^{2+} will inhibit the fast inactivation via a specific action on the N-terminal domain of Kvβ1.1 (Jow et al., 2004; see Fig. 4.9 for a similar effect involving a Kv4.2/frequenin complex). This reduction in inactivation would then lead to shorter action potentials, a reduction of Ca^{2+} influx and consequently a reduction in the magnitude or spread of Ca^{2+}-dependent processes. Thus, because Kvβ1.1 is Ca^{2+} sensitive, utilizing the Kv1.1/Kvβ1.1 complex to produce fast inactivation creates a feedback loop that can strike a balance between action potential broadening and shortening, and consequently place a limit on Ca^{2+} influx. In contrast, the inactivation rate of the Kv1.4 channel (which has an intrinsic ball and chain) is insensitive to Ca^{2+}. Consequently, if Kv1.4 were utilized as the rapidly inactivating channel, action potentials might broaden, but this process would be insensitive to Ca^{2+} influx and therefore not subject to feedback inhibition or modulation (Jow et al., 2004).

Another example of specialized function in the Kv1 family involves the Kv1.3 channel. The Kv1.3 channel is best known for its role in T-lymphocyte function, where it is involved in cytokine release. Indeed, this function of Kv1.3 is of clear clinical importance, and makes Kv1.3 a potentially important target for immunosuppression therapy (Chandy et al., 2004; Damjanovich et al., 2004; Valverde et al., 2005). But whereas Kv1.3 function is fairly straightforward in a lymphocyte, it can be a nexus for great complexity in other systems. For example, the Kv1.3 channel contributes significantly to the depolarization-activated K^+ current in olfactory neurons (Fadool and Levitan, 1998). Certainly, the simple activation of a large Kv1.3 current contributes significantly to the cell's firing behavior. However, this channel also appears to play a far more interesting and complex role in olfaction. Among the several Kv channels present in olfactory neurons, Kv1.3 is the site of modulation by multiple intracellular signals, many of which act via tyrosine kinases (Bowlby et al., 1997; Fadool and Levitan, 1998). Thus, in addition to altering action potential characteristics, deletion of Kv1.3 channels from olfactory neurons eliminates the ability of olfactory neurons to be modulated by biochemical signals that act via tyrosine kinase (Fadool et al., 2004). Moreover, elimination of Kv1.3 also results in an increased expression of several protein partners. Thus, not only have multiple cellular modulation mechanisms that converge on one particular ion channel been eliminated, but it is likely that an untold number of previously unrelated events are also indirectly influenced. For example, the elevation of expression of interacting proteins, concomitant with elimination of the Kv1.3 target, would likely lead to a greater interaction of these proteins with another target. Finally, knock-out of Kv1.3 resulted in significant structural alterations in the olfactory bulb (Fadool et al., 2004). In summary, Kv1.3 knockout changed action potential shape, duration and firing frequency, changed the structure of the olfactory bulb, eliminated the ability to modulate the olfactory neurons by a large number of modulators and changed expression of a large number of proteins that interact with both Kv1.3 and other effector proteins. Clearly, the Kv1.3 channel plays many roles in olfactory neurons.

A surprising consequence of these changes is that mice lacking Kv1.3 channels were better at detecting and discriminating odors. The opportunities for understanding channel function, and neuronal function, raised by this finding are many. One would assume that better odor detection and discrimination are good things for a mouse, yet elimination of a dominant channel in olfactory neurons apparently made olfactory function better. One might argue, from an evolutionary point of view, that the presence of Kv1.3 in the olfactory bulb, where it apparently reduces odor detection capabilities, provides an adaptive advantage. It remains to be determined what advantage has been gained, how this single ion channel influences the tradeoff of advantages and disadvantages, and indeed, how olfaction works!

4.12 Kv2 Channels

There are just two functional members of the Kv2 family, Kv2.1 and Kv2.2. To the extent that they have been compared, these channels are quite similar in both sequence and physiological characteristics. Kv2 channels have relatively long carboxy-termini for Kv channels, which suggests one or more roles for this region in channel function. These channels display differences in expression during development, apparently due to differences in the carboxy-terminal sequence of the channel (Blaine et al., 2004). Kv2.1 is present in many neuron types throughout the nervous system, as well as many peripheral organs, such as heart and pancreas. Kv2.2 channels are predominantly found in smooth muscle, but can also be found in a variety of neurons. Finally, despite the similarity of function, these two channels seem to have somewhat different roles in influencing cell excitability (cf. Malin and Nerbonne, 2002; Blaine et al., 2004).

Kv2.1 channels activate more slowly than Kv1 channels, and display a slow inactivation (Fig. 4.4). As with all other Kv channels, Kv2.1 serves a variety of purposes, depending on its relative abundance, its cellular location and the complement of other channels in the particular neuron. One unique role of Kv2.1 has been demonstrated in hippocampal neurons. The contribution of Kv2.1 channels to the overall current in hippocampal neurons is quite small (Mitterdorfer and Bean, 2002). Nonetheless, ~90% knock-down of Kv2.1 has a dramatic and somewhat surprising effect (Du et al., 2000). Consistent with its limited contribution to the total K^+

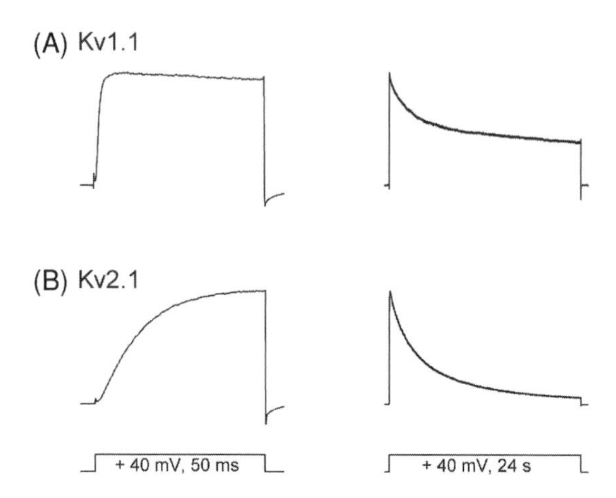

Fig. 4.4 Differences in activation and inactivation rates in Kv1.1 and Kv2.1. Kv1.1 activates rapidly and, with prolonged activation, inactivates relatively little (panel A). In contrast, Kv2.1 activates much more slowly and completely inactivates with prolonged activation (panel B). Recordings from HEK cells by Josef G. Trapani and Payam Andalib.

current, the duration of single action potentials is virtually unaffected by the absence of Kv2.1. However, at slightly elevated extracellular [K^+] (8.5 mM), which causes hippocampal neurons to fire action potentials in bursts, action potential duration is prolonged more than 10-fold in the absence of Kv2.1 channels (Fig. 4.5). Thus, it appears that the primary role of Kv2.1 in hippocampal neurons is to maintain action potential integrity when extracellular [K^+] is elevated.

Kv2.1 has an interesting and unique property that may underlie this physiological role. Upon elevation of external [K^+], both current magnitude and activation rate are increased (Fig. 4.6A; Wood and Korn, 2000; Consiglio and Korn, 2004). The effect on current magnitude is opposite to that expected given the reduction in electrochemical driving force, and opposite to that seen for most all other K^+ channels (Fig. 4.6B; the HERG K^+ channel, which is most notable for its role in heart physiology, also displays an anomalous increase in current upon [K^+] elevation, which is produced by a different mechanism). Thus, under high-frequency firing situations, when extracellular [K^+] might rise, current through Kv2.1 apparently increases while current through other channels decreases. This would serve to maintain total cellular K^+ current density during an action potential, and thus maintain action potential integrity.

The molecular mechanism that underlies this effect was described in a previous section of this chapter. Briefly, Kv2.1 can open into one of two conformations. At lower [K^+], some channels open into a high-conductance state and some open into a lower conductance state. The fraction of channels in each conformation depends on the occupancy of a specific K^+ binding site in the selectivity filter when the channel opens, which in turn is dependent on the external [K^+] (Immke et al., 1999; Immke and Korn, 2000). At higher [K^+], a larger fraction of channels open

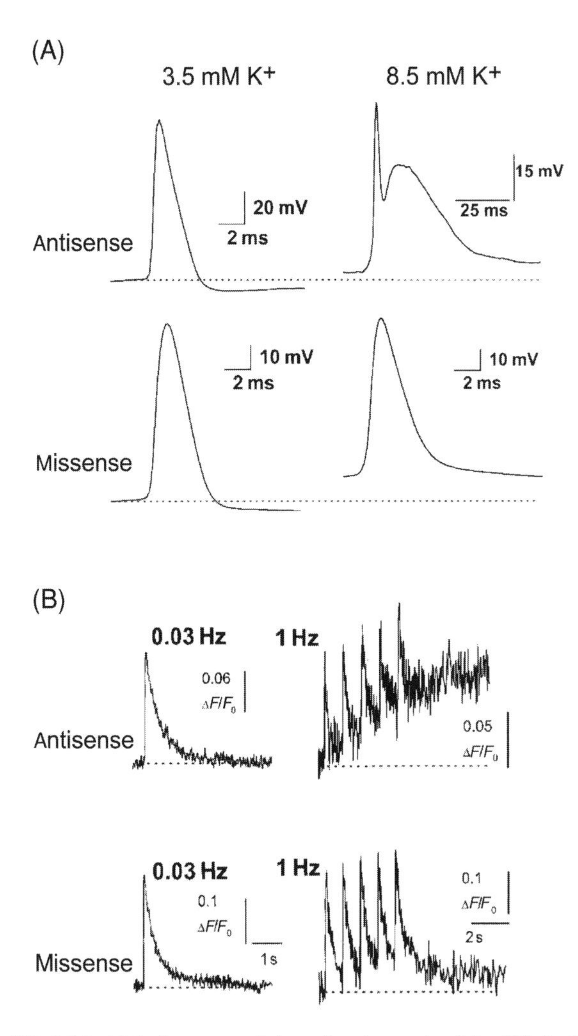

Fig. 4.5 Role of Kv2.1 in high-frequency firing of action potentials. (A) Action potentials were recorded from pyramidal cells in hippocampal slices at two external $[K^+]$. Slices were exposed either to an antisense oligonucleotide to knock down Kv2.1 expression by ~90% or a missense oligonucleotide, which did nothing to Kv2.1 expression. At normal physiological $[K^+]$ (3.5 mM), knockdown of Kv2.1 expression had little or no effect on action potential duration. However, at high external $[K^+]$ (8.5 mM), which leads to bursts of action potentials, action potential duration was greatly prolonged (note the time scale difference). (B) Intracellular Ca^{2+} transients, evoked by 0.3 Hz or 1 Hz stimulation. With Kv2.1 present, Ca^{2+} transients rise and fall with each stimulus, regardless of frequency. At higher frequency, however, the absence of Kv2.1 leads to a steady rise in $[Ca^{2+}]$ that is uncoupled from the stimulus. Figure kindly provided by Chris McBain, and adapted from Du et al. (2000).

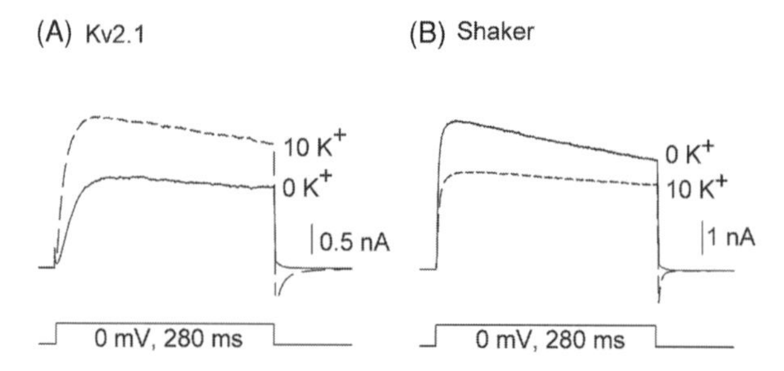

Fig. 4.6 K^+-dependent potentiation of current through the Kv2.1 channel. Elevation of external [K^+] would be expected to reduce outward current magnitude due to the reduction in electrochemical driving force. Such an effect, which happens in virtually all K^+ channels, is illustrated for *Shaker* in panel B. Interestingly, the magnitude of the reduction for a given change in driving force differs in different channels. In Kv2.1, however, the net effect of increasing external [K^+] between 0 and 10 mM is an increase in outward current (panel A). This reflects two special characteristics of Kv2.1. First, current through the channel responds very little to the change in driving force (Andalib et al., 2002). Second, the outer vestibule of the channel can open into one of two conformations, one of which has ~4-fold higher conductance than the other (Immke and Korn, 2000; Trapani et al., 2006). At higher [K^+], more channels open into the higher conductance conformation (Wood and Korn, 2000). Note also that elevation of external [K^+] speeds inactivation in Kv2.1 and slows inactivation in *Shaker*. Currents were recorded in HEK cells.

into the higher conductance state, and consequently, total macroscopic K^+ current through Kv2.1 channels is larger. Recent evidence indicates that the single-channel conductance of the lower conductance state is ~1/4 that of the higher conductance state (Trapani et al., 2006). In heterologous expression systems, only 10–20% of channels are in the low-conductance state at physiological [K^+], but conversion of these channels to the high-conductance state can have a significant impact on current magnitude. This raises an interesting issue for the in situ situation. Hippocampal neurons display a significant cell surface expression level of Kv2.1 (Misonou and Trimmer, 2004), yet, as mentioned above, Kv2.1 generates little current when single action potentials are evoked (Mitterdorfer and Bean, 2002). This may reflect two Kv2.1 properties. First, they activate and deactivate relatively slowly. Consequently, if action potentials fire more rapidly than channels deactivate, Kv2.1 current will grow with each succeeding action potential. Second, the observed function of Kv2.1 raises the intriguing possibility that, at normal physiological [K^+], a large number of Kv2.1 channels are in the low-conductance state. If so, the conformationally-based difference in conductance must be regulated by factors other than [K^+].

Kv2.1 channels have an additional pair of properties that appear to be linked: they cluster on the membrane surface and they are constitutively phosphorylated under resting conditions. Dephosphorylation results in a negative shift in the voltage dependence of activation and a dispersion of Kv2.1 (Misonou et al., 2004). Both dephosphorylation and dispersion are caused by glutamate exposure, which suggests that excitation is the responsible stimulus (Misonou et al., 2004). The negative shift in activation would tend to reduce membrane excitability. The role of clustering and dispersion are unknown. It is interesting, however, that the Kv2.1 clusters tend to be located on membranes apposed to astrocytic processes (Du et al., 1998). Astrocytes release glutamate in response to neuronal activity (cf. Araque et al., 2000), and appear to be actively involved in shaping neuronal events (cf. Fellin et al., 2004). Moreover, astrocytes have long been known to function as external [K^+] buffering systems. This raises the intriguing possibility that a dynamic set of events occur that involve high-frequency firing, external [K^+] buffering, and current magnitude through [K^+]-sensitive Kv2.1 channels.

4.13 Kv3 Channels

There are four Kv3 channels, Kv3.1–Kv3.4. Two of these channels display fast inactivation, two do not (Fig. 4.7). Kv3.1 and Kv3.2 channels have perhaps the most well-defined role in neuronal physiology (the primary difference between these two channels is their ability to be modulated by protein kinase C or A, respectively). They are designed to allow action potentials to fire rapidly and with high fidelity. As described earlier, this accurate high-frequency behavior requires Kv1 channels, which produce subthreshold currents, limit action potential firing in MNTB neurons to one per stimulus and lock action potential generation to the synaptic stimulus (Kaczmarek et al., 2005). However, another K^+ channel function is also required to ensure fidelity of rapidly firing action potentials: the channel responsible for repolarization of the action potential must (1) activate rapidly, (2) activate only at relatively positive membrane potentials, (3) carry a large current, and (4) close very quickly upon hyperpolarization. This function is supplied by Kv3 channels. Because Kv3 channels only activate at positive membrane potentials, they won't activate until well above action potential threshold, and consequently, won't interfere with action potential generation. The lack of subthreshold activation, combined with fast activation of a large current by strong depolarizations, permits the rapid activation, followed by rapid termination of the action potential. These biophysical properties keep action potentials very brief and associated temporally with the stimulus (see Fig. 4.7). Fast deactivation (channel closing) is also required because the succeeding action potential in a train would be delayed or inhibited by any residual K^+ conductance. Clearly, one would not want a channel such as Kv2.1 to be present in neurons whose function is to rapidly fire a large number of action potentials with a highly accurate temporal association with a stimulus. The slow activation of Kv2.1 would result in a broader action potential and the slow closing of Kv2.1 would prevent or delay the firing of the subsequent action potentials.

Kv3 channels are also substrates for a large number of modulators (see Rudy and McBain, 2001). As an example, Kv3.1 is phosphorylated by protein kinase C in MNTB neurons under resting conditions. Dephosphorylation can result in up to a −40 mV shift in the voltage dependence of activation, which would result in the activation of a larger current with depolarization. Modulation of Kv3.1 current magnitude by channel phosphorylation has a significant impact on both firing rate (smaller Kv3.1 currents decrease firing rate, as expected) and the fidelity with which action potentials follow high-frequency stimuli (Kaczmarek et al., 2005). The elegance of

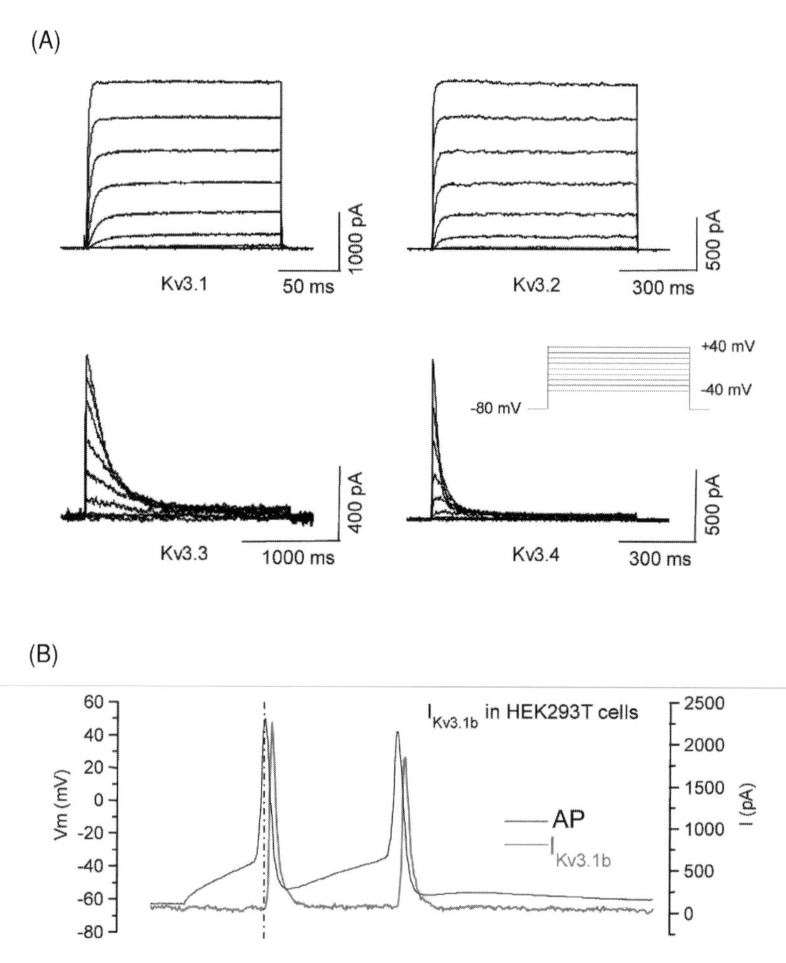

Fig. 4.7 Properties of Kv3 channels. There are four members of the Kv3 channel family, all of which open at relatively positive membrane potentials ($V_{1/2} = \sim +10$ mV). All channels activate and close rapidly. Kv3.1 and Kv3.2 don't inactivate, Kv3.3 inactivates relatively quickly and Kv3.4 inactivates even more quickly (panel A). In some neurons, Kv3.1 and Kv3.4 subunits co-assemble, resulting in a potassium channel that produces a faster hyperpolarization, decreased action potential duration and enhanced ability to fire with high frequency (Baranauskas et al., 2003). Indeed, formation of heteromeric channels with different subunit stoichiometry may provide a means to modify Kv3 channel function to optimally support action potential firing over a given frequency range (Li et al., 2001; Baranauskas et al., 2003). Panel B illustrates the activation of Kv3.1b currents by waveforms created to mimic the firing of a neocortical neuron. Note that the current activates late during the action potential (due to the positive $V_{1/2}$) and closes extremely rapidly. Figure kindly provided by Chris McBain, and adapted from Rudy and McBain (2001).

this modulation can be observed in the fine tuning of MNTB neuron physiology: high-frequency auditory stimulation causes dephosphorylation of Kv3.1 in MNTB neurons, which consequently increases Kv3.1 current magnitude and enhances the ability of the neuron to fire at high frequencies (Song et al., 2005).

The well-described roles of Kv1 and Kv3 channels in MNTB neurons provide a good example of how each kinetic component of channel behavior, together with selective modulation of a particular channel's properties, can produce an optimum result for a needed neuronal function. Additional examples of the fine-tuning of fast spiking behavior by Kv3 channels are described in an excellent older review (Rudy and McBain, 2001) and more recent papers (cf. Baranauskas et al., 2003; Lien and Jonas, 2003; Goldberg et al., 2005).

4.14 Kv4 Channels

There are three members of the Kv4 family, Kv4.1–Kv4.3, which have similar biophysical properties and pharmacology (Jerng et al., 2004a). These channels, whose hallmark is their rapid inactivation (Fig. 4.8A), are now believed to underlie the classical, transient K^+ current, I_A, first described in gastropod neurons over 30 years ago (Connor and Stevens, 1971a). Whereas they were originally shown to play a role in the timing of rhythmic firing behavior (Connor and Stevens, 1971b), they have since been shown to play an important role in a variety of electrophysiological events, including control of action potential duration, timing of responses to multiple synaptic inputs, and long-term potentiation (Schoppa and Westbrook, 1999; Mitterdorfer and Bean, 2002; Watanabe et al., 2002; Jerng et al., 2004a).

4.14.1 Influence of Kv4.2 on Hippocampal Dendrite Physiology

One function that is uniquely provided by the properties of Kv4 channels is that of regulating dendritic excitability in hippocampal pyramidal neurons. Kv4.2 is localized in dendritic membranes, with an increasing gradient of current density from proximal to distal segments (Hoffman et al., 1997). An apparently important contributor to hippocampal neuron function is the backpropagation of action potentials from the soma into the dendritic tree. The increasing gradient of Kv4.2 channels from proximal to distal dendrites causes the amplitude and number of back propagating action potentials to diminish with distance from the soma. The presence of Kv4.2 channels in distal dendrites also reduces the amplitude of excitatory postsynaptic potentials (epsps) originating from distal afferents (Hoffman et al., 1997). Most importantly, because of the rapid inactivation of Kv4.2 channels, an initial depolarization event can alter the membrane response to a second event that immediately follows. Thus, with appropriate timing, subthreshold epsps can reduce I_A and consequently increase the amplitude of closely following epsps or action potentials that are backpropagating into the distal dendrites (Johnston et al., 2000). Both epsp potentiation and the latter process, called action potential "boosting," may have a profound role in learning and memory. Long-term potentiation (LTP), which has long been studied as an electrophysiological mechanism of neuronal plasticity, is a process wherein synaptic responses to moderate stimuli are enhanced for a prolonged period by a single, preceding, powerful stimulus. Inhibition of I_A in hippocampal

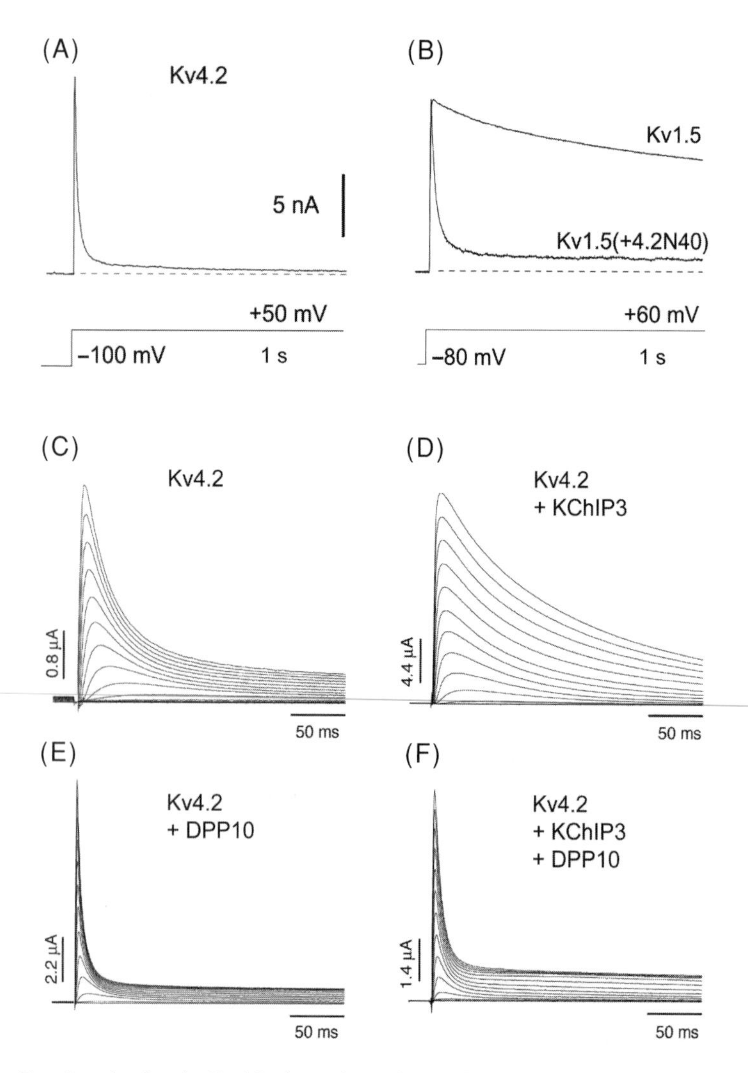

Fig. 4.8 Fast inactivation in Kv4.2 channels, and modification of Kv4.2 channel currents by KChIP3 and DPP10. (A) Illustration of a rapidly inactivating, Kv4.2 channel current. (B) Inactivation in Kv4 channel currents is complicated, potentially being caused by three or more different mechanisms. Elimination of the 40 terminal amino acids from the N-terminus of Kv4.2 slows inactivation, but still leaves a channel with a relatively fast inactivation process (see Gebauer et al., 2004; not shown). However, concatenation of the 40 N-terminal amino acids from Kv4.2 onto the slowly inactivating Kv1.5 imparts a fast inactivation mechanism onto Kv1.5. This demonstrates that the classical N-type ball and chain mechanism contributes to Kv4.2 channel inactivation. (C–D) Co-assembly with KChIP3 markedly slows Kv4.2 channel inactivation. (E) Co-assembly with DPP10 speeds Kv4.2 channel inactivation. (F) Co-assembly with both KChIP3 and DPP10 produces an intermediate inactivation rate. The rate of inactivation can be increased or decreased by changing the stoichiometry of expression of KChIP3 and DPP10 (Jerng et al., 2005). Currents in A–B were recorded in HEK cells, currents in C–F were recorded in CHO cells. Material for this figure was kindly provided by Robert Bahring (panels A–B), and Henry Jerng (panels C–F), and adapted from Gebauer et al., 2004 (panels A–B) and Jerng et al., 2005 (panels C–F).

dendrites, which results in larger epsps and action potentials in distal dendrites, reduces the threshold for LTP (Ramakers and Storm, 2002).

Reduction of Kv4.2 currents also occurs as a result of protein kinase-mediated phosphorylation of sites on the C-terminus (Anderson et al., 2000; Yuan et al., 2002). It appears that the channel is directly phosphorylated by MAP kinase, but channels can also be influenced indirectly by protein kinase A and protein kinase C, which activate MAP kinase (Yuan et al., 2002). Thus, there are many neurotransmitter-mediated pathways that can produce a reduction in I_A. As expected, this biochemically-mediated reduction of Kv4.2 current also leads to an increase in action potential amplitude in distal dendrites. Importantly, inhibition of MAP kinase reduces action potential boosting, inhibits the formation of LTP, and impairs learning (Watanabe et al., 2002; Morozov et al., 2003). Finally, Kv4.2 channels also localize Ca^{2+}-dependent events. For example, glutamate receptor activation, which would accompany excitatory synaptic input, triggers Ca^{2+}-mediated plateau potentials (Wei et al., 2001). The Ca^{2+} influx associated with these potentials would serve to activate myriad biochemical events, including second messenger cascades that modulate synaptic plasticity. With intact Kv4.2 channel function, these potentials are quickly terminated and therefore remain localized to the dendritic segment in which they were triggered (Wei et al., 2001; Cai et al., 2004). However, when Kv4.2 currents are inhibited, Ca^{2+} plateau potentials invade neighboring dendritic segments (Cai et al., 2004). In summary, depending on the role of a neuron within a circuit, dendritic membranes might contain members of the Kv1, Kv2, Kv3, Kv4, and/or K_{Ca} families (cf. Du et al., 1998; Mitterdorfer and Bean, 2002; Cai et al., 2004). Among them, Kv4.2 channels appear to be key players in the control of synaptic plasticity, and represent a common pathway by which either coincident electrophysiological stimuli or neurotransmitter-mediated biochemical cascades influence and localize events associated with neuronal plasticity and probably, learning and memory.

4.14.2 The Kv4 Channel Complex

Channels composed of just the Kv4 α subunit do not display biophysical properties identical to I_A currents in neuronal membranes. Indeed, it appears that native Kv4 channels exist in a complex with members of at least two other protein families: one called Kv channel-interacting proteins (KChIPs), a member of a larger, Ca^{2+} binding protein family, and one called dipeptidyl peptidase-like proteins (DPLs; An et al., 2000; Nadal et al., 2003; Rhodes et al., 2004; Jerng et al., 2005). The KChiIP family of proteins are cytoplasmic, and bind to the N-terminus of the Kv4 channel (cf. Scannevin et al., 2004). DPLs are integral membrane proteins that belong to a serine protease family of proteins but are without catalytic activity (Qi et al., 2003). Complexing with either KChIPs or DPLs increases Kv4 surface expression (Shibata et al., 2003; Jerng et al., 2004b), but also produces important changes in biophysical behavior. Typically, complexing with KChIPs slows inactivation (Fig. 4.8; Beck et al., 2002), but depending on KChIP and α subunit species, it can also speed inactivation (Beck et al., 2002) or eliminate inactivation (Holmqvist et al.,

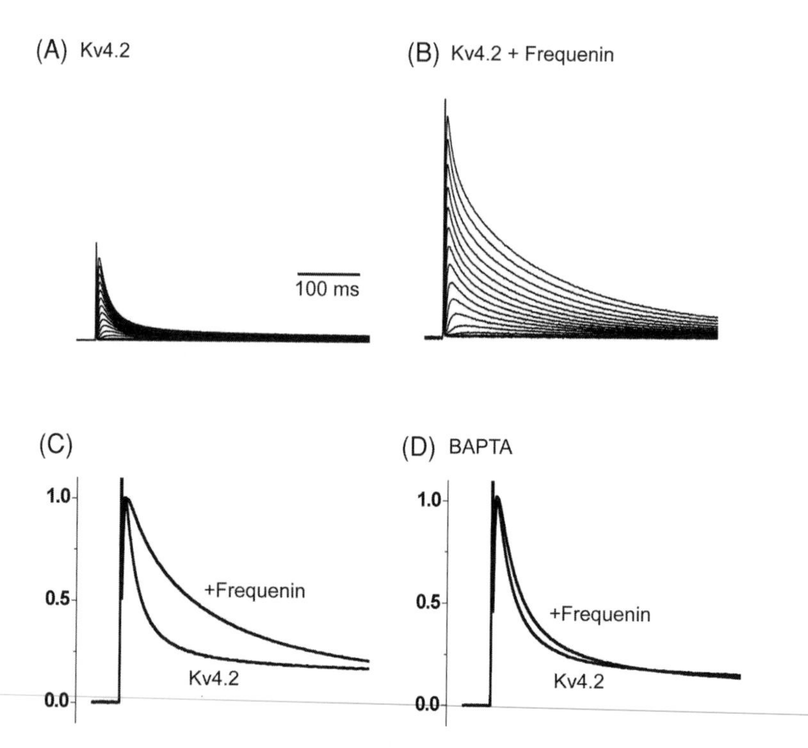

Fig. 4.9 Modulation of Kv4.2 channel currents by frequenin. (A–B) Co-expression of Kv4.2 channel subunits with frequenin, a Ca^{2+} binding protein similar to the KChIPs, increases channel expression and slows inactivation. (C) Normalized currents illustrate the slowing of Kv4.2 inactivation by co-assembly with frequenin. (D) With the Ca^{2+} chelator, BAPTA, in the cytoplasm, frequenin has little or no influence on Kv4.2 channel currents. This demonstrates that co-assembly with Ca^{2+} binding proteins can impart a Ca^{2+} dependence to Kv4.2 channel function. Currents were recorded from oocytes. Figure kindly provided by William A. Coetzee, adapted from Nakamura et al. (2001).

2002). Complexing of Kv4.2 with DPP10 speeds inactivation (Fig. 4.8; Jerng et al., 2005), and can also speed recovery from inactivation and the voltage dependence of activation and inactivation (Nadal et al., 2003). It appears that the complex formed by all three protein components represents the complete channel that underlies the I_A current (Nadal et al., 2003; Jerng et al., 2005). Moreover, it appears that varying the stoichiometry of the complex might be an approach used by cells to change I_A kinetics (Jerng et al., 2005).

The KChIP family of proteins may have an additional regulatory action on Kv4 channels. Frequenin is a member of the same Ca^{2+} binding protein family as the KChIPs. As with most other KChIP proteins, frequenin increases expression and slows inactivation of currents carried by Kv4.2 channels (Fig. 4.9A–B; Nakamura et al., 2001). However, the influence of frequenin on Kv4.2 currents is Ca^{2+}-dependent. If the rise in intracellular Ca^{2+} is prevented upon depolarization, frequenin has no impact on inactivation rate of currents carried by Kv4.2 channels

(Fig. 4.7C–D; Nakamura et al., 2001). Due to a structural difference in the N-terminal domain, frequenin influences only Kv4.2 and Kv4.3 channels; Kv4.1 channels are essentially unaffected (Nakamura et al., 2001).

In this section we presented only a small sampling of a massive and as yet incomplete literature on the role and regulation of Kv4 channels. However, we attempted in this sampling to provide insight into the enormous complexity of their involvement in neuronal physiology. Whereas Kv3 channels appear to have a rather more defined role in the accurate transmission of high-frequency input information into output information, Kv4 channels appear to have a more integrative role. Whether or not LTP, and perhaps learning and memory, occur depends on the readiness of Kv4 channels to respond to depolarization. Moreover, the size of Kv4 channel currents determine not only how large a local postsynaptic response will be but also how far it will spread down the dendrite. The sources of modulation of Kv4 current amplitude are manifold and intertwined: current amplitude is modulated by electrical activity, protein kinase activity, intracellular Ca^{2+} levels, and association with different protein partners. As a result of the fast inactivation mechanism, and the voltage range over which they are activated, Kv4 channel readiness can be modulated by changes in voltage dependence of activation or inactivation, the rate of inactivation or the rate of recovery from inactivation. Indeed, modulators not only influence one or more of all of these biophysical properties, but a number of modulators also influence cell surface expression levels. Thus, these modulators likely regulate long-term as well as short-term changes in Kv4 current amplitude.

Acknowledgments

We thank Drs. Bruce Bean, William Coetzee, Steve A.N. Goldstein, Henry Jerng, and Christopher McBain for their generosity in providing figures, some on very short notice. Much of Parts I and II of this chapter, including a color version of Fig. 4.2, were originally published in Korn, S.J. and Trapani, J.G., IEEE Trans Nanobioscience. 2005 Mar;4(1):21–33. Preparation of this chapter, and data collected in Figs. 4.4 and 4.6, were supported by NIH grants NS41090 and NS42563.

References

Abbott, G.W., and S.A. Goldstein. 2001. Potassium channel subunits encoded by the KCNE gene family: Physiology and pathophysiology of the MinK-related peptides (MiRPs). *Mol. Interv.* 1:95–107.

Abbott, G.W., and S.A. Goldstein. 2002. Disease-associated mutations in KCNE potassium channel subunits (MiRPs) reveal promiscuous disruption of multiple currents and conservation of mechanism. *FASEB J.* 16:390–400.

Abbott, G.W., M.H. Butler, S. Bendahhou, M.C. Dalakas, L.J. Ptacek, and S.A. Goldstein. 2001. MiRP2 forms potassium channels in skeletal muscle with Kv3.4 and is associated with periodic paralysis. *Cell* 104:217–231.

Abbott, G.W., F. Sesti, I. Splawski, M.E. Buck, M.H. Lehmann, K.W. Timothy, M.T. Keating, and S.A. Goldstein. 1999. MiRP1 forms IKr potassium channels with HERG and is associated with cardiac arrhythmia. *Cell* 97:175–187.

Aggarwal, S.K., and R. MacKinnon. 1996. Contribution of the S4 segment to gating charge in the Shaker K^+ channel. *Neuron* 16:1169–1177.

Ahern, C.A., and R. Horn. 2004. Stirring up controversy with a voltage sensor paddle. *Trends Neurosci.* 27:303–307.

Aldrich, R.W. 2001. Fifty years of inactivation. *Nature* 411:643–644.

An, W.F., M.R. Bowlby, M. Betty, J. Cao, H.P. Ling, G. Mendoza, J.W. Hinson, K.I. Mattsson, B.W. Strassle, J.S. Trimmer, and K.J. Rhodes. 2000. Modulation of A-type potassium channels by a family of calcium sensors. *Nature* 403:553–556.

Andalib, P., J.F. Consiglio, J.G. Trapani, and S.J. Korn. 2004. The external TEA binding site and C-type inactivation in voltage-gated potassium channels. *Biophys. J.* 87:3148–3161.

Andalib, P., M.J. Wood, and S.J. Korn. 2002. Control of outer vestibule dynamics and current magnitude in the Kv2.1 potassium channel. *J. Gen. Physiol.* 120:739–755.

Anderson, A.E., J.P. Adams, Y. Qian, R.G. Cook, P.J. Pfaffinger, and J.D. Sweatt. 2000. Kv4.2 phosphorylation by cyclic AMP-dependent protein kinase. *J. Biol. Chem.* 275:5337–5346.

Araque, A., N. Li, R.T. Doyle, and P.G. Haydon. 2000. SNARE protein-dependent glutamate release from astrocytes. *J. Neurosci.* 20:666–673.

Armstrong, C.M. 1971. Interaction of tetraethylammonium ion derivatives with the potassium channels of giant axons. *J. Gen. Physiol.* 58:413–437.

Armstrong, C.M., and F. Bezanilla. 1977. Inactivation of the sodium channel. II. Gating current experiments. *J. Gen. Physiol.* 70:567–590.

Armstrong, C.M., and B. Hille. 1972. The inner quaternary ammonium ion receptor in potassium channels of the node of Ranvier. *J. Gen. Physiol.* 59:388–400.

Bao, H., A. Hakeem, M. Henteleff, J.G. Starkus, and M.D. Rayner. 1999. Voltage-insensitive gating after charge-neutralizing mutations in the S4 segment of Shaker channels. *J. Gen. Physiol.* 113:139–151.

Baranauskas, G., T. Tkatch, K. Nagata, J.Z. Yeh, and D.J. Surmeier. 2003. Kv3.4 subunits enhance the repolarizing efficiency of Kv3.1 channels in fast-spiking neurons. *Nat. Neurosci.* 6:258–266.

Barhanin, J., F. Lesage, E. Guillemare, M. Fink, M. Lazdunski, and G. Romey. 1996. K_V LQT1 and lsK (minK) proteins associate to form the I_{Ks} cardiac potassium current. *Nature* 384:78–80.

Baukrowitz, T., and G. Yellen. 1996. Use-dependent blockers and exit rate of the last ion from the multi-ion pore of a K^+ channel. *Science* 271:653–656.

Beck, E.J., M. Bowlby, W.F. An, K.J. Rhodes, and M. Covarrubias. 2002. Remodelling inactivation gating of Kv4 channels by KChIP1, a small-molecular-weight calcium-binding protein. *J. Physiol.* 538:691–706.

Bernèche, S., and B. Roux. 2001. Energetics of ion conduction through the K^+ channel. *Nature* 414:73–77.

Bernèche, S., and B. Roux. 2005. A gate in the selectivity filter of potassium channels. *Structure (Camb)* 13:591–600.

Blaine, J.T., A.D. Taylor, and A.B. Ribera. 2004. Carboxyl tail region of the Kv2.2 subunit mediates novel developmental regulation of channel density. *J. Neurophysiol.* 92:3446–3454.

Bond, C.T., J. Maylie, and J.P. Adelman. 1999. Small-conductance calcium-activated potassium channels. *Ann. N.Y. Acad. Sci.* 868:370–378.

Bowlby, M.R., D.A. Fadool, T.C. Holmes, and I.B. Levitan. 1997. Modulation of the Kv1.3 potassium channel by receptor tyrosine kinases. *J. Gen. Physiol.* 110:601–610.

Brelidze, T.I., X. Niu, and K.L. Magleby. 2003. A ring of eight conserved negatively charged amino acids doubles the conductance of BK channels and prevents inward rectification. *Proc. Natl. Acad. Sci. USA* 100:9017–9022.

Brew, H.M., J.L. Hallows, and B.L. Tempel. 2003. Hyperexcitability and reduced low threshold potassium currents in auditory neurons of mice lacking the channel subunit Kv1.1. *J. Physiol.* 548:1–20.

Browne, D.L., S.T. Gancher, J.G. Nutt, E.R. Brunt, E.A. Smith, P. Kramer, and M. Litt. 1994. Episodic ataxia/myokymia syndrome is associated with point mutations in the human potassium channel gene, KCNA1. *Nat. Genet.* 8:136–140.

Bruening-Wright, A., M.A. Schumacher, J.P. Adelman, and J. Maylie. 2002. Localization of the activation gate for small conductance Ca^{2+}-activated K^+ channels. *J. Neurosci.* 22:6499–6506.

Buckingham, S.D., J.F. Kidd, R.J. Law, C.J. Franks, and D.B. Sattelle. 2005. Structure and function of two-pore-domain K^+ channels: Contributions from genetic model organisms. *Trends Pharmacol. Sci.* 26:361–367.

Cai, X., C.W. Liang, S. Muralidharan, S. Muralidharan, J.P. Kao, C.M. Tang, and S.M. Thompson. 2004. Unique roles of SK and Kv4.2 potassium channels in dendritic integration. *Neuron* 44:351–364.

Chanda, B., O.K. Asamoah, R. Blunck, B. Roux, and F. Bezanilla. 2005. Gating charge displacement in voltage-gated ion channels involves limited transmembrane movement. *Nature* 436:852–856.

Chandy, G.K., H. Wulff, C. Beeton, M. Pennington, G.A. Gutman, and M.D. Cahalan. 2004. K^+ channels as targets for specific immunomodulation. *Trends Pharmacol. Sci.* 25:280–289.

Chen, H., L.A. Kim, S. Rajan, S. Xu, and S.A. Goldstein. 2003a. Charybdotoxin binding in the I_{Ks} pore demonstrates two MinK subunits in each channel complex. *Neuron* 40:15–23.

Chen, H., F. Sesti, and S.A. Goldstein. 2003b. Pore- and state-dependent cadmium block of I(Ks) channels formed with MinK-55C and wild-type KCNQ1 subunits. *Biophys. J.* 84:3679–3689.

Chiara, M.D., F. Monje, A. Castellano, and J. López-Barneo. 1999. A small domain in the N terminus of the regulatory alpha-subunit Kv2.3 modulates Kv2.1 potassium channel gating. *J. Neurosci.* 19:6865–6873.

Choi, K.L., R.W. Aldrich, and G. Yellen. 1991. Tetraethylammonium blockade distinguishes two inactivation mechanisms in voltage-activated K^+ channels. *Proc. Natl. Acad. Sci. USA* 88:5092–5095.

Choi, K.L., C. Mossman, J. Aubé, and G. Yellen. 1993. The internal quaternary ammonium receptor site of Shaker potassium channels. *Neuron* 10:533–541.

Chung, S.H., T.W. Allen, and S. Kuyucak. 2002. Conducting-state properties of the KcsA potassium channel from molecular and Brownian dynamics simulations. *Biophys. J.* 82:628–645.

Connor, J.A., and C.F. Stevens. 1971a. Prediction of repetitive firing behaviour from voltage clamp data on an isolated neuron soma. *J. Physiol.* 213:31–53.

Connor, J.A., and C.F. Stevens. 1971b. Voltage clamp studies of a transient outward membrane current in gastropod neural somata. *J. Physiol.* 213:21–30.

Consiglio, J.F., P. Andalib, and S.J. Korn. 2003. Influence of pore residues on permeation properties in the Kv2.1 potassium channel. Evidence for a selective functional interaction of K^+ with the outer vestibule. *J. Gen. Physiol.* 121:111–124.

Consiglio, J.F., and S.J. Korn. 2004. Influence of permeant ions on voltage sensor function in the Kv2.1 potassium channel. *J. Gen. Physiol.* 123:387–400.

Crouzy, S., S. Bernèche, and B. Roux. 2001. Extracellular blockade of K^+ channels by TEA: Results from molecular dynamics simulations of the KcsA channel. *J. Gen. Physiol.* 118:207–218.

Cushman, S.J., M.H. Nanao, A.W. Jahng, D. DeRubeis, S. Choe, and P.J. Pfaffinger. 2000. Voltage dependent activation of potassium channels is coupled to T1 domain structure. *Nat. Struct. Biol.* 7:403–407.

Damjanovich, S., R. Gáspár, and G. Panyi. 2004. An alternative to conventional immunosuppression: Small-molecule inhibitors of kv1.3 channels. *Mol. Interv.* 4:250–254.

Decher, N., J. Chen, and M.C. Sanguinetti. 2004. Voltage-dependent gating of hyperpolarization-activated, cyclic nucleotide-gated pacemaker channels: Molecular coupling between the S4–S5 and C-linkers. *J. Biol. Chem.* 279:13859–13865.

del Camino, D., M. Holmgren, Y. Liu, and G. Yellen. 2000. Blocker protection in the pore of a voltage-gated K^+ channel and its structural implications. *Nature* 403:321–325.

del Camino, D., and G. Yellen. 2001. Tight steric closure at the intracellular activation gate of a voltage-gated K^+ channel. *Neuron* 32:649–656.

Ding, S., and R. Horn. 2002. Tail end of the s6 segment: Role in permeation in shaker potassium channels. *J. Gen. Physiol.* 120:87–97.

Ding, S., and R. Horn. 2003. Effect of S6 tail mutations on charge movement in Shaker potassium channels. *Biophys. J.* 84:295–305.

Doyle, D.A., J. Morais Cabral, R.A. Pfuetzner, A. Kuo, J.M. Gulbis, S.L. Cohen, B.T. Chait, and R. MacKinnon. 1998. The structure of the potassium channel: Molecular basis of K^+ conduction and selectivity. *Science* 280:69–77.

Du, J., L.L. Haak, E. Phillips-Tansey, J.T. Russell, and C.J. McBain. 2000. Frequency-dependent regulation of rat hippocampal somato-dendritic excitability by the K^+ channel subunit Kv2.1. *J. Physiol.* 522(Pt 1):19–31.

Du, J., J.H. Tao-Cheng, P. Zerfas, and C.J. McBain. 1998. The K^+ channel, Kv2.1, is apposed to astrocytic processes and is associated with inhibitory postsynaptic membranes in hippocampal and cortical principal neurons and inhibitory interneurons. *Neuroscience* 84:37–48.

Espinosa, F., R. Fleischhauer, A. McMahon, and R.H. Joho. 2001. Dynamic interaction of S5 and S6 during voltage-controlled gating in a potassium channel. *J. Gen. Physiol.* 118:157–170.

Fadool, D.A., and I.B. Levitan. 1998. Modulation of olfactory bulb neuron potassium current by tyrosine phosphorylation. *J. Neurosci.* 18:6126–6137.

Fadool, D.A., K. Tucker, R. Perkins, G. Fasciani, R.N. Thompson, A.D. Parsons, J.M. Overton, P.A. Koni, R.A. Flavell, and L.K. Kaczmarek. 2004. Kv1.3 channel gene-targeted deletion produces "Super-Smeller Mice" with altered glomeruli, interacting scaffolding proteins, and biophysics. *Neuron* 41:389–404.

Fellin, T., O. Pascual, S. Gobbo, T. Pozzan, P.G. Haydon, and G. Carmignoto. 2004. Neuronal synchrony mediated by astrocytic glutamate through activation of extrasynaptic NMDA receptors. *Neuron* 43:729–743.

Flynn, G.E., and W.N. Zagotta. 2001. Conformational changes in S6 coupled to the opening of cyclic nucleotide-gated channels. *Neuron* 30:689–698.

Giese, K.P., J.F. Storm, D. Reuter, N.B. Fedorov, L.R. Shao, T. Leicher, O. Pongs, and A.J. Silva. 1998. Reduced K^+ channel inactivation, spike broadening, and after-hyperpolarization in Kvbeta1.1-deficient mice with impaired learning. *Learn Mem.* 5:257–273.

Goldberg, E.M., S. Watanabe, S.Y. Chang, R.H. Joho, Z.J. Huang, C.S. Leonard, and B. Rudy. 2005. Specific functions of synaptically localized potassium channels in synaptic transmission at the neocortical GABAergic fast-spiking cell synapse. *J. Neurosci.* 25:5230–5235.

Goldstein, S.A., D. Bockenhauer, I. O'Kelly, and N. Zilberberg. 2001. Potassium leak channels and the KCNK family of two-P-domain subunits. *Nat. Rev. Neurosci.* 2:175–184.

Goldstein, S.A., L.A. Price, D.N. Rosenthal, and M.H. Pausch. 1996. ORK1, a potassium-selective leak channel with two pore domains cloned from *Drosophila melanogaster* by expression in Saccharomyces cerevisiae. *Proc. Natl. Acad. Sci. USA* 93:13256–13261.

Goldstein, S.A.N., M.T. Keating, and M.C. Sanguinetti. 2004. Cardiac arrhythmias: Inherited molecular mechanisms. *In*: Molecular Basis of Cardiovascular Disease, 2nd Ed. K.R. Chien, editor. Saunders, Philadelphia, PA, pp. 336–348.

Grissmer, S., and M. Cahalan. 1989. TEA prevents inactivation while blocking open K^+ channels in human T lymphocytes. *Biophys. J.* 55:203–206.

Gulbis, J.M., M. Zhou, S. Mann, and R. MacKinnon. 2000. Structure of the cytoplasmic beta subunit-T1 assembly of voltage-dependent K^+ channels. *Science* 289:123–127.

Gutman, G.A., K.G. Chandy, J.P. Adelman, J. Aiyar, D.A. Bayliss, D.E. Clapham, M. Covarriubias, G.V. Desir, K. Furuichi, B. Ganetzky, M.L. Garcia, S. Grissmer, L.Y. Jan, A. Karschin, D. Kim, S. Kuperschmidt, Y. Kurachi, M. Lazdunski, F. Lesage, H.A. Lester, D. McKinnon, C.G. Nichols, I. O'Kelly, J. Robbins, G.A. Robertson, B. Rudy, M. Sanguinetti, S. Seino, W. Stuehmer, M.M. Tamkun, C.A. Vandenberg, A. Wei, H. Wulff, R.S. Wymore, and International Union of Pharmacology. 2003. International Union of Pharmacology. XLI. Compendium of voltage-gated ion channels: Potassium channels. *Pharmacol. Rev.* 55:583–586.

Hackos, D.H., T.H. Chang, and K.J. Swartz. 2002. Scanning the intracellular S6 activation gate in the shaker K^+ channel. *J. Gen. Physiol.* 119:521–532.

Hartmann, H.A., G.E. Kirsch, J.A. Drewe, M. Taglialatela, R.H. Joho, and A.M. Brown. 1991. Exchange of conduction pathways between two related K^+ channels. *Science* 251:942–944.

Heginbotham, L., Z. Lu, T. Abramson, and R. MacKinnon. 1994. Mutations in the K^+ channel signature sequence. *Biophys. J.* 66:1061–1067.

Heginbotham, L., and R. MacKinnon. 1992. The aromatic binding site for tetraethylammonium ion on potassium channels. *Neuron* 8:483–491.

Hodgkin, A.L., and A.F. Huxley. 1952. A quantitative description of membrane current and its application to conduction and excitation in nerve. *J. Physiol.* 117:500–544.

Hodgkin, A.L., and B. Katz. 1949. The effect of sodium ions on the electrical activity of the giant axon of the squid. *J. Physiol.* 108:37–77.

Hoffman, D.A., J.C. Magee, C.M. Colbert, and D. Johnston. 1997. K^+ channel regulation of signal propagation in dendrites of hippocampal pyramidal neurons. *Nature* 387:869–875.

Holmes, T.C., D.A. Fadool, and I.B. Levitan. 1996. Tyrosine phosphorylation of the Kv1.3 potassium channel. *J. Neurosci.* 16:1581–1590.

Holmqvist, M.H., J. Cao, R. Hernandez-Pineda, M.D. Jacobson, K.I. Carroll, M.A. Sung, M. Betty, P. Ge, K.J. Gilbride, M.E. Brown, M.E. Jurman, D. Lawson, I. Silos-Santiago, Y. Xie, M. Covarrubias, K.J. Rhodes, P.S. Distefano, and W.F. An. 2002. Elimination of fast inactivation in Kv4 A-type potassium channels by an auxiliary subunit domain. *Proc. Natl. Acad. Sci. USA* 99:1035–1040.

Hoshi, T., W.N. Zagotta, and R.W. Aldrich. 1990. Biophysical and molecular mechanisms of Shaker potassium channel inactivation. *Science* 250:533–538.

Hoshi, T., W.N. Zagotta, and R.W. Aldrich. 1991. Two types of inactivation in Shaker K^+ channels: Effects of alterations in the carboxy-terminal region. *Neuron* 7:547–556.

Immke, D., M. Wood, L. Kiss, and S.J. Korn. 1999. Potassium-dependent changes in the conformation of the Kv2.1 potassium channel pore. *J. Gen. Physiol.* 113:819–836.

Immke, D., and S.J. Korn. 2000. Ion–ion interactions at the selectivity filter. Evidence from K^+-dependent modulation of tetraethylammonium efficacy in Kv2.1 potassium channels. *J. Gen. Physiol.* 115:509–518.

Jerng, H.H., and W.F. Gilly. 2002. Inactivation and pharmacological properties of sqKv1A homotetramers in Xenopus oocytes cannot account for behavior of the squid "delayed rectifier" K^+ conductance. *Biophys. J.* 82:3022–3036.

Jerng, H.H., K. Kunjilwar, and P.J. Pfaffinger. 2005. Multiprotein assembly of Kv4.2, KChIP3, and DPP10 produces ternary channel complexes with I_{SA}-like properties. *J. Physiol.* 767–788.

Jerng, H.H., P.J. Pfaffinger, and M. Covarrubias. 2004a. Molecular physiology and modulation of somatodendritic A-type potassium channels. *Mol. Cell Neurosci.* 27:343–369.

Jerng, H.H., Y. Qian, and P.J. Pfaffinger. 2004b. Modulation of Kv4.2 channel expression and gating by dipeptidyl peptidase 10 (DPP10). *Biophys. J.* 87:2380–2396.

Jiang, X., G.C. Bett, X. Li, V.E. Bondarenko, and R.L. Rasmusson. 2003a. C-type inactivation involves a significant decrease in the intracellular aqueous pore volume of Kv1.4 K^+ channels expressed in Xenopus oocytes. *J. Physiol.* 549:683–695.

Jiang, Y., A. Lee, J. Chen, M. Cadene, B.T. Chait, and R. MacKinnon. 2002. Crystal structure and mechanism of a calcium-gated potassium channel. *Nature* 417:515–522.

Jiang, Y., A. Lee, J. Chen, V. Ruta, M. Cadene, B.T. Chait, and R. MacKinnon. 2003b. X-ray structure of a voltage-dependent K^+ channel. *Nature* 423:33–41.

Jiang, Y., and R. MacKinnon. 2000. The barium site in a potassium channel by x-ray crystallography. *J. Gen. Physiol.* 115:269–272.

Johnson, J.P., and W.N. Zagotta. 2001. Rotational movement during cyclic nucleotide-gated channel opening. *Nature* 412:917–921.

Johnston, D., D.A. Hoffman, J.C. Magee, N.P. Poolos, S. Watanabe, C.M. Colbert, and M. Migliore. 2000. Dendritic potassium channels in hippocampal pyramidal neurons. *J. Physiol.* 525(Pt 1):75–81.

Jow, F., Z.H. Zhang, D.C. Kopsco, K.C. Carroll, and K. Wang. 2004. Functional coupling of intracellular calcium and inactivation of voltage-gated Kv1.1/Kvbeta1.1 A-type K^+ channels. *Proc. Natl. Acad. Sci. USA* 101:15535–5540.

Ju, M., L. Stevens, E. Leadbitter, and D. Wray. 2003. The roles of N- and C-terminal determinants in the activation of the Kv2.1 potassium channel. *J. Biol. Chem.* 278:12769–12778.

Kaczmarek, L.K., A. Bhattacharjee, R. Desai, L. Gan, P. Song, C.A. von Hehn, M.D. Whim, and B. Yang. 2005. Regulation of the timing of MNTB neurons by short-term and long-term modulation of potassium channels. *Hear Res.* 206:133–145.

Kang, D., C. Choe, and D. Kim. 2005. Thermosensitivity of the two-pore domain K^+ channels TREK-2 and TRAAK. *J. Physiol.* 564:103–116.

Kavanaugh, M.P., R.S. Hurst, J. Yakel, M.D. Varnum, J.P. Adelman, and R.A. North. 1992. Multiple subunits of a voltage-dependent potassium channel contribute to the binding site for tetraethylammonium. *Neuron* 8:493–497.

Kehl, S.J., C. Eduljee, D.C. Kwan, S. Zhang, and D. Fedida. 2002. Molecular determinants of the inhibition of human Kv1.5 potassium currents by external protons and Zn^{2+}. *J. Physiol.* 541:9–24.

Kerschensteiner, D., F. Monje, and M. Stocker. 2003. Structural determinants of the regulation of the voltage-gated potassium channel Kv2.1 by the modulatory alpha-subunit Kv9.3. *J. Biol. Chem.* 278:18154–18161.

Kim, D. 2003. Fatty acid-sensitive two-pore domain K^+ channels. *Trends Pharmacol. Sci.* 24:648–654.

Kiss, L., and S.J. Korn. 1998. Modulation of C-type inactivation by K^+ at the potassium channel selectivity filter. *Biophys. J.* 74:1840–1849.

Kiss, L., J. LoTurco, and S.J. Korn. 1999. Contribution of the selectivity filter to inactivation in potassium channels. *Biophys. J.* 76:253–263.

Klein, A., E. Boltshauser, J. Jen, and R.W. Baloh. 2004. Episodic ataxia type 1 with distal weakness: A novel manifestation of a potassium channelopathy. *Neuropediatrics* 35:147–149.

Klemic, K.G., G.E. Kirsch, and S.W. Jones. 2001. U-type inactivation of Kv3.1 and Shaker potassium channels. *Biophys. J.* 81:814–826.

Klemic, K.G., C.C. Shieh, G.E. Kirsch, and S.W. Jones. 1998. Inactivation of Kv2.1 potassium channels. *Biophys. J.* 74:1779–1789.

Kobertz, W.R., and C. Miller. 1999. K^+ channels lacking the 'tetramerization' domain: Implications for pore structure. *Nat. Struct. Biol.* 6:1122–1125.

Kobertz, W.R., C. Williams, and C. Miller. 2000. Hanging gondola structure of the T1 domain in a voltage-gated K^+ channel. *Biochemistry* 39:10347–10352.

Kramer, J.W., M.A. Post, A.M. Brown, and G.E. Kirsch. 1998. Modulation of potassium channel gating by coexpression of Kv2.1 with regulatory Kv5.1 or Kv6.1 alpha-subunits. *Am. J. Physiol.* 274:C1501–C1510.

Kuo, A., J.M. Gulbis, J.F. Antcliff, T. Rahman, E.D. Lowe, J. Zimmer, J. Cuthbertson, F.M. Ashcroft, T. Ezaki, and D.A. Doyle. 2003. Crystal structure of the potassium channel KirBac1.1 in the closed state. *Science* 300:1922–1926.

Li, M., Y.N. Jan, and L.Y. Jan. 1992. Specification of subunit assembly by the hydrophilic amino-terminal domain of the Shaker potassium channel. *Science* 257:1225–1230.

Li, W., and R.W. Aldrich. 2004. Unique inner pore properties of BK channels revealed by quaternary ammonium block. *J. Gen. Physiol.* 124:43–57.

Li, W., L.K. Kaczmarek, and T.M. Perney. 2001. Localization of two high-threshold potassium channel subunits in the rat central auditory system. *J. Comp. Neurol.* 437:196–218.

Lien, C.C., and P. Jonas. 2003. Kv3 potassium conductance is necessary and kinetically optimized for high-frequency action potential generation in hippocampal interneurons. *J. Neurosci.* 23:2058–2068.

Liu, Y., M. Holmgren, M.E. Jurman, and G. Yellen. 1997. Gated access to the pore of a voltage-dependent K^+ channel. *Neuron* 19:175–184.

Liu, Y., and R.H. Joho. 1998. A side chain in S6 influences both open-state stability and ion permeation in a voltage-gated K^+ channel. *Pflugers Arch.* 435:654–661.

Liu, Y., M.E. Jurman, and G. Yellen. 1996. Dynamic rearrangement of the outer mouth of a K^+ channel during gating. *Neuron* 16:859–867.

Long, S.B., E.B. Campbell, and R. MacKinnon. 2005a. Crystal structure of a mammalian voltage-dependent shaker family K^+ channel. *Science* 309:897–903.

Long, S.B., E.B. Campbell, and R. MacKinnon. 2005b. Voltage sensor of Kv1.2: Structural basis of electromechanical coupling. *Science* 309:903–908.

Loots, E., and E.Y. Isacoff. 1998. Protein rearrangements underlying slow inactivation of the Shaker K^+ channel. *J. Gen. Physiol.* 112:377–389.

Lopes, C.M., P.G. Gallagher, M.E. Buck, M.H. Butler, and S.A. Goldstein. 2000. Proton block and voltage gating are potassium-dependent in the cardiac leak channel Kcnk3. *J. Biol. Chem.* 275:16969–16978.

López-Barneo, J., T. Hoshi, S.H. Heinemann, and R.W. Aldrich. 1993. Effects of external cations and mutations in the pore region on C-type inactivation of Shaker potassium channels. *Receptors Channels* 1:61–71.

Lu, T., A.Y. Ting, J. Mainland, L.Y. Jan, P.G. Schultz, and J. Yang. 2001. Probing ion permeation and gating in a K^+ channel with backbone mutations in the selectivity filter. *Nat. Neurosci.* 4:239–246.

Lu, Z. 2004. Mechanism of rectification in inward-rectifier K^+ channels. *Annu. Rev. Physiol.* 66:103–129.

Lu, Z., A.M. Klem, and Y. Ramu. 2001. Ion conduction pore is conserved among potassium channels. *Nature* 413:809–813.

Lu, Z., A.M. Klem, and Y. Ramu. 2002. Coupling between voltage sensors and activation gate in voltage-gated K^+ channels. *J. Gen. Physiol.* 120:663–676.

MacKinnon, R., and G. Yellen. 1990. Mutations affecting TEA blockade and ion permeation in voltage-activated K^+ channels. *Science* 250:276–279.

Malin, S.A., and J.M. Nerbonne. 2002. Delayed rectifier K^+ currents, IK, are encoded by Kv2 alpha-subunits and regulate tonic firing in mammalian sympathetic neurons. *J. Neurosci.* 22:10094–10105.

Martina, M., J.H. Schultz, H. Ehmke, H. Monyer, and P. Jonas. 1998. Functional and molecular differences between voltage-gated K^+ channels of fast-spiking interneurons and pyramidal neurons of rat hippocampus. *J. Neurosci.* 18:8111–8125.

Martina, M., G.L. Yao, and B.P. Bean. 2003. Properties and functional role of voltage-dependent potassium channels in dendrites of rat cerebellar Purkinje neurons. *J. Neurosci.* 23:5698–5707.

McCrossan, Z.A., A. Lewis, G. Panaghie, P.N. Jordan, D.J. Christini, D.J. Lerner, and G.W. Abbott. 2003. MinK-related peptide 2 modulates Kv2.1 and Kv3.1 potassium channels in mammalian brain. *J. Neurosci.* 23:8077–8091.

McKay, B.E., M.L. Molineux, W.H. Mehaffey, and R.W. Turner. 2005. Kv1 K^+ channels control Purkinje cell output to facilitate postsynaptic rebound discharge in deep cerebellar neurons. *J. Neurosci.* 25:1481–1492.

Minor, D.L., Y.F. Lin, B.C. Mobley, A. Avelar, Y.N. Jan, L.Y. Jan, and J.M. Berger. 2000. The polar T1 interface is linked to conformational changes that open the voltage-gated potassium channel. *Cell* 102:657–670.

Misonou, H., D.P. Mohapatra, E.W. Park, V. Leung, D. Zhen, K. Misonou, A.E. Anderson, and J.S. Trimmer. 2004. Regulation of ion channel localization and phosphorylation by neuronal activity. *Nat. Neurosci.* 7:711–718.

Misonou, H., and J.S. Trimmer. 2004. Determinants of voltage-gated potassium channel surface expression and localization in Mammalian neurons. *Crit. Rev. Biochem. Mol. Biol.* 39:125–145.

Mitterdorfer, J., and B.P. Bean. 2002. Potassium currents during the action potential of hippocampal CA3 neurons. *J. Neurosci.* 22:10106–10115.

Morais-Cabral, J.H., Y. Zhou, and R. MacKinnon. 2001. Energetic optimization of ion conduction rate by the K^+ selectivity filter. *Nature* 414:37–42.

Morozov, A., I.A. Muzzio, R. Bourtchouladze, N. Van-Strien, K. Lapidus, D. Yin, D.G. Winder, J.P. Adams, J.D. Sweatt, and E.R. Kandel. 2003. Rap1 couples cAMP signaling to a distinct pool of p42/44MAPK regulating excitability, synaptic plasticity, learning, and memory. *Neuron* 39:309–325.

Murakoshi, H., G. Shi, R.H. Scannevin, and J.S. Trimmer. 1997. Phosphorylation of the Kv2.1 K^+ channel alters voltage-dependent activation. *Mol. Pharmacol.* 52:821–828.

Nadal, M.S., A. Ozaita, Y. Amarillo, E. Vega-Saenz de Miera, Y. Ma, W. Mo, E.M. Goldberg, Y. Misumi, Y. Ikehara, T.A. Neubert, and B. Rudy. 2003. The CD26-related dipeptidyl aminopeptidase-like protein DPPX is a critical component of neuronal A-type K^+ channels. *Neuron* 37:449–461.

Nakamura, T.Y., D.J. Pountney, A. Ozaita, S. Nandi, S. Ueda, B. Rudy, and W.A. Coetzee. 2001. A role for frequenin, a Ca^{2+}-binding protein, as a regulator of Kv4 K^+-currents. *Proc. Natl. Acad. Sci. USA* 98:12808–12813.

Neyton, J., and C. Miller. 1988. Discrete Ba^{2+} block as a probe of ion occupancy and pore structure in the high-conductance Ca^{2+}-activated K^+ channel. *J. Gen. Physiol.* 92:569–586.

Nimigean, C.M., J.S. Chappie, and C. Miller. 2003. Electrostatic tuning of ion conductance in potassium channels. *Biochemistry* 42:9263–9268.

Nishida, M., and R. MacKinnon. 2002. Structural basis of inward rectification: Cytoplasmic pore of the G protein-gated inward rectifier GIRK1 at 1.8 A resolution. *Cell* 111:957–965.

Niu, X., X. Qian, and K.L. Magleby. 2004. Linker-gating ring complex as passive spring and Ca^{2+}-dependent machine for a voltage- and Ca^{2+}-activated potassium channel. *Neuron* 42:745–756.

Ottschytsch, N., A. Raes, D. Van Hoorick, and D.J. Snyders. 2002. Obligatory heterotetramerization of three previously uncharacterized Kv channel

alpha-subunits identified in the human genome. *Proc. Natl. Acad. Sci. USA* 99:7986–7991.
Papazian, D.M., T.L. Schwarz, B.L. Tempel, Y.N. Jan, and L.Y. Jan. 1987. Cloning of genomic and complementary DNA from Shaker, a putative potassium channel gene from Drosophila. *Science* 237:749–753.
Perozo, E., D.M. Cortes, and L.G. Cuello. 1999. Structural rearrangements underlying K^+-channel activation gating. *Science* 285:73–78.
Piskorowski, R., and R.W. Aldrich. 2002. Calcium activation of BK_{Ca} potassium channels lacking the calcium bowl and RCK domains. *Nature* 420: 499–502.
Proks, P., J.F. Antcliff, and F.M. Ashcroft. 2003. The ligand-sensitive gate of a potassium channel lies close to the selectivity filter. *EMBO Rep.* 4:70–75.
Puopolo, M., B.P. Bean, and E. Raviola. 2005. Spontaneous activity of isolated dopaminergic periglomerular cells of the main olfactory bulb. *J. Neurophysiol.* 94:3618–3627.
Qi, S.Y., P.J. Riviere, J. Trojnar, J.L. Junien, and K.O. Akinsanya. 2003. Cloning and characterization of dipeptidyl peptidase 10, a new member of an emerging subgroup of serine proteases. *Biochem. J.* 373:179–189.
Ramakers, G.M., and J.F. Storm. 2002. A postsynaptic transient K^+ current modulated by arachidonic acid regulates synaptic integration and threshold for LTP induction in hippocampal pyramidal cells. *Proc. Natl. Acad. Sci. USA* 99:10144–10149.
Rettig, J., S.H. Heinemann, F. Wunder, C. Lorra, D.N. Parcej, J.O. Dolly, and O. Pongs. 1994. Inactivation properties of voltage-gated K^+ channels altered by presence of beta-subunit. *Nature* 369:289–294.
Rhodes, K.J., K.I. Carroll, M.A. Sung, L.C. Doliveira, M.M. Monaghan, S.L. Burke, B.W. Strassle, L. Buchwalder, M. Menegola, J. Cao, W.F. An, and J.S. Trimmer. 2004. KChIPs and Kv4 alpha subunits as integral components of A-type potassium channels in mammalian brain. *J. Neurosci.* 24:7903–7915.
Richter, T.A., G.A. Dvoryanchikov, N. Chaudhari, and S.D. Roper. 2004. Acid-sensitive two-pore domain potassium (K2P) channels in mouse taste buds. *J. Neurophysiol.* 92:1928–1936.
Rudy, B., A. Chow, D. Lau, Y. Amarillo, A. Ozaita, M. Saganich, H. Moreno, M.S. Nadal, R. Hernandez-Pineda, A. Hernandez-Cruz, A. Erisir, C. Leonard, and E. Vega-Saenz de Miera. 1999. Contributions of Kv3 channels to neuronal excitability. *Ann. N.Y. Acad. Sci.* 868:304–343.
Rudy, B., and C.J. McBain. 2001. Kv3 channels: Voltage-gated K^+ channels designed for high-frequency repetitive firing. *Trends Neurosci.* 24:517–526.
Sanguinetti, M.C., M.E. Curran, A. Zou, J. Shen, P.S. Spector, D.L. Atkinson, and M.T. Keating. 1996. Coassembly of K_V LQT1 and minK (IsK) proteins to form cardiac I_{Ks} potassium channel. *Nature* 384:80–83.
Scannevin, R.H., K. Wang, F. Jow, J. Megules, D.C. Kopsco, W. Edris, K.C. Carroll, Q. Lü, W. Xu, Z. Xu, A.H. Katz, S. Olland, L. Lin, M. Taylor, M. Stahl, K.

Malakian, W. Somers, L. Mosyak, M.R. Bowlby, P. Chanda, and K.J. Rhodes. 2004. Two N-terminal domains of Kv4 K^+ channels regulate binding to and modulation by KChIP1. *Neuron* 41:587–598.

Schoppa, N.E., and G.L. Westbrook. 1999. Regulation of synaptic timing in the olfactory bulb by an A-type potassium current. *Nat. Neurosci.* 2:1106–1113.

Schreiber, M., A. Wei, A. Yuan, J. Gaut, M. Saito, and L. Salkoff. 1998. Slo3, a novel pH-sensitive K^+ channel from mammalian spermatocytes. *J. Biol. Chem.* 273:3509–3516.

Schulteis, C.T., N. Nagaya, and D.M. Papazian. 1996. Intersubunit interaction between amino- and carboxyl-terminal cysteine residues in tetrameric shaker K^+ channels. *Biochemistry* 35:12133–12140.

Seoh, S.A., D. Sigg, D.M. Papazian, and F. Bezanilla. 1996. Voltage-sensing residues in the S2 and S4 segments of the Shaker K^+ channel. *Neuron* 16:1159–1167.

Sergeant, G.P., S. Ohya, J.A. Reihill, B. Perrino, G.C. Amberg, Y. Imaizumi, B. Horowitz, K.M. Sanders, and S.D. Koh. 2004. Regulation of Kv4.3 currents by Ca^{2+}-calmodulin-dependent protein kinase II. *Am. J. Physiol. Cell Physiol.* 288:C304–C313.

Sesti, F., G.W. Abbott, J. Wei, K.T. Murray, S. Saksena, P.J. Schwartz, S.G. Priori, D.M. Roden, A.L. George, and S.A. Goldstein. 2000. A common polymorphism associated with antibiotic-induced cardiac arrhythmia. *Proc. Natl. Acad. Sci. USA* 97:10613–10618.

Sesti, F., and S.A. Goldstein. 1998. Single-channel characteristics of wild-type IKs channels and channels formed with two minK mutants that cause long QT syndrome. *J. Gen. Physiol.* 112:651–663.

Shen, N.V., X. Chen, M.M. Boyer, and P.J. Pfaffinger. 1993. Deletion analysis of K^+ channel assembly. *Neuron* 11:67–76.

Shen, N.V., and P.J. Pfaffinger. 1995. Molecular recognition and assembly sequences involved in the subfamily-specific assembly of voltage-gated K^+ channel subunit proteins. *Neuron* 14:625–633.

Shepard, A.R., and J.L. Rae. 1999. Electrically silent potassium channel subunits from human lens epithelium. *Am. J. Physiol.* 277:C412–C424.

Shibata, R., H. Misonou, C.R. Campomanes, A.E. Anderson, L.A. Schrader, L.C. Doliveira, K.I. Carroll, J.D. Sweatt, K.J. Rhodes, and J.S. Trimmer. 2003. A fundamental role for KChIPs in determining the molecular properties and trafficking of Kv4.2 potassium channels. *J. Biol. Chem.* 278:36445–36454.

So, I., I. Ashmole, N.W. Davies, M.J. Sutcliffe, and P.R. Stanfield. 2001. The K^+ channel signature sequence of murine Kir2.1: Mutations that affect microscopic gating but not ionic selectivity. *J. Physiol.* 531:37–50.

Sokolova, O., A. Accardi, D. Gutierrez, A. Lau, M. Rigney, and N. Grigorieff. 2003. Conformational changes in the C terminus of Shaker K^+ channel bound to the rat Kvbeta2-subunit. *Proc. Natl. Acad. Sci. USA* 100:12607–12612.

Soler-Llavina, G.J., M. Holmgren, and K.J. Swartz. 2003. Defining the conductance of the closed state in a voltage-gated K^+ channel. *Neuron* 38:61–67.

Song, P., Y. Yang, M. Barnes-Davies, A. Bhattacharjee, M. Hamann, I.D. Forsythe, D.L. Oliver, and L.K. Kaczmarek. 2005. Acoustic environment determines phosphorylation state of the Kv3.1 potassium channel in auditory neurons. *Nat. Neurosci.* 8:1335–1342.

Starkus, J.G., L. Kuschel, M.D. Rayner, and S.H. Heinemann. 1997. Ion conduction through C-type inactivated Shaker channels. *J. Gen. Physiol.* 110:539–550.

Swensen, A.M., and B.P. Bean. 2003. Ionic mechanisms of burst firing in dissociated Purkinje neurons. *J. Neurosci.* 23:9650–9663.

Tai, K.K., and S.A. Goldstein. 1998. The conduction pore of a cardiac potassium channel. *Nature* 391:605–608.

Tarasov, A., J. Dusonchet, and F. Ashcroft. 2004. Metabolic regulation of the pancreatic beta-cell ATP-sensitive K^+ channel: A pas de deux. *Diabetes* 53(Suppl 3):S113–S122.

Trapani, J.G., P. Andalib, J.F. Consiglio, and S.J. Korn. 2006. Control of single channel conductance in the outer vestibule of the Kv2.1 potassium channel. *J. Gen. Physiol.* 128(2):231–246.

Trapani, J.G., and S.J. Korn. 2003. Control of ion channel expression for patch clamp recordings using an inducible expression system in mammalian cell lines. *BMC Neurosci.* 4:15.

Tristani-Firouzi, M., J. Chen, and M.C. Sanguinetti. 2002. Interactions between S4–S5 linker and S6 transmembrane domain modulate gating of HERG K^+ channels. *J. Biol. Chem.* 277:18994–19000.

Tzounopoulos, T., H.R. Guy, S. Durell, J.P. Adelman, and J. Maylie. 1995. Min K channels form by assembly of at least 14 subunits. *Proc. Natl. Acad. Sci. USA* 92:9593–9597.

Valverde, P., T. Kawai, and M.A. Taubman. 2005. Potassium channel-blockers as therapeutic agents to interfere with bone resorption of periodontal disease. *J. Dent. Res.* 84:488–499.

Wang, K.W., K.K. Tai, and S.A. Goldstein. 1996. MinK residues line a potassium channel pore. *Neuron* 16:571–577.

Wang, W., J. Xia, and R.S. Kass. 1998. MinK-KvLQT1 fusion proteins, evidence for multiple stoichiometries of the assembled IsK channel. *J. Biol. Chem.* 273:34069–34074.

Wang, Z., J.C. Hesketh, and D. Fedida. 2000. A high-Na^+ conduction state during recovery from inactivation in the K^+ channel Kv1.5. *Biophys. J.* 79:2416–2433.

Watanabe, S., D.A. Hoffman, M. Migliore, and D. Johnston. 2002. Dendritic K^+ channels contribute to spike-timing dependent long-term potentiation in hippocampal pyramidal neurons. *Proc. Natl. Acad. Sci. USA* 99:8366–8371.

Wei, D.S., Y.A. Mei, A. Bagal, J.P. Kao, S.M. Thompson, and C.M. Tang. 2001. Compartmentalized and binary behavior of terminal dendrites in hippocampal pyramidal neurons. *Science* 293:2272–2275.

Wood, M.J., and S.J. Korn. 2000. Two mechanisms of K^+-dependent potentiation in Kv2.1 potassium channels. *Biophys. J.* 79:2535–2546.

Xia, X.M., X. Zeng, and C.J. Lingle. 2002. Multiple regulatory sites in large-conductance calcium-activated potassium channels. *Nature* 418:880–884.

Xia, X.M., X. Zhang, and C.J. Lingle. 2004. Ligand-dependent activation of Slo family channels is defined by interchangeable cytosolic domains. *J. Neurosci.* 24:5585–5591.

Yellen, G. 1998. The moving parts of voltage-gated ion channels. *Q Rev. Biophys.* 31:239–295.

Yellen, G., D. Sodickson, T.Y. Chen, and M.E. Jurman. 1994. An engineered cysteine in the external mouth of a K^+ channel allows inactivation to be modulated by metal binding. *Biophys. J.* 66:1068–1075.

Yifrach, O., and R. MacKinnon. 2002. Energetics of pore opening in a voltage-gated K^+ channel. *Cell* 111:231–239.

Yuan, L.L., J.P. Adams, M. Swank, J.D. Sweatt, and D. Johnston. 2002. Protein kinase modulation of dendritic K^+ channels in hippocampus involves a mitogen-activated protein kinase pathway. *J. Neurosci.* 22:4860–4868.

Zagotta, W.N., T. Hoshi, and R.W. Aldrich. 1990. Restoration of inactivation in mutants of Shaker potassium channels by a peptide derived from ShB. *Science* 250:568–571.

Zerangue, N., Y.N. Jan, and L.Y. Jan. 2000. An artificial tetramerization domain restores efficient assembly of functional Shaker channels lacking T1. *Proc. Natl. Acad. Sci. USA* 97:3591–3595.

Zhang, M., M. Jiang, and G.N. Tseng. 2001. MinK-related peptide 1 associates with Kv4.2 and modulates its gating function: Potential role as beta subunit of cardiac transient outward channel? *Circ. Res.* 88:1012–1019.

Zheng, J., and F.J. Sigworth. 1997. Selectivity changes during activation of mutant Shaker potassium channels. *J. Gen. Physiol.* 110:101–117.

Zheng, J., and F.J. Sigworth. 1998. Intermediate conductances during deactivation of heteromultimeric Shaker potassium channels. *J. Gen. Physiol.* 112:457–474.

Zhou, M., and R. MacKinnon. 2004. A mutant KcsA K^+ channel with altered conduction properties and selectivity filter ion distribution. *J. Mol. Biol.* 338:839–846.

Zhou, M., J.H. Morais-Cabral, S. Mann, and R. MacKinnon. 2001a. Potassium channel receptor site for the inactivation gate and quaternary amine inhibitors. *Nature* 411:657–661.

Zhou, Y., and R. MacKinnon. 2003. The occupancy of ions in the K^+ selectivity filter: Charge balance and coupling of ion binding to a protein conformational change underlie high conduction rates. *J. Mol. Biol.* 333:965–975.

Zhou, Y., J.H. Morais-Cabral, A. Kaufman, and R. MacKinnon. 2001b. Chemistry of ion coordination and hydration revealed by a K^+ channel-Fab complex at 2.0 A resolution. *Nature* 414:43–48.

Zilberberg, N., N. Ilan, R. Gonzalez-Colaso, and S.A. Goldstein. 2000. Opening and closing of KCNKO potassium leak channels is tightly regulated. *J. Gen. Physiol.* 116:721–734.

5 BK_{Ca}-Channel Structure and Function

Daniel H. Cox

5.1 Introduction

Among ion channels, the large-conductance Ca^{2+}-activated K^+ channel (BK_{Ca} channel) is in many ways unique. It has a very large single-channel conductance—ten times that of most vertebrate K^+ channels—and yet it maintains strict K^+ selectivity. It senses as little as 200 nM Ca^{2+}, but it contains no consensus Ca^{2+}-binding motifs, and it is the only channel to be activated by both intracellular Ca^{2+} and membrane voltage. In fact, there is a synergy between these stimuli such that the higher the internal Ca^{2+} concentration ($[Ca^{2+}]$), the smaller the depolarization needed to activate the channel. Furthermore, the BK_{Ca} channel has its own brand of auxiliary subunits that profoundly affect gating. In this chapter, I will discuss what is understood about the origins of these properties in terms of allosteric models and channel structure. At the outset, however, I should say that there is not yet a crystal structure of the BK_{Ca} channel or any of its components, so much of the current thinking about BK_{Ca}-channel structure relies on analogy to other channels.

5.2 BK_{Ca}-Channel Topology

BK_{Ca} channels are formed by pore-forming α subunits and in some tissues auxiliary β subunits. Both are integral membrane proteins. Four α subunits alone form a fully functional channel (Adelman et al., 1992; Shen et al., 1994), while β subunits play a modulatory role (McManus et al., 1995; Orio et al., 2002; Fig. 5.1A). cDNAs encoding a BK_{Ca}-channel α subunit were first identified as the drosophila mutant slowpoke (*slo*) (Atkinson et al., 1991; Adelman et al., 1992). Taking advantage of sequence homologies, *slo* cDNAs were then cloned from many species including human (Butler et al., 1993; Knaus et al., 1994a; Pallanck and Ganetzky 1994; Tseng-Crank et al., 1994; McCobb et al., 1995), and three *slo*-related genes were found. None, however, encode for a Ca^{2+}-activated channel—*slo2.1* (Bhattacharjee et al., 2003) and *slo2.2* (Yuan et al., 2000) encode for Na-activated K^+ channels, and *slo3* (Schreiber et al., 1998) encodes for a pH-sensitive K^+ channel. Thus, the original *slo* gene, now termed *slo1, KCNMA1*, or K_{Ca} *1.1*, is the only BK_{Ca}-channel gene, and ignoring splice variation mammalian Slo1 proteins share greater than 95% amino acid identity. This strikingly high degree of homology suggests that there has been strong evolutionary pressure to maintain the functioning of these channels within

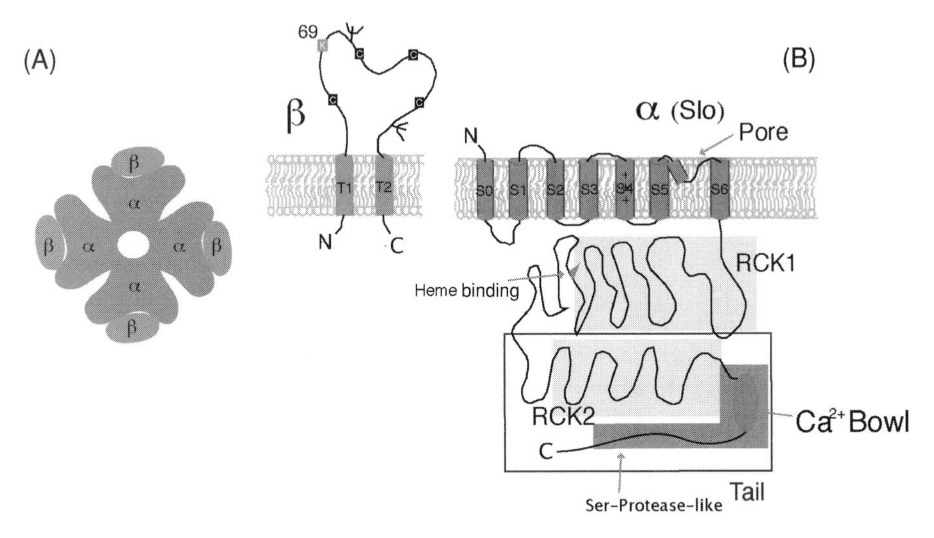

Fig. 5.1 BK_{Ca} channels are composed of α and β subunits. (A) Schematic diagram of a BK_{Ca} channel viewed from the top down. The channel contains four α and four β subunits. (B) Putative membrane topologies of the BK_{Ca} α and β subunits.

narrow tolerances. Mammalian Slo1 channels are blocked by the classic BK_{Ca}-channel blockers charybdotoxin, iberiotoxin, and tetraethylammonium (Butler et al., 1993; Dworetzky et al., 1996).

Shown in Fig. 5.1B are the proposed membrane topologies of the BK_{Ca} α and β subunits. The α subunit (~1200 amino acids) is arranged much like a purely voltage-gated (K_V-type) K^+ channel subunit, complete with an amphipathic S4 helix that likely forms the channel's voltage-sensor (Bezanilla, 2005), and a K^+-channel pore sequence (Heginbotham et al., 1992). Different from K_v channel subunits, however, the Slo1 subunit has an extra transmembrane segment at its N-terminus, termed S0, that places its N-terminus outside the cell. This was demonstrated by Meera et al. (1997) who showed that an antibody applied extracellularly can bind to an epitope placed at the N-terminus of human Slo1(hSlo1), even if cells transfected with hSlo1 are not permeabilized. The purpose of this BK_{Ca}-specific adaptation is not known, but a study of chimeric channels containing portions of drosophila Slo1 (dSlo1) and mouse Slo1 (mSlo1) has implicated S0 as the sight where α and β subunits interact (Wallner et al., 1996). Perhaps allowing for this interaction is S0's primary function.

The Slo1 subunit also contains a large (~800 amino acid) C-terminal extension that constitutes two thirds of its sequence. This extension contains four somewhat hydrophobic segments that were originally thought to be membrane spanning; however, immunohistochemical and in vitro translation experiments have shown them to be intracellular (Meera et al., 1997). For the purposes of discussion Slo1's cytoplasmic domain may be divided into proximal and distal portions, with a region of low conservation across species marking the division (Butler et al., 1993). It has been argued that the proximal portion contains a domain found in bacterial K^+ channels and transporters termed an RCK domain (Jiang et al., 2001). The distal portion, also known as the "tail", contains perhaps another RCK domain (Jiang et al., 2002) and a domain that very likely forms a Ca^{2+}-binding site termed the "Ca^{2+} bowl" (Schreiber and Salkoff, 1997; Bian et al., 2001; Bao et al., 2002, 2004; Niu et al., 2002; Xia et al., 2002). I will discuss the RCK and Ca^{2+} bowl domains later, in the context of Ca^{2+}-sensing.

The Slo1 subunit also has a region that binds heme, and nanomolar heme dramatically inhibits channel opening (Tang et al., 2003). The physiological significance of this interaction, however, has yet to be established. Nor in fact has the significance of the observation that part of Slo1's tail domain has homology to serine proteases (Moss et al., 1996). Both of these regions are indicated in Fig. 1B.

BK_{Ca} β subunits (there are now 4) are much smaller than the α subunit (191–279 amino acids). They have two membrane-spanning domains, intracellular N and C termini, and a fairly large extracellular loop (116–128 amino acids) (Knaus et al., 1994a; Lu et al., 2006). This loop contains four cysteine residues that form two disulfide bridges, the pairings of which are unclear (Hanner et al., 1998), and in $\beta 1$ a lysine residue (K69) which crosslinks to the external pore-blocker charybdotoxin (Manujos et al., 1995). Four β subunits can interact with a single BK_{Ca} channel (Wang et al., 2002). BK_{Ca} β subunits will be discussed again later in the context of their functional effects.

5.3 The Origin of the BK_{Ca} Channel's Large Conductance

The most striking property of the BK_{Ca} channel is its very large single-channel conductance. In symmetrical 150 mM K^+ the BK_{Ca} channel has a conductance of 290 pS (Cox et al., 1997a), while other vertebrate K^+ channels have conductances ranging from 2 to 50 pS (Hille, 1992). The prototypical Shaker K^+ channel, for example, has a conductance of 25 pS under these conditions (Heginbotham and MacKinnon, 1993), which means at +50 mV it passes 7.8 million ions per second, while the BK_{Ca} channel passes 11.6 times more, or 90 million ions per second. Interestingly, however, the BK_{Ca} channel is not less selective for K^+ than Shaker. For every 100 K^+ ions allowed to pass through either channel fewer that 1 Na^+ ion is allowed to pass (Blatz and Magleby, 1984; Yellen, 1984; Heginbotham and MacKinnon, 1993). How is it that the BK_{Ca} channel can maintain this high degree of selectivity and yet pass ions much more quickly that other K^+ channels?

A priori one might suppose that it has to do with the structure of the selectivity filter, the narrowest part of the pore, which interacts most closely with the permeating K^+ ions and therefore might reasonably be rate limiting. Crystal structures, however, have been determined now for five K^+ channels: KcsA (Doyle et al., 1998), KvAP (Jiang et al., 2003), MthK (Jiang et al., 2002), KirBac1.1 (Kuo et al., 2003), and Kv1.2 (Long et al., 2005), and despite varying conductances, the selectivity filters of these channels are identical. Each spans about a third of the way

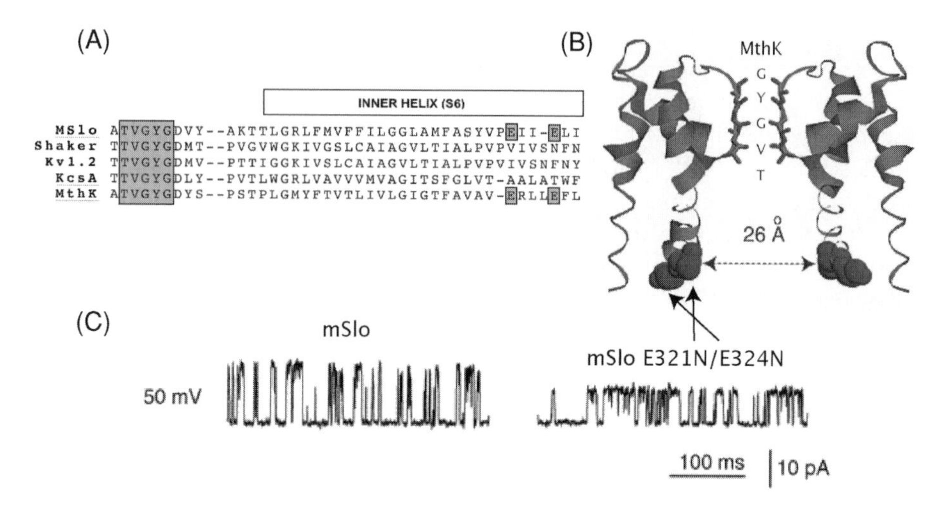

Fig. 5.2 Rings of negative charge increase the BK_{Ca} channel's conductance. (A) Pore sequences of four K^+ channels. All four have the K^+-channel signature sequence indicated in grey, and mSlo and MthK have two acidic glutamates near the end of S6 shown also in grey. (B) Ribbon diagram of the crystal structure of the pore region of MthK. Indicated in space fill are the glutamates highlighted in *A*. (C) Unitary currents from a wild-type mSlo1 channel and a mSlo1 channel that contains the double mutation E321N/E324N. The size of the channel's unitary current is reduced by half by the mutations. The figures in *A, B, C*, were adapted from Nimigean et al. (2003).

through the bilayer starting from the outside, and the same "signature sequence" TVGYG (Heginbotham et al., 1992) provides carbonyl oxygens that form four K^+ binding sites (Fig. 5.2A and B). K^+ ions pass through this filter in single file separated by water molecules (MacKinnon, 2003). The BK_{Ca} channel also contains the "TVGYG" signature sequence, and so does the Shaker channel, so it seems unlikely that it is differences in this structure that account for the BK_{Ca} channel's unusually large conductance. Similarly, all the crystallized K^+ channels have wide shallow external mouths that one would suppose would provide good diffusional access to ions entering from the outside (Fig. 5.2B). Indeed, both the BK_{Ca} channel and the Shaker channel (after a point mutation is made) can be blocked from the outside by the pore-plugging toxin charybdotoxin (MacKinnon and Miller, 1988; MacKinnon et al., 1990; Goldstein and Miller, 1992; Stocker and Miller, 1994), which makes specific contacts with the external mouth, so both channels must have similar external architectures.

Given these observations it seems likely that the large conductance of the BK_{Ca} channel arises from modifications inside the channel, on the cytoplasmic side of the selectivity filter. Recent experiments support this notion, and two mechanisms appear to be at play. First, using the crystal structure of the open MthK channel as a guide (Fig. 5.2B; Jiang et al., 2002), it appears that the BK_{Ca} channel has two rings of negatively charged glutamate residues on the inner pore-helix, just at the internal entrance of the channel, and Nimigean et al. (2003) and Brelidze et al. (2003) have shown that when these residues are neutralized to glutamine, the outward conductance of the BK_{Ca} channel decreases by half (Fig. 5.2C). Thus, part of the reason for the large conductance of the BK_{Ca} channel is rings of negative charge at the inner mouth of the channel. These rings attract K^+ creating a ~3.3-fold increase in the local $[K^+]$—from 150 mM to 500 mM in the experiments of Brelidze et al. (2003)—and this increases the rate at which K^+ ions encounter the inner mouth of the selectivity filter. In fact, at very high internal $[K^+]$ (over 1 M) neutralizing these charges has no effect on the channel's conductance (Brelidze et al., 2003) presumably because the local $[K^+]$ is very high under this condition, even without the negatively charged rings, and the rate of K^+ diffusion to the inner entrance of the selectivity filter is therefore no longer rate limiting (Brelidze et al., 2003).

These rings of negative charge, however, must not be the whole story, as their neutralization only reduces the outward K^+ flux by half (at 150 mM K^+ on both side of the membrane) (Brelidze et al., 2003; Nimigean et al., 2003), leaving a channel with still a very large 150 pS conductance. Another mechanism must also be involved, and it seems to be that the internal vestibule and internal mouth are larger in the BK_{Ca} channel than they are in other vertebrate K^+ channels. This idea rests on four observations. First, Li and Aldrich (2004) showed that large quaternary ammonium compounds diffuse much more rapidly into the BK_{Ca} channel from the inside, on their way to blocking the channel, than they do into the Shaker channel (Li and Aldrich, 2004). Second, once they are inside, the BK_{Ca} channel can close behind them, while the Shaker channel cannot (Li and Aldrich, 2004). Third, when Brelidze and Magleby (2005) increased the concentration of sucrose on the internal side of the channel to slow diffusion up to and through the inner vestibule, they found that the BK_{Ca} channel's conductance was reduced in a manner that indicated that the rate of motion of K^+ through the inner vestibule is an important determinant of conductance. And four, based on the amount of sucrose needed to make K^+ diffusion from bulk solution to the internal mouth of the channel rate limiting, Brelidze and Magleby (2005) estimated that the BK_{Ca} channel's internal mouth is twice as large (20 Å in diameter) as that of the Shaker channel (Webster et al., 2004; Brelidze and Magleby, 2005) and similar in size to the large-conductance (~2004 pS) MthK channel (Jiang et al., 2002) (Fig. 5.2B). Thus, controlling the size and shape of the portion of the pore that is internal to the selectivity filter, and strategically placing charges there to attract K^+ ions, appears to be how nature has produced K^+ channels of differing conductances but common selectivities.

The large conductance of the BK_{Ca} channel indicates that the K^+-channel selectivity filter is exquisitely designed for both selectivity and high throughput, and it suggests that the selectivity filter is not working at maximum capacity in K^+ channels with lower conductances. It is interesting to consider, however, whether as the inner mouth and vestibule of a K^+ channel becomes larger, does more of the membrane potential drop across the selectivity filter. And if so, does the increased electric-field strength drive ions through the selectivity filter more quickly than they can go through in smaller-conductance channels? Perhaps this is another reason for the enhanced conductance of the BK_{Ca} channel? As far as I am aware this remains an open question?

5.4 BK_{Ca}-Channel Gating, Studies Before Cloning

Because of their large conductance, with the advent of the patch-clamp technique BK_{Ca} channels were among the first channels to be studied at the single channel level, and throughout the 1980s several rigorous studies of their gating behavior were performed (Methfessel, and Boheim, 1982; Magleby and Pallotta, 1983; Moczydlowski, and Latorre 1983; Pallotta, 1983; McManus and Magleby, 1988, 1991; Oberhauser et al., 1988; Latorre et al., 1989; McManus, 1991; Markwardt and Isenberg, 1992). In brief they lead to the following important observations: BK_{Ca} channels can be activated both by Ca^{2+} and voltage, and they can be activated by voltage to high open probabilities over a wide range of [Ca^{2+}]. Hill coefficients for Ca^{2+}-dependent activation are usually observed to be greater than 1 and sometimes as high as 5 (Golowasch et al., 1986). More typically, however, they are between 1 and 4 suggesting that at least four Ca^{2+} ions bind to the channel and they act to some degree cooperatively when influencing opening. Analysis of the dwell times of the channel in open and closed states indicates that the channel can occupy at least five closed states and three open states, and that there are multiple paths between closed and open. These observations lead McManus and Magleby in 1991 to propose the following model of BK_{Ca}-channel Ca^{2+}-dependent gating at +30 mV.

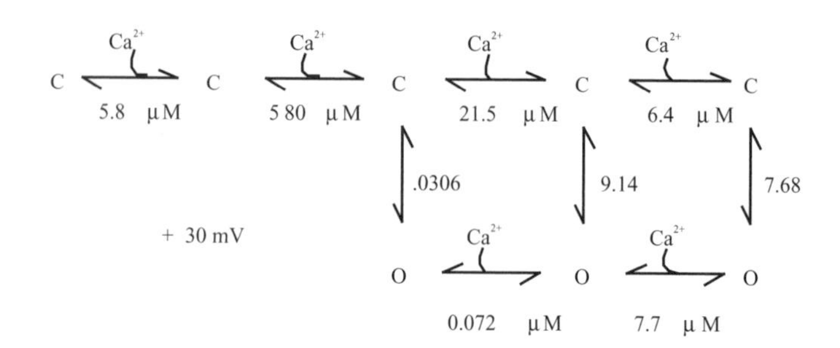

Scheme I

Here the channel binds four Ca^{2+} ions, and there is a single conformational change between closed and open. Each Ca^{2+}-binding step has unique rate constants and therefore also a unique affinity constant (as indicated). After binding the second Ca^{2+} ion, the model channel is observed to open with a detectable frequency, but it is the large difference in affinity between the open and closed conformations for the third bound Ca^{2+} that drives channel opening. As will be discussed below, although the unique importance of the third binding event is no longer widely accepted, the idea that the channel undergoes a single conformational change allosterically regulated by multiple Ca^{2+}-binding events remains an important part of all current models of BK_{Ca}-channel gating.

5.5 BK_{Ca}-Channel Gating, Macroscopic Current Properties

The cloning of *slo1* made it possible to express BK_{Ca} channels at high density in heterologous expression systems and to study their behavior as a population. Shown in Fig. 5.3A are mouse *Slo1* (mSlo1) macroscopic currents recorded from an excised inside-out patch from a *Xenopus* oocyte. The internal face of this patch was exposed to 10 μM Ca^{2+}, and the membrane potential was stepped from −80 mV, where the channels were all closed, through a series of increasingly more positive potentials, and then back again to −80 mV. From these data one can determine the

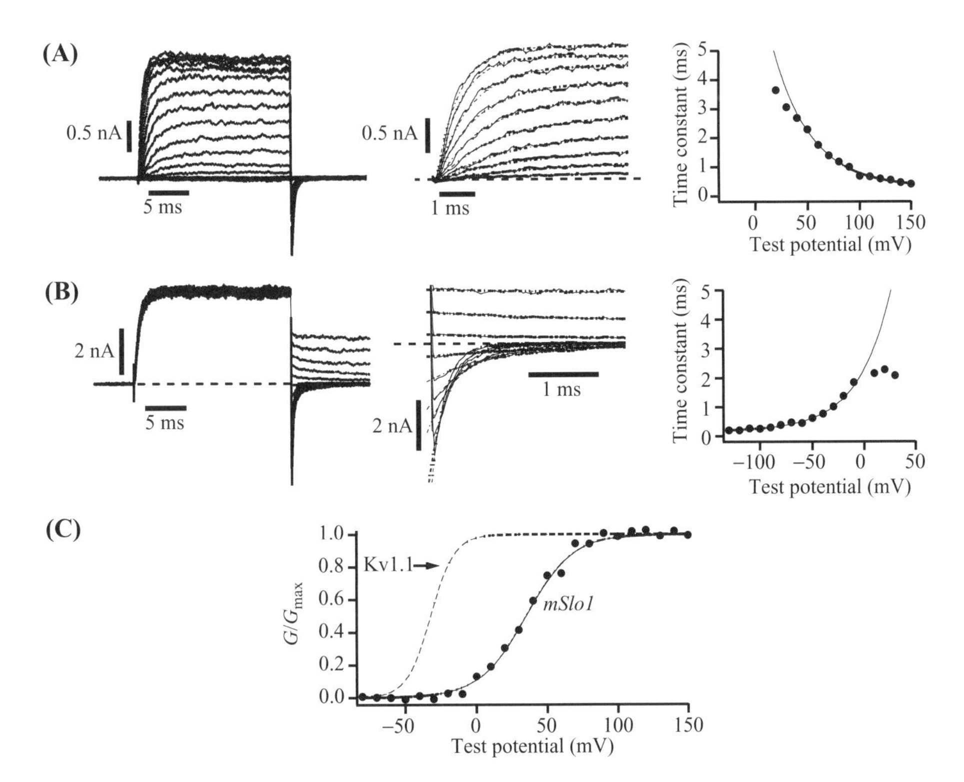

Figure 5.3 mSlo1 macroscopic currents recorded from *Xenopus* oocyte, inside-out macropatches. The internal face of each patch was exposed to 10 μM [Ca^{2+}]. (A) (left) 20 ms pulses from −80 mV to between −80 and +150 mV in 10 mV increments. (middle) The traces on the left have been expanded and fitted with exponential functions. (right) Time constants of activation plotted as a function of voltage. (B) (left) Tail currents recorded in response to repolarization to a series of potentials after depolarization to +100 mV. (middle) The traces on the left have been expanded and fitted with exponential functions. (right) Time constants of deactivation plotted as a function of voltage. (C) Normalized conductance vs. voltage curve (*G–V* curve) determined from the data in *A*. The data have been fitted with a Boltzmann function ($G/G_{\max} = 1/(1 + e^{qF(V_{1/2}-V)/RT})$ with $V_{1/2} = 24.6$, $q = 1.54$. Also indicated is the *G–V* curve of the mammalian Kv.1.1 channel (Grissmer et al. (1994). This figure was adapted from figure 1 of Cui et al. (1997).

conductance of the membrane as a function of voltage (Fig. 5.3C), and the time course of channel activation (Fig. 5.3A). Similarly, by activating the channels with a prepulse to +100 mV and then stepping back to various potentials, the time course of deactivation can be examined (Fig. 5.3B). Plotted in Fig. 5.3C is the conductance of the mSlo1 channel, relative to its maximum (G/G_{max}), as a function of test potential. As is evident the channel is significantly voltage dependent and exhibits an e-fold change in conductance per 16.5 mV. This corresponds to an apparent gating charge of ~1.54e, which is significant but substantially less than that of the prototypical voltage-dependent K^+ channel, Shaker, or its mammalian homologue Kv1.1 (dashed line).

In Fig. 5.3A (center) the traces from the left have been expanded and the activation time courses fitted with single exponentials. The surprising result here is that, over the entire voltage range, the traces appear well fitted by this simple function. Furthermore, the time course of deactivation is also well fitted by a single exponential at a series of potentials (Fig. 5.3B) (DiChiara and Reinhart, 1995; Cui et al., 1997). Given the complex behavior observed at the single channel level, and the eight states in Scheme I, this is a surprise, as in principle a kinetic scheme with eight states will relax with a time course described by seven time constants. Some of these could be small, and some could be similar to one another, so reasonably one might expect to see fewer than seven, but more than one. Even more striking, however, if one performs the same experiments at a variety of [Ca^{2+}] (Fig. 5.4), increasing [Ca^{2+}] speeds activation (Fig. 5.4A) and slows deactivation (Fig. 5.4B), but single exponential relaxations are still observed throughout. This result suggests that there is a single rate-limiting conformational change that is influenced by voltage and [Ca^{2+}] and dominates the kinetics of channel gating over a wide range of conditions (Cui et al., 1997). To be strictly correct, however, I should say that over a wide range of conditions the kinetics of activation follow an exponential time course but for a very brief delay, on the order of 200 μs (Cui et al., 1997; Stefani et al., 1997; Horrigan et al., 1999). The significance of this delay will be made more apparent below.

If one looks at the effect of [Ca^{2+}] on the steady-state gating properties of the mSlo1 channel, one sees that Ca^{2+} shifts the mSlo1 G–V curve leftward along the voltage axis, with for the most part little change in shape (Fig. 5.5A). This effect is quite dramatic. The shift starts at ~100 nM [Ca^{2+}] (Meera et al., 1996; Cox and Aldrich, 2000) and continues at concentrations as high as 10 mM, having shifted by this point over 200 mV ([Ca^{2+}] up to only 1000 μM are shown). Thus, the mSlo1 channel's response to [Ca^{2+}] spans five orders of magnitude, and one might reasonably ask how the channel is able to respond to such a wide range of [Ca^{2+}]? And what determines the magnitude of the channel's G–V shift as a function of [Ca^{2+}]? The studies described below address these questions.

If one takes the steady-state data of Fig. 5.5A and turns it on its head, plotting now G/G_{max} as a function of [Ca^{2+}] at several voltages (Fig. 5.5B), one sees that the apparent affinity of the channel for Ca^{2+} ranges from less than 1 μM to greater than 300 μM and is steeply voltage dependent. Early single-channel work suggested that this behavior arises from voltage-dependent Ca^{2+} binding (Moczydlowski and

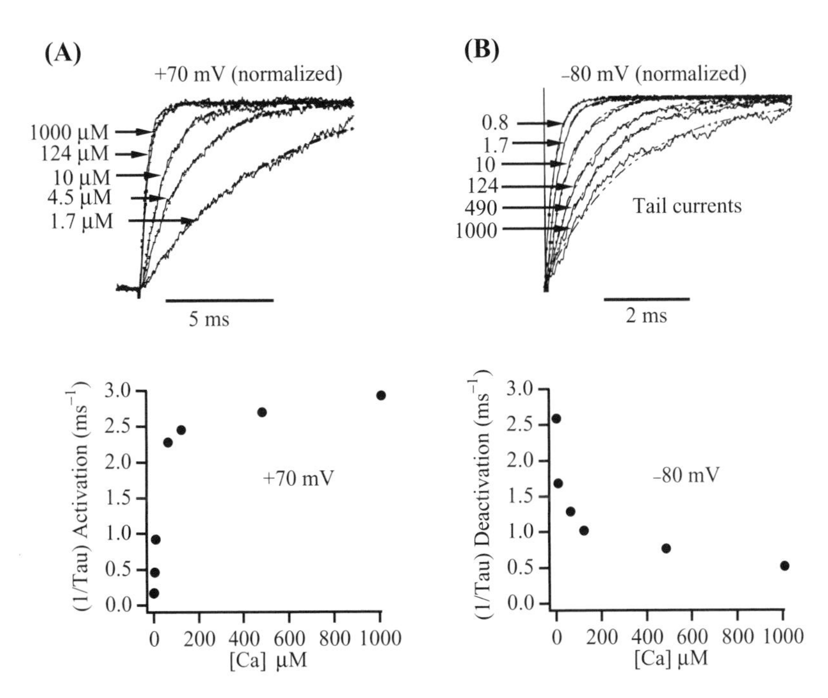

Figure 5.4 Activation rates increase and deactivation rates decrease with increasing $[Ca^{2+}]$. (A) Currents recorded after voltage steps to +70 mV were normalized to their maxima and superimposed. Each curve is fitted with an exponential function, and $1/\tau$ is plotted in the lower panel as a function $[Ca^{2+}]$. (B) Tail currents were recorded at −80 mV, after depolarizations to +100 mV. These currents were then normalized to their minima and superimposed. Each curve is fitted with an exponential function and $1/\tau$ is plotted as a function of $[Ca^{2+}]$ in the lower panel. This figure was adapted from figure 6 of Cui et al.(1997).

Latorre, 1983), however, with the cloning of *slo1* it became clear that the Slo1 channel can be near maximally activated in the essential absence of Ca^{2+} with very strong depolarizations (> +300 mV) (Meera el al., 1996; Cui et al., 1997), and that the rate of channel activation at very low $[Ca^{2+}]$ is too fast to be due to Ca^{2+}-binding with high affinity after the voltage step (Cui et al., 1997). Thus, the BK_{Ca} channel is a voltage-gated channel that is modulated by Ca^{2+} binding.

5.6 A Simple Model of BK_{Ca}-Channel Gating

The properties discussed above—a tetrameric channel (Shen et al., 1994), multiple Ca^{2+} binding sites, a single rate-limiting conformational change between open and closed influenced by both Ca^{2+} binding and membrane voltage—lead naturally too, and can be accounted for in great part by a simple model of gating known as the voltage-dependent Monaux Wyman Changeux model, or VD-MWC model (Cui et al., 1997; Cox et al., 1997b). While better and more complex models of BK_{Ca}-channel gating exist and will be discussed below, the simplicity and mathematical

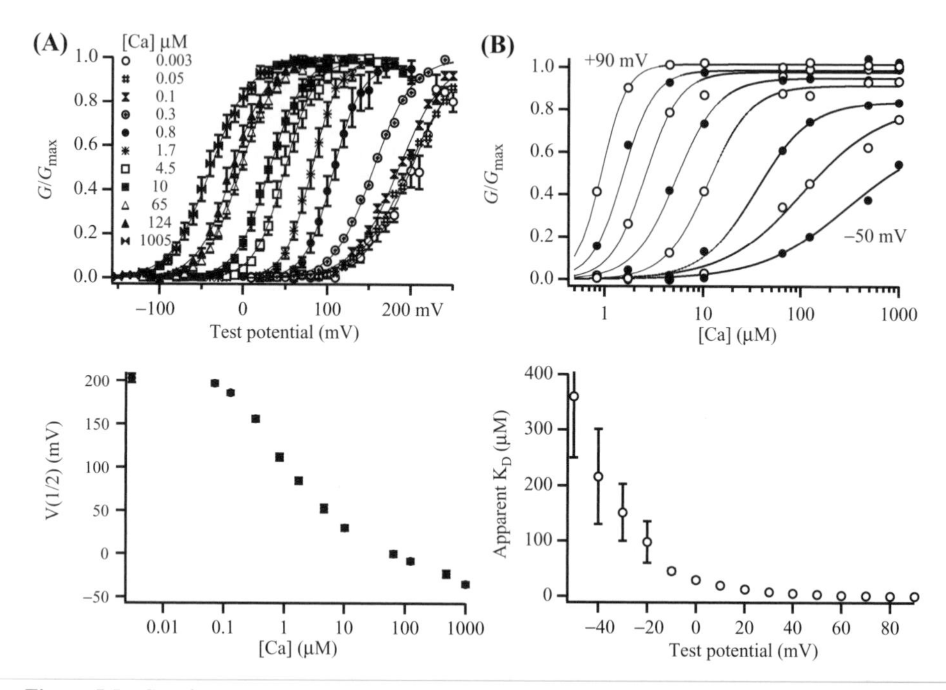

Figure 5.5 Steady-state gating properties of mSlo1 macroscopic currents as a function of voltage and $[Ca^{2+}]$. (A) mSlo1 *G–V* relations determined at the $[Ca^{2+}]$ indicated. Each curve has been fitted with a Boltzmann function. In the lower panel $V_{1/2}$ from the Boltzmann fits is plotted as a function of $\log[Ca^{2+}]$. (B) Data like that in *A* were converted to Ca^{2+} dose–response curves. The curves displayed were determined at –50 to +90 mV, in 20 mV increments. Each curve has been fitted with the Hill equation (Eq. 5.6), and the $K_{D\text{-Apparent}}$ is plotted as a function of voltage in the lower panel. The figures in *A* and *B*, were adapted from Cui et al. (1997).

tractability of the VD-MWC model make it a useful tool by which to gain and intuitive understanding of the properties of an ion channel modulated by both ligand binding and membrane voltage. The VD-MWC model is represented by Scheme II below.

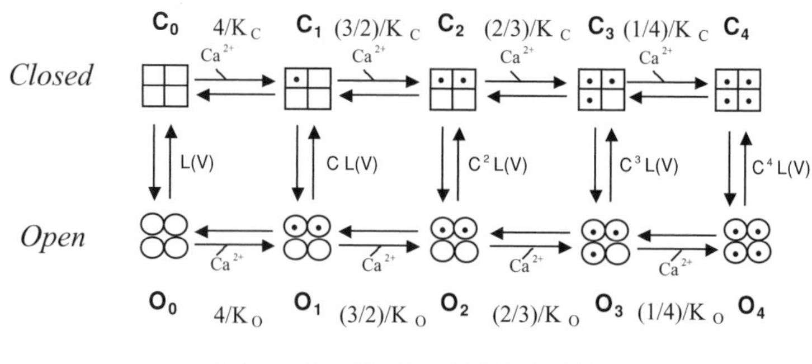

Scheme II (The VD-MWC Model)

Horizontal steps represent Ca^{2+}-binding events, each with dissociation constant K_C in the closed conformation and K_O in the open conformation. Vertical steps represent the concerted conformational change by which the channel opens. The equilibrium constant between open and closed in the absence of Ca^{2+} is referred to as L. The MWC model without voltage dependence was first formulated to describe four oxygen molecules binding to homotetrameric hemoglobin (Monod et al., 1965). It is easily adapted to the homotetrameric BK$_{Ca}$ channel supposing now four Ca^{2+}-binding sites one in each subunit (Cui et al., 1997; Cox et al., 1997b).

Important properties of the MWC model are as follows. In response to ligand-binding individual subunits do not undergo conformational changes on their own, but rather all subunits undergo a conformational change together that is coincident with channel opening. All binding sites are assumed to be identical. The binding of Ca^{2+} at one site does not affect the affinities of neighboring sites except indirectly via promoting opening. And, in order for ligand binding to promote opening the open conformation of the channel must bind Ca^{2+} more tightly than the closed conformation. We may rely on the law of detailed balance (which does apply to the BK$_{Ca}$ channel; McManus and Magleby, 1989) to see why this is so. This law states that for any cyclic gating scheme the product of equilibrium constants on any path between two given states must be equivalent. Considering then the paths between states C_0 and O_1 in Scheme II, we may write

$$\frac{4[\mathrm{Ca}]}{K_C}X = L\frac{4[\mathrm{Ca}]}{K_O}, \text{ which can be rearranged to } X = L\frac{K_C}{K_O}, \tag{5.1}$$

where X is the equilibrium constant between C_1 and O_1. Thus, in the MWC model each ligand-binding event alters the equilibrium constant between closed and open by a factor $C = K_C/K_O$, a situation with two interesting consequences. First, if $K_C = K_O$, Ca^{2+} binding will occur, but it will not produce opening, and second, the effect that Ca^{2+} binding has on channel opening can be increased by either decreasing K_O , that is making the open channel bind Ca^{2+} more tightly, or increasing K_C, making the closed channel bind Ca^{2+} more weakly. The apparent affinity of the channel, therefore, can increase as a result of a true decrease in binding affinity, a result that demonstrates that even the simplest of ligand-dependent gating systems can behave counter intuitively.

The behavior of the MWC model can be made more intuitive by considering the effect of each binding event in energetic terms. Shown in Fig. 5.6 is the MWC model plotted as an energy diagram, with energy on the vertical. Here K_C was assumed to be 10 μM and K_O 1 μM, and for simplicity L was assumed to be 1 such that the energies of the closed and open conformations of the channel are equal in the absence of bound Ca^{2+}. Each Ca^{2+}-binding event then reduces the energy of the closed channel by $\Delta G_C = RT \ln K_C$, or -28.3 KJ mol^{-1}, and it reduces the energy of the open channel by $\Delta G_O = RT \ln K_O$, or -34.0 KJ mol^{-1}, such that each Ca^{2+}-binding event tips the energetic balance toward opening by -5.7 KJ mol^{-1}. Thus,

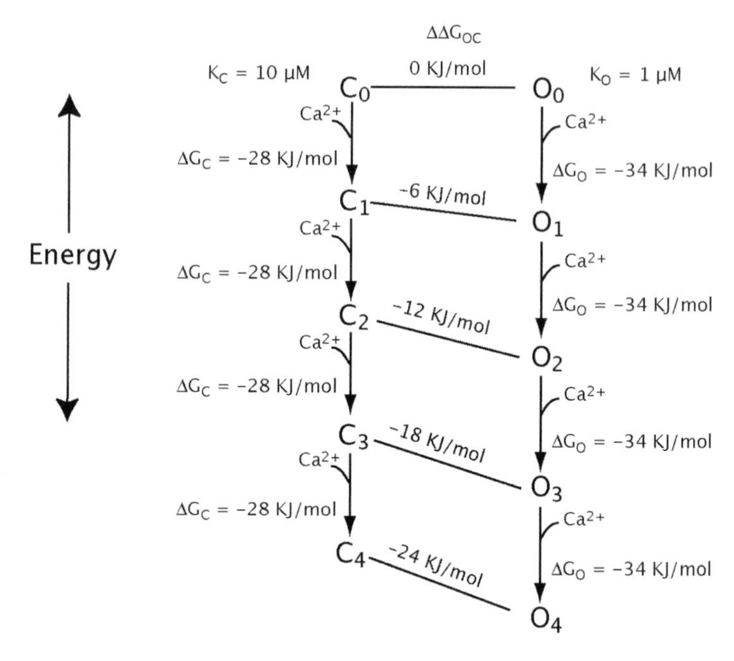

Figure 5.6 Energy diagram for the MWC model. As each Ca^{2+} ion binds, the energy of the open and closed conformations are lowered by differing amounts due to differences in binding affinity. This lowers the energy of the open channel relative to the closed. For simplicity L has been assumed here to equal 1 such that C_0 and O_0 have the same energy. This is not the case for the BK_{Ca} channel where $L(0)$ is ~0.0006.

it is necessarily the difference between ΔG_O and ΔG_C, and therefore the ratio of K_C to K_O, that drives opening, not their absolute values, and the more Ca^{2+}-binding sites the channel has the stronger the effect.

In the VD-MWC model voltage dependence is added simply by supposing that the central conformational change involves the movement of some gating charge Q through the membrane's electric field such that the equilibrium constant between closed and open becomes voltage dependent. That is

$$L(V) = \frac{[O_0]}{[C_0]} = L(0)e^{QFV/RT}, \tag{5.2}$$

and with this stipulation the open probability of the model is given by

$$P_{\text{open}} = \frac{1}{1 + \left[\dfrac{1 + \dfrac{[\text{Ca}]}{K_c}}{1 + \dfrac{[\text{Ca}]}{K_o}}\right]^4 \dfrac{1}{L(0)} e^{-QFV/RT}}. \tag{5.3}$$

Even without fitting the data it is apparent by inspection that Eq. 5.3 will display many properties of the BK_{Ca} channel. It allows the channel to be maximally activated by voltage in the absence of Ca^{2+}, and it has the form

$$P_{open} = \frac{1}{1 + Ae^{-QFV/RT}}, \tag{5.4}$$

which is a Boltzmann function, with the parameter A being related to the free energy difference between open and closed in the absence of an applied voltage. For a given Q, this parameter determines the position of the G–V curve along the voltage axis, while Q determines it steepness. In the VD-MWC model A is Ca^{2+} dependent.

$$A = \left[\frac{1 + \frac{[Ca]}{K_C}}{1 + \frac{[Ca]}{K_O}}\right]^4 \frac{1}{L(0)}. \tag{5.5}$$

As $[Ca^{2+}]$ increases, A decreases, and the model's G–V relation shifts leftward along the voltage axis. Since Q is independent of $[Ca^{2+}]$, the model's G–V relation will not change shape as $[Ca^{2+}]$ is varied—it will simply slide along the axis. While this is not true for the BK_{Ca} channel, it is true to a first approximation, particularly in the middle $[Ca^{2+}]$ range (1 – 100 μM).

In Fig. 5.7A, a series of mSlo1 G–V curves have been fitted simultaneously with Eq. 5.3, and although the fit is not exceptional, as we expected, the model does a reasonable job of capturing the Ca^{2+}-dependent shifting of the mSlo1 channel's G–V relation. The fit suggests that $K_C = 11$ μM, $K_O = 1.1$ μM, $L = 0.00061$ and $Q = 1.40\,e$. Notice, however, that the model is unable to mimic the shifting nature of the mSlo1 G–V curve at $[Ca^{2+}]$ greater than 100 μM (data and fits in grey; Cox et al., 1997b). This is now understood to be due the existence of a low-affinity set of Ca^{2+}-binding sites that are separate from the channel's higher affinity sites and not included in the model (Shi and Cui, 2001; Zhang et al., 2001).

In Fig. 5.8A and B is shown a series of Ca^{2+} dose–response curves determined at different voltages for the mSlo1 channel (*filled*) and the VD-MWC model channel (*open*). Each curve has been fitted with the Hill equation

$$\frac{G}{G_{max}} = \left[\frac{\text{Amp}}{1 + \left(\frac{K_D}{[Ca]}\right)^H}\right], \tag{5.6}$$

and the resulting parameters are plotted in Fig. 5.8C–E. As expected from Fig 5.7,

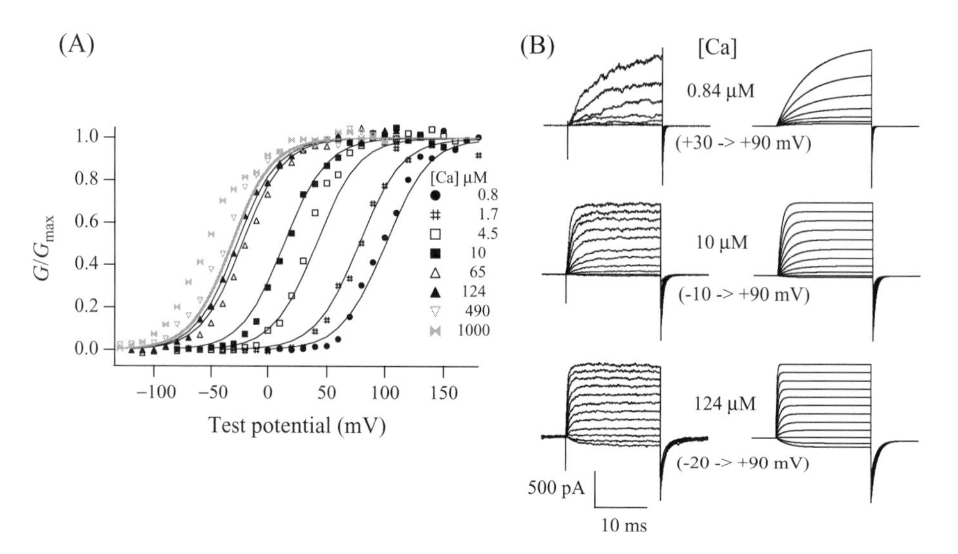

Figure 5.7 The VD-MWC model mimics mSlo1 gating. (A) mSlo1 G–V relations fitted with the VD-MWC model (solid lines). All curves were fit simultaneously. [Ca^{2+}] are as indicated. [Ca^{2+}] above 124 μM are shown in grey and were not included in the fitting. (B) Current families recorded at 0.84 μM, 10 μM, and 124 μM as indicated are shown on the left. VD-MWC-model currents for the same [Ca^{2+}] are shown on the right. For model parameters see text and Cox et al. (1997b). This figure was adapted from Cox et al. (1997b).

here too the model in many ways recapitulates the data. First, in both the data and the model the extent of activation by Ca^{2+} is limited by voltage (Fig. 5.8C), second, the Hill coefficient (H) is ∼1.5 to 2 for both the model and the channel over a fairly wide voltage range (Fig. 5.8D), and, three, although the fit is not good at low voltages, the model does predict, as appears in the data, an increase in the channel's apparent Ca^{2+} affinity (a decrease in K_D-apparent) as voltage is increased (Fig. 5.8E). Indeed, this occurs even though neither of the model's Ca^{2+}-binding constants K_C and K_O are voltage dependent.

Why is the apparent Ca^{2+} affinity of the VD-MWC model voltage-dependent even though its binding constants are not? One way to get at this question is to plot open probability (P_{open}) for the model versus the mean number of Ca^{2+} ions bound (Fig. 5.8F). Such a plot makes it clear that the model becomes more Ca^{2+} sensitive as voltage is increased, not because it binds Ca^{2+} more tightly, but because the mean number of bound Ca^{2+} ions needed to activate the channel becomes smaller. At +160 mV, for example, it takes only 1 bound Ca^{2+} to activate the channel to a P_{open} greater than 0.8, while at 0 mV, 4 bound Ca^{2+} are required. This is a graphical demonstration of the additive nature of the energies imparted to the closed-to-open conformational change by Ca^{2+} binding and membrane voltage. For the VD-MWC model P_{open} is related to the free energy difference between open and closed, ΔG,

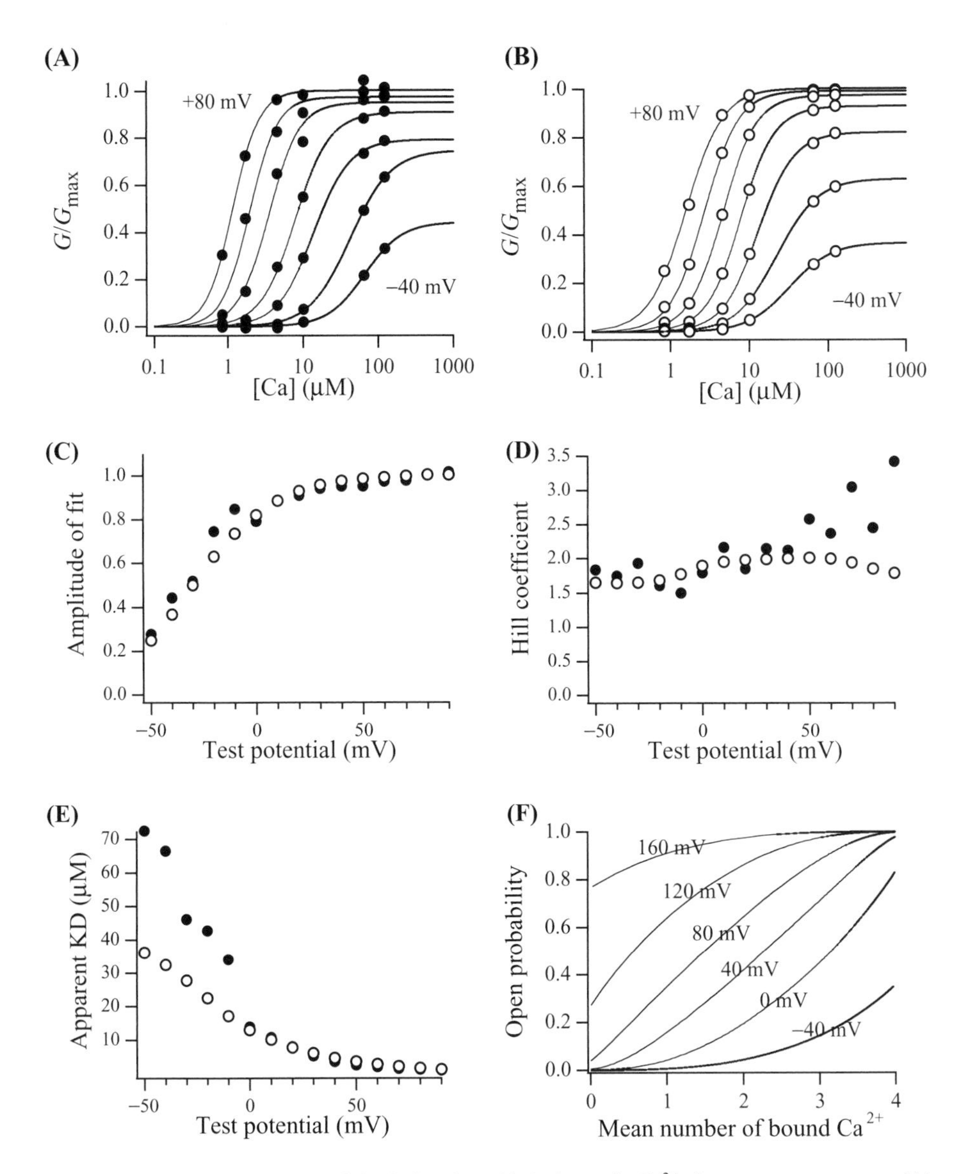

Figure 5.8 The VD-MWC model mimics the mSlo1 channel's Ca^{2+} dose–response curves. (A) Ca^{2+} dose–response curves are plotted for seven different voltages ranging from −40 to +80 mV in 20 mV steps (symbols). Each curve is and has been fitted with the Hill equation (Eq. 5.6) (solid curves), and the parameters of these fits are plotted as a function of voltage in *C–E*. (B) Simulated data from the VD-MWC model. The parameters used for these simulations are those listed in the text. (C–E) Comparison of the fit parameters from the data (closed circles) and the model (open circles). (F) Plots of open probability of the model channel as a function of the mean number of Ca^{2+} ions bound to the model channel at a series of voltages. This figure was adapted from those in Cox et al. (1997b).

by the following function

$$P_{\text{open}} = \frac{1}{1+e^{\left(\frac{\Delta G}{RT}\right)}} = \frac{1}{1+e^{\left(\frac{\Delta G_{\text{intrinsic}}}{RT}+\frac{\Delta G_{\text{voltage}}}{RT}+\frac{\Delta G_{\text{Ca}}}{RT}\right)}}$$

where

$$\begin{aligned}\Delta G_{\text{intrinsic}} &= -RT\ln(L(0)),\\ \Delta G_{\text{voltage}} &= -QFV/RT,\\ \Delta G_{\text{Ca}} &= -4RT\ln\left[\frac{1+\frac{[\text{Ca}]}{K_{\text{O}}}}{1+\frac{[\text{Ca}]}{K_{\text{C}}}}\right].\end{aligned} \tag{5.7}$$

Thus, as more energy is supplied for opening by voltage, less is required from Ca^{2+}-binding to achieve a given P_{open}. This is of course not just true for the VD-MWC model but for a large class of models in which two stimuli are regulating a central conformational change, and it is true for the more complex models to be discussed below.

One might reasonably ask, however, just how far can this go? Can an ion channel have an apparent affinity for its ligand that is higher that the true affinity of any of its binding sites in either conformation? Remarkably, the answer to this question is yes. This is illustrated in Fig. 5.9 where plotted for the model channel is P_{open} as a function of $[Ca^{2+}]$ at different values of $L(K_{\text{C}} = 10, K_{\text{O}} = 1)$(Fig. 5.9A). These curves were then normalized to have the same maximum and minimum, so that their shapes could be easily compared (Fig. 5.9B), and then the $[Ca^{2+}]$ at which the normalized curves are at half of their maximum value is plotted as a function of L in Fig. 5.9C. At $L = 1$ the model's K_{D}-apparent is 0.37 μM while K_{O} is 1 μM. Thus, at even moderate values of L the apparent K_{D} of the model can be smaller than the true K_{D} of the open channel. If one repeats this exercise, however, for a channel with only one binding site, one finds that K_{D}-apparent cannot be less than K_{O}, and it moves from K_{C}, when L is very small, to K_{O}, when L is very large. The peculiar situation, then, where an MWC system displays an apparent affinity higher than any of is real affinities, results from the condition where significantly fewer ligands need to bind to the channel to maximally activate it, than there are binding sites on the channel. For the BK_{Ca} channel this situation would only pertain at voltages greater than +100 mV and not under physiological conditions.

Another question one might ask about the BK_{Ca} channel is what governs its Hill coefficient (H)? And how should this coefficient—which is a standard measure of the cooperativity of an allosteric system—vary with voltage? The data in Fig. 5.8D tell us that the channel's Hill coefficient slowly increases with voltage from ∼1.5 to ∼ 3.0, and for the model channel it displays a fairly constant but wavy pattern around the value 1.7. What determines this value? A well known result for the MWC model is that H for ligand binding is strongly dependent on L (Segel, 1993), and

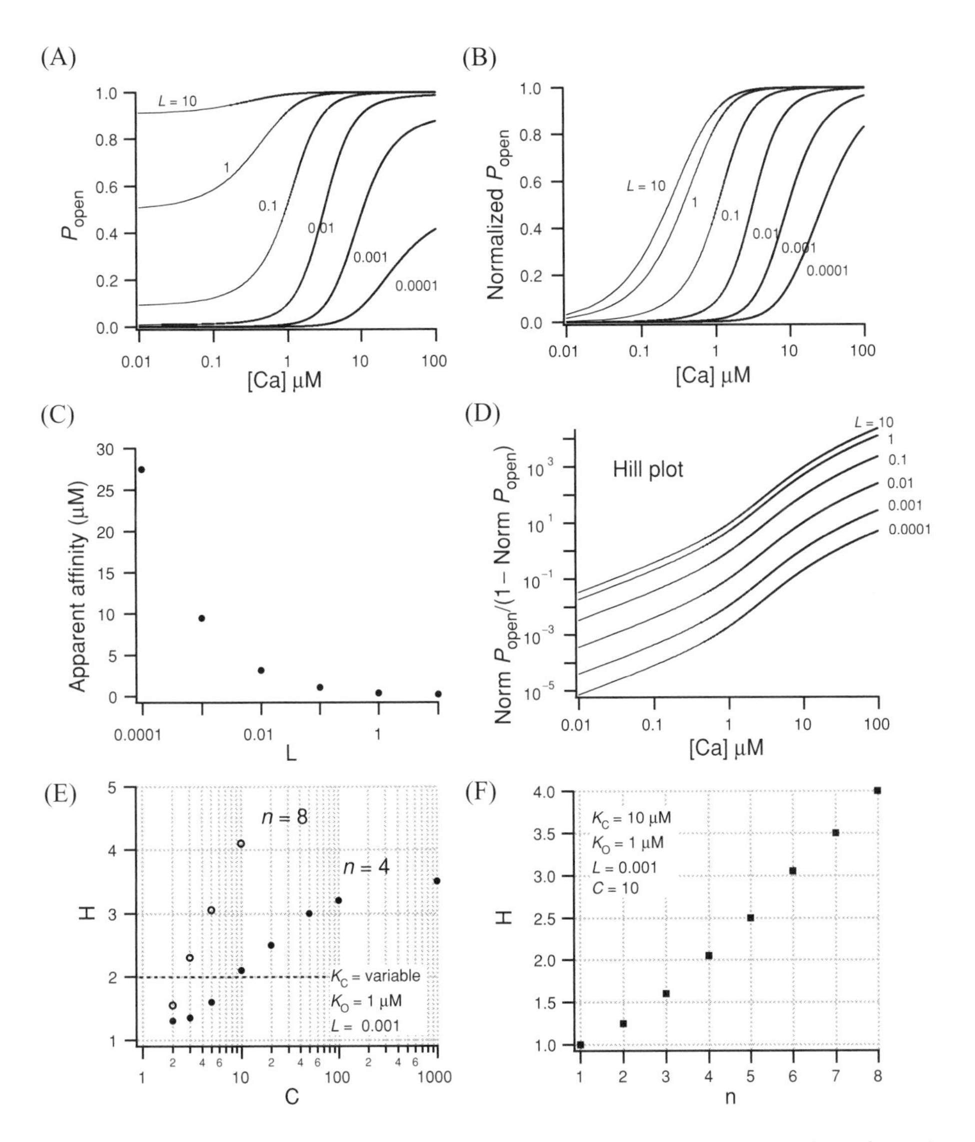

Figure 5.9 Properties of the MWC model. (A) Simulated dose–response curves for and MWC model where $K_C = 10$, $K_O = 1$, $n = 4$, and L was varied as indicated. (B) The 0 [Ca^{2+}] values of each curve in *A* were subtracted from each curve, and then they were each normalized to their maximum to yield the curves plotted *B*. (C) The [Ca^{2+}] at which the curves in *B* are at half of their maximum is plotted as a function of *L*. (D) Hill plots for the curves in *B*. The maximum slope of these relations are defined as the Hill coefficient, which here does not change with *L*. (E) Hill coefficient *(H)* plotted as a function of *C*, where $C = K_C/K_O$. (F) Hill coefficient *(H)* plotted as a function of the number of binding sites *n*. Notice *H* increases with *n* and *C*.

in the VD-MWC model it is therefore voltage dependent. Interestingly, however, if instead of measuring Ca^{2+} binding one measures channel opening, H for channel opening is independent of L. This illustrated in Fig. 5.9D were a series simulated Hill plots from an MWC model are displayed as a function of L. Each plot has the same maximum slope and thus the same Hill coefficient. According to this model, then, the BK_{Ca} channel's Hill coefficient should not vary with voltage, and indeed this is true as well for more complex models, so long as voltage sensing and Ca^{2+} binding are acting independently on channel opening, and there is only one type of Ca^{2+}-binding site.

If this is the case, however, then why does the model's Hill coefficient—and indeed the real channel's—vary in Fig. 5.8D. The answer, at least for the model, is that when the simulated data were fit with the Hill equation, as the voltage was varied, the data were spread over different parts of the P_{open}–V curve, and this created and apparent change in H as different parts of the curve were emphasized in the fitting. However, no real change is expected, and perhaps this technical problem also contributes to the variation in H observed with fits to the real data. It is now known, however —as will be discussed below—that the BK_{Ca} channel has more than one type of Ca^{2+}-binding site, a circumstance that could also give rise to a Hill coefficient that varies with voltage, if different binding sites becomes more or less important for channel opening as the membrane voltage changes.

If L does not govern the cooperativity of the MWC model, when viewed in terms of channel opening, what does? The answer is C and n, where n represents the number of binding sites the system contains, and again $C = K_C/K_O$. Fig 5.9E and F shows that H increases as C increases and as n increases, and that an H value of ~2 for the mSlo1 channel could be explained by supposing that the channel has four binding sites ($n = 4$) with a C value of ~10 ($C = K_C/K_O$), as was done by Cox et al. (1997b), or eight binding sites with a C value of ~3, which now appears closer to the truth (Bao et al., 2002).

In addition to mimicking many aspects of mSlo1 steady-state gating, the VD-MWC model can also approximate the Ca^{2+} and voltage dependence of the kinetics of channel activation and deactivation (Fig. 5.7B) (Cox et al., 1997b). That is, it can be made to activate more quickly with increasing $[Ca^{2+}]$ and voltage and deactivate more slowly. To do this rate constants must be supplied for all of the transitions in Scheme II, and two things have been found to be required. (1) The rate constant of channel opening must increase with each Ca^{2+}-binding event, while the rate constant for channel closing must decrease, and (2) the on rates and off rates for Ca^{2+} binding must be fast. In this limit the VD-MWC model approximates a two state system whose open and closing rates are a weighted average of all the vertical rate constant in Scheme II, each weighted by the percent of closed or open channels that occupy the state that precedes each rate constant (Cox et al., 1997b). For the model in Fig 5.7 the Ca^{2+} on rates were assumed to be 10^9 $M^{-1}s^{-1}$, at the very upper end of what is reasonable, and the off rates were then necessarily ~10^4s^{-1} for the closed channel and ~10^3s^{-1} for the open channel. These values produced exponential kinetics under most conditions (Cox et al., 1997b). Thus, many of the attributes of both the steady-state and kinetic properties of BK_{Ca}-channel gating can be accounted for by the simple VD-MWC scheme, and what it seems to be telling us is that, if there are four independent binding sites, then each has a K_C of ~10 μM and a K_O of ~1 μM, and the binding and unbinding of Ca^{2+} must be fast.

5.7 Interpreting Mutations

A useful aspect of the VD-MWC model and something that is not the case for models with more complicated voltage-sensing mechanisms is that is easy to write down the equation that governs its G–V position as a function of $[Ca^{2+}]$. Here in terms of the voltage of half-maximal activation $V_{1/2}$

$$V_{1/2} = \frac{4RT}{QF} \ln\left[\frac{(1+[Ca]/K_C)}{(1+[Ca]/K_O)}\right] + \frac{RT}{QF} \ln[L(0)]. \tag{5.8}$$

This equation states that $V_{1/2}$ in 0 $[Ca^{2+}]$ depends on L and Q, and the extent to which it moves with increasing $[Ca^{2+}]$ depends on K_C, K_O, and Q. An important result here is that the steepness of the $V_{1/2}$ vs. $[Ca^{2+}]$ relation depends on the channel's Ca^{2+} dissociation constants, as one might imagine, but also on Q. All other things being equal the larger Q is, the smaller the shift. Thus, if one were to make a mutation that affects the slope of the BK_{Ca} channel's $V_{1/2}$ vs. $[Ca^{2+}]$ relation, the VD-MWC model suggests that this could either be due to a change in voltage sensitivity or a change in Ca^{2+} binding. We can determine which, however, by rearranging Eq. 5.8 to

$$QFV_{1/2} = 4RT \ln\left[\frac{(1+[Ca]/K_C)}{(1+[Ca]/K_O)}\right] + RT \ln[L(0)], \tag{5.9}$$

(where $-QFV_{1/2}$ is equal to the energy difference between open and closed) and instead of plotting $V_{1/2}$ vs. $[Ca^{2+}]$, plotting $QFV_{1/2}$ vs. $[Ca^{2+}]$. On such a plot the position of the curve at 0 $[Ca^{2+}]$ is determined solely by L, while the shape of the curve is determined by K_C and K_O (Cox el al., 1997b; Cox and Aldrich, 2000; Cui and Aldrich, 2000). The VD-MWC model, therefore, provides a straightforward means by which to distinguish between mutations that effect voltage sensing (Q), Ca^{2+} binding (K_C or K_O), and the intrinsic energetics of channel opening ($L(0)$).

5.8 A Better Model of Voltage-Dependent Gating

Although the VD-MWC model provides a useful framework for thinking about a channel regulated by voltage and ligand binding, there are aspects of the BK_{Ca} channel's gating behavior that is fails to mimic. It does not predict the brief delay (200 μs) in activation mentioned above. It does not mimic well the kinetic behavior of the channel at far positive and far negative voltages, and it predicts that in the absence

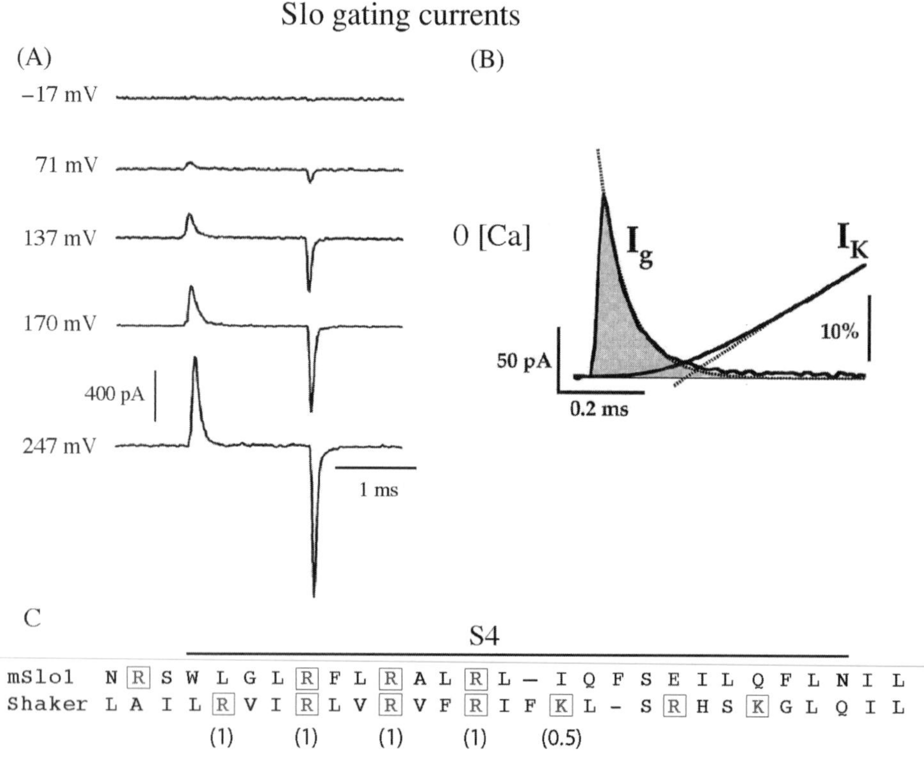

Figure 5.10 BK_{Ca}-channel gating currents are fast. (A) hSlo1 gating currents recorded with pulses to the indicated potentials in the absence of $[Ca^{2+}]$. (B) Gating current (I_g) and the time course of ionic current activation (I_K). Gating charge moves before the channels open. (C) Alignment of the S4 regions of the mSlo1 channel and the Shaker channel. *A* was adapted from Stefani et al. (1997). *B* was adapted from Horrigan and Aldrich (1999).

of Ca^{2+}, single channel recordings should reveal only a single open state and a single closed state, while the mSlo1 channel displays three closed and two open states under this condition (Talukder and Aldrich, 2000). The most damming evidence against the VD-MWC model, however, is the important observation—initially made by Stefani et al. (1997) (see Fig. 5.10A), and then Horrigan and Aldrich (1999) (see Fig. 5.10B)— that BK_{Ca}-channel gating currents are much faster than channel opening and closing (Fig. 5.10B). The VD-MWC model predicts the two processes should have the same time constant, as in the VD-MWC model voltage-sensor movement is part of the central closed-to-open conformational change.

That this is not the case for the BK_{Ca} channel, however, perhaps should not be surprising as the Shaker channel's gating currents also relax more quickly than the channel opens, and this has been explained by supposing that voltage sensors in each subunit move rapidly upon depolarization, and when all have moved to their active conformation the channel undergoes an opening conformational change (which is also weakly voltage-dependent) (Bezanilla, 2005). Three things about the BK_{Ca} channel's gating currents, however, are very different from those of Shaker. BK_{Ca}-channel gating currents are much smaller, they are much faster, and their off gating currents are not slowed by channel opening (Stefani et al., 1997; Horrigan and Aldrich, 1999). With Shaker the voltage sensors must wait for the channel to close before they can return to their resting state, while this is not the case for mSlo1 (Zagotta et al., 1994; Bezanilla, 2005).

In 1999, based on an extensive series of both gating and ionic current experiments with the mSlo1 channel, Horrigan, Cui, and Aldrich proposed a model of BK_{Ca}-channel voltage-dependent gating that does a remarkably good job of accounting quantitatively for almost all aspects of BK_{Ca}-channel gating in the absence of Ca^{2+} (Horrigan et al., 1999; Horrigan and Aldrich, 1999). This model, termed here the HCA model, is represented by Scheme III.

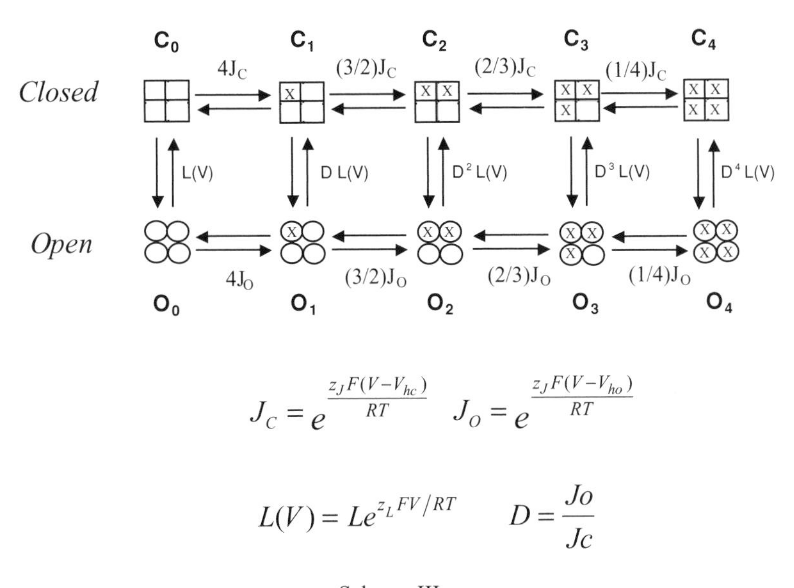

$$J_C = e^{\frac{z_J F(V - V_{hc})}{RT}} \qquad J_O = e^{\frac{z_J F(V - V_{ho})}{RT}}$$

$$L(V) = Le^{z_L FV/RT} \qquad D = \frac{Jo}{Jc}$$

Scheme III

Notice it looks very much like the MWC model except here horizontal transitions represent voltage-sensor activation, now one in each subunit, and the central closed-to-open conformational change is also proposed to weakly voltage dependent. Thus, the HCA model is an allosteric model of voltage-dependent gating, where instead of voltage-sensor movement being required for channel opening, as is the case with the VD-MWC model and models of Shaker gating, here the channel can open with any number of voltage sensor's active, but as each voltage sensor moves to its active state, the open conformation of the channel becomes energetically more favored. Also analogous to the MWC model, in the HCA model, in order for voltage-sensor movement to promote channel opening it must lower the energy of the open state of the channel more than the energy of the closed state. Practically speaking, what this means is the $V_{1/2}$ for voltage-senor movement must be more negative when the channel is open than when it is closed. Analogous to the factor C in the MWC model, Horrigan, Cui, and Aldrich defined D to represent the factor by which the movement of a voltage senor in the HCA model increases the equilibrium constant between open and closed. They estimated D to be 17 for the mSlo1 channel (Horrigan et al., 1999; Horrigan and Aldrich, 1999).

There are five parameters that govern the equilibrium behavior of the HCA model. $L(0)$ the equilibrium constant between open and closed at 0 mV when no voltage sensors are active, z_q the gating charge associated with this equilibrium constant, V_{hc} and V_{ho} the half-activation voltages of voltage-sensor movement when the channel is closed or open respectively, and z_j the gating charge associated with each voltage sensor's movement. By measuring open probabilities at very low voltages, Horrigan et al. (1999) and Horrigan and Aldrich (1999) were able to determine $L(0)$ to be ~2e–6 and z_q to be 0.4 e. By measuring gating currents and macroscopic ionic currents over a wide range of voltages, they were able to estimate V_{hc} to be +155 mV, V_{ho} to be 24 mV and z_j to be 0.55 e. With these parameters the HCA model nicely mimics the mSlo1 channel's Q–V and G–V curves in the absence of Ca^{2+}. Indeed a key feature of the data that is reproduced by the model is that at very low open probabilities the mSlo1 log(P_{open})-vs.-voltage relation reaches a limiting slope that is less than the maximum slope of this relation and reflects just the voltage dependence of the central conformational change (Horrigan et al., 1999).

Also, by studying the kinetics of gating and ionic currents Horrigan et al. (1999) and Horrigan and Aldrich (1999) were able to specify all the rate constants in Scheme III and then reproduce very well the kinetic behavior of both gating and ionic currents. In qualitative terms what their work and that of Stefani et al. (1997) suggests is that in response to changes in voltage the channel's voltage sensors move very rapidly and independently between an active and an inactive state, with the process of channel opening and closing being much slower. Indeed, in response to depolarization the voltage sensors of the closed channel reequilibrate completely during the brief delay (~200 μs) that precedes macroscopic current activation (see Fig. 5.10B), and then, as the channels open, they equilibrate again among open states. Upon repolarization the voltage sensors equilibrate very rapidly among open states, before the channels close, and then they equilibrate again through the closed states as the channels close. Combining these results, then, with observations from VD-MWC modeling, it appears that the essentially monoexponential nature of the BK_{Ca} channels macroscopic currents arises from a channel whose closed-to-open conformational change is rate limiting under nearly all conditions and is allosterically regulated by the faster processes of Ca^{2+} binding and voltage-sensor movement.

Each voltage sensor in the HCA model carries a gating charge of 0.55 e, while each voltage sensor of the Shaker channel carries a gating charge of ~3.5 e (Bezanilla, 2005). Thus the BK_{Ca} channel's voltage sensor appears to carry only 1/6 the gating charge of that of the Shaker channel. For the Shaker channel the gating charge has been shown to arise from the movement of 5 of 7 positively charges residues in the S4 segment (Aggarwal and MacKinnon, 1996; seoh et al. 1996). An alignment of the Shaker and mSlo1 S4 segments is shown in Fig 5.10C. The residues that carry gating charge along with the approximate amount of charge they carry (according to Agarwall and MacKinnon, 1996) are indicated in parentheses. The BK_{Ca} channel shares the 2nd, 3rd, and 4th positive charges that in the Shaker channel contribute 1 full charge each to the gating charge. Thus, it is surprising that the BK_{Ca} channel's voltage sensor does not contain a larger amount of gating charge, and if this is indeed the case, as it appears to be, then it suggests either that the movements that this region undergoes during gating are not the same for the two channels, or the electrical fields through which these movements occur are very different. Indeed, one might question whether S4 is involved at all in the voltage sensing of the BK_{Ca} channel. Mutagenesis experiments, however, suggest that R213 and R210 carry a small amount of gating charge (Diaz et al., 1998), so the S4 segment may form the voltage sensor of both channels, but the reason why R207, R210, and R213 do not contribute more gating charge to mSlo1 remains a mystery.

5.9 Combining HCA and MWC

Given the success of MWC-like models in describing Ca^{2+} sensing, and the HCA model in describing voltage sensing, it is natural to combine the two to produce a model of BK_{Ca}-channel gating-like that shown in Fig. 5.11A. Here the top tier of states represents the closed conformation of the channel, and the bottom tier the open conformation. The horizontal transitions along the long axis represent Ca^{2+} binding and unbinding. The horizontal transitions along the short axis represent voltage-sensor movement. There is one open state and one closed state for every possible combination of 1–4 voltage sensors active and 1–4 Ca^{2+}-binding sites occupied, and thus 25 closed and 25 open states, or 50 states in all. Rothberg and Magleby (2000) where the first to propose that a model of this form could mimic the essential aspects of the gating of the BK_{Ca} channel based on their analysis of the mSlo1 channel's single-channel behavior over a wide range of conditions.

At first glance this 50-state model might seem hopelessly complex, but it is based on simple ideas about Ca^{2+} and voltage sensing. And although it has many states, if voltage sensors and Ca^{2+}-binding sites are considered identical and independent, its open probability is determined by just seven parameters—the five HCA parameters, $L(0)$, z_q, V_{hc}, V_{ho}, and z_j, plus the MWC parameters K_C an K_O—so there is reason to hope that they could be well constrained by electrophysiological data. Indeed, Cox and Aldrich (2000), using voltage-sensing parameters close to those determined from HCA modeling (Horrigan and Aldrich, 1999), and Ca^{2+} dissociation constants close to those determined from VD-MWC modeling (Cox et al., 1997b), were able to describe the shifting nature of the mSlo1 G–V relation between 0 and 100 μM [Ca^{2+}] fairly well (Cox and Aldrich, 2000).

(A)

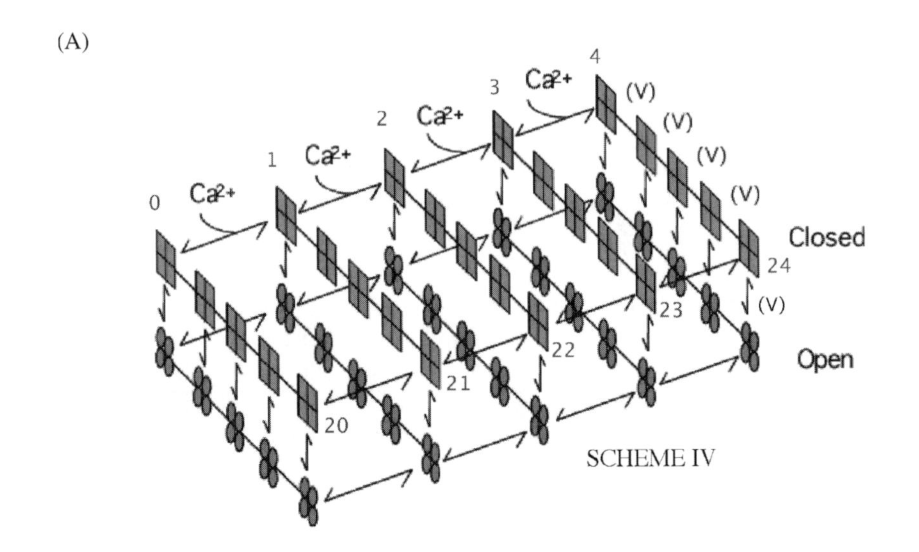

(B)

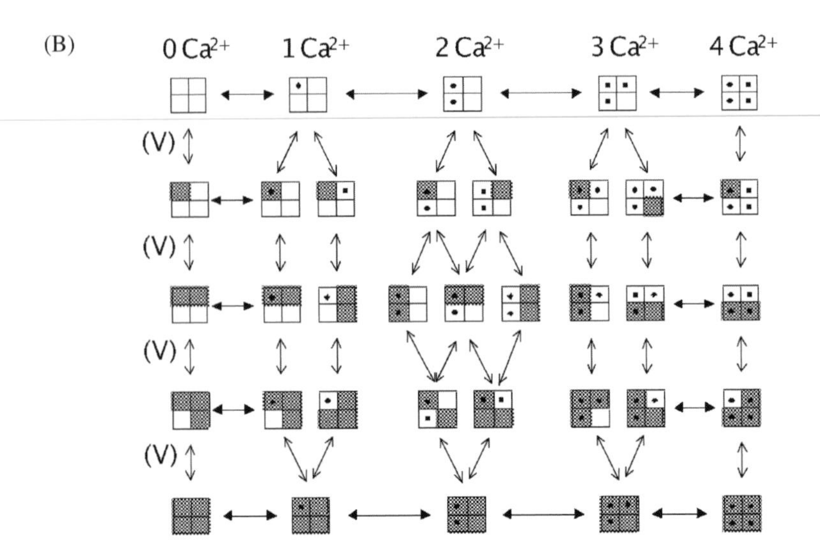

SCHEME V (closed tier)

Figure 5.11 Allosteric models of BK_{Ca}-channel gating (A) 50-state model first proposed by Rothberg and Magleby (2000). (B) The 35 states of one tier of the 70-state model of Horrigan and Aldrich (2002). *A* was adapted from Cox and Aldrich (2000). *B* was adapted from Cox et al. (1997b).

One possibility that the 50-state model ignores, however, is that Ca^{2+} binding to its site in a given subunit may directly alter the equilibrium for voltage sensor movement in that same subunit and vice versa. If one takes this possibility into account then the 50-state model must be expanded to a 70-state scheme with now 35 closed and 35 open states (Cox et al., 1997b), and one more parameter E must be added (Horigan and Aldrich, 2002). Analogous to C and D, E is the intrasubunit coupling factor between Ca^{2+} binding and voltage sensing. The 35 closed states of such a model are depicted graphically in Fig. 5.11B. One has to imagine each of these closed states being connected through an opening conformational change to a corresponding open state. The open probability of the resulting 70-state model, first proposed by Horrigan and Aldrich (2002), is given by the following equation

$$P_{\text{open}} = \frac{L(1 + K_{\text{O}} + J_{\text{O}} + J_{\text{O}}K_{\text{O}}E)^4}{L(1 + K_{\text{O}} + J_{\text{O}} + J_{\text{O}}K_{\text{O}}E)^4 + (1 + J_{\text{C}} + K_{\text{C}} + J_{\text{C}}K_{\text{C}}E)^4} \tag{5.10}$$

where

$$L = L(0)e^{\frac{z_q FV}{RT}}; \quad J_{\text{O}} = J_{\text{O}}(0)e^{\frac{z_j F(V - V_{ho})}{RT}}; \quad J_{\text{C}} = J_{\text{C}}(0)e^{\frac{z_j F(V - V_{hc})}{RT}};$$

$$K_{\text{C}} = \frac{[\text{Ca}]}{K_{\text{C}}}; \quad K_{\text{O}} = \frac{[\text{Ca}]}{K_{\text{O}}}.$$

But are the 20 extra states needed? Does Ca^{2+} binding in fact directly affect voltage sensing? Is E different from 1? Experiments with mutations that alter voltage sensing suggest it is not (Cui and Aldrich, 2000); however, Horrigan and Aldrich (2002) addressed this issue directly by measuring mSlo1 gating currents in 0 and 70 μM [Ca^{2+}]. They made the following interesting observations. mSlo1 on gating currents exhibit two exponential components, fast and slow. Q_{fast} can be assigned to the rapid equilibration of voltage sensors in the closed channel before the channel opens, and Q_{slow} to the reequilibration of the voltage sensors among open-states after the channels open. As this second component of charge movement is limited by the speed of channel opening, it appears slow even though, once a channel is open, its voltage sensors reequilibrate rapidly. The interesting result, however, is that, while increasing [Ca^{2+}] from 0 to 70 μM has a large effect on the Q_{total}–V relation (fast + slow)—it makes this curve much steeper and shifts it to the left (see Fig. 12D)—it has very little effect on the Q_{fast}–V relation. It shifts it just −20 mV, and it does not change its shape (Fig. 5.12C). Since this relation reflects the activation of voltage sensors in the closed channel, the fact that it changes little when [Ca^{2+}] is increased indicates that indeed Ca^{2+} binding does not have a large direct effect on voltage-sensor movement. That is, E is small. The −20 mV shift, however, does indicate that it is not equal to 1. Horrigan and Aldrich estimated it to be 2.4.

If the Q_{fast}–V relation does not change much with increasing [Ca^{2+}], then why does the Q_{total}–V relation change so dramatically with increasing [Ca^{2+}] (Fig. 5.12D)? It turns out this is not so strange, and it is predicted by the 70-state

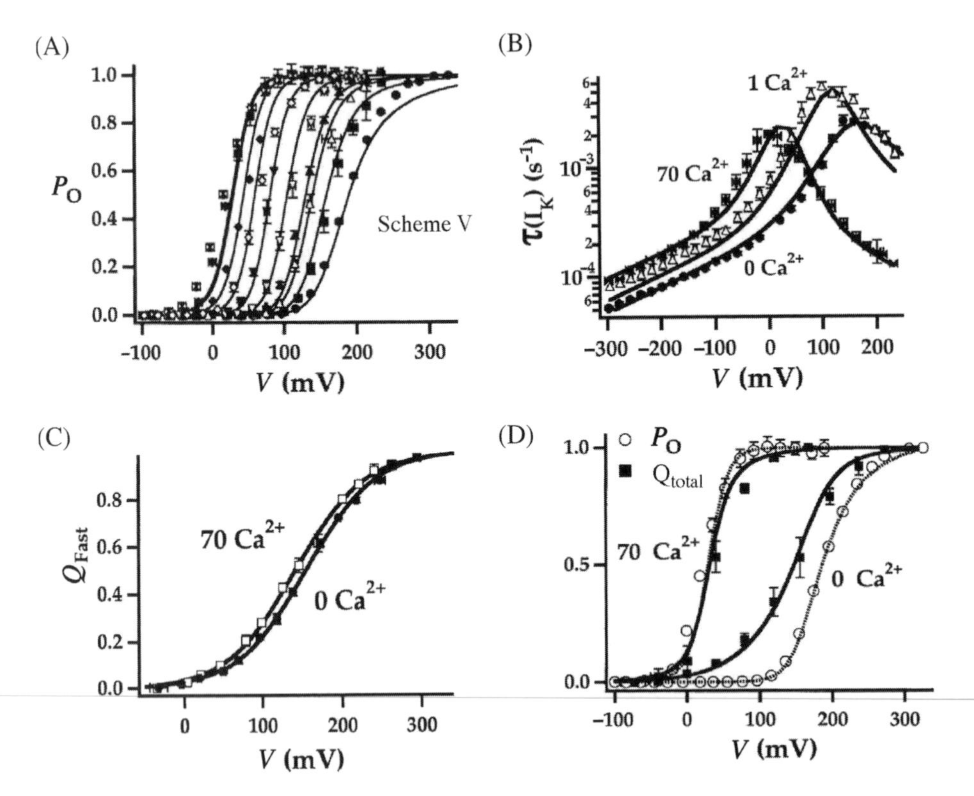

Figure 5.12 mSlo gating and ionic currents can be well described by the HA model (Scheme V see Fig. 5.11B). (A) G–V relations determined at the following [Ca^{2+}], from right to left in μM: 0.27, 0.58, 0.81, 1.8, 3.8, 8.2, 19, 68, and 99. The curves were simultaneously fit with the HA model (continuous lines) with the following parameters: $K_C = 11$ μM, $K_o = 1.4$ μM, $V_{hc} = 150$ mV, $V_{ho} = 0.7$ mV, $z_q = 0.4$, $z_J = 0.55$, $L(0) = 2 \times 10^{-6}$. (B) The HA model can also mimic the kinetic behavior of mSlo1 macroscopic currents. Plotted are macroscopic current relaxation time constants as a function of membrane voltage. Continuous lines represent fits to Scheme V. Steady-state parameters were the same as in *A*. For kinetic parameters see Horrigan and Aldrich (2002, Table III). (C) There is little change in the Q_{fast}–V relation of the mSlo1 channel when [Ca^{2+}] is raised from 0 to 70 μM. The small shift observed is −20 mV. (D) The HA model (continuous lines) can also reproduce the changes observed in the relationship between the mSlo1 Q_{total}–V relation and the P_{open}–V relation as [Ca^{2+}] is increased. This change occurs because as [Ca^{2+}] is increased fewer voltage sensors need to move to open the channel. All panels were adapted from figures in Horrigan and Aldrich (2002).

or HA model. For a detailed explanation I refer the reader to the original publication (Horrigan and Aldrich, 2002). In brief, however, the answer is that the Q_{total}–V relation of the HA model is a weighted average of the Q–V relations of the open and closed channels—with midpoints V_{hc} and V_{ho} respectively—weighted by the fraction of channels that are open and closed at a given voltage. These relations have the same shape but lie ∼140 mV apart on the voltage axis. The addition of Ca^{2+} makes the channels open at lower voltages and thereby puts greater weight on the Q_{open}–V relation than the Q_{closed}–V relation. According to the HA model, it is changes in the weighting of these two curves with increasing [Ca^{2+}] that cause the large change in shape and position of the Q_{total}–V curve.

Notice also in Fig. 5.12D that at 0 [Ca^{2+}] the Q_{total}–V relation lies to the left of the P_{open}–V relation, and it is more shallow, while at high [Ca^{2+}], the Q_{total}–V relation is very similar in shape and position to the P_{open}–V curve. This makes sense, as in 0 [Ca^{2+}] on average three voltage sensors must become active before the channel opens, while in 70 μM [Ca^{2+}] just one is required (Horrigan and Aldrich, 2002).

Not only did Horrigan and Aldrich show that their model could explain the behavior of the channel's gating currents, they also found a set or parameters that could fit the channel's G–V relation with [Ca^{2+}] between 0 and 70 μM (Horrigan and Aldrich, 2002) (Fig. 5.12A). Their estimates of K_C and K_O, assuming four Ca^{2+}-binding sites, were 11 μM and 1.4 μM, $C = 8$, similar to estimates from the VD-MWC model. And they also found rate constants for the closed-to-open transitions in the model that enabled it to reproduce the channel's macroscopic-current kinetics fairly well (Fig. 5.12B). To approximate exponential kinetics again Ca^{2+} binding was assumed to be fast. Thus, the HA model reproduces BK_{Ca}-channel behavior over an impressively wide range of conditions, and it is the best explanation of BK_{Ca}-channel gating to date. Indeed, it also allows for variations in G–V steepness as a function of [Ca^{2+}], something else observed in the data and not accounted for by the VD-MWC model.

5.10 The BK_{Ca} Channel Has Low-Affinity Ca^{2+} Binding Sites

What is wrong with the HA model? It has yet to be compared quantitatively to single-channel data, and if it were to be, it is unlikely that it would reproduce the rapid flickers observed in such recordings (McManus and Magleby, 1988; Cox et al., 1997b; Rothberg and Magleby, 1999,2000). Its fits to macroscopic-current kinetics in the middle [Ca^{2+}] range ($\sim$10 μM) could be improved, but the major problem with this model is that—like the VD-MWC model—it saturates at 100 μM [Ca^{2+}], and therefore it does not predict additional G–V shifts at [Ca^{2+}] greater than this. This drawback was acknowledged by Horrigan and Aldrich and attributed to the lack in the model of the low-affinity Ca^{2+}-binding sites described by the Cui and Lingle groups (Shi and Cui, 2001; Zhang et al., 2001; Shi et al., 2002; Xia et al., 2002).

These binding sites were discovered by studying the effects of Mg^{2+} on BK_{Ca}-channel gating. Both groups found that raising intracellular Mg^{2+} from 0 to 100 mM shifts the mSlo1 G–V curve leftward $\sim$100 mV, and that this occurs in the presence or absence of 100–300 μM Ca^{2+}, as if Mg^{2+} were acting at separate sites from the Ca^{2+} binding sites. Furthermore, high Mg^{2+} prevents Ca^{2+} from causing leftward G–V shifts at concentrations greater than 100 μM, and the mutation E399N (indicated in Fig. 5.13A) eliminates responses to Ca^{2+} above 100 μM and Mg^{2+} up to 10 mM.

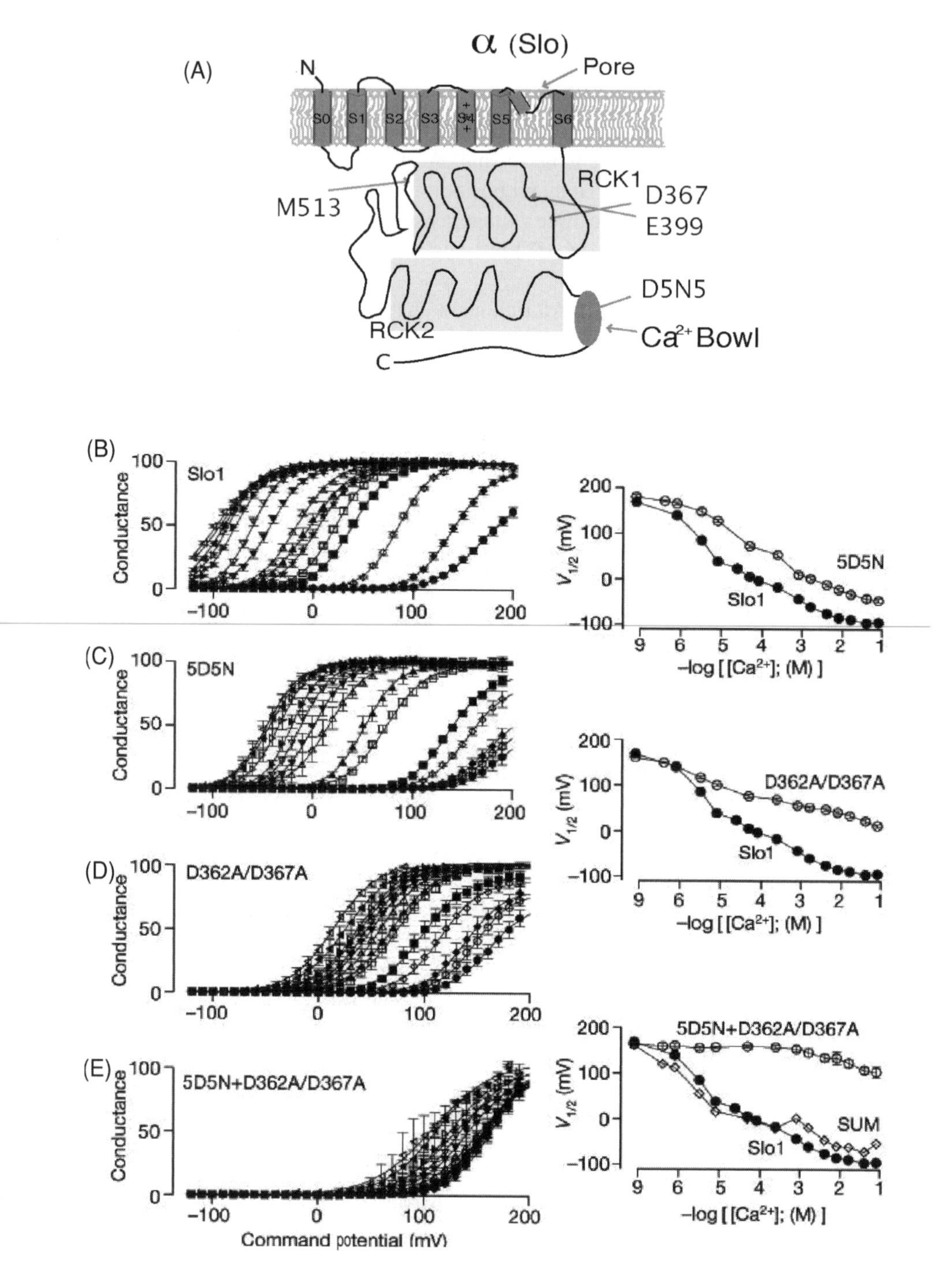
α (Slo)
(A)
N
Pore
S0
S1
S2
S3
S4
S5
S6
RCK1
M513
D367
E399
D5N5
RCK2
Ca2+ Bowl
C
(B)
Slo1
Conductance
5D5N
Slo1
(C)
5D5N
D362A/D367A
Slo1
(D)
D362A/D367A
(E)
5D5N+D362A/D367A
5D5N+D362A/D367A
SUM
Slo1
Command potential (mV)
$V_{1/2}$ (mV)
–log [[Ca2+]; (M)]

Thus, this mutation appears to eliminate a low-affinity Ca^{2+}-binding site that is responsible for the leftward *G–V* shifts cause by [Ca^{2+}] above 100 μM and is also the site of Mg^{2+} action. Indeed when Zhang et al. (2001) added a low-affinity site to the 50-state model discussed above, they found that the model could produce leftward *G–V* shifts at high [Ca^{2+}]. Thus, the BK_{Ca} channel's ability to respond to [Ca^{2+}] over five orders of magnitude is now understood to be due to the presence of both low- and high-affinity Ca^{2+}-binding sites. The low-affinity site has been estimated to have the following dissociation constants for Ca^{2+} : $K_{C\text{-}Ca}$2–3 mM (Zhang et al., 2001), $K_{O\text{-}Ca}$0.6–0.9 mM (Zhang et al., 2001), and for Mg^{2+} : $K_{C\text{-}Mg}$8–22 mM (Shi and Cui, 2001; Zhang et al., 2001), $K_{O\text{-}Mg}$2–6 mM (Shi and Cui 2001; Zhang et al., 2001). Ca^{2+}, therefore, binds weakly, but more tightly than Mg^{2+}.

Under most physiological conditions, these low-affinity sites would not be occupied by Ca^{2+}; however, as [Ca^{2+}] rises transiently into the tens of micromolar—due to Ca^{2+} entry through nearby Ca^{2+} channels or release from intracellular stores—Ca^{2+} binding at these sites may have a small effect (~10 mV *G–V* shift). Similarly, cytoplasmic [Mg^{2+}] has been estimated to be ~0.5 to 1 mM (Zhang et al., 2001), a concentration that by acting through these sites would also be expected to produce a small *G–V* shift (10–25 mV). As yet, however, what role these low-affinity sites play in the functioning of the BK_{Ca} channel in its native settings has yet to be explored.

5.11 The BK_{Ca} Channel Has Two Types of High-Affinity Ca^{2+}-Binding Sites

Another important insight that came in a part from mutagenesis work is that, rather than having a single type of high-affinity Ca^{2+}-binding site, as the HA and the VD-MWC models assume, the BK_{Ca} channel appears to have two structurally distinct, and structurally novel types of high-affinity Ca^{2+}-binding sites.

The great majority of Ca^{2+}-binding proteins thus far discovered (over 300 and counting) contain Ca^{2+}-binding sites of the EF-hand type—calmodulin, troponin C, and parvalbumin for example. In these sites the Ca^{2+} ion is coordinated in a pentagonal–bipyramidal arrangement by oxygen atoms from side chains of amino

Figure 5.13 Mutations that affect BK_{Ca}-channel Ca^{2+} sensing. (A) Diagram of the Slo1 subunit indicating the putative RCK1 and RCK2 domains, the Ca^{2+} bowl, and the positions of mutations D367A, E399A, M513I, and D5N5. (B) Wild-tye mSlo1 *G–V* curves plotted for the following [Ca^{2+}] from right to left in μM: 0, 0.5, 1.4, 10.60, 100, 300, 1000, 2000, 5000, 10,000, 20,000, 50,000, 100,000. (C) *G–V* curves as in *B* for the Ca^{2+}-bowl mutation 5D5N. (D) *G–V* curves as in *B* for the RCK1 mutation D362A/D367A.(E) *G–V* curves as in *B* for the double mutation 5D5N+ D362A/D367A. On the right are shown $V_{1/2}$ vs. [Ca^{2+}] plots for the wildtype channel and the various mutations. Notice that the double mutation in *E* eliminates Ca^{2+} sensing below 1 mM, and the sum of the $V_{1/2}$ vs. [Ca^{2+}] curves for the 5D5N mutation and the D362A/D367A mutation is very similar to the wild-type relation. This indicates that the two mutations are additive and likely acting on separate Ca^{2+}-binding sites in Panels *B–E* are from figure 2 of Xia et al. (2002).

acids spaced approximately every other residue over a loop of 12 residues (Falke et al., 1994). Typically EF-hands come in pairs with binding at one site allosterically affecting binding at the other site. Some proteins have as many as six EF hands, although two or four is more typical (Cox, 1996). The EF-hand consensus sequence is as follows: DX(D/N)X(D/N)-GXXDXXE (Falke et al., 1994). The BK_{Ca} channel has no such sequence.

The other well established high-affinity Ca^{2+}-binding motif is the C2 domain, which is found in such proteins as protein kinase C, phospholipase A2, and synaptotagmin. The C2 domain is composed of ~130 amino acids arranged in two four-stranded β sheets (Nalefski and Falke, 1996). Ca^{2+} ions, usually two or three, are coordinated by acidic residues and backbone carbonyl groups in loops that lie above the β-stranded structure (Nalefski and Falke, 1996). The C2 domain also has a consensus sequence that does not conform to any region in the Slo1 subunit. What can be learned from these motifs, however, is that Ca^{2+}-binding sites are likely to be found in loop regions containing acidic residues, often spaced every other residue.

Although Slo1 does not have any canonical Ca^{2+}-binding motifs, there is a region in the distal portion of Slo1's C-terminal domain that roughly conforms to these criteria and has been strongly implicated as a Ca^{2+} binding site. This region contains 28 amino acids, 10 of which are acidic.

T **E** L V N **D** T N V Q F L **D** Q **D** **D** **D** **D** **D** P **D** T **E** L Y L T Q

The Ca^{2+} Bowl 900

It was given the name "the Ca^{2+} bowl" (above) and proposed to be a Ca^{2+}-binding site by Schreiber and Salkoff (1997) based on their observation that mutations in this region cause rightward *G–V* shifts at moderate and high [Ca^{2+}], but no shift in the absence of Ca^{2+} (Schreiber and Salkoff, 1997). Since then a number of observations have supported this conclusion. Principally, when a portion of Slo1 that includes the Ca^{2+} bowl was transferred to the Ca^{2+}-insensitve Slo3 subunit, Ca^{2+}-sensitivity was conferred upon the previously insensitive channel (Schreiber et al., 1999). And peptides composed of portions of Slo1 that include the Ca^{2+} bowl bind Ca^{2+} in gel-overlay assays (Bian et al., 2001; Braun and Sy, 2001; Bao et al., 2004), and this binding is inhibited by the mutation of Ca^{2+}-bowl aspartic acids (Bian et al., 2001; Bao et al., 2004).

Even large mutations in the Ca^{2+} bowl, however, do not eliminate Ca^{2+} sensing, but rather they reduce the *G–V* shift induced by 100 μM Ca^{2+} by about half (Xia et al., 2002). Where is the remaining Ca^{2+} sensitivity coming from? One idea proposed by Schreiber and Salkoff (1997) is that in addition to the Ca^{2+} bowl, the BK_{Ca} channel has other high-affinity Ca^{2+} binding sites. In support of this possibility they noted that Cd^{2+} can activate the BK_{Ca} channel, and that Ca^{2+}-bowl mutations do not affect Cd^{2+} sensing as they do Ca^{2+} sensing. They proposed that Cd^{2+} was binding selectively to a second high-affinity Ca^{2+}-binding site that is unrelated to the Ca^{2+} bowl.

Also supporting this hypothesis came work from the Moczydlowski, Cox, and Lingle groups who showed that large Ca^{2+}-bowl mutants behave as if they have lost half of their high-affinity Ca^{2+}-binding sites (Bian et al., 2001; Bao et al., 2002; Xia et al., 2002). And most importantly, mutations made far upstream of the Ca^{2+} bowl in the proximal part of the intracellular domain, M513I (Bao et al., 2002) and D367A/D362A (Xia et al., 2002) (see Fig. 5.13A), also reduce the effectiveness of $[Ca^{2+}]$ in shifting the mSlo1 G–V relation (see Fig. 5.13B–D). And when either of these mutations is combined with a large Ca^{2+}-bowl mutation, the channel's Ca^{2+} sensitivity is either severely impaired (Bao et al., 2002) or eliminated up to 100 μM $[Ca^{2+}]$ (Fig. 5.13E). In fact, the Lingle group found that when they combined D367A/D362A with the Ca^{2+}-bowl mutation D857-901N (5D5N) and the low-affinity-site mutation E399A, the channel's Ca^{2+} response was completely eliminated up to 10 mM $[Ca^{2+}]$, and that D367A/D362A, unlike mutations made in the Ca^{2+} bowl, also eliminates Cd^{2+} sensing (Zeng et al., 2005)—consistent with the proposal that Cd^{2+} binds only to the second site (Schreiber and Salkoff, 1997). Together, then, these results argue that the BK_{Ca} channel has three types of Ca^{2+}-binding sites, two of high affinity and one of low affinity, and the two of high affinity are thought to lie within different subdomains (proximal vs. distal) of the channel's large intracellular domain.

There is some disagreement as to the affinities of the two types of high-affinity sites. Bao et al. (2002) estimated the Ca^{2+}-bowl-related site to have the following affinity constants for Ca^{2+}: $K_C = 3.5$ μM and $K_O = 0.8$ μM, while Xia et al. (2002) made the following estimates $K_C = 4.5$ μM and $K_O = 2.0$ μM. For the other site Bao et al.'s estimates were $K_C = 3.8$ μM and $K_O = 0.9$ μM, similar to the Ca^{2+}-bowl-related site, while the Xia et al.'s estimates (Bao et al., 2002) were considerably higher $K_C = 17.2$ μM and $K_O = 4.6$ μM. These groups, however, used different mutations and different models to make their estimates, so some disagreement is not surprising. Which are closer to the truth is as yet unclear.

When making the above estimates both groups relied on the assumption that there are four of each kind of binding site, one in each subunit. While this seems most reasonable, it is not necessarily the case. A lot depends on the symmetry of the intracellular portion of the channel and whether binding sites are contained with in subunits or formed at their interfaces. If each type of binding site is contained within a subunit, and there are three types of binding sites, then the total is 12. If some of the binding sites are formed by the interfaces between subunits, and the cytoplasmic part of the channel is fourfold rotationally symmetric, then still there must be 12. But if the cytoplasmic part of the channel is twofold rotationally symmetric—a dimer of dimers, as is the case for the small conductance Ca^{2+}-activated K^+ channel (Schumacher et al., 2001; Maylie et al., 2004)—then it could be that there are only two of each type of site. And indeed there could be combinations of sites, some at the interfaces and some within subunits, such that reasonably the channel could contain 6, 8, 10, or 12 Ca^{2+} binding sites. Which is actually the case has yet to be definitively determined, however, it appears that there are four Ca^{2+}-bowl-related sites, as when Niu and Magleby (2002) created hybrid channels that contained 1, 2,

3, or 4 subunits mutated in the Ca^{2+} bowl, they observed four phenotypes. Indeed, this work also suggested that the binding of Ca^{2+} at one Ca^{2+}-bowl does not greatly affect binding at Ca^{2+} bowls on other subunits, something that was assumed without evidence in the modeling discussed above.

5.12 Is the BK_{Ca} Channel Like the MthK Channel?

Another important question that has yet to be resolved is: how is the energy of Ca^{2+} binding is transduced into energy for opening? This question may be difficult to answer without a crystal structure, but there is an interesting hypothesis from the bacterial world. In 2002, a crystal structure of MthK, a bacterial Ca^{2+}-activated K^+ channel, was solved by the MacKinnon group (Jiang et al., 2002). Although this channel is very different from the BK_{Ca} channel in many respects—it contains only two membrane spanning helixes and is not voltage-dependent—still it was proposed that the BK_{Ca} channel's Ca^{2+} sensing mechanism may be much like that of the MthK channel (Jiang et al., 2001, 2002). The MthK channel is a fourfold symmetric tetramer with a classic K^+-channel pore sequence (TVGYG). Each monomer has an intracellular C-terminal domain whose core is an RCK domain (Jiang et al., 2001) (Fig. 5.14A). In the MthK structure eight RCK domains come together to form what is called the channel's "gating ring" (Fig. 5.14B)—four from the channel proper and four more derived from a truncated piece of the channel generated off of a second translation start site. Each RCK domain in the gating ring binds one Ca^{2+}, and this—it has been proposed—causes a shearing motion that leads to an expansion of the ring and an opening of the channel's gate (Jiang et al., 2002)(Fig. 5.14D).

Even though the homology between proteins is low (<20%), because of key regions of conservation, and the mutagenesis experiments to be discussed below, the MacKinnon group has proposed that the Slo1 subunit has an RCK domain that forms the proximal part of its cytoplasmic domain (see Fig. 5.1B). Two lines of evidence support this proposition. First, in the crystal structure of an *E. coli* RCK domain—which takes the form of a Rossman fold—there is a salt bridge that is also predicted to exist in the putative RCK domain of Slo1. When Jiang et al. (2002) mutated the residues in the mSlo1 channel that are predicted to form this salt bridge

→

Figure 5.14 The Ca^{2+}-sensing mechanism of the MthK channel. (A) Two RCK domains of the MthK channel. Spheres represent bound Ca^{2+} ions. (B) The "gating ring" of the MthK channel. This structure must be envisioned as hanging below the channel in the cytoplasm. It is compose of eight RCK domains. This crystal structure is from the work of Jiang et al. (2002). (C) Homology model of a portion of mSlo1 RCK1. Indicated are residues whose mutation disrupts low-affinity Ca^{2+} sensing and Mg^{2+} sensing. (D) Jiang et al. have proposed that as Ca^{2+} binds the gating ring expands, due to a rearrangement of its RCK domains, that leads to stress on the linkers leading to the pore helices, and this pulling action opens the channel. *A, B, D* were adapted from Jiang et al. (2002). *C* was adapted from Shi et al. (2002).

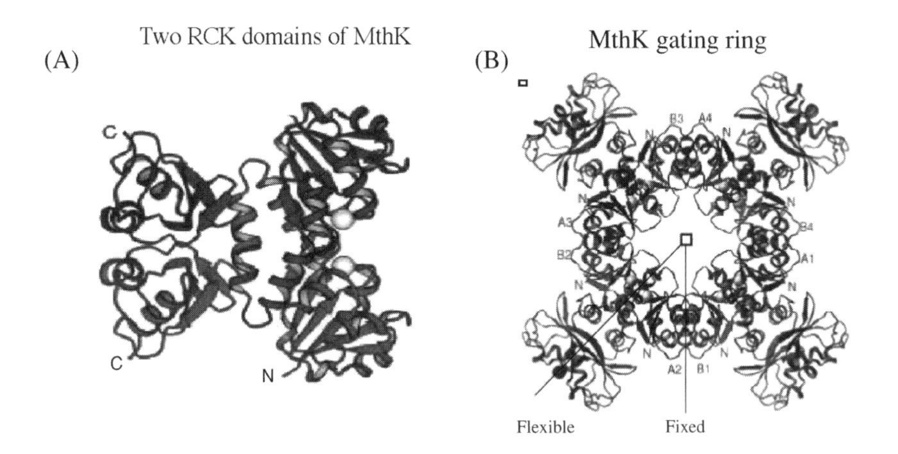

Putative low-affinity Ca^{2+}-binding site

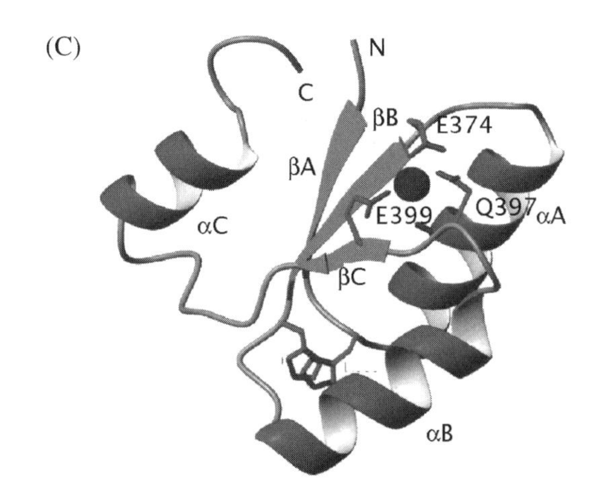

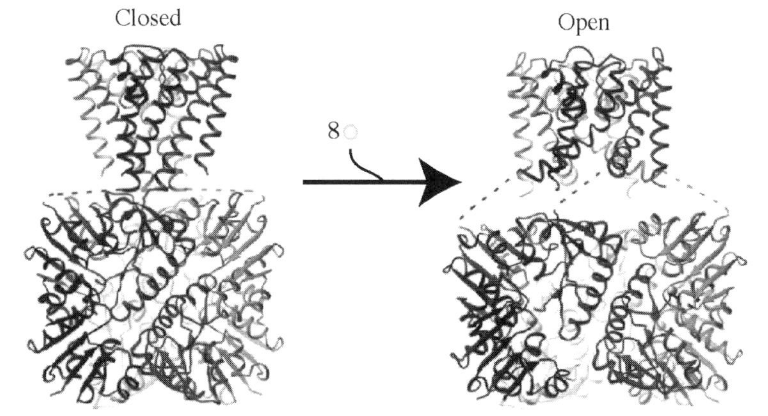

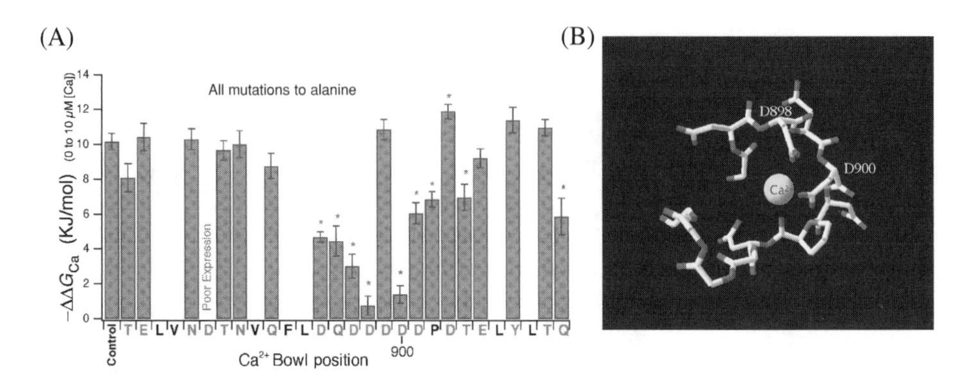

Figure 5.15 An alanine scan of the Ca^{2+}-bowl. (A) On the horizontal axis are listed the residues of the Ca^{2+} bowl. Each oxygen-containing side chain was mutated one at a time to alanine, and the resulting effect on the influence that 10 μM[Ca^{2+}] has on the free energy difference between open and closed is plotted (Bao et al., 2004). These mutations were made in a M5I3I background to eliminate Ca^{2+}-sensing through the RCK1-related high-affinity Ca^{2+}-binding site (Bao et al., 2004). (B) Model from Bao et al. (2004) of Ca^{2+} binding to the Ca^{2+} bowl, based on the mutagenesis data in *A*. This figure was adapted from those in Bao et al. (2004).

to reverse their charges, if either residue were mutated alone, the channel's Ca^{2+} sensitivity was reduced; however, if both residues were mutated together, near wild-type Ca^{2+} sensitivity was restored. This result argues that a salt bridge was broken by each single mutation and then reformed by the double mutation, and therefore that the predicted salt bridge does exist in mSlo1. Second, the Cui group found that if they supposed that the RCK domain existed in Slo1 in the position predicted by Jiang et al., then mutations that they had identified as eliminating low-affinity Ca^{2+} sensing, E399N and D374A, nicely clustered around a site where Ca^{2+} or Mg^{2+} could reasonably bind (Shi et al., 2002) (Fig. 5.14C). Thus, it seems likely that the proximal portion of Slo1's intracellular domain does form an RCK domain now referred to as RCK1.

The gating ring of the MthK channel is composed of eight RCK domains, while if each Slo1 subunit were to have one RCK domain, then that would make only four. The MacKinnon group however, has suggested that each Slo1 subunit has a second RCK domain downstream of the first, and, although they did not specify where it is, it is not difficult to identify a potential second RCK domain by homology to the first (Roosild et al., 2004). So perhaps there are two per subunit that come together to form an eight-membered gating ring like that of MthK. Indeed, consistent with this hypothesis Niu et al. (2004) found that the length of the linker between S6 and RCK1 has a large effect on the intrinsic energetics of channel opening. Lengthening the linker makes it harder to open the channel with depolarization, while shortening the linker makes it easier. Furthermore, lengthening the linker decreases the effectiveness of Ca^{2+} at opening the channel. Although not the only plausible explanation, this is what might be expected if there is a gating ring that exerts an opening force on the pore's S6 gate via these linkers, and this force increases as Ca^{2+} binds and the gating ring expands.

Some data, however, suggest that the mechanism by which the BK_{Ca} channel senses Ca^{2+} must be different from that of the MthK channel. First, the acidic residues that bind Ca^{2+} in MthK's RCK domain are not acidic in the Slo1 RCK domains. Second, the Slo1 residues M513 and D367 do not have counterparts in the MthK RCK domains. They lie in regions of sequence that are aligned as gaps with the MthK RCK sequence (Jiang et al., 2001, 2002). And third, the Ca^{2+} bowl lies outside and downstream of both putative RCK domains of Slo1. Thus, the best candidates for Ca^{2+}-binding sites in Slo1 do not exist in MthK, and the Ca^{2+}-binding sites of MthK do not exist in Slo1. It could be, however, that the gating-ring structure is used by both channels, but the BK_{Ca} channel has evolved a different mechanism for linking Ca^{2+} binding sites to it.

One such binding site is apparently the Ca^{2+} bowl, and it will be very interesting to see this domain's structure, as it is likely to reveal a novel Ca^{2+}-binding site. Indeed, while no structure is yet available, Bao et al. (2004) mutated to alanine all of the oxygen containing side chains in the Ca^{2+} bowl, and they determined that mutations at two positions eliminated Ca^{2+} sensing via the Ca^{2+}-bowl-related site, D898 and D900, while mutations at all other residues had less or no effect (Fig. 5.15A). They proposed that perhaps it is these residues and the backbone carbonyl oxygen of proline 902 that form the binding site (Fig. 5.15B). For now, however, this remains a working hypothesis.

5.13 The Discovery of β1

As discussed above, only a single BK_{Ca}-channel gene has been identified (Atkinson et al., 1991; Adelman et al., 1992; Butler et al., 1993; Pallanck and Genetzky, 1994). Native BK_{Ca} channels from various tissues, however, differ widely in their apparent Ca^{2+} sensitivities (Latorre et al., 1989; McManus, 1991). In neurons and skeletal muscle, for example, the $[Ca^{2+}]$ at which BK_{Ca} channels are half activated ($[Ca]_{1/2}$) is between 5 and 100 μM at 0 mV, whereas in smooth muscle it is usually between 0.1 and 1 μM, and in some secretory cells it even lower (McManus, 1991). Some of this phenotypic diversity may arise from alternative RNA splicing, as nine splice sites and many alternative exons have been identified in mammalian slo1 sequences (Adelman et al., 1992; Butler et al., 1993; Pallanck and Ganetzky, 1994; Tseng-Crank et al., 1994; Ferrer et al., 1996; Saito et al., 1997; Xie and McCobb, 1998; Hanaoka et al., 1999; Langer et al., 2003). However, it now appears that the most important source of functional diversity is the regulated expression of BK_{Ca} β subunits (Lu et al., 2006).

Although four α subunits form a functional channel, when Slo1 was purified from airway smooth muscle it was found to be associated with a smaller auxiliary subunit now termed β1 (Garcia-Calvo et al., 1994; Knaus et al., 1994a,b). When β1 was expressed with α, it was found to cause large leftward shifts in the channel's

G–V relation (Fig. 5.16A and B) such that at almost all voltages the channel becomes more Ca^{2+} sensitive (McManus et al., 1995). At 0 mV, for example, the β1 subunit enhances the apparent Ca^{2+} affinity of the mSlo channel by 10-fold (Bao and Cox, 2005) (Fig. 5.16C). β1 also generally slows macroscopic relaxation kinetics, particularly at hyperpolarized potentials (Dworetzky et al., 1996; Tseng-Crank et al., 1996) (Fig. 5.16D), it prolongs single-channel burst durations (Nimigean et al., 1999a,b), and it enhances the affinity of the channel for charybdotoxin (Hanner et al., 1997).

A question of recent interest has been how, in terms of the HA model, does the β1 subunit enhance the channel's Ca^{2+} sensitivity. While there is some controversy in this area (Bao and Cox, 2005; Oria and Latorre, 2005), three observations appear central. (1) β1 increases single-channel burst times even in the absence of Ca^{2+}, which indicates that β1 must be working, at least in part, on aspects of gating not involving Ca^{2+} binding (Nimigean et al., 1999b; Nimigean and Magleby, 2000); (2) β1 does not change the critical $[Ca^{2+}]$ at which the BK_{Ca} channel's *G–V* relation starts to shift leftward ~100 nM (Fig. 5.16E), which suggests that β1 does not greatly alter the affinity of the channel for $[Ca^{2+}]$ when it is open K_C (Cox and Aldrich, 2000); and (3) gating current measurements reveal that β1 shifts the closed channel's Q_{fast}–V relation leftward along the voltage axis 71 mV (Bao and Cox 2005) (Fig. 5.16F). This makes the channel's voltage sensors activate at lower voltages, an effect that lowers the free energy difference between open and closed at most voltages, and thereby lowers the apparent affinity of the channel for Ca^{2+} as well. Indeed, by analyzing mSlo1 gating and ionic currents in terms an HA-like model with eight rather than four Ca^{2+} binding sites, Bao and Cox (2005), have suggested that β1's steady-state effects are due to a −71 mV shift in V_{hc}, a −61 mV shift in V_{ho}, and an increase in K_C from 3.7 μM to 4.7 μM. Thus, counter to what one might suppose, the β1 subunit enhances the Ca^{2+} sensitivity of the BK_{Ca} channel by apparently making the channel bind Ca^{2+} with lower affinity when it is closed, and making its voltage-sensors activate more easily.

5.14 Four β Subunits Have Now Been Identified

Since the identification of β1, three β1 homologues have been identified β2–β4. Their sequences are aligned with β1 in Fig. 5.17. Each has a unique tissue distribution and unique effects on channel gating. β1 is primarily found in smooth muscle (Tseng-Crank et al., 1996; Chang et al., 1997; Jiang et al., 1999), and β1's ability to enhance the BK_{Ca} channel's Ca^{2+} sensitivity has been shown to be critical for the proper regulation of smooth muscle tone (Brenner et al., 2000a).

β2 is expressed at high levels in kidney, pancreas, ovary, and adrenal gland, at moderate levels in heart and brain, and at low levels in a wide variety of tissues (Wallner et al., 1999; Xia et al., 1999; Behrens et al., 2000; Uebele et al., 2000; Brenner et al., 2006). It has effects similar to β1 but it also has a longer N-terminus

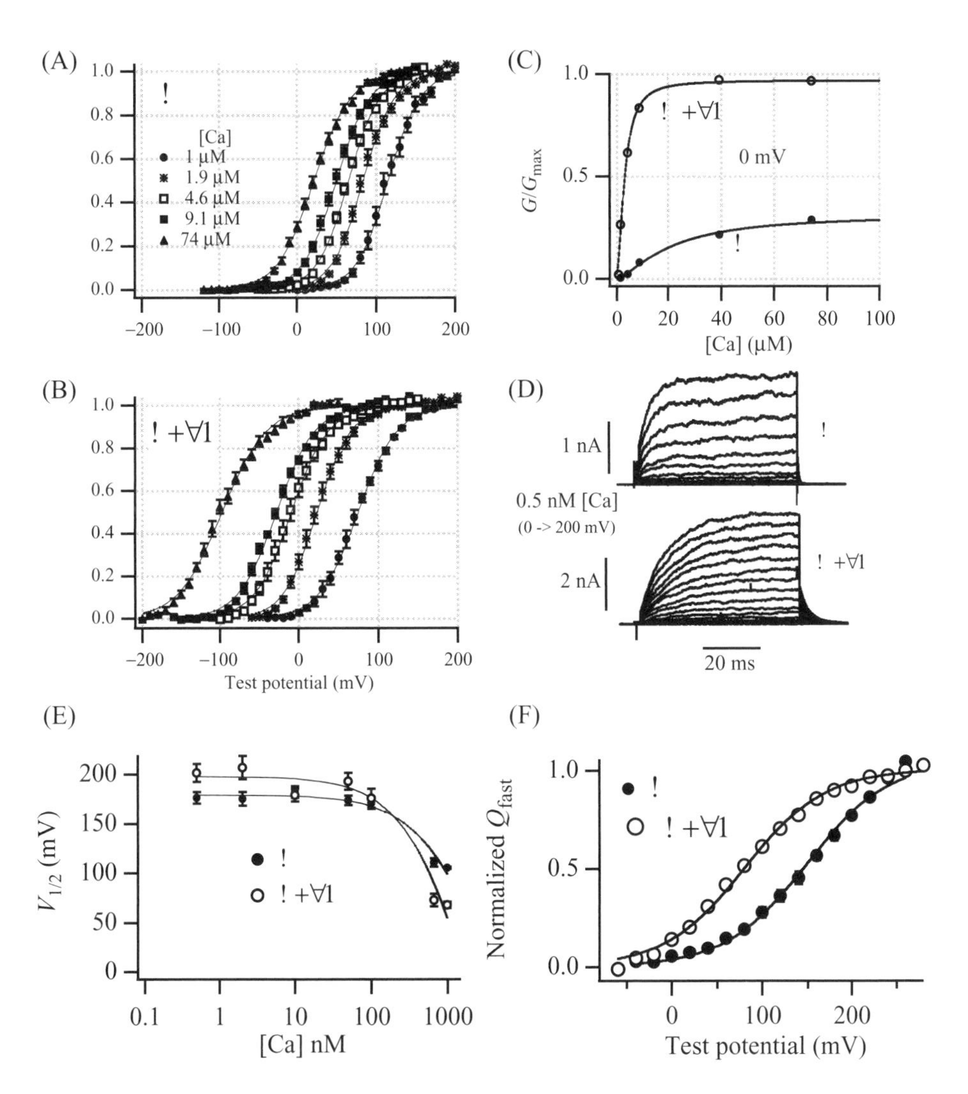

Figure 5.16 $\beta 1$ effects on BK_{Ca}-channel gating. *G–V* relations determined from excised membrane patches expressing (A) the $mSlo1_{\alpha}$ channel (α subunit alone), or (B) $mSlo1{+}\beta 1$ (bovine). (C) Ca^{2+}–dose response curves for $mSlo1_{\alpha}$ and $mSlo1_{\alpha+\beta 1}$. $[Ca^{2+}]_{1/2}$ for $mSlo1_{\alpha} = 32.8\,\mu M$; $[Ca^{2+}]_{1/2}$ for $mSlo1_{\alpha+\beta 1} = 3.4\,\mu M$ at 0 mV. (D) Macroscopic currents recorded from *Xenopus* oocyte macropatches expressing either $mSlo1_{\alpha}$ or $mSlo_{\alpha+\beta 1}$ Test potentials were between 0 and 200 mV. $[Ca^{2+}] = 0.5$ nM. Repolarizations were to -80 mV. $V_{hold} = -50$ mV. (E) Plots of $V_{1/2}$ vs. $[Ca^{2+}]$ for the $mSlo1_{\alpha}$ and $mSlo1_{\alpha+\beta 1}$ channels. Notice both channels start to respond to $[Ca^{2+}]$ at ~100 nM. (F) Q_{fast} vs. voltage curves for the mSlo1 channel with and without $\beta 1$. Notice $\beta 1$ shifts this relation 71 mV leftward without changing its slope. Panel *E* was adapted from Cox and Aldrich (2000). Panel *F* was adapted from Bao and Cox (2005).

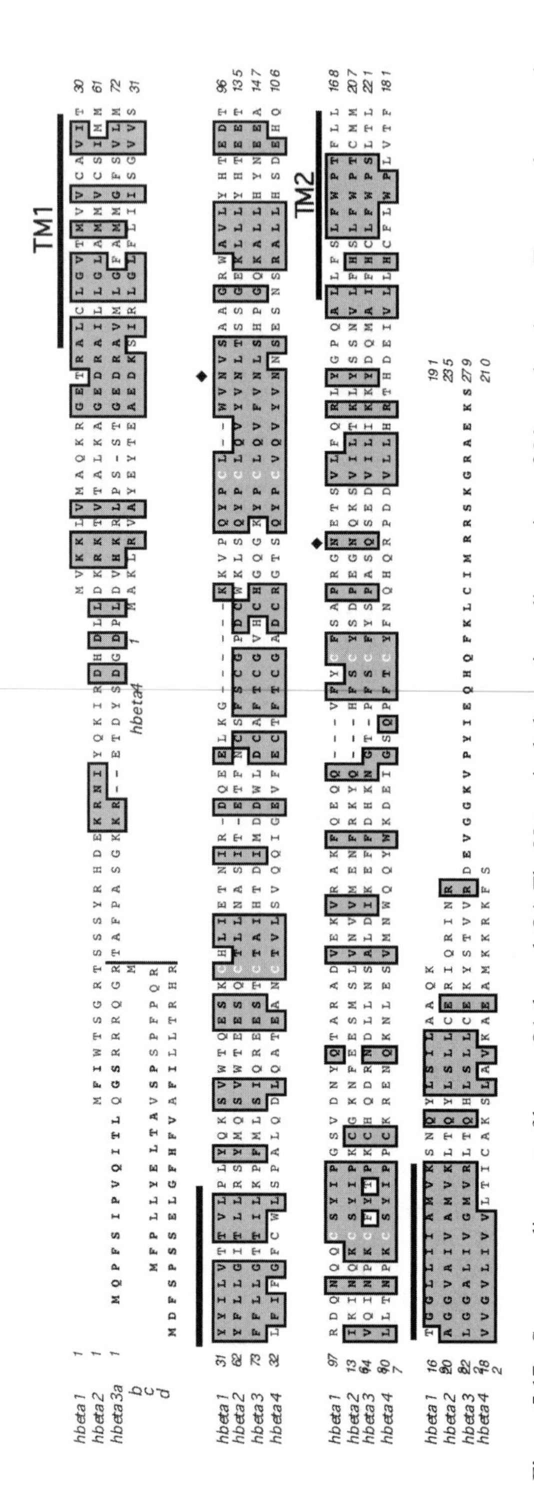

Figure 5.17 Sequence alignment of human β1 through β4. The N-terminal alternative splice versions of β3 are also shown. Transmembrane regions are overlined with dark bars. Conserved cysteines are indicated in white. N-linked glycosylation sites are indicated with ♦. Regions of sequence similarity are boxed and shaded.

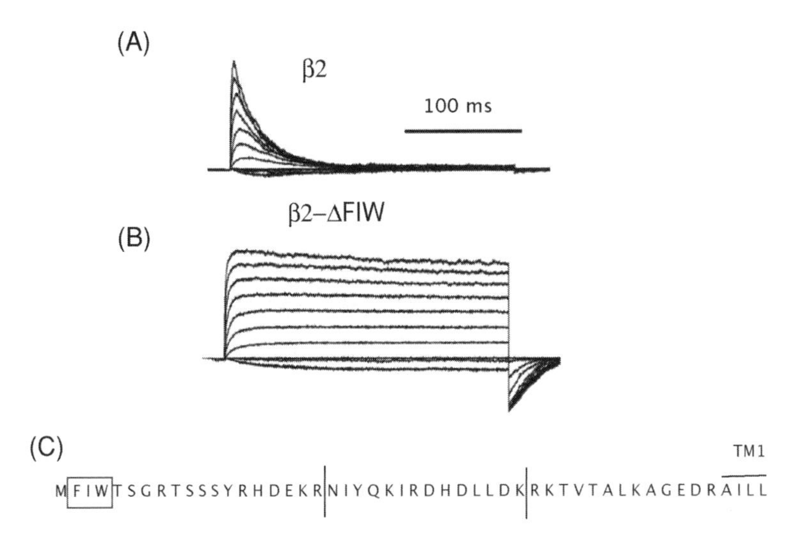

Figure 5.18 The BK$_{Ca}$ β2 subunit caused an N-type inactivation. (A and B) Current family elicited with voltage steps from –100 mV to +140 mV from mSlo1 + β2 channels. Inactivation occurs with a time constant of 20–35 ms. (B) When the N-terminal residues FIW are deleted inactivation no longer occurs. (C) Sequence of the β2 N-terminus. Boxed are the residues deleted in *B*. Panels *A* and *B* were adapted from Xia et al. (2003).

that causes rapid N-type inactivation (Wallner et al., 1999; Xia et al., 1999, 2003; Brenner et al., 2000b) (Fig. 5.18A). Indeed, when the N-terminus of β2 is removed, inactivation is eliminated (Wallner et al., 1999; Xia et al., 1999), and when it is added back as a free peptide, inactivation is again observed as the peptide binds after the channel opens (Wallner et al., 1999). Thus, a "ball and chain" mechanism does seem to apply (Aldrich, 2001). Furthermore, Xia et al. (2003) found that residues 2–4 (FIW) are important determinates of inactivation rate and extent (Fig. 5.18B), while the nature of the residues between this sequence and the first transmembrane domain are not critical, so long as there are at least 12 of them. The β2 subunit is thought to underlie the inactivation of the BK$_{Ca}$ channels of adrenal chromaffin cells (Wallner et al., 1999; Xia et al., 1999).

β3 is expressed at low levels in a wide variety of tissues as well at high levels in testis, pancreas, and heart (Behrens et al., 2000; Brenner et al., 2000b; Uebele et al., 2000). It carries an unusually long C-terminus, and four splice variants of β3 (a–d) have been identified (Uebele et al., 2000) that differ from one another at their N-termini (see Fig. 5.17). β3a–c cause rapid inactivation that is mediated in each case by the subunit's N-terminus (Uebele et al., 2000). The inactivation mediated by β3a and β3c is less complete than that mediated by β2 (Uebele et al., 2000), while β3b causes a very rapid inactivation (Xia et al., 2000; Uebele et al., 2000), so fast that it can only be observed as a decay in current at high voltages where activation is also very fast. β3d has no obvious effects on BK$_{Ca}$-channel activation or inactivation; however, all β3 subunits cause a pronounced inward rectification that is mediated by the β3 extracellular loop (Zeng et al., 2003). This loop, it has been proposed, reaches over the top of the channel to interact with ions as they pass through the channel and toxins as they block the channel. Supporting this proposal, $\alpha + \beta 3$ single-channel recordings show a rapid, flickery gating pattern that is removed when the cysteine residues of the external loop are reduced, and the disulfide bridges they form disrupted (Zeng et al., 2003).

The β4 subunit is found almost exclusively in the brain (Behrens et al., 2000; Brenner et al., 2000b; Weiger et al., 2000), and its main effect on the channel is to slow its kinetics (Behrens et al., 2000; Brenner et al., 2000b; Weiger et al., 2000; Ha et al., 2004). It also has small effects (compared to β1) on steady-state gating causing rightward G–V shifts at low [Ca^{2+}] and leftward G–V shifts at high [Ca^{2+}] (Behrens et al., 2000; Brenner et al., 2000b; Weiger et al., 2000; Ha et al., 2004). β4 knockout mice display prolonged action potentials in the dentate gyrus and seizures focused in this area (Brenner et al., 2005). The β4 subunit also renders the BK_{Ca} channel insensitive to block by charybdotoxin (Meera et al., 2000). There is still a great deal to be learned about how BK_{Ca} β subunits produce their varied effects.

5.15 Conclusions

Over the last two decades a great deal has been learned about the genetics and biophysics of the BK_{Ca} channel. The Slo1 gene was cloned. Sophisticated gating models have been produced. β subunits have been identified as an important source of functional diversity, and some progress has been made in the characterization of the channel's Ca^{2+}-sensing mechanism. In the coming decade the challenge will be to determine the structure of this channel and how this structure leads to the behaviors that have been so carefully documented and explained in energetic terms.

References

Adelman, J.P. et al. 1992. Calcium-activated potassium channels expressed from cloned complementary DNAs. *Neuron* 9:209–216.

Aggarwal, S.K., and R. MacKinnon. 1996. Contribution of the S4 segment to gating charge in the Shaker K^+ channel. *Neuron* 16:1169–1177.

Aldrich, R.W. 2001. Fifty years of inactivation. *Nature* 411:643–644.

Atkinson, N.S., G.A. Robertson, and B. Ganetzky. 1991. A.component of calcium-activated potassium channels encoded by the Drosophila slo locus. *Science* 253:551–555.

Bao, L., and D.H. Cox. 2005. Gating and ionic currents reveal how the BK_{Ca} channel's Ca^{2+} sensitivity is enhanced by its beta1 subunit. *J. Gen. Physiol.* 126:393–412.

Bao, L., C. Kaldany, E.C. Holmstrand, and D.H. Cox. 2004. Mapping the BKCa channel's "Ca2+ bowl": Side-chains essential for Ca^{2+} sensing. *J. Gen. Physiol.* 123:475–489.

Bao, L., A.M. Rapin, E.C. Holmstrand, and D.H. Cox. 2002. Elimination of the BK(Ca) channel's high-affinity Ca(2+) sensitivity. *J. Gen. Physiol.* 120:173–189.

Behrens, R. et al. 2000. hKCNMB3 and hKCNMB4, cloning and characterization of two members of the large-conductance calcium-activated potassium channel beta subunit family. *FEBS Lett.* 474:99–106.

Bezanilla, F. 2005. Voltage-gated ion channels. *IEEE Trans. Nanobiosci.* 4:34–48.

Bhattacharjee, A. et al. 2003. Slick (Slo2.1), a rapidly-gating sodium-activated potassium channel inhibited by ATP. *J. Neurosci.* 23:11681–11691.

Bian, S., I. Favre, and E. Moczydlowski. 2001. Ca^{2+}-binding activity of a COOH-terminal fragment of the Drosophila BK channel involved in Ca^{2+}-dependent activation. *Proc. Natl. Acad. Sci. USA* 98:4776–4781.

Blatz, A.L., and K.L. Magleby. 1984. Ion conductance and selectivity of single calcium-activated potassium channels in cultured rat muscle. *J. Gen. Physiol.* 84:1–23.

Braun, A.F., and L. Sy. 2001. Contribution of potential EF hand motifs to the calcium-dependent gating of a mouse brain large conductance, calcium-sensitive K(+) channel. *J. Physiol.* 533:681–695.

Brelidze, T.I., and K.L. Magleby. 2005. Probing the geometry of the inner vestibule of BK channels with sugars. *J. Gen. Physiol.* 126:105–121.

Brelidze, T.I., X. Niu, and K.L. Magleby. 2003. A ring of eight conserved negatively charged amino acids doubles the conductance of BK channels and prevents inward rectification. *Proc. Natl. Acad. Sci. USA*.

Brenner, R. et al. 2000. Vasoregulation by the β1 subunit of the calcium-activated potassium channel. *Nature* 407:870–876.

Brenner, R., T.J. Jegla, A. Wickenden, Y. Liu, and R.W. Aldrich. 2000b. Cloning and functional characterization of novel large conductance calcium-activated potassium channel beta subunits, hKCNMB3 and hKCNMB4. *J. Biol. Chem.* 275:6453–6461.

Brenner, R. et al. 2005. BK channel beta4 subunit reduces dentate gyrus excitability and protects against temporal lobe seizures. *Nat. Neurosci.* 8(12):1752–1759.

Butler, A., S. Tsunoda, D.P. McCobb, A. Wei. and L. Salkoff. 1993. mSlo, a complex mouse gene encoding "maxi" calcium-activated potassium channels. *Science* 261:221–224.

Chang, C.P., S.I. Dworetzky, J. Wang, and M.E. Goldstein. 1997. Differential expression of the alpha and beta subunits of the large-conductance calcium-activated potassium channel: Implication for channel diversity. *Brain Res. Mol. Brain Res.* 45:33–40.

Cox, D.H., and R.W. Aldrich. 2000. Role of the beta1 subunit in large-conductance Ca(2+)-activated K(+) channel gating energetics. Mechanisms of enhanced Ca(2+) sensitivity. *J. Gen. Physiol.* 116:411–432.

Cox, D.H., J. Cui, and R.W. Aldrich. 1997. Separation of gating properties from permeation and block in mslo large conductance Ca-activated K^+ channels. *J. Gen. Physiol.* 109:633–646.

Cox, D.H., J. Cui, and R.W. Aldrich. 1997. Allosteric gating of a large conductance Ca-activated K^+ channel. *J. Gen. Physiol.* 110:257–281.

Cox, J.A. 1996. Invertebrate, plant, and lower organism calcium binding proteins. *In*: Guidebook to the Calcium-Binding Proteins. M.R. Celio and T.L. Pauls, editors. Oxford University Press, Oxford, pp. 1–14.

Cui, J., and R.W. Aldrich. 2000. Allosteric linkage between voltage and Ca(2+)-dependent activation of BK-type mslo1 K(+) channels. *Biochemistry* 39:15612–15619.

Cui, J., D.H. Cox, and R.W. Aldrich. 1997. Intrinsic voltage dependence and Ca^{2+} regulation of *mslo* large conductance Ca-activated K^+ channels. *J. Gen. Physiol.* 109:647–673.

DiChiara, T.J., and P.H. Reinhart. 1995. Distinct effects of Ca^{2+} and voltage on the activation and deactivation of cloned Ca^{2+}-activated K^+ channels. *J. Physiol. (Lond.)* 489:403–418.

Diaz, L. et al. 1998. Role of the S4 segment in a voltage-dependent calcium-sensitive potassium (hSlo) channel. *J. Biol. Chem.* 273:32430–32436.

Doyle, D.A. et al. 1998. The structure of the potassium channel: Molecular basis of K^+ conduction and selectivity. *Science* 280:69–77.

Dworetzky, S.I. et al. 1996. Phenotypic alteration of a human BK (hSlo) channel by hSlobeta subunit coexpression: Changes in blocker sensitivity, activation/relaxation and inactivation kinetics, and protein kinase A modulation. *J. Neurosci.* 16:4543–4550.

Falke, J.J., S.K. Drake, A.L. Hazard, and O.B. Peersen. 1994. Molecular tuning of ion binding to calcium signaling proteins. *Q. Rev. Biophys.* 27:219–290.

Ferrer, M., M. Meyer, and G. Osol. 1996. Estrogen replacement increases beta-adrenoceptor-mediated relaxation of rat mesenteric arteries. *J. Vasc. Res.* 33:124–131.

Garcia-Calvo, M. et al. 1994. Purification and reconstitution of the high-conductance, calcium-activated potassium channel from tracheal smooth muscle. *J. Biol. Chem.* 269:676–682.

Goldstein, S.A., and C.A. Miller. 1992. A point mutation in a Shaker K^+ channel changes its charybdotoxin binding site from low to high affinity. *Biophys. J.* 62:5–7.

Golowasch, J., A. Kirkwood, and C. Miller. 1986. Allosteric effects of Mg^{2+} on the gating of Ca^{2+}-activated K^+ channels from mammalian skeletal muscle. *J. Exp. Biol.* 124:5–13.

Grissmer, S. et al. 1994. Pharmacological characterization of five cloned voltage-gated K^+ channels, types Kv1.1, 1.2, 1.3, 1.5, and 3.1, stably expressed in mammalian cell lines. *Mol. Pharmacol.* 45:1227–1234.

Ha, T.S., M.S. Heo, and C.S. Park. 2004, Functional effects of auxiliary beta4-subunit on rat large-conductance Ca(2+)-activated K(+) channel. *Biophys. J.* 86:2871–2882.

Hanaoka, K., J.M. Wright, I.B. Cheglakov, T. Morita, and , W.B. Guggino. 1999. A 59 amino acid insertion increases Ca(2+) sensitivity of rbslo1, a Ca^{2+} -activated K(+) channel in renal epithelia. *J. Membr. Biol.* 172:193–201.
Hanner, M. et al. 1997. The beta subunit of the high-conductance calcium-activated potassium channel contributes to the high-affinity receptor for charybdotoxin. *Proc. Natl. Acad. Sci. USA* 94:2853–2858.
Hanner, M. et al. 1998. The beta subunit of the high conductance calcium-activated potassium channel. Identification of residues involved in charybdotoxin binding. *J. Biol. Chem.* 273:16289–16296.
Heginbotham, L., T. Abramson, and R. MacKinnon. 1992. A functional connection between the pores of distantly related ion channels as revealed by mutant K^+ channels. *Science* 258:1152–1155.
Heginbotham, L., and R. MacKinnon. 1993. Conduction properties of the cloned Shaker K^+ channel. *Biophys. J.* 65:2089–2096.
Hille, B. 1992. Ionic Channels of Excitable Membranes. Sinauer Associates, Sunderland, MA.
Horrigan, F. T., and R.W. Aldrich. 1999. Allosteric voltage gating of potassium channels II. Mslo channel gating charge movement in the absence of Ca(2+). *J. Gen. Physiol.* 114:305–336.
Horrigan, F.T., and R.W. Aldrich. 2002. Coupling between voltage sensor activation, Ca^{2+} binding and channel opening in large conductance (BK) potassium channels. *J. Gen. Physiol.* 120:267–305.
Horrigan, F.T., J. Cui, and R.W. Aldrich. 1999. Allosteric voltage gating of potassium channels I. Mslo ionic currents in the absence of Ca(2+). *J. Gen. Physiol.* 114:277–304.
Jiang, Y., A. Pico, M. Cadene, B.T. Chait, and R. MacKinnon. 2001. Structure of the RCK domain from the *E. coli* K^+ channel and demonstration of its presence in the human BK channel. *Neuron* 29:593–601.
Jiang, Y. et al. 2002. Crystal structure and mechanism of a calcium-gated potassium channel. *Nature* 417:515–522.
Jiang, Y. et al. 2003. X-ray structure of a voltage-dependent K+ channel. *Nature* 423:33–41.
Jiang, Z., M. Wallner, P. Meera, and L. Toro. 1999. Human and rodent MaxiK channel beta-subunit genes: Cloning and characterization. *Genomics* 55:57–67.
Knaus, H.G. et al. 1994. Primary sequence and immunological characterization of beta-subunit of high conductance Ca(2+)-activated K^+ channel from smooth muscle. *J. Biol. Chem.* 269:17274–17278.
Knaus, H.G., M. Garcia-Calvo, G.J. Kaczorowski, and M.L. Garcia. 1994. Subunit composition of the high conductance calcium-activated potassium channel from smooth muscle, a representative of the mSlo and slowpoke family of potassium channels. *J. Biol. Chem.* 269:3921–3924.
Kuo, A. et al. 2003. Crystal structure of the potassium channel KirBac1.1 in the closed state. *Science* 300:1922–1926.

Langer, P., S. Grunder, and A. Rusch. 2003. Expression of Ca^{2+}-activated BK channel mRNA and its splice variants in the rat cochlea. *J. Comp. Neurol.* 455:198–209.

Latorre, R., A. Oberhauser, P. Labarca, and O. Alvarez. 1989. Varieties of calcium-activated potassium channels. *Annu. Rev. Physiol.* 51:385–399.

Li, W., and R.W. Aldrich. 2004. Unique inner pore properties of BK channels revealed by quaternary ammonium block. *J. Gen. Physiol.* 124:43–57.

Long, S.B., E.B. Campbell, and R. Mackinnon. 2005. Crystal structure of a mammalian voltage-dependent Shaker family K^+ channel. *Science* 309:897–903.

Lu, R. et al. 2006. MaxiK channel partners: Physiological impact. *J. Physiol.* 570:65–72.

MacKinnon, R. 2003. Potassium channels. *FEBS Lett.* 555:62–65.

MacKinnon, R., L. Heginbotham, and T. Abramson. 1990. Mapping the receptor site for charybdotoxin, a pore-blocking potassium channel inhibitor. *Neuron* 5:767–771.

MacKinnon, R., and C. Miller. 1998. Mechanism of charybdotoxin block of the high-conductance, Ca2+-activated K+ channel. *J. Gen. Physiol.* 91:335–349.

Magleby, K.L., and B.S. Pallotta. 1983. Burst kinetics of single calcium-activated potassium channels in cultured rat muscle. *J. Physiol. (Lond.)* 344:605–623.

Markwardt, F., and G. Isenberg. 1992. Gating of maxi K^+ channels studied by Ca^{2+} concentration jumps in excised inside-out multi-channel patches (myocytes from guinea pig urinary bladder). *J. Gen. Physiol.* 99:841–862.

Maylie, J., C.T. Bond, P.S. Herson, W.S. Lee, and J.P. Adelman. 2004. Small conductance Ca^{2+}-activated K^+ channels and calmodulin. *J. Physiol.* 554:255–261.

McCobb, D.P. et al. 1995. A human calcium-activated potassium channel gene expressed in vascular smooth muscle. *Am. J. Physiol.* 269:H767–H777.

McManus, O.B. 1991. Calcium-activated potassium channels: Regulation by calcium. *J. Bioenerg. Biomembr.* 23:537–560.

McManus, O.B., and K.L. Magleby. 1988. Kinetic states and modes of single large-conductance calcium-activated potassium channels in cultured rat skeletal muscle. *J. Physiol. (Lond.)* 402:79–120.

McManus, O.B., and K.L. Magleby. 1989. Kinetic time constants independent of previous single-channel activity suggest Markov gating for a large conductance Ca-activated K channel. *J. Gen. Physiol.* 94:1037–1070.

McManus, O.B., and K.L. Magleby. 1991. Accounting for the Ca(2+)-dependent kinetics of single large- conductance Ca(2+)-activated K^+ channels in rat skeletal muscle. *J. Physiol. (Lond.)* 443:739–777.

McManus, O.B. et al. 1995. Functional role of the beta subunit of high conductance calcium-activated potassium channels. *Neuron* 14:645–650.

Meera, P., M. Wallner, Z. Jiang, and L. Toro. 1996. A calcium switch for the functional coupling between alpha (hslo) and beta subunits (KV,Ca beta) of maxi K channels. *FEBS Lett.* 382:84–88.

Meera, P., M. Wallner, M. Song, and L. Toro. 1997. Large conductance voltage- and calcium-dependent K^+ channel, a distinct member of voltage-dependent ion channels with seven N-terminal transmembrane segments (S0–S6), an extracellular N terminus, and an intracellular (S9–S10) C terminus. *Proc. Natl. Acad. Sci. USA* 94:14066–14071.

Meera, P., M. Wallner, and L.A Toro. 2000. neuronal beta subunit (KCNMB4) makes the large conductance, voltage- and Ca^{2+}-activated K^{+} channel resistant to charybdotoxin and iberiotoxin. *Proc. Natl. Acad. Sci. USA* 97:5562–5567.

Methfessel, C., and G. Boheim. 1982. The gating of single calcium-dependent potassium channels is described by an activation/blockade mechanism. *Biophys. Struct. Mech.* 9:35–60.

Moczydlowski, E., and R. Latorre. 1983. Gating kinetics of Ca^{2+}-activated K^{+} channels from rat muscle incorporated into planar lipid bilayers. Evidence for two voltage-dependent Ca^{2+} binding reactions. *J. Gen. Physiol.* 82:511–542.

Monod, J., J. Wyman, and J.P. Changeux. 1965. On the nature of allosteric transitions: A plausible model. *J. Mol. Biol.* 12:88–118.

Moss, G.W., J. Marshall, M. Morabito, J.R. Howe, and E. Moczydlowski. 1996. An evolutionarily conserved binding site for serine proteinase inhibitors in large conductance calcium-activated potassium channels. *Biochemistry* 35:16024–16035.

Munujos, P., H.G. Knaus, G.J. Kaczorowski, and M.L. Garcia. 1995. Cross-linking of charybdotoxin to high-conductance calcium-activated potassium channels: Identification of the covalently modified toxin residue. *Biochemistry* 34:10771–10776.

Nalefski, E.A., and J.J. Falke. 1996. The C2 domain calcium-binding motif: Structural and functional diversity. *Protein Sci.* 5:2375–2390.

Nimigean, C.M., J.S. Chappie, and C. Miller. 2003. Electrostatic tuning of ion conductance in potassium channels. *Biochemistry* 42:9263–9268.

Nimigean, C.M., and K.L. Magleby. 1999. β Subunits increase the calcium sensitivity of mSlo by stabilizing bursting kinetics. *Biophys. J.* 76(2):A328.

Nimigean, C.M., and K.L. Magleby. 1999. The beta subunit increases the Ca^{2+} sensitivity of large conductance Ca^{2+}-activated potassium channels by retaining the gating in the bursting states. *J. Gen. Physiol.* 113:425–440.

Nimigean, C.M., and K.L. Magleby. 2000. Functional coupling of the beta(1) subunit to the large conductance Ca(2+)-activated K(+) channel in the absence of Ca(2+). Increased Ca(2+) sensitivity from a Ca(2+)-independent mechanism. *J. Gen. Physiol.* 115:719–736.

Niu, X., and K.L. Magleby. 2002. Stepwise contribution of each subunit to the cooperative activation of BK channels by Ca^{2+}. *Proc. Natl. Acad. Sci. USA* 99:11441–11446.

Niu, X., X. Qian, and K.L. Magleby. 2004. Linker-gating ring complex as passive spring and Ca(2+)-dependent machine for a voltage- and Ca(2+)-activated potassium channel. *Neuron* 42:745–756.

Oberhauser, A., O. Alvarez, and R. Latorre. 1988. Activation by divalent cations of a Ca^{2+}-activated K^{+} channel from skeletal muscle membrane. *J. Gen. Physiol.* 92:67–86.

Orio, P., and R. Latorre. 2005. Differential effects of beta 1 and beta 2 subunits on BK channel activity. *J. Gen. Physiol.* 125:395–411.

Orio, P., P. Rojas, G. Ferreira, and R. Latorre. 2002. New disguises for an old channel: MaxiK channel beta-subunits. *News Physiol. Sci.* 17:156–161.

Pallanck, L., and B. Ganetzky. 1994. Cloning and characterization of human and mouse homologs of the Drosophila calcium-activated potassium channel gene, slowpoke. *Hum. Mol. Genet.* 3:1239–1243.

Pallotta, B.S. 1983. Single channel recordings from calcium-activated potassium channels in cultured rat muscle. *Cell Calcium* 4:359–370.

Roosild, T.P., K.T. Le, and S. Choe. 2004. Cytoplasmic gatekeepers of K^+-channel flux: A structural perspective. Trends *Biochem. Sci.* 29:39–45.

Rothberg, B.S., and K.L. Magleby. 1999. Gating kinetics of single large-conductance Ca^{2+}-activated K^+ channels in high Ca^{2+} suggest a two-tiered allosteric gating mechanism. *J. Gen. Physiol.* 114:93–124.

Rothberg, B.S., and K.L. Magleby. 2000. Voltage and Ca^{2+} activation of single large-conductance Ca^{2+}-activated K^+ channels described by a two-tiered allosteric gating mechanism. *J. Gen. Physiol.* 116:75–99.

Saito, M., C. Nelson, L. Salkoff, and C.J. Lingle. 1997. A cysteine-rich domain defined by a novel exon in a slo variant in rat adrenal chromaffin cells and PC12 cells. *J. Biol. Chem.* 272:11710–11717.

Schreiber, M. et al. 1998. Slo3, a novel pH-sensitive K^+ channel from mammalian spermatocytes. *J. Biol. Chem.* 273:3509–3516.

Schreiber, M. and L. Salkoff. 1997. A novel calcium-sensing domain in the BK channel. *Biophys. J.* 73:1355–1363.

Schreiber, M., A. Yuan, and L. Salkoff. 1999. Transplantable sites confer calcium sensitivity to BK channels. *Nat. Neurosci.* 2:416–421.

Schumacher, M.A., A.F. Rivard, H.P. Bachinger, and J.P. Adelman. 2001. Structure of the gating domain of a Ca^{2+}-activated K^+ channel complexed with Ca^{2+}/calmodulin. *Nature* 410:1120–1124.

Segel, I. 1993. H. Enzyme Kinetics: Behavior and Analysis of Rapid Equilibrium and Steady-State Enzyme Systems. Wiley-Interscience, New York.

Seoh, S., D. Sigg, D.M. Papazian, and F. Bezanilla. 1996. Voltgae-sensing residues in the S2 and S4 segments of the *shaker* K^+ channel. *Neuron* 16:1159–1167.

Shen, K.Z. et al. 1994. Tetraethylammonium block of Slowpoke calcium-activated potassium channels expressed in Xenopus oocytes: Evidence for tetrameric channel formation. *Pflugers Arch.* 426:440–445.

Shi, J., and J. Cui. 2001. Intracellular Mg(2+) enhances the function of BK-type Ca(2+)-activated K(+) channels. *J. Gen. Physiol.* 118:589–606.

Shi, J. et al. 2002. Mechanism of magnesium activation of calcium-activated potassium channels. *Nature* 418:876–880.

Stefani, E. et al. 1997. Voltage-controlled gating in a large conductance Ca^{2+}-sensitive K^+ channel (hslo). *Proc. Natl. Acad. Sci. USA* 94:5427–5431.

Stocker, M., and C. Miller. 1994. Electrostatic distance geometry in a K^+ channel vestibule. *Proc. Natl. Acad. Sci. USA* 91:9509–9513.

Talukder, G., and R.W. Aldrich. 2000. Complex voltage-dependent behavior of single unliganded calcium-sensitive potassium channels. *Biophys. J.* 78:761–772.

Tang, X.D. et al. 2003. Haem can bind to and inhibit mammalian calcium-dependent Slo1 BK channels. *Nature* 425:531–535.

Tseng-Crank, J. et al. 1994. Cloning, expression, and distribution of functionally distinct Ca(2+)-activated K^+ channel isoforms from human brain. *Neuron* 13:1315–1330.

Tseng-Crank, J. et al. 1996. Cloning, expression, and distribution of a Ca(2+)-activated K^+ channel beta-subunit from human brain. *Proc. Natl. Acad. Sci. USA* 93:9200–9205.

Uebele, V.N. et al. 2000. Cloning and functional expression of two families of beta-subunits of the large conductance calcium-activated K^+ channel. *J. Biol. Chem.* 275:23211–23218.

Wallner, M., P. Meera, and L. Toro. 1996. Determinant for beta-subunit regulation in high-conductance voltage-activated and Ca(2+)-sensitive K^+ channels: An additional transmembrane region at the N terminus. *Proc. Natl. Acad. Sci. USA* 93:14922–14927.

Wallner, M., P. Meera, and L. Toro. 1999. Molecular basis of fast inactivation in voltage and Ca^{2+}-activated K^+ channels: A transmembrane beta-subunit homolog. *Proc. Natl. Acad. Sci. USA* 96:4137–4142.

Wang, Y.W., J.P. Ding, X.M. Xia, and C.J. Lingle. 2002. Consequences of the stoichiometry of Slo1 alpha and auxiliary beta subunits on functional properties of large-conductance Ca^{2+}-activated K^+ channels. *J. Neurosci.* 22:1550–1561.

Webster, S.M., D. Del Camino, J.P. Dekker, and G. Yellen. 2004. Intracellular gate opening in Shaker K+ channels defined by high-affinity metal bridges. *Nature* 428:864–868.

Weiger, T.M. et al. 2000. A novel nervous system beta subunit that downregulates human large conductance calcium-dependent potassium channels. *J. Neurosci.* 20:3563–3570.

Xia, X.M., J.P. Ding, and C.J. Lingle. 1999. Molecular basis for the inactivation of Ca^{2+}- and voltage-dependent BK channels in adrenal chromaffin cells and rat insulinoma tumor cells. *J. Neurosci.* 19:5255–5264.

Xia, X.M., J.P. Ding, and C.J. Lingle. 2003. Inactivation of BK channels by the NH2 terminus of the beta2 auxiliary subunit: An essential role of a terminal peptide segment of three hydrophobic residues. *J. Gen. Physiol.* 121:125–148.

Xia, X.M., J.P. Ding, X.H. Zeng, K.L.Duan, and C.J. Lingle. 2000. Rectification and rapid activation at low Ca^{2+} of Ca^{2+}-activated, voltage-dependent BK currents: consequences of rapid inactivation by a novel beta subunit. *J. Neurosci.* 20:4890–4903.

Xia, X.M., X. Zeng, and C.J. Lingle. 2002. Multiple regulatory sites in large-conductance calcium-activated potassium channels. *Nature* 418:880–884.

Xie, J., and D.P. McCobb. 1998. Control of alternative splicing of potassium channels by stress hormones. *Science* 280:443–446.

Yellen, G. 1984. Ionic permeation and blockade in Ca^{2+}-activated K^+ channels of bovine chromaffin cells. *J. Gen. Physiol.* 84:157–186.

Yuan, A. et al. 2000. SLO-2, a K^+ channel with an unusual Cl-dependence. *Nat. Neurosci.* 3:771–779. taf/DynaPage.taf?file=/neuro/journal/v3/n8/full/nn0800_771.html; taf/DynaPage.taf?file=/neuro/journal/v3/n8/abs/nn0800_771.html.

Zagotta, W.N., T. Hoshi, J. Dittman, and R.W. Aldrich. 1994. Shaker potassium channel gating. II: Transitions in the activation pathway. *J. Gen. Physiol.* 103:279–319.

Zeng, X.H., X.M. Xia, and C.J. Lingle. 2003. Redox-sensitive extracellular gates formed by auxiliary beta subunits of calcium-activated potassium channels. *Nat. Struct. Biol.* 10:448–454.

Zeng, X.H., X.M. Xia, and C.J. Lingle. 2005. Divalent cation sensitivity of BK channel activation supports the existence of three distinct binding sites. *J. Gen. Physiol.* 125:273–286.

Zhang, X., C.R. Solaro, and C.J. Lingle. 2001. Allosteric regulation of BK channel gating by Ca(2+) and Mg(2+) through a nonselective, low affinity divalent cation site. *J. Gen. Physiol.* 118:607–636.

6 Voltage-Gated Sodium Channels

Dorothy A. Hanck and Harry A. Fozzard

6.1 Introduction

Voltage-gated sodium channels subserve regenerative excitation throughout the nervous system, as well as in skeletal and cardiac muscle. This excitation results from a voltage-dependent mechanism that increases regeneratively and selectively the sodium conductance of the channel e-fold for a 4–7 mV depolarization of the membrane with time constants in the range of tens of microseconds. Entry of Na^+ into the cell without a companion anion depolarizes the cell. This depolarization, called the action potential, is propagated at rates of 1–20 meters/sec. In nerve it subserves rapid transmission of information and, in muscle cells, coordinates the trigger for contraction. Sodium-dependent action potentials depolarize the membrane to inside positive values of about 30–40 mV (approaching the electrochemical potential for the transmembrane sodium gradient). Repolarization to the resting potential (usually between −60 and −90 mV) occurs because of inactivation (closure) of sodium channels, which is assisted in different tissues by variable amounts of activation of voltage-gated potassium channels. This sequence results in all-or-nothing action potentials in nerve and fast skeletal muscle of 1–2 ms duration, and in heart muscle of 100–300 ms duration. Recovery of regenerative excitation, i.e., recovery of the ability of sodium channels to open, occurs after restoration of the resting potential with time constants of a few to several hundreds of milliseconds, depending on the channel isoform, and this rate controls the minimum interval for repetitive action potentials (refractory period).

The sodium channel that is responsible for this complex time- and voltage-dependent behavior is an integral membrane protein of greater than 200 kDa; it spans the cell's surface membrane with a large mass located both intracellularly and extracellularly. Five structural properties underlie the complex channel function described above: (1) the pore or permeation path, which connects the outside and inside salt solutions, (2) the narrowest region of the pore, which determines the selectivity for sodium over other physiologically present ions, (3) the gates, which open or shut the channel, i.e., control activation and inactivation, (4) the process that couples the gates to transmembrane voltage, and (5) the various binding sites that allow the modulation of the channel by drugs and natural toxins. Various clinical diseases of the nervous system and muscle result from subtle changes in the channel's structure and consequent function.

This chapter will describe the protein, the permeation path and its selectivity property, the gates and their control by membrane voltage, and some of the changes that result in human disease. The crucial and remarkable ability of these voltage-dependent channels to monitor transmembrane voltage has been thoroughly discussed by Francisco Bezanilla in his chapter in this volume, and we will discuss only features that are specific for the sodium channel.

The channel protein (α-subunit) is large, i.e., about 2000 amino acids. It is organized into four homologous and covalently linked domains, each resembling one subunit of the potassium channel. Some isoforms are found together with smaller β-subunits, which are single membrane-spanning subunits that modulate membrane expression and channel function. However, all of the essential elements of the channel seem to be accounted for by the α-subunit. The water-filled cation permeation path, the pore, has been located by interaction with several natural toxins and clinically used drugs. Its narrowest region is responsible for close interaction with the permeating cation, selecting sodium for permeation over all other ions physiologically present. The pore can be occluded by two gates. The activation gate, which is generally thought to lie near the inside mouth of the pore, opens in response to depolarization by coupling with part of the voltage sensor. An intracellular segment is drawn into the pore after opening; blocking permeation. The speed of this process, called fast inactivation, is coupled to the position of voltage sensors. Additional slower processes can also occlude the pore, called slow inactivation. Most diseases associated with sodium channel abnormalities are gain-of-function, although rarer loss of function mutants have also been described, which are associated with sudden cardiac death, usually from a reduction in functional channels. Obviously, loss of all sodium channels in mammalian cells is embryonically lethal. Little direct structural information is available for this large complex protein. However, the opportunity to work with cloned sodium channels expressed in heterologous cells has been a key to relating structure to function in this protein. We also benefit by analogy from insights gained from studies of other cation channel types.

6.2 The Sodium Channel as a Protein

The channel is a large, glycosylated, intrinsic membrane protein. Its primary α-subunit is usually greater than 2000 amino acids long, and this subunit forms the functional unit of the channel (Fozzard and Hanck, 1996; Catterall, 2000). Some sodium channel isoforms in some tissues are expressed with additional smaller subunits (β-subunits), which affect gating kinetics and membrane expression levels. It is possible to identify four highly homologous regions of the α-subunit that appear to form four covalently linked intramembrane domains. Each of the domains (I–IV) includes six α-helical segments that are long enough to cross the membrane (S1–S6). In addition to being similar to each other, the four domains are similar to the six-helix mammalian potassium channel subunit, four of which can assemble to form a channel. The long N- and C-termini of the protein and the three linkers between the four transmembrane domains are intracellular and they are not as homologous. Thus, the α-subunit is a four-domain protein with 24 transmembrane α-helices, organized circumferentially around a central pore. About half of the protein is intracellular, a third is intramembrane, and a sixth is extracellular.

Nine isoforms of mammalian channels have been found so far (Goldin, 2001). Three (Nav1.1, Nav1.2, and Nav1.3) are found in the central nervous system, three are in the peripheral nervous system (Nav1.7, Nav1.8, and Nav1.9), and one (Nav1.6) is seen in both regions. Nav1.4 is exclusively in skeletal muscle, and Nav1.5 is mainly in the heart, but also has been found in the central nervous system. These isoforms are 80–90% identical, but have markedly different gating kinetics and drug/toxin interactions. Many other sodium channel genes have been sequenced from nonmammalian tissues, but only an insect isoform has been expressed and studied.

Similar to the potassium channel subunit, each domain contains one α-helical segment (S4) that contains multiple positively charged residues, located at three-residue intervals, which are responsible for voltage-sensing (Bezanilla, 2000). The S5 and S6 segments of each of the four domain form the lining of the centrally located pore, with their extracellular connecting segments (P loops) folded back into the membrane to form the pore's outer vestibule. The domains are arranged around the pore in a clockwise pattern, when viewed from the outside (Dudley et al., 2000). Thus, the intramembrane part of each domain has a voltage-sensing part—the S1–S4, and a pore-forming part, the S5-P-S6.

No X-ray crystal structures have been obtained of the sodium channel α-subunit, and its size and hydrophobicity suggest that such information will not soon be available. However, its homology to the potassium channel intramembrane component, which has been crystalized, has allowed some inference as to the three-dimensional structure of that part of the protein. Pieces of the cytoplasmic regions have been structurally defined (see later). Crystalization has been successful for several bacterial channels, and a particularly interesting gene (NaChBac) has been isolated from a bacterium. It has significant homology to one domain of the mammalian channel, and expression of this bacterial gene yields a channel that is sodium selective (Ren et al., 2001).

6.3 The Pore

6.3.1 General Strategy

Ions transit the channel by diffusion down an electrochemical gradient through a water-filled path that is created by the protein structure. Although the physiological gradients are such that normally Na^+ diffuses into the cell, experimental conditions can be altered to reverse the flow. The pore is located between the intramembranous parts of the four domains. It is able to discriminate between ions as similar as $Na^+$and K^+, so it must remove some of the ions' waters of hydration, allowing direct interaction between the protein and the ion. Removal of waters of hydration would require excessive energy without surrogate interactions between the ions and the pore wall. Consequently, at least some part of the pore must be narrow enough for its walls to interact directly with ions with an unhydrated radius of ~1 Å or partially dehydrated ions. For our discussion, the pore will be divided into four parts: the outer vestibule, the selectivity ring, the inner pore, and the gating region.

The original definition of the pore was made with biophysical tools—measurements of currents carried by different ions in response to voltage changes. Since the channel has been cloned and subject to expression in cells with little or no voltage-dependent currents, chimerae and point mutations of these channels have been prime tools for unraveling the relationship between structure and function.

In nature there are a number of toxins that target the sodium channel, producing block or changes in gating kinetics (Catterall, 1980). These toxins were crucial in the biochemical purification necessary for the original channel cloning by providing tags to trace the channel in broken membrane fractions, where its characteristic electrical properties could not be measured. Some of the toxins bind with high affinity to specific sites. Two classes that have been important in resolving the pore are the guanidinium toxins, tetrodotoxin (TTX) and saxitoxin (STX), and the μ-conotoxins (μ-CTX). TTX is found mainly in organs of the pufferfish, a culinary delicacy in Japan. STX is the toxin in Red Tide. They are small, compact, and charged molecules that are only somewhat larger than hydrated sodium ions. μ-CTX is a 22-amino acid peptide that is used by the Conus snails to paralyze their prey (French and Dudley, 1999). In addition, a commonly used drug class—local anesthetics (LA)—also target the sodium channel and have been useful in defining the region of their binding.

6.3.2 Outer Vestibule

Shortly after cloning of several isoforms a stretch of amino acid residues in each extracellular connecting segment between each domain's S5 and S6 that is highly conserved across all sodium channel isoforms was predicted to be the region that might determine selectivity (Guy and Seetharamulu, 1986). On the basis of hydrophobicity and secondary structural preferences of the segment they proposed that it folded back into the membrane as a helix-strand hairpin. Because of the circumferential location of the domains, these four "P loops" then could form the outer entrance to the pore. This fourfold P-loop pattern aligned a perfectly conserved ring of amino acid residues at the innermost extent of the hairpin—aspartate, glutamate, lysine, and alanine for domains I to IV, respectively (DEKA motif), and this was proposed to contribute to the sodium selectivity of the channel (see later). Three or four residues distal (toward the C-terminus) from the putative selectivity ring was another perfectly conserved ring of carboxylates—glutamate, glutamate, aspartate, and aspartate, although its critical function was not clear at the time. Noda and colleagues found that neutralization of the glutamate in this domain I outer ring greatly diminished TTX block (Noda et al., 1989), and this finding was expanded to include the other charged ring residues (Terlau et al., 1991). This work confirmed that these

P-loops were part of the TTX binding site and likely to be the outer pore mouth. The cardiac isoform of the sodium channel (Nav1.5) is known to be dramatically less sensitive to TTX block, but only two residues in the lining of the putative outer vestibule are different. Studies of point mutations of these residues showed that the TTX sensitivity differences resulted from the amino acid difference just C-terminus to the domain I selectivity ring aspartate. Cysteine, found in this position in Nav1.5, conferred resistance to TTX (Satin et al., 1992), while tyrosine in Nav1.4 (Backx et al., 1992) or phenylalanine in Nav1.2 (Heinemann et al., 1992) was associated with sensitivity. Subsequently, Nav1.6, with a serine in this position, was found to be resistant. Lipkind and Fozzard (1994) modeled the vestibule by assembling the P-loops to form a coherent binding site for TTX and measured its predicted binding interactions. They found that the aromatic residue in domain I P-loop provided a hydrophobic interaction with TTX that added about -4 kcal mol^{-1} of interaction energy, accounting for the high affinity state. Subsequently, Penzotti et al. (2001) determined the change in TTX-channel interaction energy with mutation of the other critical vestibule residues matched that predicted by the Lipkind–Fozzard model, supporting the model as a reasonable approximation of the conformation of the outer vestibule.

The snail toxin μ-conotoxin (μ-CTX) competes with TTX binding, so its binding site is likely to overlap. Because μ-CTX is a small peptide, analogs could be synthesized with changes in individual amino acid residues, in order to determine their roles in the toxin-channel binding interaction (Becker et al., 1992). Arg-13 was found to play the largest role in μ-CTX binding energy, and mutant cycle analysis was consistent with the critical interaction of Arg-13 being with the domain II outer ring glutamate (Chang et al., 1998). However, modeling the interaction suggested that the toxin bound eccentrically in the pore and failed to occlude it. Careful study indicated that part of the blocking mechanism of Arg-13 and μ-CTX was electrostatic repulsion of sodium ions by the critically located positive charge of Arg-13 (Hui et al., 2002), rather than simple steric occlusion of the pore. In summary, interaction of the guanidinium toxins and μ-CTX with the channel created a consistent structural picture of the outer vestibule as a shallow funnel about 10 Å in diameter at the outside negatively charged ring, which tapers to a 3×5 Å selectivity ring. Its volume is sufficient for 30–40 water molecules, forming a constrained hydrophilic environment above the selectivity filter.

6.3.3 Selectivity

The channel discriminates between Na^+ and K^+ with a permeability ratio of $>10:1$, with a permeation rate under standard conditions of about 10 million Na^+ per second. The radius of Na^+ is 0.95 Å and the radius of K^+ is 1.33 Å, depending on the method of its measurement. Both have a single unit positive charge and are hydrated, with energies of -105 and of 85 kcal mol^{-1}, respectively. When hydrated, these ions are almost indistinguishable, so it was realized quite early that at least partial dehydration of the ions would be required, if they were to be identified, so that only Na^+ is allowed to permeate. However, the energies required for dehydration are huge, and there was no straightforward way for the channel to accomplish this and still allow a permeation rate of 10 million ions per second (maximal transit time of 100 ns). It seemed plausible that some component of the channel pore be able to contribute equivalent energy to the water of hydration and still permit the rapid ion transit. Therefore, it was suggested that the pore protein lining could substitute energetically for water. Since Na^+ has a higher energy of hydration than K^+, the oxygen in a carboxylate group is more likely than the lower energy of the oxygen of a carbonyl or hydroxyl group to interact with Na^+(and visa versa for K^+).

The DEKA inner ring of the vestibule formed by the P-loops has two such carboxylates, with an adjacent lysine. The glutamate in domain II and lysine in domain III were critical for Na:K selectivity, and the lysine (or arginine) was necessary to block Ca permeation (Favre et al., 1996; Schlief et al., 1996). Favre et al. suggested that in the selectivity ring the lysine acts as a tethered cation, blocking interaction of cations with the carboxylates, and Na^+ is energetically able to move it away (Lipkind and Fozzard, 2000). The lysine also lowers binding affinity of Na^+ for the selectivity filter, favoring dissociation and rapid permeation (Fig. 6.1).

The influence of the outer ring carboxylates on selectivity is small, but they play a large role in permeation rate (Terlau et al., 1991). One possible clue to the outer ring function is that the sodium current is quite sensitive to pH in the physiological range. The current can be titrated by acid solutions with single site pKa values of about 6. Free aspartate and glutamate have pKa values of 4.0–4.5, but this is shifted markedly in the alkaline direction in a negative field. The outer ring carboxylates do create a strong negative field in the vestibule, shifting pKa values of the amino acids in the vestibule lining into the physiological range. Because of the single site titration curve, they appear to create a single site in the vestibule that may contribute to the dehydration process (Schild and Moczydlowski, 1994; Khan et al., 2002). If these carboxylates are neutralized by protons at low pH, the field is reduced and they are less effective in dehydration of Na^+, reducing the single-channel conductance. The calculated field in the vestibule was −58 to −93 mV. For comparison, Hui et al. (2002) titrated a histidine mutant of μ-CTX in the pore and estimated that the vestibule potential was at least −100 mV.

6.3.4 Inner Pore Gates

Sodium channels have at least two structures that gate the channel open or closed. The activation gate is opened within microseconds by depolarization and consequent movement of the S4 voltage sensors. A fast inactivation gate then shuts, partly related to depolarization-induced movement of the S4 voltage sensors and partly related to the open conformation of the activation gate (see later). There are several slower inactivation processes that are less well understood and may result from collapse of some part of the pore.

The crystalization of a bacterial potassium channel (Doyle et al., 1998) predicts a general structure for the pore of ion selective channels. The KcsA crystal structure

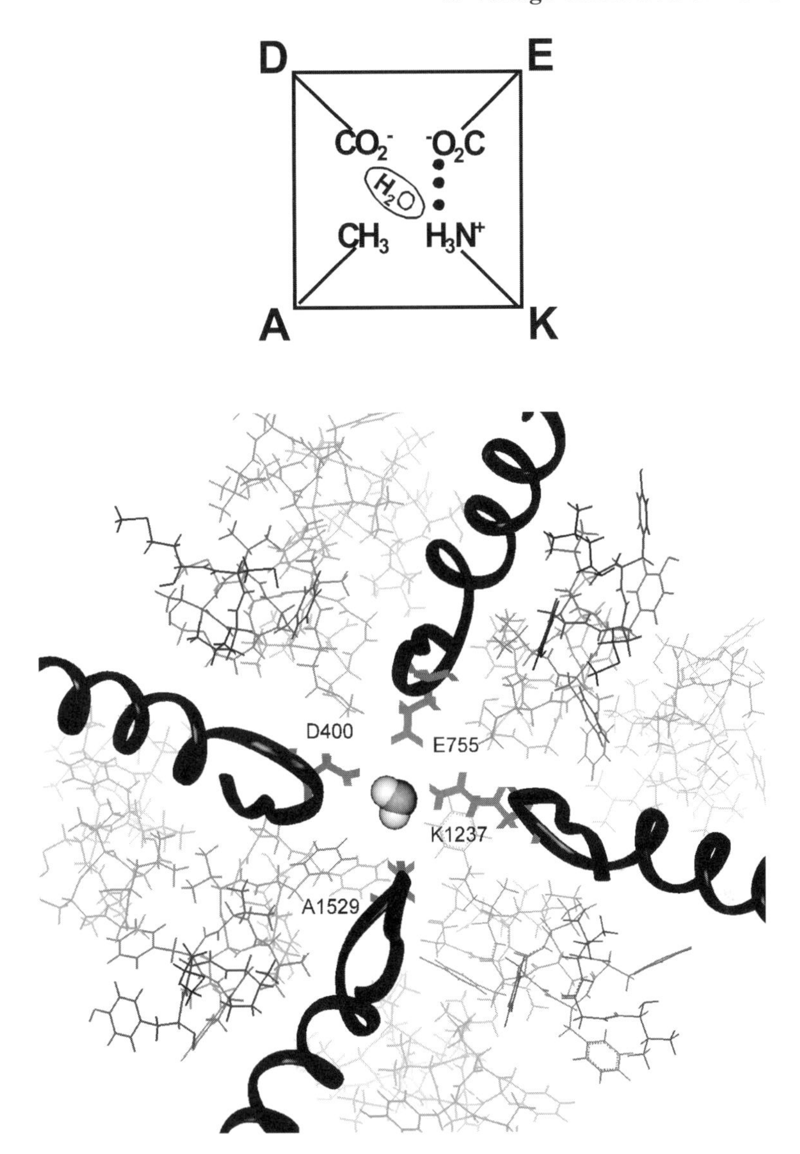

Fig. 6.1 The selectivity filter of the Na channel (using sequence of μ1) with residues of the DEKA motif shown schematically and by space-filling images. The side chains of Glu-755 and Lys-1237 of domains II and III are located at a distance compatible with a salt bridge interaction. The side chains of Asp-400 of domain I and Lys-1237 are separated by one molecule of water, which is shown in the center of the figure. The open area inside the pore is about 3 × 5 Å (in the absence of water), and there is room for the side chain of Lys-1237 to more toward Ala-1529. DEKA (Asp-400, Glu-755, Lys-1237, and Ala-1529) are located in the turns of P-loops of domain I–IV, which are shown by ribbons. The space between the P-loops is filled by S5 and S6 α-helices. This arrangement is based on available mutagenesis data and binding data for TTX and STX (Lipkind and Fozzard, 2000).

shows that the S5-P-S6 components of each domain cross the membrane at an angle, overlapping near the inside. This "teepee" structure of the KcsA channel is characteristic of a closed channel. Comparison with the crystal structure of the open MthK channel (Jiang et al., 2002) shows that for an open channel the S6 helices are bent at glycine residues and the inner halves are hinged away from the pore. Using a similar orientation of the S6 helices of the sodium channel, Sunami et al. (2004) found that the mutants F1579C and V1583C of IVS6, residues predicted to be located just below the selectivity filter were accessible to the hydrophilic, positively charged, MTSET from the inside, only if the activation gate was opened, placing the activation gate at the S6 crossover (Lipkind and Fozzard, 2000). The location of the S6 activation hinge is not entirely clear yet. For MthK the hinges were glycine residues, and for the Shaker potassium channel it is probably a proline-x-proline. The sodium channel S6 segments have no prolines and in place of a glycine in domain IV there is a serine. Using serines and glycines as hinges Lipkind and Fozzard (2005) were able to model an open pore for the sodium channel.

6.4 Gating

Prior to the molecular cloning of channels, our understanding of how voltage controls gating was almost exclusively derived from investigation of voltage-gated sodium channels in a few model organisms where channel expression was particularly high, e.g., squid giant axon. After cloning, however, potassium channels became the most common channel-type used for gating studies. DNAs were <2 kb rather than the >6 kb for sodium channels, since potassium channels could be expressed with four identical subunits rather than as a single polypeptide with four connected, nonidentical pseudo-domains. Not only is it easier to work with smaller DNAs, but it was reasonable to assume that expressed channels would be fourfold symmetric, and introduced amino acid changes would yield four substitutions per channel. Potassium channels, especially Shaker delta (amino acids 4–46 deleted, which removed the fast inactivation particle) expressed well, and the clones were freely shared in the research community. Consequently, they became the standard for studies of voltage gating. The advantage of multiple investigators working on the same isoform, often expressed in the same background (usually *Xenopus* oocytes) was a boon to biophysical investigations because time need not be spent considering isoform or background and experimental differences, and investigators could build on each others' results. Such studies, in combination with the crystallization of a voltage-gated bacterial channel, has produced a fairly detailed, although not yet complete, understanding of how voltage controls opening (activation) of voltage-gated channels. The reader is directed to the discussion of gating by Bezanilla in this volume, which focuses on activation gating largely derived from studies of these potassium channels.

Voltage-dependent sodium channels are more complicated than potassium channels because the sodium channels have four pseudo-domains coded by a single

DNA, which produces channels with four nonidentical gating structures. Simple inspection suggests that either each domain contributes to all gating processes, or more intriguingly, that domains have differentiated to become important for different aspects of gating. In fact the first structure–function studies of sodium channels (Stuhmer et al., 1989), which are principally remembered because they were the first to establish the role of S4 positive charges as the voltage sensor, also reported experiments in which DNA was cut in a region predicted to be in the intracellular linker between domains III and IV, producing channels that inactivated poorly. At the time the suggestion was made that the Hodgkin–Huxley description of sodium channel gating as m^3h (Hodgkin and Huxley, 1952) might arise because the S4s in domains I, II, and III accounted for activation while the S4 of domain IV controlled inactivation. This would contrast with potassium channels, where activation was described as n^4, with all four voltage sensors contributing to activation.

To understand why the idea of identifying the S4 segments with the Hodgkin–Huxley probability factors was not immediately seized upon with enthusiasm, one needs to consider several earlier studies beginning with the development of gating charge measurement techniques in the 1970s. Gating currents represent the membrane delimited movement of the S4 charged amino acids in response to changes in membrane potential. The suggestion by the Hodgkin–Huxley analysis that inactivation was a single, voltage-dependent process that proceeded similarly from all closed states as well as the open state (h) was one of the important predictions that could be addressed by experiments measuring gating currents. In a landmark paper, Armstrong and Bezanilla (1977) established that a voltage dependence of inactivation could not be directly demonstrated. Rather, inactivation could be appreciated only by a slow time-dependence of the recovery of activation charge, a process they described as charge immobilization. Other experimenters confirmed these results, leading to the idea that inactivation was independent of voltage, and its apparent voltage dependence arose because it was linked to voltage-dependent activation. Experimental data in sodium channels also supported the idea that voltage sensing was not the coordinated movement of identical voltage sensors (independence of gating). This meant that m^3h was a useful formalism for model calculations, but actual activation gating proceeded as a complex process with different amounts of gating charge in each step.

Recording of single channel activity made it possible for the first time to obtain direct estimation of channel gating transition rate constants. Aldrich, Corey, and Stevens recorded single-channel events of sodium channels in mammalian neuroblastoma cells, observing that mean open time was much shorter than the decay of the macroscopic current and was voltage independent over a large voltage range (Aldrich et al., 1983). They concluded, therefore, that inactivation from the open state was essentially voltage independent and the voltage dependence of macroscopic current decay arose not from a voltage-dependent inactivation process but from the distributed arrival of channels at the open state, i.e., the voltage dependence of activation. Based on these 1983 experiments and those they published later (Aldrich and Stevens, 1987), in combination with the available gating current data, the idea that inactivation was itself a voltage-independent process became accepted as axiomatic. When single-channel recordings were finally achieved in squid giant axon, Vandenberg and Bezanilla (1991) also described inactivation from the open state as voltage independent, although they required the I–O transition to be voltage dependent.

After heterologous expression of cloned sodium channels became available, mutagenesis studies in the early 90s led to the identification of a stretch of residues in the intracellular linker between domains III and IV that needed to be hydrophobic in order for inactivation from the open state to occur. The IFM motif (West et al., 1992) was supported experimentally by other similar experiments of the time, and the idea that this region of the channel bound and occluded permeation, acting as a hinged lid (Catterall, 2000), became accepted. The similarity between these data in sodium channels and the ball and chain motif identified by the Aldrich laboratory in the N-terminus of the Shaker potassium channel (Hoshi et al., 1990), where decay of the current and fast inactivation was essentially voltage independent, comfortably supported the gating current data of the late 70s and the single channel data in mammalian neuronal cells. The IFM peptide has been characterized structurally by NMR, showing a hydrophobic core hinged to an α-helical segment (Rohl et al., 1999; Kuroda et al., 2000).

In this same time period, however, several laboratories reported single-channel data for the isoform of sodium channels found in native cardiac cells. Each of these (Berman et al., 1989; Yue et al., 1989; Scanley et al., 1990) observed a biphasic dependence of channel mean open time on voltage, suggesting that inactivation from the open state might be voltage dependent in some sodium channel isoforms.

Before cloning, biophysical investigation of channels was restricted to a few cell types in which expression was particularly high. Mammalian preparations were, in general, not suitable for gating current studies. After development of techniques for the isolation of single cells from mammalian tissues, a number of cardiac preparations were found to express a high density of sodium currents, with Purkinje cells in the conduction system of the heart having a particularly high expression level (Makielski et al., 1987). Development of a large bore pipette made from theta glass allowed for recording from canine Purkinje cells with high fidelity, as well as the ability to change intracellular solutions. Using this recording method, it was possible to observe directly the gating charge of sodium channels as a nonlinear signal that added to the linear charge needed to establish voltage across the cell membrane in these cells (Hanck et al., 1990) (Fig. 6.2).

Gating charge for this cardiac isoform resembled that previously observed for other sodium channels with the exception that rather than moving over a more negative potential range than channels activated, it was quite similar to activation (Fig. 6.3) suggesting a tighter relationship of charge movement to the opening transition than for other sodium channels (Sheets and Hanck, 1999). But it was the use of a toxin that produced insight into whether inactivation was fundamentally different in this

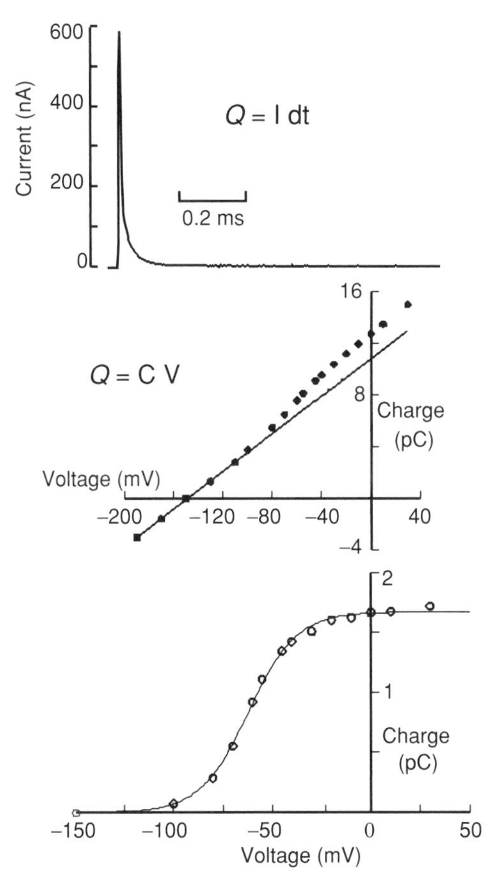

Fig. 6.2 Recording of linear and nonlinear components of capacitive currents. Upper panel shows raw currents recorded from a cardiac Purkinje cell in response to a 40 mV step from −150 mV to −110 mV. Middle panel shows total charge as a function of voltage for a series of steps in which all ionic current was eliminated by replacement of sodium with TMA and addition of 10 μM STX. Line is the best fit to data between −100 and −200 mV and represents the charging of the membrane, capacitance of this cell was 73 pF. Lower panel shows the nonlinear charge (gating charge) calculated as the difference between total charge and the linear component.

isoform. As we have noted, toxins have been important molecules for the understanding of ion channels. Certainly, the use of TTX for selective block of sodium channels is well appreciated (see earlier). Investigators searched for additional channel-specific toxins, and in the 80s the Catterall laboratory was particularly successful at discovering and characterizing toxins that produced a variety of kinetic effects (Catterall, 1980). Members of one class of toxins that was of interest in the cardiac field were the site 3 toxins. Although of diverse origin and molecular weight (isolated from both scorpions and sea anemone) they shared the property of sparing activation and selectively inhibiting inactivation in the neuronal preparations in

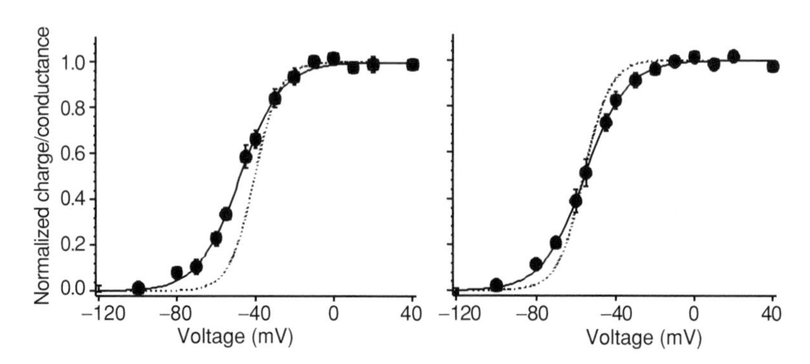

Fig. 6.3 Superimposed peak charge–voltage (•) and conductance–voltage (dotted line) relationships for rSkM1 (μ) sodium channels (left) and for hH1a Na channels (right). Note that for μ charge–voltage relationship is largely to the left of the conductance–voltage relationship whereas for the cardiac isoform the charge–voltage and conductance–voltage relationships are superimposed. Redrawn from Sheets and Hanck (1999).

which they had been studied. One of these, Anthropleura A, which was isolated from the giant green sea anemone found broadly in the pacific coastal tidepools (ApA), is particularly effective in the heart (Hanck and Sheets, 1995; Sheets and Hanck, 1995).

In voltage clamp experiments ApA indeed did not affect the time course of the onset of the current, the time course of tail currents at negative potentials (when the primary transition was channel closure from the open state), or the midpoint of activation. Instead it markedly delayed current decay and, consistent with a selective effect on inactivation from the open state, was associated with a small increase in the rate of recovery of current at negative potentials and a marked slowing of recovery at positive potentials. With respect to gating charge ApA selectively reduced gating charge only at positive potentials, and difference charges (+/− toxin) showed a voltage dependence that was itself voltage dependent. This indicated inactivation from the open state was associated with charge movement rather than deriving its voltage dependence from channel activation (Hanck and Sheets, 1995; Sheets and Hanck, 1995). Interestingly, the toxin produced similar changes in gating in both cardiac and skeletal isoforms, consistent with a similar mechanism of action, albeit different affinity binding site, i.e., inhibition of movement of the domain IV voltage sensor (DIV/S4) (Sheets and Hanck, 1999). This suggests that domain IV S4 movement is tightly coupled to inactivation from the open state in all isoforms.

Additional mutagenesis experiments in $Na_V1.5$, combining S4 charge neutralizations with toxin modification, established that channel opening largely developed before the movement of DIV/S4, i.e., its movement was not required for channels to open (Sheets et al., 1999; Sheets et al., 2000). Movement of the domain IV voltage sensor was slow, and it was closely linked to prompt closure of the inactivation lid. The time course of movement of this charge was not affected by the presence or absence of the lid structure either during activation (on-gating charge) or during repolarization (off-gating charge) (Sheets and Hanck, 2005). The conclusions of these experiments were confirmed by a second technique in which fluorescent markers were added to the channel. This fluorescent technique also demonstrated that the domain IV S4 charge moved slowly; in contrast, charge in domains I/S4 and II/S4 always moved rapidly. Domain III S4 resets slowly after hyperpolarization when fast inactivation is intact, so the presence of an intact lid retards charge recovery (immobilization).

Sodium channel recovery rates can vary significantly as a function of duration and frequency of depolarization, revealing multiple kinetically distinct components of channel inactivation. Although fast inactivation following brief and infrequent depolarization recovers in milliseconds, prolonged or repetitive depolarizations drive channels into more stable conformations from which recovery is on the order of seconds to minutes, a process termed "slow inactivation." Slow inactivation reduces cellular excitability during sustained depolarization caused, for example, by increased extracellular potassium, and during high-frequency bursts of action potentials in nerve and skeletal muscle resulting in their eventual termination. The extent of slow inactivation-induced alterations in excitability varies from tissue to tissue as a result of regional differences in action potential firing frequency and isoform-dependent differences in the extent of slow inactivation. Cardiac sodium channels, for example, do not slow inactivate as quickly or as completely as the skeletal muscle isoform (Richmond et al., 1998; O'Reilly et al., 1999), allowing for sustained firing of action potentials in cardiac tissue. In muscle, slow inactivation has been broken down further kinetically into several discrete components, called slow, intermediate, and ultra-slow. The process has been less well studied in nerve, where it is more difficult to separate discrete inactivation states and the time constants appear to be a continuum.

The structural and mechanistic details underlying slow inactivation, while not well understood, appear to be distinct from those controlling fast inactivation. Slow inactivation does not require prior development of fast inactivation as disruption of fast inactivation, either through intracellular application of proteolytic enzymes that degrade the DIII–IV linker or direct mutation of the IFM motif itself, does not abolish slow inactivation (Rudy, 1978; Cummins and Sigworth, 1996; Featherstone et al., 1996). In fact, removal of fast inactivation by mutation in the IFM motif increased the probability of channels entering the slow inactivated (Featherstone et al., 1996), suggesting that binding of the fast inactivation gate protects the channel from the conformational changes that underlie slow inactivation. Additionally, slow inactivation does not alter the binding and unbinding of the fast inactivation gate as measured by accessibility assays of the IFM motif, further supporting the idea that distinct gating mechanisms control fast and slow inactivation (Vedantham and Cannon, 1998).

A number of mutations altering slow inactivation have been identified but have yet to provide any clues as to the location of the slow inactivation gate(s). Although some mutations affecting slow inactivation have been found in the domain III–IV linker that contains the fast inactivation gate, others are widely scattered in the protein and no pattern is yet apparent. Some of the first mutations modifying slow inactivation were found in Nav1.4 from individuals with hereditary muscle diseases (Barchi, 1997; Cannon, 2000). Others were serendipitous observations on mutations made for other reasons (Balser et al., 1996). At the very least, the identification of mutations that specifically alter slow inactivation gating provides potentially useful tools to probe structure–function questions relating to slow inactivation gating. Presumably, slow inactivation involves slow and extensive, albeit reversible, rearrangement of the protein. Pore collapse, akin to the conformational changes that occur in the pore during C-type inactivation of the potassium channel, has been suggested for slow inactivation of sodium channels. Several lines of evidence are consistent with the view that during slow inactivation the pore is occluded by dramatic rearrangements of the outer vestibule (Benitah et al., 1999; Todt et al., 1999; Ong et al., 2000; Struyk and Cannon, 2002; Xiong et al., 2003), the inner vestibule (Vedantham and Cannon, 2000; Sandtner et al., 2004), or the selectivity filter (Hilber et al., 2001, 2005). The slowing of entry into slow inactivated states by extracellular metals cations (Townsend and Horn, 1997) and a pore-binding toxin peptide (Todt et al., 1999) further supports the idea that a relationship exists between the pore and conformational changes associated with slow inactivation. Although the process is clearly voltage dependent, its relation to the S4 voltage sensors has not been resolved. In summary, the structural mechanism of slow inactivation is obscure at this time but it appears to be less focused and more complex than fast inactivation.

Changes in channel gating and conformational states not only affect excitability of the cells in which they are expressed, but also influence drug sensitivity. The sodium channel is the prime target for a large class of commonly used drugs—local anesthetics (LA). They are thought to block the sodium current by binding to the inner pore of the channel, with their affinity greatly enhanced by repetitive action potentials (use-dependence). Four residues located in the inner half of domain I, domain III, and IV S6 have been identified by mutation to contribute to the LA binding site. The most important are Phe-1764 and Tyr-1771 (Nav1.2) in domain IV S6, located two helical turns apart just below the selectivity filter and near the activation gate (Yarov-Yarovoy et al., 2002). A molecular model of the inner pore and LA binding site (Fig. 6.4) has been developed (Lipkind and Fozzard, 2005). This model is of an open channel based on the suggestion from crystal structure of the potassium channel MthK that the activation gate opens by hinging of S6 segments at glycine and serine residues located just below the filter (Jiang et al., 2002). A key feature of this model is that high affinity LA binding depends on the activated open conformation but not the fast inactivated state, consistent with evidence that this fast gate is not directly affected by LA binding (Vedantham and Cannon, 1999). The mechanism of block is not yet resolved, but is probably not simply steric. One plausible proposal is that LA binding stabilizes the slow inactivated conformation (for review, see Ulbricht, 2005). Alternatives include electrostatic block by the amino head of the LA molecule positioned in the inner pore (Lipkind and Fozzard, 2005) and stabilization of the domain III S4 in its depolarized position (Sheets and Hanck, 2003).

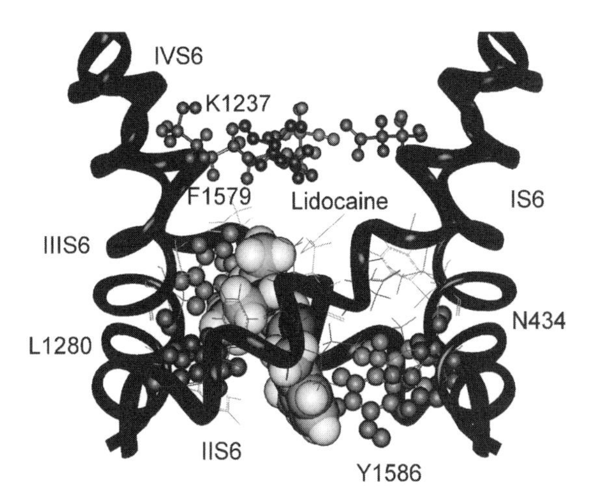

Fig. 6.4 Optimal location predicted for binding of lidocaine (shown as space filled) in the open pore of the Na channel (side view), using the μ1 Na channel sequence. The alkylamine head and the ethyl group at the linker bran produced the most optimal interactions with the side chains of Phe-1579 (DIV, S6) and Leu-1280 (DIII, S6). The dimethyl-substituted aromatic ring is in proximity to the side chain of Tyr-1586 (DIV, S6), which is adjacent to Asn-434 (DI, S6). The side chain of Phe-1579 forms the top of the binding site, and its location provides the most optimal interaction (van der Waals and electrostatic) with lidocaine. The amino acids that are predicted to make contact with lidocaine, based on available mutagenesis data, are shown as balls and sticks. Selectivity filter residues are also shown as balls and sticks in order to show the relationship of the predicted binding site to the selectivity filter. Based on the modeling of Lipkind and Fozzard (2005).

6.5 Hereditary Sodium Channel Diseases

Multiple polymorphisms have been found in the sodium channel gene, some rare and associated with high probability of disease and some common. The first sodium channel monogenic diseases were found in Nav1.4 (Cannon, 1996). Polymorphisms of Nav1.5 are found widely distributed in the population, and show some ethnic clustering (Makielski et al., 2003; Ackerman et al., 2004), and one of these seems to predispose to cardiac arrhythmia (Splawski et al., 2002). With the recent progress in genome-wide screening, it seems likely that differences in the sodium channel gene, in combination with other gene abnormalities or with environmental factors, will be found to be involved in variation in drug metabolism and in disease.

Monogenic skeletal muscle diseases include myotonias and periodic paralysis, depending on whether slow inactivation is enhanced (paralysis) or reduced (myotonia) (Cannon, 1996). The first cardiac mutation to be discovered results in one of the types of long QT syndrome—a highly lethal ventricular arrhythmia in young people (Wang et al., 1995). The mechanism of that mutation is disruption of the IFM inactivation motif. Subsequently, mutations in other locations have been found to cause long QT syndrome, especially ones in the proximal C-terminus.

Another closely related family of mutations have been associated with lethal ventricular arrhythmias in middle-aged persons (Brugada syndrome), and seem to be caused by a reduction in sodium channel expression (Chen et al., 1998). The main use for these discoveries of monogenic disease has been in diagnosis and prognosis, but directed therapy will develop in time. In the central nervous system polymorphisms have not been characterized as much. However, epilepsy associated with fevers has been associated with mutations in the β-subunit of the sodium channel (Wallace et al., 1998). These mutations not only identify disease mechanisms, but they also contribute indirectly to our understanding of the relation between structure and function of the normal sodium channel.

Acknowledgments

We thank Gregory Lipkind for help in figure preparation and Megan McNulty for editorial assistance. Supported by HL-RO1-065661.

References

Ackerman, M.J., I. Splawski, J.C. Makielski, D.J. Tester, M.L. Will, K.W. Timothy, M.T. Keating, G. Jones, M. Chadha, C.R. Burrow, J.C. Stephens, C. Xu, R. Judson, and M.E. Curran. 2004. Spectrum and prevalence of cardiac sodium channel variants among black, white, Asian, and Hispanic individuals: Implications for arrhythmogenic susceptibility and Brugada/long QT syndrome genetic testing. *Heart Rhythm* 1:600–607.

Aldrich, R.W., D.P. Corey, and C.F. Stevens. 1983. A reinterpretation of mammalian sodium channel gating based on single channel recording. *Nature* 306:436–441.

Aldrich, R.W., and C.F. Stevens. 1987. Voltage-dependent gating of single sodium channels from mammalian neuroblastoma cells. *J. Neurosci.* 7:418–431.

Armstrong, C.M., and F. Bezanilla. 1977. Inactivation of the sodium channel. II. Gating current experiments. *J. Gen. Physiol.* 70:567–590.

Backx, P.H., D.T. Yue, J.H. Lawrence, E. Marban, and G.F. Tomaselli. 1992. Molecular localization of an ion-binding site within the pore of mammalian sodium channels. *Science* 257:248–251.

Balser, J.R., H.B. Nuss, N. Chiamvimonvat, M.T. Perez-Garcia, E. Marban, and G.F. Tomaselli. 1996. External pore residue mediates slow inactivation in mu 1 rat skeletal muscle sodium channels. *J. Physiol.* 494(Pt 2):431–442.

Barchi, R.L. 1997. Ion channel mutations and diseases of skeletal muscle. *Neurobiol. Dis.* 4:254–264.

Becker, S., E. Prusak-Sochaczewski, G. Zamponi, A.G. Beck-Sickinger, R.D. Gordon, and R.J. French. 1992. Action of derivatives of mu-conotoxin GIIIA on sodium channels. Single amino acid substitutions in the toxin separately affect association and dissociation rates. *Biochemistry* 31:8229–8238.

Benitah, J.P., Z. Chen, J.R. Balser, G.F. Tomaselli, and E. Marban. 1999. Molecular dynamics of the sodium channel pore vary with gating: Interactions between P-segment motions and inactivation. *J. Neurosci.* 19:1577–1585.

Berman, M.F., J.S. Camardo, R.B. Robinson, and S.A. Siegelbaum. 1989. Single sodium channels from canine ventricular myocytes: Voltage dependence and relative rates of activation and inactivation. *J. Physiol.* 415:503–531.

Bezanilla, F. 2000. The voltage sensor in voltage-dependent ion channels. *Physiol. Rev.* 80:555–592.

Cannon, S.C. 1996. Ion-channel defects and aberrant excitability in myotonia and periodic paralysis. *Trends Neurosci.* 19:3–10.

Cannon, S.C. 2000. Spectrum of sodium channel disturbances in the nondystrophic myotonias and periodic paralyses. *Kidney Int.* 57:772–779.

Catterall, W.A. 1980. Neurotoxins that act on voltage-sensitive sodium channels in excitable membranes. *Annu. Rev. Pharmacol. Toxicol.* 20:15–43.

Catterall, W.A. 2000. From ionic currents to molecular mechanisms: The structure and function of voltage-gated sodium channels. *Neuron* 26:13–25.

Chang, N.S., R.J. French, G.M. Lipkind, H.A. Fozzard, and S. Dudley Jr. 1998. Predominant interactions between mu-conotoxin Arg-13 and the skeletal muscle Na^+ channel localized by mutant cycle analysis. *Biochemistry* 37:4407–4419.

Chen, Q., G.E. Kirsch, D. Zhang, R. Brugada, J. Brugada, P. Brugada, D. Potenza, A. Moya, M. Borggrefe, G. Breithardt, R. Ortiz-Lopez, Z. Wang, C. Antzelevitch, R.E. O'Brien, E. Schulze-Bahr, M.T. Keating, J.A. Towbin, and Q. Wang. 1998. Genetic basis and molecular mechanism for idiopathic ventricular fibrillation. *Nature* 392:293–296.

Cummins, T.R., and F.J. Sigworth. 1996. Impaired slow inactivation in mutant sodium channels. *Biophys. J.* 71:227–236.

Doyle, D.A., J. Morais Cabral, R.A. Pfuetzner, A. Kuo, J.M. Gulbis, S.L. Cohen, B.T. Chait, and R. MacKinnon. 1998. The structure of the potassium channel: Molecular basis of K^+ conduction and selectivity. *Science* 280:69–77.

Dudley, S.C., Jr., N. Chang, J. Hall, G. Lipkind, H.A. Fozzard, and R.J. French. 2000. mu-conotoxin GIIIA interactions with the voltage-gated Na(+) channel predict a clockwise arrangement of the domains. *J. Gen. Physiol.* 116:679–690.

Favre, I., E. Moczydlowski, and L. Schild. 1996. On the structural basis for ionic selectivity among Na^+, K^+, and Ca^{2+} in the voltage-gated sodium channel. *Biophys. J.* 71:3110–3125.

Featherstone, D.E., J.E. Richmond, and P.C. Ruben. 1996. Interaction between fast and slow inactivation in Skm1 sodium channels. *Biophys. J.* 71:3098–3109.

Fozzard, H.A., and D.A. Hanck. 1996. Structure and function of voltage-dependent sodium channels: Comparison of brain II and cardiac isoforms. *Physiol. Rev.* 76:887–926.

French, R.J., and S.C. Dudley Jr. 1999. Pore-blocking toxins as probes of voltage-dependent channels. *Methods Enzymol.* 294:575–605.

Goldin, A.L. 2001. Resurgence of sodium channel research. *Annu. Rev. Physiol.* 63:871–894.

Guy, H.R., and P. Seetharamulu. 1986. Molecular model of the action potential sodium channel. *Proc. Natl. Acad. Sci. USA* 83:508–512.

Hanck, D.A., and M.F. Sheets. 1995. Modification of inactivation in cardiac sodium channels: Ionic current studies with Anthopleurin-A toxin. *J. Gen. Physiol.* 106:601–616.

Hanck, D.A., M.F. Sheets, and H.A. Fozzard. 1990. Gating currents associated with Na channels in canine cardiac Purkinje cells. *J. Gen. Physiol.* 95:439–457.

Heinemann, S.H., H. Terlau, and K. Imoto. 1992. Molecular basis for pharmacological differences between brain and cardiac sodium channels. *Pflugers Arch.* 422:90–92.

Hilber, K., W. Sandtner, O. Kudlacek, I.W. Glaaser, E. Weisz, J.W. Kyle, R.J. French, H.A. Fozzard, S.C. Dudley, and H. Todt. 2001. The selectivity filter of the voltage-gated sodium channel is involved in channel activation. *J. Biol. Chem.* 276:27831–27839.

Hilber, K., W. Sandtner, T. Zarrabi, E. Zebedin, O. Kudlacek, H.A. Fozzard, and H. Todt. 2005. Selectivity filter residues contribute unequally to pore stabilization in voltage-gated sodium channels. *Biochemistry* 44:13874–13882.

Hodgkin, A.L., and A.F. Huxley. 1952. A quantitative description of membrane current and its application to conduction and excitation in nerve. *J. Physiol.* 117:500–544.

Hoshi, T., W.N. Zagotta, and R.W. Aldrich. 1990. Biophysical and molecular mechanisms of Shaker potassium channel inactivation. *Science* 250:533–538.

Hui, K., G. Lipkind, H.A. Fozzard, and R.J. French. 2002. Electrostatic and steric contributions to block of the skeletal muscle sodium channel by mu-conotoxin. *J. Gen. Physiol.* 119:45–54.

Jiang, Y., A. Lee, J. Chen, M. Cadene, B.T. Chait, and R. MacKinnon. 2002. Crystal structure and mechanism of a calcium-gated potassium channel. *Nature* 417:515–522.

Khan, A., L. Romantseva, A. Lam, G. Lipkind, and H.A. Fozzard. 2002. Role of outer ring carboxylates of the rat skeletal muscle sodium channel pore in proton block. *J. Physiol.* 543:71–84.

Kuroda, Y., K. Miyamoto, M. Matsumoto, Y. Maeda, K. Kanaori, A. Otaka, N. Fujii, and T. Nakagawa. 2000. Structural study of the sodium channel inactivation gate peptide including an isoleucine–phenylalanine–methionine motif and its analogous peptide (phenylalanine/glutamine) in trifluoroethanol solutions and SDS micelles. *J. Pept. Res.* 56:172–184.

Lipkind, G.M., and H.A. Fozzard. 1994. A structural model of the tetrodotoxin and saxitoxin binding site of the Na^+ channel. *Biophys. J.* 66:1–13.

Lipkind, G.M., and H.A. Fozzard. 2000. KcsA crystal structure as framework for a molecular model of the Na(+) channel pore. *Biochemistry* 39:8161–8170.

Lipkind, G.M., and H.A. Fozzard. 2005. Molecular modeling of local anesthetic drug binding by voltage-gated sodium channels. *Mol. Pharmacol.* 68:1611–1622.

Makielski, J.C., M.F. Sheets, D.A. Hanck, C.T. January, and H.A. Fozzard. 1987. Sodium current in voltage clamped internally perfused canine cardiac Purkinje cells. *Biophys. J.* 52:1–11.

Makielski, J.C., B. Ye, C.R. Valdivia, M.D. Pagel, J. Pu, D.J. Tester, and M.J. Ackerman. 2003. A ubiquitous splice variant and a common polymorphism affect heterologous expression of recombinant human SCN5A heart sodium channels. *Circ. Res.* 93:821–828.

Noda, M., H. Suzuki, S. Numa, and W. Stuhmer. 1989. A single point mutation confers tetrodotoxin and saxitoxin insensitivity on the sodium channel II. *FEBS Lett.* 259:213–216.

Ong, B.H., G.F. Tomaselli, and J.R. Balser. 2000. A structural rearrangement in the sodium channel pore linked to slow inactivation and use dependence. *J. Gen. Physiol.* 116:653–662.

O'Reilly, J.P., S.Y. Wang, R.G. Kallen, and G.K. Wang. 1999. Comparison of slow inactivation in human heart and rat skeletal muscle Na^+ channel chimaeras. *J. Physiol.* 515(Pt 1):61–73.

Penzotti, J.L., G. Lipkind, H.A. Fozzard, and S.C. Dudley Jr. 2001. Specific neosaxitoxin interactions with the Na^+ channel outer vestibule determined by mutant cycle analysis. *Biophys. J.* 80:698–706.

Ren, D., B. Navarro, H. Xu, L. Yue, Q. Shi, and D.E. Clapham. 2001. A prokaryotic voltage-gated sodium channel. *Science* 294:2372–2375.

Richmond, J.E., D.E. Featherstone, H.A. Hartmann, and P.C. Ruben. 1998. Slow inactivation in human cardiac sodium channels. *Biophys. J.* 74:2945–2952.

Rohl, C.A., F.A. Boeckman, C. Baker, T. Scheuer, W.A. Catterall, and R.E. Klevit. 1999. Solution structure of the sodium channel inactivation gate. *Biochemistry* 38:855–861.

Rudy, B. 1978. Slow inactivation of the sodium conductance in squid giant axons. Pronase resistance. *J. Physiol.* 283:1–21.

Sandtner, W., J. Szendroedi, T. Zarrabi, E. Zebedin, K. Hilber, I. Glaaser, H.A. Fozzard, S.C. Dudley, and H. Todt. 2004. Lidocaine: A foot in the door of the inner vestibule prevents ultra-slow inactivation of a voltage-gated sodium channel. *Mol. Pharmacol.* 66:648–657.

Satin, J., J.W. Kyle, M. Chen, P. Bell, L.L. Cribbs, H.A. Fozzard, and R.B. Rogart. 1992. A mutant of TTX-resistant cardiac sodium channels with TTX-sensitive properties. *Science* 256:1202–1205.

Scanley, B.E., D.A. Hanck, T. Chay, and H.A. Fozzard. 1990. Kinetic analysis of single sodium channels from canine cardiac Purkinje cells. *J. Gen. Physiol.* 95:411–437.

Schild, L., and E. Moczydlowski. 1994. Permeation of Na^+ through open and Zn(2+)-occupied conductance states of cardiac sodium channels modified by batrachotoxin: Exploring ion–ion interactions in a multi-ion channel. *Biophys. J.* 66:654–666.

Schlief, T., R. Schonherr, K. Imoto, and S.H. Heinemann. 1996. Pore properties of rat brain II sodium channels mutated in the selectivity filter domain. *Eur. Biophys. J.* 25:75–91.

Sheets, M.F., and D.A. Hanck. 1995. Voltage-dependent open-state inactivation of cardiac sodium channels: Gating current studies with Anthopleurin-A toxin. *J. Gen. Physiol.* 106:617–640.

Sheets, M.F., and D.A. Hanck. 1999. Gating of skeletal and cardiac muscle sodium channels in mammalian cells. *J. Physiol.* 514(Pt 2):425–436.

Sheets, M.F., and D.A. Hanck. 2003. Molecular action of lidocaine on the voltage sensors of sodium channels. *J. Gen. Physiol.* 121:163–175.

Sheets, M.F., and D.A. Hanck. 2005. Charge immobilization of the voltage sensor in domain IV is independent of sodium current inactivation. *J. Physiol.* 563:83–93.

Sheets, M.F., J.W. Kyle, and D.A. Hanck. 2000. The role of the putative inactivation lid in sodium channel gating current immobilization. *J. Gen. Physiol.* 115:609–620.

Sheets, M.F., J.W. Kyle, R.G. Kallen, and D.A. Hanck. 1999. The Na channel voltage sensor associated with inactivation is localized to the external charged residues of domain IV, S4. *Biophys. J.* 77:747–757.

Splawski, I., K.W. Timothy, M. Tateyama, C.E. Clancy, A. Malhotra, A.H. Beggs, F.P. Cappuccio, G.A. Sagnella, R.S. Kass, and M.T. Keating. 2002. Variant of SCN5A sodium channel implicated in risk of cardiac arrhythmia. *Science* 297:1333–1336.

Struyk, A.F., and S.C. Cannon. 2002. Slow inactivation does not block the aqueous accessibility to the outer pore of voltage-gated Na channels. *J. Gen. Physiol.* 120:509–516.

Stuhmer, W., F. Conti, H. Suzuki, X.D. Wang, M. Noda, N. Yahagi, H. Kubo, and S. Numa. 1989. Structural parts involved in activation and inactivation of the sodium channel. *Nature* 339:597–603.

Sunami, A., A. Tracey, I.W. Glaaser, G.M. Lipkind, D.A. Hanck, and H.A. Fozzard. 2004. Accessibility of mid-segment domain IV S6 residues of the voltage-gated Na^+ channel to methanethiosulfonate reagents. *J. Physiol.* 561:403–413.

Terlau, H., S.H. Heinemann, W. Stuhmer, M. Pusch, F. Conti, K. Imoto, and S. Numa. 1991. Mapping the site of block by tetrodotoxin and saxitoxin of sodium channel II. *FEBS Lett.* 293:93–96.

Todt, H., S.C. Dudley Jr., J.W. Kyle, R.J. French, and H.A. Fozzard. 1999. Ultra-slow inactivation in mu1 Na^+ channels is produced by a structural rearrangement of the outer vestibule. *Biophys. J.* 76:1335–1345.

Townsend, C., and R. Horn. 1997. Effect of alkali metal cations on slow inactivation of cardiac Na^+ channels. *J. Gen. Physiol.* 110:23–33.

Ulbricht, W. 2005. Sodium channel inactivation: Molecular determinants and modulation. *Physiol. Rev.* 85:1271–1301.

Vandenberg, C.A., and F. Bezanilla. 1991. Single-channel, macroscopic, and gating currents from sodium channels in the squid giant axon. *Biophys. J.* 60:1499–1510.

Vedantham, V., and S.C. Cannon. 1998. Slow inactivation does not affect movement of the fast inactivation gate in voltage-gated Na^+ channels. *J. Gen. Physiol.* 111:83–93.

Vedantham, V., and S.C. Cannon. 1999. The position of the fast-inactivation gate during lidocaine block of voltage-gated Na^+ channels. *J. Gen. Physiol.* 113:7–16.

Vedantham, V., and S.C. Cannon. 2000. Rapid and slow voltage-dependent conformational changes in segment IVS6 of voltage-gated Na(+) channels. *Biophys. J.* 78:2943–2958.

Wallace, R.H., D.W. Wang, R. Singh, I.E. Scheffer, A.L. George Jr., H.A. Phillips, K. Saar, A. Reis, E.W. Johnson, G.R. Sutherland, S.F. Berkovic, and J.C. Mulley. 1998. Febrile seizures and generalized epilepsy associated with a mutation in the Na^+-channel beta1 subunit gene SCN1B. *Nat. Genet.* 19:366–370.

Wang, Q., J. Shen, I. Splawski, D. Atkinson, Z. Li, J.L. Robinson, A.J. Moss, J.A. Towbin, and M.T. Keating. 1995. SCN5A mutations associated with an inherited cardiac arrhythmia, long QT syndrome. *Cell* 80:805–811.

West, J.W., D.E. Patton, T. Scheuer, Y. Wang, A.L. Goldin, and W.A. Catterall. 1992. A cluster of hydrophobic amino acid residues required for fast Na(+)-channel inactivation. *Proc. Natl. Acad. Sci. USA* 89:10910–10914.

Xiong, W., R.A. Li, Y. Tian, and G.F. Tomaselli. 2003. Molecular motions of the outer ring of charge of the sodium channel: Do they couple to slow inactivation? *J. Gen. Physiol.* 122:323–332.

Yarov-Yarovoy, V., J.C. McPhee, D. Idsvoog, C. Pate, T. Scheuer, and W.A. Catterall. 2002. Role of amino acid residues in transmembrane segments IS6 and IIS6 of the Na^+ channel alpha subunit in voltage-dependent gating and drug block. *J. Biol. Chem.* 277:35393–35401.

Yue, D.T., J.H. Lawrence, and E. Marban. 1989. Two molecular transitions influence cardiac sodium channel gating. *Science* 244:349–352.

7 Calcium Channels

Ben Corry and Livia Hool

7.1 Introduction

Ion channels underlie the electrical activity of cells. Calcium channels have a unique functional role, because not only do they participate in this activity, they form the means by which electrical signals are converted to responses within the cell. Calcium concentrations in the cytoplasm of cells are maintained at a low level, and calcium channels activate quickly such that the opening of ion channels can rapidly change the cytoplasmic environment. Once inside the cell, calcium acts as a "second messenger" prompting responses by binding to a variety of calcium sensitive proteins. Calcium channels are known to play an important role in stimulating muscle contraction, in neurotransmitter secretion, gene regulation, activating other ion channels, controlling the shape and duration of action potentials and many other processes. Since calcium plays an integral role in cell function, and since excessive quantities can be toxic, its movement is tightly regulated and controlled through a large variety of mechanisms.

The importance of calcium channels is highlighted by the fact many naturally occurring mutations in voltage-gated calcium channel proteins are known to underlie human disorders including childhood absence epilepsy, familial hemiplegic migraine, spinocerebellar ataxia type 6 (a severe movement disorder), hypokalemic periodic paralysis and X-linked congenital stationary night blindness (French and Zamponi, 2005). Mutations in mice have also been shown to cause a variety of disorders including seizures, atrophy, and lethargy. Knockout mice, deficient in various calcium channel pore-forming subunits (see below) are often not viable as calcium channels play an integral role in muscle function. The knockout mice exhibit abnormalities in cardiac muscle contraction, paralysis of the diaphragm, deafness, and epilepsy. Mutations of Ca^{2+} release channel genes are known to be involved in cardiac arrhythmias and malignant hypothermia (Priori and Napolitano, 2005).

The history of calcium channels dates back to soon after the discovery by Hodgkin and Huxley that Na^+ and K^+ currents were responsible for the action potential in squid giant axon. Fatt and Katz (1953) identified a form of electrical excitability in crab muscle that occurred in the absence of Na^+. In 1958, this phenomenon was correctly explained to be due to a "calcium spike," an influx of calcium during the upstroke of the action potential (Fatt and Ginsborg, 1958). Calcium spikes were also measured using radiolabeled calcium in squid axon (Hodgkin and Keynes, 1957) and in response to lowering intracellular calcium concentrations (Hagiwara and Naka, 1964). The fact that Ca^{2+} passed through pathways other than Na^+ and K^+ channels was confirmed with the use of Na^+ and K^+ channel blockers. Ca^{2+} currents are generally difficult to measure, as calcium channels are much less densely distributed than Na^+ or K^+ channels. Final verification of the hypothesis that Ca^{2+} passed through independent channels, and detailed investigations to their physiology had to wait until the development of the patch clamp and giga-seal techniques that allowed for recordings to be made from small cells and for single-channel currents to be recorded. Over the past four decades inward Ca^{2+} currents have been observed in every excitable cell, a huge variety of calcium channels have been identified and their essential biological roles are beginning to be well recognized and characterized.

In this chapter we outline the variety of different calcium channels that are utilized in biological organisms, highlighting their differences, nomenclature and function, and recent research. Next we give a more detailed discussion of the role of calcium in stimulating muscle contraction, neurotransmitter secretion and controlling electrical excitability. We then discuss a variety of properties of these channels examining such questions as how calcium channels discriminate between different ion types and how ions migrate through them. Included in this discussion is an examination of recent theoretical investigations into the nature of ion permeation and selectivity. Finally, we discuss the mechanisms by which ion flow is controlled, reviewing our knowledge of channel gating voltage- and calcium-dependent channel inactivation and the regulation of calcium channels by mechanisms such as thiol group oxidation or reduction.

It is the nature of a topic as large as this that we cannot hope to cover all the areas of current investigation. Instead we have highlighted some areas and focused our attention particularly on the family of voltage-gated calcium channels. Calcium-release, mitochondrial or nuclear channels are no less interesting and significant understanding of their structure and function at both a physiological and molecular level has been determined in recent years. We hope that these may form the focus of another review chapter in the near future. The reader is referred to a number of recent reviews for further information into voltage-gated (Jones, 1998, 2003; Catterall, 2000; Sather and McCleskey, 2003) calcium release (Sutko and Airey, 1996; George et al., 2005; Wehrens et al., 2005) and store-operated channels (Parekh and Putney, 2005).

7.2 Types of Ca^{2+} Channels

Calcium channels come in a range of different shapes and sizes, are located in different regions of the cell or different organs, carry out diverse roles and respond to different stimuli. Broadly, calcium channels are distinguished as either voltage-activated or responding to the binding of calcium or other agonists that release calcium from intracellular stores. Here we introduce the different types of Ca^{2+} channels, noting the names applied to them and the functions they fulfil.

7.2.1 Voltage-Gated Channels

Many cell types, notably excitable cells express voltage-gated Ca^{2+} channels that play an integral role in calcium influx and regulate intracellular processes such as contraction, secretion, neurotransmission, and gene expression discussed in more detail below. They are members of a gene superfamily of transmembrane ion channel proteins that includes voltage-gated K^+ and Na^+ channels. Ca^{2+} channels share structural similarities with K^+ and Na^+ channels in that they possess a pore-forming α_1 subunit in four repeats of a domain with six transmembrane-spanning segments that include the voltage sensing S4 segment and the pore forming (P) region. The α_1 subunit is large (190–250 kDa) and incorporates the majority of the known sites regulated by second messengers, toxins, and drugs. This subunit is usually complexed with at least three auxiliary subunits, α_2, δ, β and γ, with the α_2 and δ subunits always linked by a disulfide bond (Fig. 7.1).

Nomenclature for the Ca^{2+} channels has evolved over the decades since the early electrophysiology studies and has had to incorporate diverse biochemical, pharmacological, and physiological properties of the channels with the more recent identification of channel genes. Initially, Ca^{2+} channels were classified according to their ability to be activated by a large or small depolarization and whether the response was rapid or slow and persistent. For example, L-type Ca^{2+} channels are a member of the HVA or high voltage activation type because the channels are activated by strong depolarizations typically to 0 or +10 mV and are long-lasting in that they are slow to inactivate. In addition, they are blocked by the lipid-soluble 1,4-dihydropyridines such as nifedipine and BAY K 8644. T-type Ca^{2+} channels on the other hand are transiently activated at low membrane potentials and are insensitive

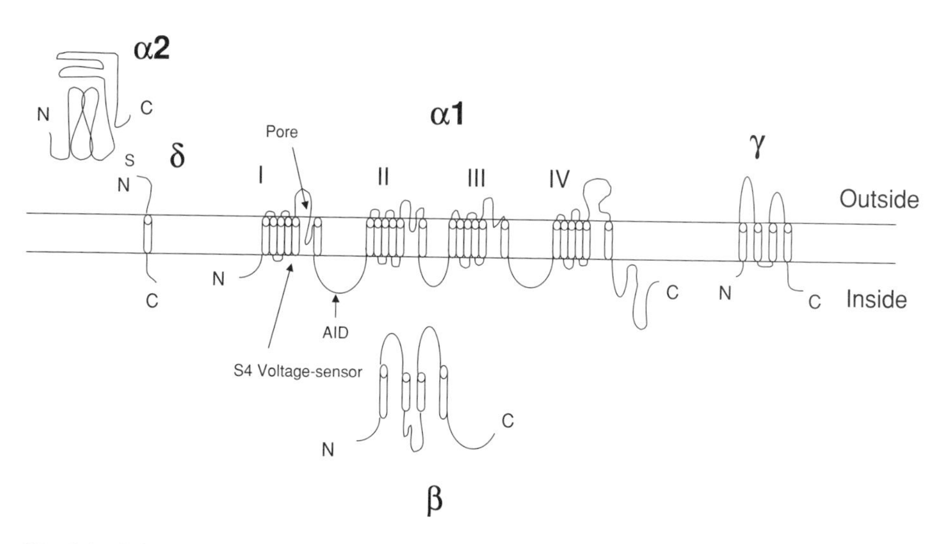

Fig. 7.1 Subunits of Ca_v channels showing the four homologous domains of the α_1 subunit, and auxiliary β, γ, and $\alpha_2\delta$ subunits. Predicted alpha-helices are depicted as cylinders. AID = alpha interacting domain where the β subunit is proposed to interact with the α subunit (see text).

to dihydropyridines and fall into the class of low voltage activated (LVA) channels. Taking into consideration the functional and pharmacological differences, Tsien and colleagues (Tsien et al., 1988) devised a nomenclature based on single letters that continues to be used today to describe the electrophysiological characteristics of expressed channels studied in their native state. In addition to the L-type (long-lasting currents) and T-type (transient currents) channels mentioned above, a number of other channel types were found with single-channel conductances between T- and L-type that were resistant to the dihydropyridines. These channels were predominant in neurons and could be blocked by the ω-Conotoxin GVIA from cone snails. They were named N-type (Nowycky et al., 1985b). A unique Ca^{2+} channel was identified in Purkinje cells by Linás' group that exhibited sensitivity to the venom of funnel web spider (FTX) but was insensitive to ω-Conotoxins and dihydropyridines (Llinas et al., 1989). The channel was named P-type. Another type of channel slightly less sensitive to ω-Conotoxins, termed Q-type, was identified in cerebellar granule neurons that otherwise has very similar physiological characteristics to P-type channels (Randall and Tsien, 1995). An R-type Ca^{2+} channel that is resistant to ω-Conotoxin, ω-Agatoxin IVA, and dihydropyridines but sensitive to ω-Agatoxin IIIA was characterized as a rapid transient current with an activation potential slightly positive to T-type channels (Randall and Tsien, 1997).

This nomenclature required revision as the molecular identities of calcium channel genes were rapidly discovered. Channel types were classified according to distinct hybridization patterns of mRNA on Northern blot analysis and grouped as A, B, C, or D (Snutch et al., 1990). Genes subsequently identified were classified E through I except the skeletal muscle isoform that was referred to as α_{1S} (Birnbaumer et al., 1994). In 2000, Ertel et al. (Ertel et al., 2000) suggested a nomenclature based on that used to classify K^+ channels with Ca as the permeating ion followed by the physiological regulator "v" for voltage in subscript and the gene subfamily in order of discovery. The Ca_v1 subfamily comprises the L-type Ca^{2+} currents, Ca_v2 subfamily comprises the P/Q, N- and R-type Ca^{2+} currents, and Ca_v3 subfamily the T-type Ca^{2+} currents as shown in Fig. 7.2. Detailed listings of voltage-dependent Ca^{2+} channel nomenclature as approved by the Nomenclature Committee of the International Union of Pharmacology including structure–function relationships are available (Catterall et al., 2003, 2005). Table 7.1 lists the nomenclature of Ca^{2+} channels as the classifications have evolved.

Although the main characteristics that define channel subtypes such as ion selectivity, voltage-dependence, and drug binding sites reside in the alpha subunit, the auxiliary subunits also play important regulatory roles. The first β subunit to be identified (now termed $Ca_v\beta1a$) was cloned from skeletal muscle and observed as a 54 kDa protein (Takahashi et al., 1987). To date four genes have been cloned encoding human β subunits ($Ca_v\beta1$–4) and there are several splice variants produced from the genes (for review, see Birnbaumer et al., 1998; Dolphin, 2003). $Ca_v\beta$ subunits are intracellularly located and bind the α subunit with high affinity via the α interaction domain (AID) on the I–II linker (see Fig. 7.1; Pragnell et al., 1994). $Ca_v\beta$ subunits promote functional expression of $Ca_v\alpha$ subunits, localization, and insertion of the

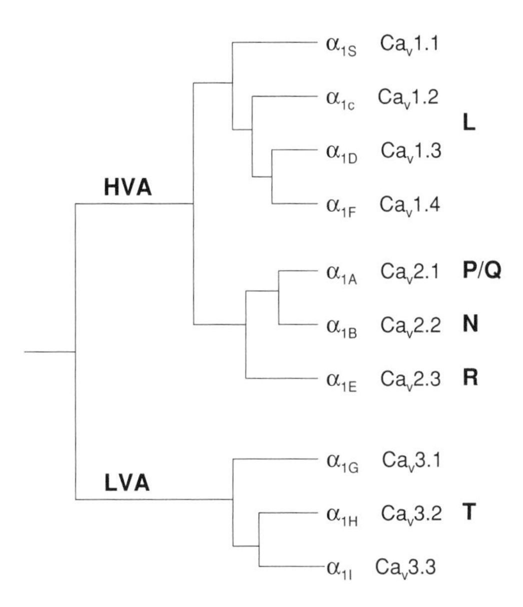

Fig. 7.2 Family tree of mammalian voltage-gated Ca^{2+} channels and nomenclature. Adapted from Jones (2003).

channel complex in the plasma membrane (Chien et al., 1995; Bichet et al., 2000). In addition, $Ca_v\beta$ subunits can modify the kinetics and amplitude of current, including increasing the channel open probability and altering the inactivation rate (Colecraft et al., 2002).

The $\alpha_2\delta$ subunit is highly glycosylated and encoded by a single gene. The δ segment spans the plasma membrane and connects to the extracellular α_2 protein via a disulfide bridge formed between a number of cysteines present on both proteins. To date four genes have been identified that encode $\alpha_2\delta1$–4 subunits (Ellis et al., 1988; Klugbauer et al., 1999; Qin et al., 2002). The α_2 protein has been shown to influence channel stimulation and the δ protein alters voltage-dependent activation and inactivation kinetics (Felix et al., 1997).

In some tissues a fourth auxiliary subunit is expressed, the γ subunit. At least eight genes are know to encode for γ subunits ($Ca_v\gamma$ 1–8), the first known subunit being cloned from skeletal muscle with a mass of 25 kDa (Jay et al., 1990). Co-expression studies reveal the subunit can modulate peak current and activation/inactivation kinetics (Singer et al., 1991).

7.2.2 Ca Release Channels

One of the critical roles served by Ca^{2+} channels in cell function is the maintenance of intracellular calcium levels and the initiation of calcium-dependent cellular processes. This is particularly poignant in cardiac muscle where calcium influx through plasma membrane $Ca_v1.2$ channels is a requirement for the release of calcium from

Table 7.1 Classification schemes for voltage-gated calcium channel proteins. The nomenclature introduced by Tsien is related to those describing the different gene types (Snutch/Birnbaumer and IUP classes). The chromosomes encoding these genes and the tissues in which they are expressed are also indicated.

HVA (high voltage activated, large conductance, persistent)				
Tsien class:	L-type			
Snutch/Birnbaumer class:	α_{1S}	α_{1C}	α_{1D}	α_{1F}
IUP classification:	$Ca_v1.1$	$Ca_v1.2$ a,b,c	$Ca_v1.3$	$Ca_v1.4$
Gene:	CACNA1S	CACNA1C	CACNA1D	CACNA1F
Human chromosome:	1q31-32	12p13.3	3p14.3	Xp11.23
Tissue:	skeletal muscle	heart, smooth muscle, brain, adrenal	brain, pancreas, kidney, cochlea, ovary	retina
HVA (high voltage activated, rapid inactivating)				
Tsien class:	P/Q-type			
Snutch/Birnbaumer class:	α_{1A}	α_{1B}	α_{1E}	
IUP class:	$Ca_v2.1$a,b	$Ca_v2.2$ a,b	$Ca_v2.3$ a,b	
Gene:	CACNA1A	CACNA1B	CACNA1E	
Human chromosome:	19p13	9q34	1q25-31	
Tissue:	brain, cochlear, pituitary	brain, NS	brain, cochlea, retina, heart, pituitary	
LVA (low voltage activated)				
Tsien class:	T-type			
Snutch/Birnbaumer class:	α_{1G}	α_{1H}	α_{1I}	
IUP class:	$Ca_v3.1$	$Ca_v3.2$	$Ca_v3.3$	
Gene:	CACNA1G	CACNA1H	CACNA1I	
Human chromosome:	17q22	16p13.3	22q12.3-13-2	
Tissue:	brain, NS	brain, heart, kidney, liver	brain	

Tsien class (Tsien et al., 1988); Snutch/Birnbaumer class (Snutch et al., 1990; Birnbaumer et al., 1994); IUP class (Ertel et al., 2000; Catterall et al., 2003, 2005).

IUP = International Union of Pharmacology; NS = nervous system.

intracellular sarcoplasmic reticulum stores that then initiates contraction. This process is known as excitation–contraction coupling and will be discussed in detail in Section 7.3. There are, broadly speaking, three families of protein, known as Ca-release channels, that are responsible for the release of calcium from intracellular stores. They are inositol triphosphate receptors (IP_3R), ryanodine receptors (RyR), and calcium-release-activated calcium (CRAC) channels. IP_3Rs and RyRs serve to release calcium into the cytoplasm from the sarcoplasmic reticulum and endoplasmic reticulum of muscle but they are also expressed in many other cell types including neurons where they play a role in calcium-dependent neurotransmission. In muscle the activation of IP_3Rs and RyRs is an important step in the process of muscle contraction, allowing intracellular calcium levels to increase and bind to troponin C that then switches on the contractile apparatus. During the course of a single muscle contraction, free intracellular calcium increases from resting concentrations (typically 10–300 nM) to micromolar levels. Equally important to the release of calcium, however, is the reuptake of calcium into intracellular stores to allow relaxation.

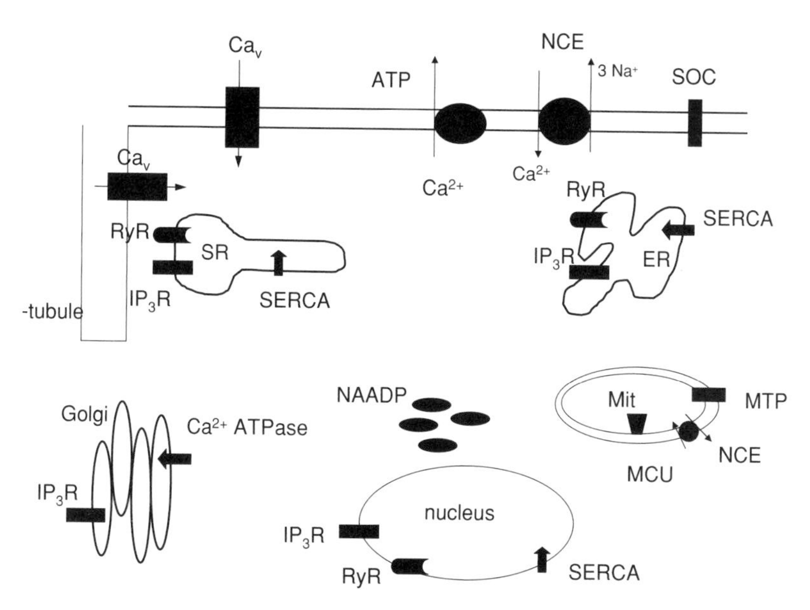

Fig. 7.3 Channels and transporters responsible for regulating calcium in a cell. Mit = mitochondria, ER = endoplasmic reticulum, SR = sarcoplasmic reticulum, SERCA = sarcoplasmic endoplasmic reticulum ATPase, NCE = Na^+/Ca^{2+} exchanger, SOC = store operated channel, MTP = mitochondrial transition pore, MCU = mitochondrial calcium uniporter, NAADP = nicotinic acid adenine dinucleotide phosphate-sensitive calcium channel depicted here on lysosomes.

The sarcoplasmic reticulum is capable of accumulating millimolar concentrations of calcium and this is returned after release by way of ATP-dependent pumps known as SERCA pumps (sarcoplasmic reticulum, endoplasmic reticulum Ca^{2+} ATPase). The SERCA family of genes encode five isoforms that are all sensitive to the inhibitor thapsigargin (Lytton et al., 1991). Like the plasma membrane Ca^{2+} ATPase and Na^+/K^+ ATPase, SERCA utilize the energy produced from the hydrolysis of ATP to drive the pumping of calcium back into sarcoplasmic reticulum stores. Other means of removing calcium from the cytoplasm and restoring resting intracellular calcium also exist such as sarcolemmal Na^+/Ca^{2+} exchange and mitochondrial calcium uptake (Fig. 7.3).

Three receptor types or isoforms of RyR and IP_3Rs have been isolated and combinations of isoforms are co-expressed in different muscle types (Nakagawa et al., 1991; Fill and Copello, 2002). RyRs are localized within membranes of organelles with stores of calcium (such as the sarcoplasmic reticulum), but IP_3Rs have been identified in the plasma membrane, endoplasmic reticulum, Golgi apparatus, and nuclear membranes (Khan et al., 1992; Leite et al., 2003). Both RyR and IP_3Rs form tetrameric complexes with the channel region located at the C-terminal end of the protein. There is good evidence that they appear to share common calcium stores and can interact with each other (McGeown, 2004). Activation of both receptor types is Ca^{2+}-dependent and varies with isoform. RyRs are activated at micromolar concentrations of calcium while IP_3Rs are generally activated at submicromolar concentrations (Bezprozvanny et al., 1991). Therefore the local release of calcium by one receptor can be amplified by activating a nearby receptor. The dependence of activation on calcium concentration, however, is often bell-shaped so that high concentrations of calcium also have inhibitory effects on channel function. This is the case for type I IP_3Rs where it has been proposed that the bell-shaped-dependence supports oscillations in cytosolic calcium, but not for type III IP_3Rs where an increase in cytosolic calcium causes a further increase in calcium that is suited to signal initiation (Hagar et al., 1998). In type I IP_3Rs the Ca^{2+}-binding sites lie within the inositol triphosphate ($InsP_3$)-binding core, and it has been proposed that calcium binding is negatively regulated by a conformational constraint imposed by $InsP_3$-binding (Bosanac et al., 2002). Both IP_3Rs and RyRs favor the conductance of divalent cations over monovalent cations exhibiting single-channel conductances of between 30 and 120 pS.

IP_3Rs mediate calcium release in response to activation of G protein or tyrosine kinase-coupled plasma membrane receptors. Agonist binding leads to stimulation of phospholipase C resulting in production of diacylglycerol, an activator of protein kinase C, and IP_3. IP_3 is a highly diffusible second messenger that binds to IP_3Rs on membranes of intracellular organelles. There is evidence, at least in neurons, that IP_3Rs may be directly activated by $G_{\beta\gamma}$ subunit following activation of a G_i protein (Zeng et al., 2003). IP_3Rs can be phosphorylated by many kinases. Protein kinase A phosphorylates IP_3Rs at serine 1589 and 1755 (Ferris et al., 1991a) as does cGMP-dependent protein kinase (Komalavilas and Lincoln, 1994). Protein kinase C and calcium calmodulin-dependent protein kinase also phosphorylate the receptor at distinct sites regulating calcium release in a spatiotemporal manner contributing to formation of calcium microdomains (Ferris et al., 1991b). ATP augments calcium release from IP_3Rs by stabilizing open channel states thereby allowing positive control of calcium release when SERCA have utilized the local ATP during reuptake or when the cell is experiencing metabolic inhibition (Mak et al., 1999). Cyclic ADP-ribose (cADPR) stimulates the association of ryanodine with RyR increasing channel opening (Guse et al., 1999). IP_3 does not bind to RyRs but caffeine and heparin potentiate RyR channel opening.

In nonexcitable cells the predominant mechanism for release of store-operated calcium (SOC) is by CRAC channels. The best characterized SOC influx channel is the CRAC channel expressed in T lymphocytes that is essential to the immune response including the regulation of gene expression and cell proliferation. The increase in intracellular calcium occurs in a biphasic manner typically with an initial transient increase due to binding of an agonist such as IP_3 and release of internal stores. The depletion of intracellular calcium stores then triggers release of calcium via CRAC channels on the plasma membrane and is entirely dependent on extracellular calcium. Recent evidence indicates that a transmembrane protein with an EF hand motif near the amino terminus located in the lumen of the ER termed STIM-1 is the sensor that migrates from the calcium store to the plasma membrane to activate CRAC channels (Zhang et al., 2005). CRAC channels have a low single-channel conductance (~25 fS for calcium and 0.2 pS for monovalent cations) with high selectivity for divalent cations and typically exhibit inward rectification. They favor opening when the cell is hyperpolarized (Zweifach and Lewis, 1993).

7.2.3 Mitochondrial and Nuclear Ion Channels

Many organelles other than the sarcoplasmic reticulum and endoplasmic reticulum are able to take up calcium and impose strict control over its movement in order to regulate organelle function. As mentioned above, the Golgi complex expresses IP_3Rs. It also contains SERCA-type pumps known as secretory pathway Ca^{2+}-ATPase that are located in close proximity to the nuclear membrane and hence may be involved in calcium signaling with the nucleus (Wootton et al., 2004). Calcium uptake into Golgi may occur by thapsigargin-sensitive and thapsigargin-insensitive mechanisms. Interestingly, only the *cis*-Golgi express IP_3Rs unlike the *trans*-Golgi suggesting that the role of the *trans*-Golgi in vesicle packaging and posttranslational modification has differential requirements for calcium compared with the *cis*-Golgi that appears to play a greater contribution to spatial and temporal calcium signals (Vanoevelen et al., 2004). Calcium release channels that do not involve IP_3 or cADPR have been identified on lysosome-related organelles. A nicotinic acid adenine dinucleotide phosphate (NAADP)-sensitive calcium channel has been located on lysosomes in pancreatic β cells that requires proton exchange to assist with calcium loading (Yamasaki et al., 2004).

It is well established that the mitochondria can accumulate calcium. Control of mitochondrial calcium is necessary for ATP production (McCormack et al., 1990) and the shaping of intracellular calcium signals (Jouaville et al., 1995). Calcium overload is associated with the processes mediating necrosis and apoptosis (programmed cell death) (Halestrap et al., 2000). In addition, several of the mitochondrial enzymes require calcium as a cofactor in higher concentration in the mitochondrial matrix than the mitochondrial cytosol, therefore transport of calcium across the mitochondria must be carefully regulated (Gunter et al., 2004). Calcium accumulation mainly occurs through the mitochondrial calcium uniporter (MCU; see Fig. 7.3) located on the inner mitochondrial membrane. Recently, the MCU was identified as a calcium channel with extremely high calcium affinity ($K_{0.5}$ of 19 mM) and biophysical characteristics that included an inwardly rectifying current suited to calcium uptake into mitochondria and inhibition by ruthenium red (Kirichok et al., 2004). Calcium may exit the mitochondria via the permeability transition pore (MPTP) that is more frequently associated with the collapse of the membrane potential and influx of proteins that lead to cell death (Halestrap et al., 1998, 2000). PTP opening is thought to occur when a protein, cyclophilin D binds to the matrix side of the PTP. Calcium is required for the binding of cyclophilin D. Calcium efflux from the mitochondria also occurs via the Na^+/Ca^{2+} exchanger that, similar to the plasma membrane exchanger transports one calcium ion out of the mitochondria in exchange for three sodium ions.

Although it is recognized that regulation of calcium within the nucleus is vital for many functions including gene transcription, little is understood about the regulatory processes involved. The nuclear membrane consists of an envelope with a luminal space capable of storing calcium. It also possesses the components necessary to regulate calcium levels such as SERCA, IP_3Rs, and RyRs (Gerasimenko et al., 1996). Both IP_3Rs and RyRs are located on inner and outer nuclear membranes. The nucleus can generate its own IP_3 but it is unclear whether it physically associates with other calcium stores such as ER and SR. However, there is evidence that the nucleus contains its own stores (termed nucleoplasmic reticulum) that may physically associate with the mitochondria suggesting a cooperative role on regulating calcium signals (Lui et al., 2003).

7.3 Roles of Ca^{2+} Channels

Calcium is essential in electrical activity in that it shapes the long plateau phase of the ventricular action potential and the upstroke and duration of the smooth muscle and atrial pacemaker action potentials. However, the influx of calcium through voltage-dependent calcium channels is also an essential component in the initiation of intracellular processes. Here we discuss in detail the role of calcium in muscle contraction, neurotransmitter secretion, and in controlling electrical excitability, highlighting recent research.

7.3.1 Muscle Contraction

Calcium is the switch that initiates the contraction of muscle fibers. These fibers are formed from many overlapping strands, comprising a thick filament made from myosin and a thin filament comprised of actin and tropomyosin. When Ca^{2+} is released from intracellular stores, it binds to troponin present on the thin filaments, and allosterically modulates the tropomyosin to unblock a series of myosin binding sites. The myosin can then be powered by hydrolyzing ATP to move along these binding sites causing muscle contraction until the cytoplasmic Ca^{2+} is removed and the myosin binding sites are once more blocked.

In the heart calcium influx across the plasma membrane through L-type Ca^{2+} channels is essential for triggering calcium release from the sarcoplasmic reticulum, a process known as calcium-induced calcium-release. This occurs as a result of a close physical association between dihydropyridine receptors (L-type Ca^{2+} channels) and transverse or T-tubules that communicate directly with RyRs on the sarcoplasmic reticulum (Fig. 7.3). Because the activation kinetics of L-type Ca^{2+} channels in skeletal muscle is 100-fold slower than in cardiac muscle, the process of calcium influx is too slow to initiate contraction. In skeletal muscle, therefore, a depolarization of the plasma membrane is the trigger for release of calcium from SR stores rather than the influx of calcium itself. Therefore, L-type Ca^{2+} channels serve as the voltage sensor used to initiate the release of Ca^{2+} from intracellular store for initiating contraction as calcium influx across the plasma membrane is not a requirement (Franzini-Armstrong and Protasi, 1997). In smooth muscle, calcium influx through L-type Ca^{2+} channels can release calcium from the SR to trigger contraction. In addition, a hormone-induced calcium release can initiate contraction through release of calcium from internal stores via binding to IP_3Rs.

The characteristics of the contractile force in muscle can also be controlled by intracellular calcium. The contractile force can be altered either by increasing the duration or amplitude of the calcium transient (the rise in intracellular calcium) or altering the sensitivity of the myofilaments to calcium. Stretch increases myofilament calcium sensitivity as it enhances actin–myosin interaction (Fukuda et al., 2001b) while acidosis reduces calcium sensitivity and enhances the length-dependence of tension (Fukuda et al., 2001a). Cytosolic calcium must be removed to allow for the relaxation of muscle fibers. The majority of cytosolic calcium uptake occurs by the SR Ca^{2+}-ATPase, with the remaining calcium being extruded from the cell either by Na^+/Ca^{2+} exchange or a small amount being taken up by the mitochondria via the MCU. Na^+/Ca^{2+} exchange is voltage-dependent and reversible with high intracellular calcium favoring calcium efflux; while a positive membrane potential and high intracellular Na^+ favors Na^+ efflux (Dipolo and Beauge, 2006). Under physiological conditions, Na^+/Ca^{2+} exchange works mostly to extrude calcium driven by the intracellular calcium transient. It is also a secondary active transport process driven by the Na^+/K^+ ATPase that extrudes three Na^+ ions in exchange for 2 K^+ ions while hydrolysing one molecule of ATP. If the Na^+/K^+ ATPase is partially inhibited, calcium influx through Na^+–Ca^{2+} exchange is increased. This is the basis for the increased force of muscle contractions induced by cardiac glycosides in the heart.

In principle, the Na^+/Ca^{2+} exchanger could be a trigger of intracellular calcium release and muscle contraction, but the influx is too slow and too small to do so (Sipido et al., 1997). A contributing factor to this is the fact that Na^+/Ca^{2+} exchangers are not positioned near the SR junctional cleft (Scriven et al., 2000). Promiscuous calcium entry through tetrodotoxin-sensitive Na^+ channels has also been proposed to trigger contraction, also termed slip-mode conductance (Santana et al., 1998). However, this remains controversial since a number of groups have been unable to trigger contraction by this means (Chandra et al., 1999; DelPrincipe et al., 2000). T-type Ca^{2+} channels are also not located near SR junctions and the current passing through them is less than for L-type channels, therefore any contribution they make to triggering contraction is thought to be small (Sipido et al., 1998; Zhou and January, 1998).

IP_3Rs are the main trigger of calcium release from SR and ER in smooth muscle. Generally, calcium release from IP_3Rs stimulates further calcium release in a cooperative fashion, but with high calcium concentrations inhibiting channel function. This creates the classical oscillatory pattern of calcium release typically seen in innervated smooth muscle such as that resulting in peristalsis in the bowel or agonist mediated calcium release pathways.

RyRs inactivate one of two ways. They either do not reopen until they recover (Schiefer et al., 1995; Sham et al., 1998), or they relax to a lower open probability but can still be reactivated at high calcium levels, a process known as adaptation and also common to IP_3Rs (Gyorke and Fill, 1993). Adaptation of RyRs enables a graded calcium-induced calcium-release for contraction in muscle.

7.3.2 Neurotransmitter Release and Neuronal Plasticity

Another important physiological process involving calcium is neurotransmission at nerve terminals. Within the presynaptic terminal of neuron hormones, neurotransmitters, or other peptides are stored in vesicles. When an action potential depolarizes the membrane, the vesicles fuse with the plasma membrane and the contents are released into the extracellular space where they can bind with receptors and initiate chemical activity in nearby cells. The process of neurotransmission is calcium-dependent in that secretion requires extracellular calcium and is inhibited by extracellular Mg^{2+} (Douglas, 1968). In addition, intracellular calcium modulates the activity of a number of ion channels involved in this process. Intracellular calcium influences the inactivation rate and inhibition of L-type Ca^{2+} channels (see above) and activates large conductance K^+ channels BK(Ca) that terminate neurotransmitter release in the presynaptic terminal by hyperpolarizing the membrane (Storm, 1990; Sah, 1996). BK(Ca) are expressed in virtually all excitable cells, have long and large unitary conductances typically 100–250 pS and can be blocked by charybdotoxin, iberiotoxin, and tetraethylammonium compounds. The BK(Ca) channels are derived from the *slo* family of genes originally cloned from the Drosophila mutant *slowpoke*. Intermediate (IK) and small (SK) conductance channels of K(Ca) family have also been cloned and characterized (Vergara et al., 1998). Calcium-dependent chloride channels also contribute to membrane potential hyperpolarization along with the BK(Ca) channels in neurons (Mayer, 1985).

The release of neurotransmitters is steeply calcium dependent and a decrease in calcium channel activity contributes to presynaptic inhibition (Dunlap and Fischbach, 1978). An example of presynaptic inhibition is the block of transmission of sensory fibers that occurs with opiates such as morphine. This occurs as a result of rapid coupling of G protein $\beta\gamma$ subunits to Ca_v2 family of calcium channels causing a slowing of activation of the channels (Dolphin, 1998). Several modulatory transmitters such as acetylcholine, norepinephrine, adenosine, and GABA contribute to depression of calcium channels at presynaptic terminals through the G protein-dependent mechanism.

Both presynaptic and postsynaptic calcium stores are involved in long-term depression in the developing hippocampus (Caillard et al., 2000). However, presynaptic RyRs appear to be important in determining the strength of synaptic transmission induced by NMDA receptors (Unni et al., 2004). Activation of presynaptic NMDA, kainate, and nicotinic acetylcholine receptors can lead to calcium-induced calcium-release from RyR stores similar to that recorded in muscle that acts to inhibit or regulate secretion. Although still controversial, calcineurin may interact with IP_3Rs and RyRs to modulate intracellular calcium release and synaptic plasticity (Bultynck et al., 2003).

Calcium can also play a role in building synaptic connections. Rapid changes in synaptic strength are mediated by posttranslational modifications of preexisting protein but enduring changes are dependent on gene expression and the synthesis of a variety of proteins required for synaptic transmission. One of the key mediators of gene expression involved in synaptic plasticity is the transcription factor cAMP response element binding protein (CREB). This transcription factor plays an important role in behavioral adaptations to changes in the environment and complex processes such as learning and memory. It is also important in proliferation and differentiation in developing vertebrates. Calcium influx through L-type Ca^{2+} channels or NMDA receptors is required for phosphorylation and activation of CREB (Kornhauser et al., 2002). Another transcription factor responsible for shaping long-term changes in neurons is the NFAT family, and similar to regulation of CREB, activation of NFAT is dependent upon calcium influx through L-type Ca^{2+} channels. Interactions of calcium with PDZ domain proteins (PSD-95, Discs-large, ZO-1 domain proteins that often function as scaffolding proteins and have been shown to play important roles in signal transduction) appear to be necessary for the coupling of L-type Ca^{2+} channels to NFAT as does CREB activation (Weick et al., 2003). Neurotrophins, such as the brain derived neurotrophic factor, have also been implicated in the activation of NFAT-dependent transcription. Initiating such transcriptions leads to further increase in brain derived neurotrophic factor mRNA and protein as positive feedback (Groth and Mermelstein, 2003).

7.3.3 Excitability and the Action Potential

Calcium plays an important role in shaping the action potential of muscle and neurons. One example where the shape and duration of the action potential must be carefully controlled is in cardiac pacemaker activity. This originates from either specialized cells located in the right atrium of the heart known as sinoatrial node cells or from secondary pacemaker cells located within the atrioventricular node and also through the Purkinje fibers of the ventricles. Pacemaker action potentials are designed to originate from the sinoatrial node cells so that the resting heartbeat is maintained at 60–80 beats min^{-1}. If the pacemaker fails in its duty, secondary pacemakers in the atrioventricular node or in the purkinje fibers are activated to take over and maintain ventricular function and ultimately cardiac output, albeit at a lower heart rate. The spontaneous activity of pacemaker cells is attributed to a phase of the action potential known as the slow diastolic depolarization. During this phase, the membrane potential slowly depolarizes following termination of an action potential until the threshold for a new action potential is reached. It is well accepted that the electrical activity of cardiac pacemakers is thought to depend exclusively on voltage-dependent ion channels within the plasma membrane of nodal cells. These include the hyperpolarization-activated inward current (I_f), the time-dependent decay of K^+ conductance, inward L-type and T-type Ca^{2+} currents, and the lack of background K^+ conductance. The hyperpolarization-activated channel carries Na^+ and K^+ ions inwards and activates upon hyperpolarization. However, interventions that affect intracellular Ca^{2+} can also affect pacemaker activity. A rise in intracellular Ca^{2+} results in activation of the Na^+/Ca^{2+} exchanger. This produces an inward current that is sufficient to induce diastolic depolarization and pacemaker activity (Ju and Allen, 1998). The importance of Ca^{2+} in contributing to pacemaker firing is further supported by studies that show that embryonic stem cell-derived cardiomyocytes do not show increased beating rate with differentiation when they lack ryanodine receptors (Yang et al., 2002).

The rate of change in membrane potential during the upstroke is much slower in sinoatrial node cells than in ventricular cells. This is because depolarization of the membrane during phase 0 (the rapid upstroke) of a slow response action potential in pacemaker cells is caused by an increase in the Ca^{2+} conductance due to activation of L-type Ca^{2+} channels. Nodal cells have fewer Na^+ channels than ventricular cells and because the membrane potential in phase 0 is depolarized, changes in Na^+ conductance do not contribute. Pacemaker cells do not have a pronounced plateau phase. Action potential duration is determined by a balance between Ca^{2+} and K^+ conductances.

In ventricular myocytes the prolonged phase 2 (plateau phase) of the action potential is characterized by slow inward current produced by L-type Ca^{2+} channels. This is a distinguishing feature of the cardiac action potential and is the necessary trigger for calcium-induced calcium-release and contraction. Similarly in smooth muscle and skeletal muscle calcium influx through L-type Ca^{2+} channels are important in shaping the duration of the action potential, however, the duration is much shorter than that of ventricular myocytes. In smooth muscle and neurons the calcium-dependent large conductance K^+channels contribute to hyperpolarization of the membrane.

7.4 Ion Selectivity and Permeation

It is clear that to carry out their biological role, calcium channels have to be selective, allowing Ca^{2+} ions to pass while blocking Na^+, K^+, Cl^-, and other ions. A high degree of specificity is important for these channels as Na^+ ions are more than 100-fold more numerous in the extracellular solution than Ca^{2+}. Voltage-gated calcium channels are extremely discriminating, selecting calcium over sodium at a ratio of over 1000:1 (Hess et al., 1986), yet the picoampere currents they carry require over one million ions to pass through a single channel every second (Tsien et al., 1987). A central problem in understanding the function of these channels is to determine how they can be both highly selective while still passing so many ions. A narrow pore can block the passage of large ions, but, Na^+ and Ca^{2+} are similar in size and much larger ions are known to permeate the channel (McCleskey and Almers, 1985). Thus, these proteins must distinguish between ions using more than just size. Instead, it appears that calcium channels utilize the different strengths of interaction between ions and the protein as well as a multi-ion conduction process to obtain ion selectivity. Here we focus the discussion of selectivity on L-type voltage-gated calcium channels as these have been most widely studied. It is hoped that many of the mechanisms of this channel will apply to a variety of proteins in this family, particularly as the important residues required to obtain calcium specificity are highly conserved.

7.4.1 High-Affinity Binding of Permeant Ions

A surprising characteristic of voltage-gated calcium channels is that monovalent ions conduct through the channel at much higher rates than any divalent ions when no divalent ions are present (Kostyuk et al., 1983; Almers and McCleskey, 1984; Fukushima and Hagiwara, 1985; Hess et al., 1986; Kuo and Hess, 1993a). But, these monovalent currents are blocked when the calcium concentration reaches only 1 μM (Kostyuk at al., 1983; Almers et al., 1984). Indeed, the ability of some ion species to block currents carried by others gave the first clues to the origin of ion selectivity in these channels, suggesting that it relied on high-affinity binding of ions within a single file pore. Those ions that bound most strongly in a single file pore would block the passage of other ions.

A large number of mono- and divalent ion types pass through the pore, but a characteristic sequence in which some ion types block the passage of others has been observed. In addition to Ca^{2+} blocking monovalent cations, Sr^{2+} current is also blocked by Ca^{2+} (Vereecke and Carmeliet, 1971) while Cd^{2+}, Co^{2+}, and $La3^{+}$ block currents carried by Ca^{2+} even though all of these species were known to pass through the channel at some rate (Hagiwara et al., 1974; Lansman et al., 1986; Chow, 1991). Furthermore, Co^{2+} was found to block Ca^{2+} currents less effectively than Ba^{2+} currents, even though Ba^{2+} currents were found to be larger than Ca^{2+} currents (Hagiwara et al., 1974). This property was explained by positing that Ca^{2+} binds more strongly to the pore than Ba^{2+} but not as strongly as Co^{2+}. Together these data suggested a high-affinity binding site in the pore, with an order of affinity estimated from bi-ionic reversal potentials to be $La3^{+} > Cd^{2+} > Co^{2+} > Ca^{2+} > Sr^{2+} > Ba^{2+} > Li^{+} > Na^{+} > K^{+} > Cs^{+}$ (Reuter and Scholz, 1977; Fenwick et al., 1982; Lee and Tsien, 1984; Hess et al., 1986; Taylor, 1988).

One complication, however, has been observed. Although Ca^{2+} was found to prevent monovalent currents, Na^{+} was also found to attenuate currents carried by Ca^{2+} in N-type channels (Polo-Parada and Korn, 1997). Although this appears to contradict the earlier results, it can easily be explained if these Na^{+} ions compete for entry into the pore rather than permanently occupying the high-affinity binding site (Polo-Parada and Korn, 1997; Corry et al., 2001).

Single-channel conductance measurements on the other hand, have the exact reverse order of conductance values than the binding affinity: $La3^{+} < Cd^{2+} < Co^{2+} < Ca^{2+} < Sr^{2+} < Ba^{2+} < Li^{+} < Na^{+} < K^{+} < Cs^{+}$ (Hess et al., 1986; Kuo and Hess, 1993a,b). The inverse relationship between binding affinity and permeation were first explained with the so-called "sticky-pore" hypothesis. In this, ions that are bound with higher affinity pass through the channel more slowly and so have a lower conductance (Bezanilla and Armstrong, 1972). For example, the lower current carried by Ca^{2+} than by Ba^{2+} can be understood if Ca^{2+} binds more strongly in the pore, and thus takes longer to dissociate once bound. Indeed it has been observed that permeabilities for monovalent and divalent ions are ordered according to an Eisenman sequence that models selectivity between ions as a balance between the dehydration and binding energy of each (Eisenman, 1962). According to Eisenman, the permeability sequence observed in calcium channels can be explained by an electrostatic binding site of strong field strength.

A paradox arises, however, when explaining ion selectivity with high-affinity binding. How can ions conduct through the pore at picoampere rates if they are bound so strongly? This paradox disappears if one considers the implications of evidence that calcium channels are multi-ion pores.

7.4.2 Calcium Channels Have a Multi-Ion Pore

A number of lines of evidence suggest that calcium channels must bind multiple ions. Multi-ion conduction is a natural consequence of the sticky-pore model as otherwise there should not be a difference in conduction rates between ion types. In a single ion pore, ions of all binding affinities conduct equally well. Those that bind more effectively pass more slowly through the pore, but they are also more likely to block the passage of lower affinity ions (Bezanilla and Armstrong, 1972).

Other strands of evidence also suggest a multi-ion pore. In many cases when two permeating species are mixed the current is lower than for either species on its own. This phenomenon is seen clearly for mixtures of Ca^{2+} and Ba^{2+} and also Ca^{2+} and Na^{+} (Almers and McCleskey, 1984; Hess and Tsien, 1984) in calcium channels and has also been observed in gramicidin (Neher, 1975) and potassium channels (Hagiwara et al., 1977). This so-called "anomalous permeability" or "anomalous mole fraction effect" is most easily explained supposing the pore contains multiple ions. If ions moved independently through the pore, then the current in mixed solutions should always lie between the currents observed with either permeating ion on its own. If, however, the channel holds multiple ions then it is possible that unfavorable interactions between the ions in the pore could reduce the current. As noted by Hille (2001), it is possible to explain the anomalous permeability by supposing some form of conformational or chemical alteration of the channel protein caused by binding of ions exterior to the pore, rather than by a multi-ion pore. But, such an explanation lacks the simplicity of the multi-ion model that is also supported by other data.

Another important strand of evidence for multiple ions lies in the measurement of two differing affinities for Ca^{2+} binding. Ca^{2+} blocks Na^{+} currents with an affinity of $\sim 1\ \mu M$, while calcium currents saturate with an affinity closer to 14 mM. This can easily be explained by a multi-ion pore if the first affinity represents binding of a calcium ion to either an empty pore or a pore containing one Na^{+}. However, if one Ca^{2+} ion is already present in the pore, then electrostatic repulsion between the two ions means that the second ion is bound more weakly giving rise to the second affinity measured.

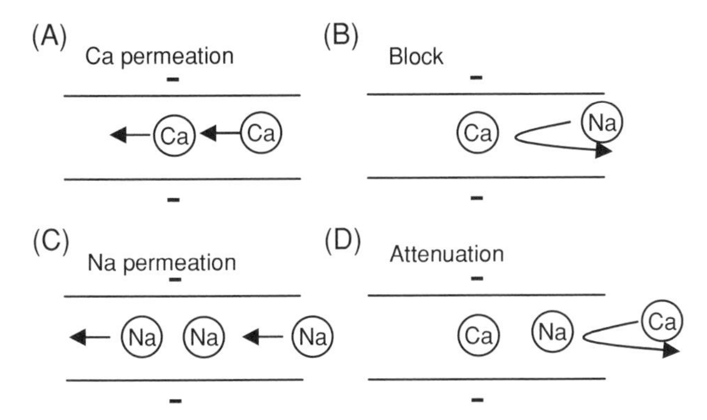

Fig. 7.4 "Electrostatic" model of permeation and selectivity in calcium channels. (A) When one Ca^{2+} ion is electrostatically bound to the negative charge of the EEEE locus, it can only be removed with the aid of Coulomb repulsion from another Ca^{2+}ion. (B) The lesser repulsion from Na^{+} is unable to displace a resident Ca^{2+} ion, thus Ca^{2+} blocks Na^{+} currents. (C) Na^{+} is able to pass through the channel in the absence of Ca^{2+} in a "knock on" mechanism involving three ions. (D) Na^{+} can attenuate Ca^{2+} currents by slowing the entry of the second Ca^{2+} ion required for conduction.

The presence of multiple ions in the pore is also supported by the fact that calcium block of Na^{+} currents is dependent on membrane voltage (Fukushima and Hagiwara, 1985; Lux et al., 1990) and on the direction of ion movement (Kuo and Hess, 1993a,b). Further evidence that two divalent ions can inhabit the pore comes from a detailed study of the blocking of Ba^{2+} currents by Cd^{2+} (Lansman et al., 1986). Raising the concentration of Ba^{2+} was found to increase the unbinding rate of Cd^{2+} as if increasing the probability of Ba^{2+} also being present in the pore enhances the exit of Cd^{2+}.

A multi-ion pore provides a simple explanation of the permeation and selectivity properties of the channel as depicted schematically in Fig. 7.4. The selectivity—conductivity paradox is avoided by utilizing ion–ion interactions inside the selectivity filter. As soon as one Ca^{2+} ion occupies the channel, it binds tightly to the pore and thereby blocks the passage of Na^{+}. However, if the second site in the pore is occupied by a calcium ion then the strong electrostatic repulsion between the two ions allows one to overcome its interaction with the protein and exit the channel at a relatively fast rate (Almers and McCleskey, 1984; Hess and Tsien, 1984). Further details of the mechanisms of selectivity are described below in the context of various theoretical models.

As noted by Hille, the concepts of binding and repulsion described here need not refer to specific details (Hille, 2001). Binding sites need not be specifically shaped or involve stereochemical fit such as in the active sites of enzymes. Rather it is more likely to refer to regions of space that form local energy minima created by broad electrostatic interactions between the ion and the protein. Furthermore, the locations of such minima are likely to change when, for example, other ions are nearby. Thus a multi-ion pore need not contain two or more specific localized sites, but could just have one or more regions of negative potential. Similarly, the repulsion between ions in the pore may be direct Coulomb repulsion between like charged particles or may involve interaction through the protein, conformational changes or restrictions on the entropic freedom of nearby ions.

7.5 Channel Structure

It has long been recognized that voltage-gated sodium, potassium, and calcium channels are related. Not only do they share many functional similarities, the amino acid sequences indicate many structural similarities as well. This similarity is no longer surprising as we now know that many proteins have evolved in families and that all these channel types most likely evolved from a common ancestor. This means that we can learn much about one member of the family from findings made about other members. As no atomic resolution images of calcium channel structures exist, much has been learnt about their structure since the recent determination of crystal structures of a number of potassium channels (Doyle et al., 1998; Jiang et al., 2003; Long et al., 2005).

7.5.1 Subunit Structure and Function

Voltage-gated calcium channels are formed by the association of separate subunits, which come together to form an active channel. As noted previously the variety of voltage-gated calcium channels arises from differences in the sequences of the individual subunits comprising the channel in particular the $\alpha 1$ subunit. As with voltage-gated sodium channels, but unlike the potassium channels, the pore-forming domain of the calcium channel protein is comprised of four regions of the one subunit rather than by separate subunits.

The first step in determining the structural characteristics of these channels was the purification, cloning, and sequencing of the protein. Purification was first achieved from transverse tubule membranes of skeletal muscle (Curtis and Catterall, 1984). As described previously, this study and further analysis (Hosey et al., 1987, Leung et al., 1987, Striessnig et al., 1987, Takahashi et al., 1987) indicated that this channel was comprised of five subunits: $\alpha 1$ weighing 190 kDa and later shown to form the pore; a disulphide linked $\alpha 2$-δ dimer of 170 kDa; and intracellular phosphorylated β subunit weighing 55 kDa; and a transmembrane γ subunit of 33 kDa as indicated in Fig. 7.5.

Subunits of similar sizes were soon purified form cardiac L-type channels (Chang and Hosey, 1988; Schneider and Hofmann, 1988) whereas neuronal L-type channels revealed $\alpha 1$, $\alpha 2$-δ, and β but no γ subunit (Ahlijanian et al., 1990). The similarity between these various L-type channels also extends to neuronal N-type (McEnery et al., 1991; Witcher et al., 1993) as well as P/Q-type channels (Martin-Moutot et al., 1995, 1996; Liu et al., 1996) that have also been purified and contain the $\alpha 1$, $\alpha 2$-δ, and β subunits. A possible γ subunit has also been found in P/Q

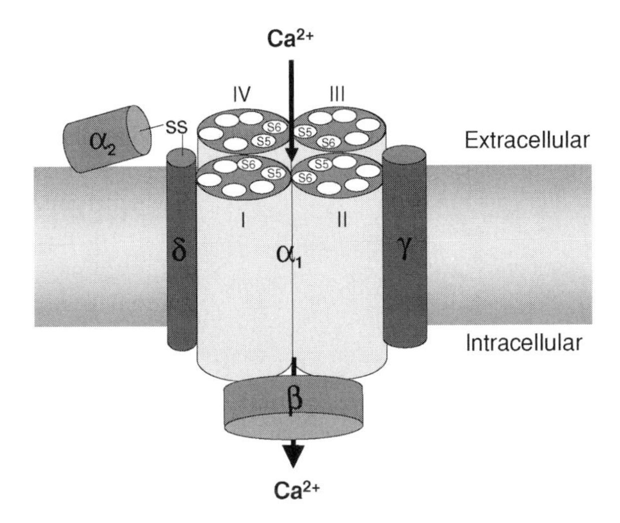

Fig. 7.5 The likely structure and arrangement of the L-type channel subunits. The pore is formed by the S5 and S6 helices and linker of the α1 subunit. The β subunit interacts with the linker between domains I and II of α1 and modifies gating and inactivation kinetics. The δ and γ subunits lie in the membrane while α2 is accessible to the extracellular solution.

channels (Letts et al., 1998) that plays a role in modulating the voltage dependence of the channels, yielding even more similarity with the L-type channels. Cloning and sequencing of the five subunits from L-type channels has since been achieved in skeletal (Tanabe et al., 1987; Ellis et al., 1988) and cardiac muscle (Mikami et al., 1989).

The α1 subunit has a structure similar to the previously cloned pore forming a subunit of the voltage-gated Na^+ channels. It contains four repeated domains (I–IV) each of which includes six transmembrane segments (S1–S6) and a membrane-associated loop (The "P-loop") between segments S5 and S6 as indicated in Fig. 7.5. That the α1 subunit forms the ion-conducting pore is supported by the fact that expression of just this subunit is enough to produce function channels, albeit with unusual kinetics and voltage dependence (Perez-Reyes et al., 1989). The four repeated domains are also remarkably similar to the four subunits known to form the voltage-gated potassium channels. With the aid of the crystal structure of the potassium channel the basic structure of the α1 subunit is clear. However, prior to the existence of this crystal structure, a large number of structural and functional channels in the voltage-gated Na^+, K^+, and Ca^{2+} channels had already characterized the organization and role of many parts of the protein (see the review by Hofmann et al., 1999, for example). Each of the six transmembrane segments most likely form α-helices. The four repeated domains surround the ion-conducting pore, with the S5, S6, and P-loops lining the conduit. Much evidence discussed below, and analogy with the potassium channels implies that ion selectivity takes place in a region surrounded by the four P-loops. The S4 helix contains a number of arginine residues and is believed to act as the voltage sensor whose action is described in more detail below.

7.5.2 Pore Structure

Many lines of reasoning suggest that the pore must be narrow enough to force ions to pass in single file. Firstly, the blockage of currents carried by some ion types by other permeant ions is indicative of single file binding within the pore. The discrimination between different like-charged ion types requires close range interaction between the protein and the permeating ions. Next, the largest known ion to pass through the pore of L-type channels is tetramethylammonium,with a radius of ~2.8 Å (McCleskey and Almers, 1985), suggesting that the pore acts as a sieve that is too narrow for larger ions to pass. As discussed below, the selectivity properties of the channel can best be explained assuming a single file pore in which ions cannot pass each other. Finally, the similarity to K^+ channels that have a known single file pore structure is persuasive.

Surprisingly, the first evidence for the regions of the pore that determine ion selectivity came from a study of sodium channels. Replacement of particular lysine or alanine residues in the P-loop of a voltage-gated sodium channel with glutamate residues changed the channel from being sodium selective to calcium selective (Heinemann et al., 1992). Not only did this suggest that this region of the pore was critical for ion selectivity, the fact that all calcium channels have conserved glutamate residues at this position suggested that it was these glutamates that are crucial for Ca^{2+} selectivity.

Further, site-directed mutagenesis indeed determined that four glutamate residues, one from each of the P-loop of the channel (often called the "EEEE locus"), were responsible for the high-affinity calcium binding site. Mutation of one or more of these had a significant effect on the channel selectivity (Kim et al., 1993; Yang et al., 1993; Ellinor et al., 1995; Parent and Gopalakrishnan, 1995; Bahinski et al., 1997). Removing any one of these residues was found to decrease the relative permeability and binding affinity of Ca^{2+}. Replacing all of these glutamate residues with either glutamine or alanine residues was found to remove the specificity of the channel for Ca^{2+} over Na^+ altogether and left only weak affinity for Ca^{2+} in the pore as shown in Fig. 7.6 (Ellinor et al., 1995). As no evidence of other high-affinity binding sites was apparent for divalent ions entering either end of the channel (Cibulsky and Sather, 2000) the EEEE locus was posited to be the sole origin of ion selectivity. Mutation of any one or any pair of the glutamate residues affected the binding affinity significantly, indicating that all four residues participated in the binding of even a single ion. Importantly, not all the glutamates are identical as the mutation of each has a slightly different effect on the blockage rate of ions in the pore. Presumably, this is a consequence of structural differences between the four repeated domains, although the functional importance of this nonequivalence is yet to be fully understood.

It is easier to make sense of how all four glutamate residues can contribute to creating a high-affinity binding region that can hold more than one ion by remembering that these binding sites can represent broad regions of local energy minimum. In this way, for example, the negative charge of the glutamate residues can create a

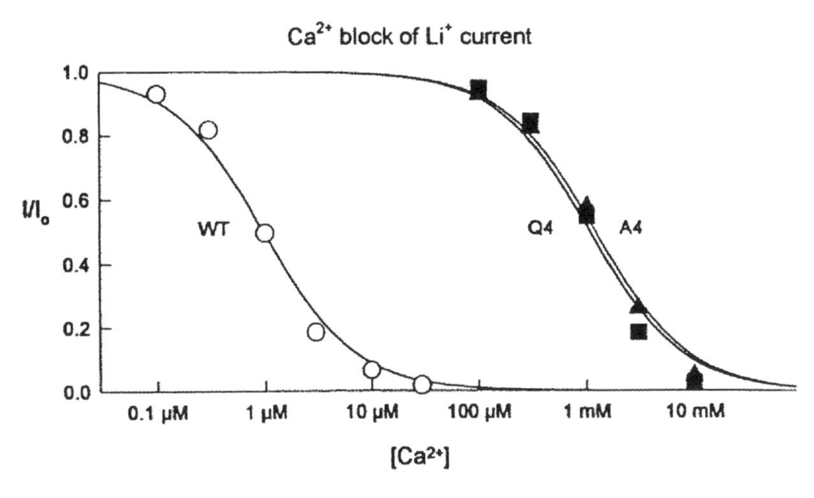

Fig. 7.6 The influence of quadruple mutants of the EEEE locus on high-affinity binding of Ca^{2+}. The fraction of Li^+ current remaining is plotted against Ca^{2+} concentration for the wild-type (WT), quadruple glutamine (Q4, squares), and quadruple alanine (A4, triangles) mutants.

single attractive region of the pore that can hold multiple ions in a semistable equilibrium. In this model, mutation of any one of the charged residues would influence the binding affinity of both the first and second ion entering the pore.

The determination of the atomic structure of a bacterial potassium channel by X-ray crystallography (Doyle et al., 1998) also aided the understanding of calcium channel structures. The organization of the membrane spanning domains of voltage-gated calcium, potassium, and sodium channels are all quite similar, and the discovery that the P-loop formed the selectivity filter of potassium channels gave credence to the view that this region would also be important in calcium channels.

An important difference is evident in the structures of the potassium and calcium channel selectivity filters. Whereas ions are stabilized by the backbone carbonyl groups in voltage-gated potassium channels, most people believe that Ca^{2+} ions interact with the side chains of the glutamate residues in calcium channels. This belief is supported by a number of tests. Protons are found to be able to block Ca^{2+} currents (Root and MacKinnon, 1994), presumably by weakening electrostatic attraction of the pore. Point mutations within the EEEE locus suggested that multiple glutamate residues interacted with a single proton (Chen et al., 1996; Chen and Tsien, 1997; Klockner et al., 1996), suggesting both that these residues were in close proximity and that their charged side chains were accessible to the interior of the pore. Also, when individual residues in the EEEE locus were replaced with cysteine, the side chains of the cysteine residues appeared to be accessible to the solvent as they could become ionized to react with methanethiosulfonate (Koch et al., 2000; Wu et al., 2000). The bulky methanethiosulfonate groups also blocked the pore upon reaction, supporting the conclusion that the EEEE locus lay inside the pore.

7.6 Theoretical Models of Permeation and Selectivity

7.6.1 Rate Theory

As the understanding of calcium channel permeation and selectivity was progressing through innovative experimental techniques, a number of theoretical models were also proposed to help elucidate the mechanisms underlying these properties. For a long time, especially when detailed structural information was lacking, the main approach utilized was rate theory. In this, the channel is represented as a series of energy wells and barriers, with each well representing an ion-binding site and its depth being proportional to the binding affinity. Ion conduction is described as the hopping of ions between the wells, with the rate of flux over each barrier related to the exponential of its height. All rate models include at least one binding site for calcium with an energy well depth of about 14 kT to account for the micro molar dissociation constant determined since micro molar concentrations of Ca^{2+} block monovalent currents.

When the multi-ion nature of permeation was determined the first models depicted this with two separate energy wells representing two binding sites, surrounded by two larger barriers as depicted in Fig. 7.7. A single ion in the pore could move between these sites, but not over the barrier to exit the channel (arrow in Fig. 7.7A). When a second ion entered and occupied the second site, however, it was proposed

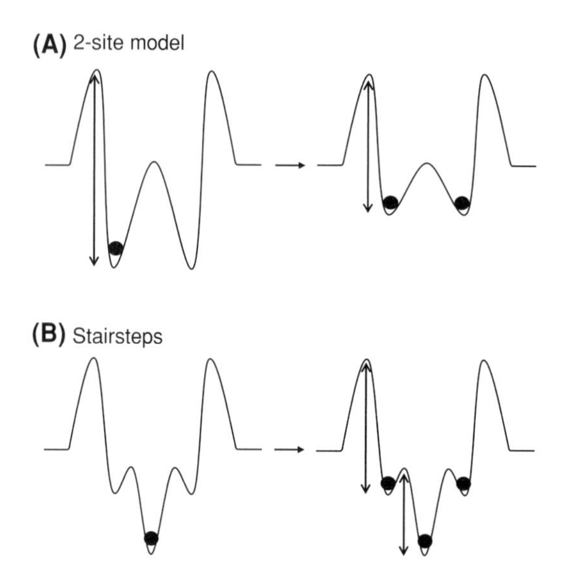

Fig. 7.7 Rate models of ion conduction. The energy landscape in the pore is represented by a series of energy wells and barriers. Ion conduction is slow for a single ion in the pore as it has to overcome a large energy barrier to move out of the channel (arrows). In the two site model (A), ion transit is increased due to interactions between ions in each site altering the energy landscape (B). In a "stair step" model (C and D) the ion can climb out of the deep energy well into low-affinity sites on either side.

that the repulsion between them alters the energy landscape and reduces the barrier for dissociation as shown in Fig. 7.7B (Almers and McCleskey, 1984; Hess and Tsien, 1984). As noted by Sather and McCleskey (2003), the nature of the ion–ion repulsion is ambiguous in this model as it deals only with the shape of the energy landscape not the source.

When it was discovered that there was only one binding region in the channel created by the EEEE locus, the idea of two similar binding sites had to be modified. A new model emerged in which there was only a single high-affinity binding site, but this was flanked by low-affinity sites that helped ions to step out of the large well (Kuo and Hess, 1993a; Dang and McCleskey, 1998) and the role of ion–ion interaction was relegated (Fig. 7.7C and D). Other mechanisms involving single sites have also been developed involving competition between calcium ions for the binding charges (Armstrong and Neyton, 1991; Yang et al., 1993). However, it is difficult within rate theory to accurately model a single binding region that is capable of holding many ions.

The rate theory models are surprisingly simple, yet they capture a number of the salient features of permeation and selectivity. In particular they helped to provide a simple explanation of how a multi-ion pore can utilize repulsion between ions to achieve both specificity and high throughput (cf. Fig. 7.4). However, these models can impart only limited information about the origins of specificity as the energy landscape is usually derived from measurements of binding affinities and is not directly related to the structure of the pore. For example, these models cannot impart information about which residues are responsible for ion binding and as no physical distances are used there is no direct connection between energy minima used in the theory and physical sites within the pore.

7.6.2 Physical Models

More recently there have been a large number of theoretical studies involving physical models of the channel protein. Information about the functional characteristics of the model is then determined by applying either a continuum theory, Brownian dynamics simulations, Monte-Carlo calculations, or molecular dynamics simulations.

In the first of these, Nonner and Eisenberg (1998) modeled the channel as a simple cylinder with conical vestibules and included the charge of the EEEE locus in the channel walls. By adjusting the diffusion coefficients and including an excess chemical potential parameter that helps ions to bind in the pore they are able to use the drift-diffusion equations (or Poisson–Nernst–Planck theory as it is often known) to reproduce the conductance and selectivity of the channel. They find that Ca^{2+} ions bind to a broad region in the interior of the pore due to the Coulomb potential of the glutamate residues as well as the excess chemical potential, and that the multi-ion repulsion may be important in speeding flux. Questions raised about the validity of the continuum description of ions in narrow pores (Moy et al., 2000; Corry et al., 2000a,b), the difficulty in treating ion–ion interactions in this model, and the arbitrary nature of the adjustable diffusions and chemical parameters limit the impact of these conclusions.

A different explanation of the origins of ion selectivity than that proposed in the rate representation was proposed in two later studies employing different calculation schemes (Monte-Carlo simulations and a Mean Spherical Approximation method; Boda et al., 2000; Nonner et al., 2000). In these models eight partially charged oxygens comprising the glutamate side chains protrude into a cylindrical pore and can freely diffuse within a limited region of the channel. Cation binding is achieved due to the electrostatic attraction of the negatively charged oxygens. Selectivity of Ca^{2+} ions is proposed to be a result of ions competing to achieve charge neutrality in a selectivity filter having finite space. Ca^{2+} ions are preferred to Na^{+} in the model as they have the same charge neutralizing effect as two Na^{+} ions while occupying less of the limited volume of the filter. Essentially, not enough Na^{+} ions can be squeezed in to the filter to achieve charge neutrality. This charge–space competition model was analyzed further by Yang et al. (2003), who conducted nonequilibrium molecular dynamics simulations of a model pore containing half-charged oxygens in an atomistic cylinder. In these simulations a second Ca^{2+} ion is required to release a Ca^{2+} already bound in the channel, but cation binding was found to be nonselective.

A simple version of the single file ion–ion repulsion model of permeation and selectivity has been presented by Corry and coworkers who simulate the trajectories of ions moving in and around models of the channel protein (Corry et al., 2000c, 2001, 2005). For this purpose, Brownian dynamics simulations are used in which the ions pass through the channel in a random walk subject to electrical forces. In these studies a rigid model of the channel is derived from analogy to other channels and a variety of experimental data. The most important aspect of the channel model is that it includes a relatively narrow region in which ions cannot pass each other, surrounded by the four glutamate residues as shown in Fig. 7.8. Unlike the previous models, the glutamate charges do not compete for space within the pore, but the concentration of negative charge does attract cations within this region. Indeed the electrostatic attraction of the protein is all that is required to account for ion permeation and selectivity in this model. The charge of the glutamate residues creates a deep energy well that strongly attracts multiple ions. In a process akin to that suggested by earlier rate models, repulsion between two resident Ca^{2+} ions is found to speed their exit as illustrated in Fig. 7.4. Because the divalent Ca^{2+} ions are more strongly attracted by the channel, they can displace Na^{+} to occupy this region. Once there, the Ca^{2+} can only be moved by the repulsion from another divalent ion and not by the lesser repulsion from Na^{+} (Fig. 7.4). As well as providing a simple mechanism for ion selectivity depicted schematically in Fig. 7.4, this model was also used to calculate the conductance of the channel in a variety of situations and replicate and explain the observed I–V curves, current concentration curves, anomalous mole fraction behavior between Na^{+} and Ca^{2+} (Fig. 7.8B) as well as the attenuation of Ca^{2+} currents by Na^{+}.

All the models described thus far assume simplified pore geometries, but one goal of ion channel modeling is to include all the atomistic detail of the pore in

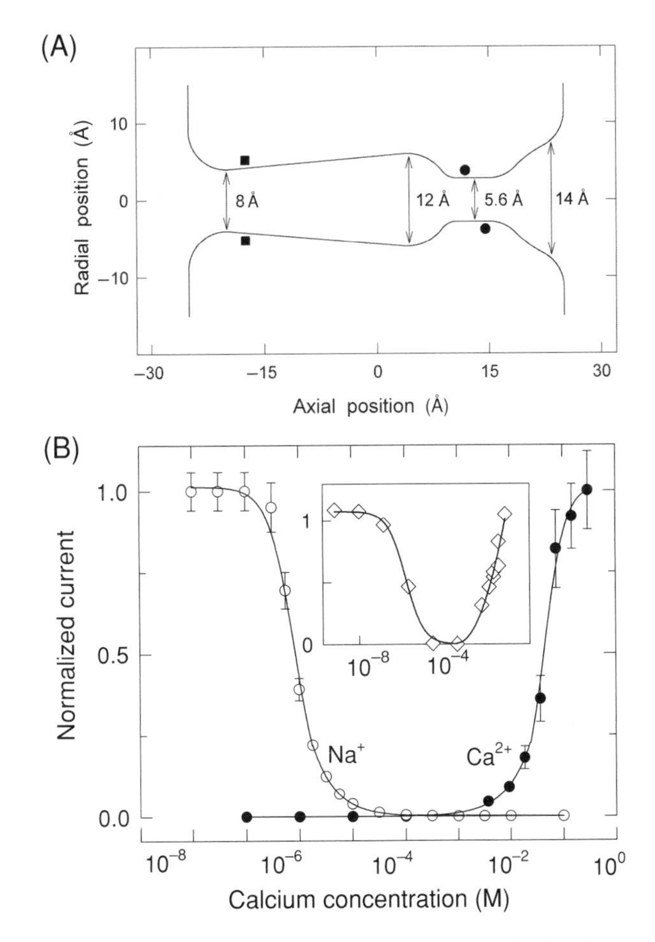

Fig. 7.8 Brownian dynamics model of an L-type calcium channel. (A) The channel shape contains a narrow portion of the pore surrounded by the charge from the four glutamate residues (circles). (B) Results from Brownian dynamics simulations in mixed Na^+ and Ca^{2+} solution reproduce the "anomalous mole fraction" behavior seen in the experimental data (Almers et al., 1984). Figures from Corry et al. (2001).

molecular simulations. Some caution should be applied when taking this approach, however, as the accuracy of an atomistic model cannot be determined until an atomic resolution structure is determined experimentally. A first attempt to make a molecular model of the outer vestibule of an N-type calcium channel was presented by Doughty et al. (1995), who suggested a β-hairpin motif for the P-loop segment of the protein. In contrast, more recent models suggest that the one half of the P-loop actually forms an α-helix (Guy and Durrel, 1995; Lipkind and Fozzard, 2001; Barreiro et al., 2002). Lipkind and Fozzard (2001) created an atomic resolution homology model of the outer vestibule and filter of an L-type channel and determined the likely positions of ions, and the resulting electrostatic fields, within it. The model, based on the KcsA crystal structure, included the P-loops as α-helix-turn-β-strand motifs, with the glutamate residues of the EEEE motif located at the turns. The glutamate side chains point into the pore and form a 6 Å wide conduit in the absence of Ca^{2+}. With Ca^{2+} present the glutamate residues move inward and form an electrostatic trap that firmly holds a single calcium ion. At millimolar concentrations they find that two additional ions can be bound either side of the first Ca^{2+}, and this enables Ca^{2+} to escape the electrostatic trap and permeate through the channel.

A second model of the P-loop has also been made that also places the important glutamate residues of the EEEE locus at a bend at the end of an α helical segment, but places them asymmetrically in two distinguishable planes to account for the asymmetries seen in site-directed mutagenesis (Barreiro et al., 2002). Molecular dynamics simulations are carried out on this model pore containing either two Ca^{2+} or one Ca^{2+} and one Na^{+}. In a 100-ps simulation they find that the two Ca^{2+} move apart but the Ca^{2+} and Na^{+} do not. It is hypothesized that Na^{+} cannot displace Ca^{2+} but that another Ca^{2+} can, similar to the results seen in the Brownian dynamics model, although the simulations may be too short to assess the significance of these conclusions. This mechanism of permeation and selectivity is supported by a second study into the potential of mean force of Na^{+} or Ca^{2+} in this channel model (Barreiro et al., 2003).

Ramakrishnan et al. (2004) also carry out nonequilibrium molecular dynamics simulations of a much simpler model pore containing glutamate residues in a β-barrel scaffold. Apparent off rates when the pore was preloaded with three Na^{+} ions are much higher than when the channel contained one Ca^{2+} and two Na^{+}, consistent with Ca^{2+} blockage. As in the other models, the presence of multiple Ca^{2+} ions was able to release this block.

The rate theory and physical models all share some common elements. Permeation involves multiple ions, most likely two ions for divalent ion conduction or three ions for monovalent ion conduction. Divalent ions are strongly attracted by the glutamates in the P-loops and this results in divalent blockage of monovalent ion currents.

These models present two similar but distinct physical mechanisms underlying ion selectivity. In both, the electrostatic attraction of the negatively charged glutamate residues is essential for attracting multiple ions into the channel and the size of the pore is such that ions cannot pass each other. In this way, the glutamate residues create an attractive binding region. We avoid using the term "binding site" here as the physical models all suggest a broader electrostatic attraction rather than any kind of chemosteric binding. The difference between the explanations of selectivity lies in whether this electrostatic attraction is of itself enough to create selectivity or whether a specific volume and flexibility of the pore is also necessary.

In the simpler "electrostatic" model of ion selectivity, Ca^{2+} is bound more strongly by the channel due to its greater charge and can only be removed by the repulsion of a similarly charged ion as represented in Fig. 7.4. In the "charge–volume" model, the volume of the pore must be such that competition for space as well as electrostatic attraction is important. Given the limited structural information available it is difficult to say exactly which, if either, of these mechanisms is correct.

An advantage of the simpler model is that less is required; it sets much weaker limits on the size and flexibility of the pore. The only geometric limitation required is that permeation through the selectivity filter occurs in single file. In both cases the charge surrounding this region needs to be appropriate to hold two divalent ions in a semistable equilibrium. The simpler electrostatic model also stresses a common theme that has arisen out of rate theory, continuum and simulation models, some of which are able to elegantly reproduce and explain a wealth of physiological data for a variety of different channel types. It appears that the extra complication of setting limits on the channel volume and flexibility, and appealing to charge–space competition to explain selectivity is not necessary.

It is of interest in this context to compare models of selectivity in calcium channel with explanations of valence selectivity in Na^+ and K^+ channels. As noted in the models of the calcium channel, the discrimination between monovalent and divalent cations cannot just rely on presence of a net negative charge as this acts to attract both kinds of ion. Rather, selectivity between monovalent and divalent ions appears to rely on the exact strength of attraction to the protein (or binding affinity) of the permeating ions. A divalent ion has a stronger electrostatic interaction with any charge on the protein than a monovalent ion due to its larger charge. A recent study suggests that this fact on its own appears to be enough to explain how potassium channels are blocked by Ba^{2+}. When Ba^{2+} enters the selectivity filter it is electrostatically bound such that it is unlikely to leave, even with the aid of repulsion from nearby cations (Corry et al., 2005). Electrostatic calculations in model calcium channels suggest that the binding of divalent ions is not as strong as in the potassium channel (Corry et al., 2000c, 2001, 2005). In this case, once a divalent ion enters it can only be forced out with the aid of the Coulomb repulsion from a second divalent ion, but not with the weaker repulsion from a monovalent ion. Divalent ions are known to block sodium channels, but with a much lower affinity and duration than in potassium channels (Taylor et al., 1976; French et al., 1994). Electrostatic calculations on a model sodium channel (Vora et al., 2004; Corry et al., 2005) find that divalent binding is much weaker than in either the calcium or potassium channels. In this case once a divalent ion enters the channel, additional ions are not attracted into the channel to help relieve the channel blockage through ion–ion repulsion. But also, being less strongly bound, the divalent ion dwell time is much shorter than for potassium channels and so divalent block is less effective.

7.7 Channel Gating

7.7.1 Voltage-Activated Gating

As the name suggests, voltage-dependent calcium channels open in response to depolarization. The open probability rises to a maximum at large depolarizations and is generally faster in muscle than in neuron. Fitting the open probability with a Boltzmann factor, it can be estimated that three to five gating charges move across

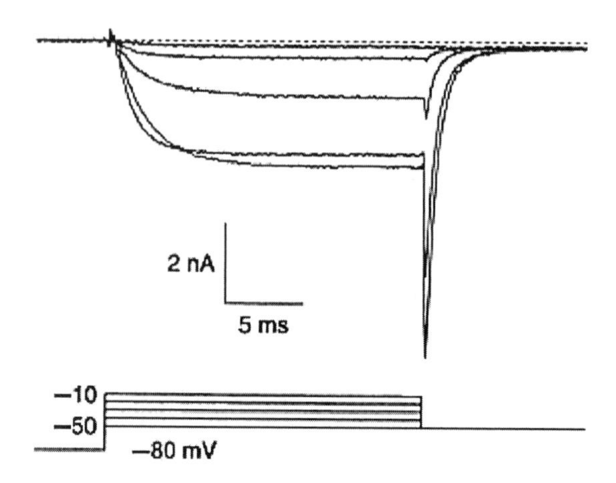

Fig. 7.9 Activation kinetics of whole cell inward calcium currents in frog sympathetic neurons. Measurements are made with 2 mM Ba^{2+}, and the largest currents are at −20 mV and the largest tail currents at −50 mV. Figure reproduced with permission from Jones (1998).

the electric field of the membrane during activation (Kostyuk et al., 1988; Coulter et al., 1989). The opening of calcium channels does not happen instantaneously after depolarization, but rather after a delay of one or more milliseconds depending on the channel type and membrane potential as illustrated in Fig. 7.9. This is generally taken to mean that the channel protein must move through many closed conformations before channel opening occurs.

Open channels are normally deactivated immediately upon repolarization of the membrane. Notably, a transient tail current, observed as a spike in the current immediately after repolarization, occurs before conduction ceases. As noted by Jones (1998), this tail current may serve an important functional role, increasing the calcium intake upon hyperpolarization following an action potential. Measured tail currents typically follow either a single or double exponential time course, however the double exponential behavior has been shown in many cases to be the result of two or more types of Ca channel closing at different rates (Matteson and Armstrong, 1986; Swandulla and Armstrong, 1988).

There also appear to be a number of means by which calcium currents into cells can be increased in a process known as facilitation. An increase in calcium currents created by depolarizing prepulses was first noticed in bovine chromaffin cells by Fenwick et al. (1982). More recent examinations have demonstrated an almost doubling of Ca^{2+} currents created by such prepulses (Artalejo et al., 1990). These enhanced currents may have a functional role in creating enhanced neurotransmitter secretion or muscle contraction in response to danger or stress. A possible contributor to this effect is delayed channel deactivation that has been noted for some HVA channels previously subjected to large depolarization (Pietrobon and Hess, 1990), in particular for P-type (McFarlane, 1997) and L-type channels (Sculptoreanu et al., 1993) where it is believed to intensify muscle contraction. In extreme cases, continual calcium channel activity has been seen for many seconds after a series of large depolarizations (Cloues et al., 1997). It has been suggested that such behavior could be a result of voltage-dependent phosphorylation of the channel protein or of changing the kinetics of the voltage sensor, and may be related to the "mode switching" behavior described below.

As described in the chapter by Bezanilla, the S4 segment of the $\alpha 1$ subunit is believed to be the voltage sensor for channel gating, moving across the electric field of the membrane and causing a conformational change of the protein to open the pore. There have been many fewer studies of gating of Ca^{2+} channels than K^+ or Na^+ channels, however, the sequence similarities between the S4 regions means that many of the insights gleaned from studies on the K^+ or Na^+ channels are likely to apply to Ca channels as well. By similarity with K^+ channels it is expected that the channel gate lies at the internal end of the S6 helices, which most likely form a narrowing of the pore in the closed state.

This internal location is also supported by studying blockage of Ca^{2+} currents by Cd^{2+}, which suggest an internal gate as well as a possible gate at the external end of the pore. Cd^{2+} is thought to tightly bind in the selectivity filter in the same high-affinity binding site utilized by Ca^{2+}. Cd^{2+} blockage of the open channel is voltage-dependent: at large hyperpolarizing or depolarizing voltages block becomes alleviated as if Cd^{2+} can be forced out either end of the pore (Thévenod and Jones, 1992; Swandulla and Armstrong, 1989). Block of the resting closed channel is not alleviated at large hyperpolarizing voltages as if Cd^{2+} cannot easily exit the internal end of the pore. But, when Cd^{2+} is applied to closed channels, blockage is much slower than in open channels, and relief of block is much slower if Cd^{2+} is washed out while the channels are closed. Together these data suggest that Cd^{2+} cannot easily enter either end of the pore in the closed state (Thévenod and Jones, 1992). Jones (2003) raises an interesting question in regard to these data. If Cd^{2+} can still exit a closed channel, albeit at a much slower rate than an open channel, is the closed channel really closed? Could for example small Ca^{2+} currents also pass through a closed pore? The question is not as absurd as it sounds, as a small leakage current of only around 1% of the open channel current would be difficult to detect in single channel recordings, but could have important physiological consequences.

7.7.2 Mode Switching

L-type channels have also been observed to have another interesting gating behavior known as "mode switching." Typically, a single channel will open and close many times during a prolonged depolarization, with each opening lasting less than 1 ms on average. But, on occasion there are times when there is a series of much longer channel openings as if the channel deactivation was 100 times slower. The typical behavior has been termed "mode 1", while the long channel openings is called "mode 2" (Nowycky et al., 1985a; Hess et al., 1986). At other times channel opening appears to cease altogether ("mode 0"). Most likely, there is some physical change taking place to create this mode switching behavior, such as a conformational change or chemical modification of important amino acid residues. This is supported by evidence that antagonistic drugs such as dihydropyridines appear to bind more strongly to channels in mode 0, and help stabilize that state, whereas agonists bind most favorably during mode 2. It has been demonstrated that phosphorylation (most likely near the C-terminus of the α1 subunit) can increase the open probability (Bean et al., 1984) and the mean open time of channel opening (Yue et al., 1990) and this may be related to the mode switching behavior (Dolphin, 1996).

7.8 Inactivation of HVA Channels

The influx of calcium is essential for initiating cellular responses; however, these responses must be controlled within a narrow range of calcium concentrations as intracellular calcium is toxic in excessive levels. For this reason calcium entry is tightly controlled in a number of feedback mechanisms. During prolonged periods of depolarization the initial channel activity is lost in a process known as inactivation that prevents the excess buildup of intracellular calcium. Inactivation can be either voltage-dependent, or alternatively calcium influx can also be controlled by the concentrations of internal calcium that can inactivate channels, or create increased currents (facilitation).

7.8.1 Voltage-Dependent Inactivation

Ca channels can be inactivated in a Ca^{2+}-independent, voltage-dependent manner. For L-type calcium channels of muscle the rate of voltage-dependent inactivation is much slower than Ca-dependent inactivation (Kass and Sanguinetti, 1984; Giannattasio et al., 1991), but this is not necessarily the case for other types of HVA channel (Werz et al., 1993; Zhang et al., 1993). Some N-type channels, for example, show no voltage-dependent inactivation (Artalejo et al., 1992), while others convert between inactivating and noninactivating modes (Plummer and Hess, 1991). Most likely these modes are caused by structural or chemical alterations in the protein.

Repeated short depolarizing bursts can also induce rapid inactivation (Patil et al., 1998). It has been proposed that inactivation occurs preferentially from a partially activated state in which some of the voltage sensors have moved to the activated position, but the channel remains closed (Klemic et al., 1998; Patil et al., 1998). While long depolarizing pulses tend to drive all the channels to the fully open state, short repeated pulses might place more channels in this partially activated state and increase the likelihood of inactivation. This possibility could also explain the inactivating and noninactivating modes of the N-type channel and why N-type channel inactivation is maximal at voltages that also produce maximal inward currents (Jones and Marks, 1989).

In many K^+ and Na^+ channels, inactivation has been demonstrated to take place via a "ball and chain" mechanism in which the amino terminus of the α subunit acts as a tethered ball that can enter the channel, occluding the pore and blocking conduction. Recent evidence suggests that a similar mechanism might form the basis of voltage-dependent inactivation in HVA calcium channels (Stotz et al., 2004). Considerable evidence suggests that both the intracellular ends of the S6 helices (see for example, Kraus et al., 2000; Stotz et al., 2000; Stotz and Zamponi, 2001) and the linker between domains I and II (the AID region; Herlitz et al., 1997; Cens et al., 1999; Berrou et al., 2001) are involved in voltage-dependent inactivation. The simplest way to reconcile these facts is that the S6 helices form a docking site for the I–II linker that when bound partially occludes the pore. Further evidence suggests that the channel β subunit, as well as the N, C, and III–IV linkers can influence inactivation kinetics by interacting with the I–II domain (Stotz et al., 2004). Recent crystal structures of the β subunit–AID complex represent the first detailed pictures of calcium channel components and indicate that binding occurs by the formation of a deep hydrophobic groove in the β subunit that can associate with $\alpha 1$ (Chen et al., 2004; Van Petegem et al., 2004). Furthermore, the structure of the β subunit itself indicates how cellular signaling molecules could interact with the β subunit and therefore influence the kinetics of channel inactivation by altering the interaction with $\alpha 1$.

7.8.2 Ca-Dependent Inactivation and Facilitation

Ca^{2+}-dependent inactivation (CDI) was initially observed in L-type calcium channels from muscle where it was found that inactivation was faster in solutions containing Ca^{2+} than Ba^{2+}. In muscle, CDI is rapid ($\tau \sim 20$ ms). In neurons, CDI of L-type channels can be either fast (Kohr and Mody, 1991) or slow (Heidelberger and Matthews, 1992) depending on the situation. For many years non-L-type HVA channels were believed to not undergo CDI. It recently became apparent, however, that one of the reasons that CDI was not observed was that most recordings were made in solutions containing high levels of the calcium buffer EGTA. When this was dropped to more physiological conditions calcium-dependent regulation has been observed in P/Q- type channels (Lee et al., 1999; DeMaria et al., 2001; Chaudhuri et al., 2005) and more recently N- and R-type channels (Liang et al., 2003). The reasons for EGTA masking CDI in non-L-type channels, but not in L-type channels is due to the differing molecular mechanisms involved as described below. In addition to CDI, it has been shown that non-L-type HVA calcium channels also have a mechanism of calcium-dependent facilitation (CDF), in which calcium influx is amplified by internal calcium concentrations. The study of CDI has progressed significantly in the last few years, and it is beyond the scope of this chapter to cover all the history of the area. For an introduction to the field, the reader is referred to an excellent review (Budde et al., 2002).

Early attempts to understand the mechanisms of CDI revolved around physiological studies of L-type channels. CDI was observed in single channel recordings, suggesting that Ca^{2+} can inactivate the same channel through which it enters the cell (Yue et al., 1990). Buffering of internal Ca^{2+} concentrations by BAPTA reduces, but does not eliminate inactivation (Ginnattasio et al., 1991) as if calcium from the same or nearby channels (Yue et al., 1990) is inactivating the channels before being chelated by the buffer.

A detailed understanding of the phenomenon of CDI and CDF only took place when calmodulin (CaM) was identified as the Ca^{2+} sensor (Lee et al., 1999; Peterson et al., 1999; Qin et al., 1999; Zühlke et al., 1999). The basic principle of both CDI and CDF appears to involve the specific binding characteristics of CaM in its different Ca^{2+} loaded states. Inactivation and facilitation are inhibited when calmodulin is bound to the channel in its unloaded form. Local or global changes in calcium concentration load the CaM in different ways, which initiate different binding patterns between the CaM and the $\alpha 1$ subunit C-terminus which promote either CDI or CDF.

The location of CaM binding has now been pinpointed to two regions of the C-terminal domain of the $\alpha 1$ subunit: a calmodulin binding domain (CBD) and an isoleucine-glutamine (IQ) or similar motif as mutations of these regions influence CDI and CDF (Qin et al., 1999; Zühlke et al., 2000; DeMaria et al., 2001; Lee et al., 2003; Kobrinsky et al., 2005). The details of CaM binding are beginning to reveal the mechanism of Ca^{2+} regulation and the differences between L- and non-L-type inactivation. CDI and CDF are believed to be regulated independently by the binding of Ca^{2+} to the N-terminal or C-terminal CaM lobes (N-lobe and C-lobe respectively). Binding of Ca^{2+} to the C-lobe takes place at high affinity while the binding affinity of Ca^{2+} to the N-lobe is much weaker (Wang, 1985; Johnson et al., 1996). For this reason, C-lobe binding can reflect local changes in Ca^{2+} concentrations created by local influx through calcium channels, while N-lobe binding is likely to take place only once the global concentration, resulting from the influx through many channels, has increased significantly.

Inactivation is triggered by the binding of Ca^{2+} to the opposite lobes in the two classes of channel. In L-type channels, CDI is governed by binding of Ca^{2+} to the C-lobe (Peterson et al., 1999), whereas in non-L-type channels it is regulated by binding to the N-lobe (DeMaria et al., 2001; Lee et al., 2003; Liang et al., 2003). CDF in non-L-type channels, on the other hand, appears to involve binding to the C-lobe (Lee et al., 2003). Although CDF is not usually detected in L-type channels, a CaM mutation that limits Ca^{2+} binding to the C-lobe promotes clear CDF as if it arises through Ca^{2+} binding via the N-lobe (Van Petegem et al., 2005). The presence of the fast acting Ca^{2+} chelator BAPTA influences both CDI and CDF in all channel types. The slower acting by high-affinity chelator EGTA, on the other hand, prevents CDF in L-type channels and CDI in non-L-type channels. The sensitivities to these chelators support the conclusion that the C-lobe processes are triggered by local Ca^{2+} which can be rapidly intercepted by BAPTA but not by EGTA. N-lobe processes appear to be the result of global Ca^{2+} influx (Lee et al., 2003; Liang et al., 2003). The different responses of CaM to local and global Ca^{2+} concentrations allow the voltage-gated calcium channels to accurately control calcium influx.

The exact details of CaM binding to the $\alpha 1$ subunit and the differences in binding between the unloaded, C-lobe loaded, and fully loaded CaM are still being worked out, but an interesting mechanism for regulating CDI and CDF in P/Q-type

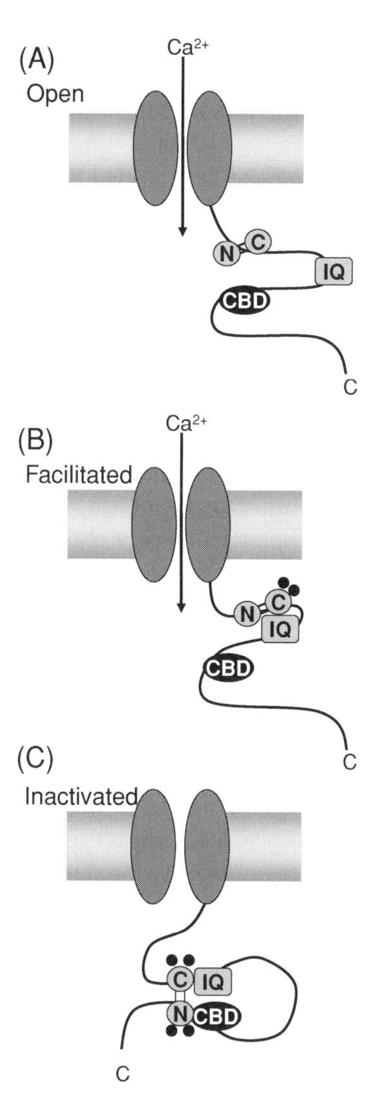

Fig. 7.10 Model of Ca^{2+} binding and the initiation of facilitation and inactivation in $Ca_v2.1$ channels. (A) At rest, no Ca^{2+} is bound to the calmodulin lobes N and C. (B) Low levels of Ca^{2+} prompt binding to the C-lobe which strengthens or initiates the interaction between calmodulin and the IQ domain creating facilitation. (C) High levels of intracellular Ca^{2+} lead to additional binding to the N-lobe and a conformational change enabling interactions of calmodulin with both the IQ and CBD domains and initiates inactivation.

channels has been suggested as depicted in Fig. 7.10 (Lee et al., 2003). In the resting state, with low intracellular calcium, CaM is believed to be preassociated with the α_1 C-terminus (Fig. 7.10A; Erickson et al., 2001). Local influx through the channel promotes Ca^{2+} binding to the C-lobe of the preassociated CaM which either strengthens or initiates binding of CaM to the IQ domain and initiates CDF

(Fig. 7.10B). A global increase in Ca^{2+} encourages Ca^{2+} binding to the CaM N-lobe. The fully loaded CaM then binds to the CBD domain (while possibly remaining bound to the IQ domain) creating a conformational change that promotes inactivation (Fig. 7.10C).

A recent structure of $Ca2^{+}$/CaM bound to the calcium channel IQ domain indicates the details of CaM–α1 interactions (Van Petegem et al., 2005). CaM is seen to bind to the IQ domain primarily through anchoring of a set of aromatic residues in the C-terminus into hydrophobic pockets of CaM. As predicted, C-lobe CaM interactions appeared to be of higher affinity than N-lobe interactions. It is suggested that the difference in CDI and CDF between channel subtypes may be the result of differing CaM anchor positions and the interplay between CDI and CDF appears to arise from competition for binding between Ca^{2+}-loaded N and C-lobes.

One of the big remaining questions in this area is how the binding or unbinding of calmodulin is linked to changes in the channel pore. What are the conformational changes induced by partially or fully loaded CaM? Does CDI utilize the same inactivation gate as voltage-dependent inactivation? If so, what is the link between the C-terminus and the I–II linker of the α1 subunit? What are the differences between L-type and non-L-type channels that are responsible for the opposite effects of Ca^{2+} binding to the CaM C-lobes? How is the calcium influx enhanced by CDF? It should also be noted that in addition to the calmodulin-dependent inactivation mechanisms, it has been suggested that calcium inhibits the channel via activation of protein phosphatase 2B and/or by inhibition of adenylate cyclase (Schuhmann et al., 1997a; You et al., 1997).

7.8.3 Gating of LVA Channels

LVA channels have different activation, deactivation, and inactivation kinetics than HVA channels. As described earlier, activation arises at more negative potentials. Inactivation is very rapid in these channels ($\tau \sim 10$ to 50 ms) and also arises at more negative potentials (Huguenard, 1996). In contrast, LVA channels are slower to activate than HVA channels and are also 10-fold slower to deactivate upon hyperpolarization (Armstrong and Matteson, 1985).

7.9 Regulation of Channel Function

Many factors influence the function of voltage-gated calcium channels so that biochemical pathways can influence intracellular calcium levels. The electrochemical gradient associated with varying concentrations of ions across the plasma membrane, notably calcium ions and H^{+} ions, influences channel gating, activation, and inactivation as described previously. It is well recognized that voltage-gated Ca^{2+} channel function is also regulated by many serine–threonine and tyrosine kinases as a result of direct phosphorylation of the channel protein. However, more recently there is good evidence that the cell's reduction–oxidation state can also influence channel function. In this section the factors that influence channel function will be discussed with particular emphasis on the effects of oxygen and oxygen metabolites on the redox state of the channel protein and its function.

Acidosis has been reported to inhibit L-type Ca^{2+} channels by reducing channel availability whereas alkalosis increases channel availability (Klockner and Isenberg, 1994). The β subunit of the channel complex may be required for this effect (Schuhmann et al., 1997b), as if chemical modifications in the β subunit may affect its interaction with α1.

The α_1 subunit is phosphorylated by cAMP-dependent protein kinase A at Ser-1928 near the carboxy terminus and this results in an increase in the macroscopic peak inward current while slowing the rate of inactivation (McDonald et al., 1994). It is now recognized that this response requires a direct association of protein kinase A with the channel via a leucine zipper interaction with a kinase anchoring protein (Hulme et al., 2003). Protein kinase G inhibits both basal channel activity and the isoproterenol stimulated current in a voltage-dependent manner that may involve competition with cAMP (Wahler et al., 1990). Tyrosine kinase activation can increase channel activity in smooth muscle (Wijetunge and Hughes, 1995) and mediates α-adrenergic receptor inhibition of β-adrenergically stimulated current at the level of the β-adrenergic receptor in cardiac myocytes probably by phosphorylation of the β-adrenergic receptor (Hool et al., 1998; Belevych et al., 2001).

Activators of protein kinase C (PKC) have been reported to have no effect (Lacerda et al., 1988; Mamas and Terrar, 2001), inhibit (Tseng and Boyden, 1991; Zhang et al., 1997; McHugh et al., 2000), or stimulate (Dosemeci et al., 1988; He et al., 2000; Aiello and Cingolani, 2001; Alden et al., 2002; Blumenstein et al., 2002) basal channel activity while decreasing β-adrenergic receptor sensitivity of the channel (Schwartz and Naff, 1997; Belevych et al., 2004). The variable responses with pharmacological activators of PKC (such as phorbol esters) may be due to nondiscrimatory activation of all PKC isoforms. PKC is a large family of lipid-derived serine/threonine kinases that are classified according to their mode of activation and homology in the regulatory domain. The classical isoforms (cPKC) are activated by Ca^{2+} and diacylglycerol (DAG). This group are the most abundant and comprise the α, β_I, β_{II}, and γ isozymes. β_I and β_{II} are C-terminal splice variants of the same gene that differ by 50 amino acids. The novel PKCs (nPKC) comprise ε, η, δ, and θ PKC and are activated by DAG but not Ca^{2+}. The atypical PKCs (aPKC) are dependent on lipids but do not require DAG or Ca^{2+} for activation. Isoforms appear to be restricted to particular cell sites before stimulation and stimulation results in the localization of isoforms to specific subcellular sites. For example, stimulation of cardiac myocytes with an α-adrenergic receptor agonist causes the translocation of β_I PKC from the cytosol to the nucleus and β_{II} PKC from fibrillar structures outside the nucleus to the membrane (Mochly-Rosen et al., 1990; Disatnik et al., 1994; Mochly-Rosen, 1995). The translocation is then assumed to be associated with specific function at the site. The localization of activated isoforms requires protein–protein interactions between the isoforms and anchoring proteins. These anchoring proteins bind unique sites on each isoforms and thereby regulate the activity of individual isoforms. It would appear that different isoforms of PKC produce variable responses depending on the level of expression and activity in tissue. The use of specific peptide inhibitors of isoforms has helped to elucidate the role of PKC in physiological and pathological processes. Zhang et al. (1997) exploited the fact that peptides can be delivered into myocytes via the patch-pipette, and recorded the effect of phorbol esters and peptide inhibitors of cPKC isozymes on basal and isoproterenol-stimulated L-type Ca^{2+} channel currents. The authors found that phorbol esters inhibit basal L-type Ca^{2+} current and dialysis of the cell with a peptide inhibitor of cPKC isozymes (βC2-2) partially attenuated the basal inhibition by PMA. In addition, PMA inhibited the isoproterenol-stimulated L-type Ca^{2+} current. Addition of βC2-2 and βC2-4 peptides to the patch pipette resulted in 89% attenuation of the effect of PMA. These results suggest that C2 containing PKC isozymes are involved in the regulation of basal L-type Ca^{2+} channel current and isoproterenol-stimulated L-type Ca^{2+} currents.

Many hormones and neurotransmitters regulate channel function by binding to G proteins that themselves regulate channel function via activation of kinases and second messengers. However, it is well recognized that P/Q-type and N-type Ca^{2+} channels are directly coupled to and regulated by G proteins. This occurs as a result of direct interaction between the G protein $\beta\gamma$ complex and the α_1 subunit at the α_1 interaction domain AID that may disturb the interaction with the β subunit (Herlitze et al., 1996; Ikeda, 1996; De Waard et al., 1997). The small G protein kir/Gem inhibits calcium channels by interacting directly with the β subunit and decreasing α_1-subunit expression (Beguin et al., 2001). This occurs with the binding of Ca^{2+}/CaM. The direct coupling of L-type Ca^{2+} channels to G_s proteins has been reported (Yatani et al., 1987). Of the T-type Ca^{2+} channels, only α_{1H} but not α_{1G} can be inhibited by activation of G protein $\beta_2\gamma_2$ subunits that bind to the intracellular loop connecting domains II and III suggesting unique control of channel function by G protein $\beta\gamma$ subunits in this class of channel (Wolfe et al., 2003).

More recently, it has been demonstrated that the channel can respond to changes in oxygen (O_2) tension raising the possibility that the channel itself is an O_2 sensor. It is assumed that acute changes in O_2 tension will result in adaptive responses that are designed to restore O_2 to tissue and maintain normal cellular function in mammals. This is true of cardiorespiratory reflexes in the brain that respond rapidly to discharge levels from afferent chemosensory fibers in the carotid body as a result of alterations in blood levels of O_2, CO_2, or pH. However, in the heart, an acute change in cellular oxygen tension represents a trigger for cardiac arrhythmia where an appropriate substrate such as myocardial infarction or a defect in a gene encoding an ion channel exists (Keating and Sanguinetti, 2001; Sanguinetti, 2002). Therefore, it would appear at least for cardiac myocytes that the cellular response to changes in oxygen tension is not always adaptive. Understanding how the myocyte senses changes in oxygen tension is important in determining treatment strategies to prevent life-threatening arrhythmias.

Ion channels have been considered O_2 sensors because their modulation by hypoxia is rapid and occurs in excised membrane patches where cytosolic variables such as second messengers, ATP, and Ca^{2+} are absent. A number of studies have reported modulations in ion channel activity during hypoxia, the first being the demonstration of O_2-sensitive K^+ channels in glomus or type I cells of the rabbit carotid body (Lopez-Barneo et al., 1988). The closure of K^+ channels by hypoxia results in membrane depolarization and Ca^{2+} influx, transmitter release to the innervated organ and activation of afferent sensory fibers. This is not the case however in some pulmonary resistance vessel myocytes where hypoxia increases Ca^{2+} channel conductance (Franco-Obregon and Lopez-Barneo, 1996a). In fact, ion channel responses to hypoxia appear to vary depending on the cell type and the functional response. In arterial smooth muscle cells, hypoxia appears to inhibit L-type Ca^{2+} channels resulting in relaxation of vessels (Franco-Obregon et al., 1995; Franco-Obregon and Lopez-Barneo, 1996b).

In cardiac myocytes, there is good evidence that hypoxia inhibits the L-type Ca^{2+} channel. This has been demonstrated in the α_{1C} subunit of the human cardiac L-type Ca^{2+} channel recombinant in HEK 293 cells (Fearon et al., 1997) and in native channels in guinea-pig cardiac myocytes (Fig. 7.11; Hool, 2000, 2001) suggesting that there is a requirement for the α_{1C} subunit of the channel in the hypoxia response. In search of the identity of the O_2-sensing site, Fearon et al. (2000) produced splice variants of the α_{1C} subunit of the human cardiac L-type Ca^{2+} channel and examined the O_2 sensitivity of the variants. They were able to isolate a 39 amino acid segment of the C-terminal domain of the subunit as the O_2-sensitive component. However, a direct effect of changes in O_2 tension on the channel protein has not been demonstrated. In addition, the varied responses by ion channels to changes in O_2 tension has made it difficult to assign a universal O_2-sensing component to the channel.

A clue to the mechanism of O_2 sensing arises from the fact that the function of the α_{1C} subunit can be modulated by reduction or oxidation of critical thiol groups (Chiamvimonvat et al., 1995; Hu et al., 1997; Fearon et al., 1999). This has also been demonstrated in native channels (Campbell et al., 1996; Hool, 2000, 2001; Hool and Arthur, 2002). In addition, the reactive oxygen species superoxide and H_2O_2 can modulate ion channel function (Anzai et al., 2000; Hool and Arthur, 2002; Liu and Gutterman, 2002; Lebuffe et al., 2003). Since the channel does not possess an O_2 binding domain it is more likely that the O_2-sensitive region on the C-terminal domain of the α_{1C} subunit contains a number of critical thiol groups that can be modified in response to alterations in cellular redox state occurring during hypoxia. Consistent with this, we have shown that exposure of myocytes to thiol-specific reducing agent dithiothreitol mimics the effect of hypoxia on the channel while the oxidizing agent 5,5'-dithio-bis[2-nitrobenzoic acid](DTNB) attenuates the effect of hypoxia. In addition to decreasing basal channel activity, hypoxia increases the sensitivity of the channel to β-adrenergic receptor stimulation (Hool, 2000, 2001; Hool and Arthur, 2002). In the presence of hypoxia, the $K_{0.5}$ for activation of the channel by the β-adrenergic receptor agonist isoproterenol is decreased from 5.1 nM to 1.6 nM. We used the response of the channel to the β-adrenergic receptor agonist

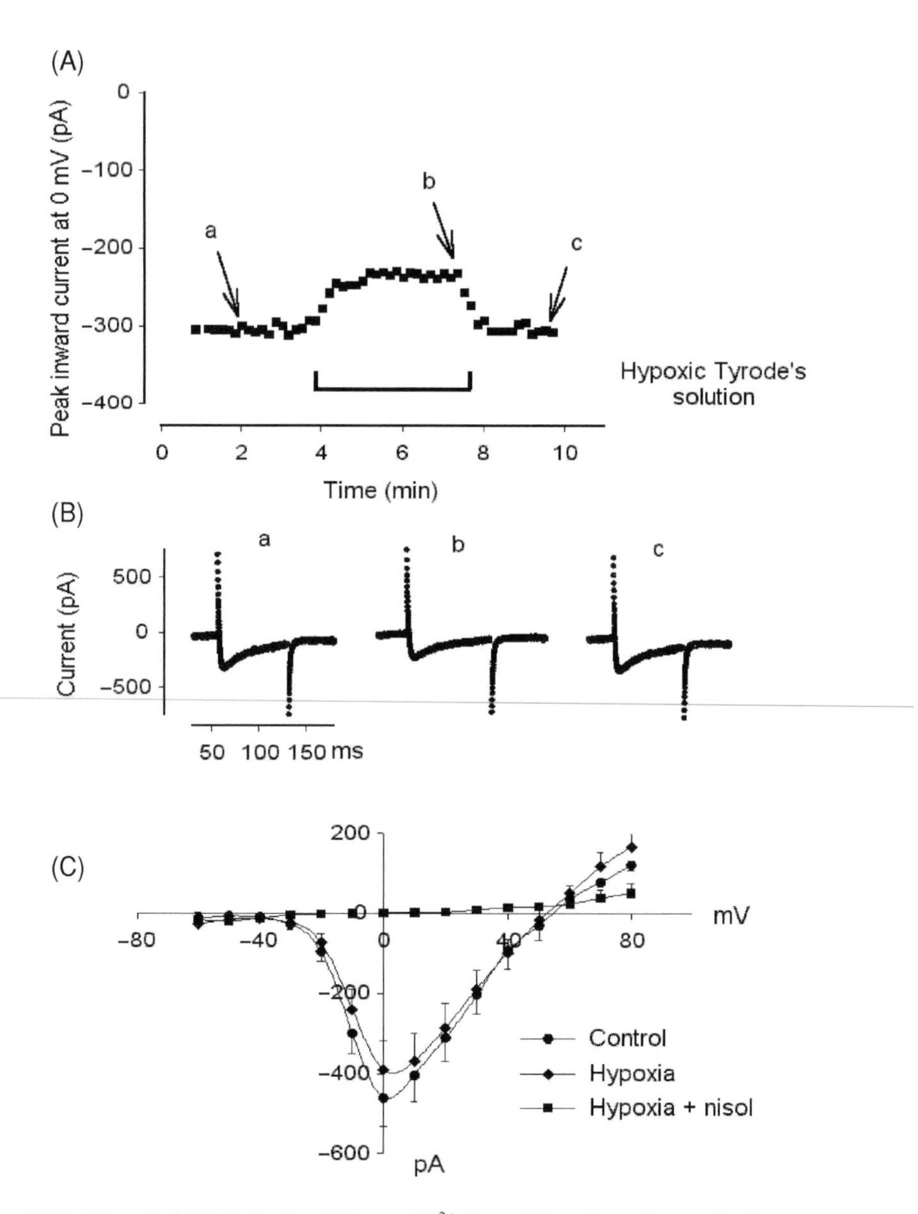

Fig. 7.11 Hypoxia inhibits the basal L-type Ca^{2+}channel in guinea-pig ventricular myocytes. (A) Time course of changes in membrane current recorded during a 75 ms test pulse to 0 mV applied once every 10 s. Basal current recorded during exposure to hypoxic Tyrode's solution that is reversed upon switching the superfusate to Tyrode's solution containing room oxygen tension. (B) Membrane currents recorded at time points in protocol illustrated in (A). (C) Mean ± SE current–voltage (I–V) relationship for five cells exposed to hypoxia and hypoxia + nisoldipine (nisol). Reproduced with permission from Hool (2000).

isoproterenol as a functional reporter of changes in cellular redox state. In cardiac myocytes, hypoxia is associated with a decrease in cellular superoxide (derived from the mitochondria) that then signals a change in the redox state of the channel resulting in altered function. Perfusing myocytes intracellularly with catalase (that specifically converts H_2O_2 to $H_2O$and O_2) mimics the effect of hypoxia on channel function. In addition, preexposing myocytes to H_2O_2 attenuates the effect of catalase and the effect of hypoxia (Fig. 7.12). Given that the channel responds to alterations in cellular redox state, it would appear that O_2-sensing is not intrinsic to ion channels. The challenge now is to determine how a change in channel thiol group redox state alters the conformational state of the channel and ion conduction.

7.10 Conclusions and Outlook

Our knowledge of calcium channels has come a long way since the first calcium currents were recorded in the 1950s. At that time the importance and complexity of the calcium signaling network could not have been appreciated. We now have a much better idea of the role of calcium in translating electrical stimuli into responses and the many functions of Ca^{2+} within the cell such as stimulating neurotransmitter secretion and muscle contraction. Because Ca^{2+} is used to initiate a vast range of responses, a huge variety of biochemical pathways exist to control its movement. Although some of the pathways by which Ca^{2+} can interact with components of the cell, and by which the cell can influence the passage of Ca^{2+} through ion channels are being clarified, many still remain to be elucidated.

Given that there is still much to be learnt about the biochemical properties of Ca^{2+} channels, it is probably not surprising that we know comparatively little about the physical mechanisms underlying their function. Indeed there are many aspects of calcium channel functions that remain to be elucidated. Using channel inactivation as just one example, although the domain I–II linker is implicated as being directly involved in channel inactivation, we still do not have a clear answer as to how this inactivation takes place at a structural level. Does this region physically block the pore along the lines of the ball and chain model? What are the relationships between the mechanisms of calcium- and voltage-dependent inactivation? How do interactions with the β subunits or the calmodulin bound C-terminus affect the process of inactivation, and how does the binding of other regulatory factors to the β subunit get passed on to alter the current passing through the channel pore? Other questions also remain to be answered. How do the structures of calcium channels respond to phosphorylation, binding to G proteins, pharmaceutical agents, or changes in oxidation state? How accurate are our models of ion permeation and selectivity?

Much progress has been made towards understanding the mechanisms of voltage sensing and gating in K^+ channels, however, there is still much that is unknown. Indeed one important question that must be asked, given that so much of our knowledge of Ca^{2+} channel structure and function is derived from the study of bacterial K^+ channels, is whether they really make good models of Ca^{2+} channel pores. It

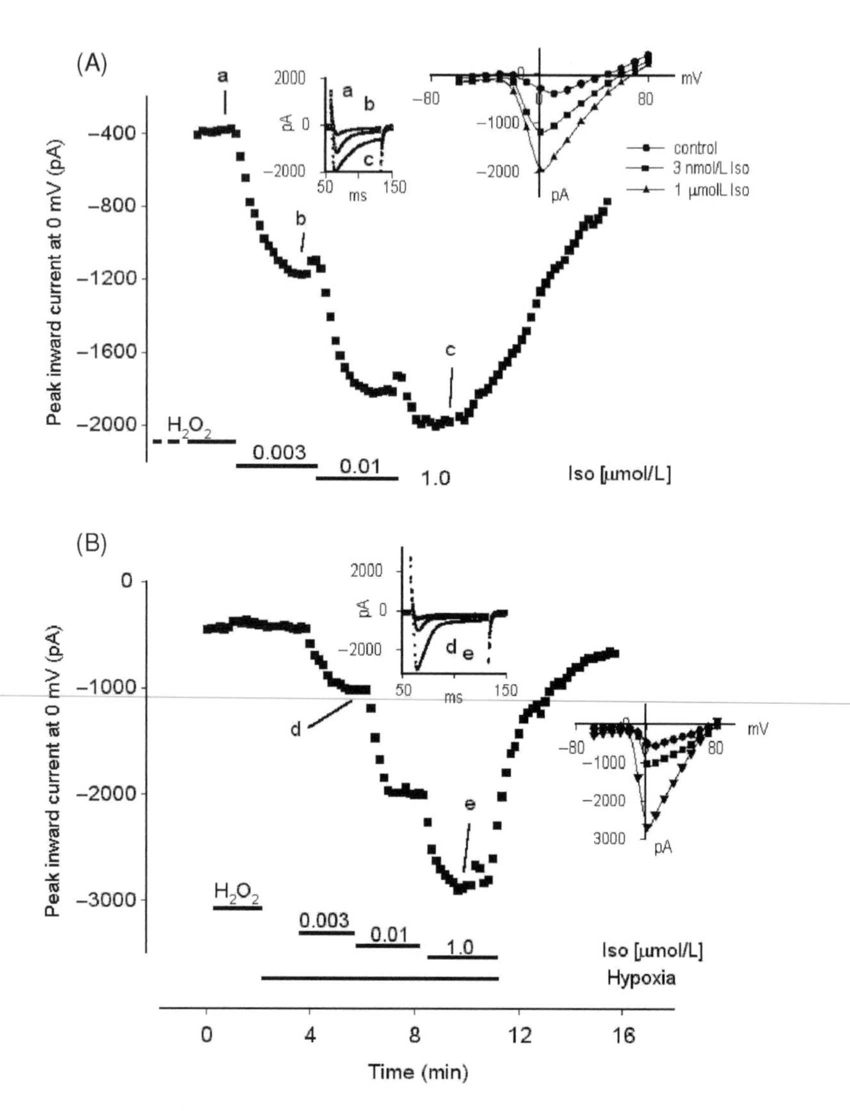

Fig. 7.12 Preexposure of guinea-pig ventricular myocytes to H_2O_2 attenuates the effect of hypoxia and the effect of catalase on the sensitivity of the L-type Ca^{2+} channel to beta-adrenergic receptor stimulation. During hypoxia (or when the cell is perfused intracellularly with catalase) the current produced in response to 3 nM isoproterenol would typically represent approximately 80% of the response elicited by a maximally stimulating concentration of the agonist (1 µM) within the same cell. (A) Time course of changes in membrane current recorded in a cell during exposure to isoproterenol while being dialyzed with catalase. The cell was preexposed to 8.8 μmol L^{-1} H_2O_2 for at least 5 min prior to superfusion with isoproterenol in the absence of H_2O_2. Membrane currents recorded at the time points indicated are shown inset at left. I–V relationship in the same cell shown inset at right. (B) Time course of changes in membrane current recorded in a cell during exposure to isoproterenol and hypoxia. The cell was also preexposed to 8.8 μmol L^{-1} H_2O_2 for at least 5 min prior to superfusion with isoproterenol in the absence of H_2O_2. Membrane currents recorded at the time points indicated are shown inset at left. I–V relationship in the same cell shown inset at right. Adapted from Hool and Arthur (2002).

is known that the selectivity filter of K^+ channels are lined by the backbone carbonyl oxygen atoms, but it is believed that glutamate side chains perform that role in calcium channels. Whether there are significant differences elsewhere in the structure can only really be determined from an atomic resolution structure of a calcium channel.

Obtaining a detailed structure of a bacterial channel was a major step in K^+ channel studies, and has advanced the field significantly. The study of calcium channels would also benefit significantly from such a structure. This would, for example, enable the details of the selectivity filter to be determined, as well as the likely channel gates. Furthermore, one hopes that the interactions between the pore and the inactivation gate on the $\alpha 1$ I–II linker and calcium sensors on the $\alpha 1$ C-terminus could be deduced from such information. Although crystal structures of the β subunit and of CaM bound to the C-terminus of the $\alpha 1$ subunit have been obtained recently, a complete structure of the entire calcium channel appears a long way off.

Surprisingly, one of the most promising leads in detailed structural studies of calcium channels has come from the successful expression of a bacterial sodium channel NaChBac (Ren et al., 2001). Being more closely related to the family of calcium channels than K^+ channels, deductions on the basic structure of NaChBac will be particularly pertinent to Ca^{2+} channel studies. In particular, NaChBac is found to contain an "EEEE" locus, and using appropriate mutations the NaChBac channel can be made selective for Ca^{2+} (Yue et al., 2002). Although many difficulties and uncertainties will no doubt have to be overcome, if a crystal structure of NaChBac is determined it will undoubtedly provide important information about the selectivity and ion conduction properties of the family of Ca^{2+} channels.

References

Ahlijanian, M.K., R.E. Westenbroek, and W.A. Catterall. 1990. Subunit structure and localization of dihydropyridine-sensitive calcium channels in mammalian brain, spinal cord, and retina. *Neuron* 4:819–832.

Aiello, E.A., and H.E. Cingolani. 2001. Angiotensin II stimulates cardiac L-type Ca(2+) current by a Ca(2+)- and protein kinase C-dependent mechanism. *Am. J. Physiol. Heart Circ. Physiol.* 280:H1528–H1536.

Alden, K.J., P.H. Goldspink, S.W. Ruch, P.M. Buttrick, and J. Garcia. 2002. Enhancement of L-type Ca(2+) current from neonatal mouse ventricular myocytes by constitutively active PKC-betaII. *Am. J. Physiol. Cell Physiol.* 282:C768–C774.

Almers, W., E.W. McCleskey, and P.T. Palade. 1984. A non-selective cation conductance in frog muscle membrane blocked by micromolar external calcium ions. *J. Physiol.* 353:565–583.

Almers, W., and E.W. McCleskey. 1984. Non-selective conductance in calcium channels of frog muscle: Calcium selectivity in a single file pore. *J. Physiol.* 353:585–608.

Anzai, K., K. Ogawa, T. Ozawa, and H. Yamamoto. 2000. Oxidative modification of ion channel activity of ryanodine receptor [Review] [28 refs]. *Antioxidants Redox Signaling* 2:35–40.

Armstrong, C.M., and J. Neyton. 1991. Ion permeation through calcium channels—a one site model. *Ann. N.Y. Acad. Sci.* 635:18–25.

Armstrong, C.M., and D.R. Matteson. 1985. Two distinct populations of calcium channels in a clonal line of pituitary cells. *Science* 227:65–67.

Artalejo, C.R., M.A. Ariano, R.L. Perlman, and A.P. Fox. 1990. Activation of facilitation calcium channels in chromaffin cells by D_1 dopamine receptors through cAMP/protein kinase A-dependent mechanism. *Nature* 348:239–242.

Artalejo, C.R., R.L. Perlman, and A.P. Fox. 1992. ω-conotoxin GVIA blocks a Ca^{2+} current in bovine chromaffin cells that is not the "classic" N-type. *Neuron* 8: 85–95.

Bahinski, A., A. Yatani, G. Mikala, S. Tang, S. Yamamoto, and A. Schwartz. 1997. Charged amino acids near the pore entrance influence ion-conduction of a human L-type cardiac calcium channel. *Mol. Cell. Biochem.* 166:125–134.

Barreiro, G., C.R. Guimaraes, and R.B. de Alencastro. 2002. A molecular dynamics study of an L-type calcium channel model. *Protein Eng.* 15:109–122.

Barreiro, G., C.R. Guimaraes, and R.B. de Alencastro. 2003. Potential of mean force calculations on an L-type calcium channel model. *Protein Eng.* 16:209–215.

Bean, B.P., M.C. Nowycky, and R.W. Tsien. 1984. Beta-adrenergetic modulation of calcium channels in frog ventricular heart-cells. *Nature* 307:371–375.

Beguin, P., K. Nagashima, T. Gonoi, T. Shibasaki, K. Takahashi, Y. Kashima, N. Ozaki, K. Geering, T. Iwanaga, and S. Seino. 2001. Regulation of Ca^{2+} channel expression at the cell surface by the small G-protein kir/Gem. *Nature* 411:701–706.

Belevych, A.E., I. Juranek, and R.D. Harvey. 2004. Protein kinase C regulates functional coupling of beta1adrenergic receptors to Gi/o-mediated responses in cardiac myocytes. *FASEB J.* 18:367–369.

Belevych, A.E., A. Nulton-Persson, C. Sims, and R.D. Harvey. 2001. Role of tyrosine kinase activity in alpha-adrenergic inhibition of the beta-adrenergically regulated L-type Ca(2+) current in guinea-pig ventricular myocytes. *J Physiol.* 537:779–792.

Berrou, L., G. Bernatchez, and L. Parent. 2001. Molecular determinants of inactivation within the I–II linker of alpha1E (CaV2.3) calcium channels. *Biophys. J.* 80:215–228.

Bezanilla, F., and C.M. Armstrong. 1972. Negative conductance caused by entry of sodium and cesium ions into the potassium channels of squid giant axons. *J. Gen. Physiol.* 60:588–608.

Bezprozvanny, I., J. Watras, and B.E. Ehrlich. 1991. Bell-shaped calcium-response curves of Ins(1,4,5)P3- and calcium-gated channels from endoplasmic reticulum of cerebellum. *Nature* 351:751–754.

Bichet, D., V. Cornet, S. Geib, E. Carlier, S. Volsen, T. Hoshi, Y. Mori, and M. De Waard. 2000. The I–II loop of the Ca^{2+} channel alpha1 subunit contains an endoplasmic reticulum retention signal antagonized by the beta subunit. *Neuron* 25:177–190.

Birnbaumer, L., K.P. Campbell, W.A. Catterall, M.M. Harpold, F. Hofmann, W.A. Horne, Y. Mori, A. Schwartz, T.P. Snutch, T. Tanabe, et al. 1994. The naming of voltage-gated calcium channels. *Neuron* 13:505–506.

Birnbaumer, L., N. Qin, R. Olcese, E. Tareilus, D. Platano, J. Costantin, and E. Stefani. 1998. Structures and functions of calcium channel beta subunits. *J. Bioenerg. Biomembr.* 30:357–375.

Blumenstein, Y., N. Kanevsky, G. Sahar, R. Barzilai, T. Ivanina, and N. Dascal. 2002. A novel long N-terminal isoform of human L-type Ca^{2+} channel is up-regulated by protein kinase C. *J. Biol. Chem.* 277:3419–3423.

Boda, D., D. Busath, D. Henderson, and S. Sokolowski. 2000. Monte Carlo simulations of the mechanism for channel selectivity: The competition between volume exclusion and charge neutrality. *J. Phys. Chem.* 104:8903–8910.

Bosanac, I., J.R. Alattia, T.K. Mal, J. Chan, S. Talarico, F.K. Tong, K.I. Tong, F. Yoshikawa, T. Furuichi, M. Iwai, T. Michikawa, K. Mikoshiba, and M. Ikura. 2002. Structure of the inositol 1,4,5-trisphosphate receptor binding core in complex with its ligand. *Nature* 420:696–700.

Budde, T., S. Meuth, and H.C. Pape. 2002. Calcium-dependent inactivation of neuronal calcium channels. *Nat. Rev. Neurosci.* 3:873–883.

Bultynck, G., E. Vermassen, K. Szlufcik, P. De Smet, R.A. Fissore, G. Callewaert, L. Missiaen, H. De Smedt, and J.B. Parys. 2003. Calcineurin and intracellular Ca^{2+}-release channels: Regulation or association? *Biochem. Biophys. Res. Commun.* 311:1181–1193.

Caillard, O., Y. Ben-Ari, and J.L. Gaiarsa. 2000. Activation of pre-synaptic and postsynaptic ryanodine-sensitive calcium stores is required for the induction of long-term depression at GABA-ergic synapses in the neonatal rat hippocampus. *J. Neurosci.* 20(17):RC94.

Campbell, D.L., J.S. Stamler, and H.C. Strauss. 1996. Redox modulation of L-type calcium channels in ferret ventricular myocytes. Dual mechanism regulation by nitric oxide and S-nitrosothiols. *J. Gen. Physiol.* 108:277–293.

Catterall, W.A. 2000. From ionic currents to molecular mechanisms: The structure and function of voltage-gated sodium channels. *Neuron* 26:13–25.

Catterall, W.A., J. Striessnig, T.P. Snutch, and E. Perez-Reyes. 2003. International Union of Pharmacology. XL. Compendium of voltage-gated ion channels: Calcium channels. *Pharmacol. Rev.* 55:579–581.

Catterall, W.A., E. Perez-Reyes, T.P. Snutch, and J. Striessnig. 2005. International Union of Pharmacology. XLVIII. Nomenclature and structure–function relationships of voltage-gated calcium channels. *Pharmacol. Rev.* 57:411–425.

Cens, T., S. Restituito, S. Galas, and P. Charnet. 1999. Voltage and calcium use the same molecular determinants to inactivate calcium channels. *J. Biol. Chem.* 274:5483–5490.

Chandra, R., V.S. Chauhan, C.F. Starmer, and A.O. Grant. 1999. beta-Adrenergic action on wild-type and KPQ mutant human cardiac Na^{+} channels: Shift in gating but no change in $Ca2^+$:Na^+ selectivity. *Cardiovasc. Res.* 42:490–502.

Chang, F.C., and M.M. Hosey. 1988. Dihydropyridine and phenylalkylamine receptors associated with cardiac and skeletal muscle calcium channels are structurally different. *J. Biol. Chem.* 263:18929–18937.

Chaudhuri, D., B.A. Alseikhan, S.Y. Chang, T.W. Soong, and D.T. Yue. 2005. Developmental activation of calmodulin dependent facilitation of cerebellar P-type Ca^{2+} current. *J. Neurosci.* 25:8282–8294.

Chen, X.H., I. Bezprozvanny, and R.W. Tsien. 1996. Molecular basis of proton block of L-type Ca^{2+} channels. *J. Gen. Physiol.* 108:363–374.

Chen, X.H., and R.W. Tsien. 1997. Aspartate substitutions establish the concerted action of p-region glutamates in repeats I and III in forming the protonation site of L-type Ca^{2+} channels. *J. Biol. Chem.* 272:30002–30008.

Chen, Y.H., M.H. Li, Y. Zhang, L.L. He, Y. Yamada, A. Fitzmaurice, Y. Shen, H. Zhang, L. Tong, and J. Yang. 2004. Structural basis of the α_1-β subunit interaction of voltage gated Ca^{2+} channels. *Nature* 429:675–680.

Chiamvimonvat, N., B. O'Rourke, T.J. Kamp, R.G. Kallen, F. Hofmann, V. Flockerzi, and E. Marban. 1995. Functional consequences of sulfhydryl modification in the pore-forming subunits of cardiovascular Ca^{2+} and Na^+ channels. *Circ. Res.* 76:325–334.

Chien, A.J., X. Zhao, R.E. Shirokov, T.S. Puri, C.F. Chang, D. Sun, E. Rios, and M.M. Hosey. 1995. Roles of a membrane-localized beta subunit in the formation and targeting of functional L-type Ca^{2+} channels. *J. Biol. Chem.* 270:30036–30044.

Chow, R.H. 1991. Cadmium block of squid calcium currents. Macroscopic data and a kinetic model. *J. Gen. Physiol.* 98:751–770.

Cibulsky, S.M., and W.A. Sather. 2000. The EEEE locus is the sole high-affinity Ca^{2+} binding structure in the pore of a voltage-gated Ca^{2+} channel–Block by Ca^{2+} entering from the intracellular pore entrance. *J. Gen. Physiol.* 116:569–585.

Cloues, R.K., S.J. Tavalin, and N.V. Marrion. 1997. Beta-adrenergic stimulation selectively inhibits long-lasting L-type calcium channel facilitation in hippocampal pyramidal neurons. *J. Neurosci.* 17:6493–6503.

Colecraft, H.M., B. Alseikhan, S.X. Takahashi, D. Chaudhuri, S. Mittman, V. Yegnasubramanian, R.S. Alvania, D.C. Johns, E. Marban, and D.T. Yue. 2002. Novel functional properties of Ca(2+) channel beta subunits revealed by their expression in adult rat heart cells. *J. Physiol.* 541:435–452.

Corry, B., S. Kuyucak, and S.H. Chung. 2000a. Invalidity of continuum theories of electrolytes in nanopores. *Chem. Phys. Lett.* 320:35–41.

Corry, B., S. Kuyucak, and S.H. Chung. 2000b. Tests of continuum theories as models of ion channels: II. Poisson–Nernst–Planck theory versus brownian dynamics. *Biophys. J.* 78:2364–2381.

Corry, B., T.W. Allen, S. Kuyucak, and S.H. Chung. 2000c. A model of calcium channels. *Biochim. Biophys. Acta* 1509:1–6.

Corry, B., T.W. Allen, S. Kuyucak, and S.H. Chung. 2001. Mechanisms of permeation and selectivity in calcium channels. *Biophys. J.* 80:195–214.

Corry, B., T. Vora, and S.H. Chung. 2005. Electrostatic basis of valence selectivity in cationic channels. *Biochim. Biophys. Acta* 1711:72–86.

Coulter, D.A., J.R. Huguenard, and D.A. Prince. 1989. Calcium currents in rat thalamocortical relay neurons: Kinetic properties of the transient, low threshold current. *J. Physiol.* 414:587–604.

Curtis, B.M., and W.A. Catterall. 1984. Purification of the calcium antagonist receptor of the voltage-sensitive calcium channel from skeletal muscle transverse tubules. *Biochemistry* 23:2133–2138.

Dang, T.X., and E.W. McCleskey. 1998. Ion channel selectivity through stepwise changes in binding affinity. *J. Gen. Physiol.* 111:185–193.

DeMaria, S.D., T.W. Soong, B.A. Alseikhan, R.S. Alvania, and D.T. Yue. 2001. Calmodulin bifurcates the local Ca^{2+} signal that modulates P/Q-type Ca^{2+} channels. *Nature* 411:484–489.

De Waard, M., H. Liu, D. Walker, V.E. Scott, C.A. Gurnett, and K.P. Campbell. 1997. Direct binding of G-protein betagamma complex to voltage-dependent calcium channels. *Nature* 385:446–450.

DelPrincipe, F., M. Egger, and E. Niggli. 2000. L-type Ca^{2+} current as the predominant pathway of Ca^{2+} entry during I(Na) activation in beta-stimulated cardiac myocytes. *J. Physiol.* 527(Pt 3):455–466.

Dipolo, R., and L. Beauge. 2006. Sodium/calcium exchanger: Influence of metabolic regulation on ion carrier interactions. *Physiol. Rev.* 86:155–203.

Disatnik, M.H., G. Buraggi, and D. Mochly-Rosen. 1994. Localization of protein kinase C isozymes in cardiac myocytes. *Exp. Cell Res.* 210:287–297.

Dolphin, A.C. 1996. Facilitation of Ca^{2+} current in excitable cells. *Trends Neurosci.* 19:35–43.

Dolphin, A.C. 1998. Mechanisms of modulation of voltage-dependent calcium channels by G proteins. *J. Physiol.* 506(Pt 1):3–11.

Dolphin, A.C. 2003. Beta subunits of voltage-gated calcium channels. *J. Bioenerg. Biomembr.* 35:599–620.

Dosemeci, A., R.S. Dhallan, N.M. Cohen, W.J. Lederer, and T.B. Rogers. 1988. Phorbol ester increases calcium current and simulates the effects of angiotensin II on cultured neonatal rat heart myocytes. *Circ. Res.* 62:347–357.

Doughty, S.W., F.E. Blaney, and W.G. Richards. 1995. Models of ion pores in N-type voltage gated calcium channels. *J. Mol. Graph.* 13:342–348.

Douglas, W.W. 1968. Stimulus-secretion coupling: The concept and clues from chromaffin and other cells. *Br. J. Pharmacol.* 34:453–474.

Dunlap, K., and G.D. Fischbach. 1978. Neurotransmitters decrease the calcium component of sensory neuron action potentials. *Nature* 276:837–839.

Doyle, D.A., J.M. Cabral, R.A. Pfuetzner, A. Kuo, J.M. Gulbis, S.L. Cohen, B.T. Chait, and R. MacKinnon. 1998. The structure of the potassium of K^+ conduction and channel: Molecular basis selectivity. *Science* 280:69–77.

Eisenman, G. 1962. Cation selective glass electrodes and their mode of operation. *Biophys. J.* 2(Pt 2):259–323.

Ellinor, P.T., J. Yang, W.A. Sather, J.F. Zhang, and R.W. Tsien. 1995. Ca^{2+} channel selectivity at a single locus for high-affinity Ca^{2+} interactions. *Neuron* 15:1121–1132.

Ellis, S.B., M.E. Williams, N.R. Ways, R. Brenner, A.H. Sharp, A.T. Leung, K.P. Campbell, E. McKenna, W.J. Koch, A. Hui, et al. 1988. Sequence and expression of mRNAs encoding the alpha 1 and alpha 2 subunits of a DHP-sensitive calcium channel. *Science* 241:1661–1664.

Erickson, M.G., B.A. Alselkhan, B.Z. Peterson, and D.T. Yue. 2001. Preassociation of calmodulin with voltage-gated Ca^{2+} channels revealed by FRET in single living cells. *Neuron* 31:973–985.

Ertel, E.A., K.P. Campbell, M.M. Harpold, F. Hofmann, Y. Mori, E. Perez-Reyes, A. Schwartz, T.P. Snutch, T. Tanabe, L. Birnbaumer, R.W. Tsien, and W.A. Catterall. 2000. Nomenclature of voltage-gated calcium channels. *Neuron* 25:533–535.

Fatt, P., and B. Katz. 1953. The electrical properties of crustacean muscle fibres. *J. Physiol.* 120:374–389.

Fatt, P., and B.L. Ginsborg. 1958. The ionic requirements for the production of action potentials in crustacean muscle fibres. *J. Physiol.* 142:516–543.

Fearon, I.M., A.C. Palmer, A.J. Balmforth, S.G. Ball, G. Mikala, A. Schwartz, and C. Peers. 1997. Hypoxia inhibits the recombinant alpha 1C subunit of the human cardiac L-type Ca^{2+} channel. *J. Physiol.* 500:551–556.

Fearon, I.M., A.C. Palmer, A.J. Balmforth, S.G. Ball, G. Varadi, and C. Peers. 1999. Modulation of recombinant human cardiac L-type Ca^{2+} channel alpha1C subunits by redox agents and hypoxia. *J. Physiol.* 514:629–637.

Fearon, I.M., G. Varadi, S. Koch, I. Isaacsohn, S.G. Ball, and C. Peers. 2000. Splice variants reveal the region involved in oxygen sensing by recombinant human L-type Ca^{2+} channels. *Circ. Res.* 87:537–539.

Felix, R., C.A. Gurnett, M. De Waard, and K.P. Campbell. 1997. Dissection of functional domains of the voltage-dependent Ca^{2+} channel alpha2delta subunit. *J. Neurosci.* 17:6884–6891.

Ferris, C.D., A.M. Cameron, D.S. Bredt, R.L. Huganir, and S.H. Snyder. 1991a. Inositol 1,4,5-trisphosphate receptor is phosphorylated by cyclic AMP-dependent protein kinase at serines 1755 and 1589. *Biochem. Biophys. Res. Commun.* 175:192–198.

Fenwick, E.M., A. Marty, and E. Neher. 1982. Sodium and calcium channels in bovine chromaffin cells. *J. Physiol.* 331:599–635.

Ferris, C.D., R.L. Huganir, D.S. Bredt, A.M. Cameron, and S.H. Snyder. 1991b. Inositol trisphosphate receptor: Phosphorylation by protein kinase C and calcium calmodulin-dependent protein kinases in reconstituted lipid vesicles. *Proc. Natl. Acad. Sci. USA* 88:2232–2235.

Fill, M., and J.A. Copello. 2002. Ryanodine receptor calcium release channels. *Physiol. Rev.* 82:893–922.

Franco-Obregon, A., and J. Lopez-Barneo. 1996a. Differential oxygen sensitivity of calcium channels in rabbit smooth muscle cells of conduit and resistance pulmonary arteries [published erratum appears in J Physiol (Lond) 1996 Jun 15;493(Pt 3):923]. *J. Physiol.* 491:511–518.

Franco-Obregon, A., and J. Lopez-Barneo. 1996b. Low PO2 inhibits calcium channel activity in arterial smooth muscle cells. *Am. J. Physiol.* 271:H2290–H2299.

Franco-Obregon, A., J. Urena, and J. Lopez-Barneo. 1995. Oxygen-sensitive calcium channels in vascular smooth muscle and their possible role in hypoxic arterial relaxation. *Proc. Natl. Acad. Sci. USA* 92:4715–4719.

Franzini-Armstrong, C., and F. Protasi. 1997. Ryanodine receptors of striated muscles: A complex channel capable of multiple interactions. *Physiol. Rev.* 77:699–729.

French, R.J., J.F. Worley III, W.F. Wonderlin, A.S. Kularatna, and B.K. Krueger 1994. Ion permeation, divalent ion block, and chemical modification of single sodium channels. *J. Gen. Physiol.* 103:447–470.

French, R.J., and G.W. Zamponi. 2005. Voltage-gated sodium and calcium channels in nerve, muscle and heart. *IEEE Trans. Nanobiosci.* 4:58–69.

Fukuda, N., O.U. Jin, D. Sasaki, H. Kajiwara, S. Ishiwata, and S. Kurihara. 2001a. Acidosis or inorganic phosphate enhances the length dependence of tension in rat skinned cardiac muscle. *J. Physiol.* 536:153–160.

Fukuda, N., D. Sasaki, S. Ishiwata, and S. Kurihara. 2001b. Length dependence of tension generation in rat skinned cardiac muscle: Role of titin in the Frank-Starling mechanism of the heart. *Circulation* 104:1639–1645.

Fukushima, A., and S. Hagiwara. 1985. Currents carried by monovalent cations through calcium channels in mouse neoplastic B lymphocytes. *J. Physiol.* 358:255–284.

George, C.H., C.C. Yin, and F.A. Lai. 2005. Toward a molecular understanding of the structure-function of ryanodine receptor Ca^{2+} release channels–perspectives from recombinant expression systems. *Cell Biochem. Biophys.* 42:197–222.

Gerasimenko, O.V., J.V. Gerasimenko, A.V. Tepikin, and O.H. Petersen. 1996. Calcium transport pathways in the nucleus. *Pflugers Arch.* 432:1–6.

Giannattasio, B., S.W. Jones, and A. Scarpa. 1991. Calcium currents in the A7r5 smooth muscle-derived cell line. Calcium-dependent and voltage-dependent inactivation. *J. Gen .Physiol.* 98:987–1003.

Groth, R.D., and P.G. Mermelstein. 2003. Brain-derived neurotrophic factor activation of NFAT (nuclear factor of activated T-cells)-dependent transcription: A role for the transcription factor NFATc4 in neurotrophin-mediated gene expression. *J. Neurosci.* 23:8125–8134.

Gunter, T.E., D.I. Yule, K.K. Gunter, R.A. Eliseev, and J.D. Salter. 2004. Calcium and mitochondria. *FEBS Lett.* 567:96–102.

Guse, A.H., C.P. da Silva, I. Berg, A.L. Skapenko, K. Weber, P. Heyer, M. Hohenegger, G.A. Ashamu, H. Schulze-Koops, B.V. Potter, and G.W. Mayr. 1999. Regulation of calcium signalling in T lymphocytes by the second messenger cyclic ADP-ribose. *Nature* 398:70–73.

Guy, H.R., and S.R. Durrel. 1995. Structural models of Na^+, Ca^{2+}, and K^+ channels.*Soc. Gen. Physiol. Ser.* 50:1–16.

Gyorke, S., and M. Fill. 1993. Ryanodine receptor adaptation: Control mechanism of Ca(2+)-induced Ca^{2+} release in heart. *Science* 260:807–809.

Hagar, R.E., A.D. Burgstahler, M.H. Nathanson, and B.E. Ehrlich. 1998. Type III InsP3 receptor channel stays open in the presence of increased calcium. *Nature* 396:81–84.

Hagiwara, A., and K.I. Naka. 1964. The initiation of spike potential in barnacle muscle fibres under low intracellular Ca^{++}. *J. Gen. Physiol.* 48:141–161.

Hagiwara, S., J. Fukuda, and D.C. Eaton. 1974. Membrane currents carried by Ca, Sr, and Ba in barnacle muscle fiber during voltage clamp. *J. Gen. Physiol.* 63:565–578.

Hagiwara, S., S. Miyazaki, S. Krasne, and S. Ciani. 1977. Anomalous permeabilities of the egg cell membrane of star fish in K^+–Ti^+ mixtures. *J. Gen. Physiol.* 70:269–281.

Halestrap, A.P., E. Doran, J.P. Gillespie, and A. O'Toole. 2000. Mitochondria and cell death. *Biochem. Soc. Trans.* 28:170–177.

Halestrap, A.P., P.M. Kerr, S. Javadov, and K.Y. Woodfield. 1998. Elucidating the molecular mechanism of the permeability transition pore and its role in reperfusion injury of the heart. *Biochim. Biophys. Acta.* 1366:79–94.

He, J.Q., Y. Pi, J.W. Walker, and T.J. Kamp. 2000. Endothelin-1 and photoreleased diacylglycerol increase L-type Ca^{2+} current by activation of protein kinase C in rat ventricular myocytes. *J. Physiol.* 524:807–820.

Heinemann, S.H., H. Terlan, W. Stühmer, K. Imoto, and S. Numa. 1992. Calcium channel characteristics conferred on the sodium channel by single mutations. *Nature* 356:441–443.

Herlitze, S., D.E. Garcia, K. Mackie, B. Hille, T. Scheuer, and W.A. Catterall. 1996. Modulation of Ca^{2+} channels by G-protein beta gamma subunits. *Nature* 380:258–262.

Herlitze, S., G.H. Hockerman, T. Scheuer, and W.A. Catterall. 1997. Molecular determinants of inactivation and G protein modulation in the intracellular loop connecting domains I and II of the calcium channel alpha1A subunit. *Proc. Natl. Acad. Sci. USA* 94:1512–1516.

Hess, P., and R.W. Tsien. 1984. Mechanism of ion permeation through calcium channels. *Nature* 309:453–456.

Hess, P., J.B. Lansman, and R.W. Tsien. 1986. Calcium channel selectivity for divalent and monovalent cations: Voltage and concentration dependence of single channel current in ventricular heart cells. *J. Gen. Physiol.* 88:293–319.

Heidelberger, R., and G. Matthews. 1992. Calcium influx and calcium current in single synaptic terminals of goldfish retinal bipolar neurons. *J. Physiol.* 447:235–256.

Hille, B. 2001. Ion Channels of Excitable Membranes, 3rd Ed. Sinauer Associates, Sunderland, MA.

Hodgkin, A.L., and R.D. Keynes. 1957. Movements of labelled calcium in squid giant axons. *J. Physiol.* 138:253–281.

Hofmann, F., L. Lacinová, and N. Klugbauer. 1999. Voltage-dependent calcium channels: From structure to function. *Rev. Physiol. Biochem. Pharmacol.* 139:33–87.

Hool, L.C. 2000. Hypoxia increases the sensitivity of the L-type Ca(2+) current to beta-adrenergic receptor stimulation via a C2 region-containing protein kinase C isoform. *Circ. Res.* 87:1164–1171.

Hool, L.C. 2001. Hypoxia alters the sensitivity of the L-type Ca(2+) channel to alpha-adrenergic receptor stimulation in the presence of beta-adrenergic receptor stimulation. *Circ. Res.* 88:1036–1043.

Hool, L.C., and P.G. Arthur. 2002. Decreasing cellular hydrogen peroxide with catalase mimics the effects of hypoxia on the sensitivity of the L-type Ca^{2+} channel to beta-adrenergic receptor stimulation in cardiac myocytes. *Circ. Res.* 91:601–609.

Hool, L.C., L.M. Middleton, and R.D. Harvey. 1998. Genistein increases the sensitivity of cardiac ion channels to beta-adrenergic receptor stimulation. *Circ. Res.* 83:33–42.

Hosey, M.M., J. Barhanin, A. Schmid, S. Vandaele, J. Ptasienski, C. Ocallahan, C. Cooper, and M. Lazdunski. 1987. Photoaffinity labelling and phosphorylation of a 165 kilo-dalton peptide associated with dihydropyridine and phenylalkylamine-sensitive calcium channels. *Biochem. Biophys. Res. Commun.* 147:1137–1145.

Huguenard, J.R. 1996. Low-threshold calcium currents in central nervous system neurons. *Annu. Rev. Physiol.* 58:329–348.

Hu, H., N. Chiamvimonvat, T. Yamagishi, and E. Marban. 1997. Direct inhibition of expressed cardiac L-type Ca^{2+} channels by S-nitrosothiol nitric oxide donors. *Circ. Res.* 81:742–752.

Hulme, J.T., T.W. Lin, R.E. Westenbroek, T. Scheuer, and W.A. Catterall. 2003. Beta-adrenergic regulation requires direct anchoring of PKA to cardiac CaV1.2 channels via a leucine zipper interaction with A kinase-anchoring protein 15. *Proc. Natl. Acad. Sci. USA* 100:13093–13098.

Ikeda, S.R. 1996. Voltage-dependent modulation of N-type calcium channels by G-protein beta gamma subunits. *Nature* 380:255–258.

Jay, S.D., S.B. Ellis, A.F. McCue, M.E. Williams, T.S. Vedvick, M.M. Harpold, and K.P. Campbell. 1990. Primary structure of the gamma subunit of the DHP-sensitive calcium channel from skeletal muscle. *Science* 248:490–492.

Jiang, Y.X., A. Lee, J.Y. Chen, V. Ruta, M. Cadene, B.T. Chait, and R. MacKinnon. 2003. X-ray structure of a voltage-dependent K^+ channel. *Nature* 423:33–41.

Johnson, J.D., C. Snyder, M. Walsh, and M. Flynn. 1996. Effects of myosin light chain kinase and peptides on Ca^{2+} exchange with the N- and C-terminal Ca^{2+} binding sites of calmodulin. *J. Biol. Chem.* 271:761–767.

Jones, S.W., and T.N. Marks. 1989. Calcium currents in bullfrog sympathetic neurons. II. Inactivation. *J. Gen. Physiol.* 94:169–182.

Jones, SW. 1998. Overview of voltage-dependent calcium channels. *J. Bioenerg. Biomembr.* 30:299–312.

Jones, SW. 2003. Calcium channels: Unanswered questions. *J. Bioenerg. Biomembr.* 35:461–475.

Jouaville, L.S., F. Ichas, E.L. Holmuhamedov, P. Camacho, and J.D. Lechleiter. 1995. Synchronization of calcium waves by mitochondrial substrates in Xenopus laevis oocytes. *Nature* 377:438–441.

Ju, Y.K., and D.G. Allen. 1998. Intracellular calcium and Na^+–Ca^{2+} exchange current in isolated toad pacemaker cells. *J. Physiol.* 508:153–166.

Kass, R.S., and M.C. Sanguinetti. 1984. Inactivation of calcium channel current in the calf cardiac Purkinje fibre. Evidence for voltage- and calcium-mediated mechanisms. *J. Gen. Physiol.* 84:705–726.

Keating, M.T., and M.C. Sanguinetti. 2001. Molecular and cellular mechanisms of cardiac arrhythmias. *Cell* 104:569–580.

Khan, A.A., J.P. Steiner, M.G. Klein, M.F. Schneider, and S.H. Snyder. 1992. IP3 receptor: Localization to plasma membrane of T cells and cocapping with the T cell receptor. *Science* 257:815–818.

Kim, M.S., T. Morii, L.X. Sun, K. Imoto, and Y. Mori. 1993. Structural determinants of ion selectivity in brain calcium channel. *FEBS Lett.* 318:145–148.

Kirichok, Y., G. Krapivinsky, and D.E. Clapham. 2004. The mitochondrial calcium uniporter is a highly selective ion channel. *Nature* 427:360–364.

Klemic, K.G., C.C. Shieh, G.E. Kirsch, and S.W. Jones. 1998. Inactivation of Kv2.1 potassium channels. *Biophys. J.* 74:1779–1789.

Klockner, U., and G. Isenberg. 1994. Intracellular pH modulates the availability of vascular L-type Ca^{2+} channels. *J. Gen. Physiol.* 103:647–663.

Klockner, U., G. Mikala, A. Schwartz, and G. Varadi. 1996. Molecular studies of the asymmetric pore structure of the human cardiac voltage-dependent Ca^{2+} channel. Conserved residue, Glu-1086, regulates proton-dependent ion permeation. *J. Biol. Chem.* 271:22293–22296.

Klugbauer, N., L. Lacinova, E. Marais, M. Hobom, and F. Hofmann. 1999. Molecular diversity of the calcium channel alpha2delta subunit. *J. Neurosci.* 19:684–691.

Kobrinsky, E., S. Tiwari, V.A. Maltsev, J.B. Harry, E. Lakatta, D.R. Abernethy, and N.M. Soldatov. 2005. Differential role of the alpha1C subunit tails in regulation of the Cav1.2 channel by membrane potential, beta subunits, and Ca^{2+} ions. *J. Biol. Chem.* 280:12474–12485.

Koch, S.E., I. Bodi, A. Schwartz, and G. Varadi. 2000. Architecture of Ca^{2+} channel pore lining segments revealed by covalent modification of substituted cysteines. *J. Biol. Chem.* 275:34493–34500.

Kohr G., and I. Mody. 1991. Endogenous intracellular calcium buffering and the activation/inactivation of HVA calcium currents in rat dentate gyrus granule cells. *J. Gen. Physiol.* 98:941–967.

Komalavilas, P., and T.M. Lincoln. 1994. Phosphorylation of the inositol 1,4,5-trisphosphate receptor by cyclic GMP-dependent protein kinase. *J. Biol. Chem.* 269:8701–8707.

Kornhauser, J.M., C.W. Cowan, A.J. Shaywitz, R.E. Dolmetsch, E.C. Griffith, L.S. Hu, C. Haddad, Z. Xia, and M.E. Greenberg. 2002. CREB transcriptional activity in neurons is regulated by multiple, calcium-specific phosphorylation events. *Neuron* 34:221–233.

Kostyuk, P.G., S.L. Mironov, and Y.M. Shuba. 1983. Two ion-selecting filters in the calcium channel of the somatic membrane of mollusc neurons. *J. Membr. Biol.* 76:83–93.

Kostyuk, P.G., Y.M. Shuba, A.N. Savchenko, and V.I. Teslenko. 1988. Kinetic characteristics of different calcium channels in the nueronal membrane. *In*: Bayer Centenary Symposium, the Calcium Channel: Structure, Function and Implications. M. Morad, W. Nayler, S. Kazda, and M. Schramm, editors. Springer-Verlag, Berlin, pp. 442–464.

Kraus, R.L., M.J. Sinnegger, A. Koschak, H. Glossmann, S. Stenirri, P. Carrera, and J. Striessnig. 2000. Three new familial hemiplegic migraine mutants affect P/Q-type Ca(2+) channel kinetics. *J. Biol. Chem.* 275:9239–9243.

Kuo, C.C., and P. Hess. 1993a. Ion permeation through the L-type Ca^{2+} channel in rat phaeochromocytoma cells: Two sets of ion binding sites in the pore. *J. Physiol.* 466:629–655.

Kuo, C.C., and P. Hess. 1993b. Characterization of the high-affinity Ca^{2+} binding sites in the L-type Ca^{2+} channel pore in rat phaeochromocytoma cells. *J. Physiol.* 466:657–682.

Lacerda, A.E., D. Rampe, and A.M. Brown. 1988. Effects of protein kinase C activators on cardiac Ca^{2+} channels. *Nature.* 335:249–251.

Lansman, J.B., P. Hess, and R.W. Tsien. 1986. Blockade of current through single calcium channels by Cd^{2+}, Mg^{2+}, and Ca^{2+}: Voltage and concentration dependence of calcium entry into the pore. *J. Gen. Physiol.* 88:321–347.

Lebuffe, G., P.T. Schumacker, Z.H. Shao, T. Anderson, H. Iwase, and T.L. vanden Hoek. 2003. ROS and NO trigger early preconditioning: Relationship to mitochondrial KATP channel. *Am. J. Physiol. Heart Circ. Physiol.* 284:H299–H308.

Lee, K.S., and R.W. Tsien. 1984. High selectivity of calcium channels in single dialysed heart cells of guinea-pig. *J. Physiol.* 354:253–272.

Lee, A., S.T. Wong, D. Gallagher, B. Li, D.R. Storm, T. Scheur, and W.A. Catterall. 1999. Ca^{2+}/calmodulin binds to and modulates P/Q-type calcium channels. *Nature* 399:155–159.

Lee, A., H. Zhou, T. Scheuer, and W.A. Catterall. 2003. Molecular determinants of Ca^{2+}/calmodulin-dependent inactivation of L-type calcium channels. *Proc. Natl. Acad. Sci. USA* 100:16059–16064.

Leite, M.F., E.C. Thrower, W. Echevarria, P. Koulen, K. Hirata, A.M. Bennett, B.E. Ehrlich, and M.H. Nathanson. 2003. Nuclear and cytosolic calcium are regulated independently. *Proc. Natl. Acad. Sci. USA* 100:2975–2980.

Letts, V.A., R. Felix, G.H. Biddlecome, J. Arikkath, C.L. Mahaffey, A. Valenzuela, F.S. Bartlett II, Y. Mori, K.P. Campbell, and W.N. Frankel. 1998. The mouse stargazer gene encodes a neuronal Ca^{2+}-channel γ subunit. *Nat. Genet.* 19:340–347.

Leung, A.T., T. Imagawa, and K.P. Campbell. 1987. Structural characterization of the 1,4-dihydropyridine receptor of the voltage dependent Ca^{2+} channel from rabbit skeletal muscle. Evidence for two distinct molecular weight subunits. *J. Biol. Chem.* 262:7943–7946.

Lipkind, G.M., and H.A. Fozzard. 2001. Modeling of the outer vestibule and selectivity filter of the L-type Ca^{2+} channel. *Biochemistry* 40:6786–6794.

Liang, H., C.D. DeMaria, M.G. Erickson, M.X. Mori, B.A. Alseikhan, and D.T. Yue. 2003. Unified mechanism of Ca(2+) regulation across the Ca(2+) channel family. *Neuron* 39:951–960.

Liu, H., M. De Waard, V.E.S. Scott, C.A. Gurnett, V.A. Lennon, and K.P. Campbell. 1996. Identification of three subunits of the high affinity ω-conotoxin MVIIC-sensitive Ca^{2+} channel. *J. Biol. Chem.* 271:13804–13810.

Liu, Y., and D.D. Gutterman. 2002. Oxidative stress and potassium channel function. *Clin. Exp. Pharmacol. Physiol.* 29:305–311.

Llinas, R., M. Sugimori, J.W. Lin, and B. Cherksey. 1989. Blocking and isolation of a calcium channel from neurons in mammals and cephalopods utilizing a toxin fraction (FTX) from funnel-web spider poison. *Proc. Natl. Acad. Sci. USA* 86:1689–1693.

Long, S.B., E.B. Campbell, and R. MacKinnon. 2005. Crystal structure of a mammalian voltage-dependent shaker family K^+ channel. *Science* 309:897–903.

Lopez-Barneo, J., J.R. Lopez-Lopez, J. Urena, and C. Gonzalez. 1988. Chemotransduction in the carotid body: K^+ current modulated by PO2 in type I chemoreceptor cells. *Science* 241:580–582.

Lui, P.P., F.L. Chan, Y.K. Suen, T.T. Kwok, and S.K. Kong. 2003. The nucleus of HeLa cells contains tubular structures for Ca^{2+} signaling with the involvement of mitochondria. *Biochem. Biophys. Res. Commun.* 308:826–833.

Lux, H. D., E. Carbone, and H. Zucker. 1990. Na^+ currents through low-voltage-activated Ca^{2+} channels of chick sensory neurons: Block by external Ca^{2+} and Mg^{2+}. *J. Physiol.* 430:159–188.

Lytton, J., M. Westlin, and M.R. Hanley. 1991. Thapsigargin inhibits the sarcoplasmic or endoplasmic reticulum Ca-ATPase family of calcium pumps. *J. Biol. Chem.* 266:17067–17071.

Mak, D.O., S. McBride, and J.K. Foskett. 1999. ATP regulation of type 1 inositol 1,4,5-trisphosphate receptor channel gating by allosteric tuning of Ca(2+) activation. *J. Biol. Chem.* 274:22231–22237.

Mamas, M.A., and D.A. Terrar. 2001. Inotropic actions of protein kinase C activation by phorbol dibutyrate in guinea-pig isolated ventricular myocytes. *Exp. Physiol.* 86:561–570.

Martin-Moutot, N., C. Leveque, K. Sato, R. Kato, M. Takahashi, and M. Seagar. 1995. Properties of omega conotoxin MVIIC receptors associated with α_{1A} calcium channel subunits in rat brain. *FEBS Lett.* 366:21–25.

Martin-Moutot, N., N. Charvin, C. Leveque, K. Sato, T. Nishi, S. Kozaki, M. Takahashi, and M. Seagar. 1996. Interaction of SNARE complexes with P/Q-type calcium channels in rat cerebellar synaptosomes. *J. Biol. Chem.* 271:6567–6570.

Matteson, D.R., and C.M. Armstrong. 1986. Properties of two types of channels in clonal pituitary cells. *J. Gen. Physiol.* 87:161–182.

Mayer, M.L. 1985. A calcium-activated chloride current generates the after-depolarization of rat sensory neurons in culture. *J. Physiol.* 364:217–239.

McCleskey, E.W., and W. Almers. 1985. The Ca channel in skeletal muscle is a large pore. *Proc. Natl. Acad. Sci. USA* 82:7149–7153.

McCormack, J.G., A.P. Halestrap, and R.M. Denton. 1990. Role of calcium ions in regulation of mammalian intramitochondrial metabolism. *Physiol. Rev.* 70:391–425.

McDonald, T.F., S. Pelzer, W. Trautwein, and D.J. Pelzer. 1994. Regulation and modulation of calcium channels in cardiac, skeletal, and smooth muscle cells. [Review] [1868 refs]. *Physiol. Rev.* 74:365–507.

McEnery, M.W., A.M. Snowman, A.H. Sharp, M.E. Adams, and S.H. Snyder. 1991. Purified ω-conotoxin GVIA receptor of rat brain resembles a dihydropyridine-sensitive L-type calcium channel. *Proc. Natl. Acad. Sci. USA* 88:11095–11099.

McFarlane, M.B. 1997. Depolarization-induced slowing of Ca^{2+} channel deactivation in squid neurons. *Biophys. J.* 72:1607–1621.

McGeown, J.G. 2004. Interactions between inositol 1,4,5-trisphosphate receptors and ryanodine receptors in smooth muscle: One store or two? *Cell Calcium* 35:613–619.

McHugh, D., E.M. Sharp, T. Scheuer, and W.A. Catterall. 2000. Inhibition of cardiac L-type calcium channels by protein kinase C phosphorylation of two sites in the N-terminal domain. *Proc. Natl. Acad. Sci. USA* 97:12334–12338.

Mikami, A., K. Imoto, T. Tanabe, T. Niidome, Y. Mori, H. Takeshima, S. Narumiya, and S. Numa. 1989. Primary structure and functional expression of the cardiac dihydropyridine-sensitive calcium channel. *Nature* 340:230–233.

Mochly-Rosen, D. 1995. Localization of protein kinases by anchoring proteins: A theme in signal transduction. *Science* 268:247–251.

Mochly-Rosen, D., C.J. Henrich, L. Cheever, H. Khaner, and P.C. Simpson. 1990. A protein kinase C isozyme is translocated to cytoskeletal elements on activation. *Cell Reg.* 1:693–706.

Moy, G., B. Corry, S. Kuyucak, and S.H. Chung. 2000. Tests of continuum theories as models of ion channels: I. Poisson–Boltzmann theory versus Brownian dynamics. *Biophys. J.* 78:2349–2363.

Nakagawa, T., H. Okano, T. Furuichi, J. Aruga, and K. Mikoshiba. 1991. The subtypes of the mouse inositol 1,4,5-trisphosphate receptor are expressed in a tissue-specific and developmentally specific manner. *Proc. Natl. Acad. Sci. USA* 88:6244–6248.

Neher, E. 1975. Ionic specificity of the gramicidin channel and the thallous ion. *Biochim. Biophys. Acta* 401:540–544.

Nonner, W., and B. Eisenberg. 1998. Ion permeation and glutamate residues linked by Poisson–Nernst–Planck theory in L-type calcium channels. *Biophys. J.* 75:1287–1305.

Nonner, W., L. Catacuzzeno, and B. Eisenberg. 2000. Binding and selectivity in L-type Ca channels: A mean spherical approximation. *Biophys. J.* 79:1976–1992.

Nowycky, M.C., A.P. Fox, and R.W. Tsien. 1985a. Long opening mode of gating of neuronal calcium channels and its promotion by the dihydropyridine calcium agonist Bay K8644. *Proc. Natl. Acad. Sci. USA* 82:2178–2182.

Nowycky, M.C., A.P. Fox, and R.W. Tsien. 1985b. Three types of neuronal calcium channel with different calcium agonist sensitivity. *Nature* 316:440–443.

Parent, L., and M. Gopalakrishnan. 1995. Glutamate substitution in repeat IV alters divalent and monovalent cation permeation in the heart Ca^{2+} channel. *Biophys. J.* 69:1801–1813.

Parekh, A.B., and J.W. Putney. 2005. Store-operated calcium channels. *Physiol. Rev.* 85:757–810.

Patil, P.G., D.L. Brody, and D.T. Yue. 1998. Preferential closed-state inactivation of nueronal calcium channels. *Neuron* 20:1027–1038.

Peterson, B.Z., C.D. DeMaria, J.P. Adelman, and D.T. Yue. 1999. Calmodulin is the Ca^{2+} sensor for Ca^{2+}-dependent inactivation of L-type calcium channels. *Neuron* 22:549–558.

Perez-Reyes, E., H.S. Kim, A.E. Lacerda, W. Horne, X.Y. Wei, D. Rampe, K.P. Campbell, A.M. Brown, and L. Birnbaumer. 1989. Induction of calcium currents by the expression of the alpha 1-subunit of the dihydropyridine receptor from skeletal muscle. *Nature* 340:233–236.

Pietrobon, D., and P. Hess. 1990. Novel mechanism of voltage-dependent gating in L-type calcium channels. *Nature* 346:651–655.

Plummer, M.R., and P. Hess. 1991. Reversible uncoupling of inactivation in N-type calcium channels. *Nature* 351:657–659.

Polo-Parada, L., and S.J. Korn. 1997. Block of N-type calcium channels in check sensory neurons by external sodium. *J. Gen. Physiol.* 109:693–702.

Pragnell, M., M. De Waard, Y. Mori, T. Tanabe, T.P. Snutch, and K.P. Campbell. 1994. Calcium channel beta-subunit binds to a conserved motif in the I–II cytoplasmic linker of the alpha 1-subunit. *Nature* 368:67–70.

Priori, S.G., and C. Napolitano. 2005. Cardiac and skeletal muscle disorders caused by mutations in the intracellular Ca^{2+} channels. *J. Clin. Invest.* 115:2033–2038.

Qin, N., R. Olcese, M. Bransby, T. Lin, and L. Birnbaumer. 1999. Ca^{2+}-induced inhibition of the cardiac Ca^{2+} channel depends on calmodulin. *Proc. Natl. Acad. Sci. USA* 96:2435–2438.

Qin, N., S. Yagel, M.L. Momplaisir, E.E. Codd, and M.R. D'Andrea. 2002. Molecular cloning and characterization of the human voltage-gated calcium channel alpha(2)delta-4 subunit. *Mol. Pharmacol.* 62:485–496.

Ramakrishnan, V., D. Hendersen, and D.D. Busath. 2004. Applied field nonequilibrium molecular dynamics simulations of ion exit from a β-barrel model of the L-type calcium channel. *Biochim. Biophys. Acta* 1664:1–8.

Randall, A.D., and R.W. Tsien. 1995. Pharmacological disection of multiple types of Ca^{2+} channel currents in rat cerebellar granule neurons. *Neuroscience* 15:2995–3012.
Randall, A.D., and R.W. Tsien. 1997. Contrasting biophysical and pharmacological properties of T-type and R-type calcium channels. *Neuropharmacology* 36:879–893.
Ren, D., B. Navarro, H. Xu, L. Yue, Q. Shi, and D.E. Clapham. 2001. A prokaryotic voltage-gated sodium channel. *Science* 294:2372–2375.
Reuter, H., and H. Scholz. 1977. A study of the ion selectivity and the kinetic properties of the calcium dependent slow inward current in mammalian cardiac muscle. *J. Physiol.* 264:17–47.
Root, M.J., and R. MacKinnon. 1994. Two identical noninteracting sites for an ion channel revealed by proton transfer. *Science* 265:1852–1856.
Sah, P. 1996. Ca(2+)-activated K^+ currents in neurons: Types, physiological roles and modulation. *Trends Neurosci.* 19:150–154.
Sanguinetti, M.C. 2002. When the KChIPs are down. *Nature Med.* 8:18–19.
Santana, L.F., A.M. Gomez, and W.J. Lederer. 1998. Ca^{2+} flux through promiscuous cardiac Na^+ channels: Slip-mode conductance. *Science* 279:1027–1033.
Sather, W.A., and E.W. McCleskey. 2003. Permeation and selectivity in calcium channels. *Annu. Rev. Physiol.* 65:133–159.
Schiefer, A., G. Meissner, and G. Isenberg. 1995. Ca^{2+} activation and Ca^{2+} inactivation of canine reconstituted cardiac sarcoplasmic reticulum Ca(2+)-release channels. *J. Physiol.* 489(Pt 2):337–348.
Schneider, T., and F. Hofmann. 1988. The bovine cardiac receptor for calcium channel blockers is a 195-kDa protein. *Eur. J. Biochem.* 174:369–375.
Schuhmann, K., C. Romanin, W. Baumgartner, and K. Groschner. 1997a. Intracellular Ca^{2+} inhibits smooth muscle L-type Ca^{2+} channels by activation of protein phosphatase type 2B and by direct interaction with the channel. *J. Gen. Physiol.* 110:503–513.
Schuhmann, K., C. Voelker, G.F. Hofer, H. Pflugelmeier, N. Klugbauer, F. Hofmann, C. Romanin, and K. Groschner. 1997b. Essential role of the beta subunit in modulation of C-class L-type Ca^{2+} channels by intracellular pH. *FEBS Lett.* 408:75–80.
Sculptoreanu, A., T. Scheuer, and W.A. Catterall. 1993. Voltage-dependent potentiation of L-type Ca^{2+} channels due to phosphorylation by cAMP-dependent protein kinase. *Nature* 364:240–243.
Schwartz, D.D., and B.P. Naff. 1997. Activation of protein kinase C by angiotensin II decreases beta 1-adrenergic receptor responsiveness in the rat heart. *J. Cardiovasc. Pharmacol.* 29:257–264.
Scriven, D.R., P. Dan, and E.D. Moore. 2000. Distribution of proteins implicated in excitation-contraction coupling in rat ventricular myocytes. *Biophys. J.* 79:2682–2691.

Sham, J.S., L.S. Song, Y. Chen, L.H. Deng, M.D. Stern, E.G. Lakatta, and H. Cheng. 1998. Termination of Ca^{2+} release by a local inactivation of ryanodine receptors in cardiac myocytes. *Proc. Natl. Acad. Sci. USA* 95:15096–15101.

Singer, D., M. Biel, I. Lotan, V. Flockerzi, F. Hofmann, and N. Dascal. 1991. The roles of the subunits in the function of the calcium channel. *Science* 253:1553–1557.

Sipido, K.R., E. Carmeliet, and F. Van de Werf. 1998. T-type Ca^{2+} current as a trigger for Ca^{2+} release from the sarcoplasmic reticulum in guinea-pig ventricular myocytes. *J. Physiol.* 508(Pt 2):439–451.

Sipido, K.R., M. Maes, and F. Van de Werf. 1997. Low efficiency of Ca^{2+} entry through the Na(+)–Ca^{2+} exchanger as trigger for Ca^{2+} release from the sarcoplasmic reticulum. A comparison between L-type Ca^{2+} current and reverse-mode Na(+)–Ca^{2+} exchange. *Circ. Res.* 81:1034–1044.

Snutch, T.P., J.P. Leonard, M.M. Gilbert, H.A. Lester, and N. Davidson. 1990. Rat brain expresses a heterogeneous family of calcium channels. *Proc. Natl. Acad. Sci. USA* 87:3391–3395.

Storm, J.F. 1990. Potassium currents in hippocampal pyramidal cells. *Prog. Brain Res.* 83:161–187.

Stotz, S.C., J. Hamid, R.L. Spaetgens, S.E. Jarvis, and G.W. Zamponi. 2000. Fast inactivation of voltage-dependent calcium channels. A hinged-lid mechanism? *J. Biol. Chem.* 275:24575–24582.

Stotz, S.C., and G.W. Zamponi. 2001. Structural determinants of fast inactivation of high voltage-activated Ca^{2+} channels. *Trends Neurosci.* 24:176–182.

Stotz, S.C., S.E. Jarvis, and G.W. Zamponi. 2004. Functional roles of cytoplasmic loops and pore lining transmembrane helices in the voltage-dependent inactivation of HVA calcium channels. *J. Physiol.* 554(Pt 2):263–273.

Striessnig, J., H.G. Knaus, M. Grabner, K. Moosburger, W. Steitz, H. Lietz, and H. Glossmann. 1987. Photoaffinity labelling of the phenylalkylamine receptor of the skeletal muscle transverse-tubule calcium channel. *FEBS Lett.* 212:247–253.

Sutko, J.L., and J.A. Airey. 1996. Ryanodine receptor Ca^{2+} release channels: Does diversity in form equal diversity in function. *Physiol. Rev.* 76:1027–1071.

Swandulla, D., and C.M. Armstrong. 1988. Fast deactivating calcium channels in chick sensory neurons. *J. Gen. Physiol.* 92:197–218.

Swandulla, D., and C.M. Armstrong. 1989. Calcium channel block by cadmium in chick sensory neurons. *Proc. Natl. Acad. Sci. USA* 86:1736–1740.

Takahashi, M., M.J. Seagar, J.F. Jones, B.F. Reber, and W.A. Catterall. 1987. Subunit structure of dihydropyridine-sensitive calcium channels from skeletal muscle. *Proc. Natl. Acad. Sci. USA* 84:5478–5482.

Tanabe, T., H. Takeshima, A. Mikami, V. Flockerzi, H. Takahashi, K. Kangawa, M. Kojima, H. Matsuo, T. Hirose, and S. Numa. 1987. Primary structure of the receptor for calcium channel blockers from skeletal muscle. *Nature* 328:313–318.

Taylor, R.E., C.M. Armstrong, and F. Bezanilla. 1976. Block of sodium channels by external calcium ions. *Biophys. J.* 16:27.

Taylor, W.R. 1988. Permeation of barium and cadmium through slowly inactivating calcium channels in cat sensory neurons. *J. Physiol.* 407:433–452.

Thévenod, F., and S.W. Jones. 1992. Cadmium block of calcium current in frog sympathetic neurons. *Biophys. J.* 63:162–168.

Tseng, G.N., and P.A. Boyden. 1991. Different effects of intracellular Ca and protein kinase C on cardiac T and L Ca currents. *Am. J. Physiol.* 261:H364–H379.

Tsien, R.W., P. Hess, E.W. McCleskey, and R.L. Rosenberg. 1987. Calcium channels: Mechanisms of selectivity, permeation and block. *Ann. Rev. Biophys. Chem.* 16:265–290.

Tsien, R.W., D. Lipscombe, D.V. Madison, K.R. Bley, and A.P. Fox. 1988. Multiple types of neuronal calcium channels and their selective modulation. *Trends Neurosci.* 11:431–438.

Unni, V.K., S.S. Zakharenko, L. Zablow, A.J. DeCostanzo, and S.A. Siegelbaum. 2004. Calcium release from presynaptic ryanodine-sensitive stores is required for long-term depression at hippocampal CA3–CA3 pyramidal neuron synapses. *J. Neurosci.* 24:9612–9622.

Vanoevelen, J., L. Raeymaekers, J.B. Parys, H. De Smedt, K. Van Baelen, G. Callewaert, F. Wuytack, and L. Missiaen. 2004. Inositol trisphosphate producing agonists do not mobilize the thapsigargin-insensitive part of the endoplasmic-reticulum and Golgi Ca^{2+} store. *Cell Calcium* 35:115–121.

Van Petegem, F., F.C. Chatelain, and D.L. Minor Jr. 2005. Insights into the voltage-gated calcium channel regulation from the structure of the $Ca_V1.2$ IQ domain-Ca^{2+}/calmodulin complex. *Nat. Struct. Mol. Biol.* 12:1108–1115.

Van Petegem, F., K.A. Clark, F.C. Chatelain, and D.L. Minor Jr. 2004. Structure of a complex between a voltage-gated calcium channel β-subunit and an α-subunit domain. *Nature* 429:671–675.

Vereecke, J., and E. Carmeliet. 1971. Sr action potentials in cardiac Purkinje fibres. II Dependence of the Sr conductance on the external Sr concentration and Sr–Ca antagonism. *Pfluegers Arch.* 322:565–578.

Vergara, C., R. Latorre, N.V. Marrion, and J.P. Adelman. 1998. Calcium-activated potassium channels. *Curr. Opin. Neurobiol.* 8:321–329.

Vora, T., B. Corry, and S.H. Chung. 2004. A model of sodium channels. *Biochim. Biophys. Acta* 1668:106–116.

Wang, Cl. 1985. A note on Ca^{2+} binding to calmodulin. *Biochem. Biophys. Res. Commun.* 130:426–430.

Wahler, G.M., N.J. Rusch, and N. Sperelakis. 1990. 8-Bromo-cyclic GMP inhibits the calcium channel current in embryonic chick ventricular myocytes. *Can. J. Physiol. Pharmacol.* 68:531–534.

Wehrens, X.H.T., S.E. Lenhart, and A.R. Marks. 2005. Intracellular calcium release and cardiac disease. *Annu. Rev. Physiol.* 67:69–98.

Weick, J.P., R.D. Groth, A.L. Isaksen, and P.G. Mermelstein. 2003. Interactions with PDZ proteins are required for L-type calcium channels to activate cAMP response element-binding protein-dependent gene expression. *J. Neurosci.* 23:3446–3456.

Werz, M.A., K.S. Elmslie, and S.W. Jones. 1993. Phosphorylation enhances inactivation of N-type calcium current in bullfrog sympathetic neurons. *Plügers. Arch.* 424:538–545.

Wijetunge, S., and A.D. Hughes. 1995. pp60c-src increases voltage-operated calcium channel currents in vascular smooth muscle cells. *Biochem. Biophys. Res. Commun.* 217:1039–1044.

Witcher, D.R., M. De Waard, J. Sakamoto, C. Franzini-Armstrong, M. Pragnell, S.D. Kahl, and K.P. Campbell. 1993. Subunit identification and reconstitution of the N-type Ca^{2+} channel complex purified from brain. *Science* 261:486–489.

Wolfe, J.T., H. Wang, J. Howard, J.C. Garrison, and P.Q. Barrett. 2003. T-type calcium channel regulation by specific G-protein betagamma subunits. *Nature* 424:209–213.

Wootton, L.L., C.C. Argent, M. Wheatley, and F. Michelangeli. 2004. The expression, activity and localisation of the secretory pathway Ca^{2+}-ATPase (SPCA1) in different mammalian tissues. *Biochim. Biophys. Acta.* 1664:189–197.

Wu, X.S., H.D. Edwards, and W.A. Sather. 2000. Side chain orientation of the selectivity filter of a voltage-gated Ca^{2+} channel. *J. Biol. Chem.* 275:31778–31785.

Yamasaki, M., R. Masgrau, A.J. Morgan, G.C. Churchill, S. Patel, S.J. Ashcroft, and A. Galione. 2004. Organelle selection determines agonist-specific Ca^{2+} signals in pancreatic acinar and beta cells. *J. Biol. Chem.* 279:7234–7240.

Yang, J., P.T. Ellinor, W.A. Sather, J.F. Zhang, and R.W. Tsien. 1993. Molecular determinants of Ca^{2+} selectivity and ion permeation in L-type Ca^{2+} channels. *Nature* 366:158–161.

Yang, H.T., D. Tweedie, S. Wang, A. Guia, T. Vinogradova, K. Bogdanov, P.D. Allen, M.D. Stern, E.G. Lakatta, and K.R. Boheler. 2002. The ryanodine receptor modulates the spontaneous beating rate of cardiomyocytes during development. *Proc. Natl. Acad. Sci. USA* 99:9225–9230.

Yang, Y., D. Hendersen, and D. Busath. 2003. Applied field molecular dynamics study of a model calcium channel selectivity filter. *J. Chem. Phys.* 118:4213–4220.

Yatani, A., J. Codina, Y. Imoto, J.P. Reeves, L. Birnbaumer, and A.M. Brown. 1987. A G protein directly regulates mammalian cardiac calcium channels. *Science* 238:1288–1292.

You, Y., D.J. Pelzer, and S. Pelzer. 1997. Modulation of L-type Ca^{2+} current by fast and slow Ca^{2+} buffering in guinea pig ventricular cardiomyocytes. *Biophys. J.* 72:175–187.

Yue, D.T., P.H. Backx, and J.P. Imredy. 1990. Calcium-sensitive inactivation in the gating of single calcium channels. *Science* 250:1735–1738.

Yue, L.X., B. Navarro, D.J. Ren, A. Ramos, and D.E. Clapham. 2002. The cation selectivity filter of the bacterial sodium channel, NaChBac. *J. Gen. Physiol.* 120:845–853.

Zeng, W., D.O. Mak, Q. Li, D.M. Shin, J.K. Foskett, and S. Muallem. 2003. A new mode of Ca^{2+} signaling by G protein-coupled receptors: Gating of IP3 receptor Ca^{2+} release channels by Gbetagamma. *Curr. Biol.* 13:872–876.

Zhang, J.F., A.D. Randall, P.T. Ellinor, W.A. Horne, W.A. Sather, T. Tanabe, T.L. Scwarz, and R.W. Tsien. 1993. Distinctive pharmacology and kinetics of cloned neuronal Ca^{2+} channels and their possible counterparts in mammalian CNS neurons. *Nueropharmacology* 32:1075–1088.

Zhang, S.L., Y. Yu, J. Roos, J.A. Kozak, T.J. Deerinck, M.H. Ellisman, K.A. Stauderman, and M.D. Cahalan. 2005. STIM1 is a Ca^{2+} sensor that activates CRAC channels and migrates from the Ca^{2+} store to the plasma membrane. *Nature* 437:902–905.

Zhang, Z.H., J.A. Johnson, L. Chen, N. El-Sherif, D. Mochly-Rosen, and M. Boutjdir. 1997. C2 region-derived peptides of beta-protein kinase C regulate cardiac Ca^{2+} channels. *Circ. Res.* 80:720–729.

Zhou, Z., and C.T. January. 1998. Both T- and L-type Ca^{2+} channels can contribute to excitation–contraction coupling in cardiac Purkinje cells. *Biophys. J.* 74:1830–1839.

Zühlke, R.D., G.S. Pitt, K. Deisseroth, R.W. Tsien, and H. Reuter. 1999. Calmodulin supports both inactivation and facilitation of L-type calcium channels. *Nature* 399:159–162.

Zühlke, R.D., G.S. Pitt, R.W. Tsien, and H. Reuter. 2000. Ca^{2+}-sensitive inactivation and facilitation of L-type Ca^{2+} channels both depend on specific amino acid residues in a consensus calmodulin-binding motif in the α1C subunit. *J. Biol. Chem.* 275:21121–21129.

Zweifach, A., and R.S. Lewis. 1993. Mitogen-regulated Ca^{2+} current of T lymphocytes is activated by depletion of intracellular Ca^{2+} stores. *Proc. Natl. Acad. Sci. USA* 90:6295–6299.

8 Chloride Transporting CLC Proteins[1]

Michael Pusch

8.1 Introduction

In the early 1980s, Chris Miller and colleagues described a curious "double-barreled" chloride channel from the electric organ of *Torpedo* fish reconstituted in planar lipid bilayers (Miller and White, 1980). Single-channel openings occurred in "bursts" separated by long closures. A single burst was characterized by the presence of two open conductance levels of equal size and the gating (i.e., openings and closings) during a burst could be almost perfectly described as a superposition of two identical and independent conductances that switched between open and closed states with voltage-dependent rates α and β (Hanke and Miller, 1983) (Fig. 8.1).

$$\mathbf{c} \underset{\beta}{\overset{\alpha}{\longleftrightarrow}} \mathbf{o}$$

These relatively fast openings and closing events gave rise to the name "fast gate" for these gating transitions during a burst. But how could it be excluded that these events just represent the presence of two identical channels in the bilayer? It was the presence of the long inter-burst closed events, during which no channel activity was observed, that demonstrated that the two "protopores" were tied together in a molecular complex, and could be inactivated by the so-called slow gate. From these results Miller formulated the "double-barreled" model according to which the channel consisted of two physically distinct, identical protopores, each with a proper fast gate and an additional common gate that acts simultaneously on both protopores (Miller and White, 1984) (Fig. 8.1).

For a relatively long time this double-barreled Cl^- channel remained a somewhat unique curiosity with little physiological relevance. This situation changed dramatically with the molecular cloning of the *Torpedo* channel by Jentsch and co-workers (Jentsch et al., 1990) in 1990, and immediately afterwards with the identification of mammalian homologues. Numerous novel physiological functions of the various CLC homologues were discovered (Jentsch et al., 2005; Pusch and Jentsch, 2005). We now know that CLC proteins are a large structurally defined family of Cl^- ion channels and Cl^-/H^+ antiporters with nine distinct genes in mammals. The

[1] Portions reprinted, with permission, from Pusch, M., and T.J. Jentsch. 2005. Unique structure and function of chloride transporting CLC proteins. IEEE Trans. Nanobioscience 4:49–57.

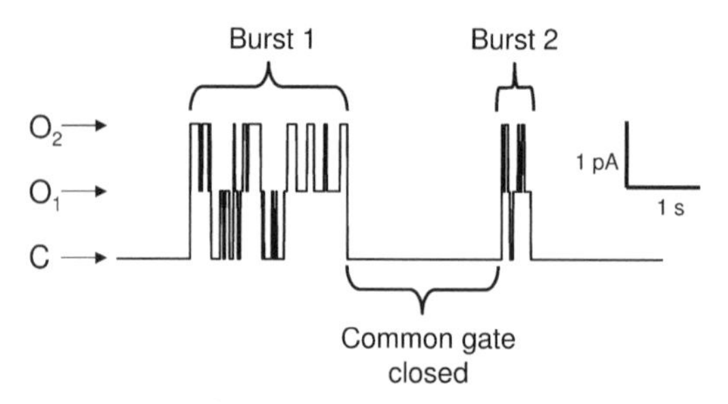

Fig. 8.1 The double-barreled ClC-0 channel. A simulated single channel trace of ClC-0 is shown in which three conductance levels (C = closed; O_1, O_2 open conductance levels) are seen. Activity occurs in bursts and during a burst the probability of observing one of the conductance levels is distributed binomial as if activity arises from two independent and identical pores. The slow common gate acts on both pores simultaneously.

membrane-embedded part of CLC proteins bears no obvious similarity to any other class of membrane proteins, while the cytoplasmic C-terminus of most eukaryotic and some prokaryotic CLCs contains two regions with homology to cystathionine beta synthase (CBS) domains that are found in other proteins as well. Different members serve a broad range of physiological roles including stabilization of the membrane potential, transepithelial ion transport, and vesicular acidification. Their physiological importance is underscored by the causative involvement in at least four different human genetic diseases. The homodimeric architecture with two physically separate ion conduction pathways that was anticipated from functional studies of the *Torpedo* homologue ClC-0 was fully confirmed by solving the crystal structure of prokaryotic CLC homologues. The determination of the crystal structure of bacterial homologues marked at breakthrough for the structure-function analysis of these proteins (Dutzler et al., 2002, 2003). The structure revealed a complex fold of 18 α-helices per subunit with at least two Cl^- ions bound in the center of each protopore. A critical glutamic acid residue was identified whose side-chain seems to occupy a third Cl^- ion binding site in the closed state and that moves away to allow Cl^- binding.

A big surprise was then the recent discovery that the bacterial CLC homologue that was used for crystallization is not a Cl^- ion channel but a Cl^-/H^+ antiporter (Accardi and Miller, 2004). Subsequently, the mammalian CLC-4 and CLC-5 proteins were shown to be Cl^-/H^+ exchangers as well, and not Cl^- ion channels, as previously assumed (Picollo and Pusch, 2005; Scheel et al., 2005). Thus, CLC proteins are either high-throughput voltage- and substrate-gated Cl^- ion channels or strictly coupled, low throughput secondary active electrogenic Cl^-/H^+ transporters. It will be interesting to find out, how the same basic molecular architecture allows for these seemingly different functions. Also, the physiological and pathophysiological role of Cl^-/H^+ exchange activity is still an unexplored territory. In the present chapter, the history and the present knowledge about CLC chloride channels and transporters will be covered including the recent developments.

8.2 Overview Over the Family of CLC Proteins

Expression cloning was used to isolate the first member of the CLC family, the *Torpedo* channel ClC-0 (Jentsch et al., 1990). This channel is highly expressed in the electric organ of electric rays. It localizes to the noninnervated membrane of the electrocyte, where it serves to stabilize the voltage across that membrane and to pass large currents that are generated by the depolarizing influx of sodium through acetylcholine receptor channels located in the opposite, innervated membrane.

The sequence of ClC-0 (Jentsch et al., 1990) revealed a protein of about 90 kDa with a large hydrophobic core that was predicted to be able to span the membrane at least 10 times and a large predicted cytoplasmic tail. The channel properties of the Cl^- current measured in oocytes after expressing the single cDNA coding for the ClC-0 protein were virtually identical to those described for the native, in bilayers reconstituted *Torpedo* channel (Bauer et al., 1991). In particular, the channel showed the typical double-barreled behavior (see Fig. 8.1). This showed that no additional subunits are required to form a completely and functional ClC-0 channel. Of note, the protein sequence of ClC-0 showed no significant homology to any other known ion channels, including the cAMP (cyclic adenosine monophosphate)-activated Cl^- channel called cystic fibrosis transmembrane conductance regulator (CFTR) and the ligand-gated $GABA_A$ (gamma amino butyric acid) and glycine receptor Cl^- channels. Thus, ClC-0 represented an entirely new channel class.

The isolation of ClC-0 opened the gate for cloning an entire family of CLC chloride channels by homology. Eventually, CLC homologues were found in all phyla. In mammals, the CLC family comprises nine distinct members. Based on sequence similarity, they can be grouped into three branches (Fig. 8.2). Members of the first

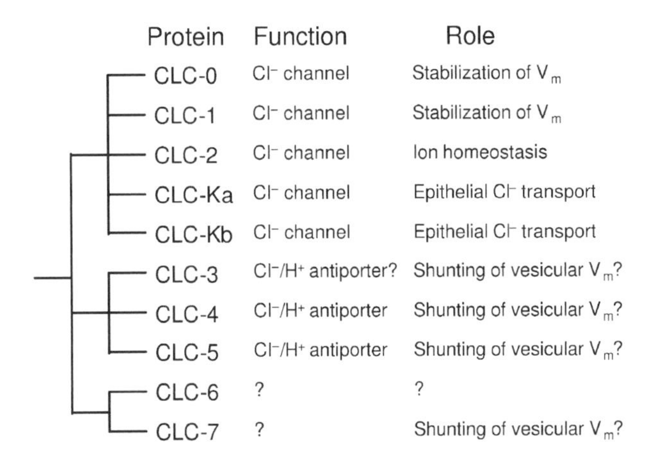

Fig. 8.2 The CLC family of Cl^- channels and Cl^-/H^+ transporters. The schematic dendrogram highlights the sequence similarity among the three groups in which the nine human CLC homologues can be classified. The *Torpedo* channel ClC-0 is included in the figure because it serves as an important model channel for functional studies. The *E. coli* homologue, ClC-ec1 (not shown in the figure), is most closely related to the ClC-3–ClC-5 branch.

branch (ClC-1,-2, -Ka, and -Kb) perform their function in the plasma membrane. Also, the *Torpedo* channel ClC-0 belongs to this class. In contrast, the members of the two other branches (comprising ClC-3, -4, -5, and ClC-6 and -7, respectively) are predominantly found in membranes of intracellular vesicles, in particular the endosomal–lysosomal pathway. Despite their predominant intracellular localization in vivo, ClC-3, ClC-4, and ClC-5 mediated currents can be measured in the plasma membrane in heterologous expression systems like *Xenopus* oocytes or transfected cells (Steinmeyer et al., 1995; Friedrich et al., 1999; Li et al., 2000). These three proteins, ClC-3, ClC-4, and ClC-5, share about 80% sequence identity, and have very similar functional properties. In contrast, no reproducible currents have been reported for ClC-6 and ClC-7 (Brandt and Jentsch, 1995; Buyse et al., 1997; Jentsch et al., 2002) and consequently, nothing is known about the biophysical properties of ClC-6 and ClC-7.

The diverse and important physiological functions of mammalian CLC proteins were impressively demonstrated by the pathologies resulting from their mutational inactivation in human genetic diseases, in mouse models, and in other representative model organisms, like *Caenorhabditis elegans* (Schriever et al., 1999; Strange, 2003) or *Arabidopis thaliana* (Hechenberger et al., 1996; Barbier-Brygoo et al., 2000). CLC proteins are also present in many prokaryotes (Mindell and Maduke, 2001), and so far the only structural data available are actually from bacterial CLCs (Mindell et al., 2001; Dutzler et al., 2002, 2003). At striking difference to what had been expected a priori, the bacterial CLCs whose crystal structure had been determined, turned out to be secondary active Cl^-/H^+ exchangers, and definitely not Cl^- ion channels (Accardi and Miller, 2004). We now know that also some mammalian CLCs are Cl^-/H^+ exchangers (Picollo et al., 2004; Scheel et al., 2005). However, it is not known if any prokaryotic CLC behaves as a Cl^- channel. Yet, the molecular and functional diversity in the microbe world is probably much larger than that anticipated from our human-centered viewpoint (Kung and Blount, 2004). Nevertheless, the current chapter focuses mainly on mammalian CLC proteins, with the *Torpedo* ClC-0 as the model CLC Cl^- channel.

The electric organ of *Torpedo* is derived from skeletal muscle, and skeletal muscle is known to possess a large background chloride conductance (Bryant and Morales-Aguilera, 1971; Bretag, 1987). In fact, the first CLC homologue to be cloned after the *Torpedo* channel was ClC-1, the major skeletal muscle chloride channel (Steinmeyer et al., 1991b). Similar to ClC-0, its function is to stabilize the plasma membrane voltage of skeletal muscle fibers. Its disruption in a natural mouse mutant (Steinmeyer et al., 1991a) and mutations in humans (Koch et al., 1992; George et al., 1993; Pusch, 2002) and defects in ClC-1 splicing (Charlet et al., 2002; Mankodi et al., 2002) lead to myotonia, a form of muscle stiffness. The Cl^- equilibrium potential in skeletal muscle is close to the K^+ equilibrium potential, and the lack of repolarizing current through ClC-1 leads to muscle hyperexcitability. Thus, stimuli that normally elicit just one action potential give rise to trains of action potentials, so-called myotonic runs (Adrian and Bryant, 1974), resulting in impairment of muscle relaxation. Interestingly, myotonia in humans can be inherited as a recessive or a dominant trait. Many recessive myotonic mutations, like early stop codons, lead to a complete loss of function of the affected allele. The fact that heterozygous carriers of such mutations do not show myotonia demonstrates that a 50% gene-dosage is not sufficient to cause myotonia. In dominant myotonia, the mutant protein can still associate with wild-type (WT) ClC-1 subunits to form dimeric channels. Frequently, dominant mutations exert a dominant negative effect by changing the voltage-dependence of the dimer through the common gate (Pusch et al., 1995b; Saviane et al., 1999). As expected, dominant myotonia is clinically less severe as 25% of the channels are composed entirely of WT subunits in heterozygous patients.

In contrast to ClC-1, which is nearly exclusively expressed in skeletal muscle, ClC-2 is broadly, if not ubiquitously expressed (Thiemann et al., 1992). Its currents may be activated by hyperpolarization, cell swelling, and moderate extracellular acidification (Gründer et al., 1992; Jordt and Jentsch, 1997; Arreola et al., 2002). Many functions were attributed to ClC-2. Those roles, however, were not confirmed by a ClC-2 knockout (KO) mouse, which unexpectedly revealed testicular and retinal degeneration that were attributed to defects in transepithelial transport across Sertoli cells and the retinal pigment epithelium, respectively (Bösl et al., 2001; Nehrke et al., 2002). Although ClC-2 might play some role in regulating the intracellular chloride in neurons (Staley et al., 1996) and thereby affect inhibitory neurotransmission, ClC-2 KO mice lacked signs of epilepsy (Bösl et al., 2001). In contrast to these clear results from KO mice, in humans, heterozygous ClC-2 mutations were found to be associated with epilepsy in a few families (Haug et al., 2003). However, several key findings of the functional analysis of these mutations (Haug et al., 2003) could not be reproduced (Niemeyer et al., 2004). Additional families with mutations that have a more clear-cut functional effect are probably necessary to firmly establish the gene coding for ClC-2 as being associated with epilepsy. Thus, while ClC-2 is expressed in neurons, its precise role in these cells is not clear. Currents that are probably mediated by ClC-2 are also observed in astrocytes (Ferroni et al., 1997) and immunocytochemistry localized ClC-2 to astrocytic endfeet (Sik et al., 2000). It is thus conceivable that ClC-2 is involved in the Cl^- ion homeostasis mediated by astrocytes.

For some time a controversy existed regarding the polarized expression of ClC-2 in epithelial cells. Transepithelial, i.e., vectorial ion and water transport depends critically on the expression of the specific transporter and channel proteins on either the apical, i.e., lumen or "outside-world" side, or the basolateral, or "blood facing" side of the epithelial cells. These different membrane compartments are separated by tight junctions and the correct targeting of membrane proteins depends on specific signal sequences that are recognized in the biosynthetic and/or the recycling pathways that deliver the membrane proteins via the fusion of small vesicles (Muth and Caplan, 2003). Several studies had reported an apical localization of ClC-2 (Blaisdell et al., 2000; Gyömörey et al., 2000) while others favored a basolateral localization (Lipecka et al., 2002; Catalán et al., 2004). Assuming an apical localization it was proposed that activation of endogenous ClC-2 channels might provide an alternative Cl^- conductance pathway in epithelia of patients with cystic fibrosis (Thiemann et al., 1992; Schwiebert et al., 1998), a genetic disease caused by the dysfunction of another, non-CLC Cl^- channel called CFTR (Riordan et al., 1989). However, results from double KO mice lacking both, CFTR and ClC-2, demonstrated that this is not a valid hypothesis (Zdebik et al., 2004). In agreement with this negative result, it has recently been unequivocally demonstrated that ClC-2 is localized basolaterally in intestinal epithelia (Peña-Münzenmayer et al., 2005). Interestingly, in this study a putative di-leucine motif in the C-terminal cytoplasmic CBS2 domain was identified as being important for the correct targeting of ClC-2 (Peña-Münzenmayer et al., 2005).

ClC-2 has also been proposed to be an apical Cl^- channel that is important for gastric acid secretion (see Jentsch et al., 2005 for discussion), but also this role could not be confirmed (Hori et al., 2004).

Thus, while ClC-2 is broadly expressed, its elimination in KO mice leads to a very specific and limited phenotype in testis and retina (Bösl et al., 2001). However, its precise cellular role in these tissues and in all other tissues where its function can be demonstrated, is not yet clearly established.

ClC-Ka and ClC-Kb are two highly homologous channels that are both expressed in certain epithelial cells of the kidney, as well as in epithelia of the inner ear (Uchida et al., 1993; Kieferle et al., 1994; Estévez et al., 2001). It is now known that these plasma membrane Cl^- channels need a small accessory β-subunit, barttin, for their transport to the plasma membrane (Estévez et al., 2001). Both channels are important for transepithelial transport in kidney and in the inner ear. Mutational inactivation of ClC-Kb in humans leads to Bartter syndrome, a disease associated with severe renal salt wasting, because this channel plays an important role in NaCl reabsorption in a certain nephron segment (the thick ascending limb, TAL) (Simon et al., 1997). In this segment a NaK2Cl-cotransporter takes up Na^+, K^+, and Cl^- ions at the apical membrane, in an electroneutral manner. Potassium ions are "recycled" back to the tubular lumen through the ROMK K^+ channel, generating a negative intracellular membrane potential, helping to extrude Cl^- ions at the basolateral membrane through the CLC-Kb/barttin channel. Na^+ is extruded through the Na–K–ATPase at the basolateral membrane, that provides the energy for this vectorial NaCl transport. Interestingly, also mutations in the apical NaK2Cl co-transporter and mutations in the apical K^+ channel ROMK lead to Bartter's syndrome, with slightly different phenotypes (Simon et al., 1996a,b). The disruption of ClC-K1 in mice (probably equivalent to ClC-Ka in humans) leads to renal water loss as its expression in the thin limb is important for the establishment of a high osmolarity in kidney medulla (Matsumura et al., 1999). The "counter-current system" in the kidney serves to retain water and to produce urine of high osmolarity in situations of limited water supply (Greger and Windhorst, 1996) and, in particular, a large Cl^- conductance in the thin ascending limb of the loop of Henle appears to be necessary, a conductance possibly mediated by ClC-Ka/barttin (Matsumura et al., 1999). The precise localization of the channel (only apical or apical and basolateral) is, however, not yet clearly resolved (Uchida et al., 1995; Vandewalle et al., 1997). It must be said that the precise role of CLC-Ka in the human kidney is, still, not fully resolved. It may be that the murine

CLC-K1 channel is not the exact functional homologue of the human CLC-Ka. Indeed, the two human CLC-K homologues are more closely related to each other than each one is two either murine CLC-K (Kieferle et al., 1994). The two CLC-K genes are localized very close to each other on chromosome 1, raising the possibility that they arose from a recent gene-duplication, after the separation of the human and mice lineages. Alternatively, the evolution of the two human genes might have been associated with convergent recombination events, reducing the sequence difference between them, and resulting in slightly different localization and function compared to the respective murine genes.

In humans, the loss of barttin, the common β-subunit of ClC-Ka and ClC-Kb, leads to sensineural deafness in addition to renal salt loss (Birkenhäger et al., 2001). This disease is also called type IV Bartter's syndrome, while mutations in CLC-Kb lead to type III Bartter's syndrome. The hearing loss in type IV Bartter's syndrome has been attributed to a defect in potassium secretion by inner ear epithelia. In these cells, ClC-Ka/barttin and ClC-Kb/barttin are needed for the basolateral recycling of chloride that is taken up by a basolateral sodium–potassium-two-chloride cotransporter (Estévez et al., 2001). The basolateral localization of ClC-Kb and ClC-Ka in the inner ear suggests that both channels might be localized exclusively in the basolateral membrane also in the kidney. Loss of barttin impairs the function of both ClC-Ka and ClC-Kb. Consequently, the renal disease phenotype in type IV Bartter's syndrome is more severe than that for patients having mutations only in ClC-Kb. However, many mechanistic aspects of the interaction between ClC-K channels and barttin are still poorly understood. Recently, patients with mutations in genes coding for both ClC-Ka and ClC-Kb, but not in the gene coding for barttin, have been described (Schlingmann et al., 2004). The disease phenotype was similarly severe as that found for patients with a loss of function of barttin. This result demonstrates that CLC-Ka is of functional relevance in humans and not e.g., a pseudogene. Barttin seems to associate exclusively with CLC-K channels but not other CLC proteins (Estévez et al., 2001). Furthermore, no close homologs of barttin seem to be present in the human genome. Thus, barttin is the only true β-subunit of CLC proteins, even though other cellular proteins have been shown to interact with CLC-proteins (Ahmed et al., 2000; Furukawa et al., 2002; Rutledge et al., 2002; Zheng et al., 2002; Gentzsch et al., 2003; Hryciw et al., 2003; Embark et al., 2004; Hryciw et al., 2004; Hinzpeter et al., 2005). The determination of the precise physiological relevance of these interactions remains an interesting and important future objective of research (Dhani and Bear, 2005).

The physiological roles of intracellular CLC proteins are best understood for ClC-5 and ClC-7. In humans, the mutational loss of ClC-5 leads to Dent's disease, an inherited kidney stone disorder which is also associated with the loss of proteins into the urine (Lloyd et al., 1996). ClC-5 is predominantly expressed in the proximal tubule of the kidney, where it localizes to apical endosomes (Günther et al., 1998). The knock-out of ClC-5 in mice has revealed that the lack of this channel impairs endocytosis of protein and fluid-phase markers (Piwon et al., 2000; Wang et al., 2000). Further, the loss of phosphate and calcium into the urine, which ultimately leads to kidney stones, may be explained by the decreased renal endocytosis and processing of calciotropic hormones (Piwon et al., 2000). The decrease in endocytosis is associated with an impairment of the luminal acidification of endosomes (Piwon et al., 2000; Günther et al., 2003; Hara-Chikuma et al., 2005). The simplest model to explain a decrease in the acidification rate would be that ClC-5 functions as a Cl^- channel and that this channel activity is important to neutralize the electric current of the H^+-ATPase that acidifies these vesicles (Piwon et al., 2000; Jentsch et al., 2005). Without such a neutralization, the voltage over the endosomal membrane would inhibit further H^+-pumping, severely limiting endosomal acidification. However, as outlined in more detail below, recently it has been demonstrated that ClC-4 and ClC-5 (and probably also ClC-3) are actually not Cl^- channels but secondary active Cl^-/H^+ exchangers (Picollo and Pusch, 2005; Scheel et al., 2005), similar as the bacterial *E. coli* homologue ClC-ec1 (Accardi and Miller, 2004). The precise physiological role of the ClC-5 is thus still relatively unclear.

Vesicular acidification is apparently also the major role of ClC-3, a Cl^- channel or Cl^-/H^+ transporter expressed on synaptic vesicles in addition to endosomes (Stobrawa et al., 2001; Li et al., 2002). The genetic knock-out of ClC-3 in mice led to blindness and to a severe degeneration of the hippocampus (Stobrawa et al., 2001). The mechanism of this degeneration, however, is incompletely understood. ClC-3 has also been proposed to act as a "volume regulated" Cl^- channel by several authors (see Jentsch et al., 2002 for review), but these findings could not be reproduced by several laboratories. In fact, ClC-3 is about 80% identical in protein sequence to ClC-4 and ClC-5, and these proteins have very similar biophysical properties (Li et al., 2000; Picollo and Pusch, 2005).

ClC-7 is the only mammalian CLC protein that is prominently expressed on lysosomes in addition to late endosomes (Kornak et al., 2001; Kasper et al., 2005). Surprisingly, its disruption in mice led to a severe osteopetrotic phenotype (Kornak et al., 2001). It was shown that in bone-resorbing osteoclasts, ClC-7 can be inserted together with the H^+-ATPase into its ruffled border, a specialized plasma membrane that faces the so-called resorption lacuna. Osteoclasts attach to the bone material, forming a shielded environment, that is also called extracellular lysosome. An acidification of the resorption lacuna is necessary for the degradation of bone. On the one hand, the acid directly dissolves the inorganic bone structure. On the other hand, an acidic pH is needed for specialized proteases that degrade organic material. A steady balance between bone formation mediated by osteoblasts and bone degradation mediated by osteoclasts guarantees the maintenance of bone mass and provides the exquisitely stable structure of bone. Osteoclasts obtained from ClC-7 KO mice fail to acidify the resorption lacuna, readily explaining the osteopetrotic phenotype of the mice (Kornak et al., 2001). Bones in these mice are denser but also more fragile. This phenotype suggested that ClC-7 might also be mutated in human osteopetrosis. Indeed, ClC-7 mutations were found in autosomal recessive malignant infantile osteopetrosis (Kornak et al., 2001) and later also in an autosomal dominant form of the disease (Cleiren et al., 2001). Like with myotonia due to ClC-1 mutations, autosomal dominant osteopetrosis is clinically more benign since 25%

of all ClC-7 dimers are expected to consist entirely of WT subunits, while recessive osteopetrosis is equivalent to a complete loss of ClC-7 function. In addition to osteopetrosis, knock-out of ClC-7 led to blindness and neurodegeneration (Kornak et al., 2001; Kasper et al., 2005). The blindness was caused directly by retinal degeneration and not indirectly by the obstruction of the optic canal (Kornak et al., 2001; Kasper et al., 2005). ClC-7 KO mice survived only for about 6 weeks and it was thus difficult to study in detail the neurodegenerative phenotype of these mice (Kornak et al., 2001). To overcome this problem Kasper et al. (2005) rescued the osteopetrotic phenotype of the ClC-7 mouse by reintroducing the ClC-7 gene under the control of an osteoclast specific promoter. These mice had severe neurodegeneration with the accumulation of electron-dense material in lysosomes, characteristic of a lysosomal storage disease such as neuronal ceroid lipofuscinosis. Surprisingly, however, the pH of the lysosomes was not different in these animal compared to that from WT mice. Thus, the precise role ClC-7 for lysosomal function is not yet fully understood (Kasper et al., 2005).

ClC-4 and ClC-6 remain the only CLC channels for which no disease or mouse knock-out phenotype has been described as yet. The physiological functions of CLC channels, and the pathologies resulting from their disruption, are nonetheless impressive and reveal the previously unsuspected importance of chloride channels. Furthermore, the biophysical effects of CLC mutations found in human disease have often been studied and have greatly increased our understanding of the structure–function relationship of this channel class.

We have learnt that several CLC proteins fulfill their physiological roles in specific intracellular membranes while others are directed to the plasma membrane, either “alone” or in collaboration with the barttin subunit in the case of CLC-K channels. The correct targeting is, of course, of crucial importance. Nevertheless, practically nothing is known about the signals that must be contained in the structure of the various CLC proteins that target them to different membrane compartments within the eukaryotic cell. Several research groups have tried to re-direct intracellular CLCs like ClC-7 to the plasma-membrane using chimeric constructs, without success (e.g., Traverso and Pusch, unpublished results). Thus, the signals that address the intracellular CLC transporters to specific endomembranes are probably not simple “motifs” that are coded by a few amino acids, but represent more complex structural entities that involve several parts of the protein.

8.3 Architecture of CLC Proteins

The double-barreled appearance of the single-channel data obtained after reconstitution into lipid bilayers (Miller and White, 1980) strongly suggests by itself at least a homodimeric architecture because it is difficult to imagine how a single polypeptide can form two independent and equal conductance pathways. However, a direct test of this hypothesis is difficult with functional data on native proteins alone. After the cloning of ClC-0 (Jentsch et al., 1990), several important questions regarding the double-barreled structure of CLC proteins could be answered using site-directed mutagenesis and functional electrophysiological analysis of mutated channels expressed in "heterologous" systems like frog oocytes, lipid bilayers, and mammalian cell lines. Early on, a multimeric structure was suggested on the basis of dominant negative disease-causing mutations of the muscle channel ClC-1 (Steinmeyer et al., 1994; Pusch et al., 1995b). Such mutants are able to suppress the currents of WT subunits when co-expressed. Using mutations that altered fast gating, slow gating, and/or channel conductance of ClC-0 and combining altered and unaltered channel subunits, a homodimeric architecture could be established (Ludewig et al., 1996; Middleton et al., 1996). For example, the cDNA coding for a single subunit can be genetically linked to another copy, that might carry a mutation or not, such that the N-terminus of one subunit is attached to the C-terminus to another subunit (both termini are cytoplasmic). It was then found that when mutations that altered pore properties like single-channel conductance or ion selectivity, or mutations that altered the fast gate, were introduced in one of the two subunits, the resulting single channels behaved like a superposition of an unaltered pore and a mutated pore (Ludewig et al., 1996). In this respect, the two subunits appeared completely independent. However, the presence of a mutation (S123T) in one of the subunits was sufficient to abolish the voltage-dependence of the "slow gate" that closes both pores simultaneously (Ludewig et al., 1996). This shows that the two subunits functionally interact. However, the nature of the slow gate is still one of the biggest secrets of CLC proteins (see below).

It could also be shown that each pore of the double-barreled channel is formed entirely from one subunit (Ludewig et al., 1996; Weinreich and Jentsch, 2001). This architecture is fundamentally different from that found, e.g., in K^+ channels that are fourfold symmetric tetramers in which the pore is formed in the central symmetry axis (Doyle et al., 1998). The K^+ channel architecture poses restrictions on the topology of the pore: it has to be straight and perpendicular to the membrane. No such restrictions apply to CLC-channels and indeed the crystal structure of bacterial CLC homologues did not reveal a clear straight ion-conducting pore (Dutzler et al., 2002). The crystal structure of two bacterial homologues, a *Salmonella* and an *E. coli* homologue have been determined (Dutzler et al., 2002). The two structures are very similar and thus only the *E. coli* CLC-ec1 structure, for which more recently a higher resolution was obtained (Dutzler et al., 2003), will be discussed here. The overall architecture of CLC proteins revealed by the crystal structure is strikingly consistent with the double-barreled shotgun cartoon: several nearby Cl^- ion binding sites were identified in each subunit, indicating probably the most selective permeation points, but the binding sites of the two subunits are far from each other ($>\sim 40$ Å) consistent with the independence of the permeation process in the two pores.

Overall, the crystal structure displays a complex fold with 18 α-helices per subunit. Unexpectedly, the structure revealed an internal pseudo-symmetry within each subunit: the N-terminal half and the C-terminal half have a very similar fold and also some spurious sequence similarity but are oppositely oriented in the membrane. The two halves of each subunit "sandwich" around the central Cl^- ion binding sites

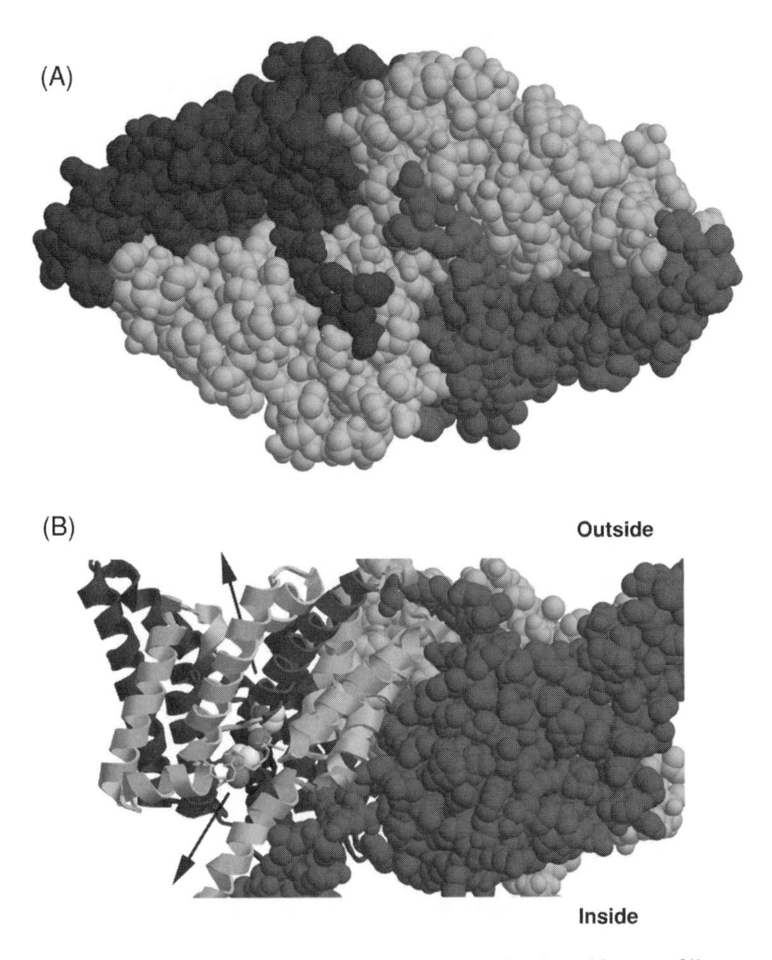

Fig. 8.3 Overall architecture of ClC-ec1. In (A) the protein is viewed in spacefill representation from the extracellular side. The pseudo-twofold internal symmetry for each subunit is highlighted with the different gray shades for each half-subunit. In (B) the protein is rotated such that the view is more lateral. One subunit is shown in cartoon with a small stretch of helix J shown as backbone-trace for clarity. Chloride ions are shown in light gray, and the amino acids S107 (between the two Cl^- ions) and E148 (above the top Cl^- ion) are shown in spacefill. Arrows indicate the probable extracellular and intracellular ion entry/exit pathways. This figure was prepared with the RasTop program that is based on Rasmol (Sayle and Milner-White, 1995) using the pdb-entry 1OTS from which water and Fab fragments have been removed for clarity.

with helix-ends pointing into the center of the membrane. Thus, CLC proteins have possibly evolved by gene duplication of an ancient protein with half of the length of the "modern" version and assembled as a homotetramer in a dimer-of-dimers arrangement (Fig. 8.3).

The dimeric architecture of CLC proteins could enable, in principle, the formation of heterodimeric complexes with two different subunits. In vitro, the formation of heterodimers could indeed be observed between the plasma membrane channels

ClC-1 and ClC-2 (Lorenz et al., 1996) and ClC-0 and ClC-2 (Weinreich and Jentsch, 2001). Such dimers could be obtained either by co-injecting the respective subunits into *Xenopus* oocytes or by linking two subunits in a single polypeptide in a head-to-tail fashion. The dimeric interface between the two subunits is extensive (Fig. 8.3) and the formation of heterodimeric CLC proteins is probably only possible if both subunits are sufficiently similar in sequence and structure to mimic a homodimeric configuration. Indeed, no functional dimers could be obtained by linking, for example, ClC-7 or ClC-6 to ClC-0 (Brandt, Pusch, and Jentsch, unpublished observation), even though other reasons, like improper plasma membrane targeting, may underlie these negative results. Physiologically, ClC-1/ClC-2 heterodimers are probably of little relevance as suggested by the nonoverlapping phenotype of the respective knock-out mouse models (Steinmeyer et al., 1991a; Bösl et al., 2001). The members of the ClC-3–5 branch are highly homologous and, in particular, ClC-4 and CLC-5 show considerable overlap in their expression profiles. In fact, a physiologically relevant hetero-dimerization of ClC-4 and ClC-5 was suggested by co-immunoprecipitation experiments (Mohammad-Panah et al., 2003).

Before describing in more detail the atomic structure of CLC-ec1, basic functional properties of mammalian CLC-channels will be discussed.

8.4 Gating of CLC-0 and Mammalian CLC Channels

As described above, functionally and structurally, animal CLC proteins can be divided into three classes. The first class consists of plasma-membrane channels (ClC-1, ClC-2, ClC-Ka, and ClC-Kb) and also the *Torpedo* channel ClC-0 belongs to this class. These proteins are clearly Cl^- ion channels in contrast to some bacterial homologues and ClC-4 and ClC-5, that are instead secondary active Cl^-/H^+ antiporters (see below). Features of these plasma membrane proteins that identify them as channels are a directly measurable single-channel conductance (≥1 pS) (Miller, 1982; Bauer et al., 1991; Pusch et al., 1994; Weinreich and Jentsch, 2001) (Alessandra Picollo and M.P., unpublished results); a reversal potential measured in Cl^- gradients that follows the Nernstian prediction of passive diffusion (Ludewig et al., 1997a; Rychkov et al., 1998; Thiemann et al., 1992); and relatively slow macroscopic gating relaxations that depend on voltage, Cl^- concentration and pH. These channels will be discussed in more detail below. The second class comprises the highly homologous proteins ClC-3, ClC-4, and ClC-5. In vivo, these proteins are mostly expressed in intracellular organelles but at least some functional plasma membrane expression can be achieved in heterologous systems (Steinmeyer et al., 1995; Friedrich et al., 1999; Li et al., 2000) allowing an electrophysiological characterization. Currents induced by these proteins are characterized by an extreme outward rectification—corresponding to movement of Cl^- ions from the extracellular side into the cytoplasm or correspondingly to a movement from the lumen of intracellular vesicles into the cytosol. Currents are measurable only for voltages >∼ +20–40 mV, such that a true reversal potential cannot be measured (Steinmeyer et al., 1995;

Friedrich et al., 1999). Current activation at positive voltages has apparently two kinetic components. A minor component shows relatively slow kinetics while the major component of the current is associated with very fast activation and deactivation kinetics that cannot be resolved using standard electrophysiological techniques. These fast kinetics may actually represent an "instantaneous" activation, i.e., a rectification of the open pore. Noise analysis supports this hypothesis (Hebeisen et al., 2003), however, the single-channel properties of ClC-3, -4, or -5 are still unclear (Vanoye and George, 2002; Hebeisen et al., 2003). Also, the relationship of the double-barreled structure and the functional properties is unknown. The functional properties of ClC-3–ClC-5 are unusual for an ion channel. Indeed, it was recently found that these proteins are in fact not ion channels but Cl^-/H^+ antiporters (Picollo and Pusch, 2005; Scheel et al., 2005), like the bacterial ClC-ec1 (see below).

The third group of CLC proteins is composed of ClC-6 and ClC-7. Both proteins are mostly found in membranes of intracellular organelles (Buyse et al., 1998; Kornak et al., 2001). No functional electrophysiological data have so far been obtained for these proteins, and it is thus unclear if they are Cl^- channels, Cl^-/H^+ antiporters, or even other transport proteins (Fig. 8.2).

The plasma membrane localized channels ClC-1 and ClC-2, and in particular the "prototype" ClC-0 channel, have been studied most extensively from a functional point of view. A fundamental difference between classical voltage-gated cation channels and the CLC Cl^- channels regards their voltage-dependence. Both type of gates (i.e., the single protopore gate and the common gate) of the *Torpedo* channel ClC-0 are voltage-dependent. The apparent gating valence of the protopore gate is around one and that of the common gate around two, while they exhibit an opposite voltage-dependence (Hanke and Miller, 1983; Pusch et al., 1995a, 1997). However, no clear "voltage-sensor" such as the S4-segment of K^+ and Na^+ channels is evident from the primary or from the 3D-structure. In fact, the voltage-dependence of the fast gate seems to arise, at least partly, indirectly from the coupling of a transmembrane movement of the permeant anion to channel opening (Pusch et al., 1995a; Chen and Miller, 1996). Lowering the extracellular Cl^- concentration "shifts" the voltage-dependence of the open-probability to more positive voltages, i.e., renders opening more difficult (Pusch et al., 1995a). Also, intracellular Cl^- affects the open-probability of the protopore gate mainly by altering the closing rate constant (Chen and Miller, 1996; Ludewig et al., 1997a). This strong coupling of gating and "down-hill" permeation renders the gating an intrinsically irreversible process. Furthermore, it implies that altering the permeation process (e.g., by mutation) also affects channel gating and vice versa. In fact, many results from mutagenesis studies, and attempts to establish a transmembrane topology have been difficult to interpret in the absence of a crystal structure (see Jentsch et al., 2002 for review). An irreversible coupling of the single protopore gate and the common gate of ClC-0 has been observed by Richard and Miller at the single-channel level: bursts of channel activity (see Fig. 8.1) started more often with both protopores gate open and ended more likely with only one of the protopore gates open (Richard and Miller, 1990). This imbalance, that on principal grounds requires an external energy input, was indeed more pronounced in stronger electrochemical gradients, demonstrating a coupling of permeation and gating and a coupling of the two types of gates (Richard and Miller, 1990). The protopore gate of ClC-0 depends also strongly on intracellular and extracellular pH (Hanke and Miller, 1983; Chen and Chen, 2001). Acidification on either side favors channel opening, however, with different qualitative effects (see Pusch, 2004 for a review). The pH dependence of fast gating is probably related to the Cl^-/H^+ antiport activity of other CLC proteins because the same central glutamate (E166 in ClC-0; E148 in ClC-ec1; E211 in ClC-5) is centrally involved in the pH dependence.

Based on a large temperature-dependence of the kinetics of the common gate it has been suggested that it is accompanied by a substantial conformational rearrangement (Pusch et al., 1997). This is consistent with the finding that mutations in many protein regions affect the common gate (Ludewig et al., 1996, 1997a,b; Fong et al., 1998; Maduke et al., 1998; Lin et al., 1999; Saviane et al., 1999; Accardi et al., 2001; Duffield et al., 2003; Traverso et al., 2003; Estévez et al., 2004). However, the molecular mechanism of the common gate is still obscure. Gating of ClC-1 largely resembles that of ClC-0 (Rychkov et al., 1996; Accardi and Pusch, 2000; Accardi et al., 2001), however, its small single-channel conductance (Pusch et al., 1994) renders a detailed analysis difficult. Qualitatively, also ClC-1 shows a double-barreled appearance (Saviane et al., 1999). However, the two gates seem to be more strongly coupled in this channel compared to ClC-0 (Accardi and Pusch, 2000). ClC-2 shows a quite complicated and slow gating-behavior that, in addition, depends on the expression system, and will not be discussed here in detail (Gründer et al., 1992; Thiemann et al., 1992; Jordt and Jentsch, 1997; Pusch et al., 1999; Arreola et al., 2002; Niemeyer et al., 2003; Zúñiga et al., 2004).

A pair of highly homologous kidney and inner ear-specific channels (ClC-K1 and ClC-K2 in rodents; ClC-Ka and ClC-Kb humans) has been cloned more than a decade ago (Uchida et al., 1993; Adachi et al., 1994; Kieferle et al., 1994). However, these channels expressed only little in heterologous systems despite their elevated sequence similarity to ClC-0. The low expression was also surprising because in vivo immunocytochemistry showed a plasma membrane localization (Uchida et al., 1995; Vandewalle et al., 1997). Furthermore, a basolateral plasma membrane localization of ClC-Kb was strongly suggested by the disease phenotype of Bartter's syndrome (Simon et al., 1997). The lack of expression rendered a detailed biophysical analysis impossible. Only recently it became clear that ClC-K channels need the associated small transmembrane protein barttin for efficient plasma membrane expression (Estévez et al., 2001). The barttin gene was identified by positional cloning of the locus leading to a particular type of Bartter's syndrome associated with deafness (Birkenhäger et al., 2001). Gating of ClC-K channels co-expressed with barttin is slightly voltage-dependent (Estévez et al., 2001; Liantonio et al., 2002, 2004; Picollo et al., 2004). ClC-K1 and ClC-Ka are similarly dependent on voltage being activated at negative voltages, while ClC-Kb shows an opposite voltage-dependence. The voltage-dependence of ClC-K channels is somewhat surprising, because they lack a glutamate in a highly conserved stretch of amino acids, having instead a valine residue. Charge neutralizing mutations of the corresponding glutamate in ClC-0 and

ClC-1 completely abolish any voltage-dependent gating relaxations, while in ClC-2 the neutralization of the glutamate almost completely abolishes gating (Dutzler et al., 2003; Estévez et al., 2003; Niemeyer et al., 2003; Traverso et al., 2003). In fact, re-introducing a glutamate into ClC-K1 confers much more pronounced gating into the channel (Estévez et al., 2001). However, the mechanisms of ClC-K gating and the relationship with the double-barreled structure is not understood.

Interestingly, ClC-K activity can be markedly regulated by the extracellular pH and by the extracellular Ca^{2+} concentration in the millimolar concentration range (Uchida et al., 1995; Waldegger and Jentsch, 2000; Estévez et al., 2001). Since the fast protopore-gate is probably mostly open in these channels, the pH and Ca^{2+} regulation could act via the common gate, but the underlying mechanisms remain to be identified.

8.5 Permeation of CLC-0 and Mammalian CLC Channels

Cl^- channels do not generally discriminate strongly among halides, except for fluoride, that is quite impermeant in most Cl^- channels probably due to its strong hydration. Based on anomalous mole fraction effects it has been suggested that CLC pores can accommodate more than one ion at a time (Pusch et al., 1995a; Rychkov et al., 1998) in agreement with the presence of multiple Cl^- ions in the crystal of ClC-ec1 (Dutzler et al., 2003). Ion permeation has been most extensively studied in the muscle channel ClC-1 at the macroscopic level. The channel exhibits an almost perfect selectivity of anions over cations (Rychkov et al., 1998). Among various anions tested the permeability sequence obtained from reversal potential measurements was $SCN^- \sim ClO_4^- > Cl^- > Br^- > NO_3^- \sim ClO_3^- > I^- \gg BrO_3^- > HCO_3^- \sim F^- \gg$ methanesulfonate $\sim$ cyclamate $\sim$ glutamate, where the latter three organic anions are practically impermeable (Rychkov et al., 1998). Furthermore, ClC-1 is blocked in a voltage-dependent manner by various organic anions and permeation and block appear to involve strong hydrophobic interactions (Rychkov et al., 1998; Rychkov et al., 2001). Recently, the permeation properties of ClC-0 and ClC-1 have been modeled using Brownian dynamics based on the crystal structure of ClC-ec1 (Corry et al., 2004). Overall these results predict a "knock-off" mechanism of permeation in which two relatively strongly bound Cl^- ions are destabilized by the arrival of a third ion. The predictions of these simulations remain to be tested experimentally.

8.6 The X-ray Structure and Its Functional Implication: A Pivot Glutamate Controls the Protopore Gate

The X-ray structure of ClC-ec1 marked a breakthrough for the structure–function analysis of CLC proteins (Dutzler et al., 2002, 2003). As shown in Fig. 8.4, in each subunit two Cl^- ions could be resolved. The central binding site (S_{cent}) is completely

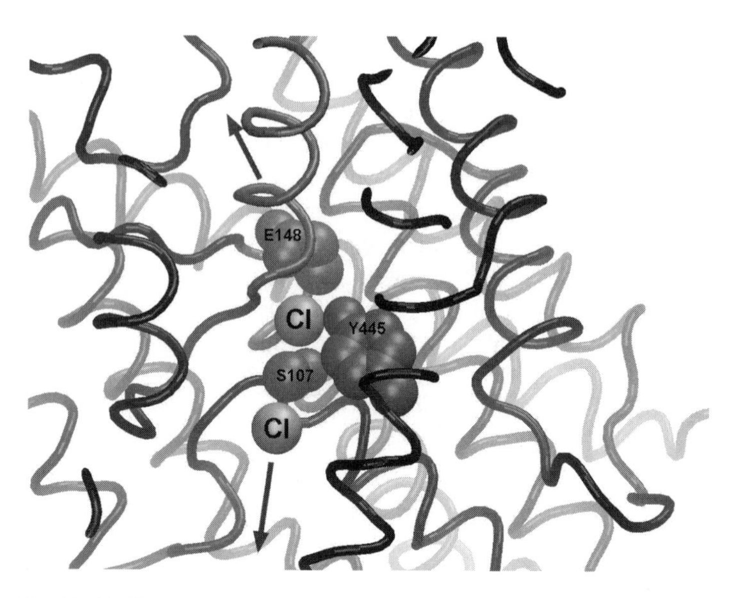

Fig. 8.4 The Cl^- binding sites in ClC-ec1. Only one subunit is shown with the intracellular side at the bottom. Cl^- ions (light gray) and the amino acids S107 (between the two Cl^- ions), E148 (above the top Cl^- ion) and the coordinating Y445 are shown in spacefill (dark gray). This figure was prepared using the VMD program (Humphrey et al., 1996).

buried in the protein. Major direct protein–ion interactions involve the amide NH groups of some amino acids and the side chains of two residues (Y445 and S107). The directly Cl^- interacting residues are well-conserved among CLC proteins and most of them are located in loops connecting two helices. In fact, several helices have their N-terminal end within the membrane. This topology reflects the pseudo-symmetrical sandwich structure described above (Fig. 8.3). It is interesting to note that none of the directly interacting residues is positively charged. Direct salt-bridging of a Cl^- ion would probably lead to a too strong binding and less efficient conduction. However, simulation studies have shown that several more distantly located positively charged residues make significant contributions to the free energy of Cl^- ion binding (Corry et al., 2004; Miloshevsky and Jordan, 2004; Cohen and Schulten, 2004; Bostick and Berkowitz, 2004). The more intracellularly located binding site, S_{int}, is in direct contact with the intracellular solution (Fig. 8.4) indicating the most likely intracellular pore entrance. In agreement with this, and confirming that the ClC-ec1 crystal structure is relevant for eukaryotic CLC channels, mutations in many residues that delineate the putative entrance alter typical pore properties like rectification and single channel conductance in an expected way (Pusch et al., 1995a; Ludewig et al., 1996, 1997a; Middleton et al., 1996; Rychkov et al., 1996; Pusch et al., 1999; Accardi and Pusch, 2003; Chen et al., 2003; Lin and Chen, 2003; Engh and Maduke, 2005).

In contrast, the exit of the central Cl^- ion toward the intracellular side seem to be impeded by the highly conserved serine 107 that is located between S_{int} an S_{cent} (Fig. 8.4) (Corry et al., 2004; Miloshevsky and Jordan, 2004). Probably, some conformational rearrangement of the intracellular entrance has to occur for an efficient Cl^- conduction. In agreement with the necessity of a conformational change of the inner pore during channel opening strong state-dependent binding of small organic inhibitors from the inside has been described (Pusch et al., 2001; Accardi and Pusch, 2003; Traverso et al., 2003). However, as discussed by Yin et al. (2004), the state-dependent block might be caused by an altered occupancy of Cl^- binding sites in the closed compared to the open state. The exit of the central Cl^- ion toward the extracellular side is clearly impeded by the negatively charged side-chain of a conserved glutamate (E148) (Fig. 8.4) (Dutzler et al., 2002). In order to study the role of E148, Dutzler and colleagues solved the structure of ClC-ec1 in which the glutamate was mutated to alanine (E148A) or glutamine (E148Q). Since at that time direct functional analysis of ClC-ec1 was not yet possible (Maduke et al., 1999), Dutzler et al. used the well-studied *Torpedo* ClC-0 as a model to investigate the functional effect of the mutations (Dutzler et al., 2003). The mutant structures were almost identical to that of WT ClC-ec1 with a single difference: where the negatively charged E148 side-chain in WT CLC-ec1 blocked ion movement was now sitting a third crystallographically identified, third, Cl^- ion, Cl^-_{ext} (Dutzler et al., 2003). This structure appears to be an opened confirmation, at least concerning movement of the central Cl^- toward the outside. Consistent with this simple structural result, both ClC-0 mutants, E166A and E166Q, showed a constitutively open phenotype: they appeared to have lost the voltage- and Cl^--dependence of the open probability (Dutzler et al., 2003; Traverso et al., 2003). Interestingly, WT ClC-0 can also be opened by lowering the extracellular pH, probably by protonation of the corresponding glutamate (Dutzler et al., 2003), mimicking the E166Q mutation. These findings clearly establish E148 as a major player in the gating of CLC proteins. As already discussed above, it is noteworthy in this respect that ClC-K channels are only little voltage-dependent showing almost constitutive activity at all voltages (Estévez et al., 2001; Picollo et al., 2004). The wimpy voltage-dependence is related to the fact that ClC-K channels, and only these, carry a valine residue at the glutamate-position. Also for the strongly outwardly rectifying channels, ClC-3–ClC-5, the conserved glutamate is of prime importance: mutating E211 in ClC-5 (or its equivalents in ClC-3 or ClC-4) to alanine leads to a loss of the strong rectification and to an almost linear current–voltage relationship (Friedrich et al., 1999; Li et al., 2002).

8.7 The Function as a Cl^-/H^+ Antiporter

Accardi and Miller succeeded recently in obtaining a high-yield and extremely pure protein preparation of ClC-ec1 that allowed them to study macroscopic currents of ClC-ec1 proteins reconstituted in lipid bilayers (Accardi et al., 2004). Currents were voltage-independent but activated by low pH; the apparent single-channel conductance was too low to be reliably estimated even from noise analysis (Accardi et al., 2004). Currents carried by the glutamate mutant E148A were independent of pH (Accardi et al., 2004), in agreement with analogous behavior for ClC-0. A puzzling finding was however, that the reversal potential measured for WT ClC-ec1 did not follow the Nernstian prediction for a purely Cl^- selective channel but was significantly smaller, indicating the permeation of some cationic species (Accardi et al., 2004; Accardi and Miller, 2004). After detailed analysis, Accardi and Miller clearly demonstrated that ClC-ec1 is actually not a diffusive Cl^--selective channel, but instead a secondary active strictly coupled Cl^-/H^+ exchanger (Accardi and Miller, 2004). The apparent stoichiometry was found to be an exchange of two Cl^- ions for each transported H^+ (Accardi and Miller, 2004). Given the opposite charges of Cl^- and H^+, this transporter is highly electrogenic with a net transfer of three elementary charges per transport cycle.

This result came as a big surprise for the ClC-field as no previous piece of evidence, and in particular not even the atomic resolution structure of the protein, had indicated such a function. The basic mechanism of function of a coupled transporter is a priori quite different from that of a passively diffusive channel and it will be very interesting to decipher the mechanism of this transport in the future. This result does not imply that all CLC proteins are Cl^-/H^+ antiporters. Certain well-known ClC proteins such as ClC-0, ClC-1, and ClC-2 are clearly Cl^- ion channels: they have a rather large conductance and they show almost perfect Nernstian $[Cl^-]$-dependence of the reversal potential. Their gating, however, depends on pH_{int} and pH_{ext} (Pusch, 2004) and this dependence is likely related to the transport mechanism of ClC-ec1. In contrast, the function of several intracellular ClC proteins, like ClC-3–ClC-7, has never been clearly related to single Cl^- ion channel activity. For ClC-6 and ClC-7 no functional electrophysiological measurement could be performed so far, leaving open the question if these are channels or transporters. However, for the endosomal proteins ClC-3–ClC-5, it became recently clear that they are indeed not Cl^- channels as assumed previously. In parallel studies, the groups of Jentsch (Scheel et al., 2005) and Pusch (Picollo and Pusch, 2005) demonstrated that the highly homologous proteins ClC-4 and ClC-5 are Cl^-/H^+ antiporters, exactly like the bacterial ClC-ec1. Because the electrical currents carried by ClC-3 are very similar to those of ClC-4 and ClC-5, by analogy also ClC-3 is most likely a Cl^-/H^+ antiporter (Fig. 8.2). The currents carried by these transporters are so strongly rectifying, however, such that it is impossible to measure a true reversal potential (Steinmeyer et al., 1995; Friedrich et al., 1999). Thus a “simple” verification that these proteins are not channels was not possible and this explains why it had not been discovered previously but only after the example provided by Accardi and Miller for ClC-ec1 (Accardi and Miller, 2004). Scheel et al. (2005) and Picollo and Pusch (2005) showed that ClC-4 and ClC-5 transport protons, coupled to the inward movement of chloride ions. Physiologically, the role of a Cl^-/H^+ exchange, that is highly electrogenic, is not at all clear. If it operates as a shunt in acidifying vesicles, the counter-transport of H^+ would lead to a waste of pumped protons. It might also be that the chloride concentration itself is an important factor in the physiology of intracellular vesicles, and the regulation of endo- and exocytosis. A major unresolved issue regards the extreme rectification of ClC-3–ClC-5 that is very difficult to reconcile with any physiological role, because the physiological membrane voltages assumed to be present in intracellular compartments are opposite to the voltages needed to activate ClC-3–ClC-5 in heterologous expression systems.

Apart from the need to understand in more detail the physiological role of these transporters, one of the most fascinating problems remaining for CLC proteins is to understand how the same basic architecture can produce transport proteins that are either "simple" Cl^- selective channels or stoichiometrically coupled Cl^-/H^+ antiporters.

8.8 Pharmacology

Small organic molecules and peptides that interact with ion channel proteins have been extremely useful tools for the studying of voltage-gated cation channels and ligand-gated channels (Hille, 2001). In this respect it suffices to recall the potent neurotoxins α-bungarotoxin or tetrodotoxin that act on the nicotinic acetylcholine receptor and the voltage-gated Na^+ channel respectively with high affinity (Hille, 2001). Also, many medically useful drugs act on various types of ion channels. Unfortunately, the situation is worse for Cl^- channels in general and for CLC proteins in particular, even though some progress has been made recently. Typical, "classical", Cl^- channel blockers include molecules like 4,4′-diisothiocyanato-2,2′-stilbenedisulfonic acid (DIDS), 4-acetamido-4′-isothiocyanato-2,2′-stilbenedisulfonic acid (SITS), 9-anthracenecarboxylic acid (9-AC), niflumic acid (NFA), 5-nitro-2-(3-phenylpropylamino)benzoic acid (NPPB) and others. Most of these drugs block, i.e., reduce the currents, of many Cl^- channels. They are thus quite unspecific. Furthermore, block occurs often with a low affinity ($<100\ \mu M^{-1}$). There are, however, exceptions. For example, some CLC proteins are particularly resistant to any blocker: no sub 100 μM organic blocker of ClC-2 and no blocker at all of ClC-5 has been identified so far (Pusch et al., 2002). On the other hand, the muscle ClC-1 channel is so far the most "sensitive" CLC protein. It is blocked by 9-AC and by derivatives of p-chloro-phenoxy propionic acid (collectively abbreviated as CPP) at concentrations below 10 μM (Bryant and Morales-Aguilera, 1971; Conte-Camerino et al., 1988; Liantonio et al., 2002). The block of ClC-0 and ClC-1 by these classes of substances (i.e., 9-AC and CPP) has been studied in considerable detail and they have proved to be useful tools to investigate the structure–function relationship of these channels. First, it has been found that both types of drugs can directly access their binding site only from the intracellular side and that they bind to the pore region (Pusch et al., 2000, 2001, 2002; Estévez et al., 2003). Their action is strongly state-dependent in that they bind much more tightly to the closed state than to the open state of the channel leading to a marked apparent voltage-dependence of block (Pusch et al., 2000, 2001, 2002; Accardi and Pusch, 2003). Starting from a critical amino acid identified independently from the crystal structure, Estévez and co-workers then used the bacterial structure as a guide to identify the crucial amino acids involved in blocking the muscle ClC-1 channel by 9-AC and CPP (Estévez et al., 2003). When mapped onto the bacterial structure these critical amino acids clustered in a region around the central Cl^- ion binding site (Estévez et al., 2003). On the one hand, this rough identification of the binding site opens the way for further analysis of the interaction of these molecules with ClC-0 and ClC-1. On the other hand the results of Estévez et al. demonstrate that the overall structure of ClC-0 and ClC-1 must be very similar to that of the bacterial ClC-ec1, and that, despite the quite different functional properties, the bacterial crystal structure is a good guide for the mammalian homologues.

Intracellularly applied DIDS has been found to irreversibly inhibit the *Torpedo* channel ClC-0, acting on individual protopores of the double-barreled channel (Miller and White, 1984). Irreversible action of DIDS is thought to be mediated by the covalent modification of lysine residues or other free amines. The target of DIDS in ClC-0 has, however, not yet been identified. Interestingly also the bacterial ClC-ec1 is inhibited by intracellular, but not extracellular DIDS (Matulef and Maduke, 2005) in a slowly reversible manner with an apparent K_D of about 30 μM. In an elegant study, Matulef and Maduke showed that the DIDS block could be potentially used to "functionally orient" ClC-ec1 transporters incorporated into lipid bilayers (Matulef and Maduke, 2005). In fact, a general limitation of the reconstitution of purified ClC-ec1 is that the orientation of the transporter in the membrane is practically completely random (Matulef and Maduke, 2005). Inclusion of a saturating DIDS concentration on one side of the bilayers leaves functional only those transporters that face the DIDS solution with their extracellular side. However, the mechanism of DIDS block is still unclear and block was not complete even at concentrations of more than 10-fold larger than the apparent K_D (Matulef and Maduke, 2005). It remains to be seen if there is overlap between the 9-AC/CPA site of ClC-1 and the DIDS site of ClC-ec1.

A different blocker-binding site was identified for certain kidney ClC-K channels. Some derivatives of CPP with two phenyl-groups (abbreviated here as 3-phenyl-CPP) and also DIDS were found to block ClC-K1 and CLC-Ka channels but not ClC-Kb and not other CLC channels (Liantonio et al., 2002; Liantonio et al., 2004; Picollo et al., 2004). ClC-K1, ClC-Ka, and ClC-Kb are about 80% identical to each other, and block by 3-phenyl-CPP and DIDS occurs from the extracellular side in competition with Cl^- ions (Liantonio et al., 2004). Comparing the primary sequence between these channels, and using the ClC-ec1 crystal structure as a guide, two amino acids could be identified that are critically important for the differences in inhibitor sensitivity (Picollo et al., 2004). Both residues (N68/D68 and G72/E72 in CLC-Ka/CLC-Kb, respectively) are located in helix B, and in ClC-ec1 the corresponding side-chains point toward the extracellular channel entrance. The N/D68 residue is the most sensitive residue involved in drug binding and it is highly conserved: practically all CLC proteins, including ClC-ec1, present a negatively charged amino acid at the corresponding position. Mutating this residue in ClC-0 or ClC-1 leads to profound alterations in the gating behavior (Fahlke et al., 1995; Ludewig et al., 1997b). Also, the ClC-Kb D68N mutation drastically alters the gating behavior (Picollo et al., 2004). The precise role of this residue will be interesting to study in the future.

Inhibitors of CLC-K channels might be interesting from a pharmaceutical point of view because they could be used as diuretics (Fong, 2004). Another pharmacologically interesting target is ClC-7 because its blockage might be beneficial to treat osteoporosis (Schaller et al., 2004; Karsdal et al., 2005). Unfortunately, heterologous ClC-7 expression does not induce electrical currents in the plasma membrane of transfected cells such that a direct block cannot be assayed. However, indirect functional assays have been developed that led to the identification of putative ClC-7 blockers (Schaller et al., 2004; Karsdal et al., 2005). It remains to be shown, however, that these substances directly act on ClC-7.

8.9 CBS Domains

The crystallized bacterial CLC homologues have short cytoplasmic N- and C-termini. In contrast eukaryotic CLCs have longer N and especially C-termini. These long C-termini are of significant functional importance (Schwappach et al., 1998; Estévez and Jentsch, 2002) and they contain two conserved so-called CBS domains (Ponting, 1997). Probably, the two CBS domains of each subunit interact and they seem to play a role in regulating the common gate (Estévez et al., 2004). However, none of the CBS domains is strictly essential for channel function (Hryciw et al., 1998; Estévez et al., 2004; Hebeisen et al., 2004) and their precise role is still unclear (see Babini and Pusch, 2004 for review). A possibly important clue about the role of CBS domains is that (as isolated peptides) they bind intracellular nucleotides like AMP and ATP (Scott et al., 2004). Some disease causing mutations in CBS domains diminish the binding (Scott et al., 2004) suggesting that it plays an important functional role. Among these mutations the amino acid change G715E that is located between CBS1 and CBS2 in ClC-2 reduced the affinity for ATP 10-fold (Scott et al., 2004). This amino acid change was reported to cause epilepsy in humans and to alter the $[Cl^-]_{int}$-sensitivity of the channel (Haug et al., 2003). These gating-effects could, however, not be reproduced (Niemeyer et al., 2004). Nevertheless, Niemeyer et al. found that the G715E-mutation altered a small kinetic effect of intracellular AMP on the gating (Niemeyer et al., 2004). Recently, also the gating of the muscle channel ClC-1 was shown to be dependent on intracellular ATP, AMP, and adenosine (Bennetts et al., 2005). These nucleotides "shifted" the voltage-dependence of the common gating process to more positive voltages compared to the situation in the absence of nucleotides. Modeling a CBS1/CBS2 dimer revealed a putative ATP binding site and mutating amino acids of the site indeed reduced or even abolished nucleotide regulation (Bennetts et al., 2005). These results provide strong evidence for a functional regulation of CLC proteins by intracellular nucleotides. Bennetts et al. speculated that a reduction of ATP levels in skeletal muscle caused by exercise could lead to a significant activation of ClC-1 mediated conductance and a larger threshold for action potential firing, contributing to fatigue (Bennetts et al., 2005). A failure of skeletal muscle action potential generation and/or propagation upon nerve stimulation due to an increased Cl^- conductance remains, however, to be verified experimentally. The role of intracellular nucleotides in the regulation of CLC proteins and the involvement of the CBS domains are an exciting new aspect of their physiology.

8.10 Conclusion

CLC proteins have been full of surprises right from their first revelation as funny double-barreled, irreversibly gating Cl^- channels. They challenge the classical distinction of ion channel and active transporter. Unexpected physiological roles have been discovered. They are involved in many human genetic diseases. They display an extremely complex three-dimensional structure, with a two-pore architecture of unknown relevance. These fascinating proteins will certainly occupy scientists from disciplines ranging from molecular medicine to molecular dynamics for several years.

Acknowledgment

The financial support by Telethon Italy (grant GGP04018) is gratefully acknowledged.

References

Accardi, A., L. Ferrera, and M. Pusch. 2001. Drastic reduction of the slow gate of human muscle chloride channel (ClC-1) by mutation C277S. *J. Physiol.* 534:745–752.

Accardi, A., L. Kolmakova-Partensky, C. Williams, and C. Miller. 2004. Ionic currents mediated by a prokaryotic homologue of CLC Cl^- channels. *J. Gen. Physiol.* 123:109–119. Epub 2004 Jan 12.

Accardi, A., and C. Miller. 2004. Secondary active transport mediated by a prokaryotic homologue of ClC Cl^- channels. *Nature* 427:803–807.

Accardi, A., and M. Pusch. 2000. Fast and slow gating relaxations in the muscle chloride channel CLC-1. *J. Gen. Physiol.* 116:433–444.

Accardi, A., and M. Pusch. 2003. Conformational changes in the pore of CLC-0. *J. Gen. Physiol.* 122:277–293.

Adachi, S., S. Uchida, H. Ito, M. Hata, M. Hiroe, F. Marumo, and S. Sasaki. 1994. Two isoforms of a chloride channel predominantly expressed in thick ascending limb of Henle's loop and collecting ducts of rat kidney. *J. Biol. Chem.* 269:17677–17683.

Adrian, R.H., and S.H. Bryant. 1974. On the repetitive discharge in myotonic muscle fibres. *J. Physiol.* 240:505–515.

Ahmed, N., M. Ramjeesingh, S. Wong, A. Varga, E. Garami, and C.E. Bear. 2000. Chloride channel activity of ClC-2 is modified by the actin cytoskeleton. *Biochem. J.* 352:789–794.

Arreola, J., T. Begenisich, and J.E. Melvin. 2002. Conformation-dependent regulation of inward rectifier chloride channel gating by extracellular protons. *J. Physiol.* 541:103–112.

Babini, E., and M. Pusch. 2004. A two-holed story: Structural secrets about CLC proteins become unraveled? *Physiology* 19:293–299. DOI:10.1152/physiol.00019.2004.

Barbier-Brygoo, H., M. Vinauger, J. Colcombet, G. Ephritikhine, J. Frachisse, and C. Maurel. 2000. Anion channels in higher plants: Functional characterization, molecular structure and physiological role. *Biochim. Biophys. Acta* 1465:199–218.

Bauer, C.K., K. Steinmeyer, J.R. Schwarz, and T.J. Jentsch. 1991. Completely functional double-barreled chloride channel expressed from a single Torpedo cDNA. *Proc. Natl. Acad. Sci. USA* 88:11052–11056.

Bennetts, B., G.Y. Rychkov, H.-L. Ng, C.J. Morton, D. Stapleton, M.W. Parker, and B.A. Cromer. 2005. Cytoplasmic ATP-sensing domains regulate gating of skeletal muscle ClC-1 chloride channels. *J. Biol. Chem.* 280:32452–32458.

Birkenhäger, R., E. Otto, M.J. Schurmann, M. Vollmer, E.M. Ruf, I. Maier-Lutz, F. Beekmann, A. Fekete, H. Omran, D. Feldmann, D.V. Milford, N. Jeck, M. Konrad, D. Landau, N.V. Knoers, C. Antignac, R. Sudbrak, A. Kispert, and F. Hildebrandt. 2001. Mutation of BSND causes Bartter syndrome with sensorineural deafness and kidney failure. *Nat. Genet.* 29:310–314.

Blaisdell, C.J., R.D. Edmonds, X.T. Wang, S. Guggino, and P.L. Zeitlin. 2000. pH-regulated chloride secretion in fetal lung epithelia. *Am. J. Physiol. Lung Cell. Mol. Physiol.* 278:L1248– L1255.

Bösl, M.R., V. Stein, C. Hübner, A.A. Zdebik, S.E. Jordt, A.K. Mukhopadhyay, M.S. Davidoff, A.F. Holstein, and T.J. Jentsch. 2001. Male germ cells and photoreceptors, both dependent on close cell–cell interactions, degenerate upon ClC-2 Cl(−) channel disruption. *EMBO J.* 20:1289–1299.

Bostick, D.L., and M.L. Berkowitz. 2004. Exterior site occupancy infers chloride-induced proton gating in a prokaryotic homolog of the ClC chloride channel. *Biophys. J.* 87:1686–1696.

Brandt, S., and T.J. Jentsch. 1995. ClC-6 and ClC-7 are two novel broadly expressed members of the CLC chloride channel family. *FEBS Lett.* 377:15–20.

Bretag, A.H. 1987. Muscle chloride channels. *Physiol. Rev.* 67:618–724.

Bryant, S.H., and A. Morales-Aguilera. 1971. Chloride conductance in normal and myotonic muscle fibres and the action of monocarboxylic aromatic acids. *J. Physiol.* 219:367–383.

Buyse, G., D. Trouet, T. Voets, L. Missiaen, G. Droogmans, B. Nilius, and J. Eggermont. 1998. Evidence for the intracellular location of chloride channel (ClC)-type proteins: Co-localization of ClC-6a and ClC-6c with the sarco/endoplasmic-reticulum Ca^{2+} pump SERCA2b. *Biochem. J.* 330:1015–1021.

Buyse, G., T. Voets, J. Tytgat, C. De Greef, G. Droogmans, B. Nilius, and J. Eggermont. 1997. Expression of human pICln and ClC-6 in xenopus oocytes induces an identical endogenous chloride conductance. *J. Biol. Chem.* 272:3615–3621.

Catalán, M., M.I. Niemeyer, L.P. Cid, and F.V. Sepúlveda. 2004. Basolateral ClC-2 chloride channels in surface colon epithelium: Regulation by a direct effect of intracellular chloride. *Gastroenterology* 126:1104–1114.

Charlet, B.N., R.S. Savkur, G. Singh, A.V. Philips, E.A. Grice, and T.A. Cooper. 2002. Loss of the muscle-specific chloride channel in type 1 myotonic dystrophy due to misregulated alternative splicing. *Mol. Cell* 10:45–53.

Chen, M.F., and T.Y. Chen. 2001. Different fast-gate regulation by external Cl(−) and H(+) of the muscle-type ClC chloride channels. *J. Gen. Physiol.* 118:23–32.

Chen, T.Y., M.F. Chen, and C.W. Lin. 2003. Electrostatic control and chloride regulation of the fast gating of ClC-0 chloride channels. *J. Gen. Physiol.* 122:641–651.

Chen, T.Y., and C. Miller. 1996. Nonequilibrium gating and voltage dependence of the ClC-0 Cl^- channel. *J. Gen. Physiol.* 108:237–250.

Cleiren, E., O. Benichou, E. Van Hul, J. Gram, J. Bollerslev, F.R. Singer, K. Beaverson, A. Aledo, M.P. Whyte, T. Yoneyama, M.C. deVernejoul, and W. Van Hul. 2001. Albers-Schonberg disease (autosomal dominant osteopetrosis, type II) results from mutations in the ClCN7 chloride channel gene. *Hum. Mol. Genet.* 10:2861–2867.

Cohen, J., and K. Schulten. 2004. Mechanism of anionic conduction across ClC. *Biophys. J.* 86:836–845.

Conte-Camerino, D., M. Mambrini, A. DeLuca, D. Tricarico, S.H. Bryant, V. Tortorella, and G. Bettoni. 1988. Enantiomers of clofibric acid analogs have opposite actions on rat skeletal muscle chloride channels. *Pflugers Arch.* 413:105–107.

Corry, B., M. O'Mara, and S.H. Chung. 2004. Conduction mechanisms of chloride ions in ClC-type channels. *Biophys. J.* 86:846–860.

Dhani, S.U., and C.E. Bear. 2005. Role of intramolecular and intermolecular interactions in ClC channel and transporter function. *Pflügers Arch.* 16:16.

Doyle, D.A., J. Morais Cabral, R.A. Pfuetzner, A. Kuo, J.M. Gulbis, S.L. Cohen, B.T. Chait, and R. MacKinnon. 1998. The structure of the potassium channel: Molecular basis of K^+conduction and selectivity. *Science* 280:69–77.

Duffield, M., G. Rychkov, A. Bretag, and M. Roberts. 2003. Involvement of helices at the dimer Interface in ClC-1 common gating. *J. Gen. Physiol.* 121:149–161.

Dutzler, R., E.B. Campbell, M. Cadene, B.T. Chait, and R. MacKinnon. 2002. X-ray structure of a ClC chloride channel at 3.0 Å reveals the molecular basis of anion selectivity. *Nature* 415:287–294.

Dutzler, R., E.B. Campbell, and R. MacKinnon. 2003. Gating the selectivity filter in ClC chloride channels. *Science* 300:108–112.

Embark, H.M., C. Bohmer, M. Palmada, J. Rajamanickam, A.W. Wyatt, S. Wallisch, G. Capasso, P. Waldegger, H.W. Seyberth, S. Waldegger, and F. Lang. 2004. Regulation of CLC-Ka/barttin by the ubiquitin ligase Nedd4-2 and the serum-and glucocorticoid-dependent kinases. *Kidney Int.* 66:1918–1925.

Engh, A.M., and M. Maduke. 2005. Cysteine accessibility in ClC-0 supports conservation of the ClC intracellular vestibule. *J. Gen. Physiol.* 125:601–617.

Estévez, R., T. Boettger, V. Stein, R. Birkenhäger, E. Otto, F. Hildebrandt, and T.J. Jentsch. 2001. Barttin is a Cl^- channel beta-subunit crucial for renal Cl^- reabsorption and inner ear K^+ secretion. *Nature* 414:558–561.

Estévez, R., and T.J. Jentsch. 2002. CLC chloride channels: Correlating structure with function. *Curr. Opin. Struct. Biol.* 12:531–539.

Estévez, R., M. Pusch, C. Ferrer-Costa, M. Orozco, and T.J. Jentsch. 2004. Functional and structural conservation of CBS domains from CLC channels. *J. Physiol.* 557:363–378.

Estévez, R., B.C. Schroeder, A. Accardi, T.J. Jentsch, and M. Pusch. 2003. Conservation of chloride channel structure revealed by an inhibitor binding site in ClC-1. *Neuron* 38:47–59.

Fahlke, C., R. Rüdel, N. Mitrovic, M. Zhou, and A.L. George Jr. 1995. An aspartic acid residue important for voltage-dependent gating of human muscle chloride channels. *Neuron* 15:463–472.

Ferroni, S., C. Marchini, M. Nobile, and C. Rapisarda. 1997. Characterization of an inwardly rectifying chloride conductance expressed by cultured rat cortical astrocytes. *Glia* 21:217–227.

Fong, P. 2004. CLC-K channels: If the drug fits, use it. *EMBO Rep.* 5:565–566.

Fong, P., A. Rehfeldt, and T.J. Jentsch. 1998. Determinants of slow gating in ClC-0, the voltage-gated chloride channel of Torpedo marmorata. *Am. J. Physiol.* 274:C966–C973.

Friedrich, T., T. Breiderhoff, and T.J. Jentsch. 1999. Mutational analysis demonstrates that ClC-4 and ClC-5 directly mediate plasma membrane currents. *J. Biol. Chem.* 274:896–902.

Furukawa, T., T. Ogura, Y.J. Zheng, H. Tsuchiya, H. Nakaya, Y. Katayama, and N. Inagaki. 2002. Phosphorylation and functional regulation of ClC-2 chloride channels expressed in Xenopus oocytes by M cyclin-dependent protein kinase. *J. Physiol.* 540:883–893.

Gentzsch, M., L. Cui, A. Mengos, X.B. Chang, J.H. Chen, and J.R. Riordan. 2003. The PDZ-binding chloride channel ClC-3B localizes to the Golgi and associates with cystic fibrosis transmembrane conductance regulator-interacting PDZ proteins. *J. Biol. Chem.* 278:6440–6449. Epub 2002 Dec 5.

George, A.L., Jr., M.A. Crackower, J.A. Abdalla, A.J. Hudson, and G.C. Ebers. 1993. Molecular basis of Thomsen's disease (autosomal dominant myotonia congenita). *Nat. Genet.* 3:305–310.

Greger, R., and U. Windhorst. 1996. Comprehensive Human Physiology. Springer, Berlin.

Gründer, S., A. Thiemann, M. Pusch, and T.J. Jentsch. 1992. Regions involved in the opening of CIC-2 chloride channel by voltage and cell volume. *Nature* 360:759–762.

Günther, W., A. Luchow, F. Cluzeaud, A. Vandewalle, and T.J. Jentsch. 1998. ClC-5, the chloride channel mutated in Dent's disease, colocalizes with the proton pump in endocytotically active kidney cells. *Proc. Natl. Acad. Sci. USA* 95:8075–8080.

Günther, W., N. Piwon, and T.J. Jentsch. 2003. The ClC-5 chloride channel knock-out mouse—an animal model for Dent's disease. *Pflügers Arch.* 445:456–462. Epub 2002 Nov 29.

Gyömörey, K., H. Yeger, C. Ackerley, E. Garami, and C.E. Bear. 2000. Expression of the chloride channel ClC-2 in the murine small intestine epithelium. *Am. J. Physiol. Cell Physiol.* 279:C1787–C1794.

Hanke, W., and C. Miller. 1983. Single chloride channels from Torpedo electroplax. Activation by protons. *J. Gen. Physiol.* 82:25–45.

Hara-Chikuma, M., Y. Wang, S.E. Guggino, W.B. Guggino, and A.S. Verkman. 2005. Impaired acidification in early endosomes of ClC-5 deficient proximal tubule. *Biochem. Biophys. Res. Commun.* 329:941–946.

Haug, K., M. Warnstedt, A.K. Alekov, T. Sander, A. Ramirez, B. Poser, S. Maljevic, S. Hebeisen, C. Kubisch, J. Rebstock, S. Horvath, K. Hallmann, J.S. Dullinger, B. Rau, F. Haverkamp, S. Beyenburg, H. Schulz, D. Janz, B. Giese, G. Muller-Newen, P. Propping, C.E. Elger, C. Fahlke, H. Lerche, and A. Heils. 2003. Mutations in CLCN2 encoding a voltage-gated chloride channel are associated with idiopathic generalized epilepsies. *Nat. Genet.* 33:527–532.

Hebeisen, S., A. Biela, B. Giese, G. Müller-Newen, P. Hidalgo, and C. Fahlke. 2004. The role of the carboxyl terminus in ClC chloride channel function. *J. Biol. Chem.* 279:13140–13147. Epub 2004 Jan 12.

Hebeisen, S., H. Heidtmann, D. Cosmelli, C. Gonzalez, B. Poser, R. Latorre, O. Alvarez, and C. Fahlke. 2003. Anion permeation in human ClC-4 channels. *Biophys. J.* 84:2306–2318.

Hechenberger, M., B. Schwappach, W.N. Fischer, W.B. Frommer, T.J. Jentsch, and K. Steinmeyer. 1996. A family of putative chloride channels from Arabidopsis and functional complementation of a yeast strain with a CLC gene disruption. *J. Biol. Chem.* 271:33632–33638.

Hille, B. 2001. Ion Channels of Excitable Membranes. Sinauer, Sunderland, MA.

Hinzpeter, A., J. Lipecka, F. Brouillard, M. Baudouin-Legros, M. Dadlez, A. Edelman, and J. Fritsch. 2005. Association between Hsp90nand the ClC-2 chloride channel upregulates channel function. *Am. J. Physiol. Cell Physiol.* 00209.2005.

Hori, K., Y. Takahashi, N. Horikawa, T. Furukawa, K. Tsukada, N. Takeguchi, and H. Sakai. 2004. Is the ClC-2 chloride channel involved in the Cl-secretory mechanism of gastric parietal cells? *FEBS Lett.* 575:105–108.

Hryciw, D.H., J. Ekberg, A. Lee, I.L. Lensink, S. Kumar, W.B. Guggino, D.I. Cook, C.A. Pollock, and P. Poronnik. 2004. Nedd4-2 functionally interacts with ClC-5: Involvement in constitutive albumin endocytosis in proximal tubule cells. *J. Biol. Chem.* 279:54996–55007. Epub 2004 Oct 15.

Hryciw, D.H., G.Y. Rychkov, B.P. Hughes, and A.H. Bretag. 1998. Relevance of the D13 region to the function of the skeletal muscle chloride channel, ClC-1. *J. Biol. Chem.* 273:4304–4307.

Hryciw, D.H., Y. Wang, O. Devuyst, C.A. Pollock, P. Poronnik, and W.B. Guggino. 2003. Cofilin interacts with ClC-5 and regulates albumin uptake in proximal tubule cell lines. *J. Biol. Chem.* 278:40169–40176. Epub 2003 Aug 6.

Humphrey, W., A. Dalke, and K. Schulten. 1996. VMD: Visual molecular dynamics. *J. Mol. Graph.* 14:33–38.

Jentsch, T.J., M. Poet, J.C. Fuhrmann, and A.A. Zdebik. 2005. Physiological functions of CLC Cl channels gleaned from human genetic disease and mouse models. *Ann. Rev. Physiol.* 67:779–807.

Jentsch, T.J., V. Stein, F. Weinreich, and A.A. Zdebik. 2002. Molecular structure and physiological function of chloride channels. *Physiol. Rev.* 82:503–568.

Jentsch, T.J., K. Steinmeyer, and G. Schwarz. 1990. Primary structure of Torpedo marmorata chloride channel isolated by expression cloning in Xenopus oocytes. *Nature* 348:510–514.

Jordt, S.E., and T.J. Jentsch. 1997. Molecular dissection of gating in the ClC-2 chloride channel. *EMBO J.* 16:1582–1592.

Karsdal, M.A., K. Henriksen, M.G. Sorensen, J. Gram, S. Schaller, M.H. Dziegiel, A.M. Heegaard, P. Christophersen, T.J. Martin, C. Christiansen, and J. Bollerslev. 2005. Acidification of the osteoclastic resorption compartment provides insight into the coupling of bone formation to bone resorption. *Am. J. Pathol.* 166:467–476.

Kasper, D., R. Planells-Cases, J.C. Fuhrmann, O. Scheel, O. Zeitz, K. Ruether, A. Schmitt, M. Poet, R. Steinfeld, M. Schweizer, U. Kornak, and T.J. Jentsch. 2005. Loss of the chloride channel ClC-7 leads to lysosomal storage disease and neurodegeneration. *EMBO J.* 24:1079–1091. Epub 2005 Feb 10.

Kieferle, S., P. Fong, M. Bens, A. Vandewalle, and T.J. Jentsch. 1994. Two highly homologous members of the ClC chloride channel family in both rat and human kidney. *Proc. Natl. Acad. Sci. USA* 91:6943–6947.

Koch, M.C., K. Steinmeyer, C. Lorenz, K. Ricker, F. Wolf, M. Otto, B. Zoll, F. Lehmann-Horn, K.H. Grzeschik, and T.J. Jentsch. 1992. The skeletal muscle chloride channel in dominant and recessive human myotonia. *Science* 257:797–800.

Kornak, U., D. Kasper, M.R. Bösl, E. Kaiser, M. Schweizer, A. Schulz, W. Friedrich, G. Delling, and T.J. Jentsch. 2001. Loss of the ClC-7 chloride channel leads to osteopetrosis in mice and man. *Cell* 104:205–215.

Kung, C., and P. Blount. 2004. Channels in microbes: So many holes to fill. *Mol. Microbiol.* 53:373–380.

Li, X., K. Shimada, L.A. Showalter, and S.A. Weinman. 2000. Biophysical properties of ClC-3 differentiate it from swelling-activated chloride channels in Chinese hamster ovary-K1 cells. *J. Biol. Chem.* 275:35994–35998.

Li, X., T. Wang, Z. Zhao, and S.A. Weinman. 2002. The ClC-3 chloride channel promotes acidification of lysosomes in CHO-K1 and Huh-7 cells. *Am. J. Physiol. Cell Physiol.* 282:C1483–C1491.

Liantonio, A., A. Accardi, G. Carbonara, G. Fracchiolla, F. Loiodice, P. Tortorella, S. Traverso, P. Guida, S. Pierno, A. De Luca, D.C. Camerino, and M. Pusch. 2002.

Molecular requisites for drug binding to muscle CLC-1 and renal CLC-K channel revealed by the use of phenoxy-alkyl derivatives of 2-(p-chlorophenoxy)propionic acid. *Mol. Pharmacol.* 62:265–271.

Liantonio, A., M. Pusch, A. Picollo, P. Guida, A. De Luca, S. Pierno, G. Fracchiolla, F. Loiodice, P. Tortorella, and D. Conte Camerino. 2004. Investigations of pharmacologic properties of the renal CLC-K1 chloride channel co-expressed with barttin by the use of 2-(p-Chlorophenoxy)propionic acid derivatives and other structurally unrelated chloride channels blockers. *J. Am. Soc. Nephrol.* 15:13–20.

Lin, C.W., and T.Y. Chen. 2003. Probing the pore of ClC-0 by substituted cysteine accessibility method using methane thiosulfonate reagents. *J. Gen. Physiol.* 122:147–159.

Lin, Y.W., C.W. Lin, and T.Y. Chen. 1999. Elimination of the slow gating of ClC-0 chloride channel by a point mutation. *J. Gen. Physiol.* 114:1–12.

Lipecka, J., M. Bali, A. Thomas, P. Fanen, A. Edelman, and J. Fritsch. 2002. Distribution of ClC-2 chloride channel in rat and human epithelial tissues. *Am. J. Physiol. Cell Physiol.* 282:C805–C816.

Lloyd, S.E., S.H. Pearce, S.E. Fisher, K. Steinmeyer, B. Schwappach, S.J. Scheinman, B. Harding, A. Bolino, M. Devoto, P. Goodyer, S.P. Rigden, O. Wrong, T.J. Jentsch, I.W. Craig, and R.V. Thakker. 1996. A common molecular basis for three inherited kidney stone diseases. *Nature* 379:445–449.

Lorenz, C., M. Pusch, and T.J. Jentsch. 1996. Heteromultimeric CLC chloride channels with novel properties. *Proc. Natl. Acad. Sci. USA* 93:13362–13366.

Ludewig, U., T.J. Jentsch, and M. Pusch. 1997a. Analysis of a protein region involved in permeation and gating of the voltage-gated Torpedo chloride channel ClC-0. *J. Physiol.* 498:691–702.

Ludewig, U., T.J. Jentsch, and M. Pusch. 1997b. Inward rectification in ClC-0 chloride channels caused by mutations in several protein regions. *J. Gen. Physiol.* 110:165–171.

Ludewig, U., M. Pusch, and T.J. Jentsch. 1996. Two physically distinct pores in the dimeric ClC-0 chloride channel. *Nature* 383:340–343.

Maduke, M., D.J. Pheasant, and C. Miller. 1999. High-level expression, functional reconstitution, and quaternary structure of a prokaryotic ClC-type chloride channel. *J. Gen. Physiol.* 114:713–722.

Maduke, M., C. Williams, and C. Miller. 1998. Formation of CLC-0 chloride channels from separated transmembrane and cytoplasmic domains. *Biochemistry.* 37:1315–1321.

Mankodi, A., M.P. Takahashi, H. Jiang, C.L. Beck, W.J. Bowers, R.T. Moxley, S.C. Cannon, and C.A. Thornton. 2002. Expanded CUG repeats trigger aberrant splicing of ClC-1 chloride channel pre-mRNA and hyperexcitability of skeletal muscle in myotonic dystrophy. *Mol.Cell* 10:35–44.

Matsumura, Y., S. Uchida, Y. Kondo, H. Miyazaki, S.B. Ko, A. Hayama, T. Morimoto, W. Liu, M. Arisawa, S. Sasaki, and F. Marumo. 1999. Overt nephrogenic

diabetes insipidus in mice lacking the CLC-K1 chloride channel. *Nat. Genet.* 21:95–98.
Matulef, K., and M. Maduke. 2005. Side-dependent inhibition of a prokaryotic ClC by DIDS. *Biophys. J.* 89:1721–1730.
Middleton, R.E., D.J. Pheasant, and C. Miller. 1996. Homodimeric architecture of a ClC-type chloride ion channel. *Nature* 383:337–340.
Miller, C. 1982. Open-state substructure of single chloride channels from Torpedo electroplax. *Philos. Trans. R. Soc. Lond. B Biol. Sci.* 299:401–411.
Miller, C., and M.M. White. 1980. A voltage-dependent chloride conductance channel from Torpedo electroplax membrane. *Ann. N.Y. Acad. Sci.* 341:534–551.
Miller, C., and M.M. White. 1984. Dimeric structure of single chloride channels from Torpedo electroplax. *Proc. Natl. Acad. Sci. USA* 81:2772–2775.
Miloshevsky, G.V., and P.C. Jordan. 2004. Anion pathway and potential energy profiles along curvilinear bacterial ClC Cl^- pores: Electrostatic effects of charged residues. *Biophys. J.* 86:825–835.
Mindell, J.A., and M. Maduke. 2001. ClC chloride channels. *Genome Biol.* 2:Reviews3003. Epub 2001 Feb 7.
Mindell, J.A., M. Maduke, C. Miller, and N. Grigorieff. 2001. Projection structure of a ClC-type chloride channel at 6.5 A resolution. *Nature* 409:219–223.
Mohammad-Panah, R., R. Harrison, S. Dhani, C. Ackerley, L.J. Huan, Y. Wang, and C.E. Bear. 2003. The chloride channel ClC-4 contributes to endosomal acidification and trafficking. *J. Biol. Chem.* 278:29267–29277.
Muth, T.R., and M.J. Caplan. 2003. Transport protein trafficking in polarized cells. *Ann. Rev. Cell Dev. Biol.* 19:333–366.
Nehrke, K., J. Arreola, H.V. Nguyen, J. Pilato, L. Richardson, G. Okunade, R. Baggs, G.E. Shull, and J.E. Melvin. 2002. Loss of hyperpolarization-activated Cl(−) current in salivary acinar cells from Clcn2 knockout mice. *J. Biol. Chem.* 277:23604–23611. Epub 2002 Apr 25.
Niemeyer, M.I., L.P. Cid, L. Zúñiga, M. Catalán, and F.V. Sepúlveda. 2003. A conserved pore-lining glutamate as a voltage- and chloride-dependent gate in the ClC-2 chloride channel. *J. Physiol.* 553:873–879. Epub 2003 Nov 14.
Niemeyer, M.I., Y.R. Yusef, I. Cornejo, C.A. Flores, F.V. Sepúlveda, and L.P. Cid. 2004. Functional evaluation of human ClC-2 chloride channel mutations associated with idiopathic generalized epilepsies. *Physiol. Genomics* 13:13.
Peña-Münzenmayer, G., M. Catalán, I. Cornejo, C.D. Figueroa, J.E. Melvin, M.I. Niemeyer, L.P. Cid, and F.V. Sepúlveda. 2005. Basolateral localization of native ClC-2 chloride channels in absorptive intestinal epithelial cells and basolateral sorting encoded by a CBS-2 domain di-leucine motif. *J. Cell. Sci.* 118:4243–4252.
Picollo, A., A. Liantonio, M.P. Didonna, L. Elia, D.C. Camerino, and M. Pusch. 2004. Molecular determinants of differential pore blocking of kidney CLC-K chloride channels. *EMBO Rep.* 5:584–589. Epub 2004 May 28.
Picollo, A., and M. Pusch. 2005. Chloride/proton antiporter activity of mammalian CLC proteins ClC-4 and ClC-5. *Nature* 436:420–423.

Piwon, N., W. Günther, M. Schwake, M.R. Bösl, and T.J. Jentsch. 2000. ClC-5 Cl^--channel disruption impairs endocytosis in a mouse model for Dent's disease. *Nature* 408:369–373.
Ponting, C.P. 1997. CBS domains in CIC chloride channels implicated in myotonia and nephrolithiasis (kidney stones). *J. Mol. Med.* 75:160–163.
Pusch, M. 2002. Myotonia caused by mutations in the muscle chloride channel gene CLCN1. *Hum. Mutat.* 19:423–434.
Pusch, M. 2004. Structural insights into chloride and proton-mediated gating of CLC chloride channels. *Biochemistry* 43:1135–1144.
Pusch, M., A. Accardi, A. Liantonio, L. Ferrera, A. De Luca, D.C. Camerino, and F. Conti. 2001. Mechanism of block of single protopores of the Torpedo chloride channel ClC-0 by 2-(p-chlorophenoxy)butyric acid (CPB). *J. Gen. Physiol.* 118:45–62.
Pusch, M., A. Accardi, A. Liantonio, P. Guida, S. Traverso, D.C. Camerino, and F. Conti. 2002. Mechanisms of block of muscle type CLC chloride channels (Review). *Mol. Membr. Biol.* 19:285–292.
Pusch, M., and T.J. Jentsch. 2005. Unique structure and function of chloride transporting CLC proteins. *IEEE Trans. Nanobiosci.* 4:49–57.
Pusch, M., S.E. Jordt, V. Stein, and T.J. Jentsch. 1999. Chloride dependence of hyperpolarization-activated chloride channel gates. *J. Physiol.* 515:341–353.
Pusch, M., A. Liantonio, L. Bertorello, A. Accardi, A. De Luca, S. Pierno, V. Tortorella, and D.C. Camerino. 2000. Pharmacological characterization of chloride channels belonging to the ClC family by the use of chiral clofibric acid derivatives. *Mol. Pharmacol.* 58:498–507.
Pusch, M., U. Ludewig, and T.J. Jentsch. 1997. Temperature dependence of fast and slow gating relaxations of ClC-0 chloride channels. *J. Gen. Physiol.* 109:105–116.
Pusch, M., U. Ludewig, A. Rehfeldt, and T.J. Jentsch. 1995a. Gating of the voltage-dependent chloride channel CIC-0 by the permeant anion. *Nature* 373:527–531.
Pusch, M., K. Steinmeyer, and T.J. Jentsch. 1994. Low single channel conductance of the major skeletal muscle chloride channel, ClC-1. *Biophys. J.* 66:149–152.
Pusch, M., K. Steinmeyer, M.C. Koch, and T.J. Jentsch. 1995b. Mutations in dominant human myotonia congenita drastically alter the voltage dependence of the CIC-1 chloride channel. *Neuron* 15:1455–1463.
Richard, E.A., and C. Miller. 1990. Steady-state coupling of ion-channel conformations to a transmembrane ion gradient. *Science* 247:1208–1210.
Riordan, J.R., J.M. Rommens, B. Kerem, N. Alon, R. Rozmahel, Z. Grzelczak, J. Zielenski, S. Lok, N. Plavsic, J.L. Chou, M.L. Drumm, M.C. Iannuzzi, F.S. Collins, and L.-C. Tsui. 1989. Identification of the cystic fibrosis gene: Cloning and characterization of complementary DNA. *Science* 245:1066–1073.
Rutledge, E., J. Denton, and K. Strange. 2002. Cell cycle- and swelling-induced activation of a Caenorhabditis elegans ClC channel is mediated by CeGLC-7alpha/beta phosphatases. *J. Cell. Biol.* 158:435–444. Epub 2002 Aug 5.

Rychkov, G., M. Pusch, M. Roberts, and A. Bretag. 2001. Interaction of hydrophobic anions with the rat skeletal muscle chloride channel ClC-1: Effects on permeation and gating. *J. Physiol.* 530:379–393.

Rychkov, G.Y., M. Pusch, D.S. Astill, M.L. Roberts, T.J. Jentsch, and A.H. Bretag. 1996. Concentration and pH dependence of skeletal muscle chloride channel ClC-1. *J. Physiol.* 497:423–435.

Rychkov, G.Y., M. Pusch, M.L. Roberts, T.J. Jentsch, and A.H. Bretag. 1998. Permeation and block of the skeletal muscle chloride channel, ClC-1, by foreign anions. *J. Gen. Physiol.* 111:653–665.

Saviane, C., F. Conti, and M. Pusch. 1999. The muscle chloride channel ClC-1 has a double-barreled appearance that is differentially affected in dominant and recessive myotonia. *J. Gen. Physiol.* 113:457–468.

Sayle, R.A., and E.J. Milner-White. 1995. RASMOL: Biomolecular graphics for all. *Trends Biochem. Sci.* 20:374–376.

Schaller, S., K. Henriksen, C. Sveigaard, A.M. Heegaard, N. Helix, M. Stahlhut, M.C. Ovejero, J.V. Johansen, H. Solberg, T.L. Andersen, D. Hougaard, M. Berryman, C.B. Shiodt, B.H. Sorensen, J. Lichtenberg, P. Christophersen, N.T. Foged, J.M. Delaisse, M.T. Engsig, and M.A. Karsdal. 2004. The chloride channel inhibitor NS3736 [corrected] prevents bone resorption in ovariectomized rats without changing bone formation. *J. Bone Miner. Res.* 19:1144–1153.

Scheel, O., A.A. Zdebik, S. Lourdel, and T.J. Jentsch. 2005. Voltage-dependent electrogenic chloride/proton exchange by endosomal CLC proteins. *Nature* 436:424–427.

Schlingmann, K.P., M. Konrad, N. Jeck, P. Waldegger, S.C. Reinalter, M. Holder, H.W. Seyberth, and S. Waldegger. 2004. Salt wasting and deafness resulting from mutations in two chloride channels. *N. Engl. J. Med.* 350:1314–1319.

Schriever, A.M., T. Friedrich, M. Pusch, and T.J. Jentsch. 1999. CLC chloride channels in Caenorhabditis elegans. *J. Biol. Chem.* 274:34238–34244.

Schwappach, B., S. Stobrawa, M. Hechenberger, K. Steinmeyer, and T.J. Jentsch. 1998. Golgi localization and functionally important domains in the NH2 and COOH terminus of the yeast CLC putative chloride channel Gef1p. *J. Biol. Chem.* 273:15110–15118.

Schwiebert, E.M., L.P. Cid-Soto, D. Stafford, M. Carter, C.J. Blaisdell, P.L. Zeitlin, W.B. Guggino, and G.R. Cutting. 1998. Analysis of ClC-2 channels as an alternative pathway for chloride conduction in cystic fibrosis airway cells. *Proc. Natl. Acad. Sci. USA* 95:3879–3884.

Scott, J.W., S.A. Hawley, K.A. Green, M. Anis, G. Stewart, G.A. Scullion, D.G. Norman, and D.G. Hardie. 2004. CBS domains form energy-sensing modules whose binding of adenosine ligands is disrupted by disease mutations. *J. Clin. Invest.* 113:274–284.

Sik, A., R.L. Smith, and T.F. Freund. 2000. Distribution of chloride channel-2-immunoreactive neuronal and astrocytic processes in the hippocampus. *Neuroscience* 101:51–65.

Simon, D.B., R.S. Bindra, T.A. Mansfield, C. Nelson-Williams, E. Mendonca, R. Stone, S. Schurman, A. Nayir, H. Alpay, A. Bakkaloglu, J. Rodriguez-Soriano, J.M. Morales, S.A. Sanjad, C.M. Taylor, D. Pilz, A. Brem, H. Trachtman, W. Griswold, G.A. Richard, E. John, and R.P. Lifton. 1997. Mutations in the chloride channel gene, CLCNKB, cause Bartter's syndrome type III. *Nat. Genet.* 17:171–178.

Simon, D.B., F.E. Karet, J.M. Hamdan, A.D. Pietro, S.A. Sanjad, and R.P. Lifton. 1996a. Bartter's syndrome, hypokalaemic alkalosis with hypercalciuria, is caused by mutations in the Na-K-2CI cotransporter NKCC2. *Nat. Genet.* 13:183–188.

Simon, D.B., F.E. Karet, J. Rodriguez-Soriano, J.H. Hamdan, A. DiPietro, H. Trachtman, S.A. Sanjad, and R.P. Lifton. 1996b. Genetic heterogeneity of Barter's syndrome revealed by mutations in the K^+ channel, ROMK. *Nat. Genet.* 14:152–156.

Staley, K., R. Smith, J. Schaack, C. Wilcox, and T.J. Jentsch. 1996. Alteration of $GABA_A$ receptor function following gene transfer of the CLC-2 chloride channel. *Neuron* 17:543–551.

Steinmeyer, K., R. Klocke, C. Ortland, M. Gronemeier, H. Jockusch, S. Gründer, and T.J. Jentsch. 1991a. Inactivation of muscle chloride channel by transposon insertion in myotonic mice. *Nature* 354:304–308.

Steinmeyer, K., C. Lorenz, M. Pusch, M.C. Koch, and T.J. Jentsch. 1994. Multimeric structure of ClC-1 chloride channel revealed by mutations in dominant myotonia congenita (Thomsen). *EMBO J.* 13:737–743.

Steinmeyer, K., C. Ortland, and T.J. Jentsch. 1991b. Primary structure and functional expression of a developmentally regulated skeletal muscle chloride channel. *Nature* 354:301–304.

Steinmeyer, K., B. Schwappach, M. Bens, A. Vandewalle, and T.J. Jentsch. 1995. Cloning and functional expression of rat CLC-5, a chloride channel related to kidney disease. *J. Biol. Chem.* 270:31172–31177.

Stobrawa, S.M., T. Breiderhoff, S. Takamori, D. Engel, M. Schweizer, A.A. Zdebik, M.R. Bösl, K. Ruether, H. Jahn, A. Draguhn, R. Jahn, and T.J. Jentsch. 2001. Disruption of ClC-3, a chloride channel expressed on synaptic vesicles, leads to a loss of the hippocampus. *Neuron* 29:185–196.

Strange, K. 2003. From genes to integrative physiology: Ion channel and transporter biology in Caenorhabditis elegans. *Physiol. Rev.* 83:377–415.

Thiemann, A., S. Gründer, M. Pusch, and T.J. Jentsch. 1992. A chloride channel widely expressed in epithelial and non-epithelial cells. *Nature* 356:57–60.

Traverso, S., L. Elia, and M. Pusch. 2003. Gating competence of constitutively open CLC-0 mutants revealed by the interaction with a small organic Inhibitor. *J. Gen. Physiol.* 122:295–306.

Uchida, S., S. Sasaki, T. Furukawa, M. Hiraoka, T. Imai, Y. Hirata, and F. Marumo. 1993. Molecular cloning of a chloride channel that is regulated by dehydration and expressed predominantly in kidney medulla. *J. Biol. Chem.* 268:3821–3824.

Uchida, S., S. Sasaki, K. Nitta, K. Uchida, S. Horita, H. Nihei, and F. Marumo. 1995. Localization and functional characterization of rat kidney-specific chloride channel, ClC-K1. *J. Clin. Invest.* 95:104–113.

Vandewalle, A., F. Cluzeaud, M. Bens, S. Kieferle, K. Steinmeyer, and T.J. Jentsch. 1997. Localization and induction by dehydration of ClC-K chloride channels in the rat kidney. *Am. J. Physiol.* 272:F678–F688.

Vanoye, C.G., and A.L. George Jr. 2002. Functional characterization of recombinant human ClC-4 chloride channels in cultured mammalian cells. *J. Physiol.* 539:373–383.

Waldegger, S., and T.J. Jentsch. 2000. Functional and structural analysis of ClC-K chloride channels involved in renal disease. *J. Biol. Chem.* 275:24527–24533.

Wang, S.S., O. Devuyst, P.J. Courtoy, X.T. Wang, H. Wang, Y. Wang, R.V. Thakker, S. Guggino, and W.B. Guggino. 2000. Mice lacking renal chloride channel, CLC-5, are a model for Dent's disease, a nephrolithiasis disorder associated with defective receptor-mediated endocytosis. *Hum. Mol. Genet.* 9:2937–2945.

Weinreich, F., and T.J. Jentsch. 2001. Pores formed by single subunits in mixed dimers of different CLC chloride channels. *J. Biol. Chem.* 276:2347–2353.

Yin, J., Z. Kuang, U. Mahankali, and T.L. Beck. 2004. Ion transit pathways and gating in ClC chloride channels. *Proteins* 57:414–421.

Zdebik, A.A., J.E. Cuffe, M. Bertog, C. Korbmacher, and T.J. Jentsch. 2004. Additional disruption of the ClC-2 Cl^- channel does not exacerbate the cystic fibrosis phenotype of cystic fibrosis transmembrane conductance regulator mouse models. *J. Biol. Chem.* 279:22276–22283.

Zheng, Y.J., T. Furukawa, T. Ogura, K. Tajimi, and N. Inagaki. 2002. M phase-specific expression and phosphorylation-dependent ubiquitination of the ClC-2 channel. *J. Biol. Chem.* 277:32268–32273. Epub 2002 Jun 24.

Zúñiga, L., M.I. Niemeyer, D. Varela, M. Catalán, L.P. Cid, and F.V. Sepúlveda. 2004. The voltage-dependent ClC-2 chloride channel has a dual gating mechanism. *J. Physiol.* 555:671–682. Epub 2004 Jan 14.

9 Ligand-Gated Ion Channels: Permeation and Activation[1]

Joseph W. Lynch and Peter H. Barry

9.1 Introduction

Ligand-gated ion channels (LGICs) are fast-responding channels in which the receptor, which binds the activating molecule (the ligand), and the ion channel are part of the same nanomolecular protein complex. This chapter will describe the properties and functions of the nicotinic acetylcholine LGIC superfamily, which play a critical role in the fast chemical transmission of electrical signals between nerve cells at synapses and between nerve and muscle cells at endplates. All the processing functions of the brain and the resulting behavioral output depend on chemical transmission across such neuronal interconnections. To describe the properties of the channels of this LGIC superfamily, we will mainly use two examples of this family of channels: the excitatory nicotinic acetylcholine receptor (nAChR) and the inhibitory glycine receptor (GlyR) channels. In the chemical transmission of electrical signals, the arrival of an electrical signal at the synaptic terminal of a nerve causes the release of a chemical signal—a neurotransmitter molecule (the ligand, also referred to as the agonist). The neurotransmitter rapidly diffuses across the very narrow 20–40 nm synaptic gap between the cells and binds to the LGIC receptors in the membrane of the target (postsynaptic) cell and generates a new electrical signal in that cell (e.g., Kandel et al., 2000). How this chemical signal is converted into an electrical one depends on the fundamental properties of LGICs and the ionic composition of the postsynaptic cell and its external solution.

The LGICs are small highly specialized protein complexes, about 12 nm long and 8 nm in diameter, which span the 3 nm or so lipid bilayer membranes of the nerve or muscle cells (see Fig. 9.1). Many of these fast neurotransmitter channels belong to the nicotinic acetylcholine superfamily of LGICs, which are also referred to as the Cys-loop superfamily of LGICs (because of the presence of a conserved signature Cys-loop, loop 7, which is present in all family members, as shown for the GlyR in Fig. 9.2, in Section 9.2). Although other ligand-gated ion channels exist, we will be confining ourselves to this superfamily, whose members are all closely

---

[1] This chapter is a revised and expanded version of Barry and Lynch (2005), and portions of that paper are reprinted, with permission, from *IEEE Trans. Nanobiosci.*, vol. 4, no. 1, pp. 70–80, Mar. 2005 IEEE.

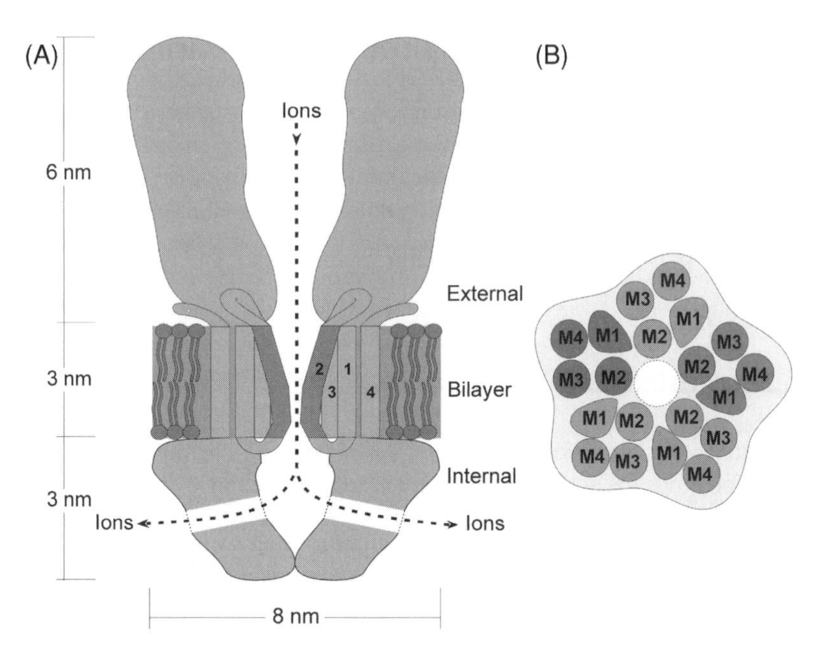

Fig. 9.1 Schematic longitudinal and cross-sectional diagrams of a typical LGIC. (Originally modified from Fig. 1 of Keramidas et al., 2004, and based on Unwin, 1993; Miyazawa et al., 1999; and Brejc et al., 2001). Panel A shows two of the five subunits and the pathway for ions entering the exterior end of the channel and moving into the cell interior via lateral portals at the cytoplasmic end of the channel, as in the nAChR channel. The numbers 1–4 refer to the M1–M4 segments for each subunit. Panel B shows a cross-sectional view of the LGIC with the four transmembrane segments of each of the five subunits. It also illustrates the five M2 segments, lining the pore region of the channel within the membrane lipid bilayer. The figure has been reproduced in monochrome from Fig. 1 of Barry and Lynch (2005), with copyright permission of [2005] IEEE.

related genetically and are very similar structurally and mechanistically, though their precise physiological function can be quite different.

When the appropriate chemical neurotransmitter, the ligand, binds to the LGIC, which incorporates both the receptor and the ion channel, the action of the ligand binding to its docking site on the receptor can somehow then cause the channel to open. The open channel selectively allows certain species of ions to pass from one side of the cell membrane to the other, through the channel. The magnitude and direction of the resulting current depends on the signs of the permeating ions and their electrochemical potential energy gradient across the membrane. The LGICs can exist in three main states. They are normally closed in the absence of the binding of any ligand, will very rapidly open within about 20 μs of the ligand binding and will close when the ligand dissociates from the receptor. However, in the continued presence of a relatively high concentration of ligand, they can also go into a third desensitized, nonconducting, state, while the ligand is still bound. This is functionally similar to the closed state except that it cannot be opened by the further addition of ligand molecules. In the desensitized state, the LGIC is unable to be reactivated until the ligand has dissociated from the receptor and the channel has returned to the closed state.

The four main LGIC types found in vertebrates, and named according to their endogenous agonist, are: the nicotinic acetylcholine receptor (nAChR) channel (so named, because the subgroup of acetylcholine LGICs can selectively be also activated by nicotine), the 5-hydroxytryptamine type 3 receptor (5-HT_3R) channel, the γ-aminobutyric acid receptor channels, types A and C ($GABA_AR$ and $GABA_CR$, respectively), and the strychnine-sensitive glycine receptor (GlyR) channel.

To understand how the current passing through an LGIC is generated, consider, for example, either the nAChR or the 5-HT_3R channels, which are predominantly selective to monovalent cations, mainly Na^+ and K^+ ions. For further general details on some of these electrophysiological principles see, for example, Aidley and Stanfield (1996) and Hille (2001). The 'driving force' per mole of ions for ion movement through a membrane is proportional to the electrochemical potential energy difference across that membrane. The electrochemical potential energy ($\bar{\mu}$), which is the free energy per mole of ions, at a particular point in a solution or channel is given by:

$$\bar{\mu} = \mu_o + RT \ \ln a + zFV, \tag{9.1}$$

where μ_o represents the standard state potential, which will be the same in the two aqueous solutions on either side of the membrane and where a hydrostatic term has been omitted, since it is usually negligible for ions in comparison to the electrical and activity terms. R, T, and F have their usual significance as gas constant, temperature in K, and Faraday constant; z and a represent the valency and activity ($\approx$ concentration in dilute solutions) of the appropriate ion under consideration, and V represents the electrical potential at that point.

In the absence of any large hydrostatic pressure difference across the membrane, the electrochemical potential energy difference across a membrane, between two aqueous solutions, for a mole of ions of species 'j', may readily be shown to be given by:

$$\Delta\bar{\mu}_j = \bar{\mu}_j^i - \bar{\mu}_j^o = zF[V_m - V_j], \tag{9.2}$$

where superscripts 'i' and 'o' refer to the internal and external solutions respectively, V_m is defined by electrophysiologists as the membrane potential, with the convention that it is the potential of the cell interior relative to the outside solution, and V_j is defined as:

$$V_j = (RT/z_jF) \ \ln\left(a_j^o/a_j^i\right), \tag{9.3}$$

V_j has the dimensions of electrical potential and represents the membrane potential that would just balance the energy due to the ionic activity difference across the membrane for those ions. Hence, V_j is referred to as the equilibrium (Nernst) potential for ion j and represents the potential at which those ions would be in equilibrium. The driving force for the movement of ion j across a membrane would be given by $(V_m - V_j)$.

For example, for Na^+, the equilibrium potential, V_{Na}, would be given by:

$$V_{Na} = (RT/F)\ \ln\left(a^{o}_{Na}/a^{i}_{Na}\right), \tag{9.4}$$

since $z_{Na} = 1$. There will be a corresponding equilibrium potential for K^+ in terms of its activities similar to Eq. 9.4 with "Na" replaced by "K". The driving force for Na^+ ion movement is therefore given by $(V_m - V_{Na})$. Nerve and muscle cells, like most other cells under *in vivo* conditions, have a very large inward driving force on Na^+ ions ($V_m \ll 0$, $V_{Na} \gg 0$; so that $V_m - V_{Na}$ is very much less than 0) and a small outward driving force on K^+ ions (V_m is normally just a little greater than V_K, so that $V_m - V_K$ is only just a little greater than 0) at typical membrane potentials (–70 to –90 mV). Hence, when nAChR or 5-HT$_3$R channels open, there is a net influx of cations (inward current), due to a large influx of Na^+ ions and only a small efflux of K^+ ions. This net influx of positive ions tends to drive the membrane potential positive (to depolarize the cell) and, if large enough, can exceed the threshold for initiating an electrical impulse, an action potential, in the postsynaptic cell. The nAChR and 5-HT$_3$R channels are therefore excitatory. In particular, the nAChR channel in muscle cells plays a major role in neuromuscular transmission. In response to a nerve impulse traveling down a motor nerve, there is a release of the neurotransmitter, acetylcholine (ACh) molecules, from the nerve ending. The ACh molecules bind to the nAChRs on the muscle cell membrane, depolarize the muscle cell and normally initiate an action potential in that cell, which, in turn, leads to the muscle contracting. One interesting competitive blocker of the nAChR is the plant alkaloid, curare (D-tubocurarine), used as a paralytic poison in the Amazon and also as a muscle relaxant in surgery (e.g., Hille, 2001). Also, neuronal nAChRs are involved in nicotine addiction (Cordero-Erausquin et al., 2000). The 5-HT$_3$R is known to be involved in sensory processing, including pain reception and aspects of motor control. Selective 5-HT$_3$R antagonists are also used in clinical situations as anti-emetic agents to reduce nausea and vomiting (Conley, 1996).

In contrast to the above two excitatory LGICs, the GABA$_A$R, GABA$_C$R, and GlyR channels are predominantly selective to anions, such as Cl^- ions, the major anion in the external solution bathing animal cells. In most, but not all cells, because of the usual very low internal concentration of Cl^- ions, with its concentration gradient more than compensating for the negative membrane potential, there is a net inward driving force $[(V_m - V_{Cl}) > 0]$. Therefore, when GABA$_A$R and GlyR channels open, there is a net influx of Cl^- ions (outward current), which drives the membrane potential more negative (or provides a conductance shunt), and as a result the effectiveness of any concurrent excitatory signals would be reduced, so that such an excitatory signal may be unable to depolarize the membrane potential sufficiently to initiate an action potential. The GABA$_A$Rs, GABA$_C$Rs, and GlyRs are therefore normally inhibitory. These inhibitory channels play a major role in the processing of sensory and motor signals in the central nervous system. Bicucculine and picrotoxin, which block these inhibitory $GABA_AR$ channels, can cause convulsions. In contrast, benzodiazepines (e.g., valium) or barbiturates, which increase the response of $GABA_AR$ to GABA, can have a sedative or calming effect (Mohler et al., 2000). Similarly, GlyR channels are essential, in addition to other functions, for modulating reflex responses important in the maintenance of posture. For example, low doses of the plant alkaloid, strychnine, which can antagonize the GlyR response to glycine, can produce an increased responsiveness to sensory stimuli, whereas high doses can result in exaggerated reflexes, an absence of muscle control, convulsions, and then death (Schofield et al., 1996). There is one particular genetic disease, startle disease, or hyperekplexia, in which the GlyR channel is defective (Schofield et al., 1996). If sufferers of hyperekplexia are suddenly startled, the reduction of this inhibitory modulation can lead to a greatly exaggerated reflex response, which can result in them becoming rigid and falling over.

9.2 Physicochemical Structure

9.2.1 Overall Structure of LGICs

The above LGIC complexes are all pentameric, comprised of five similar subunits, which form a barrel-like structure (Fig. 9.1), with a channel pore running through the middle of the complex. Each subunit is comprised of four transmembrane segments (M1, M2, M3, and M4), as shown in Fig. 9.2, and as indicated in Fig. 9.1B, the channel pore is known to be lined by the M2 segments. There is a large N-terminal hydrophilic extracellular domain, which contains the main ligand-binding region in some subunits. This extracellular domain is connected to the M1 domain. There is a short intracellular loop connecting M1 to M2 and a short extracellular loop connecting M2 to M3 (Fig. 9.2). A long intracellular loop connecting M3 to M4 is thought to be associated with cytoskeletal proteins within the cell (such as rapsyn in the nAChR and gephyrin in the GlyR) that acts as a support structure to control the clustering of the receptors at appropriate regions of the membrane. The five nAChR subunits are α, β, γ, δ, and ε, with some subunits possessing a number of variants ($\alpha1$–$\alpha10$, $\beta1$–$\beta4$; Alexander et al., 2004). The embryonic muscle nAChR and the *Torpedo* organ nAChR have a stoichiometry of 2α:β:γ:δ (with two α and one each of β, γ, and δ subunits; Alexander et al., 2004; Miyazawa et al., 2003). In addition, the ligand-binding region is generally located at the interfaces between the α subunit and the γ and δ subunits. Similarly, GlyRs can be heteromeric with a stoichiometry of 2α:3β subunits (Grudzinska et al., 2005), or homomeric with five identical α subunits. Other LGICs known to exist as homomers are some neuronal $\alpha7$ nAChRs and the $\rho1$ $GABA_CR$.

The complete amino acid sequences of each of the subunits of these LGICs are known and there is a high degree of sequence identity and similarity between them in the different members of this LGIC family.

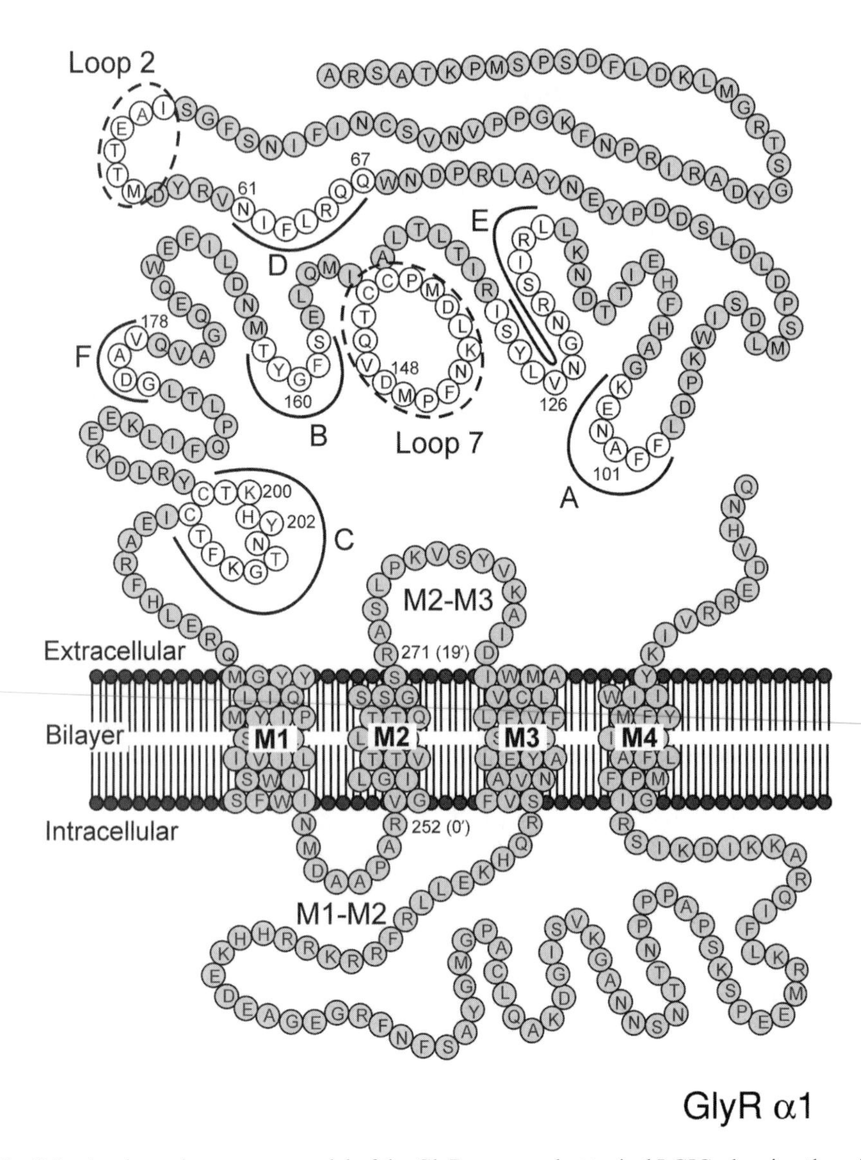

Fig. 9.2 A schematic sequence model of the GlyR, a somewhat typical LGIC, showing the principal ligand-binding domains (A–C) and complementary ligand-binding domains (D–F). These binding domains are indicated with a solid curved line and the residues involved are shown in white. Two of the extracellular loops, considered to be important in signal transduction (loops 2 and 7), are also shown with dashed ellipses, with included resides shown in white. It should be noted that the physical position of some of these extracellular loops, in particular, is likely to be different from that in this illustration, with, for example, loops 2 and 7 probably both close to the transmembrane domain. The transmembrane segments M1–M4 are shown, with the M2–M3 and M1–M2 loops labeled. Redrawn and modified from Fig. 2 of Schofield et al. (1996) to incorporate later data from Brejc et al. (2001) on ligand-binding domains and loops 2 and 7.

In spite of the fact that until recently (see below) there has been no information about the crystal structure of the nAChR, much work has been done to build up a 3-D physical picture of these LGICs using other methods. Nigel Unwin and his colleagues have been extremely successful in such efforts with the nAChRs, using special freezing techniques to drop membrane tubes of densely-packed helically arranged nAChRs from the electric ray *Torpedo* into liquid ethane at below $-160°C$ to "freeze" the nAChRs prior to making electron micrographic measurements (Unwin, 1993). In later experiments, the tubes were sometimes sprayed with a mist of ACh as they were being dropped into the liquid ethane to freeze the channels in the open state prior to the electron microscopy (Unwin, 2003). In both cases, the electron micrographs were scanned with a densitometer, averaged and used to build up optical density arrays of the nAChR channel. The differences between the images in the open and closed state have given some very useful information about the mechanism of channel opening, which will be discussed further in Section 9.4. Computational Fourier techniques were then used to build up a 3-D structure of the nAChR from the different views of the above arrays (Unwin, 2000; Miyazawa et al., 2003). Figure 9.3A shows a reconstructed 2-D image from 0.9 nm resolution electron micrographs, with the suggestion of a kinked α-helical M2 region lining the channel pore within the membrane bilayer. Figure 9.3B and C shows sections of a higher 0.46 nm resolution 3-D image of the bottom cytoplasmic region of the nAChR channel showing the lateral portals or windows between the channel pore and the cell cytoplasm, which have been suggested to at least affect channel conductance in some LGICs.

Recently, a soluble acetylcholine-binding protein (AChBP) from a freshwater snail has been found which shares similar pharmacological properties to the $\alpha 7$ homomeric nAChR. Although lacking the transmembrane (TM) domains, it does share an approximately 20% amino acid sequence identity with the ligand-binding domain of the nAChR and incorporates the LGIC signature Cys-loop (loop 7 in Fig. 9.2). A recent study that replaced the ligand-binding domain of a 5-HT_3R with AChBP revealed that AChBP does indeed replicate the function of a real ligand-binding domain (Bouzat et al., 2004). The crystal structure of AChBP, solved by Sixma and colleagues in 2001 (Brejc et al., 2001), reconciles many years of biochemical and electrophysiological investigations into the LGIC family and is thus considered an accurate template of the LGIC ligand-binding domain. As shown in Fig. 9.4, it consists of five identical subunits arranged symmetrically around a central water-filled vestibule. Each subunit comprises 10 β-sheets arranged in a novel immunoglobulin fold with no significant match to any known protein structure. The dimensions of the AChBP are similar to those previously determined from electron diffraction images of *Torpedo* nAChRs (Miyazawa et al., 1999). Agonist-binding pockets are present at the subunit interfaces, approximately midway between the top and bottom of the protein, and abundant evidence (reviewed in Corringer et al., 2000) identifies these as ligand-binding sites. Three loops (Domains A, B, and C) form the "principal" ligand-binding surface on the one side of the interface and three β-strands (Domains D, E, and F) from the adjacent subunit comprise the "complementary"

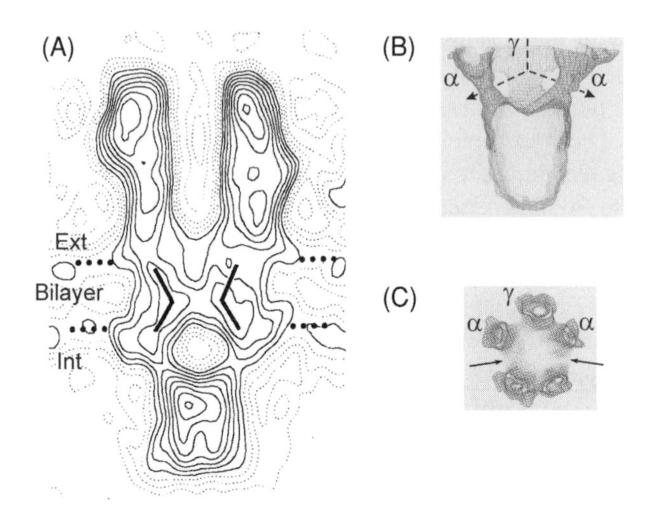

Fig. 9.3 Reconstructed images derived from cryo-electron micrographic images of an nAChR channel. Panel A shows an early reconstruction, at 0.9 nm resolution, of a longitudinal view of the channel, with the black lines shown to depict two kinked-regions of the M2 α-helical parts of two subunits. The position of the membrane bilayer is shown as between the two rows of filled circles and 'Ext' and 'Int' refer to the extracellular and intracellular bilayer interfaces respectively. Panel B, with the higher level of resolution now available, shows part of the very bottom cytoplasmic region of the nAChR channel, at 0.46 nm resolution. This illustrates two of the five lateral portals (tunnels/windows) in the walls of the cavity at the end of the channel pore between the rod-like protrusions at the bottom of the subunits. The rods from the α and γ subunits are shown and the added dashed lines indicate two of the pathways for current flow through the lateral portals. Panel C shows the five rod-like structures in cross-section. The positions of the two largest lateral portals are shown with arrows. Panel A was originally reprinted from Fig. 13a of Unwin (1993) and Panels B and C from part of Fig. 12a and Fig. 12b of Miyazawa et al. (1999) with permission from Elsevier, and the whole figure is now reproduced from Fig. 2 of Barry and Lynch (2005), with copyright permission of [2005] IEEE.

face. Figure 9.4B shows a side-on view of one interface formed between adjacent subunits. The ligand-binding domains are shown for the GlyR in Fig. 9.2.

The structure of the nAChR TM domains was determined by cryo-electron microscopy to a resolution of 0.4 nm by Miyazawa et al. (2003). This structure, depicted later in Fig. 9.11 (in Section 9.4), confirms the long-held view that the TM domains comprise a cluster of four α-helical domains. A water-filled crevice, continuous with the extracellular fluid, is formed between the M1, M2, and M3 domains of each subunit. Abundant evidence identifies this crevice as an alcohol and volatile anesthetic binding site in the $GABA_AR$ and GlyR (Lynch, 2004). The Miyazawa paper did not resolve the structure of the large intracellular loop linking M3 and M4, although it has now been resolved to some degree by Unwin (2005). This domain, which varies considerably in amino acid identity and length among LGIC members, contains phosphorylation sites and other sites responsible for mediating interactions with cytoplasmic factors.

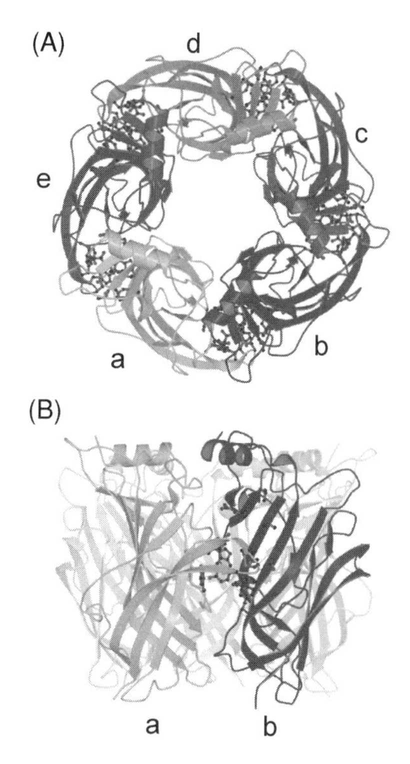

Fig. 9.4 AChBP viewed along the fivefold axis of symmetry toward the membrane. (A) Each subunit is identical, earlier shown in a different color and labeled as a–e. (B) Side view of the same structure with only two subunits (a and b) displayed. The principal ligand-binding domains are on the left side (a) of this interface and the likely ligand-binding residues are shown in ball-and-stick representation. This figure was originally reproduced from Fig. 2 of Brejc et al. (2001), by copyright permission from the Nature Publishing Group, and is now reproduced in monochrome from Fig. 3 of Barry and Lynch (2005), with copyright permission of [2005] IEEE.

9.2.2 Structure of the Channel Pathway

Figure 9.1 shows the pathway for ion movement through a cation-selective channel, like the nAChR, which has opened as a result of the binding of ACh. For an nAChR at a normal resting membrane potential, Na^+ ions will enter the channel vestibule from outside the cell, pass down through a narrow region, which is known as the selectivity filter, before entering a cavity and then passing out into the cytoplasm of the cell via lateral portals (see Fig. 9.3B). K^+ ions will normally move in the opposite direction, because of their oppositely directed driving force.

9.2.3 Other Information from Amino Acid Sequences

In addition to the above structural data, a significant amount of information has been determined from the known amino acid sequences of the subunits of the LGICs.

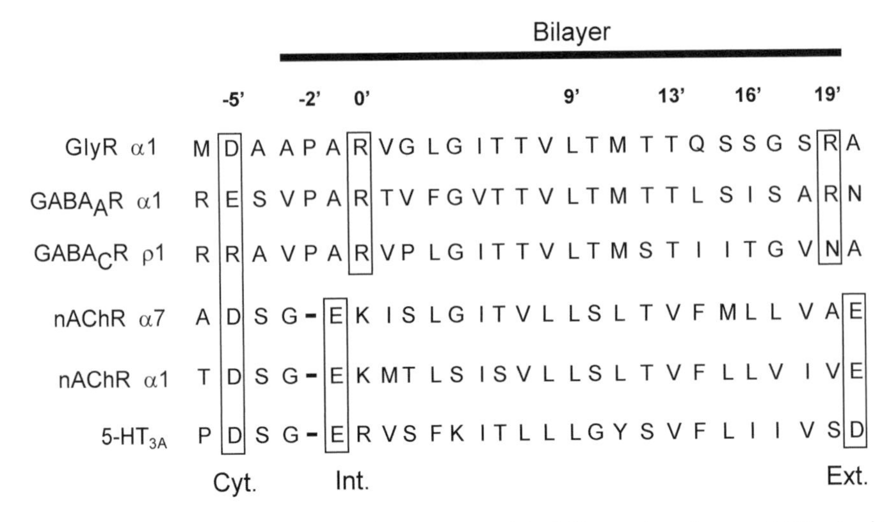

Fig. 9.5 A comparison of the amino acid sequences of the M2 domains of some LGIC subunits, giving their single-letter amino acid codes. The top bar shows the approximate position of the bilayer in relation to this region of the pore. The boxed regions outline many of the critical charged amino acids, the aspartate (D) and glutamate (E) residues being negative and the arginine (R) and the lysine (K) residues being positive. Cyt., Int., and Ext. refer to the cytoplasmic, intracellular and extracellular rings of charge respectively. This is a modified version of Fig. 3 of Keramidas et al. (2004), where details of the references for the sequences of each of the LGICs may be found. The set of numbers with primes represents a convenient standard relative numbering system for the M2 regions of the various subunits, arbitrarily defined with respect to the arginine or lysine residues, in the region of the intracellular charged ring, as being at position 0′. Based on Fig. 3A of Keramidas et al. (2004). This figure has been reproduced from Fig. 4 of Barry and Lynch (2005), with permission of [2005] IEEE.

It was from hydropathy plots of the amino acid sequence of each subunit protein that the presence of the TM domains (M1–M4; see Fig. 9.2) was identified on the basis of the nonaqueous solubility of these regions (e.g., Hille, 2001). Together with our knowledge of the properties of the amino acid residues, data from crystal structure inferred from that of the AChBP (discussed earlier), and structure–function studies to be discussed in detail later, our picture of the physical structure of these LGICs is being considerably extended. For example, if we compare the amino acid sequences of the M2 domains of the different LGICs, as shown in Fig. 9.5, it is clear that the amino acid sequences of the M2 pore regions for each of the main LGIC members are very similar, particularly with respect to the charged regions. Both the cation-selective nAChR and 5-HT$_3$R channels have three rings of negative charge. Each ring of charge arises because there is a negative amino acid residue at the same position in the M2 domains of each of the five subunits (e.g., Fig. 9.5). When the subunits assemble to form a channel, these charged residues from each subunit face in toward the channel pore and form a ring of charged residues in the pore (see Fig. 9.7, later). However, the middle, intracellular, ring also has adjacent positive amino acids, but presumably these are set back somewhat from the surface of the channel pore and therefore do not contribute significantly to the electrical potential in the pore. In contrast to the above LGICs, the anion-selective GlyR and $GABA_A$R channels have two rings (intracellular and extracellular) of positive charge, and a negatively charged cytoplasmic ring, which evidence suggests is buried in the protein (Section 9.3.2). In addition, from other amino acid sequence evidence for the nAChR, there are also known to be negative charges lining the cytoplasmic portals (Fig. 9.3), whereas for the GlyR channel there are positive charges in the portal region.

An obvious question is: "What role do all these charged amino acids play in determining ion permeation and selectivity?" This will be discussed in detail in Section 9.3.

9.2.4 Structure–Function Studies

The development of two major technologies has enabled very significant advances to be made in our knowledge of the relationship between the molecular structure of these channels and their functional properties, which in turn has led to a greater understanding of their underlying functional mechanisms.

The first of these technologies has been the patch-clamp technique (Hamill et al., 1981; see the discussion in Jordan, 2006; Chapter 1 in this volume). With such techniques, the current of a single ion channel can be directly measured in pA (10^{-12} A), the single-channel conductance determined in pS (10^{-12} S), and the duration of the average open time of the channel determined in ms. Alternatively, using such techniques, the total current of all the open channels in a very small cell can also now be accurately determined.

The second of these technologies has been molecular biology, and, in particular, site-directed mutagenesis, so that individual (or groups of) amino acids can be deleted, inserted or mutated, with the mutant channels then being expressed in tissue-cultured cells. The effect of such mutations on the electrophysiological properties of these specific channels can then be investigated.

The main part of this chapter will now concentrate on how such techniques have been used to determine how ion permeation and selectivity is controlled in these LGICs (Section 9.3) and how these channels are opened in response to the binding of ligands (Section 9.4).

9.3 Ion Conductances, Permeation and Selectivity

As noted earlier, the cation selective nAChRs and 5-HT_3Rs have three negative rings lining their channel pores, whereas the anion selective $GABA_A$Rs, $GABA_C$Rs, and GlyRs have two positively charged rings lining their pores. The effect of such charged rings on single-channel conductances, ion permeation, and selectivity will now be discussed.

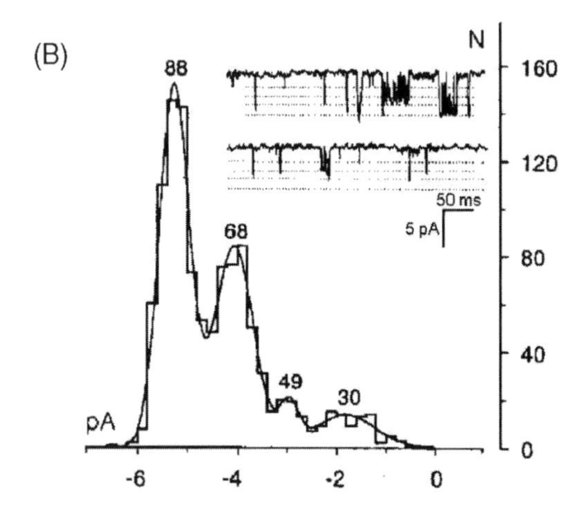

Fig. 9.6 Panel A shows a current record of a toad muscle nAChR channel showing clearly defined transitions between closed and open states (originally modified from a current record of *N*. Quartararo in Fig. 7 of Barry and Gage, 1984) at a potential of −70 mV. Panel B shows current records of normal (wild type; WT) α1 GlyR channels at a potential of −55 mV with a histogram of the current amplitude distribution in pA of the current records (*N* being the frequency for the ordinate), to give the multiple conductance levels in pS at each fitted Gaussian peak (originally a modified version of Fig. 2A of Rajendra et al., 1995). This figure has been reproduced from Fig. 5 of Barry and Lynch (2005), with copyright permission of [2005] IEEE.

9.3.1 Conductances

Patch clamp measurements indicated typical single-channel conductances in the range of about 25–50 pS for neuronal nAChRs (e.g., Conley, 1996) with clear well-defined channel openings with a predominant main conductance level, infrequent sub-conductance levels (e.g., Fig. 9.6A), and typical open times of 5–10 ms or longer. Maximum conductance levels of about 50 pS were observed for heteromeric $\alpha_1\beta$ GlyRs (in human embryonic kidney, HEK, cells) with much shorter (flickery) openings, tending to occur in bursts, and displaying multiple sub-conductance states (Bormann et al., 1993). The main conductance level was dependent on subunit composition and increased to around 90 pS in homomeric α_1 GlyRs (Bormann et al., 1993; cf. Fig. 9.6B, and Fig. 2A of Rajendra et al., 1995).

To investigate the role of such charged rings on conductance in the nAChR channel, site-directed mutagenesis was used to change the charge on each ring, by mutating the appropriate residues in the different subunits (e.g., neutralizing a negative charge on the two α subunits, changes the total ring charge by +2 charges, whereas neutralizing it on the single β subunit changes it by +1 charge; Imoto

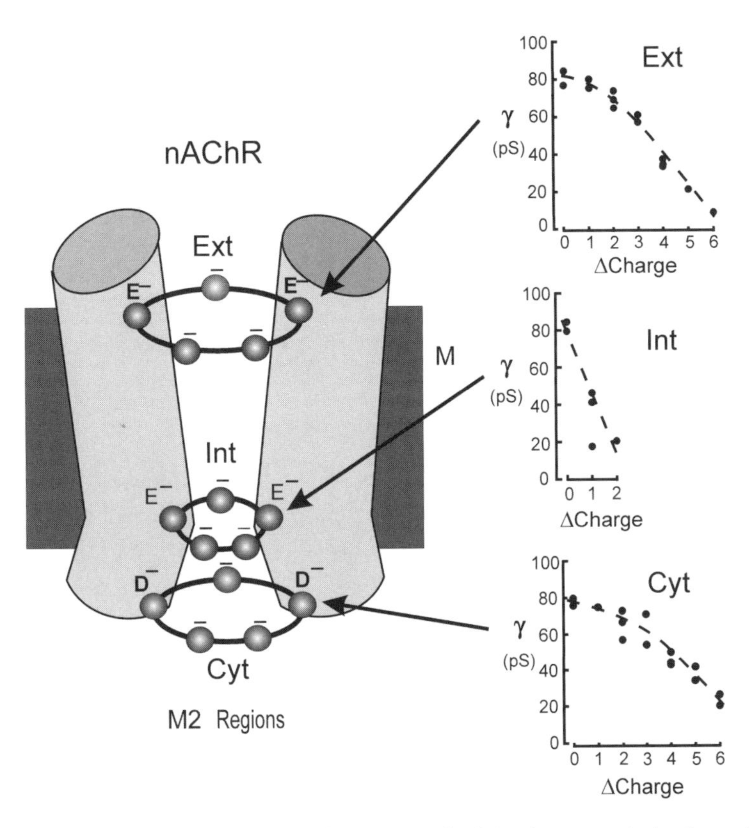

Fig. 9.7 The effect of changing the net charge on each of the three negatively charged rings in the nAChR channel on the single-channel conductance, γ, in pS (10^{-12} S). ΔCharge represents the change in units of electronic charge, the zero value representing the WT nAChR channel. Ext., Int., and Cyt. represent the extracellular, intracellular, and cytoplasmic rings of negative charge, contributed by each charged residue on each of the five subunits (only two of the five M2 helical domains are shown in the schematic diagram), and M indicates the position of the membrane. The three graphs were originally redrawn from Figs. 2d, 2e, and 3b of Imoto et al. (1988) and the figure has been reproduced in monochrome from Fig. 6 of Barry and Lynch (2005), with copyright permission of [2005] IEEE.

et al., 1988). Figure 9.7 shows the graphs of single-channel conductance against the change in charge for various combinations of mutations. The conductance changes are seen to be especially sensitive to changes in net charge at the intermediate ring, where a change of just +2 charges is enough to drop the conductance to 20% of its normal (wild type; WT) value (Fig. 9.7). Similar reductions in single-channel conductance were seen for homomeric α1 GlyR channels, when the five positive charges at the extracellular ring were neutralized (Langosch et al., 1994; Rajendra et al., 1995), though the GlyR channel remained anion-selective. Later experiments (Section 9.3.2) to change the charge on the extracellular ring of a cation-selective GlyR mutant channel showed that the charge on these rings also controlled rectification. A positive extracellular ring of charge produced outward rectification and a negative one inward rectification (Moorhouse et al., 2002). These charge effects on rectification are discussed further in Keramidas et al. (2004).

It has also been suggested that charges lining the lateral portals at the cytoplasmic end of the channels (Fig. 9.3) may contribute to the channel conductance. In support of this, a series of positive arginines were identified in this region (in the M3–M4 loop) in the 5-HT_3R, which when neutralized or replaced by a negative aspartate, radically increased the inward cation conductance from below 1 pS to about 20 pS (Kelley et al., 2003).

In the $GABA_A$R channel, diazepam and pentobarbitone can also act by increasing single-channel conductance (Eghbali et al., 1997).

9.3.2 Permeation and Selectivity

In the nAChR channel, it has been shown that selectivity between alkali cations could be altered by changing one specific polar residue in the M2 region of the α subunit (Position 2′; see Fig. 9.5) with residues of varying volume (Villarroel et al., 1991). For example, decreasing the volume of the residue to glycine increased the permeability of the channel to the larger Rb^+ relative to the smaller Na^+.

The question still remained: What factors actually determine anion-to-cation selectivity? In an extensive series of mutations, following a comparison of the M2 sequences between the different LGIC members (Fig. 9.5), it has been shown that a minimum of three mutations were able to convert the homomeric α7 nAChR from being cation permeable to being anion permeable (Galzi et al., 1992). The valine at position 13′ was mutated to a polar threonine (V13′T), the glutamate at −1′ was neutralized with an alanine (E-1′A) and a proline was inserted at −2′ (−2′P) (cf. Fig. 9.5). By itself, the glutamate neutralization did remove the calcium permeability (P_{Ca}/P_{Na} decreased from about 10 to <0.02), although the channel was still cation selective. Also by itself, the mutation V13′T actually increased P_{Ca}/P_{Na} and decreased P_{Na}/P_K, and seemed to suggest that an additional desensitized state had become conducting. The precise position of the proline insertion was not absolutely critical, since, for example, if it was inserted at position −4′ instead of at −2′ (along with E-1′A and V13′T), the channel again became anion selective (Corringer et al., 1999). The minimum requirement of the above three mutations, including a proline insertion, suggested a conformational change. The need for such a conformational change seemed to make the precise mechanism for selectivity conversion somewhat unclear (see Galzi et al., 1992; Corringer et al., 1999).

The next LGIC to have its ion charge selectivity inverted was the GlyR (Keramidas et al., 2000). The reverse of the three mutations in the nAChR was introduced at the equivalent locations in the GlyR. The mutations were P-2′Δ, A-1′E, and T13′V (where Δ = deletion). These mutations did convert the GlyR channel from being anion-selective ($P_{Cl}/P_{Na} = 25$) to being cation-selective ($P_{Cl}/P_{Na} = 0.27$). However, the resultant currents were extremely small and brief (requiring noise analysis to determine their single-channel conductance values of 3 pS for inward and 11 pS for outward currents; Keramidas et al., 2000). In addition, the amount of glycine required as an agonist to ensure a maximum current response was very high (100 mM compared to <0.1 mM for WT GlyRs; Keramidas et al., 2002). The relative permeabilities were determined by doing experiments to dilute the external NaCl concentrations to approximately 50% and 25% of their control values and using the Goldman–Hodgkin–Katz equation to determine the permeability ratios from the voltage required for zero agonist generated current (see discussion in Keramidas et al., 2004). The cation selectivity sequence was $Cs^+ > K^+ > Na^+ > Li^+ \gg Ca^{2+}$, with Ca^{2+} being impermeant (this is a low field strength Eisenman selectivity sequence I or II; see discussion in Jordan, 2006). In such a sequence, the ions with the smaller hydration shell (but larger ionic radius; e.g., Cs^+) are more permeant than the ions with the larger hydration shell (but smaller ionic radius; e.g., Li^+).

Further measurements in the GlyR indicated that the T13′V mutation was actually counterproductive, with the selectivity double mutation (SDM; P-2′Δ and A-1′E) producing a more cation-selective channel ($P_{Cl}/P_{Na} = 0.13$), which had now become permeable to Ca^{2+} ($P_{Ca}/P_{Na} = 0.29$) (Keramidas et al., 2002), and required less glycine to activate it than the selectivity triple mutant (STM) GlyR channel. Although the conductances were similar in magnitude and rectification to the triple mutant GlyR channel, the longer channel openings in the double mutant channel could now be measured directly from current records (Moorhouse et al., 2002). Dissecting the effect of these mutations further, a single proline deletion (P-2′Δ) was found to be unable to invert the ion charge selectivity, though this mutant GlyR channel was less anion selective ($P_{Cl}/P_{Na} = 3.8$) than the WT GlyR ($P_{Cl}/P_{Na} = 25$) (Keramidas et al., 2002). However, a single glutamate mutation (A-1′E) was able to invert the selectivity, giving this cation-selective mutant GlyR a P_{Cl}/P_{Na} value of 0.34 and clarifying a major role for charged residues in determining the ion charge selectivity of these LGICs (Keramidas et al., 2002).

In addition to measurements of anion–cation permeability ratios, measurements were made of the minimum pore diameters of these and other mutant GlyRs. The aim was to investigate whether there were any additional structural changes in the channel, which could be contributing to its ion selectivity. To determine minimum pore diameters, the permeabilities of a range of large organic cations (or anions) were measured for cation-selective (or anion-selective) mutant GlyR channels, in order to determine the maximum dimension of ions, which could just permeate through them (Keramidas et al., 2002; Lee et al., 2003). Such measurements indicated values of ~0.54 nm for the WT GlyR channel (Rundström et al., 1994), ~0.69 nm for the single P-2′Δ GlyR channel (Lee et al., 2003), ~0.65 nm for the A-1′E GlyR channel, and ~0.97 nm for the SDM GlyR channel (Keramidas et al., 2002). Further mutations to change the charge of the external charged ring in the cation-selective SDM GlyR channel gave some interesting results. Neutralizing the external charged ring (P-2′Δ, A-1′E, and R19′A) changed rectification from outward to 'linear,' whereas making it negatively charged (P-2′Δ, A-1′E, and R19′E), made it inwardly rectifying (Moorhouse et al., 2002). Both of the above results were consistent with the cytoplasmic charged ring being buried in the protein, which is not unreasonable, since this charged ring in the WT GlyR channel is of the inappropriate sign

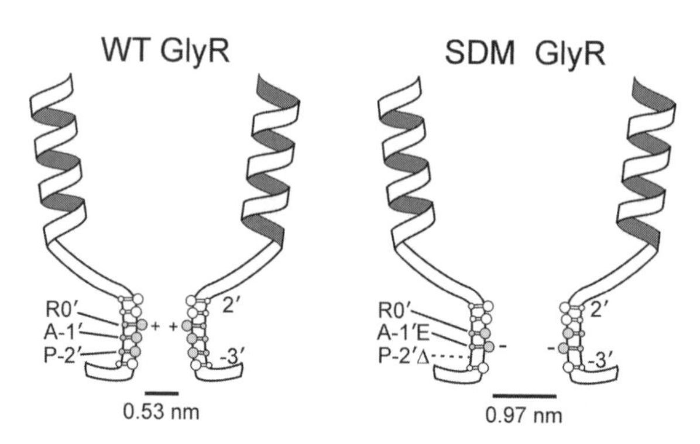

Fig. 9.8 A schematic diagram depicting the selectivity filter region of the GlyR channel at the cytoplasmic (internal) end of two of the M2 segments, where it meets the M1–M2 loop, and the suggested local structural changes which take place when the anion-selective WT GlyR channel is mutated to the cation-selective SDM (A-1′E, P-2′Δ) mutant GlyR channel, with experimentally determined minimum pore diameters shown (see Keramidas et al., 2002; Keramidas et al., 2004). This figure has been reproduced from Fig. 7 of Barry and Lynch (2005), with copyright permission of [2005] IEEE.

for an anion-selective channel. The SDM GlyR channel with the negatively charged external ring (P-2′Δ, A-1′E, and R19′E) also had an increased relative calcium permeability compared to the SDM GlyR (Keramidas et al., 2002).

The results in this section indicated that there were two factors determining anion–cation selectivity in these LGICs. The dominant factor was the presence of an effective charged residue in the selectivity filter region, a negative residue making the channel cation-selective and a positive one making it anion-selective. Presumably, the presence of the negative glutamate in the A-1′E mutation caused the adjacent positive arginine (R0′) to move back from the channel surface and so contribute less to the electrical potential in the selectivity filter region of the channel, as indicated in Fig. 9.8. This is supported by the observation in the nAChR channel that the mutations to the K0′ position in the β and γ subunits had no significant effect on cation conductances (Imoto et al., 1988). Ideally, it would have been very instructive to have directly done the mutation R0′E, but mutations of this residue in α1 GlyRs (R0′Q, R0′E, R0′N) failed to express effective channels (Langosch et al., 1994; Rajendra et al., 1995). In addition to the effect of residue charge, the data indicate that the size of the minimum pore diameter of the channel also plays a role, with smaller diameters tending to increase P_{Cl}/P_{Na} and larger ones tending to decrease it (Keramidas et al., 2004). The suggestion was made that in the smaller channel the ions have to be dehydrated to permeate through the filter region (Fig. 9.9). Since it is easier for the larger Cl^- to shed its hydration shell compared to the smaller Na^+ with its much larger hydration shell, this would tend to increase P_{Cl} relative to P_{Na} (Keramidas et al., 2002). In contrast in the larger negatively charged cation-selective channels, Na^+ ions could pass through in a more hydrated state (Keramidas et al., 2002) (Fig. 9.9).

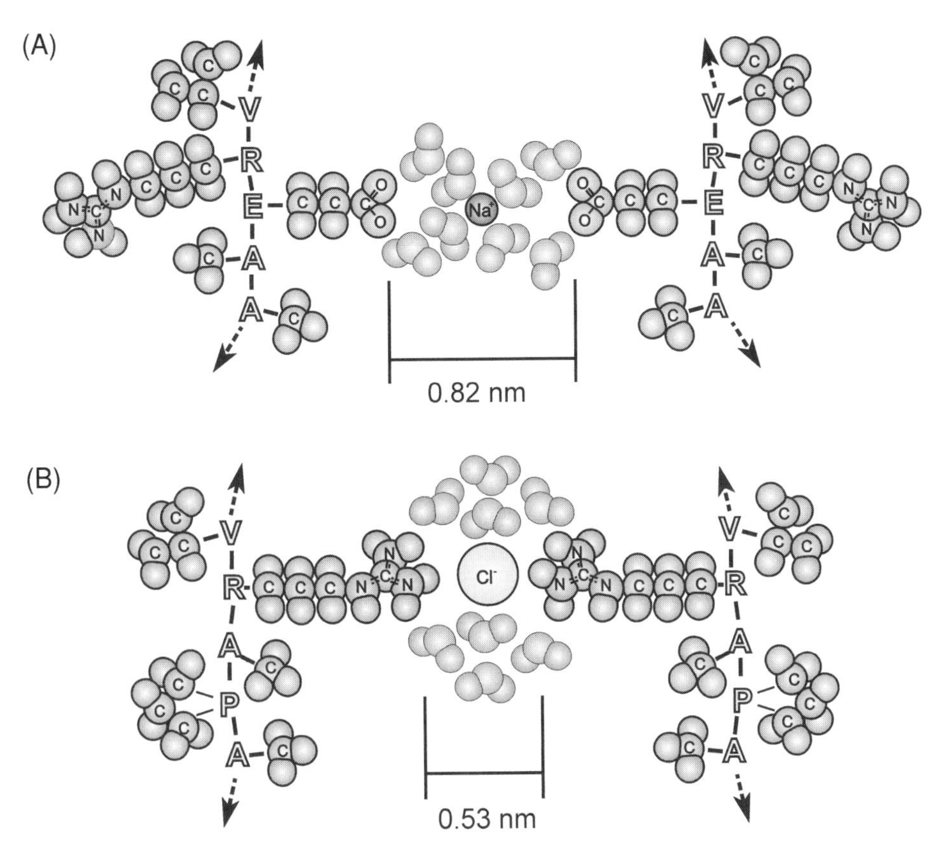

Fig. 9.9 A schematic representation of the selectivity filter region of cation- and anion-selective LGICs between positions 1′ to −3′ and based on data from WT and mutant α1 GlyR channels. Only parts of two M2 domains are shown. Atoms and molecules are drawn approximately to scale and for simplicity only the side chains of the amino acid residues are shown, with the remainder of the residues (the peptide backbone) being represented by the single-letter code for the amino acids. Panel A is intended to depict the situation for the cation-selective (SDM + R19′A) GlyR with the Na^+ ion permeating through the larger diameter filter region without having to be completely dehydrated, whereas for the WT α1 GlyR channel, the Cl^- ion is more readily able to permeate through the smaller diameter filter region in its dehydrated form. This figure has been modified from Fig. 9 of Keramidas et al. (2002) and Fig. 5 of Keramidas et al. (2004).

The relationship between the effect of residue charge and pore diameter is illustrated for WT and mutant GlyRs in Fig. 9.10, where it can be seen that an increase in pore diameter is correlated with a decrease in the relative anion to cation permeability of the channel. However, it may also be seen from this figure that the switch from the channel being predominantly anion-selective to being predominantly cation-selective requires a change in the sign of the effective charge in its selectivity filter region.

Recent measurements have supported similar mechanisms underlying selectivity in the other members of the LGIC family and have added further support to the

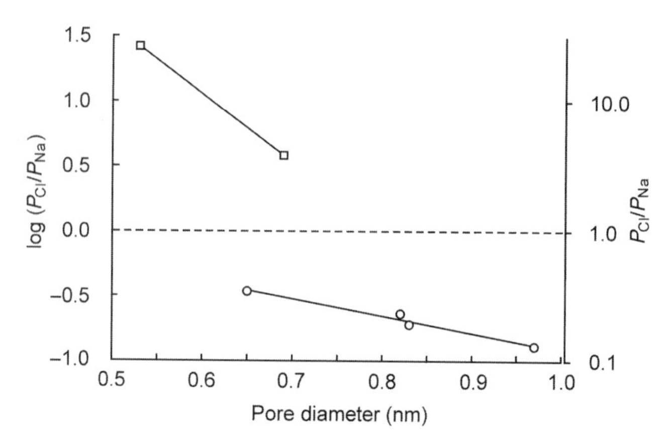

Fig. 9.10 This graph illustrates the relationship between anion–cation permeability ratio and pore diameter, together with the sign of the residue charge in the selectivity filter region, in homomeric α1 GlyRs. The anion-selective GlyR data (open squares) and cation-selective data (open circles) are fitted to two separate lines. The figure shows that increasing pore diameter is correlated with a decrease in the anion–cation permeability ratio, but that a predominantly anion-selective channel requires an effective positive residue charge in the selectivity filter region and a predominantly cation-selective channel an effective negative residue charge. The minimum pore diameter values for the WT GlyR and P-2′Δ GlyR channels were taken from Rundström et al. (1994) and Lee et al. (2003). All other values were taken from Keramidas et al. (2002). The figure has been redrawn from Fig. 4A of Keramidas et al. (2004).

role of the residues in the region from at least −2′ to 2′ as making up the selectivity filter region (see Fig. 9.8). Firstly, a set of triple mutations in the cationic 5-HT_3R, similar to those in the nAChR channel, also made it anion selective (Gunthorpe and Lummis, 2001). In the homomeric ρ1 $GABA_C$R, a single A-1′E mutation made it seemingly nonselective ($P_{Cl}/P_{Na} = 1.3$), whereas a double mutation (P-2′Δ and A-1′E), made it cation-selective ($P_{Cl}/P_{Na} = 0.31$; HEK cell data). Also, in this channel it was possible to directly change the charge at position 0′. Neutralizing this charge made it slightly less anion-selective than the WT ρ1 $GABA_C$R channel, but it still remained anion-selective, whereas replacing the positive charge with a negative one (R0′E), did make the channel weakly cation selective ($P_{Cl}/P_{Na} = 0.4$) (Wotring et al., 2003; see also discussion in Keramidas et al., 2004). In 2004, it was also shown that placing a negative charge in the 2′ position in the ρ1 $GABA_C$R channel (a single mutation of the 2′ proline to a glutamate, P2′E) did invert its charge selectivity from being anion-selective ($P_{Cl}/P_{Na} = 7.1$) to cation-selective (with $P_{Cl}/P_{Na} = 0.08$) (Carland et al., 2004). These results further support the important role of residue charge in controlling the ion selectivity of these LGICs.

9.3.3 Modeling Ion Permeation

To precisely determine the mechanisms underlying ion permeation, it is very instructive to be able to derive a mathematical permeation model to fit to the experimental data. Permeation models can be classified into three broad classes. These are (1) the classical type of continuum electrodiffusion models, such as the Goldman–Hodgkin–Katz (GHK) equation, the Planck equation, or the Poisson–Nernst–Planck model; (2) kinetic rate theory models, with ions hopping between discrete energy wells, and (3) molecular models such as Brownian dynamics and molecular dynamics, which require knowledge of the precise molecular structure of a channel (see Keramidas et al., 2004, for references).

Clearly, the easiest models to use for determining relative permeability ratios are the electrodiffusion ones, like the GHK equation, and this equation is the one used by most electrophysiologists for this purpose. It is reasonably argued by theoreticians that the underlying Goldman equation, of which the GHK equation is its zero-current form, is dependent on fundamental assumptions, which are likely to be invalid for very narrow channels of sub nm dimensions, such as the LGICs (e.g., 0.6–0.9 nm). However, it has been shown that for membrane potentials, in certain situations and *under zero current conditions*, the permeability ratios, determined by a range of different models, are very similar, in spite of the fact that each model is based on totally different assumptions (Keramidas et al., 2004). These certain situations are the so-called "bi-ionic potentials", where membranes separate two different electrolytes with a common ion (e.g., NaCl:KCl at the same concentration, where P_{Na} and $P_{\mathrm{K}} \gg P_{\mathrm{Cl}}$) and so-called "dilution potentials" where the membrane separates the same electrolyte at two different concentrations (e.g., $\mathrm{NaCl}_{\mathrm{C1}}$:$\mathrm{NaCl}_{\mathrm{C2}}$, at concentrations C1 and C2). Hence, such derived permeability ratios for these situations seem to be essentially model-independent (Keramidas et al., 2004), and to be similar to parameters which might result from a consideration of the irreversible thermodynamics of anion and cation fluxes through an ion channel (Barry, 2006).

Nevertheless, it is clearly important to be able to fully understand the mechanisms underlying permeation. To achieve this we need to accurately model ion permeation and current–voltage relationships through such ion channels, taking into account the substantial amount of information that is becoming available about the 3-D molecular structure of the channels. The best practical approach currently is a combination of Brownian dynamics (BD) with other information determined from molecular dynamics (MD). In BD, the total force acting on each ion is calculated from (A) all the electrical forces acting on the ion and from (B) a random fluctuating force due to thermal motion and collisions between ions and water molecules, together with a related frictional viscosity term due to the movement of the ion in the solution. This force acting on the ion is used to calculate the new position and velocity of each ion at a particular time in the channel. This needs to be done in femtosecond (10^{-15} s) time steps for all the ions in the channel and the calculations repeated millions of times, to determine the trajectories of the ions and subsequent ionic fluxes, for different voltages across the channel. O'Mara et al. (2003) were able to run such simulations for WT and some cation-selective mutant $\alpha 1$ GlyR channels and were able to simulate the key permeation features of these channels and particularly the basic role of charged residues in determining ion charge selectivity (O'Mara et al., 2003). Nevertheless, that study was unable to explain the presence of permeant counterions in these LGIC channels, and their BD models of the above channels only seemed to be able to allow ions of one sign to permeate (i.e., P_{Cl}/P_{Na} was either ∞ for anion-selective WT GlyRs or 0 for cation-selective mutant GlyRs), in spite of the fact that the reversal potentials measured experimentally did not support such conclusions. However, more recently a BD study by Cheng et al. (2005) of both WT and cation-selective mutant GlyRs, based on the same experimental data, but slightly different channel parameters and simulation conditions was able to observe counterion permeation through both sets of channels, with magnitudes consistent both with their reversal potential data and with the experimental anion/cation permeability ratios estimated from reversal potential measurements in those studies (Keramidas et al., 2000, 2002, 2004). It seems most likely that a critical factor in the relationship between relative permeabilities and consistent reversal potentials must lie in the way in which the boundary algorithms, for the ion concentrations on either side of the channel, are implemented.

9.4 Ion Channel Gating

All LGICs contain 2–5 agonist-binding sites. Agonist binding initiates a conformational change that is propagated throughout the protein, culminating in the opening of the channel 'gate.' The gate is the physical barrier that stops ions from traversing the pore in the unliganded state. This section describes the molecular basis of agonist binding, and the nature of the structural changes that occur once the agonist has bound.

The activation of multi-subunit proteins can be described by two starkly contrasting models: the Monod–Wyman–Changeux (MWC) model (Changeux and Edelstein, 1998, 2005) or the Koshland–Nemethy–Filmer (KNF) model (Koshland et al., 1966). The MWC model proposes that all subunits change conformation simultaneously so the receptor can exist only in either the closed or entirely activated states. Alternatively, the KNF model proposes that each subunit can independently adopt a specific conformation change depending on the number of bound agonist molecules, leading to a series of intermediate states between fully closed and fully open. The KNF model also predicts that different agonists may activate the receptor via different conformational changes. On the other hand, MWC theory predicts a single activated state, although different agonists may stabilize this state to differing degrees. A recent study on the $\alpha 1$ homomeric GlyR found that a full agonist, glycine, and a partial agonist, taurine, both induced similar conformational changes in the external loop linking the M2 and M3 domains (Han et al., 2004). This provides strong evidence in favor of the MWC model.

One of the difficulties in experimentally discriminating between the MWC and KNF models stems from the fact that many oligomeric protein channels are comprised of different subunits. As structurally different subunits would not be expected to respond to agonist binding via identical conformational changes, it can be difficult to experimentally determine whether dissimilar subunits undergo concerted conformational changes. Taking into account such difficulties, the weight of evidence to date favors an MWC model over a KNF model for the LGIC family (Auerbach, 2003; Changeux and Edelstein, 2005) and most researchers accept that LGIC pore opening is accompanied by the simultaneous activation of all five subunits.

9.4.1 Agonist Binding

Since the ligand-binding domains A–D (see Fig. 9.2 in Section 9.2) of most LGIC members each contain highly conserved aromatic residues, it is likely that all LGIC agonist-binding sites are lined by aromatic rings (Lester et al., 2004). There is abundant evidence that the ACh-nAChR binding reaction is mediated largely by noncovalent 'cation-π' electrostatic interactions (Zhong et al., 1998; Celie et al., 2004). In this system, the side chains of the aromatic amino acids (phenylalanine, tyrosine or tryptophan) contribute a negatively charged π surface, while the cation is provided by the agonist. The highly conserved nature of the aromatic groups suggests that this binding mechanism may be broadly applicable across the LGIC family. Indeed, the homologous conserved aromatic residues have been shown to be involved in agonist binding in the $GABA_AR$ (Amin and Weiss, 1993), $GABA_CR$ (Lummis et al., 2005), $5\text{-}HT_3R$ (Spier and Lummis, 2000), and GlyR (Schmieden et al., 1993; Grudzinska et al., 2005). Of course, different LGIC members are highly selective for particular agonists, implying that other receptor-specific binding interactions are also necessary to confer agonist specificity.

The crystal structures of nicotinic agonists (lobeline and epibatidine) and antagonists (α-Conotoxin ImI and methyllycaconitine) complexed with AChBP have recently been published (Hansen et al., 2005). These structures (cf. Fig. 9.2 for the GlyR) confirm that aromatic residues from the principal ligand-binding domains A–C (Fig. 9.2) and complementary ligand-binding domain D form an aromatic 'nest' that largely engulfs the ligands. As anticipated from functional studies (Karlin, 2002), the vicinal disulfide between Cys 190 and Cys 191 (in domain C; cf. Fig. 9.2 for the GlyR) and residues in the complementary ligand-binding domains E and F also provide important agonist-binding determinants. The Hansen et al. (2005) study showed that the large antagonists, α-Conotoxin ImI and methyllycaconitine, were also coordinated by the aromatic nest on the principal ligand-binding side, but that there was more variability in the contact sites on the complementary side of the interface (Hansen et al., 2005). A similar study using the peptide antagonist, α-Conotoxin PnIA, revealed a similar picture (Celie et al., 2005). Antagonist binding was not associated with significant movements of domain C, whereas agonist binding resulted in this domain wrapping tightly around the bound agonist (Hansen et al., 2005).

9.4.2 Conformational Changes in the Ligand-Binding Domain

As noted in Section 9.2.1, Unwin obtained low-resolution electron diffraction images of the *Torpedo* nAChR in both the closed and open states (Unwin et al., 1995). Since

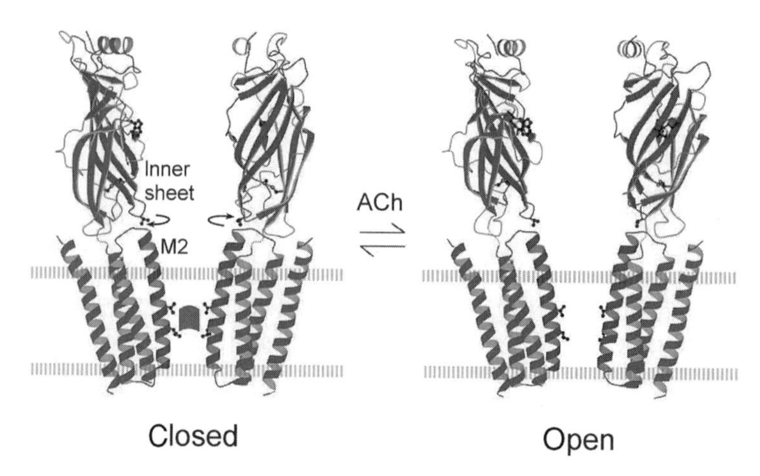

Fig. 9.11 Conformations of the closed (left) and open (right) nAChR, as determined by fitting the polypeptide chains to the electron image density maps of Unwin and colleagues (Unwin et al., 2002; Miyazawa et al., 2003). Moving parts are shown in blue in the original figure. Only two subunits per receptor are shown for clarity. The gate is shown as the curved rectangle (dark pink in the original figure). The TM domain reflects the Miyazawa cryo-EM structure (Miyazawa et al., 2003) and the ligand-binding domain is modeled on AChBP (Brejc et al., 2001). This figure was originally reprinted in color by permission of the Federation of the European Biochemical Societies from Fig. 3 of Unwin (2003) and has been reproduced in monochrome from Fig. 8 of Barry and Lynch (2005), with copyright permission of [2005] IEEE.

only the open state structure provided a close match with AChBP, he was able to deduce how the nAChR extracellular domains move upon agonist binding (Unwin et al., 2002). The Unwin model separates the extracellular domain into inner and outer parts. The inner, vestibule-lining part (comprising seven β-sheets) contains most of the inter-subunit contact points plus agonist-binding domain A (see Fig. 9.4, and Fig. 9.2 for the GlyR in Section 9.2). The outer part (which comprises three β-sheets) includes the agonist-binding domains B and C (cf. Fig. 9.2 for the GlyR). Upon agonist binding, the outer part was hypothesized to undergo an upwards tilt around an axis parallel with the membrane plane, while the inner part rotated $\sim 15^{\circ}$ in a clockwise direction (when viewed from the synapse) around an axis perpendicular to the membrane plane (Fig. 9.11). The movement of the outer part has the effect of clasping binding domain C around the agonist, virtually burying it in the binding site. A variety of evidence, including direct crystallographic analysis (Celie et al., 2004; Hansen et al., 2005), tryptophan fluorescence (Gao et al., 2005), agonist structure–function relationships (Wagner and Czajkowski, 2001) and molecular structure–function studies (Grutter et al., 2003; Gao et al., 2005), provides strong support for this mechanism. There is, as yet, no independent evidence to support a rotation of the inner sheets, although channel activation has long been considered to be mediated by some kind of relative movement of residues on each side of the subunit interface (Corringer et al., 2000). As noted previously, the recent crystal structures of AChBP complexed with a variety of nicotinic agonists and antagonists reveal that different surfaces of the complementary binding domain interact with agonists relative to antagonists (Hansen et al., 2005). This implies that agonist binding induces substantial movements in the complementary domain, which is located on the inner sheets. However, no rotation of the inner sheets was observed. Nevertheless, as noted by the authors, it is unclear whether these conformational changes are related to gating since the AChBP lacks the residues that functionally couple ligand-binding with channel activation (Hansen et al., 2005).

The inner sheets (Fig. 9.11) contain two loops that protrude from the bottom of the extracellular part of the structure toward the TM domains (Brejc et al., 2001; Unwin et al., 2002). These loops, numbered 2 and 7, are therefore prime candidates for transmitting agonist-binding information to the activation gate. Loop 7 is also known as the conserved cysteine loop, as mentioned in Section 9.1. Detailed investigations, described below, have begun to unravel the molecular interactions that mediate the information transfer from the agonist-binding site to the activation gate. Studies have focused on the interactions between ligand-binding domain loops 2 and 7, the (extracellular) M2–M3 linker domain, and the (extracellular) pre-M1 domain (see Fig. 9.2 for the GlyR).

9.4.3 Conformational Changes in the Membrane-Spanning Domains

A signal transduction role for the entire M2–M3 domain was first suggested by a systematic site-directed mutagenesis study on the α1 GlyR (Lynch et al., 1997). Although this study showed that mutations to this domain uncoupled the agonist-binding site from the channel activation gate, it provided no information as to whether this domain moved upon channel activation. The substituted cysteine accessibility method (SCAM) was subsequently employed to address the question of domain movement (Karlin and Akabas, 1998). This technique entails introducing cysteine residues one-by-one into the domain of interest. The surface accessibility of cysteines is then probed by highly water-soluble methanethiosulfonate reagents (Karlin and Akabas, 1998). If a functional property of the channel is irreversibly changed by the reagent, then it is assumed that the cysteine lies on the protein surface. Changes in the cysteine modification rate between open and closed states may provide information about the movement of the domain relative to its surroundings. Lynch et al. (2001) employed this approach to show that the surface accessibility of six contiguous residues in the M2–M3 loop of the α1 GlyR subunit experienced an increased surface accessibility in the open state. Surprisingly, a subsequent study by the same group using a similar approach showed that the homologous residues in the GlyR β subunit did not appear to be exposed at the protein–water interface (Shan et al., 2003).

An even more direct measure of protein conformational change is to combine electrophysiology with the quantization of state-dependent changes in the fluorescence of small labels attached to the domain of interest (Gandhi and Isacoff, 2005). Because small fluorophores are often sensitive to the hydrophobicity of their environment, they may report local structural reorganizations. Such an approach has been employed to monitor changes in the fluorescence of a small fluorophore (rhodamine) tethered to cysteine side chains inserted to the 19′ residue at the N-terminal end of the nAChR M2–M3 domain (Dahan et al., 2004). As expected, this method indeed identified a state-dependent movement in this domain. It also showed that transitions at the 19′ site were not tightly coupled to activation, suggesting that sequential rather than fully concerted transitions occur during channel activation. These results bode well for the use of this technique in characterizing conformational changes in surface domains associated with channel activation. However, it must be acknowledged that we currently know little about the structural reorganization of the M2–M3 domain that occurs during channel activation.

The Auerbach group has pioneered the use of linear free-energy relationships (LFERs) of nAChRs incorporating mutations in various positions (Auerbach, 2003). They have shown that the energy transitions experienced by M2–M3 domain residues are midway between those experienced by residues at the agonist-binding site and the activation gate. The authors conclude that the M2–M3 domain lies at the midpoint of an agonist-initiated conformational 'wave' that proceeds from the agonist-binding site to the activation gate (Grosman et al., 2000).

A variety of approaches have been employed to characterize the structural basis of the interactions between the ligand-binding domain and the M2–M3 domain. Research has focused on loops 2 and 7 of the ligand-binding domain as these intercalate directly with the M2–M3 domain. The Miyazawa et al. (2003) nAChR TM structure suggests that a loop 2 hydrophobic side chain fits into the end of the M2 α-helix like a 'pin in a socket'. Movements of the ligand-binding domain inner sheets would thus rotate the M2 domain to the open state (Figs. 9.11 and 9.12). Functional approaches have also identified strong interactions between the M2–M3 domain and loops 2 and 7 of the ligand-binding domains of GlyRs (see Fig. 9.2) and $GABA_A$Rs (Absalom et al., 2003; Kash et al., 2003, 2004). Of particular note, Kash et al. (2003) used mutant cycle analysis to identify an electrostatic interaction between positively charged lysine K279 in the M2–M3 domain and negatively charged D149 in loop 7. The same study also found that cysteines substituted into these positions were able to crosslink in the open state. Together, these results suggest that $GABA_A$R channel activation is mediated by a decreased distance and hence an increased electrostatic interaction between D149 and K279 (Kash et al., 2003). However, the electrostatic interaction between the corresponding residues in the GlyR is weaker (Absalom et al., 2003), and it is not known whether this mechanism is applicable to other members of the nAChR family. There is also evidence for molecular interactions between loops 2 and 7 and the pre-M1 domain (Kash et al., 2004). However, evidence to date is insufficient to identify any common linkage mechanisms that may pertain to the whole LGIC family.

One way of establishing whether LGIC activation occurs via a universal linkage mechanism between the ligand-binding domain and the M2–M3 loop is to test the functionality of chimeric receptors. It has been shown that the nAChR ligand-binding domain or AChBP attached to the 5-HT_3R TM domains results in

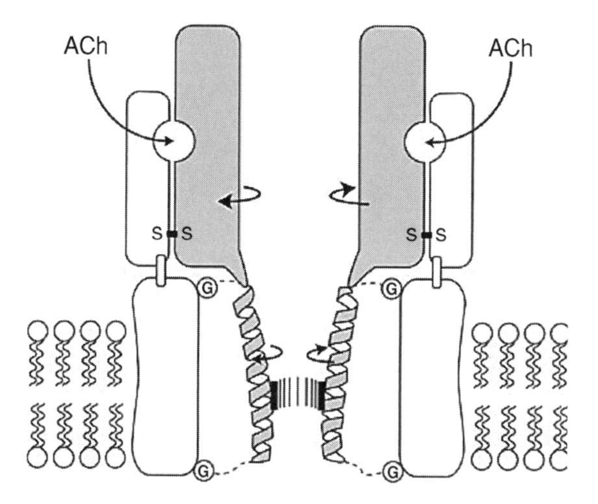

Fig. 9.12 A schematic model for the gating mechanism, depicted in Fig. 9.11. ACh binding induces a rotation in the α subunits, which is transmitted to a hydrophobic barrier or a restriction in the channel through the M2 helices. The helices are linked by flexible loops to the outer protein wall containing glycine residues (G). The two S–S represent disulphide bridge pivots and the moving parts are shown in grey. Originally reprinted from Fig. 6 of Miyazawa et al. (2003) with copyright permission from the Nature Publishing Group, this figure has been reproduced from Fig. 9 of Barry and Lynch (2005), with copyright permission of [2005] IEEE.

functional acetylcholine-gated currents (Bouzat et al., 2004). Similarly, the $GABA_C R$ ligand-binding domain coupled to the GlyR TM domains (Mihic et al., 1997) produces functional channels. However, until functional activation can be demonstrated in chimeras comprising components of both anion- and cation-gated LGICs it is premature to conclude that LGIC activation occurs via a common linkage mechanism.

Low-resolution electron diffraction images of *Torpedo* nAChR originally indicated that the M2 domain incorporated a centrally located kink at the 9′ position (Unwin, 1995) (cf. Fig. 9.3A). The 9′ leucine residue is highly conserved across all LGICs, implying a critical role in channel function. Mutating the 9′ leucines to small polar residues had an equal effect on the ACh sensitivity regardless of which subunit was mutated (Labarca et al., 1995). As binding sites exist at only two of the five subunit interfaces, the implication is that neighboring nAChR subunits interact via their respective 9′ residues. The Miyazawa TM domain structure (Miyazawa et al., 2003) reveals the existence of hydrophobic bonds between the 9′ and 10′ residues of adjacent subunits. These bonds probably maintain the central part of the pore into a fivefold radially symmetrical arrangement that holds the channel closed. It is likely that agonist-induced conformational changes asymmetrically disrupt some of these bonds, leading to a collapse of symmetry and a simultaneous conversion of all M2 domains to the activated state. At this stage it is uncertain how the M2 domains might move during activation. The current prevailing view, shown in Figs. 9.11 and 9.12, is that the domains rotate about their long axes (Unwin, 1995, 2003; Horenstein et al., 2001; Goren et al., 2004; Taly et al., 2005).

9.4.4 The Gate

As mentioned earlier in this section, the gate is a physical barrier or restriction within the channel, which blocks current flow when the channel is in the closed state. There is currently some uncertainty as to the precise location of this structure. We have also noted that in the open state of LGICs there is a selectivity filter region, which controls the type of ions, which flow through the channel. The question arises: Are the gate and selectivity filter regions physically separate or are they co-localized?

One proposal, originally suggested by Unwin and colleagues (Unwin, 1995; Miyazawa et al., 1999), is that the M2 domains are kinked inwards (see Fig. 9.3A) to form a centrally located hydrophobic gate near the highly conserved 9′ leucine residue. This view is supported by a molecular modeling study (Kim et al., 2004) and by experiments designed to probe the surface accessibility of cysteines introduced into the pore of the 5-HT_3R (Panicker et al., 2002). However, similar experiments on the $GABA_AR$, $GABA_CR$, and nAChR have delimited the gate to the same narrow pore region ($-2'$ to $+2'$) that houses the selectivity filter (Akabas et al., 1994; Xu and Akabas, 1996; Wilson and Karlin, 1998; Filippova et al., 2004).

At present there is no obvious way of reconciling the two sets of observations. One possibility is that the position of the gate may vary among LGIC members.

9.4.5 Modulation of LGIC Receptors

LGIC receptors are modulated by a wide variety of molecules, including endogenous substances of physiological or pathological relevance and exogenous pharmacological probes. In addition to direct protein–protein interactions, LGICs are modulated by post-translational modifications, notably phosphorylation. This involves the covalent attachment of a phosphate group to serine, threonine or tyrosine side chains. A classic means of ion channel modulation is direct channel block: a process whereby a molecule binds to a site in the pore and directly prevents the passage of ions. The molecular basis of quinacrine block of the nAChR has recently been characterized in detail (Yu et al., 2003). Another classic modulatory mechanism is competitive antagonism. This description applies to a molecule that binds to an agonist site but cannot initiate the conformational change required to activate the channel. Strychnine, which binds in the glycine pocket, is a classic competitive antagonist of the glycine receptor (Lynch, 2004). The final category considered here is the 'allosteric modulator'. This somewhat loose term describes any molecule that alters the conformation of a receptor. By doing so, such modulators can change the function of the receptor in a wide variety of ways. Two examples are the effects of diazepam and pentobarbitone on increasing the single-channel conductance of $GABA_A$, mentioned at the end of Section 9.3.2. Many classical blockers and competitive antagonists also exert allosteric effects.

9.5 Conclusions and Some Questions Still Pending

9.5.1 Ion Conductances, Permeation and Selectivity

It has been shown that changing the net charge on the rings of charge in the LGICs alters their single-channel conductance and that the most sensitive response results from changes at the intracellular ring. For example, decreasing the negative charge in that ring in the cation-selective nAChR radically reduces channel conductivity.

Structure–function selectivity experiments in all of the LGICs have implicated very similar underlying mechanisms with a major role being due to the sign of the charge residues in the ion selectivity filter region (from at least position $-2'$ to $+2'$) close to the intracellular ring. Data for the mutant cation-selective GlyRs also suggest a secondary contribution from changes in the minimum pore diameter in these mutations, with larger diameters tending to increase the cation/anion permeability ratio.

Four pending ion permeation questions follow. (1) What role, if any, do the negative or positive charges lining the cytoplasmic portals play in determining ion selectivity? The experimental evidence outlined in this chapter for the GlyR channel suggests that this may well be a very minor role, but this needs to be fully investigated both for the GlyR channel and other LGICs. (2) Do other ion selectivity mutant LGICs also display equivalent shifts in minimum pore diameter similar to those demonstrated for the GlyR mutant channels? (3) What is the precise physical mechanism for counterion permeation through the selectivity filter region and can this be satisfactorily simulated by Brownian and molecular dynamics studies? (4) Can ion hydration factors be feasibly incorporated into present and future Brownian and molecular dynamics studies in order to explain selectivity between different ions of the same sign in such channels?

9.5.2 Ion Channel Gating

The recently resolved structures of AChBP and the *Torpedo* nAChR TM domains together provide an excellent basis for modeling the structure of LGIC family members. Because these models make precise predictions about the relative spatial positioning of residues, they permit the design of more specific experiments aimed at understanding how LGICs open and close. Hopefully, these new high-resolution models will provide new ways of addressing the following three critical questions. (1) How does the ligand-binding domain move upon agonist binding, and how is this movement transferred to the TM domains? (2) How do the TM domains move as the channel opens and closes? (3) Is the gate located at the same position in all LGIC members?

Acknowledgments

We thank Dr. Trevor M. Lewis for reading this manuscript and acknowledge his very helpful comments and suggestions, and thank both him and Dr. Andrew Moorhouse for their comments on the previous IEEE manuscript. We also thank the IEEE for permission to use much of the text and the figures of Barry and Lynch (2005).

References

Absalom, N.L., T.M. Lewis, W. Kaplan, K.D. Pierce, and P.R. Schofield. 2003. Role of charged residues in coupling ligand binding and channel activation in the extracellular domain of the glycine receptor. *J. Biol. Chem.* 278:50151–50157.

Akabas, M.H., C. Kaufmann, P. Archdeacon, and A. Karlin. 1994. Identification of acetylcholine receptor channel-lining residues in the entire M2 segment of the alpha subunit. *Neuron* 13:919–927.

Aidley, D.J., and P.R. Stanfield. 1996. Ion Channels: Molecules in Action. Cambridge University Press, Cambridge, UK.

Alexander, S.P., A. Mathie, and J.A. Peters. 2004. Guide to receptors and channels. *Br. J. Pharmacol.* 141:S1–S126.

Amin, J., and D.S. Weiss. 1993. $GABA_A$ receptor needs two homologous domains of the beta-subunit for activation by GABA but not by pentobarbital. *Nature* 366:565–569.

Auerbach, A. 2003. Life at the top: The transition state of AChR gating. *Sci. STKE.* 2003:re11.

Barry, P.H. 2006. The reliability of relative anion-cation permeabilities deduced from reversal (dilution) potential measurements in ion channel studies. *Cell Biochem. Biophys.* 46(2) (October, *in press*).

Barry, P.H., and P.W. Gage. 1984. Ionic selectivity of channels at the end plate. *Curr. Top. Membr. Transp.* 21:1–51.

Barry, P.H., and J.W. Lynch. 2005. Ligand-gated channels. *IEEE Trans. Nanobiosci.* 4:70–80.

Bormann, J., N. Rundstrom, H. Betz, and D. Langosch. 1993. Residues within transmembrane segment M2 determine chloride conductance of glycine receptor homo- and hetero-oligomers. *EMBO J.* 12:3729–3737.

Bouzat, C., F. Gumilar, G. Spitzmaul, H.L. Wang, D. Rayes, S.B. Hansen, P. Taylor, and S.M. Sine. 2004. Coupling of agonist binding to channel gating in an ACh-binding protein linked to an ion channel. *Nature* 430:896–900.

Brejc, K., W.J. van Dijk, R.V. Klaassen, M. Schuurmans, J. van der Oost, A.B. Smit, and T.K. Sixma. 2001. Crystal structure of an ACh-binding protein reveals the ligand-binding domain of nicotinic receptors. *Nature* 411:269–276.

Carland, J.E., A.J. Moorhouse, P.H. Barry, G.A.R. Johnston, and M. Chebib. 2004. Charged residues at the 2′ position of human GABAC rho 1 receptors invert ion selectivity and influence open state probability. *J. Biol. Chem.* 279:54153–54160.

Celie, P.H., S.E. van Rossum-Fikkert, W.J. van Dijk, K. Brejc, A.B. Smit, and T.K. Sixma. 2004. Nicotine and carbamylcholine binding to nicotinic acetylcholine receptors as studied in AChBP crystal structures. *Neuron* 41:907–914.

Celie, P.H., I.E. Kasheverov, D.Y. Mordvintsev, R.C. Hogg, P. van Nierop, R. van Elk, S.E. van Rossum-Fikkert, M.N. Zhmak, D. Bertrand, V. Tsetlin, T.K. Sixma, and A.B. Smit. 2005. Crystal structure of nicotinic acetylcholine receptor homolog AChBP in complex with an alpha-conotoxin PnIA variant. *Nat. Struct. Mol. Biol.* 2:582–588.

Changeux, J.P., and S.J. Edelstein. 1998. Allosteric receptors after 30 years. *Neuron* 21:959–980.

Changeux, J.P., and S.J. Edelstein. 2005. Allosteric mechanisms of signal transduction. *Science* 308:1424–1428.

Cheng, M.H., M. Cascio, and R.D. Coalson. 2005. Theoretical studies of the M2 transmembrane segment of the glycine receptor: Models of the open pore structure and current–voltage characteristics. *Biophys. J.* 89:1669–1680.

Conley, E.C. 1996. The Ion Channels Factbook I—Extracellular Ligand-Gated Channels. Academic Press, London, UK.

Cordero-Erausquin, M., L.M. Marubio, R. Klink, and J.P. Changeux. 2000. Nicotinic receptor function: New perspectives from knockout mice. *Trends Pharmacol. Sci.* 21:211–217.

Corringer, P.J., S. Bertrand, J.-L. Galzi, A. Devillers-Thiery, J.P. Changeux, and D. Bertrand. 1999. Mutational analysis of the charge selectivity filter of the alpha7 nicotinic acetylcholine receptor. *Neuron* 22:831–843.

Corringer, P.J., N. Le Novere, and J.P. Changeux. 2000. Nicotinic receptors at the amino acid level. *Annu. Rev. Pharmacol. Toxicol.* 40:431–458.

Dahan, D.S., M.I. Dibas, E.J. Petersson, V.C. Aeyeung, B. Chanda, F. Bezanilla, D.A. Dougherty, and H.A. Lester. 2004. A fluorophore attached to nicotinic acetylcholine receptor beta M2 detects productive binding of agonist to the alpha delta site. *Proc. Natl. Acad. Sci. USA* 101:10195–10200.

Eghbali, M., J.P. Curmi, B. Birnir, and P.W. Gage. 1997. Hippocampal GABA(A) channel conductance increased by diazepam. *Nature* 388:71–75.

Filippova, N., V.E. Wotring, and D.S. Weiss. 2004. Evidence that the TM1-TM2 contributes to the rho1 GABA receptor pore. *J. Biol. Chem.* 279:20906–20914.

Galzi, J.-L., A. Devillers-Thiery, N. Hussy, S. Bertrand, J.P. Changeux, and D. Bertrand. 1992. Mutations in the channel domain of a neuronal nicotinic receptor convert ion selectivity from cationic to anionic. *Nature* 359:500–505.

Gandhi, C.S., and E.Y. Isacoff. 2005. Shedding light on membrane proteins. *Trends Neurosci.* 28:472–479.

Gao, F., N. Bren, T.P. Burghardt, S. Hansen, R.H. Henchman, P. Taylor, J.A. McCammon, and S.M. Sine. 2005. Agonist-mediated conformational changes in acetylcholine-binding protein revealed by simulation and intrinsic tryptophan fluorescence. *J. Biol. Chem.* 280:8443–8451.

Goren, E.N., D.C. Reeves, and M.H. Akabas. 2004. Loose protein packing around the extracellular half of the GABA(A) receptor beta1 subunit M2 channel-lining segment. *J. Biol. Chem.* 279:11198–11205.

Grosman, C., M. Zhou, and A. Auerbach. 2000. Mapping the conformational wave of acetylcholine receptor channel gating. *Nature* 403:773–776.

Grudzinska, J., R. Schemm, S. Haeger, A. Nicke, G. Schmalzing, H. Betz, and B. Laube. 2005. The beta subunit determines the ligand binding properties of synaptic glycine receptors. *Neuron* 45:727–739.

Grutter, T., L. Prado de Carvalho, N. Le Novere, P.J. Corringer, S. Edelstein, and J.P. Changeux. 2003. An H-bond between two residues from different loops of the acetylcholine binding site contributes to the activation mechanism of nicotinic receptors. *EMBO J.* 22:1990–2003.

Gunthorpe, M.J., and S.C. Lummis. 2001. Conversion of the ion selectivity of the 5-HT(3A) receptor from cationic to anionic reveals a conserved feature of the ligand-gated ion channel superfamily. *J. Biol. Chem.* 276:10977–10983.

Hamill, O.P., A. Marty, E. Neher, B. Sakmann, and F.J. Sigworth. 1981. Improved patch-clamp techniques for high-resolution current recording from cells and cell-free membrane patches. *Pflügers Arch.* 391:85–100.

Han, N.L., J.D. Clements, and J.W. Lynch. 2004. Comparison of taurine- and glycine-induced conformational changes in the M2-M3 domain of the glycine receptor. *J. Biol. Chem.* 279:19559–19565.

Hansen, S.B., G. Sulzenbacher, T. Huxford, P. Marchot, P. Taylor, and Y. Bourne. 2005. Structures of *Aplysia* AChBP complexes with nicotinic agonists and antagonists reveal distinctive interfaces and conformations. *EMBO J.* 24:3635–3646.

Hille, B. 2001. Ionic Channels of Excitable Cells, 3rd Ed. Sinauer Associates, Sunderland, MA.

Horenstein, J., D.A. Wagner, C. Czajkowski, M.H. Akabas. 2001. Protein mobility and GABA-induced conformational changes in GABA(A) receptor pore-lining M2 segment. *Nat. Neurosci.* 4:477–485.

Imoto, K., C. Busch, B. Sakmann, M. Mishina, T. Konno, J. Nakai, H. Bujo, Y. Mori, K. Fukuda, and S. Numa. 1988. Rings of negatively charged amino acids determine the acetylcholine receptor channel conductance. *Nature* 335:645–648.

Jordan, P.C. 2006. Ion channels from fantasy to facts in fifty years. *In*: Handbook of Ion Channels: Dynamics, Structure and Applications. S-H. Chung, O.S. Andersen, and V. Krishnamurthy, editors. Springer-Verlag, New York, pp. 3–29 (Chapter 1, this volume).

Kandel, E.R., and S.A. Siegelbaum. 2000. Overview of synaptic transmission and signalling at the nerve–muscle synapse: Direct-gated transmission. *In*: Principles of Neural Science, 4th Ed. E.R. Kandel, J.H. Schwartz, and T.M. Jessell, editors. McGraw-Hill, New York, pp. 175–206.

Karlin, A. 2002. Emerging structure of the nicotinic acetylcholine receptors. *Nat. Rev. Neurosci.* 3:102–114.

Karlin, A., and M.H. Akabas. 1998. Substituted-cysteine accessibility method. *Meth. Enzymol.* 293:123–145.

Kash, T.L., M.J. Dizon, J.R. Trudell, and N.L. Harrison. 2004. Charged residues in the beta2 subunit involved in GABAA receptor activation. *J. Biol. Chem.* 279:4887–4893.

Kash, T.L., A. Jenkins, J.C Kelley, J.R. Trudell, and N.L. Harrison. 2003. Coupling of agonist binding to channel gating in the GABA(A) receptor. *Nature* 421:272–275.

Kelley, S.P., J.I. Dunlop, E.F. Kirkness, J.J. Lambert, and J.A. Peters. 2003. A cytoplasmic region determines single-channel conductance in 5-HT_3 receptors. *Nature* 424:321–324.

Keramidas, A., A.J. Moorhouse, C.R. French, P.R. Schofield, and P.H. Barry. 2000. M2 pore mutations convert the glycine receptor channel from being anion- to cation-selective. *Biophys. J.* 79:247–259.

Keramidas, A., A.J. Moorhouse, K.D. Pierce, P.R. Schofield, and P.H. Barry. 2002. Cation-selective mutations in the M2 domain of the inhibitory glycine receptor channel reveal determinants of ion-charge selectivity. *J. Gen. Physiol.* 119:393–410.

Keramidas, A., A.J. Moorhouse, P.R. Schofield, and P.H. Barry. 2004. Ligand-gated ion channels: Mechanisms underlying ion selectivity. *Prog. Biophys. Mol. Biol.* 86:161–204.

Kim, S., A.K. Chamberlain, and J.U. Bowie. 2004. A model of the closed form of the nicotinic acetylcholine receptor M2 channel pore. *Biophys. J.* 87:792–799.

Koshland, D.E., Jr., G. Nemethy, and D. Filmer. 1966. Comparison of experimental binding data and theoretical models in proteins containing subunits. *Biochemistry* 5:365–385.

Labarca, C., M.W. Nowak, H. Zhang, L. Tang, P. Deshpande, and H.A. Lester. 1995. Channel gating governed symmetrically by conserved leucine residues in the M2 domain of nicotinic receptors. *Nature* 376:514–516.

Langosch, D., B. Laube, N. Rundstrom, V. Schmieden, J. Bormann, and H. Betz. 1994. Decreased agonist affinity and chloride conductance of mutant glycine receptors associated with human hereditary hyperekplexia. *EMBO J.* 13:4223–4228.

Lee, D.J.-S., A. Keramidas, A.J. Moorhouse, P.R. Schofield, and P.H. Barry. 2003. The contribution of proline 250 (P-2′) to pore diameter and ion selectivity in the human glycine receptor channel. *Neurosci. Lett.* 351:196–200.

Lester, H.A., M.I. Dibas, D.S. Dahan, J.F. Leite, and D.A. Dougherty. 2004. Cys-loop receptors: New twists and turns. *Trends Neurosci.* 27:329–336.

Lummis, S.C., D. Beene, N.L. Harrison, H.A. Lester, and D.A. Dougherty. 2005. A cation-pi binding interaction with a tyrosine in the binding site of the GABA(C) receptor. *Chem. Biol.* 12:993–997.

Lynch, J.W. 2004. Molecular structure and function of the glycine receptor chloride channel. *Physiol. Rev.* 84:1051–1095.

Lynch, J.W., N.L. Han, J.L. Haddrill, K.D. Pierce, and P.R. Schofield. 2001. The surface accessibility of the glycine receptor M2-M3 loop is increased in the channel open state. *J. Neurosci.* 21:2589–2599.

Lynch, J.W., S. Rajendra, K.D. Pierce, C.A. Handford, P.H. Barry, and P.R. Schofield. 1997. Identification of intracellular and extracellular domains mediating signal transduction in the inhibitory glycine receptor chloride channel. *EMBO J.* 16:110–120.

Mihic, S.J., Q. Ye, M.J. Wick, V.V. Koltchine, M.D. Krasowski, S.E. Finn, M.P. Mascia, C.F. Valenzuela, K.K. Hansen, E.P. Greenblatt, R.A. Harris, and N.L. Harrison. 1997. Sites of alcohol and volatile anaesthetic action on GABA(A) and glycine receptors. *Nature* 389:385–389.

Miyazawa, A., Y. Fujiyoshi, M. Stowell, and N. Unwin. 1999. Nicotinic acetylcholine receptor at 4.6 A resolution: Transverse tunnels in the channel wall. *J. Mol. Biol.* 288:765–786.

Miyazawa, A., Y. Fujiyoshi, and N. Unwin. 2003. Structure and gating mechanism of the acetylcholine receptor pore. *Nature* 423:949–955.

Mohler, H., D. Benke, J.M. Fritschy, and J. Benson. 2000. The benzodiazepine site of $GABA_A$ receptors. *In*: GABA in the Nervous System: The View at 50 Years. D.L. Martin and R.W. Olsen, eds. Lippincott, Williams and Wilkins, Philadelphia, pp. 97–112.

Moorhouse, A.J., A. Keramidas, A. Zaykin, P.R. Schofield, and P.H. Barry. 2002. Single channel analysis of conductance and rectification in cation-selective, mutant glycine receptor channels. *J. Gen. Physiol.* 119:411–425.

O'Mara, M., P.H. Barry, and S.-H. Chung. 2003. A model of the glycine receptor deduced from Brownian dynamics studies. *Proc. Natl. Acad. Sci. USA* 100:4310–4315.

Panicker, S., H. Cruz, C. Arrabit, and P.A. Slesinger. 2002. Evidence for a centrally located gate in the pore of a serotonin-gated ion channel. *J. Neurosci.* 22:1629–1639.

Rajendra, S., J.W. Lynch, K.D. Pierce, C.R. French, P.H. Barry, and P.R. Schofield. 1995. Mutation of an arginine residue in the human glycine receptor transforms beta-alanine and taurine from agonists into competitive antagonists. *Neuron* 14:169–175.

Rundström, N., V. Schmieden, H. Betz, J. Bormann, and D. Langosch. 1994. Cyanotriphenylborate: Subtype specific blocker of glycine receptor chloride channels. *Proc. Natl. Acad. Sci. USA* 91:8950–8954.

Schmieden, V., J. Kuhse, and H. Betz. 1993. Mutation of glycine receptor subunit creates beta-alanine receptor responsive to GABA. *Science* 262:256–258.

Schofield, P.R., J.W. Lynch, S. Rajendra, K.D. Pierce, C.A. Handford, and P.H. Barry. 1996. Molecular and genetic insights into ligand binding and signal transduction at the inhibitory glycine receptor. *Cold Spring Harb. Symp. Quant. Biol.* 61:333–342.

Shan, Q., S.T. Nevin, J.L. Haddrill, and J.W. Lynch. 2003. Asymmetric contribution of alpha and beta subunits to the activation of alphabeta heteromeric glycine receptors. *J. Neurochem.* 86:498–507.

Spier, A.D., and S.C. Lummis. 2000. The role of tryptophan residues in the 5-hydroxytryptamine(3) receptor ligand binding domain. *J. Biol. Chem.* 275:5620–5625.

Taly, A., M. Delarue, T. Grutter, M. Nilges, N. Le Novere, P.J. Corringer, and J.P. Changeux. 2005. Normal mode analysis suggests a quaternary twist model for the nicotinic receptor gating mechanism. *Biophys. J.* 88:3954–3965.

Unwin, N. 1993. Nicotinic acetylcholine receptor at 9 A resolution. *J. Mol. Biol.* 229:1101–1124.

Unwin, N. 1995. Acetylcholine receptor channel imaged in the open state. *Nature* 373:37–43.

Unwin, N. 2000. The Croonian Lecture 2000. Nicotinic acetylcholine receptor and the structural basis of fast synaptic transmission. *Philos. Trans. R. Soc. Lond. B. Biol. Sci.* 355:1813–1829.

Unwin, N. 2003. Structure and action of the nicotinic acetylcholine receptor explored by electron microscopy. *FEBS Lett.* 555:91–95.

Unwin, N. 2005. Refined structure of the nicotinic acetylcholine receptor at 4 angstrom resolution. *J. Mol. Biol.* 346:967–989.

Unwin, N., A. Miyazawa, J. Li, and Y. Fujiyoshi. 2002. Activation of the nicotinic acetylcholine receptor involves a switch in conformation of the alpha subunits. *J. Mol. Biol.* 319:1165–1176.

Villarroel, A., S. Herlitze, M. Koenen, and B. Sakmann. 1991. Location of a threonine residue in the alpha-subunit M2 transmembrane segment that determines the ion flow through the acetylcholine receptor channel. *Proc. R. Soc. Lond. B. Biol. Sci.* 243:69–74.

Wagner, D.A., and C. Czajkowski. 2001. Structure and dynamics of the GABA binding pocket: A narrowing cleft that constricts during activation. *J. Neurosci.* 21:67–74.

Wotring, V.E., T.S. Miller, and D.S. Weiss. 2003. Mutations at the GABA receptor selectivity filter: A possible role for effective charges. *J. Physiol.* 548:527–540.

Wilson, G.G., and A. Karlin. 1998. The location of the gate in the acetylcholine receptor channel. *Neuron* 20:1269–1281.

Xu, M., and M.H. Akabas. 1996. Identification of channel-lining residues in the M2 membrane-spanning segment of the GABA(A) receptor alpha1 subunit. *J. Gen. Physiol.* 107:195–205.

Yu, Y., L. Shi, and A. Karlin. 2003. Structural effects of quinacrine binding in the open channel of the acetylcholine receptor. *Proc. Natl. Acad. Sci. USA* 100:3907–3912.

Zhong, W., J.P. Gallivan, Y. Zhang, L. Li, H.A. Lester, and D.A. Dougherty. 1998. From ab initio quantum mechanics to molecular neurobiology: A cation-pi binding site in the nicotinic receptor. *Proc. Natl. Acad. Sci. USA* 95:12088–12093.

10 Mechanosensitive Channels

Boris Martinac

Living cells are exposed to a variety of mechanical stimuli acting throughout the biosphere. The range of the stimuli extends from thermal molecular agitation to potentially destructive cell swelling caused by osmotic pressure gradients. Cellular membranes present a major target for these stimuli. To detect mechanical forces acting upon them cell membranes are equipped with mechanosensitive (MS) ion channels. Functioning as molecular mechanoelectrical transducers of mechanical forces into electrical and/or chemical intracellular signals these channels play a critical role in the physiology of mechanotransduction. Studies of prokaryotic MS channels and recent work on MS channels of eukaryotes have significantly increased our understanding of their gating mechanism, physiological functions, and evolutionary origins as well as their role in the pathology of disease.

10.1 Introduction

The idea of mechano-gated or mechanosensitive (MS) ion channels arose originally from studies of specialized mechanosensory neurons (Katz, 1950; Loewenstein, 1959; Detweiler, 1989). By converting mechanical stimuli exerted on the cell membrane into electrical or biochemical signals these channels function in a variety of physiological processes ranging from regulation of cellular turgor and growth in bacteria and fungi to touch, hearing, salt and fluid balance, and blood pressure regulation in mammals (García-Añovernos and Corey, 1997; Sachs and Morris, 1998; Hamill and Martinac, 2001; Gillespie and Walker, 2001; Corey, 2003a,b; Martinac, 2004; Lin and Corey, 2005). After their original discovery in embryonic chick skeletal muscle (Guharay and Sachs, 1984) and frog muscle (Brehm et al., 1984) MS channels have been documented in various types of nonspecialized cells (Sachs, 1988; Morris, 1990; Hamill and Martinac, 2001; Martinac, 2004) as well as in specialized mechanoreceptor neurons (Tavernarakis and Driscoll, 1997; Sukharev and Corey, 2004). Although limited compared to what we know about voltage- and ligand-gated channels our knowledge of the structure and function of MS channels has increased significantly over the last 20 years. The discovery of MS channels in bacteria (Martinac et al., 1987) has ultimately led to molecular identification and structural determination of the MS type of ion channels. The cloning of the bacterial MscL and MscS (Sukharev et al., 1994; Levina et al., 1999), the elucidation of their 3D crystal structures (Chang et al., 1998; Bass et al., 2002) and the demonstration of their physiological role in bacterial osmoregulation (Booth and Louis, 1999; Levina et al., 1999) have provided basis for intensive research and rapid progress in studies of the structure and function in this class of ion channels (Hamill and Martinac, 2001; Martinac, 2004). Later on, the cloning and genetic analysis of the *mec* genes in *Cenorhabditis elegans* (Tavernarakis and Driscoll, 1997; Goodman et al., 2002; Goodman et al., 2003), genetic and functional studies of the TRP-type MS channels (Montell et al., 2002; Zhou et al., 2003; Corey et al., 2004; Maroto et al., 2005) as well as molecular biological and functional studies of the TREK-1 family of 2P-type potassium channels (Patel et al., 1998, 2001) have further contributed to our understanding of the role of MS channels in the physiology of mechanosensory transduction.

10.2 Evolutionary Origins of MS Channels

MS channels exist in all three domains of living organisms indicating their early evolutionary origins (Fig. 10.1). It is likely that very early on MS channels evolved as cellular osmoregulators designed to measure small changes in the concentration of water across membranes of primordial cells (e.g., bacteria or archaea) (Sachs, 1988; Kung et al., 1990; Martinac, 1993; Kung and Saimi, 1995; Kung, 2005). At some later stage these osmoregulators could have been employed in regulation of cell size and volume (Christensen, 1987; Ubl et al., 1988) or in specialized forms of mechanotransduction, such as gravitropism in plants (Pickard and Ding, 1992), contractility of the heart (Sigurdson et al., 1992; Kohl and Sachs, 2001; Sachs,

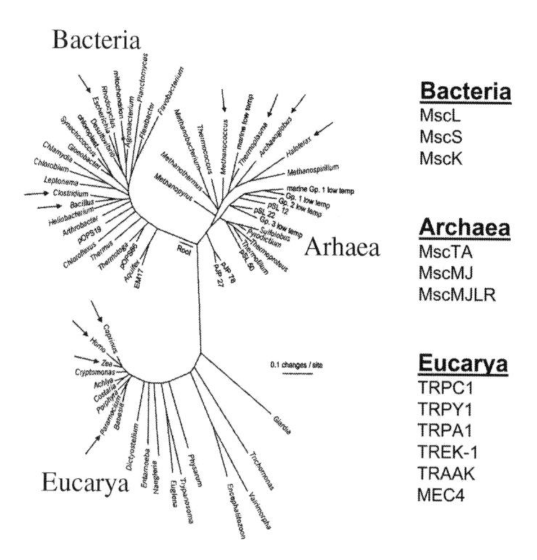

Fig. 10.1 *Ubiquity of MS channels*. Universal phylogenetic tree showing the life on Earth organized in three kingdoms of living organisms based on small subunit tRNA sequences (modified from Pace, 1997; with permission). Organisms, in which MS channels have been identified, are indicated by arrows. MS channels that have been identified at the molecular level are shown on the right.

2004a), or hearing (Hamill and Martinac, 2001; Gillespie and Walker, 2001). A basic question that remains to be answered is about the extent to which the mechanism of gating by mechanical force characteristic of prokaryotic (bacterial and archaeal) MS channels has been conserved and adapted to gating of MS channels in eukaryotes.

10.3 Bilayer and Tethered Model of MS Channel Gating by Mechanical Force

MS channels respond to membrane tension caused by mechanical force stretching membranes of living cells (Gustin et al., 1988; Sokabe and Sachs, 1990; Sokabe et al., 1991). Membrane tension required for half activation of most of the known MS channels is usually around several dynes/cm (10^{-3}N/m) (Sachs, 1988). The forces of that magnitude can be produced, for example, by differences in transmembrane (TM) osmolarity of a few milliosmols (Martinac, 1993; Sachs, 2004b).

Currently, there are two models which describe gating of MS channels by mechanical force: the bilayer and the more speculative tethered model (Hamill and Martinac, 2001) (Fig. 10.2). According to the bilayer model the tension in the lipid

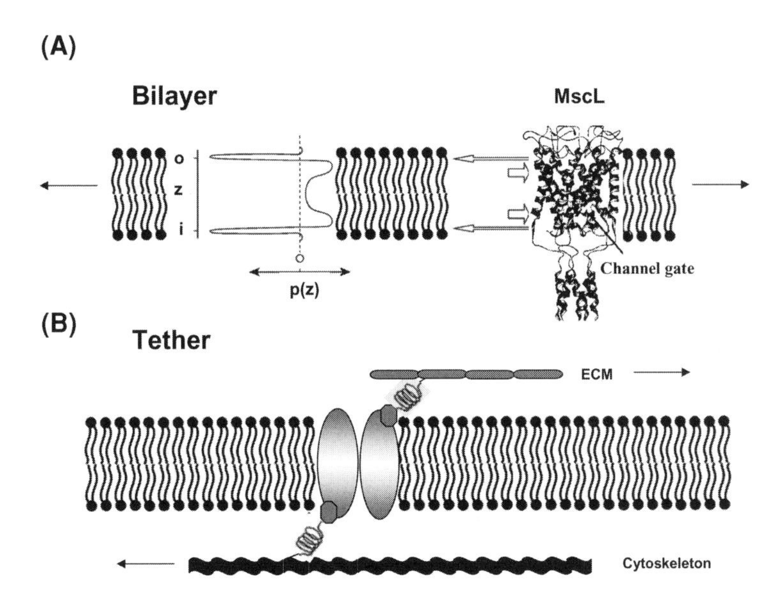

Fig. 10.2 *Bilayer and tethered models of gating MS channels by mechanical force.* (A) The internal pressure profile $p(z)$ plotted along the depth z of the bilayer (left) (Cantor, 1997), and a cartoon of the MscL channel (right). Tension (narrow arrows) pulling on the channel near the lipid head groups is balanced by positive pressure (broad arrows) exerted on the channel–lipid interface. Stretching (black arrows) or bending the bilayer causes a change in the internal pressure profile, which in turn causes change in the channel conformation leading to opening of MS channels, such as MscL (bilayer model). (B) Tension exerted on the cytoskeleton and/or ECM directly gates MS channels without involvement of the lipid bilayer (tethered model).

bilayer alone is sufficient to gate directly the MS channels. Since prokaryotic cells lack a cytoskeleton it is most likely the lipid bilayer is a tension-bearing element transmitting the mechanical force to the MS channels (Martinac, 2004; Kung, 2005). First proposed for the gating of bacterial MS channels (Martinac et al., 1990; Markin and Martinac, 1991), the bilayer model has been documented in various types of MS channels. Recent studies demonstrated that changes in the trans-bilayer tension profile, which gate bacterial MS channels, are caused either by protein–lipid bilayer hydrophobic mismatch and/or membrane curvature (Perozo et al., 2002a,b) (Fig. 10.2). Purified MscL, MscS, and other prokaryotic MS channels remain MS when reconstituted into liposomes (Sukharev et al., 1994; Häse et al., 1995; Kloda and Martinac, 2001a,b,c; Martinac, 2001; Perozo and Rees, 2003) (Fig. 10.3).

A number of eukaryotic MS channels including 2P-type potassium channels TREK-1 and TRAAK (Patel et al., 2001), TRP-type channels TRPC1 (Maroto et al., 2005), TRPY1 (Zhou et al., 2003), TRPY2 and TRPY3 (Zhou et al., 2005) as well as some BK-type potassium channels (Qi et al., 2003, 2005) have also been shown to be gated by bilayer deformation forces. The tethered model invokes direct connections between MS channels and cytoskeletal proteins or extracellular matrix (ECM) and requires relative displacement of the channel gate with respect to the cytoskeleton or ECM for channel gating (Hamill and McBride, 1997; Gillespie and Walker, 2001) (Fig. 10.2). This model was originally proposed for gating of MS channels in hair cells (Corey and Hudspeth, 1983; Corey, 2003a) and chick skeletal muscle (Guharay and Sachs, 1984) and should apply to eukaryotic MS channels in specialized mechanoreceptor cells (Lin and Corey, 2005). A summary of MS channels and their gating mechanism is given in Table 10.1.

The evidence showing that lipids play essential role in opening and closing not only of prokaryotic MS channels but also of the MS channels of fungi, plants, and animals has recently led to a proposal of a possible unifying principle for mechanosensation based on the bilayer mechanism (Kung, 2005). The main idea of the unifying principle is that forces from lipids gate MS channels independently of their evolutionary origin and type of cells in which they are found. According to this principle tethering of MS channels to rigid elements (cytoskeleton or ECM) does not necessarily imply mechanical force transmission. Tethers could serve instead to station the channels close to the cell surface involved in mechanotransduction and to attenuate or amplify the mechanical force and thus adjust the dynamic range within which MS channels operate in a particular cellular setting. Recent mutational and proteolytic cleavage studies have shown that bacterial MscL can be made as MS as any of the eukaryotic MS channels studied in patch-clamp experiments (Yoshimura et al., 1999; Ajouz et al., 2000). Furthermore, experiments in *Xenopus* oocytes seem to indicate that all necessary forces required to activate MS channels are transmitted through the lipids (Zhang et al., 2000). Consequently, the function of the cytoskeleton and/or ECM would consist in altering the forces within the lipid bilayer by absorbing mechanical stresses and modifying the time dependence of MS channel adaptation (Hamill and Martinac, 2001).

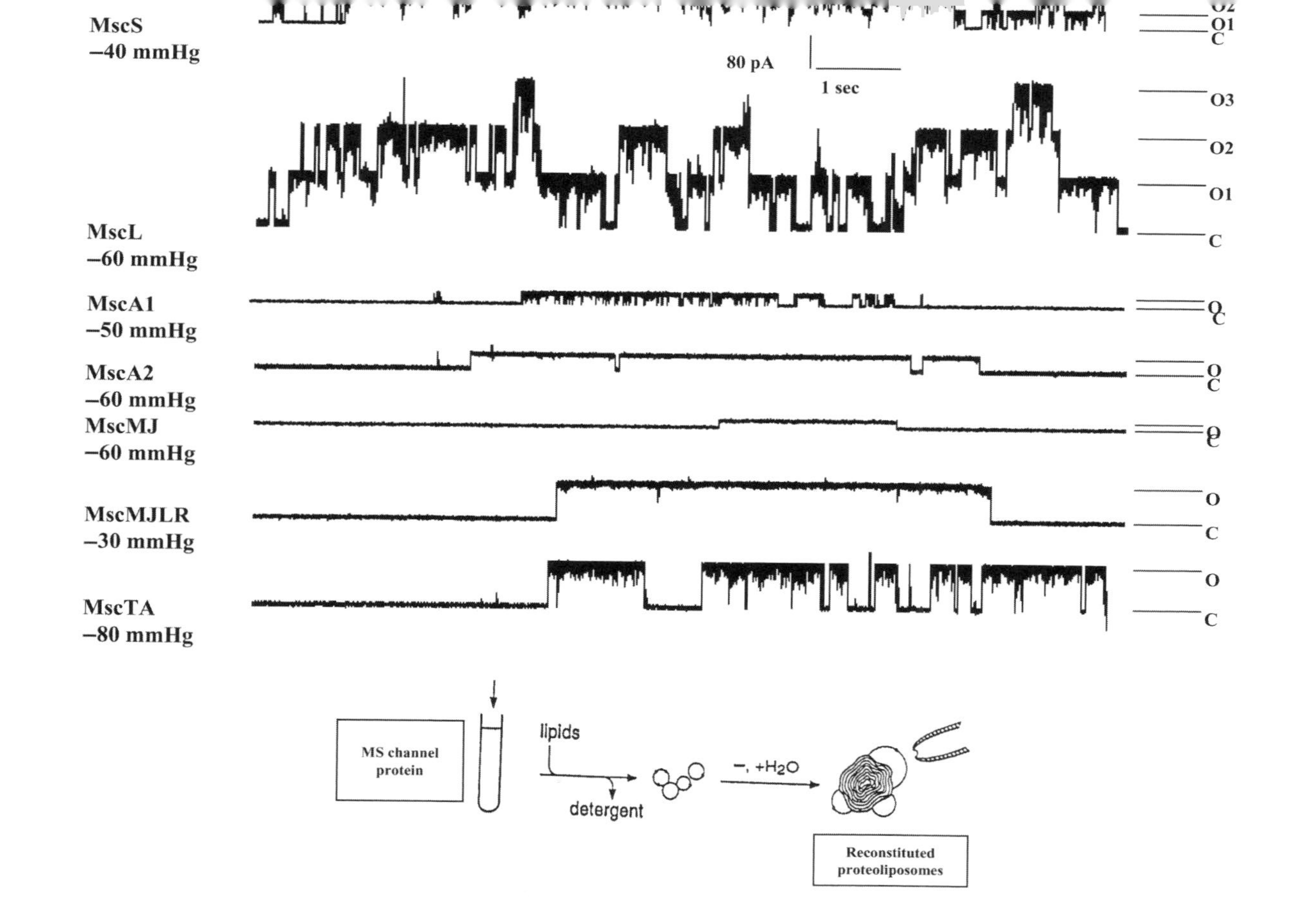

Fig. 10.3 *Current traces of prokaryotic MS channels reconstituted into liposomes*. MS channels found in bacteria and archaea. Current traces of MscL and MscS of *E. coli* are followed by traces of MscA1 and MscA2 of *H. volcanii*, MscMJ and MscMJLR of *M. jannashii* and MscTA of *T. acidophilum*. Currents were recorded at pipette voltage of +40 mV. C denotes the closed state and O_n denotes open state of *n* channels. Note: 1 mm Hg = 133 Pa. *Bottom*: A diagram showing dehydration/rehydration method of MS channel reconstitution into liposomes.

Table 10.1 Summary of cloned MS channels, whose gating mechanism and/or physiological function have been characterized.

MS channel	Source	Gating mechanism	Amphipaths	Physiological function	References
MscL	Bacteria	Bilayer	Yes	Cellular turgor and growth	Häse et al., 1995; Sukharev et al., 1999; Levina et al., 1999
MscS	Bacteria	Bilayer	Yes	Cellular turgor and growth	Martinac et al., 1990; Sukharev et al.,1993 Levina et al., 1999
MscA1	Archaea	Bilayer	NT	Cellular turgor*	Le Dain et al., 1998
MscA2	Archaea	Bilayer	NT	Cellular turgor*	Le Dain et al., 1998
MscMJ	Archaea	Bilayer	Yes	Cellular turgor*	Kloda and Martinac, 2001a
MscMJLR	Archaea	Bilayer	Yes	Cellular turgor*	Kloda and Martinac, 2001b
MscTA	Archaea	Bilayer	Yes	Cellular turgor*	Kloda and Martinac, 2001c
MEC4	*C. elegans*	Tether*	NT	Touch	Tavernarakis and Driscoll, 1997; Hamill and Martinac, 2001
TREK-1	Brain, heart	Bilayer	Yes	Resting membrane potential	Patel et al., 1998, 2001
TRAAK	Brain, spinal chord	Bilayer	Yes	Resting membrane potential	Patel et al., 1998, 2001
EnaC	Rat, human, *C. elegans*	Bilayer/tether*	Yes	Touch	Awayda et al., 2004; Ronan and Gillespie, 2005
TRPC1	*Xenopus* oocytes	Bilayer	NT	Unknown	Maroto et al., 2005
TRPY	Fungi	Bilayer*	NT	Cellular turgor	Zhou et al., 2003, 2005
TRPA1	Hair cells	Tether*	NT	Hearing	Corey et al., 2004; Lin and Corey, 2005
TRPN	Drosophila, zebrafish	Tether*	NT	Touch, hearing	Lin and Corey, 2003; Sidi et al., 2003; Sotomayor et al., 2005
SAKCa	Chick heart	Bilayer/tether	Yes	Myogenic tone	Kawakubo et al., 1999; Qi et al., 2005

Note however, that in contrast to the bilayer mechanism there is no single experimental result that provides unequivocal support for the tethered model of MS channel gating (Hamill and Martinac, 2001). Asterisk indicates a likely gating mechanism or physiological function that has not yet been firmly established. NT indicates that the effect of amphipaths has not been tested in the particular type of MS channels.

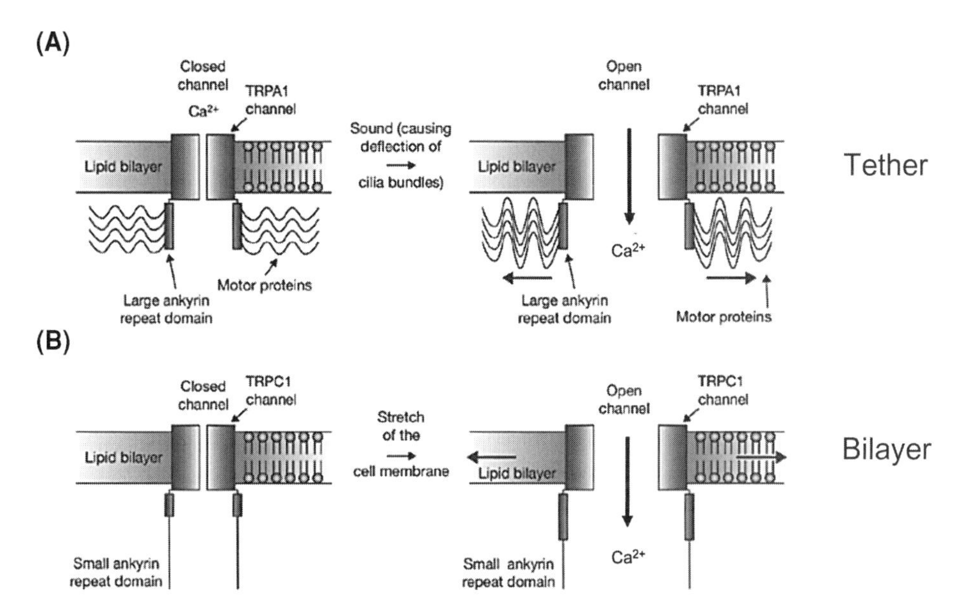

Fig. 10.4 *TRPA1 and TRPC1 acting as mechanosensitive channels*. (A) The functional TRPA1 is probably a homotetramer made of TRPA1 monomers or a heterotetramer composed of TRPA1 monomers and subunits of another channel. Each TRPA1 monomer has 17 ankyrin repeats at the cytoplasmic N-terminal end. It is proposed that the ankyrin repeats function as a spring gating the channel (tethered model). (B) The functional TRPC1 is probably also a homotetramer composed of TRPC1 monomers or a heterotetramer made of TRPC1 monomers and subunits of another channel protein. Each TRPC1 monomer contains three or four ankyrin repeats at the N-terminal cytoplasmic end, which may not be directly involved in channel gating. It is proposed that TRPC1 like MscL, is directly gated by bilayer tension (bilayer model) (modified from Barritt and Rychkov, 2005; with permission).

Of particular interest in regard to the preservation and adaptation of the bilayer mechanism to the gating of eukaryotic MS channels is the TRP superfamily of ion channels that have also been associated with mechanosensation (Fig. 10.4). Among these channels the TRPC1 comes closest to showing that certain TRP channels are gated by mechanical force directly from lipids (Maroto et al., 2005). In contrast, the role of tethers seems to be well established for the hair-cell transduction channel identified to be most likely the TRPA1 protein (Corey et al., 2004).* A recent study by Kwan et al., 2006 indicates that although TRPA1 is critical for the transduction of noxious cold and mechanical stimuli, it is not essential for hair-cell transduction. Although the familiar "trapdoor" model of hair-cell mechanotransduction had to be modified the long N-terminal ankyrin repeats of TRPA1 are compatible with the elastic properties of the gating spring (Howard and Bechstedt, 2004; Lin and Corey,

* Note added in proof:
A recent study by Kwan et al., 2006 indicates that although TRPA1 is critical for the transduction of noxious cold and mechanical stimuli, it is not essential for hair-cell transduction.

2005; Sotomayor et al., 2005). Nevertheless, it cannot be excluded that TRPA1 may also "sense" the force from lipids given its activation by lipid-like compounds such as mustard oils, cannabinoids or bradykinin (Bandell et al., 2004; Jordt et al., 2004) and requirement of PIP_2 for hair-cell mechanotransduction (Hirono et al., 2004).

10.4 MS Channels of Bacteria and Archaea

The advent of the patch-clamp technique (Hamill et al., 1981) made possible studies of ion channels in bacteria despite their minute size (Martinac et al., 1994). Bacterial MS channels have been discovered and extensively studied in giant spheroplasts of *E. coli* (Martinac et al., 1987, 1992; Zoratti and Ghazi, 1993; Martinac, 2001; Stokes et al., 2003; Martinac, 2004), which harbors three types of MS channels in its cellular membrane: MscL (*L*arge), MscS/MscK (*S*mall/*K*alium, i.e., potassium), and MscM (*M*ini). The channels are named according to their single-channel conductance, which is ~3 nS for MscL, ~1 nS for MscS and MscK, and ~0.3 nS for MscM (Berrier et al., 1996). MS channels of either MscL- or MscS-type have also been found in other Gram-negative and Gram-positive bacteria as well as in archaea (Le Dain et al., 1998; Kloda and Martinac, 2002; Martinac and Kloda, 2003).

10.4.1 MscL

MscL, cloned and sequenced by Kung and co-workers (Sukharev et al., 1993, 1994), was the first among the MS-type of ion channels characterized at the molecular level. It comprises 136 residues and shares no significant sequence similarity with known voltage- or ligand-gated ion channels. A few years after its cloning Rees and co-workers (Chang et al., 1998) resolved the 3D oligomeric structure of the MscL from *Mycobacterium tuberculosis* (Tb-MscL) by X-ray crystallography. The structure of the channel resolved at 3.5 Å was obtained in the closed state and showed that the channel is a homopentamer whose subunits have two α-helical TM domains, TM1 and TM2, cytoplasmic N- and C-terminal domains, and a central periplasmic domain. The structure of the open MscL channel has recently been solved by a combination of cysteine-scanning mutagenesis, site-directed spin labeling, and EPR spectroscopy (Perozo et al., 2002a). The open state of MscL has a water-filled pore of >25 Å in diameter which is lined by the TM1 helices from the five subunits indicating an overall large conformational change in the channel structure during the channel opening (Perozo et al., 2002a). Another recent MscL study employing FRET spectroscopy confirmed these findings by showing a difference of 16 Å in diameter between the closed and the open conformations of the MscL channel (Corry et al., 2005), a result in a close agreement with the EPR spectroscopic study (Perozo et al., 2002a). Several independent studies also showed that the channel undergoes large conformational changes during opening and closing (Cruickshank et al., 1997; Sukharev et al., 2001; Biggin and Sansom, 2001; Gullingsrud et al., 2001; Betanzos et al., 2002; Colombo et al., 2003; Gullingsrud and Schulten, 2003).

MscL was also the first MS channel unambiguously shown to be gated by the mechanical force transmitted directly to the channel through the lipid bilayer (Sukharev et al., 1994; Häse et al., 1995) (Fig. 10.2). A recent study evaluated two potential triggers of MscL gating by the bilayer mechanism: (i) protein–lipid bilayer hydrophobic mismatch and (ii) membrane curvature (Perozo et al., 2002b). The study demonstrated that hydrophobic mismatch is not the driving force that triggers MscL opening, although specific mismatch levels could stabilize intermediate conformational states along the kinetic path toward the open state. According to this study the mechanism of mechanotransduction in MscL and possibly other MS channels is defined by both local and global asymmetries in the trans-bilayer tension profile at the lipid–protein interface. Lipid composition effects on MscL gating were also investigated by performing molecular dynamics simulations (Elmore and Dougherty, 2003). The results of this study showed that protein–lipid interactions were clearly altered by the headgroup changes, leading to conformational differences in the C-terminal region of MscL. In addition, all MD simulations showed evidence of hydrophobic matching between MscL and the lipid membrane and indicated further that protein–lipid interactions could be more important for proper MscL function and assembly than the protein–protein interactions. In accordance with this notion an MscL mutagenesis study showed that the disturbance of the hydrophobic interaction between the membrane lipid and the periplasmic rim of the channel's funnel impaired the function of MscL (Yoshimura et al., 2004). In further support of the significant role that hydrophobic matching could play for the function of MS channels, an independent study using gramicidin A (gA), which exhibits mechanosensitivity in lipid bilayers (Goulian et al., 1998; Hamill and Martinac, 2001), demonstrated that the gA channel could behave as a stretch-activated or stretch-inactivated channel depending on the bilayer thickness (Martinac and Hamill, 2002). These key findings in bilayer-controlled functional properties of mechanotransducer channels emphasize further that bilayer is much more than a neutral solvent by actively modulating the specificity and fidelity of signaling by membrane proteins.

10.4.2 MscS

Among bacterial MS channels MscS was discovered first (Martinac et al., 1987). Besides being activated by membrane tension MscS is also regulated by voltage (Martinac et al., 1987; Cui et al., 1995). It is encountered in 100% of spheroplast membrane patches characterized by a large number of channels that inactivate rapidly upon sustained application of suction to the patch pipette (Levina et al., 1999). Furthermore, MscS exhibits an anion selectivity (P_{Cl}/P_K) of $\sim$1.5–3.0 (Kloda and Martinac, 2002; Sukharev, 2002). MscS and its close relative MscK were also cloned in *E. coli* (Levina et al., 1999). MscS, encoded by the *yggB* locus is a small, 286-residue membrane protein. By contrast, MscK, encoded by *kefA*, is a large, multi-domain 120-kDa protein comprising 1120 residues. Originally, MscS and MscK were considered to represent a single type of bacterial MS channel, since in patch-clamp experiments they exhibit similar sensitivities to activation by pressure and have conductance of ~1 nS (Martinac et al., 1987; Sukharev et al., 1993). However, their activities can clearly be distinguished (Li et al., 2002). MscK activity is found in about 70% of spheroplast patches. These are characterized by fewer channels, which do not inactivate upon continuous application of suction. Another distinguishing property of MscK is its sensitivity to the extracellular ionic environment (Li et al., 2002). MscK was reported to show some anionic preference (Li et al., 2002), although mutational analysis has suggested that it might actually be cation specific (McLaggan et al., 2002). Nevertheless, the overall similarities in their conductance, selectivity, and sensitivity to membrane tension seem to reflect structural similarity between the two channels, since MscK contains an MscS-like domain at its C-terminus (Levina et al., 1999).

MscS was the second MS channel, whose 3D structure was solved by X-ray crystallography (Bass et al., 2002). The crystal structure resolved at 3.9 Å resolution shows that unlike MscL the MscS channel folds as a homoheptamer with each of the seven subunits consisting of three TM domains. The crystal structure also shows that the TM3 helices line the channel pore whereas the TM1 and TM2 helices constitute the sensors for membrane tension and voltage (Bass et al., 2002). The fact that MscS is not only a tension sensor but also a likely voltage sensor might provide insight into the structural changes induced by voltage in membrane proteins (Bezanilla and Perozo, 2002).

10.4.3 Archaeal MS Channels

MS channel activities have also been documented in several archaeal species including *Haloferax volcanii* (Le Dain et al., 1998), *Thermoplasma acidophilum* (Kloda and Martinac, 2001c), and *Methanococcus jannashii* (Kloda and Martinac, 2001a,b). Similar to MscL and MscS these channels are gated by the bilayer mechanism (Fig. 10.3). They all have large conductances and low selectivity for ions. They are weakly voltage dependent, can be activated by amphipaths and are blocked by submillimolar concentrations of gadolinium (Kloda and Martinac, 2002; Martinac and Kloda, 2003). It is important however, to remember that gadolinium (Gd^{3+}) has a limited use as a tool for MS channel studies because an increasing number of reports indicated that Gd^{3+}might not directly block MS channels, but instead should affect the nature of the lipid bilayer and thus indirectly alter the mechanosensitivity of MS channels (Ermakov et al., 2001; Tanaka et al., 2002).

10.4.4 Families of Prokaryotic MS Channels

Sequence alignments of MscL and MscS homologues obtained from a large number of recently cloned genomes of various bacteria and archaea revealed that these channels form families, which have common evolutionary origins. MscL relatives form a separate family of MS channels, encompassing Gram-negative and Gram-positive bacteria, as well as a few representatives from archaea and fungi (Kumánovics et al., 2003; Martinac and Kloda, 2003; Pivetti et al., 2003). In contrast, the MscS relatives are much more diverse and include representatives from bacteria, archaea, fungi, and plants (Kloda and Martinac, 2002; Martinac and Kloda, 2003; Pivetti et al., 2003). It is worth noting that the TM1 domain of MscL was successfully used as a genetic probe for molecular identification of MscMJ and MscMJLR of *M. jannashi*, which belong to the MscS channel family. This is of possible interest because it seems to suggest that all prokaryotic MS channels might have a common evolutionary ancestry (Kloda and Martinac, 2002). This proposal has however, been questioned, because of the lack of statistical evidence for a link between the MscL and MscS families. Instead it has been proposed that the MscL and MscS families of MS channels evolved independently (Okada et al., 2002; Pivetti et al., 2003). Nonetheless, sequence similarity between the highly conserved pore-lining helices in the two types of MS channels, i.e., TM1 of MscL and TM3 of MscS might indicate a possible evolutionary link between MscS and MscL families (Kloda and Martinac, 2001a; Pivetti et al., 2003).

10.5 MS Channels of Eukaryotes

MS ion channels have first been documented in eukaryotes (Brehm et al., 1984; Guharay and Sachs, 1984). Despite their extensive electrophysiological characterization in a large variety of eukaryotic cells (Sachs and Morris, 1998; Hamill and Martinac, 2001), the structural characterization and elucidation of the roles that MS channels play in mechanosensory transduction in eukaryotes have been slow compared to the progress made in understanding of the structure and function of MS channels in prokaryotes. Nonetheless, significant new developments toward understanding of the structure and function of eukaryotic MS ion channels have occurred through recent work on TREK-1 and TRAAK, two MS channels belonging to a new family of two-pore-domain (2P-domain) weakly inward-rectifying K^+ channels, and mutagenesis studies in *Caenorhabditis elegans*, zebrafish and *Drosophila*, which revealed that some of the MEC/DEG (mechanosensory abnormal/degenerins) and TRP (*t*ransient *r*eceptor *p*otential) proteins function as MS channels (Tavernarakis and Driscoll, 1997; Hamill and Martinac, 2001; Patel et al., 2001; Minke and Cook, 2002; Clapham, 2003).

10.5.1 TREK and TRAAK

The 2P domain K^+ channels function as dimers in which both N and C termini face the cytoplasm. To date, 15 human members of this channel superfamily have been identified. Most of the channels behave as pure leak or background K^+ channels, whose main function is to maintain the resting level of membrane potential. Two of them, TREK-1 and TRAAK belong to a subfamily of MS 2P-domain potassium channels (Patel et al., 1998; Maingret et al., 1999a). TREK channels are polymodal (i.e., gated by a variety of chemical and physical stimuli) K^+ channels. They are expressed in a variety of tissues, but are particularly abundant in the brain and in the heart (Patel et al., 1999). They are opened by both physical (stretch, cell swelling, intracellular acidosis, heat, and voltage) (Maingret et al., 1999b, 2000a, 2002) and chemical stimuli (polyunsaturated fatty acids, lysophospholipids, membrane crenators, and volatile general anesthetics) (Patel et al., 1998, 1999; Maingret et al., 2000b; Terrenoire et al., 2001). Characteristic for TREK channels is that they are preferentially activated by positive curvature of the membrane (induced by suction applied to the patch pipette) similar to MS channels in astrocytes (Bowman et al., 1992; Bowman and Lohr, 1996). In addition, their activity is downmodulated by phosphorylation of a C-terminal serine residue by cAMP-dependent protein kinase (Bockenhauer et al., 2001). The C-terminal domain is essential for mechanosensitivity and acid sensing in these channels. Partial deletion of this region impairs activation by membrane tension or lipids (Patel et al., 2001). Protonation of E306 in the C-terminus is responsible for acidic activation, and an E306A substitution locks channel in an open configuration and prevents the PKA-mediated downmodulation (Honoré et al., 2002). The polymodality of the TREK channels indicates that in neurons of the central nervous system they play an important role in physiological (electrogenesis), pathophysiological (ischemia), and pharmacological conditions (anesthesia).

TRAAK is similar to TREK in that it can be activated by arachidonic acid and membrane tension (Maingret et al., 1999a). Similar to TREK, TRAAK is also activated by external but not internal application of the negatively charged amphipath trinitrophenol (TNP), which presumably partitions into and expands the less negative external monolayer to induce a convex curvature (Maingret et al., 1999a). However, unlike TREK, TRAAK is not opened by intracellular acidosis in the absence of membrane stretch. Also unlike TREK, it can be activated by the cytoskeletal inhibitors colchicine and cytochalasin, which indicates that the cytoskeleton might constrain tension development in the bilayer (Patel et al., 1998). It is worth mentioning that stretch activation of both TREK and TRAAK is also observed in excised membrane patches, indicating that cell integrity is not required for activation by membrane tension. In common with TREK and MscL, the C-terminal domain of TRAAK contains a charged cluster that is critical for both arachidonic acid activation and mechanosensitivity. TRAAK is widely expressed in the brain, spinal cord, and retina, which indicates that it has a function wider than mechanotransduction in neuronal excitability (Patel et al., 1999).

10.5.2 DEG/ENaC

Genetic screens of *C. elegans* have identified a number of membrane proteins being required for touch sensitivity in this nematode. Four of these proteins, MEC-2, MEC-4, MEC-6, and MEC-10, are candidates for a mechanically gated ion channel complex (Driscoll and Chalfie, 1991; Huang et al., 1995). These proteins belong to a superfamily of amiloride-sensitive Na^+ channels of the transporting epithelia and the degenerins (DEG/ENaC channels) with many of them suspected to be directly gated by mechanical stimuli (Tavernarakis and Driscoll, 1997; Sukharev and Corey, 2004). A recent electrophysiological recording from *C. elegans* touch receptor neurons showed that external mechanical force activated mechanoreceptor currents carried by Na^+ blocked by amiloride thus suggesting that a DEG/ENaC channel is mechanically gated (O'Hagan et al., 2005; Ronan and Gillespie, 2005). ENaC is a hetero-oligomer of unknown stoichiometry composed of α, β, and γ subunits, which share 30% sequence identity (Canessa et al., 1994; Garty and Palmer, 1997; Schild et al., 1997). A fourth, δ-subunit is mainly expressed in the testis and ovaries (Darboux et al., 1998). The secondary structure and membrane topology of ENaC are similar to those of the bacterial MscL channel and the ATP-gated P2X receptor channels (North, 1996). Each subunit has two TM domains, TM1 and TM2, intracellular N- and C-termini, and a large extracellular loop.

MEC/DEG superfamily of ion channels have diverse functions and include acid-sensing ion channels ASICs (Waldmann and Lazdunski, 1998), molluscan FMRFamide-gated channels (Lingueglia et al., 1995), and Drosophila Na^+ channels expressed in gonads (Adams et al., 1998). The MEC/DEG subfamily of degenerins are responsible for swelling-induced neuronal degeneration in *C. elegans* and includes the MEC-2, MEC-4, MEC-6, and MEC-10 proteins, which have been shown to underlay mechanoreceptor currents in this nematode (O'Hagan et al., 2005). A role for MEC/DEG proteins in mechanotransduction in *C. elegans* has also been indicated by the finding that mutations in MEC-4 resulted in touch insensitivity and that dominant mutations in the same gene resulted in swelling-induced degeneration and lysis of the mechanosensory neurons, which seems consistent with continuously open channels. Other DEG/ENaC channels including UNC-8 in *C. elegans* (Tavernarakis et al., 1997) and pickpocket in *D. melanogaster* (Ainsley et al., 2003) have also been implicated in mechanosensation.

A hypothetical model of the MEC-mechanotransduction complex proposes that MEC-4 and MEC-10 form the MS transduction channel, which is intracellularly attached via MEC-2 linker to β-tubulin (MEC-7) and α-tubulin (MEC-12) (Du et al., 1996; Ronan and Gillespie, 2005). According to the model the MEC-6 protein should interact with the MEC-4/MEC-10 channel to regulate the activity of the channel, which may also be attached to the ECM. Interestingly, mechanoreceptor currents became reduced but not eliminated in mutants of *C. elegans* affecting MEC-7 β-tubulin found only in touch-receptor neurons (O'Hagan et al., 2005). Therefore, forces transmitted through microtubules and/or ECM proteins might not directly gate the channel but instead produce tension in the lipid bilayer and indirectly activate the MS channels in neuronal membranes (Hamill and Martinac, 2001). In support of this view a recent study provided some evidence for alteration of the ENaC activity by changes in membrane bilayer properties (Awayada et al., 2004). Whatever the mechanism of gating in DEG/ENaC ion channels might be, i.e., bilayer, tether or both, a combination of the patch-clamp electrophysiology and *C. elegans* genetics should significantly advance our understanding of the role of MEC/DEG-type MS ion channels in mechanosensory neurons.

10.5.3 TRPs

Another large and diverse family of ion channels comprises more than 20 TRP (transient receptor potential) proteins organized in six subfamilies of cation-selective channels (Clapham, 2003). Each of the channels has six TM segments and a membrane topology similar to that of some voltage- and cyclic-nucleotide-gated channels (Hartneck et al., 2000; Montell et al., 2002). Expressed in many tissues in numerous organisms the TRP channels function as cellular sensors mediating responses to a variety of physical (e.g., light, osmolarity, temperature, and pH) and chemical stimuli (e.g., odors, pheromones, and nerve growth factor) (Minke and Cook, 2002; Voets et al., 2005). TRP channels can be activated by sensory stimuli either directly or by a variety of second messengers. They function as specialized biological sensors that are essential in processes such as vision, taste, tactile sensation, and hearing. Several TRP channels may be inherently MS, such as *Drosophila* NompC (now TRPN) identified to have a clear role in mechanosensation (Montell et al., 2002; Montell, 2003; Lin and Corey, 2005), OSM-9 channel, which is a member of the TRPV subfamily involved in touch and osmosensing in *C. elegans* (Colbert et al., 1997), PKD-1 and PKD-2, members of the TRPP subfamily, which mediate mechanosensation in kidney cilia (Nauli et al., 2003), TRPA1 (previously ANKTM1), which is the most likely candidate for the hair-cell mechanotransduction channel (Corey et al., 2004),* TRPY1 (formerly Yvc1p), the MS channel found in yeast vacuole (Zhou et al., 2003), and TRPC1 channel, a member of the canonical TRP subfamily, which has been identified as MscCa, the Ca^{2+} permeable MS channel in *Xenopus* oocytes (Maroto et al., 2005).

TRPN (NompC, i.e. *no m*echanosensory *p*otential) was identified in uncoordinated *Drosophila* mutants that also show an absence or reduction in the mechanoreceptor potentials recorded from external sensory bristles (Walker et al., 2000). It is a large protein containing 29 ankyrin (ANK) repeats in its N-terminal domain, which is a unique feature that it shares with TRPA1 implicated in hair-cell function (Corey et al., 2004; Lin and Corey, 2005). ANK repeats may couple the TRPN channel to cytoskeletal proteins, which could gate or anchor the channel. TRPN has however, not been conclusively shown to function as an MS channel in patch-clamp experiments. The best evidence for the TRPN functioning as a mechanotransduction channel comes from experiments showing that flies carrying a missense mutation between the third and fourth TM domains of TRPN showed more rapid adaptation to sustained bristle deflections than the wild-type flies (Walker et al., 2000; Sukharev and Corey, 2004). In contrast to the MEC/DEG proteins, which are expressed only in nonciliated touch cells (Tavernarakis and Driscoll, 1997; Adams et al., 1998), TRPN and its homologue in *C. elegans* are selectively expressed in ciliated mechanoreceptors (Hartneck et al., 2000) suggesting that TRPN could also play a role in hearing given that Drosophila TRPN mutants have partially defective auditory responses (Lin and Corey, 2005). In zebrafish TRPN has been shown to underlie the "microphonic"

* However, see the note added in proof on p. 375.

current usually generated when the hair bundle is mechanically deflected. Inhibiting the synthesis of TRPN by morpholino nucleotides not only inhibited the microphonic current, but also caused behavioral abnormalities characteristic of vestibular dysfunction (Sidi et al., 2003). As in Drosophila TRPN may constitute the MS channel necessary for auditory hair-cell transduction in zebrafish. A TRPN orthologue in mammalian vertebrates has not been identified (Lin and Corey, 2005).

Another MS channel involved in hearing in flies is a member of the TRPV subfamily, named Nanchung. It is a Ca^{2+}-permeant cation channel localized to the ciliated endings of the *Drosophila* auditory organ (Kim et al., 2003). Deletion of Nanchung causes deafness and a lack of coordination, thus strongly indicating that Nanchung is a component of the auditory mechanosensor (Corey, 2003b). Among the members of the TRPV subfamily two relatives of the vanilloid receptor VR1 (neuronal membrane receptor for capsaicin and related irritant compounds; Szallasi and Blumberg, 1990, 1999), i.e., SIC (for *sic* transfected Chinese hamster ovary cells) (Suzuki et al., 1999) and VR-OAC (for *v*anilloid *r*eceptor-related *o*smotically *a*ctivated ion *c*hannel) (Liedtke et al., 2000), function as osmotically gated ion channels. SIC is a mammalian Ca^{2+}-permeable channel activated by cell shrinkage and inhibited by cell swelling. VR-OAC was cloned from *C. elegans* as a large protein, which like VR1 has three ANK repeats in its N-terminal region (Minke and Cook, 2002).

PKD2, a member of the TRPP subfamily, functions as a Ca^{2+}-permeable ion channel when co-expressed in Chinese hamster ovary (CHO) cells with PKD1, another much larger (10–12 TM helices) member of the same subfamily (Hanaoka et al., 2000). The cytosolic carboxy terminal domain of PKD1 includes a coiled-coil domain which is implicated in protein–protein interactions. The coiled-coil domain of PKD1 interacts with the carboxy terminal domain of PKD2 (Delmas, 2004; Nauli and Zhou, 2004). Localized in the apical cilium of the kidney tubule epithelia, the PKD1/PKD2 complex plays a mechanosensory role in renal epithelia by sensing fluid flow in ciliated epithelial cells (Nauli et al., 2003; Nauli and Zhou, 2004).

Deflection of stereocilia of vertebrate hair cells leads to the opening of MS ion channels at the tips of these cells. Identification of the mechanotransduction channel protein underlying the detection of sound by the ear has been difficult because of the low copy number of a few hundred channels per hair cell. By screening all 33 TRP channels of the mouse using in situ hybridization in the inner ear Corey et al. (2004) have recently provided evidence that the TRPA1 (formerly ANKTM1) constitutes, or is a component of the MS transduction channel of vertebrate hair cells. Similar to other TRP channels TRPA1 is a nonselective cation channel highly permeable to Ca^{2+}. The channel has 17 ankyrin repeats in its N-terminal domain, which led to a proposal that ankyrin repeats are key elements acting as a spring-like gating structure in the activation mechanism of the channel (Fig. 10.4). This proposal has found support in a study examining the elastic properties of the ankyrin repeats by molecular dynamic simulations showing that the extension and stiffness of the large ankyrin-repeat structures matched those predicted by the gating-spring model of the mechanotransduction channel (Sotomayor et al., 2005). Nevertheless, it remains unclear how the channel is activated by the ankyrin spring (Barritt and Rychkov, 2005; Lin and Corey, 2005). Given that TRPA1 is also expressed in neurons of the dorsal root ganglia, the trigeminal nerve and photoreceptors leaving open what the function of the ankyrin repeats in these other locations might be, another proposal has been put forward suggesting that the polyankyrin tether could pass the mechanical force to the channel indirectly through the lipids (Kung, 2005).

That a member of the TRP family functions as an MS channel was directly demonstrated by the patch-clamp recording from TRPY1, a 300–400 pS channel present in the vacuole of the yeast *Saccharomyces cerevisiae* (Zhou et al., 2003). TRPY1 is a Ca^{2+}-permeable channel that mediates Ca^{2+} release from the yeast vacuole induced by hyperosmotic shock. It is functionally analogous to bacterial MscL and MscS, since a temporary osmotic imbalance is sufficient to activate it. Two distant fungal homologues of TRPY1, TRPY2 from *Kluyveromyces lactis* and TRPY3 from *Candida albicans* have also been shown to function as MS channels in patch-clamp experiments (Zhou et al., 2005). When heterologously expressed in *S. cerevisiae* they retained their mechanosensitivity and could restore the ability to release vacuolar Ca^{2+} upon hypertonic stimulation in the yeast cells having deletion in the *trpy1* gene. Although the conclusive evidence that TRPY channels are gated by bilayer deformation forces is still lacking, because to date they have not been reconstituted and activated in liposomes, it is likely that these channels could respond directly to bilayer tension given their microbial origin and evolutionary distance to TRP channels in higher organisms.

Hamill and colleagues used detergent solubilization of frog oocyte membrane proteins, followed by liposome reconstitution and functional examination by the patch-clamp and identified TRPC1, a member of the canonical TRP subfamily (TRPC1-TRPC7), as MscCa, the verterbrate MS cation channel (Maroto et al., 2005). The evidence came from SDS-PAGE analysis of a membrane protein fraction showing an abundance of 80 kDa protein band that correlated with high MscCa activity in liposomes. The protein band was identified by immunological techniques as TRPC1. Further evidence was provided by injecting a TRPC1-specific antisense RNA in *Xenopus* oocytes, which abolished endogenous MscCa, and heterologous expression of human TRPC1 into CHO-K1 cells, which significantly increased MscCa expression. The role of TRPC1 in mechanosensation seems consistent with the general roles of TRP proteins as sensory detectors (Clapham, 2003).

TRPC1 and TRPA1 present an interesting case of MS channels in terms of their gating by mechanical force. Although definitive evidence for the gating of TRPC1 by the bilayer mechanism is lacking, because the purified TRPC1 protein has not yet been reconstituted and activated in liposomes, the strategy used for its identification as an MS channel, which was similar to that leading to the molecular cloning of MscL (Sukharev et al., 1994, 1997), strongly suggest that these channels could respond directly to bilayer tension (Kung, 2005; Lin and Corey, 2005; Maroto et al., 2005) (Fig. 10.4). In contrast, according to the present evidence TRPA1 is most likely activated directly via the elastic ankyrin spring located in its N-terminal domain (Corey et al., 2004; Lin and Corey, 2005; Sotomayor et al., 2005) (Fig. 10.4). The question is whether the number of ankyrin repeats (17 in TRPA1 and 3 in TRPC1) is what determines how mechanical force is conveyed to a TRP MS channel. The answer to this question would help to resolve and reconcile the current dichotomy between the bilayer and tethered mechanism of MS channel gating.

10.5.4 SAKCa

It is worth considering here mechanosensitivity of the BK_{Ca} channels. These channels have been found in a variety of cells having function in processes, which include firing in neurons, secretion in endocrine and exocrine cells, and myogenic tone in arterial smooth muscle (Gribkoff et al., 2001). Activation of BK_{Ca} by membrane stretch in the absence of calcium has been reported in several types of cells including the heart cells of the chick, which thus led to renaming the channel to SAKCa for stretch-activated Ca-dependent potassium channel (Kawakubo et al., 1999). It has been found that the sequence ERA present in a Stress-axis Regulated Exon (STREX) segment of the chick and human relatives of the SAKCa channels is essential for the channel mechanosensitivity. A single amino acid substitution A674T in STREX or deletion of STREX itself in a cloned SAKCa channel from the chick apparently eliminated the mechanosensitivity of this channel (Tang et al., 2003). A recent study demonstrated that the activity of the SAKCa channels could be modulated by amphipaths chlorpromazine (CPZ) and trinitrophenol (TNP) in a manner strikingly similar to their effect on bacterial MS channels (Qi et al., 2005). These results seem to indicate that the SAKCa channels may respond to the tension in the lipid bilayer. Interestingly, the study reported residual sensitivity to TNP in the STREX deletion mutant channel, but no sensitivity to CPZ. This suggests that the SAKCa channel without STREX may be sensitive to bilayer deformation forces by responding preferentially to tension in one leaflet of the bilayer. An explanation put forward by the study suggested that STREX might interact directly or indirectly via a membrane associated protein, which could sense or transmit the force in the bilayer more efficiently. Consequently, STREX would act as an intermediate structure that could indirectly convey and amplify the mechanical force to the gating mechanism of the SAKCa channels. Future work is expected to resolve the issue of interaction between the bilayer and tether mechanism by focusing on determination of the specific mechanisms of gating eukaryotic MS channels by mechanical force.

10.6 The Role of MS Channels in Cell Physiology and Pathology of Disease

Bacteria possess multiple adaptation mechanisms, which enables them to grow in a wide range of external osmolarities (Wood, 1999; Sleator and Hill, 2001). To release osmotic stress caused by a sudden change in external osmolarity they are equipped with MS channels serving as "emergency valves." It has been well established that

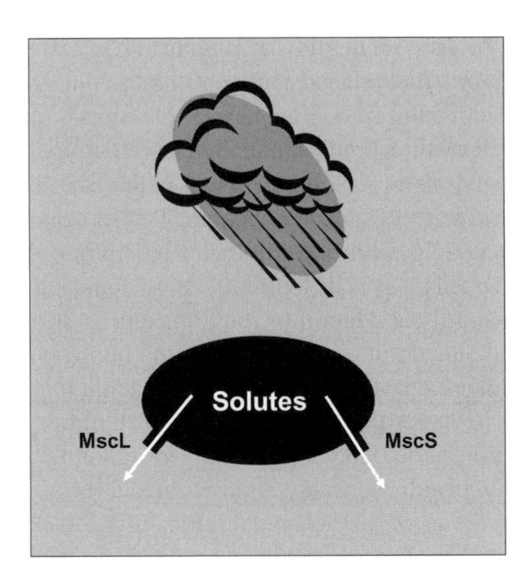

$$\Delta\Pi = 24.1\ \Delta C$$

Fig. 10.5 *MS channels in bacteria are essential to maintain cell integrity*. Osmotic stress caused by a sudden drop in external osmolarity (hypo-osmotic shock) opens MscL and MscS in *E. coli* to release excessive turgor pressure. Normally, cell turgor of a bacterial cell is of the order of 4–6 atm. Depending on the magnitude of the hypo-osmotic shock the turgor pressure may increase $\gg$10 atm, which without MS channel opening would cause cell death. $\Delta\Pi$ is osmotic pressure difference in atm (at 22°C), and ΔC is concentration gradient in mol/liter (osmolarity).

bacterial MS channels regulate the cell turgor by opening upon hypo-osmotic shock (Booth and Louis, 1999; Levina et al., 1999) (Fig. 10.5). The channels have large conductances and mostly lack ionic specificity, so that they could well serve such a function. Mutants of *E. coli* lacking both MscL and MscS channel proteins die upon transfer from a medium of high to a medium of low osmolarity. The mini MS channel, MscM, is insufficient alone to protect them. Cells with only MscS or MscL deleted are however, fully functional, which suggest that the redundancy of MS channels provides a safeguard against the deleterious effects of sudden changes in external osmolarity. Another role that MS channels could play in physiology of bacterial cells is in sensing changes in turgor pressure during cell division and cell growth (Csonka and Epstein, 1996). Increase in turgor is required for stretching the cell envelope and increase in cell volume, which may trigger the synthesis and the assembly of cell wall components. This notion has been supported by recent evidence showing that the expression of MscS and MscL is regulated by the stress sigma factor, RpoS (σ^S) (RNA polymerase holoenzyme containing σ^S) (Stokes et al., 2003). During entry into stationary phase the cells undergo cell wall re-modeling, which is accompanied by an increase in the number of MS channels due to RpoS.

The role of MS channels in archaea is most likely very similar to those of their bacterial counterparts. Although not much is known about cellular turgor in archaea, cell turgor is essential for growth and cell wall synthesis in prokaryotes, as stretch of the cellular envelope resulting from turgor is required for enlargement of the envelope and consequently for growth of prokaryotic cells (Csonka and Epstein, 1996). Changes in external osmolarity due to flood, drought, or volcanic activity can also be expected to occur in the extreme habitats of archaea making MS channels indispensable as emergency valves in cellular osmoregulation in these microbes (Kloda and Martinac, 2001d).

Examples of physiological processes in which MS channels could play a role include touch, hearing, proprioception, osmotic gradients, control of cellular turgor, and gravitropism (Hamill and Martinac, 2001). As demonstrated by genetic studies in worms, flies and zebrafish mechano- and osmosensation in these organisms are regulated by members of the TRP channel superfamily (Minke and Cook, 2002; Corey, 2003b), There is increasing evidence that this ion channel family plays a major role in mechanotransduction in eukaryotic cells (Lin and Corey, 2005). Different types of conductances have been identified that have a physiological function in mechanotransduction in eukaryotic microbes (Martinac, 1993). Examples include the Ca^{2+}-selective and K^{+}-selective MS conductances of Paramecium, which regulate the direction of the ciliary beating during the "avoidance response" and the "escape response" of this ciliate upon mechanical stimulation of the cell anterior and cell posterior, respectively (Machemer and Ogura, 1979; Naitoh, 1984). Neither of the two conductances has to date been characterized at the single channel level.

Abnormalities of MS channel function cause neuronal (Driscoll and Chalfie, 1991; Hong and Driscoll, 1994) and muscular degeneration (Franco and Lansman, 1990; Franco-Obregon and Lansman, 1994), cardiac arrhythmias (Hansen et al., 1990; Franz et al., 1992), hypertension (Kohler et al., 1999), and polycystic kidney disease (Chen et al., 1999). A large number of patients affected by polycystic kidney disease suffer also from serious cardiovascular lesions (Devuyst et al., 2003; Rossetti et al., 2003). Another disease linked to defective MS channels is atrial fibrillation, which is the most common cardiac arrhythmia to occur in humans (Kohl and Sachs, 2001; Sachs, 2004a,b). A role for SA-CAT channels has been implicated in this heart disorder. The blockade of SA-CAT channels by the spider venom peptide GsMtx-4 was shown to reduce the extent of the abnormalities of the heart beat induced by atrial fibrillation in the rabbit heart (Bode et al., 2001). GsMtx-4 blocks MS channels in dystrophic muscle as well (Yeung et al., 2005). Since GsMtx-4 was also found to inhibit TRPC1 channels expressed in the CHO cells (F. Sachs and P. Gottlieb, personal communication), this suggests that SA-CAT channels may be TRPC1 or closely related channels.

10.7 Conclusion

The investigation of MS channels went from a serendipitous discovery to disgrace of artifacts, to be fully re-established over the last decade through molecular identification, structural determination, and functional analysis of a number of MS channels from prokaryotes and eukaryotes. The cloning of MscL and MscS and elucidation of their 3D crystal structure together with the unambiguous demonstration of their physiological role in bacterial osmoregulation have provided a solid basis for further research in this class of ion channels. The cloning and genetic analysis of the *mec* genes in *Cenorhabditis elegans* as well as genetic and functional studies of the TREK and TRP families of MS channels have greatly contributed to our understanding of the role of MS channels in physiology of mechanosensory transduction. We may expect to see significant future developments in this exciting research field.

Acknowledgments

I would like to thank Drs. F. Sachs, P. Gottlieb, and M. Sokabe for sharing unpublished information and Dr. O. P. Hamill for critical reading of the manuscript and helpful suggestions. This work was supported by the Australian Research Council.

References

Adams, C.M., M.G. Anderson, D.G. Motto, M.P. Price, W.A. Johnson, and M.J. Welsh. 1998. Ripped pocket and pickpocket, novel Drosophila DEG/ENaC subunits in early development and mechanosensory neurons. *J. Cell Biol.* 140:143–152.

Ainsley, J.A. et al. 2003. Enhanced locomotion caused by loss of the Drosophila DEG/ENaC protein Pickpocket1. *Curr. Biol.* 13:1557–1563.

Ajouz, B., C. Berrier, M. Besnard, B. Martinac, and A. Ghazi. 2000. Contributions of the different extramembranous domains of the mechanosensitive ion channel MscL to its response to membrane tension. *J. Biol. Chem.* 275:1015–1022.

Awayda, M.S., W. Shaom, F. Guo, M. Zeidel, and W.G. Hill. 2004. ENaC-membrane interactions: Regulation of channel activity by membrane order. *J. Gen. Physiol.* 123(6):709–727.

Bandell, M. et al. 2004. Noxious cold ion channel TRPA1 is activated by pungent compounds and bradykinin. *Neuron* 41:849–857.

Barritt, G., and G. Rychkov. 2005. TRPs as mechanosensitive channels. *Nat. Cell Biol.* 7:105–107.

Bass, R.B., P. Strop, M. Barclay, and D. Rees. 2002. Crystal structure of *Escherichia coli* MscS, a voltage-modulated and mechanosensitive channel. *Science* 298:1582–1587.

Berrier, C., M. Besnard, B. Ajouz, A. Coulombe, and A. Ghazi. 1996. Multiple mechanosensitive ion channels from *Escherichia coli*, activated at different thresholds of applied pressure. *J. Membr. Biol.* 151:175–187.

Betanzos, M., C.-S. Chiang, H.R. Guy, and S. Sukharev. 2002. A large iris-like expansion of a mechanosensitive channel protein induced by membrane tension. *Nat. Struct. Biol.* 9(9):704–710.

Bezanilla, F., and E. Perozo. 2002. Force and voltage sensors in one structure. *Science* 298:1562–1563.
Biggin, P.C., and M.S.P. Sansom. 2001. Channel gating: Twist to open. *Curr. Biol.* 11(9):R364–R366.
Bockenhauer, D., N. Zilberberg, and S.A.N. Goldstein. 2001. KCNK2: Reversible conversion of a hippocampal potassium leak into a voltage-dependent channel. *Nat. Neurosci.* 4(5):486–491.
Bode, F., F. Sachs, and M.R. Franz. 2001. Tarantula peptide inhibits atrial fibrillation during stretch. *Nature* 409:35–36.
Booth, I.R., and P. Louis. 1999. Managing hypoosmotic stress: Aquaporins and mechano-sensitive channels in *Escherichia coli. Curr. Opin. Microbiol.* 2:166–169.
Bowman, C.B., J.P. Ding, F. Sachs, and M. Sokabe. 1992. Mechanotransducing ion channels in astrocytes. *Brain Res.* 584:272–286.
Bowman, C.B., and J.W. Lohr. 1996. Mechanotransducing ion channels in C6 glioma cells. *Glia* 18:161–176.
Brehm, P., R. Kullberg, and F. Moody-Corbet. 1984. Properties of nonjunctional acetylcholine receptor channels on innervated muscle of Xenopus laevis. *J. Physiol.* 350:631–648.
Canessa, C.M., A.M. Merillat, and B.C. Rossier. 1994. Membrane topology of the epithelial sodium channel in intact cells. *Am. J. Physiol. Cell Physiol.* 267:C1682–C1690.
Cantor, R.S. 1997. Lateral pressures in cell membranes: A mechanism for modulation of protein function. *J. Phys. Chem.* 101:1723–1725.
Chang, G., R. Spencer, A. Lee, M. Barclay, and C. Rees. 1998. Structure of the MscL homologue from Mycobacterium tuberculosis: A gated mechanosensitive ion channel. *Science* 282:2220–2226.
Chen, X.Z., P.M. Vassilev, N. Basora, J.B. Peng, H. Nomura, Y. Segal, E.M. Brown, S.T. Reeders, M.A. Hediger, and J. Zhou. 1999. Polycystin-L is a calcium-regulated cation channel permeable to calcium ions. *Nature* 401:383–386.
Christensen, O. 1987. Mediation of cell volume regulation by Ca^{2+} influx through stretch-activated channels. *Nature* 330:66–68.
Clapham, D.E. 2003. TRP channels as cellular sensors. *Nature* 426:517–524.
Colbert, H.A., T.L. Smith, and C.I. Bargmann. 1997. Osm-9, a novel protein with structural similarity to channels, is required for olfaction, mechanosensation, and olfactory adapation in *Caenorhabditis elegans*. *J. Neurosci.* 17:8259–8269.
Colombo, G., S.J. Marrink, and A.E. Mark. 2003. Simulation of MscL gating in a bilayer under stress. *Biophys. J.* 84:2331–2337.
Corey, D. 2003a. Sensory transduction in the ear. *J. Cell Sci.* 116:1–3.
Corey, D.P. 2003b. New TRP channels in hearing and mechanosensation. *Neuron* 39:585–588.
Corey, D.P., J. Garcia-Añoveros, J.R. Holt, K.Y. Kwan, S.Y. Lin, M.A. Vollrath, A. Amalfitano, E.L. Cheung, B.H. Derfler, A. Duggan, et al. 2004. TRPA1 is

a candidate for the mechanosensitive channel of vertebrate hair cells. *Nature* 432:723–730.

Corey, D.P., and A.J. Hudspeth. 1983. Kinetics of the receptor current in bullfrog saccular hair cells. *J. Neurosci.* 3:962–976.

Corry, B., P. Rigby, P., Z.-W. Liu, Z.-W. and B. Martinac, B. 2005. Conformational changes involved in MscL channel gating measured using FRET spectroscopy. *Biophys. J.* 89:L49–L51.

Cruickshank, C., R. Minchin, A. LeDain, and B. Martinac. 1997. Estimation of the pore size of the large-conductance mechanosensitive ion channel of *Escherichia coli*. *Biophys. J.* 73:1925–1931.

Csonka, L.N., and W. Epstein. 1996. Osmoregulation. In *Escherichia coli* and Salmonella: Cellular and Molecular Biology, 2nd Ed. F.C. Neidhardt, R. Curtis III, J.L. Ingraham, E.C.C. Lin, K. Brooks Low, B. Magasanik, W.S. Reznikoff, M. Riley, M. Schaechter, and H.E. Umbarger, editors. ASM Press, Washington, DC, pp.1210–1225.

Cui, C., D.O. Smith, and J. Adler. 1995. Characterization of mechanosensitive channels in *Escherichia coli* cytoplasmic membranes by whole cell patch clamp recording. *J. Membr. Biol.* 144:31–42.

Darboux, I., E. Lingueglia, G. Champigny, S. Coscoy, P. Barbry, and M. Lazdunski. 1998. dGNaCl, a gonad-specific amiloride sensitive Na^+ channel. *J. Biol. Chem.* 273:9424–9429.

Delmas, P. 2004. Polycystins: From mechanosensation to gene regulation. *Cell* 118(2):145–148.

Detweiler, P.B. 1989. Sensory transduction. In *Textbook of Physiology: Excitable Cells and Neurophysiology*. Patton, H.D., Fuchs, A.F., Hille, B., Sher, A.M. and Streiner, R. (eds.), Saunders Company, Philadelphia, London, Toronto, Montreal, Sydney, Tokyo, pp. 98–129.

Devuyst, O. et al. 2003. Autosomal dominant polycystic kidney disease: modifier genes and endothelial dysfunction. *Nephrol. Dial. Transplant.* 18(11);2211–2215.

Driscoll, M., and M. Chalfie. 1991. The mec-4 gene is a member of a family of *Caenorhabditis elegans* genes that can mutate to induce neuronal degeneration. *Nature* 349:588–593.

Du, H., G. Gu, C.M. William, and M. Chalfie. 1996. Extracellular proteins needed for *C. elegans* mechanosensation. *Neuron* 16:183–194.

Elmore, D.E., and D.A. Dougherty. 2003. Investigating lipid composition effects on the mechanosensitive channel of large conductance (MscL) using molecular dynamics simulations. *Biophys. J.* 85(3):1512–1524.

Ermakov, Y.A., A.Z. Averbakh, A.I. Yusipovich, and S.I. Sukharev. 2001. Dipole potentials indicate restructuring of the membrane interface induced by gadolinium and beryllium ions. *Biophys. J.* 80:1851–1862.

Franco, A., and J.B. Lansman. 1990. Calcium entry through stretch-inactivated channels in mdx myotubes. *Nature* 344:670–673.

Franco-Obregon, A., Jr., and J.B. Lansman. 1994. Mechanosensitive ion channels from normal and dystrophic mice. *J. Physiol.* 481:299–309.
Franz, M.R., R. Cima, D. Wang, D. Profitt, and R. Kurz. 1992. Electrophysiological effects of myocardial stretch and mechanical determinants of stretch-activated arrhythmias. *Circulation* 86:968–978.
García-Añovernos, J., and D.P. Corey. 1997. The molecules of mechanosensation. *Ann. Rev. Neurosci.* 20:567–594.
Garty, H., and L.G. Palmer. 1997. Epithelial sodium channels: Function, structure, and regulation. *Physiol. Rev.* 77:359–396.
Gillespie, P.G., and R.G. Walker. 2001. Molecular basis of mechanotransduction. *Nature* 413:194–202.
Goodman, M.B., G.G. Ernstrom, D.S. Chelur, R.C. O'Hagan, A. Yao, and M. Chalfie. 2002. MEC-2 regulates *C. elegans* DEG/EnaC channels needed for mechanosensation. *Nature* 415:1039–1042.
Goodman, M.B., and E.M. Schwarz. 2003. Transducing touch in *Caenorhabditis elegans*. *Annu. Rev. Physiol.* 65:429–452.
Goulian, M., O.N. Mesquita, D.K. Fygenson, C. Neilsen, O.S. Andersen, and A. Libchaber. 1998. Gramicidin channel kinetics under tension. *Biophys. J.* 74:328–337.
Gribkoff, V.K., J.E. Atarrett Jr., and S.I. Dworetzky. 2001. Maxi-K potassium channels: Form, function, and modulation of a class of endogenous regulators of intracellular calcium. *Neuroscientist* 7:166–177.
Guharay, F., and F. Sachs. 1984. Stretch-activated single ion channel currents in tissue cultured embryonic chick skeletal muscle. *J. Physiol.* 352:685–701.
Gullingsrud, J., D. Kosztin, and K. Schulten. 2001. Structural determinants of MscL gating studied by molecular dynamics simulations. *Biophys. J.* 80:2074–2081.
Gullingsrud, J., and K. Schulten. 2003. Gating of MscL studied by steered molecular dynamics. *Biophys. J.* 85:2087–2099.
Gustin, M.C., X.-L. Zhou, B. Martinac, and C. Kung. 1988. A mechanosensitive ion channel in the yeast plasma membrane. *Science* 242:762–765.
Häse, C.C., A.C. Le Dain, and B. Martinac. 1995. Purification and functional reconstitution of the recombinant large mechanosensitive ion channel (MscL) of *Escherichia coli*. *J. Biol. Chem.* 270:18329–18334.
Hamill, O.P., A. Marty, E. Neher, B. Sackmann, and F.J. Sigworth. 1981. Improved patch-clamp techniques for high-resolution current recording from cells and cell-free membrane patches. *Pflügers Arch. Eur. J. Physiol.* 391:85–100.
Hamill, O.P., and D.W. McBride Jr. 1997. Induced membrane hypo/hyper-mechanosensitivity: A limitation of patch-clamp recording. *Ann. Rev. Physiol.* 59:621–631.
Hamill, O.P., and B. Martinac. 2001. Molecular basis of mechanotransduction in living cells. *Physiol. Rev.* 81:685–740.

Hanaoka, K., F. Qian, A. Boletta, A.K. Bhunia, K. Piontek, L. Tsiokas, V.P. Sukhatme, W.B. Guggino, and G.G. Germino. 2000. Co-assembly of polycystin-1 and -2 produces unique cation-permeable currents. *Nature* 408:990–994.

Hansen, D.E., C.S. Craig, and L.M. Hondeghem. 1990. Stretch induced arrhythmias in the isolated canine ventricle. Evidence for the importance of mechanoelectrical feedback. *Circulation* 81:1094–1105.

Hartneck, C., T.D. Plant, and G. Schultz. 2000. From worm to man: Three subfamilies of TRP channels. *TINS* 23(4):159–166.

Hirono, M., C.S. Denis, G.P. Richardson, and P.G. Gillespie. 2004. Hair cells require phophatidylinositol 4,5-bisphosphate for mechanical transduction and adaptation. *Neuron* 44:309–320.

Hong, K., and M. Driscoll. 1994. A transmembrane domain of the putative channel subunit MEC-4 influences mechanotransduction and neurodegeneration in *C. elegans*. *Nature* 367:470–473.

Honoré, E., F. Maingret, M. Lazdunski, A.J. Patel. 2002. An intracellular proton sensor commands lipid- and mechano-gating of the K(+) channel TREK-1. *EMBO J.* 21(12):2968–2976.

Howard, J., and S. Bechstedt. 2004. Hypothesis: A helix of ankyrin repeats of the NOMPC-TRP ion channel is the gating spring of mechanoreceptors. *Curr. Biol.* 14:R224–R226.

Huang, M., G. Gu, E.L. Ferguson, and M. Chalfie. 1995. A stomatin-like protein is needed for mechano-sensation in *C. elegans*. *Nature* 378:292–295.

Jordt, S.E. et al. 2004. Mustard oils and cannabinoids excite sensory nerve fibres through the TRP channel ANKTM1. *Nature* 427:260–265.

Katz, B. 1950. Depolarization of sensory terminals and the initiation of impulses in the muscle spindle. *J. Physiol. (Lond.)* 111:261–282.

Kawakubo, T., K. Naruse, T. Matsubara, N. Hotta, and M. Masahiro Sokabe. 1999. Characterization of a newly found stretch-activated KCa,ATP channel in cultured chick ventricular myocytes. *Am. J. Physiol.* 276:H1827–H1838.

Kim, J., Y.D. Chung, D.-Y. Park, S. Choi, D.W. Shin, H. Soh, H.W. Lee, W. Son, J. Yim, C.-S. Park, M.J. Kernan, and C. Kim. 2003. A TRPV family ion channel required for hearing in Drosophila. *Nature* 424:81–84.

Kloda, A., and B. Martinac. 2001a. Molecular identification of a mechanosensitive ion channel in archaea. *Biophys. J.* 80:229–240.

Kloda, A., and B. Martinac. 2001b. Structural and functional similarities and differences between MscMJLR and MscMJ, two homologous MS channels of *M. jannashii*. *EMBO J.* 20:1888–1896.

Kloda, A., and B. Martinac. 2001c. Mechanosensitive channel in Thermoplasma a cell wall-less archaea: Cloning and molecular characterization. *Cell Biochem. Biophys.* 34:321–347.

Kloda, A., and B. Martinac. 2001d. Mechanosensitive channels in archaea. *Cell Biochem. Biophys.* 34:349–381.

Kloda, A., and B. Martinac. 2002. Mechanosensitive channels of bacteria and archaea share a common ancestral origin. *Eur. Biophys. J.* 31:14–25.

Kohl, P., and F. Sachs. 2001. Mechanoelectric feedback in cardiac cells. *Philos. Trans. R. Soc. Lond. B Biol. Sci.* 359:1173–1185.

Kohler, R., A. Distler, and J. Hoyer. 1999. Increased mechanosensitive currents in aortic endothelial cells from genetically hypertensive rats. *J. Hypertens.* 17:365–371.

Kumánovics, A., G. Levin, and P. Blount. 2003. Family ties of gated pores: Evolution of the sensor module. *FASEB J.* 16:1623–1629.

Kung, C. 2005. A possible unifying principle for mechanosensation. *Nature* 436:647–654.

Kung, C., and Y. Saimi. 1995. Solute sensing vs. solvent sensing, a speculation. *J. Eukaryot. Microbiol.* 42:199–200.

Kwan K.Y., A.J. Allchorne, M.A. Vollrath, A.P. Christensen, D.S. Zhang, C.J. Woolf and D.P. Corey. 2006. TRPA1 contributes to cold, mechanical, and chemical nociception but is not essential for hair-cell transduction. *Neuron* 50(2):177–180.

Kung, C., Y. Saimi, and B. Martinac. 1990. Mechanosensitive ion channels in microbes and the early evolutionary origin of solvent sensing. In Protein Interactions, Current Topics in Membranes and Transport. Academic Press, New York, pp. 145–153.

Le Dain, A.C., N. Saint, A. Kloda, A. Ghazi, and B. Martinac. 1998. Mechanosensitive ion channels of the archaeon *Haloferax volcanii*. *J. Biol. Chem.* 273:12116–12119.

Levina, N., S. Totemeyer, N.R. Stokes, P. Louis, M.A. Jones, and I.R. Booth. 1999. Protection of *Escherichia coli* cells against extreme turgor by activation of MscS and MscL mechanosensitive channels: Identification of genes required for MscS activity. *EMBO J.* 18:1730–1737.

Li, Y., P.C. Moe, S. Chandrasekaran, I.R. Booth, and P. Blount. 2002. Ionic regulation of MscK, a mechanosensitive channel from *Escherichia coli*. *EMBO J.* 21:5323–5330.

Liedtke, W., Y. Choe, M.A. Marti-Renom, A.M. Bell, C.S. Denis, A. Sali, A.J. Hudspeth, J.M. Friedman, and S. Heller. 2000. Vanilloid receptor-related osmotically activated channel (VR-OAC), a candidate vertebrate osmoreceptor. *Cell* 103:525–535.

Lin, S.-Y., and D.P. Corey. 2005. TRP channels in mechanosensation. *Curr. Opin. Neurobiol.* 15:350–357.

Lingueglia, E., G. Champigny, M. Lazdunski, and P. Barbry. 1995. Cloning of the amiloride-sensitive FMRFamide peptide-gated sodium channel. *Nature* 378(6558):730–733.

Loewenstein, W.R. 1959. The generation of electric activity in a nerve ending. *Ann. N.Y. Acad. Sci.* 81:367–387.

Machemer, H., and A. Ogura. 1979. Ionic conductances of membranes in ciliated and deciliated *Paramecium*. *J. Physiol.* 296:49–60.

Maingret, F., M. Fosset, F. Lesage, M. Lazdunski, and E. Honoré. 1999a. Traak is a mammalian neuronal mechano-gated K^+ channel. *J. Biol. Chem.* 274:1381–1387.

Maingret, F., E. Honore, M. Lazdunski, and A.J. Patel. 2002. Molecular basis of the voltage-dependent gating of TREK-1, a mechano-sensitive K(+) channel. *Biochem. Biophys. Res. Commun.* 292(2):339–346.

Maingret, F., I. Lauritzen, A.J. Patel, C. Heurteaux, R. Reyes, F. Lesage, M. Lazdunski, and E. Honore. 2000a. TREK-1 is a heat-activated background K(+) channel. *EMBO J.* 19(11):2483–2491.

Maingret, F., A.J. Patel, F. Lesage, M. Lazdunski, and E. Honore. 1999b. Mechano- or acid stimulation, two interactive modes of activation of the TREK-1 potassium channel. *J. Biol. Chem.* 274(38):26691–26696.

Maingret, F., A.J. Patel, F. Lesage, M. Lazdunski, and E. Honore. 2000b. Lysophospholipids open the two-pore domain mechano-gated K^+ channels TREK-1 and TRAAK. *J. Biol. Chem.* 275:10128–10133.

Markin, V.S., and B. Martinac. 1991. Mechanosensitive ion channels as reporters of bilayer expansion. A theoretical model. *Biophys. J.* 60:1120–1127.

Maroto, R., A. Raso, T.G. Wood, A. Kurosky, B. Martinac, and O.P. Hamill. 2005. TRPC1 forms the stretch-activated cation channel in vertebrate cells. *Nat. Cell Biol.* 7(2):179–185.

Martinac, B. 1993. Mechanosensitive ion channels: biophysics and physiology. *In*: *Thermo-dynamics of Membrane Receptors and Channels*. M.B. Jackson (ed.) CRC Press, Boca Raton, FL, pp. 327–351.

Martinac, B. 2001. Mechanosensitive channels in prokaryotes. *Cell. Physiol. Biochem.* 11:61–76.

Martinac, B. 2004. Mechanosensitive ion channels: Molecules of mechanotransduction. *J. Cell Sci.* 117:2449–2460.

Martinac, B., J. Adler, and C. Kung. 1990. Mechanosensitive ion channels of *E. coli* activated by amphipaths. *Nature* 348:261–263.

Martinac, B., M. Buechner, A.H. Delcour, J. Adler, and C. Kung. 1987. Pressure-sensitive ion channel in *Escherichia coli*. *Proc. Natl. Acad. Sci. USA* 84:2297–2301.

Martinac, B., A.H. Delcour, M. Buechner, J. Adler, and C. Kung. 1992. Mechanosensitive ion channels in bacteria. *In*: Comparative Aspects of Mechanoreceptor Systems. F. Ito, editor. Springer-Verlag, Berlin, pp. 3–18.

Martinac, B., and O.P. Hamill. 2002. Gramicidin A channels switch between stretch activation and stretch inactivation depending on bilayer thickness. *Proc. Natl. Acad. Sci. USA* 99:4308–4312.

Martinac, B., and A. Kloda. 2003. Evolutionary origins of mechanosensitive ion channels. *Prog. Biophys. Mol. Biol.* 82:11–24.

Martinac, B., X.-L. Zhou, A. Kubalski, S. Sukharev, and C. Kung. 1994. Microbial channels. *In*: Handbook of Membrane Channels: Molecular and Cellular Physiology. C. Perrachia, editor. Academic Press, New York, pp. 447–459.

McLaggan, D., M.A. Jones, G. Gouesbet, N. Levina, S. Lindey, W. Epstein, and I.R. Booth. 2002. Analysis of the kefA2 mutation suggests that KefA is a cation-specific channel involved in osmotic adaptation in *Escherichia coli*. *Mol. Microbiol.* 43(2):521–536.

Minke, B., and B. Cook. 2002. TRP channel proteins and signal transduction. *Physiol. Rev.* 82:429–472.

Montell, C. 2003. Thermosensation: Hot findings make TRPNs very cool. *Curr. Biol.* 13:R476–R478.

Montell, C., L. Birnbaumer, and V. Flockerzi. 2002. The TRP channels, a remarkably functional family. *Cell* 108:595–598.

Morris, C.E. 1990. Mechanosensitive ion channels. *J. Membr. Biol.* 113:93–107.

Naitoh, Y. 1984. Mechanosensory transduction in protozoa. *In*: Membranes and Sensory Transduction. G. Colombetti and F. Lenci, editors. Plenum Press, New York.

Nauli, S.M., F.J. Alenghat, Y. Luo, E. Williams, P. Vassilev, X. Li, A.E.H. Elia, W. Lu, E.M. Brown, S.J. Quinn, D.E. Ingber, and J. Zhou. 2003. Polycystins 1 and 2 mediate mechanosensation in the primary cilium of kidney cells. *Nat. Genet.* 33:129–137.

Nauli, S.M., and J. Zhou. 2004. Polycystins and mechanosensation in renal and nodal cilia. *Bioassays* 26(8):844–856.

North, R.A. 1996. Families of ion channels with two hydrophobic segments. *Curr. Opin. Cell Biol.* 8:474–483.

O'Hagan, R., M. Chalfie, and M. Goodman. 2005. The MEC-4 DEG/ENaC channel of *Caenorhabditis elegans* touch receptor neurons transduces mechanical signals. *Nat. Neurosci.* 8:43–50.

Okada, K., P.C. Moe, and P. Blount. 2002. Functional design of bacterial mechanosensitive channels. *J. Biol. Chem.* 277:27682–27688.

Pace, N.R. 1997. A molecular view of microbial diversity and the biosphere. *Science* 276:734–740.

Patel, A., E. Honoré, F. Maingret, F. Lesage, M. Fink, F. Duprat, and M. Lazdunski. 1998. A mammalian two pore domain mechano-gated S-like K^+ channel. *EMBO J.* 17:4283–4290.

Patel, A.J., E. Honore, F. Lesage, M. Fink, G. Romey, M. Lazdunski. 1999. Inhalational anesthetics activate two-pore-domain background K^+ channels. *Nat. Neurosci.* 2:422–426.

Patel, A.J., E. Honore, and M. Lazdunski. 2001. Lipid and mechano-gated 2P domain K^+ channels. *Curr. Opin. Cell Biol.* 13:422–428.

Perozo, E., D.M. Cortes, P. Sompornpisut, A. Kloda, and B. Martinac. 2002a. Structure of MscL in the open state and the molecular mechanism of gating in mechanosensitive channels. *Nature* 418:942–948.

Perozo, E., A. Kloda, D.M. Cortes, and B. Martinac. 2002b. Physical principles underlying the transduction of bilayer deformation forces during mechanosensitive channel gating. *Nat. Struct. Biol.* 9:696–703.

Perozo, E., and D.C. Rees. 2003. Structure and mechanism in prokaryotic mecahnosensitive channels. *Curr. Opin. Struct. Biol.* 13:432–442.

Pickard, B.G., and J.P. Ding. 1992. Gravity sensing by higher plants. *In*: Comparative Aspects of Mechanoreceptor Systems. F. Ito, editor. Springer-Verlag, Berlin, pp. 81–110.

Pivetti, C.D., M.-R. Yen, S. Mille, W. Busch, Y.-H. Tseng, I.R. Booth, and M.H. Saier Jr. 2003. Two families of mechanosensitive channel proteins. *Microbiol. Mol. Biol. Rev.* 67(1):66–85.

Qi, Z., S. Chi, X. Su, K. Naruse, and M. Sokabe. 2005. Activation of a mechanosensitive BK channel by membrane stress created with amphipaths. *Mol. Membr. Biol.* 22(6):519–527.

Qi, Z., K. Naruse, and M. Sokabe. 2003. Ionic amphipaths affect gating of a stretch activated BK channel (SAKCa) cloned from chick heart. *Biophys. J.* 84:A234.

Ronan, D., and P. Gillespie. 2005. Metazoan mechanotransduction mystery finally solved. *Nat. Neurosci.* 8:7–8.

Rossetti, S. et al. 2003. Association of mutation position in polycystic kidney disease 1 (PKD1) gene and development of a vascular phenotype. *Lancet* 361(9376):2196–2201.

Sachs, F. 1988. Mechanical transduction in biological systems. *CRC Crit. Revs. Biomed. Eng.* 16:141–169.

Sachs, F. 2004a. Heart Mechanoelectric Transduction. In *Cardiac electrophysiology; From Cell to Bedside* (eds.), Jalife, J. and Zipes, D., pp. 96–102. Saunders (Elsevier), Philadelphia.

Sachs, F. 2004b. Stretch-activated channels in the heart. In *Cardiac Mechano-Electric Feedback and Arrhythmias: From Pipette to Patient* (eds.) Kohl, P., Franz, M.R. and Sachs, F. Saunders (Elsevier), Philadelphia.

Sachs, F., and C.E. Morris. 1998. Mechanosensitive ion channels in nonspecialized cells. *Revs. Physiol. Biochem. Pharmacol.* 132:1–77.

Schild, L., E. Schneeberger, I. Gautschi, and D. Firsov. 1997. Identification of amino acid residues in the alpha, beta, and gamma subunits of the epithelial sodium channel (ENaC) involved in amiloride block and ion permeation. *J. Gen. Physiol.* 109:15–26.

Sidi, S., R.W. Friedrich, and T. Nicolson. 2003. NompC TRP channel required for vertebrate sensory hair cell mechanotransduction. *Science* 301:96–99.

Sigurdson, W.J., A. Ruknudin, and F. Sachs. 1992. Calcium imaging of mechanically induced fluxes in tissue cultured chick heart: The role of stretch-activated ion channels. *Am. J. Physiol.* 262:H1110–H1115.

Sleator, R.D., and C. Hill. 2001. Bacterial osmoadaptation: The role of osmolytes in bacterial stress and virulence. *FEMS Microbiol. Rev.* 26:49–71.

Sokabe, M., and F. Sachs. 1990. The structure and dynamics of patch clamped membrane: A study using differential interference contrast microscopy. *J. Cell Biol.* 111:599–606.

Sokabe, M., F. Sachs, and Z. Jing. 1991. Quantitative video microscopy of patch clamped membranes: Stress, strain, capacitance and stretch channel activation. *Biophys. J.* 59:722–728.

Sotomayor, M., D.P. Corey, and K. Schulten. 2005. In search of the hair-cell gating spring: Elastic properties of ankyrin and cadherin repeats. *Structure* 13:669–682.

Stokes, N.R., H.D. Murray, C. Subramaniam, R.L. Gourse, P. Louis, W. Bartlett, S. Miller, and I.R. Booth. 2003. A role for mechanosensitive channels in survival of stationary phase: Regulation of channels expression by RpoS. *Proc. Natl. Acad. Sci. USA* 100:15959–15964.

Sukharev, S. 2002. Purification of the small mechanosensitive channel of *Escherichia coli* (MscS): The subunit structure, conduction, and gating characteristics in liposomes. *Biophys. J.* 83:290–298.

Sukharev, S.I., P. Blount, B. Martinac, F.R. Blattner, and C. Kung. 1994. A large mechanosensitive channel in *E. coli* encoded by mscL alone. *Nature* 368:265–268.

Sukharev, S.I., P. Blount, B. Martinac, and C. Kung. 1997. Mechanosensitive channels of *Escherichia coli*: The MscL gene, protein, and activities. *Annu. Rev. Physiol.* 59:633–657.

Sukharev, S., M. Betanzos, C.S. Chiang, and H.R. Guy. 2001. The gating mechanism of the large mechanosensitive channel MscL. *Nature* 409:720–724.

Sukharev, S.I., and D. Corey. 2004. Mechanosensitive channels: Multiplicity of families and gating paradigms. Science's STKE www.stke.org/cgi/content/full/sigtrans;2004/219/re4, pp. 1–24.

Sukharev, S.I., B. Martinac, V.Y. Arshavsky, and C. Kung. 1993. Two types of mechanosensitive channels in the *E. coli* cell envelope: Solubilization and functional reconstitution. *Biophys. J.* 65:177–183.

Suzuki, M., J. Sato, K. Kutsuwada, G. Ooki, and M. Imai. 1999. Cloning of a stretch-inhibitable nonselective cation channel. *J. Biol. Chem.* 274:6330–6335.

Szallasi, A., and P.M. Blumberg. 1990. Resiniferatoxin and analogs provide novel insights into the pharmacology of vanilloid (capsaicin) receptors. *Life Sci.* 47:1399–1408.

Szallasi, A., and P.M. Blumberg. 1999. Vanilloid (capsaicin) receptors and mechanisms. *Pharmacol. Rev.* 51:159–211.

Tanaka, T., Y. Tamba, S.Md. Masum, Y. Yamashita, and M. Yamazaki. 2002. La3+ and Gd3+ induce shape change of giant unilamellar vesicles of phosphatidylcholine. *Biochim. Biophys. Acta* 1564(1):173–182.

Tang, Q.Y., Z. Qi, K. Naruse, and M. Sokabe. 2003. Characterization of a functionally expressed stretch-activated BK_{Ca} channel cloned from chick ventricular myocytes. *J. Membr. Biol.* 196:185–200.

Tavernarakis, N., and M. Driscoll. 1997. Molecular modelling of mechanotransduction in the nematode *Caenorhabditis elegans*. *Annu. Rev. Physiol.* 59:659–689.

Tavernarakis, N., W. Shreffler, S. Wang, and M. Driscoll. 1997. unc-8, a DEG/ENaC family member, encodes a subunit of a candidate mechanically-gated channel that modulates locomotion in *C. elegans*. *Neuron* 18:107–119.

Terrenoire, C., I. Lauritzen, F. Lesage, G. Romey, M. Lazdunski. 2001. A TREK-1-like potassium channel in atrial cells inhibited by beta-adrenergic stimulation and activated by volatile anesthetics. *Circ. Res.* 89(4):336–342.

Ubl, J., H. Maurer, and H.-A. Kolb. 1988. Ion channels activated by osmotic and mechanical stress in membranes of opposum kidney cells. *J. Membr. Biol.* 104:223–232.

Voets, T., K. Talavera, G. Owsianik, and B. Nilius. 2005. Sensing with TRP channels. *Nat. Chem. Biol.* 1:85–92.

Waldmann, R., and M. Lazdunski. 1998. H^+-gated cation channels; neuronal acid sensors in the NaC/Deg family of ion channels. *Curr. Opin. Neurobiol.* 8:418–424.

Walker, R.G., A.T. Willingham, and C.S. Zuker. 2000. A Drosophilia mechanosensory transduction channel. *Science* 287:2229–2234.

Wood, J.M. 1999. Osmosensing by bacteria: Signals and membrane-based sensors. *Microbiol. Mol. Biol. Rev.* 63:230–262.

Yeung, E.W., N.P. Whitehead, T.M. Suchyna, P.A. Gottlieb, S. Sachs and D.G. Allen. 2005. Effects of stretch-activated channel blockers on $[Ca^{2+}]i$ and muscle damage in the *mdx* mouse. *J. Physiol.* 562:367–380.

Yoshimura, K., A. Batiza, M. Schroeder, P. Blount, and C. Kung. 1999. Hydrophilicity of a single residue within MscL correlates with increased mechanosensitivity. *Biophys. J.* 77:1960–1972.

Yoshimura, K., T. Nomura, and M. Sokabe. 2004. Loss-of-function mutations at the rim of the funnel of mechanosensitive channel MscL. *Biophys. J.* 86:2113–2120.

Zhang, Y., G. Gao, V.L. Popov, and O.P. Hamill. 2000. Mechanically gated channel activity in cytoskeleton deficient plasma membrane blebs and vesicles. *J. Physiol. (Lond.)* 523:117–129.

Zhou, X.-L., A.F. Batiza, S.H. Loukin, C.P. Palmer, C. Kung, and Y. Saimi. 2003. The transient receptor potential channel on the yeast vacuole is mechanosensitive. *Proc. Natl. Acad. Sci. USA* 100(12):7105–7110.

Zhou, X.-L., S.H. Loukin, C. Coria, C. Kung, and Y. Saimi. 2005. Heterologously expressed fungal transient receptor potential channels retain mechanosensitivity *in vitro* and osmotic response *in vivo*. *Eur. Biophys. J.* 34:413–422.

Zoratti, M., and A. Ghazi. 1993. Stretch-activated channels in prokaryotes. In *Alkali Transport Systems in Prokaryotes*. E.P. Bakker (ed.), pp. 349–358, Boca Raton, CRC Press.

11 TRP Channels

Thomas Voets, Grzegorz Owsianik, and Bernd Nilius

11.1 Introduction

The TRP superfamily represents a highly diverse group of cation-permeable ion channels related to the product of the *Drosophila trp* (*t*ransient *r*eceptor *p*otential) gene. The cloning and characterization of members of this cation channel family has experienced a remarkable growth during the last decade, uncovering a wealth of information concerning the role of TRP channels in a variety of cell types, tissues, and species. Initially, TRP channels were mainly considered as phospholipase C (PLC)-dependent and/or store-operated Ca^{2+}-permeable cation channels. More recent research has highlighted the sensitivity of TRP channels to a broad array of chemical and physical stimuli, allowing them to function as dedicated biological sensors involved in processes ranging from vision to taste, tactile sensation, and hearing. Moreover, the tailored selectivity of certain TRP channels enables them to play key roles in the cellular uptake and/or transepithelial transport of Ca^{2+}, Mg^{2+}, and trace metal ions. In this chapter we give a brief overview of the TRP channel superfamily followed by a survey of current knowledge concerning their structure and activation mechanisms.

11.2 TRP Channel History

TRP history started in 1969, when Cosens and Manning performed a screening for *Drosophila* mutants with impaired vision. They identified a mutant that exhibited a transient instead of sustained response to bright light (Cosens and Manning, 1969). Analysis of the photoreceptor cells of the mutant fly strain revealed that sustained light induced a transient rather than the normal sustained, plateau-like receptor potential. The mutant was baptized *trp*, for transient receptor potential. Two decades later, the *trp* gene was cloned (Montell and Rubin, 1989) and subsequently its product, TRP, was characterized as a Ca^{2+}-permeable cation channel, the founding member of the TRP superfamily (Hardie and Minke, 1992). Additional close homologues of TRP, named TRPL (or TRP-like) (Phillips et al., 1992) and TRPγ (Xu et al., 2000) were identified in *Drosophila* photoreceptor cells, and all three proteins were found to contribute to the light-induced currents (Phillips et al., 1992; Reuss et al., 1997; Xu et al., 2000). It is now well established that *Drosophila* TRP is not directly involved in the detection of light, but rather functions as a receptor-operated channel that is activated downstream of the light-induced, PLC-mediated hydrolysis of phosphatidylinositol 4,5-bisphosphate (PIP_2) (Hardie and Raghu, 2001). The debate whether TRP opens in response to the reduced PIP_2 levels, or whether it is activated by diacyl glycerol (DAG) or by polyunsatured fatty acids derived from DAG (Chyb et al., 1999) is still unsettled.

The first mammalian TRP homologues were identified in 1995 (Petersen et al., 1995; Wes et al., 1995; Birnbaumer et al., 1996; Zhu et al., 1996; Zitt et al., 1996). It was quickly realized that they were members of a larger family of cation channels, which could form the molecular counterparts of the different nonselective cation channels that had been functionally described in many cell types. The quest for *trp*-related genes resulted in the identification and characterization of more than 50 TRP channels in yeast, worms, insects, fish, and mammals (Vriens et al., 2004a; Montell, 2005). More recent analysis of published genomes reveals that there are 28 *trp*-related genes in mice, 27 *trp*-related genes in humans, 17 *trp*-related genes in the worm *C. elegans*, and 13 *trp*-related genes in *Drosophila*.

11.3 Classification

The identification of a large number of TRP-related channels in a short time span led to considerable confusion as to channel nomenclature. This resulted in a situation where several TRP channels had either multiple names or shared the same name with distinct TRPs. A functional classification, like the one used to classify for example K^+ channels, was not really feasibly for TRP channels, due to the high diversity of activating mechanisms and permeability properties. In 2002, a unified nomenclature was adopted, classifying part of the TRP superfamily into three subfamilies based on amino acid sequence homology (Montell et al., 2002). The *TRPC subfamily* (C stands for canonical or classical) contains proteins with the highest homology to founding member, *Drosophila* TRP. Two other subfamilies with significant homology to *Drosophila* TRP were named after their first identified members: the *TRPV subfamily* after the vanilloid receptor 1 (VR-1, now TRPV1), and the *TRPM subfamily* after the tumor suppressor melastatin (now TRPM1). However, several more divergent members of the TRP superfamily cannot be classified into any of these three subfamilies. Therefore, there is now a general consensus that these members belong to one of four additional subfamilies (Clapham, 2003; Montell, 2005). The *TRPA subfamily* is named after the protein ankyrin-like with transmembrane domains 1 (ANKTM1, now TRPA1), the *TRPN subfamily* after the *no mechanoreceptor potential C* gene (*nompC*) from *Drosophila*, the *TRPP subfamily* after the polycystic kidney disease-related protein 2 (PKD2, now TRPP2), and the *TRPML subfamily* after mucolipin 1 (TRPML1). Figure 11.1 presents a phylogenetic tree of the mammalian members of the TRP superfamily. Note that the TRPN subfamily does not contain any mammalian member, and that the TRPC2 gene in humans is a pseudogene. A detailed phylogenetic analysis of TRP channels in nonmammals can be found elsewhere (Vriens et al., 2004a).

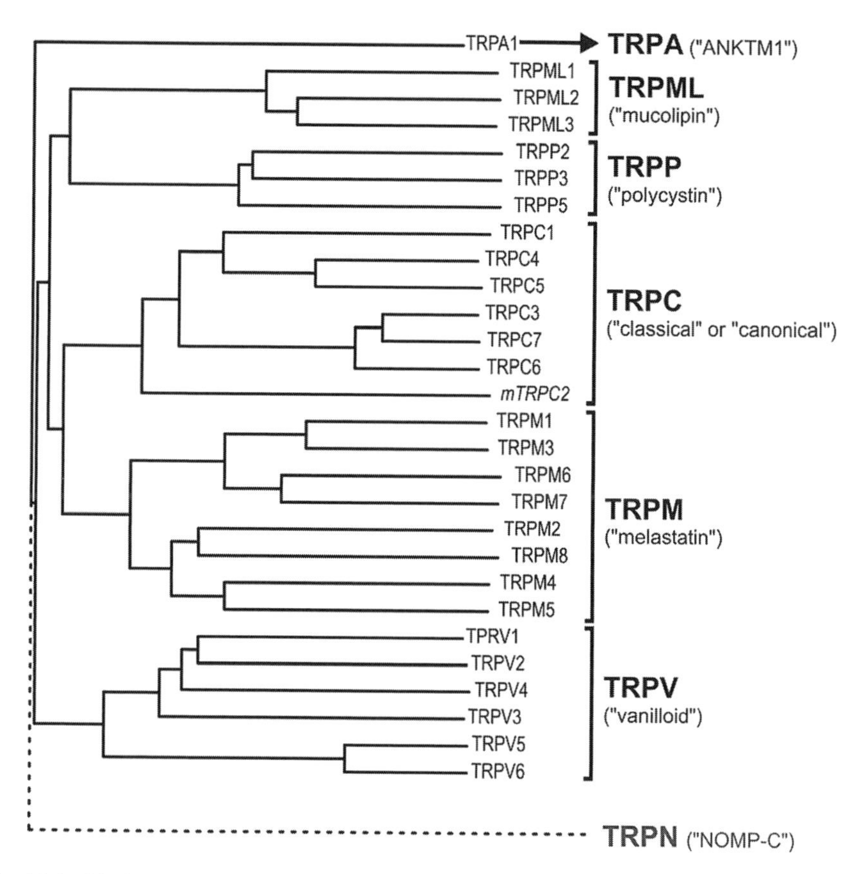

Fig. 11.1 Phylogenetic tree of human members of the TRP superfamily. All human TRP channels are classified into six subfamilies: TRPC, TRPV, TRPM, TRPML, TRPP, and TRPA. The TRPN (in grey) subfamily contains no mammalian members, hence the dashed line. The human TRPC2 gene is a pseudogene and therefore the mouse homologue (*mTRPC2*) was used for phylogenetic analysis.

11.4 Structural Aspects

11.4.1 TRP Channels Are Tetramers

All TRP channel proteins have a predicted transmembrane topology characterized by six transmembrane segments (S1–S6) and cytoplasmic N- and C-terminal tails (Fig. 11.2A). This transmembrane configuration is similar to the topology of the pore-forming subunits of other six-transmembrane (6TM) channels such as voltage-gated and Ca^{2+}-activated K^+ channels, cyclic nucleotide-gated (CNG) cation channels, hyperpolarization-activated cyclic-nucleotide-gated (HCN) channels and to each of the four repeats in the α-subunit of voltage-gated Ca^{2+} and Na^+ channels (Swartz, 2004). Results obtained for members of the TRPV subfamily convincingly demonstrated that, similar to other 6TM channels, four identical or similar TRP channel

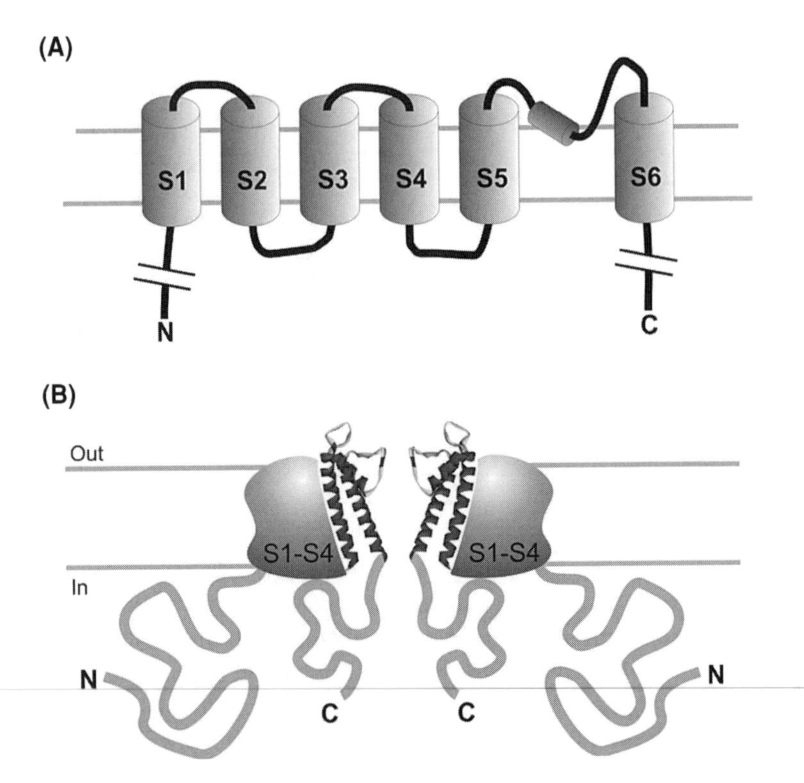

Fig. 11.2 Structure of TRP channels. (A) Transmembrane topology of TRP channels. (B) Proposed architecture of a functional TRP channel. The structure of the selectivity filter is based on SCAM analysis of TRPV6 (Voets et al., 2004b). The structure of S5 and S6 is based on the crystal structure of KvAP (Doyle et al., 1998).

subunits tetramerize to form a functional channel (Kedei et al., 2001; Hoenderop et al., 2003), and that S5, S6, and the connecting pore loop form the central cation-conducting pore (Dodier et al., 2004; Voets et al., 2004b) (Fig. 11.2B). In analogy to other 6TM channels, it is likely that S1–S4 and the cytoplasmic N- and C-terminal parts regulate the opening and closing of the pore and/or interact with other proteins, but mechanistic insight in the gating process of TRP channels is in most cases very limited.

11.4.2 Pore Region and Selectivity Filter

The permeability properties within the TRP superfamily display a tremendous variation. Roughly speaking, four main permeability profiles can be distinguished, ranging from exquisite Ca^{2+} selectivity over permeability to mono- and divalent cations alike, to selective permeability for monovalent cations and finally broad permeability to different divalent cations including Ca^{2+}, Mg^{2+}, and trace divalent cations such as Zn^{2+}, Fe^{2+} or Co^{2+} (Table 11.1). This variability contrasts with most other

Table 11.1 Permeability properties of members of the TRP superfamily.

Permeability properties	TRP channel
Highly Ca^{2+}-selective ($P_{Ca}/P_{Na} > 100$)	TRPV5, TRPV6
Relatively nonselective for mono- and divalent cations	TRPCs, TRPV1-4, TRPM2, TRPM3, TRPM8, TRPAs, TRPPs, TRPMLs
Nonselective for monovalent cations, impermeable to divalent cations	TRPM4, TRPM5
Selective for divalent cations including Mg^{2+}, Ca^{2+}, and trace metal ions	TRPM6, TRPM7

families of ion channels, where the differences in permeation properties within one family are generally small. It is therefore not too surprising that a consensus pore sequence, such as the well-described "signature" sequence that delineates the selectivity filter of K^+ channels (TXXTXGYGD) (Yellen, 2002), cannot be defined in TRP channels. At present, data concerning the localization and structure of the pore region and selectivity filter are scarce and limited to the TRPV, TRPC, and TRPM subfamilies (Owsianik et al., 2006b).

11.4.2.1 TRPV

Structure–function analysis of the pore region is most advanced for the TRPV subfamily. Here, sequence comparison revealed that the loop between S5 and S6 of the different TRPVs shows significant sequence homology with the outer pore (P-loop) of potassium channels including KcsA (Voets et al., 2004b), for which a detailed crystal structure has been provided thanks to the pioneering work of the MacKinnon lab (Doyle et al., 1998). Neutralization of negatively charged aspartate residues in the corresponding region in TRPV1, TRPV4, TRPV5, and TRPV6 resulted in altered permeability properties and lowered sensitivity to voltage-dependent block by ruthenium red, Mg^{2+} or Cd^{2+} (Garcia-Martinez et al., 2000; Nilius et al., 2001b; Voets et al., 2002, 2003; Hoenderop et al., 2003), establishing the contribution of this region to the TRPV pore. Most strikingly, neutralization of Asp^{542} in TRPV5 and Asp^{541} in TRPV6 converted these channels from highly Ca^{2+}-selective to nonselective for monovalent cations (Nilius et al., 2001b; Voets et al., 2003).

TRPV5 and TRPV6 share many permeation properties with voltage-gated Ca^{2+} channels (Tsien et al., 1987; McDonald et al., 1994; Varadi et al., 1999; Hille, 2001; Sather and McCleskey, 2003). A characteristic feature of these highly Ca^{2+}-selective channels is the anomalous mole fraction effect. At sub-micromolar Ca^{2+} concentrations, these channels are highly permeable to monovalent cations. With increasing Ca^{2+} concentrations, the channels first become blocked and then start to conduct Ca^{2+}. The high ion transfer rate of the Ca^{2+} channel pore in the case of L-type voltage-gated Ca^{2+} channels has been explained by two models: the "repulsion model" (Almers and McCleskey, 1984; Hess and Tsien, 1984; Hess et al., 1986) and the "step model" (Dang and McCleskey, 1998). With moderate changes in parameters, such models can also be applied to TRPV5 and TRPV6 (Vennekens et al., 2001). The crucial aspartate residues in the pore region of TRPV5 and TRPV6 (Asp^{542} and Asp^{541}, respectively) can be considered as the functional analogues of the ring of four negative residues (aspartates and/or glutamates) in the pore of voltage-gated Ca^{2+} channels (Heinemann et al., 1992; Yang et al., 1993; Talavera et al., 2001).

Further details of the pore structure of the Ca^{2+}-selective TRPV5 and TRPV6 were obtained using the substituted cysteine accessibility method (SCAM) (Dodier et al., 2004; Voets et al., 2004b). Cysteines introduced in a region preceding Asp^{542} and Asp^{541}, respectively, displayed a cyclic pattern of reactivity, indicating that these residues form a pore helix, similar as in the KcsA crystal structure. The pore helix is followed by the selectivity filter, which has a diameter of approximately 5.4 Å at its narrowest point as determined by measurements of permeability to cations of increasing size (Voets et al., 2004b). This value is between those reported for the L-type (6.2 Å; McCleskey and Almers, 1985) and T-type (5.1 Å; Cataldi et al., 2002) Ca^{2+} channels. The TRPV6 pore diameter was significantly increased when D^{541} was replaced by amino acids with a shorter side chain, indicating that this aspartate lines the narrowest part of the selectivity filter and contributes to the sieving properties of the pore. Based on these experimental data and on homology to KcsA, the first TRP channel pore structures have been presented (Dodier et al., 2004; Voets et al., 2004b). As illustrated in Fig. 11.2B, the external vestibule in TRPV5/6 encompasses two main structural domains: a pore helix of about 15 amino acids followed by a nonhelical loop. In contrast to KcsA (Doyle et al., 1998), the selectivity filter of TRPV6 appears to be lined by amino acid side chains rather than by backbone carbonyls.

In a recent study, Yeh and colleagues nicely demonstrated the role of the pore helix of TRPV5 in the regulation of the channel by pH (Yeh et al., 2003, 2005). They found that increased proton concentrations diminish the apparent diameter of the TRPV5 pore. Most interestingly, using SCAM they provided evidence that acidic intracellular conditions lead to a clockwise rotation of the pore helix, which then leads to restriction of the selectivity filter (Yeh et al., 2005). It is tempting to speculate that rotation of the pore helix is a more general mechanism for gating of TRP channel pores. Such a mechanism has already previously been proposed for gating of other types of 6TM channels, such as CNG channels.

11.4.2.2 TRPC

The number of studies aimed at the identification of the pore region of TRPC channels is very limited. Moreover, unlike TRPV channels, the region between S5 and S6 does not share significant sequence homology to the pore segment of bacterial K^+ channels such as KscA. This loss of homology to the bacterial archetype pore might signify that TRPC are phylogenetically younger than TRPV channels, and thus that the "classical" TRPCs were not the actual starting point for the evolution of the TRP superfamily. Nevertheless, mutagenesis studies on TRPC1 and TRPC5 indicate that the loop between S5 and S6 contains residues that determine the pore properties of these channels. Expression of a TRPC1 mutant in which all seven negatively charged residues in the region between S5 and S6 were neutralized resulted in a channel with decreased Ca^{2+} but normal Na^{+} currents (Liu et al., 2003). A systematic mutagenesis of all negatively charged residues localized in the putative extracellular loops of TRPC5 revealed that three of the five glutamates in the loop between S5 and S6 are involved in channel activation by La^{3+} and determine single-channel properties (Jung et al., 2003). Negatively charged residues in the other extracellular loops did not play a significant role (Jung et al., 2003). Taken together, these data suggest that the pore-forming region in TRPC is also localized between S5 and S6. However, current data do not allow drawing any conclusion as to the position of the pore helix or selectivity filter in TRPCs. Moreover, it seems that the negatively charged residues identified in these studies are located at the periphery of the S5–S6 loop.

11.4.2.3 TRPM

Two recent studies have addressed the structural determinants of the pore of TRPM channels. The first study focused on the loop between S5 and S6 of TRPM4, which shows only very limited homology to the P-loop of K^{+} channels. Substitution of residues Glu^{981} to Ala^{986} in this region with the known selectivity filter of TRPV6 yielded a functional channel that combined the gating hallmarks of TRPM4 with TRPV6-like pore properties such as sensitivity to block by extracellular Ca^{2+} and Mg^{2+} (Nilius et al., 2005a). Substitution of Glu^{981} by an alanine strongly reduced the channel's affinity to block by intracellular spermine, indicating that E^{981} is located in the inner part of the pore (Nilius et al., 2005a). Substitution of Gln^{977} by a glutamate altered the monovalent cation permeability sequence and resulted in a pore with moderate Ca^{2+} permeability. Taken together, these results suggest that the loop between S5 and S6 indeed contains the selectivity filter of TRPM4.

Another recent study describes several splice variants of TRPM3, which differ in the length of the putative pore region since one splice site is located in the loop between S5 and S6 (Oberwinkler et al., 2005). Two splice variants, termed TRPM3α1 and TRPM3α2, which only differ by the presence or absence of a stretch of 12 amino acids in the S5–S6 loop display >10-fold difference in relative permeability to Ca^{2+} and Mg^{2+} and in the sensitivity to block by extracellular monovalent cations (Oberwinkler et al., 2005). This is probably the first example of an ion channel whose selectivity is regulated through alternative splicing.

11.4.2.4 TRP Pores: Many Open Questions

From the above, it is clear that structure–function research concerning the pore of TRP channels is still in its infancy. We are only starting to understand the structural basis of the different permeability profiles in TRP channels. The structure of the inner pore, which in the case of K^{+} channels is lined by residues of the S6 domain, has not yet been addressed. Moreover, in the case of TRPP, TRPML, TRPA, and TRPN channels, not a single study has addressed the localization or structure of the pore region. High-resolution crystal structures of the pore region of different TRP channels would be needed to give definitive answers to these questions.

11.4.3 Transmembrane Segments S1–S4

In other 6TM channels such as the voltage-gated K^+ and HCN channels, the first four transmembrane segments (S1–S4) form a sensor that detects changes in transmembrane voltage (Swartz, 2004; Bezanilla, 2005). In particular, a cluster of positively charged residues in S4 is known to sense changes in the transmembrane electrical field and to move relative to it (Bezanilla, 2005). Notably, TRP channels lack the clustering of basic residues in S4. According to the latest structural information obtained for voltage-gated K^+ channels, the S1–S4 region acts as an independent domain inside the membrane and its movement is translated into opening of the pore region via an interaction between the S4–S5 linker and the C-terminal part of S6 (Long et al., 2005a,b).

Only few reports describe the functional impact of the S1–S4 segments in regulation of the TRP channel function. Yet, it is conceivable that the S1–S4 domain of TRP channels is similarly involved in channel gating. Indeed, a growing number of studies indicate that the S2–S4 region of TRP channels plays a key role in ligand-dependent channel gating. Jordt and Julius first identified a YS motif in the S2–S3 linker of TRPV1 ($Tyr^{511}Ser^{512}$) that is crucial for binding of the activating ligand capsaicin (Jordt and Julius, 2002). A further study revealed additional residues in S2 and S4 involved in vanilloid sensitivity of TRPV1 (Gavva et al., 2004). Following up on these results, Vriens et al. identified Tyr^{555} in S3 of TRPV4 as a determinant of the sensitivity to the synthetic phorbol ester 4α-phorbol 12, 13-dideconoate (4α-PDD) and heat (Vriens et al., 2004b). Finally, Chuang et al. identified residues in S3 and the S2–S3 linker of TRPM8 that are involved in channel activation by the synthetic super-cooling agent icilin (Chuang et al., 2004).

11.4.4 Cytoplasmic Tails

Compared to other 6TM channels, TRP channels can have extremely long cytoplasmic N- and C-terminal tails containing a plethora of potential structural and/or regulatory modules (Clapham, 2003; Montell, 2005; Nilius et al., 2005b). In at least three instances, the C-terminal tails encompass entire functional enzymes: TRPM2 contains a Nudix hydrolase domain in its C terminus, which functions as an ADP-ribose (ADPR) pyrophosphatase (Perraud et al., 2001). TRPM6 and TRPM7 contain a C-terminal atypical α-kinase domain (Nadler et al., 2001; Runnels et al., 2001). Other notable domains are ankyrin repeats, 33-residue motifs consisting of pairs of antiparallel α-helices connected by β-hairpin motifs, which can be identified in the N terminus of most subfamilies of TRP channels. The number of ankyrin repeats is highly variable, ranging from 3–4 in TRPCs and TRPVs, to 14–15 in TRPAs and approximately 29 (!) in TRPNs. The so-called TRP box in TRPCs, a stretch of six amino acids in the proximal part of the C terminus, has initially been proposed as a signature sequence for all TRP channels. However, the TRP box consensus sequence (EWKFAR) is only poorly conserved in TRPVs and TRPMs, and fully absent in the TRPP, TRPML, TRPA, and TRPN subfamilies. The role of these different intracellular domains in TRP channel functioning is mostly unclear. For a more detailed account of the structure of TRP channels, we refer to recent reviews focusing on this aspect (Owsianik et al., 2006a,b).

11.5 Activation Mechanisms

Given the functional properties of *Drosophila* TRP, mammalian TRP homologues were initially mainly envisaged as the potential molecular correlates of PLC-dependent and/or store-operated cation channels (Petersen et al., 1995; Wes et al., 1995; Birnbaumer et al., 1996; Clapham, 1996; Zhu et al., 1996; Zitt et al., 1996). This view has changed completely, especially since the cloning of VR1 (vanilloid receptor 1; now TRPV1), a Ca^{2+}-permeable cation channel activated by vanilloid compounds such as capsaicin, by protons and by noxious heat (>43°C) (Caterina et al., 1997). Now, one of the trademarks of the TRP superfamily is the huge variability in gating mechanisms. Below we give an overview of TRP channel gating mechanisms. Most likely, this list of activating mechanisms is incomplete. Moreover, it should be kept in mind that many TRP channels respond to different types of activating stimuli, or even require the simultaneous presence of two distinct stimuli to open.

11.5.1 PLC- and/or Store-Dependent Gating

In most nonexcitable cell types, receptor-induced Ca^{2+} signals are composed of a rapid, transient release of Ca^{2+} from internal stores followed by a slower, sustained Ca^{2+} entry from the extracellular medium (Venkatachalam et al., 2002). Both G protein- and tyrosine kinase-coupled plasma membrane receptors lead to the activation of PLC, which catalyze the production of inositol 3,4,5,-triphosphate (IP_3) and DAG from PIP_2 (Venkatachalam et al., 2002). It is well established that IP_3 directly activates IP_3 receptors, which form Ca^{2+}-permeable channels located on the membrane of intracellular Ca^{2+} stores. Members of the TRP superfamily, in particular TRPC channels, have been put forward as the main pathways for the slower, sustained Ca^{2+} entry from the extracellular medium that follows PLC activation (Petersen et al., 1995; Wes et al., 1995; Birnbaumer et al., 1996; Zhu et al., 1996; Zitt et al., 1996). Several mechanisms have been put forward to explain the opening of these TRPC channels by PLC activation. Firstly, it has been convincingly shown that TRPC3, TRPC6, and TRPC7 are activated by DAG and DAG-analogues. The DAG-dependent gating mechanism occurs in a membrane-delimited manner, suggesting a direct interaction between DAG and the channel protein (Hofmann et al., 1999). Secondly, for all TRPCs and also for TRPV6 it has been proposed that they are gated in a store-dependent manner, meaning that they become activated whenever the intracellular Ca^{2+} stores are depleted. Unfortunately, in the case of all these channels, subsequent reports have contested their store dependence (Putney, 2005). A further complication is that several reports indicate that the level of expression determines whether a specific TRP channel is activated in a store-dependent or store-independent manner.

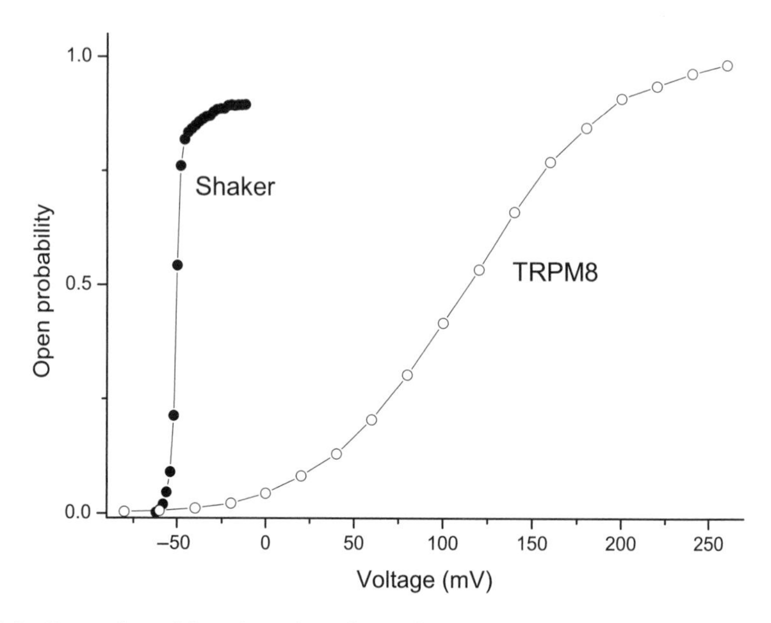

Fig. 11.3 Comparison of the voltage dependence of TRPM8 with that of the archetypical voltage-gated K^+ channel Shaker. Note that the activation curve for TRPM8 mainly extends into the nonphysiological voltage range (>50 mV) and has a much shallower voltage dependence.

For a detailed account of the knowledge and controversies concerning PLC- and/or store-dependent TRPC activation we refer to recent reviews (Putney, 2004, 2005).

11.5.2 Voltage-Gated TRP Channels

For a long time, TRP channels were considered to be channels with little or no voltage dependence (Clapham, 2003). This view was mainly founded on the absence of the cluster of positively charged residues in S4, which was known to constitute the voltage sensor of other 6TM channels. Moreover, the first observations of voltage dependence in members of the TRP superfamily revealed voltage-dependent activation curves that extended mainly into the nonphysiological positive voltage range and were characterized by very low gating valences (Fig. 11.3), suggesting that voltage dependence was an epiphenomenona of little physiological relevance (Gunthorpe et al., 2000; Nilius et al., 2003, 2005b). However, recent studies have demonstrated that factors such as temperature or chemical ligands can drastically shift the voltage-dependent activation curves toward the physiological voltage range, and that this constitutes an important mechanism for TRP channel gating (see below).

11.5.3 Temperature-Sensitive TRP Channels

Since the discovery of the TRPV1 as a heat-activated channel, seven additional mammalian temperature-sensitive TRP channels [also called thermoTRPs (Patapoutian et al., 2003)] have been described. TRPV1 and its closest homologues TRPV2 (Caterina et al., 1999), TRPV3 (Peier et al., 2002b; Smith et al., 2002; Xu et al., 2002), and TRPV4 (Güler et al., 2002; Watanabe et al., 2002b), as well as the Ca^{2+}-dependent monovalent cation channels TRPM4 and TRPM5 (Talavera et al., 2005) are activated upon heating. In contrast, TRPM8 (McKemy et al., 2002; Peier et al., 2002a) and TRPA1 (Story et al., 2003) open upon cooling, although cold activation of TRPA1 could not be reproduced by some groups (Jordt et al., 2004; Nagata et al., 2005). Together, these temperature-sensitive TRP channels have the potential to detect changes in temperature from <10 to >50°C, which corresponds to the physiological range of temperatures that humans can discriminate.

All types of enzymes, including ion channels, display some degree of temperature dependence. It is, therefore, important to briefly consider how specialized the temperature sensitivity of these thermoTRP channels actually is. In general, the temperature dependence of a reaction rate can be quantified using the 10 degree temperature coefficient or Q_{10}, which is defined as $Q_{10} = \text{rate}(T + 10)/\text{rate}(T)$ (Hille, 2001). For example, the single-channel conductance of any channel increases with temperature, with typical Q_{10} values between 1.2 and 1.4, which can be accounted for by the temperature dependence of ionic diffusion (Hille, 2001). Voltage-dependent gating of classical voltage-gated channels displays Q_{10} values that are typically between 2 and 4 (Hille, 2001). Quantification of the temperature dependence of the ionic current through heat-activated TRPs has yielded Q_{10} values between 6 and 30 (Caterina et al., 1997, 1999; Peier et al., 2002b; Smith et al., 2002; Watanabe et al., 2002b; Xu et al., 2002; Talavera et al., 2005). Moreover, in the case of the cold-activated TRP channels TRPM8 and TRPA1, ionic currents decrease upon heating (McKemy et al., 2002; Peier et al., 2002a; Story et al., 2003), which in theory corresponds to Q_{10} values <1.

Several mechanisms can be envisaged to explain the remarkable temperature sensitivity of temperature-sensitive TRP channels (Clapham, 2003). First, changes in temperature could lead to the production of channel-activating ligands. In such a model, the ligand-producing enzyme rather than the TRP channel itself would be temperature dependent. In most thermoTRPs thermal sensitivity is well preserved in cell-free membranes, indicating that temperature-dependent ligand binding is unlikely to be a general mechanism for thermosensation. In the case of TRPV4, however, heat activation no longer occurs in cell-free inside-out patches, which could point toward loss of some soluble messenger (Watanabe et al., 2002b). Second, channel activation might result from a temperature-dependent phase transition of the lipid membrane or a conformational transition (or denaturation) of the channel protein. Phase transitions of lipid membrane or conformational transitions in proteins usually occur over a narrow temperature range, which is in line with the steep temperature dependence of thermoTRP activation.

In a recent study we presented a fundamental thermodynamic principle to explain cold activation of TRPM8 and heat activation of TRPV1, not necessitating diffusible messengers or conformational transitions (Voets et al., 2004a). It was revealed that TRPM8 and TRPV1 are voltage-gated channels activated upon membrane depolarization. Thermal activation reflects a robust but graded shift of the voltage dependence of activation from strongly depolarized potentials toward the physiological potential range. This finding implies that the thermal sensitivity of these channels depends on voltage, and that temperature-dependent activation represents a gradual increase in the open probability of the channel rather than a threshold phenomenon. Moreover, it strongly argues against temperature-dependent phase transition of the lipid membrane or conformational transitions of the channel protein as a mechanism for thermal activation, as such processes would predict a single sharp thermal threshold (Voets et al., 2004a).

A two-state gating model was found to accurately approximate the temperature-dependent activation of TRPV1, TRPM8, TRPM4, and TRPM5 (Voets et al., 2004a; Talavera et al., 2005):

$$\text{Closed} \underset{\beta(V,T)}{\overset{\alpha(V,T)}{\rightleftarrows}} \text{Open}.$$

For TRPM8, the temperature dependence of channel opening (α) is much less steep than that of channel closing (α), which leads to channel activation upon cooling. In the case of TRPV1, TRPM4, and TRPM5, α displays a much steeper temperature dependence than β, leading to channel activation upon heating (Voets et al., 2004a; Talavera et al., 2005). According to Eyring rate theory, α and β are related to membrane voltage and temperature according to

$$\alpha = \kappa \frac{kT}{h} \cdot e^{\frac{-\Delta H_{\text{open}} + T\Delta S_{\text{open}} + \delta z F V}{RT}}$$
$$\beta = \kappa \frac{kT}{h} \cdot e^{\frac{-\Delta H_{\text{close}} + T\Delta S_{\text{close}} - (1-\delta) z F V}{RT}},$$

where ΔH_{open} and ΔH_{close} represent the enthalpies and ΔS_{open} and ΔS_{close} the entropies associated with channel opening and closing, respectively. R represents the gas constant (8.31 J mol^{-1} K^{-1}), T the absolute temperature, z the effective charge associated with voltage-dependent gating, δ the fraction of z moved in the outward direction, F the Faraday constant (9.65×10^4 C mol^{-1}), k Boltzmann's constant (1.381×10^{23} J K^{-1}), and h Planck's constant (6.626×10^{-34} J s). κ is the transmission coefficient, which is usually assumed to equal 1. Half-maximal activation occurs when $\alpha = \beta$, which refers to membrane potential $V_{1/2}$ given by the expression:

$$V_{1/2} = \frac{1}{zF}(\Delta H - T\Delta S),$$

where ΔH and ΔS represent the difference in enthalpy and entropy between the open and closed state, respectively ($\Delta H = \Delta H_{\text{open}} - \Delta H_{\text{close}}$; $\Delta S = \Delta S_{\text{open}} - \Delta S_{\text{close}}$). Note that the shallow voltage dependence of TRP channels ($z < 1$) is highly relevant, as $V_{1/2}$ changes by a factor of $-\Delta S/zF$ when temperature increases by 1 degree (Voets et al., 2005). Thus, ΔS is negative for a cold-activated channel and positive for a heat-activated channel. In all examined thermoTRPs channels (TRPV1, TRPM8, TRPM4, TRPM5), the $|\Delta S|$ amounts to approximately 500 J mol^{-1} K^{-1} (Talavera et al., 2005). The factors that determine this significant difference in entropy between open and closed states are presently not known. Moreover, it is not at all understood which structural elements determine that TRPM8 is cold-activated ($\Delta S < 0$), whereas the closely related TRPM4 and TRPM5 are heat-activated ($\Delta S > 0$).

11.5.4 Ligand-Gated TRP Channels

11.5.4.1 Exogenous Ligands

The vanilloid receptor TRPV1 was identified in an expression cloning approach that used capsaicin, the pungent extract of hot peppers, as an agonist (Caterina et al., 1997). After this discovery, a growing number of structurally unrelated botanical compounds have been identified as potent activators of TRP channels from different subfamilies. In particular, it seems that temperature-sensitive TRPs have evolved as the favorite targets for plant-derived chemicals. The finding that a single receptor can respond to both thermal and chemical stimuli explains why we sometimes attribute inherent thermal features to chemical compounds (e.g., "hot" chilli pepper or "cool" mint), a phenomenon known as chemesthesis. Other well-known examples of plant-derived compounds that act on TRPV1 are resiniferatoxin, an active compound from the cactus *Euphorbia resinifera* (Szallasi et al., 1999), and by piperine, the pungent component in black pepper (McNamara et al., 2005). Camphor, the waxy substance with penetrating odor extracted from *Cinnamomum camphora*, acts as an agonist of both TRPV1 and TRPV3 (Moqrich et al., 2005; Xu et al., 2005). TRPM8 is directly activated by menthol and eucalyptol, two cooling compound extracted from the mint plant *Mentha piperita* and the tree *Eucalyptus globulus*, respectively (McKemy et al., 2002; Peier et al., 2002a). TRPA1 acts as a receptor for isothiocyanates (the pungent component in mustard, horseradish, and wasabi), cinnamaldehyde (an active compound in cinnamon oil) and Δ9-tetrahydrocannabinol (the psychoactive compound in marijuana (*Cannabis sativa*)) (Bandell et al., 2004; Jordt et al., 2004). Allicin, an unstable component of fresh garlic (*Allium sativum*), is an agonist for TRPA1 (Bautista et al., 2005; Macpherson et al., 2005) and TRPV1 (Macpherson et al., 2005).

Additionally, more and more synthetic TRP channel ligands are becoming available. Some of these synthetic ligands display relatively high selectivity for distinct TRP channels, such as olvanil (TRPV1) and 4α-phorbol-12,13-didecanoate (4α-PDD; TRPV4) (Watanabe et al., 2002a), which makes them important and widely used pharmacological tools. Many others, such as 2-aminoethyl diphenylborinate (TRPV1, TRPV2, TRPV3) (Chung et al., 2004; Hu et al., 2004) or icilin (TRPM8, TRPA1) (McKemy et al., 2002; Story et al., 2003) are activators of more than one TRP channel. At present, determination of the role of most TRP channels in cells, tissues, and living animals is still strongly hampered by the lack of specific (ant)agonists.

Interestingly, we recently demonstrated that menthol (TRPM8) and capsaicin (TRPV1) induce channel activation by shifting the voltage dependence of activation, such that the channels open at physiologically relevant voltages (Voets et al., 2004a). Thus, these compounds should not be considered as activating agonists that can switch the channels on, but rather of modulators of the intrinsic voltage dependence of these channels. Identification of the voltage sensor of TRP channels would therefore be instrumental to understand the exact *modus operandi* of these chemical ligands.

11.5.4.2 Endogenous Ligands

Several TRP channels are known to act as receptors for endogenous compounds. The activation of different TRPC by DAG was already discussed above. Other well-known examples are the gating of TRPV1 and TRPV4 by a number of compounds related to arachidonic acid (AA). The amide anandamide (arachidonoylethanolamide) (Zygmunt et al., 1999) as well as lipoxygenase metabolites of AA such as 12,15-(*S*)-hydroperoxyeicosatetraenoic acid and leukotriene B_4 (Hwang et al., 2000) activate TRPV1 at low micromolar concentrations. 5′,6′-epoxieicosatrienoic acid (5′,6′-EET), a cytochrome P450 epoxygenase-dependent metabolite of AA, activates TRPV4 at submicromolar concentrations in a membrane-delimited manner (Watanabe et al., 2003). TRPM3 was recently shown to be activated by sphingosine, a metabolite that is formed during the de novo synthesis of cellular sphingolipids (Grimm et al., 2005). Finally, TRPM4 and TRPM5 are directly gated by increases in intracellular Ca^{2+} (Launay et al., 2002; Hofmann et al., 2003; Liu and Liman, 2003; Prawitt et al., 2003). Unfortunately, structural information about the binding site for all these ligands or mechanistic insight in how they cause opening of TRP channels is currently lacking.

11.5.5 Mechanosensitive TRP Channels

The involvement of TRP channels in mechanosensation was first deduced from genetic studies in *C. elegans*. Worms with mutations in the *osm-9* gene, which encodes a TRPV channel expressed in sensory ASH neurons, were found to exhibit strong defects in their avoidance reaction to noxious odors, high osmolality, and nose touch (Colbert et al., 1997). Subsequently, the mammalian OSM-9 homologue TRPV4 as a cation channel gated by hypotonic cell swelling (Liedtke et al., 2000; Strotmann et al., 2000; Wissenbach et al., 2000; Nilius et al., 2001a). Interestingly, mouse TRPV4 expressed in ASH neurons from *C. elegans osm-9* mutants is able to restore the response to hypertonicity and nose touch, but not that to noxious odors, indicating that OSM-9 and TRPV4 exhibit similar mechanosensitive but distinct chemosensing properties (Liedtke et al., 2003). Activation by cell swelling has also been reported for mammalian TRPV2 (Muraki et al., 2003) and TRPM3 (Grimm et al., 2003), and for Nanchung, a TRPV channel required for hearing in *Drosophila* (Kim et al., 2003). Further research is required to establish whether these channels have a physiological role as mechano- or osmosensors.

Recently, Maroto and colleagues observed a high abundance of TRPC1 in the protein fraction that reconstitutes the mechanosensitive cation channel (MscCa) from *Xenopus laevis* oocytes in liposomes (Maroto et al., 2005). Antisense inhibition of TRPC1 expression in *Xenopus* abolished endogenous MscCa, whereas heterologous TRPC1 expression led to a dramatic increase in MscCa activity. They concluded that TRPC1 forms all or part of vertebrate MscCa (Maroto et al., 2005), a hypothesis that strongly challenges the general view of TRPC channels as store- and/or PLC-dependent channels (Clapham, 2003).

Loss-of-function mutations in the *no mechanoreceptor potential C* (*nompC*) gene, which codes for the founding member of the TRPN subfamily, abolishes mechanotransduction in *Drosophila* (Walker et al., 2000). Similarly, knock-down of the expression of the zebrafish orthologue leads to larval deafness and imbalance (Sidi et al., 2003). Similarly, vertebrate TRPA1 was implicated in mechanosensation, as a molecular candidate for the transduction channel in vertebrate hair cells (Corey et al., 2004). This conclusion was mainly based on three observations: TRPA1 is highly expressed hair bundles of the hair cells, TRPA1 messenger expression coincides with the onset of hair cell mechanosensitivity, and, most importantly, disruption of TRPA1 expression in zebrafish using morpholino oligonucleotides and in mice using small interference RNA strongly inhibited mechanotransduction in the hair cells (Corey et al., 2004). However, the most recent studies (Bautista et al., 2006; Kwan et al., 2006) conducted with *trpa*1$^{(-/-)}$ mice revealed no obvious deficits in auditory function. Clearly, TRPA1 can no longer be regarded as an essential component of the transduction channel in hearing, at least in adult animals (Bautista et al., 2006; Kwan et al., 2006). As far as mechano-activation of TRPA1 is concerned, in one study of *trpa*1$^{(-/-)}$ mice a reduction in sensitivity to mechanical stimulation of the hindpaws was reported (Kwan et al., 2006). It remains to be demonstrated that they act as mechanosensitive channels.

Three possible mechanisms have been put forward to explain for the transduction of a mechanical signal to the opening of the channel gate. First, mechanical stimuli may lead to changes in the tension in the lipid bilayer, which can then be sensed by the transmembrane segments of the channel resulting in gating of the pore. Such a mechanism is well established for MscL (Perozo et al., 2002), a large conductance mechanosensitive cation channel ubiquitously expressed in bacteria but with no significant homology to TRPs. It was recently shown that the mechanosensitivity of TRPC1 is conserved after incorporation in artificial liposomes, suggesting that it may also directly sense changes in bilayer tension (Maroto et al., 2005). Second, mechanical stress may be directly transduced to the channel, for example, via a connection between cytosolic tails and cytoskeletal elements. Such a mechanism has been proposed to explain the mechanosensitivity of TRPA1 and TRPN1 (Corey et al., 2004; Howard and Bechstedt, 2004; Sotomayor et al., 2005). In particular, it was found that these channels contain a large number of ankyrin repeats in their N terminus, which from crystallographic studies are known to form a spring-like helical structure (Howard and Bechstedt, 2004; Sotomayor et al., 2005). Molecular dynamics simulations and theoretical calculations of the properties of such an ankyrin helix indicates that its extension and stiffness match with those of the gating spring in vertebrate hair cells (Howard and Bechstedt, 2004; Sotomayor et al., 2005). Third, an enzyme whose activity induces channel gating rather than the channel itself can be mechanosensitive. We have shown that such a mechanism undelies the swelling-dependent activation of TRPV4 (Vriens et al., 2004b). Blockers of phospholipase A2 (PLA_2) and cytochrome P450 epoxygenases inhibit activation of TRPV4 by hypotonic solution but not by the direct ligand 4α-PDD. Cell swelling is known to directly activate PLA_2. Thus, it was concluded that activation of TRPV4 proceeds via the PLA_2-dependent formation of AA and its subsequent metabolization to 5′,6′-EET (Vriens et al., 2004b), which then acts as a channel-activating ligand (Watanabe et al., 2003).

11.5.6 Constitutively Open TRP Channels

Several TRP channels display significant open probability under "control conditions," i.e., in the absence of any of the above-described activatory stimuli. Striking examples are the Ca^{2+}-selective TRPV5 and TRPV6, which act as gatekeepers for Ca^{2+} entry in the apical membrane of Ca^{2+}-transporting epithelia in kidney and intestine (Hoenderop et al., 2005). Both channels undergo prominent Ca^{2+}-dependent inhibition, which acts as a negative feedback brake mechanism to prevent Ca^{2+} overload (Vennekens et al., 2000; Hoenderop et al., 2001; Yue et al., 2001).

An analogous mechanism appears to regulate TRPM6 and TRPM7. These two closely related channels function as influx pathways for Mg^{2+}, and are involved in the cellular Mg^{2+} homeostasis and in Mg^{2+} reabsorption in the kidney (Schlingmann et al., 2002; Walder et al., 2002; Schmitz et al., 2003; Voets et al., 2004c). Both channels undergo feedback inhibition by intracellular Mg^{2+}, with half-maximal inhibition occurring at a free Mg^{2+} concentration of $\sim$0.5 mM (Nadler et al., 2001; Voets et al., 2004c), which is close to the normal resting free Mg^{2+} concentration in mammalian cells.

11.6 Concluding Remarks

The discovery of the TRP superfamily meant a great leap forward in our molecular understanding of the cation channels in nonexcitable and excitable cells. From the above account it is clear that we are only beginning to understand how these fascinating channels work. Given the extremely diverse and complex permeation and gating mechanisms, we foresee that TRP channels will entertain and bother ion channel biophysicists and structural biologists for the coming decades.

Acknowledgments

Work in the authors' laboratory is supported by the Human Frontiers Science Programme (HFSP Research Grant Ref. RGP 32/2004), the Belgian Federal Government, the Flemish Government, and the research council of the K.U. Leuven (GOA 2004/07, F.W.O. G.0214.99, F.W.O. G.0136.00; F.W.O. G.0172.03, Interuniversity Poles of Attraction Program, Prime Ministers Office IUAP, Excellentic financiering EF1951010).

References

Almers, W., and E.W. McCleskey. 1984. Non-selective conductance in calcium channels of frog muscle: Calcium selectivity in a single-file pore. *J. Physiol.* 353:585–608.

Bandell, M., G.M. Story, S.W. Hwang, V. Viswanath, S.R. Eid, M.J. Petrus, T.J. Earley, and A. Patapoutian. 2004. Noxious cold ion channel TRPA1 is activated by pungent compounds and bradykinin. *Neuron* 41:849–857.

Bautista, D.M., Jordt, S.E., Nikai, T., Tsuruda, P.R., Read, A.J., Poblete, J., Yamoah, E.N., Basbaum, A.I., and Julius, D. 2006. TRPA1 Mediates the Inflammatory Actions of Environmental Irritants and Proalgesic Agents. *Cell* 124:1269–1282.

Bautista, D.M., P. Movahed, A. Hinman, H.E. Axelsson, O. Sterner, E.D. Hogestatt, D. Julius, S.E. Jordt, and P.M. Zygmunt. 2005. Pungent products from garlic activate the sensory ion channel TRPA1. *Proc. Natl. Acad. Sci. USA* 102:12248–12252.

Bezanilla, F. 2005. Voltage-gated ion channels. *IEEE Trans. Nanobiosci.* 4:34–48.

Birnbaumer, L., X. Zhu, M. Jiang, G. Boulay, M. Peyton, B. Vannier, D. Brown, D. Platano, H. Sadeghi, E. Stefani, and M. Birnbaumer. 1996. On the molecular basis and regulation of cellular capacitative calcium entry: Roles for Trp proteins. *Proc. Natl. Acad. Sci. USA* 93:15195–15202.

Cataldi, M., E. Perez-Reyes, and R.W. Tsien. 2002. Differences in apparent pore sizes of low and high voltage-activated Ca^{2+} channels. *J. Biol. Chem.* 277:45969–45976.

Caterina, M.J., T.A. Rosen, M. Tominaga, A.J. Brake, and D. Julius. 1999. A capsaicin-receptor homologue with a high threshold for noxious heat. *Nature* 398:436–441.

Caterina, M.J., M.A. Schumacher, M. Tominaga, T.A. Rosen, J.D. Levine, and D. Julius. 1997. The capsaicin receptor: A heat-activated ion channel in the pain pathway. *Nature* 389:816–824.

Chuang, H.H., W.M. Neuhausser, and D. Julius. 2004. The super-cooling agent icilin reveals a mechanism of coincidence detection by a temperature-sensitive TRP channel. *Neuron* 43:859–869.

Chung, M.K., H. Lee, A. Mizuno, M. Suzuki, and M.J. Caterina. 2004. 2-aminoethoxydiphenyl borate activates and sensitizes the heat-gated ion channel TRPV3. *J. Neurosci.* 24:5177–5182.

Chyb, S., P. Raghu, and R.C. Hardie. 1999. Polyunsaturated fatty acids activate the Drosophila light-sensitive channels TRP and TRPL. *Nature* 397:255–259.

Clapham, D.E. 1996. TRP is cracked but is CRAC TRP? *Neuron* 16:1069–1072.

Clapham, D.E. 2003. TRP channels as cellular sensors. *Nature* 426:517–524.

Colbert, H.A., T.L. Smith, and C.I. Bargmann. 1997. OSM-9, a novel protein with structural similarity to channels, is required for olfaction, mechanosensation, and olfactory adaptation in Caenorhabditis elegans. *J. Neurosci.* 17:8259–8269.

Corey, D.P., J. Garcia-Anoveros, J.R. Holt, K.Y. Kwan, S.Y. Lin, M.A. Vollrath, A. Amalfitano, E.L. Cheung, B.H. Derfler, A. Duggan, G.S. Geleoc, P.A. Gray, M.P. Hoffman, H.L. Rehm, D. Tamasauskas, and D.S. Zhang. 2004. TRPA1 is a candidate for the mechanosensitive transduction channel of vertebrate hair cells. *Nature* 432:723–730.

Cosens, D.J., and A. Manning. 1969. Abnormal electroretinogram from a *Drosophila* mutant. *Nature* 224:285–287.

Dang, T.X., and E.W. McCleskey. 1998. Ion channel selectivity through stepwise changes in binding affinity. *J. Gen. Physiol.* 111:185–193.

Dodier, Y., U. Banderali, H. Klein, O. Topalak, O. Dafi, M. Simoes, G. Bernatchez, R. Sauve, and L. Parent. 2004. Outer pore topology of the ECaC-TRPV5 channel by cysteine scan mutagenesis. *J. Biol. Chem.* 279:6853–6862.

Doyle, D.A., J. Morais Cabral, R.A. Pfuetzner, A. Kuo, J.M. Gulbis, S.L. Cohen, B.T. Chait, and R. MacKinnon. 1998. The structure of the potassium channel: Molecular basis of K^+ conduction and selectivity. *Science* 280:69–77.

Garcia-Martinez, C., C. Morenilla-Palao, R. Planells-Cases, J.M. Merino, and A. Ferrer-Montiel. 2000. Identification of an aspartic residue in the P-loop of the vanilloid receptor that modulates pore properties. *J. Biol. Chem.* 275:32552–32558.

Gavva, N.R., L. Klionsky, Y. Qu, L. Shi, R. Tamir, S. Edenson, T.J. Zhang, V.N. Viswanadhan, A. Toth, L.V. Pearce, T.W. Vanderah, F. Porreca, P.M. Blumberg, J. Lile, Y. Sun, K. Wild, J.C. Louis, and J.J. Treanor. 2004. Molecular determinants of vanilloid sensitivity in TRPV1. *J. Biol. Chem.* 279:20283–20295.

Grimm, C., R. Kraft, S. Sauerbruch, G. Schultz, and C. Harteneck. 2003. Molecular and functional characterization of the melastatin-related cation channel TRPM3. *J. Biol. Chem.* 278:21493–21501.

Grimm, C., R. Kraft, G. Schultz, and C. Harteneck. 2005. Activation of the melastatin-related cation channel TRPM3 [corrected] by D-erythro-sphingosine. *Mol. Pharmacol.* 67:798–805.

Güler, A.D., H. Lee, T. Iida, I. Shimizu, M. Tominaga, and M. Caterina. 2002. Heat-evoked activation of the ion channel, TRPV4. *J. Neurosci.* 22:6408–6414.

Gunthorpe, M.J., M.H. Harries, R.K. Prinjha, J.B. Davis, and A. Randall. 2000. Voltage- and time-dependent properties of the recombinant rat vanilloid receptor (rVR1). *J. Physiol. (Lond.)* 525:747–759.

Hardie, R.C., and B. Minke. 1992. The trp gene is essential for a light-activated Ca^{2+} channel in Drosophila photoreceptors. *Neuron* 8:643–651.
Hardie, R.C., and P. Raghu. 2001. Visual transduction in Drosophila. *Nature* 413:186–193.
Heinemann, S.H., H. Terlau, W. Stuhmer, K. Imoto, and S. Numa. 1992. Calcium channel characteristics conferred on the sodium channel by single mutations. *Nature* 356:441–443.
Hess, P., J.B. Lansman, and R.W. Tsien. 1986. Calcium channel selectivity for divalent and monovalent cations. Voltage and concentration dependence of single channel current in ventricular heart cells. *J. Gen. Physiol.* 88:293–319.
Hess, P., and R.W. Tsien. 1984. Mechanism of ion permeation through calcium channels. *Nature* 309:453–456.
Hille, B. 2001. Ionic Channels of Excitable Membranes. Sinauer Associates, Sunderland, MA.
Hoenderop, J.G., B. Nilius, and R.J. Bindels. 2005. Calcium absorption across epithelia. *Physiol. Rev.* 85:373–422.
Hoenderop, J.G.J., R. Vennekens, D. Müller, J. Prenen, G. Droogmans, R.J.M. Bindels, and B. Nilius. 2001. Function and expression of the epithelial Ca^{2+} channel family: Comparison of the epithelial Ca^{2+} channel 1 and 2. *J. Physiol. Lond.* 537:747–761.
Hoenderop, J.G.J., T. Voets, S. Hoefs, F. Weidema, J. Prenen, B. Nilius, and R.J.M. Bindels. 2003. Homo- and heterotetrameric architecture of the epithelial Ca^{2+} channels, TRPV5 and TRPV6. *EMBO J.* 22:776–785.
Hofmann, T., V. Chubanov, T. Gudermann, and C. Montell. 2003. TRPM5 is a voltage-modulated and Ca(2+)-activated monovalent selective cation channel. *Curr. Biol.* 13:1153–1158.
Hofmann, T., A.G. Obukhov, M. Schaefer, C. Harteneck, T. Gudermann, and G. Schultz. 1999. Direct activation of human TRPC6 and TRPC3 channels by diacylglycerol. *Nature* 397:259–263.
Howard, J., and S. Bechstedt. 2004. Hypothesis: A helix of ankyrin repeats of the NOMPC-TRP ion channel is the gating spring of mechanoreceptors. *Curr. Biol.* 14:R224–R226.
Hu, H.Z., Q. Gu, C. Wang, C.K. Colton, J. Tang, M. Kinoshita-Kawada, L.Y. Lee, J.D. Wood, and M.X. Zhu. 2004. 2-aminoethoxydiphenyl borate is a common activator of TRPV1, TRPV2, and TRPV3. *J. Biol. Chem.* 279:35741–35748.
Hwang, S.W., H. Cho, J. Kwak, S.Y. Lee, C.J. Kang, J. Jung, S. Cho, K.H. Min, Y.G. Suh, D. Kim, and U. Oh. 2000. Direct activation of capsaicin receptors by products of lipoxygenases: Endogenous capsaicin-like substances. *Proc. Natl. Acad. Sci. USA* 97:6155–6160.
Jordt, S.E., D.M. Bautista, H.H. Chuang, D.D. McKemy, P.M. Zygmunt, E.D. Hogestatt, I.D. Meng, and D. Julius. 2004. Mustard oils and cannabinoids excite sensory nerve fibres through the TRP channel ANKTM1. *Nature* 427:260–265.
Jordt, S.E., and D. Julius. 2002. Molecular basis for species-specific sensitivity to "hot" chili peppers. *Cell* 108:421–430.

Jung, S., A. Muhle, M. Schaefer, R. Strotmann, G. Schultz, and T.D. Plant. 2003. Lanthanides potentiate TRPC5 currents by an action at extracellular sites close to the pore mouth. *J Biol Chem*. 278:3562–3571.

Kedei, N., T. Szabo, J.D. Lile, J.J. Treanor, Z. Olah, M.J. Iadarola, and P.M. Blumberg. 2001. Analysis of the native quaternary structure of vanilloid receptor 1. *J. Biol. Chem*. 276:28613–28619.

Kim, J., Y.D. Chung, D.Y. Park, S. Choi, D.W. Shin, H. Soh, H.W. Lee, W. Son, J. Yim, C.S. Park, M.J. Kernan, and C. Kim. 2003. A TRPV family ion channel required for hearing in Drosophila. *Nature* 424:81–84.

Kwan, K.Y., A.J. Allchorne, M.A. Vollrath, A.P. Christensen, D.S. Zhang, C.J. Woolf, and D.P. Corey. 2006. TRPA1 Contributes to Cold, Mechanical, and Chemical Nociception but Is Not Essential for Hair-Cell Transduction. *Neuron* 50:277–289.

Launay, P., A. Fleig, A.L. Perraud, A.M. Scharenberg, R. Penner, and J.P. Kinet. 2002. TRPM4 is a Ca^{2+}-activated nonselective cation channel mediating cell membrane depolarization. *Cell* 109:397–407.

Liedtke, W., Y. Choe, M.A. Marti-Renom, A.M. Bell, C.S. Denis, A. Sali, A.J. Hudspeth, J.M. Friedman, and S. Heller. 2000. Vanilloid receptor-related osmotically activated channel (VR-OAC), a candidate vertebrate osmoreceptor. *Cell* 103:525–535.

Liedtke, W., D.M. Tobin, C.I. Bargmann, and J.M. Friedman. 2003. Mammalian TRPV4 (VR-OAC) directs behavioral responses to osmotic and mechanical stimuli in Caenorhabditis elegans. *Proc. Natl. Acad. Sci. USA* 100(Suppl 2):14531–14536.

Liu, D., and E.R. Liman. 2003. Intracellular Ca^{2+} and the phospholipid PIP2 regulate the taste transduction ion channel TRPM5. *Proc. Natl. Acad. Sci. USA* 100:15160–15165.

Liu, X., B.B. Singh, and I.S. Ambudkar. 2003. TRPC1 is required for functional store-operated Ca^{2+} channels. Role of acidic amino acid residues in the S5-S6 region. *J. Biol. Chem*. 278:11337–11343.

Long, S.B., E.B. Campbell, and R. Mackinnon. 2005a. Crystal structure of a mammalian voltage-dependent Shaker family K^+ channel. *Science* 309:897–903.

Long, S.B., E.B. Campbell, and R. Mackinnon. 2005b. Voltage sensor of Kv1.2: Structural basis of electromechanical coupling. *Science* 309:903–908.

Macpherson, L., B.H. Geierstanger, V. Viswanath, M. Bandell, S.R. Eid, and A. Patapoutian. 2005. The pungency of garlic: Activation of TRPA1 and TRPV1 in response to allicin. *Curr. Biol.* 15:929–934.

Maroto, R., A. Raso, T.G. Wood, A. Kurosky, B. Martinac, and O.P. Hamill. 2005. TRPC1 forms the stretch-activated cation channel in vertebrate cells. *Nat. Cell Biol*. 7:179–185.

McCleskey, E.W., and W. Almers. 1985. The Ca channel in skeletal muscle is a large pore. *Proc. Natl. Acad. Sci. USA* 82:7149–7153.

McDonald, T.F., S. Pelzer, W. Trautwein, and D.J. Pelzer. 1994. Regulation and modulation of calcium channels in cardiac, skeletal, and smooth muscle cells. *Physiol. Rev*. 74:365–507.

McKemy, D.D., W.M. Neuhäusser, and D. Julius. 2002. Identification of a cold receptor reveals a general role for TRP channels in thermosensation. *Nature* 416:52–58.

McNamara, F.N., A. Randall, and M.J. Gunthorpe. 2005. Effects of piperine, the pungent component of black pepper, at the human vanilloid receptor (TRPV1). *Br. J. Pharmacol.* 144:781–790.

Montell, C. 2005. The TRP superfamily of cation channels. *Sci. STKE* 2005:re3.

Montell, C., L. Birnbaumer, V. Flockerzi, R.J. Bindels, E.A. Bruford, M.J. Caterina, D.E. Clapham, C. Harteneck, S. Heller, D. Julius, I. Kojima, Y. Mori, R. Penner, D. Prawitt, A.M. Scharenberg, G. Schultz, N. Shimizu, and M.X. Zhu. 2002. A unified nomenclature for the superfamily of TRP cation channels. *Mol. Cell.* 9:229–231.

Montell, C., and G.M. Rubin. 1989. Molecular characterization of the Drosophila trp locus: A putative integral membrane protein required for phototransduction. *Neuron* 2:1313–1323.

Moqrich, A., S.W. Hwang, T.J. Earley, M.J. Petrus, A.N. Murray, K.S. Spencer, M. Andahazy, G.M. Story, and A. Patapoutian. 2005. Impaired thermosensation in mice lacking TRPV3, a heat and camphor sensor in the skin. *Science* 307:1468–1472.

Muraki, K., Y. Iwata, Y. Katanosaka, T. Ito, S. Ohya, M. Shigekawa, and Y. Imaizumi. 2003. TRPV2 is a component of osmotically sensitive cation channels in murine aortic myocytes. *Circ. Res*. 93:829–838.

Nadler, M.J., M.C. Hermosura, K. Inabe, A.L. Perraud, Q. Zhu, A.J. Stokes, T. Kurosaki, J.P. Kinet, R. Penner, A.M. Scharenberg, and A. Fleig. 2001. LTRPC7 is a Mg.ATP-regulated divalent cation channel required for cell viability. *Nature* 411:590–595.

Nagata, K., A. Duggan, G. Kumar, and J. Garcia-Anoveros. 2005. Nociceptor and hair cell transducer properties of TRPA1, a channel for pain and hearing. *J. Neurosci*. 25:4052–4061.

Nilius, B., J. Prenen, G. Droogmans, T. Voets, R. Vennekens, M. Freichel, U. Wissenbach, and V. Flockerzi. 2003. Voltage dependence of the Ca^{2+}-activated cation channel TRPM4. *J. Biol. Chem.* 278:30813–30820.

Nilius, B., J. Prenen, A. Janssens, G. Owsianik, C. Wang, M.X. Zhu, and T. Voets. 2005a. The selectivity filter of the cation channel TRPM4. *J. Biol. Chem*. 280:22899–22906.

Nilius, B., J. Prenen, U. Wissenbach, M. Bodding, and G. Droogmans. 2001a. Differential activation of the volume-sensitive cation channel TRP12 (OTRPC4) and volume-regulated anion currents in HEK-293 cells. *Pflugers Arch*. 443:227–233.

Nilius, B., K. Talavera, G. Owsianik, J. Prenen, G. Droogmans, and T. Voets. 2005b. Gating of TRP channels: A voltage connection? *J. Physiol. (Lond.)* 567:35–44.

Nilius, B., R. Vennekens, J. Prenen, J.G. Hoenderop, G. Droogmans, and R.J. Bindels. 2001b. The single pore residue Asp542 determines Ca^{2+} permeation and Mg^{2+} block of the epithelial Ca^{2+} channel. *J. Biol. Chem.* 276:1020–1025.

Oberwinkler, J., A. Lis, K.M. Giehl, V. Flockerzi, and S.E. Philipp. 2005. Alternative splicing switches the divalent cation selectivity of TRPM3 channels. *J. Biol. Chem.* 280:22540–22548.

Owsianik, G., D. D'hoedt, T. Voets, and B. Nilius. 2006a. Structure–function relationships of the TRP channel superfamily. *Rev. Physiol. Biochem. Pharmacol.* 156:61–90.

Owsianik, G., K. Talavera, T. Voets, and B. Nilius. 2006b. Permeation and selectivity of TRP channels. *Annu. Rev. Physiol.* 68:4.1–4.33.

Patapoutian, A., A.M. Peier, G.M. Story, and V. Viswanath. 2003. ThermoTRP channels and beyond: Mechanisms of temperature sensation. *Nat. Rev. Neurosci.* 4:529–539.

Peier, A.M., A. Moqrich, A.C. Hergarden, A.J. Reeve, D.A. Andersson, G.M. Story, T.J. Earley, I. Dragoni, P. McIntyre, S. Bevan, and A. Patapoutian. 2002a. A TRP channel that senses cold stimuli and menthol. *Cell* 108:705–715.

Peier, A.M., A.J. Reeve, D.A. Andersson, A. Moqrich, T.J. Earley, A.C. Hergarden, G.M. Story, S. Colley, J.B. Hogenesch, P. McIntyre, S. Bevan, and A. Patapoutian. 2002b. A heat-sensitive TRP channel expressed in keratinocytes. *Science* 296:2046–2049.

Perozo, E., D.M. Cortes, P. Sompornpisut, A. Kloda, and B. Martinac. 2002. Open channel structure of MscL and the gating mechanism of mechanosensitive channels. *Nature* 418:942–948.

Perraud, A.L., A. Fleig, C.A. Dunn, L.A. Bagley, P. Launay, C. Schmitz, A.J. Stokes, Q. Zhu, M.J. Bessman, R. Penner, J.P. Kinet, and A.M. Scharenberg. 2001. ADP-ribose gating of the calcium-permeable LTRPC2 channel revealed by Nudix motif homology. *Nature* 411:595–599.

Petersen, C.C., M.J. Berridge, M.F. Borgese, and D.L. Bennett. 1995. Putative capacitative calcium entry channels: Expression of Drosophila trp and evidence for the existence of vertebrate homologues. *Biochem. J.* 311(Pt 1):41–44.

Phillips, A.M., A. Bull, and L.E. Kelly. 1992. Identification of a Drosophila gene encoding a calmodulin-binding protein with homology to the trp phototransduction gene. *Neuron* 8:631–642.

Prawitt, D., M.K. Monteilh-Zoller, L. Brixel, C. Spangenberg, B. Zabel, A. Fleig, and R. Penner. 2003. TRPM5 is a transient Ca^{2+}-activated cation channel responding to rapid changes in [Ca2+]i. *Proc. Natl. Acad. Sci. USA* 100:15166–15171.

Putney, J.W. 2005. Physiological mechanisms of TRPC activation. *Pflugers Arch.* 451:29–34.

Putney, J.W., Jr. 2004. The enigmatic TRPCs: Multifunctional cation channels. *Trends Cell Biol.* 14:282–286.

Reuss, H., M.H. Mojet, S. Chyb, and R.C. Hardie. 1997. In vivo analysis of the Drosophila light-sensitive channels, TRP and TRPL. *Neuron* 19:1249–1259.

Runnels, L.W., L. Yue, and D.E. Clapham. 2001. TRP-PLIK, a bifunctional protein with kinase and ion channel activities. *Science* 291:1043–1047.

Sather, W.A., and E.W. McCleskey. 2003. Permeation and selectivity in calcium channels. *Annu. Rev. Physiol.* 65:133–159.

Schlingmann, K.P., S. Weber, M. Peters, L. Niemann Nejsum, H. Vitzthum, K. Klingel, M. Kratz, E. Haddad, E. Ristoff, D. Dinour, M. Syrrou, S. Nielsen, M. Sassen, S. Waldegger, H.W. Seyberth, and M. Konrad. 2002. Hypomagnesemia with secondary hypocalcemia is caused by mutations in TRPM6, a new member of the TRPM gene family. *Nat. Genet.* 31:166–170.

Schmitz, C., A.L. Perraud, C.O. Johnson, K. Inabe, M.K. Smith, R. Penner, T. Kurosaki, A. Fleig, and A.M. Scharenberg. 2003. Regulation of vertebrate cellular Mg^{2+} homeostasis by TRPM7. *Cell* 114:191–200.

Sidi, S., R.W. Friedrich, and T. Nicolson. 2003. NompC TRP channel required for vertebrate sensory hair cell mechanotransduction. *Science* 301:96–99.

Smith, G.D., M.J. Gunthorpe, R.E. Kelsell, P.D. Hayes, P. Reilly, P. Facer, J.E. Wright, J.C. Jerman, J.P. Walhin, L. Ooi, J. Egerton, K.J. Charles, D. Smart, A.D. Randall, P. Anand, and J.B. Davis. 2002. TRPV3 is a temperature-sensitive vanilloid receptor-like protein. *Nature* 418:186–190.

Sotomayor, M., D.P. Corey, and K. Schulten. 2005. In search of the hair-cell gating spring elastic properties of ankyrin and cadherin repeats. *Structure (Camb.)* 13:669–682.

Story, G.M., A.M. Peier, A.J. Reeve, S.R. Eid, J. Mosbacher, T.R. Hricik, T.J. Earley, A.C. Hergarden, D.A. Andersson, S.W. Hwang, P. McIntyre, T. Jegla, S. Bevan, and A. Patapoutian. 2003. ANKTM1, a TRP-like channel expressed in nociceptive neurons, is activated by cold temperatures. *Cell* 112:819–829.

Strotmann, R., C. Harteneck, K. Nunnenmacher, G. Schultz, and T.D. Plant. 2000. OTRPC4, a nonselective cation channel that confers sensitivity to extracellular osmolarity. *Nat. Cell Biol.* 2:695–702.

Swartz, K.J. 2004. Towards a structural view of gating in potassium channels. *Nat. Rev. Neurosci.* 5:905–916.

Szallasi, A., P.M. Blumberg, L.L. Annicelli, J.E. Krause, and D.N. Cortright. 1999. The cloned rat vanilloid receptor VR1 mediates both R-type binding and C-type calcium response in dorsal root ganglion neurons. *Mol. Pharmacol.* 56:581–587.

Talavera, K., M. Staes, A. Janssens, N. Klugbauer, G. Droogmans, F. Hofmann, and B. Nilius. 2001. Aspartate residues of the Glu-Glu-Asp-Asp (EEDD) pore locus control selectivity and permeation of the T-type Ca(2+) channel alpha(1G). *J. Biol. Chem.* 276:45628–45635.

Talavera, K., K. Yasumatsu, T. Voets, G. Droogmans, N. Shigemura, Y. Ninomiya, R.F. Margolskee, and B. Nilius. 2005. Heat-activation of TRPM5 underlies the thermal sensitivity of sweet taste. *Nature* 438:1022–1025.

Tsien, R.W., P. Hess, E.W. McCleskey, and R.L. Rosenberg. 1987. Calcium channels: Mechanisms of selectivity, permeation, and block. *Annu. Rev. Biophys. Biophys. Chem.* 16:265–290.

Varadi, G., M. Strobeck, S. Koch, L. Caglioti, C. Zucchi, and G. Palyi. 1999. Molecular elements of ion permeation and selectivity within calcium channels. *Crit. Rev. Biochem. Mol. Biol.* 34:181–214.

Venkatachalam, K., D.B. van Rossum, R.L. Patterson, H.T. Ma, and D.L. Gill. 2002. The cellular and molecular basis of store-operated calcium entry. *Nat. Cell Biol.* 4:E263–E272.

Vennekens, R., J.G. Hoenderop, J. Prenen, M. Stuiver, P.H. Willems, G. Droogmans, B. Nilius, and R.J. Bindels. 2000. Permeation and gating properties of the novel epithelial Ca^{2+} channel. *J. Biol. Chem.* 275:3963–3969.

Vennekens, R., J. Prenen, J.G. Hoenderop, R.J. Bindels, G. Droogmans, and B. Nilius. 2001. Pore properties and ionic block of the rabbit epithelial calcium channel expressed in HEK 293 cells. *J. Physiol. (Lond.)* 530:183–191.

Voets, T., G. Droogmans, U. Wissenbach, A. Janssens, V. Flockerzi, and B. Nilius. 2004a. The principle of temperature-dependent gating in cold- and heat-sensitive TRP channels. *Nature* 430:748–754.

Voets, T., A. Janssens, G. Droogmans, and B. Nilius. 2004b. Outer pore architecture of a Ca^{2+}-selective TRP channel. *J. Biol. Chem.* 279:15223–15230.

Voets, T., A. Janssens, J. Prenen, G. Droogmans, and B. Nilius. 2003. Mg^{2+}-dependent gating and strong inward rectification of the cation channel TRPV6. *J. Gen. Physiol.* 121:245–260.

Voets, T., B. Nilius, S. Hoefs, A.W. van der Kemp, G. Droogmans, R.J. Bindels, and J.G. Hoenderop. 2004c. TRPM6 forms the Mg^{2+} influx channel involved in intestinal and renal Mg^{2+} absorption. *J. Biol. Chem.* 279:19–25.

Voets, T., J. Prenen, J. Vriens, H. Watanabe, A. Janssens, U. Wissenbach, M. Bodding, G. Droogmans, and B. Nilius. 2002. Molecular determinants of permeation through the cation channel TRPV4. *J. Biol. Chem.* 277:33704–33710.

Voets, T., K. Talavera, G. Owsianik, and B. Nilius. 2005. Sensing with TRP channels. *Nat. Chem. Biol.* 1:85–92.

Vriens, J., G. Owsianik, T. Voets, G. Droogmans, and B. Nilius. 2004a. Invertebrate TRP proteins as functional models for mammalian channels. *Pflugers Arch.* 449:213–226.

Vriens, J., H. Watanabe, A. Janssens, G. Droogmans, T. Voets, and B. Nilius. 2004b. Cell swelling, heat, and chemical agonists use distinct pathways for the activation of the cation channel TRPV4. *Proc. Natl. Acad. Sci. USA* 101:396–401.

Walder, R.Y., D. Landau, P. Meyer, H. Shalev, M. Tsolia, Z. Borochowitz, M.B. Boettger, G.E. Beck, R.K. Englehardt, R. Carmi, and V.C. Sheffield. 2002. Mutation of TRPM6 causes familial hypomagnesemia with secondary hypocalcemia. *Nat. Genet.* 31:171–174.

Walker, R.G., A.T. Willingham, and C.S. Zuker. 2000. A Drosophila mechanosensory transduction channel. *Science* 287:2229–2234.

Watanabe, H., J.B. Davis, D. Smart, J.C. Jerman, G.D. Smith, P. Hayes, J. Vriens, W. Cairns, U. Wissenbach, J. Prenen, V. Flockerzi, G. Droogmans, C.D. Benham, and B. Nilius. 2002a. Activation of TRPV4 channels (hVRL-2/mTRP12) by phorbol derivatives. *J. Biol. Chem.* 277:13569–13577.

Watanabe, H., J. Vriens, J. Prenen, G. Droogmans, T. Voets, and B. Nilius. 2003. Anandamide and arachidonic acid use epoxyeicosatrienoic acids to activate TRPV4 channels. *Nature* 424:434–438.

Watanabe, H., J. Vriens, S.H. Suh, C.D. Benham, G. Droogmans, and B. Nilius. 2002b. Heat-evoked activation of TRPV4 channels in a HEK293 cell expression system and in native mouse aorta endothelial cells. *J. Biol. Chem.* 277:47044–47051.

Wes, P.D., J. Chevesich, A. Jeromin, C. Rosenberg, G. Stetten, and C. Montell. 1995. TRPC1, a human homolog of a Drosophila store-operated channel. *Proc. Natl. Acad. Sci. USA* 92:9652–9656.

Wissenbach, U., M. Bodding, M. Freichel, and V. Flockerzi. 2000. Trp12, a novel Trp related protein from kidney. *FEBS Lett.* 485:127–134.

Xu, H., N.T. Blair, and D.E. Clapham. 2005. Camphor activates and strongly desensitizes the transient receptor potential vanilloid subtype 1 channel in a vanilloid-independent mechanism. *J. Neurosci.* 25:8924–8937.

Xu, H., I.S. Ramsey, S.A. Kotecha, M.M. Moran, J.A. Chong, D. Lawson, P. Ge, J. Lilly, I. Silos-Santiago, Y. Xie, P.S. DiStefano, R. Curtis, and D.E. Clapham. 2002. TRPV3 is a calcium-permeable temperature-sensitive cation channel. *Nature* 418:181–186.

Xu, X.Z., F. Chien, A. Butler, L. Salkoff, and C. Montell. 2000. TRPgamma, a Drosophila TRP-related subunit, forms a regulated cation channel with TRPL. *Neuron* 26:647–657.

Yang, J., P.T. Ellinor, W.A. Sather, J.F. Zhang, and R.W. Tsien. 1993. Molecular determinants of Ca^{2+} selectivity and ion permeation in L-type Ca^{2+} channels. *Nature* 366:158–161.

Yeh, B.I., Y.K. Kim, W. Jabbar, and C.L. Huang. 2005. Conformational changes of pore helix coupled to gating of TRPV5 by protons. *EMBO J.* 24:3224–3234.

Yeh, B.I., T.J. Sun, J.Z. Lee, H.H. Chen, and C.L. Huang. 2003. Mechanism and molecular determinant for regulation of rabbit transient receptor potential type 5 (TRPV5) channel by extracellular pH. *J. Biol. Chem.* 278:51044–51052.

Yellen, G. 2002. The voltage-gated potassium channels and their relatives. *Nature* 419:35–42.

Yue, L., J.B. Peng, M.A. Hediger, and D.E. Clapham. 2001. CaT1 manifests the pore properties of the calcium-release-activated calcium channel. *Nature* 410:705–709.

Zhu, X., M. Jiang, M. Peyton, G. Boulay, R. Hurst, E. Stefani, and L. Birnbaumer. 1996. trp, a novel mammalian gene family essential for agonist-activated capacitative Ca^{2+} entry. *Cell* 85:661–671.

Zitt, C., A. Zobel, A.G. Obukhov, C. Harteneck, F. Kalkbrenner, A. Luckhoff, and G. Schultz. 1996. Cloning and functional expression of a human Ca^{2+}-permeable cation channel activated by calcium store depletion. *Neuron* 16:1189–1196.

Zygmunt, P.M., J. Petersson, D.A. Andersson, H. Chuang, M. Sorgard, V. Di Marzo, D. Julius, and E.D. Hogestatt. 1999. Vanilloid receptors on sensory nerves mediate the vasodilator action of anandamide. *Nature* 400:452–457.

12 Ion Channels in Epithelial Cells

Lawrence G. Palmer

12.1 Ion Channels and Epithelial Function

Ion channels in epithelial cells serve to move ions, and in some cases fluid, between compartments of the body. This function of the transfer of *material* is fundamentally different from that of the transfer of *information*, which is the main job of most channels in excitable cells. Nevertheless the basic construction of the channels is similar in many respects in the two tissue types. This chapter reviews the nature of channels in epithelia and discusses how their functions have evolved to accomplish the basic tasks for which they are responsible. I will focus on three channel types: epithelial Na^+ channels, inward-rectifier K^+ channels, and CFTR Cl^- channels.

To appreciate the biological roles of these channels, it is necessary to consider the basic structure of an epithelial cell. These cells separate the major body fluid compartments from those which are essentially in contact with the outside world, including those fluids of the urinary tract, GI tract, and sweat. In the lung, epithelial cells separate the body from an air space, with a thin layer of fluid in between. All epithelia form layers of cells that are joined by so-called tight junctions. These junctions consist of an extracellular caulking material that helps to seal the layer and prevent material from leaking between the cells. The cells therefore have two very distinct membrane surfaces: one in contact with the internal body compartments (i.e., the interstitial fluid) and the other in contact with the external fluids. These are called the basolateral and apical membranes, respectively. A basic tenet of epithelial biology, originally emphasized by Ussing and colleagues (Koefoed-Johnsen and Ussing, 1958), is that the transport properties of these two membranes are very different. It turns out that many of the channels and other transport proteins that are specialized to epithelial cells reside in the apical cell membrane in contact with the urine, the intestinal contents, etc. while the transporters that are shared with other cells in the body are often expressed in the basolateral membranes.

Epithelial Na^+ channels reside in the apical membranes of epithelia in the kidney (mainly in the so-called distal nephron segments), the colon, the sweat and salivary ducts, and the lung. Their main job is to *reabsorb* Na^+ ions from the fluid within these organs. This term is used to denote the fact that most of the Na^+ in these fluids originates within the body and is either filtered into the urine or secreted by glands into the GI tract, sweat or saliva. Thus their importance is in the conservation of Na^+. (In one special but historically important case, the skin of frogs, toads, and other amphibians, the channels may actually be used to obtain Na^+ from the environment.) This conservation process is critical, as in its absence the body would become depleted of Na^+ and of fluid, especially from the plasma and other extracellular compartments.

The basic arrangement is shown in Fig. 12.1A, which is a modern representation of the model of Koefoed-Johnsen and Ussing (1958). Na^+ ions enter the cells through the channels by the process of electrodiffusion. This movement is driven by a concentration difference for Na^+, which is lower within the cytoplasm than in the extracellular fluids, and by the electrical potential across the membrane, which is generally more negative within the cell. The membrane on the other side of the cell contains a "pump" which drives Na^+ out of the cell (toward a higher electrochemical potential) using ATP as an energy source. The net result is a transfer of Na^+ across the cell layer from outside to inside the body. The power underlying this process comes from the ATP-driven pump (or ATPase). It turns out, however, that the rate at which Na^+ is transported is determined, and regulated, by the channels. One other essential element in this system is the set of K^+ channels that operates in parallel with the pump in the basolateral membrane. These K^+ channels have two important functions with respect to epithelial Na^+ transport. First, they recycle the K^+ that enters the cells as part of the pump process as Na^+ is moved out. Furthermore, this conductance to K^+, together with the high concentration of K^+ inside the cells and the low concentration in the extracellular fluids, establishes the negative electrical potential within the cell. This provides part of the driving force for Na^+ to enter through the apical channels. As suggested above, neither the pump nor the K^+ channels in the basolateral membrane are special to epithelial cells; they are found in nearly every cell in the body and in each case their fundamental job is to keep intracellular Na^+ concentration low and the resting cell membrane voltage negative. In contrast, the Na^+ channels are found almost exclusively in epithelia and are always located in the apical membranes. This mechanism is predicated on the segregation of transport functions, with the channels on one side of the cell and the pumps on the other side. How these membrane proteins are sorted by the cell's biosynthetic machinery and maintained in different places within the cell is an important problem in epithelial biology that has not been completely solved (Caplan, 1997; Mostov et al., 2000; Campo et al., 2005). A discussion of this issue is beyond the scope of this chapter.

The net reabsorption of Na^+ shown in Fig. 12.1A would by itself lead to a separation of charge and the development of a large voltage difference that in turn would halt the process rather quickly. For Na^+ transport to continue the charge translocation must be balanced either by the movement of an anion (usually Cl^-) in the same direction or of a cation (usually K^+) in the opposite direction. To some extent this movement takes place between the cells, through the intercellular caulking which is not absolutely tight. This so-called paracellular transport is not very selective among small ions. There are, however, specific processes, some involving other ion channels, which mediate the transepithelial transport of both Cl^- and of K^+.

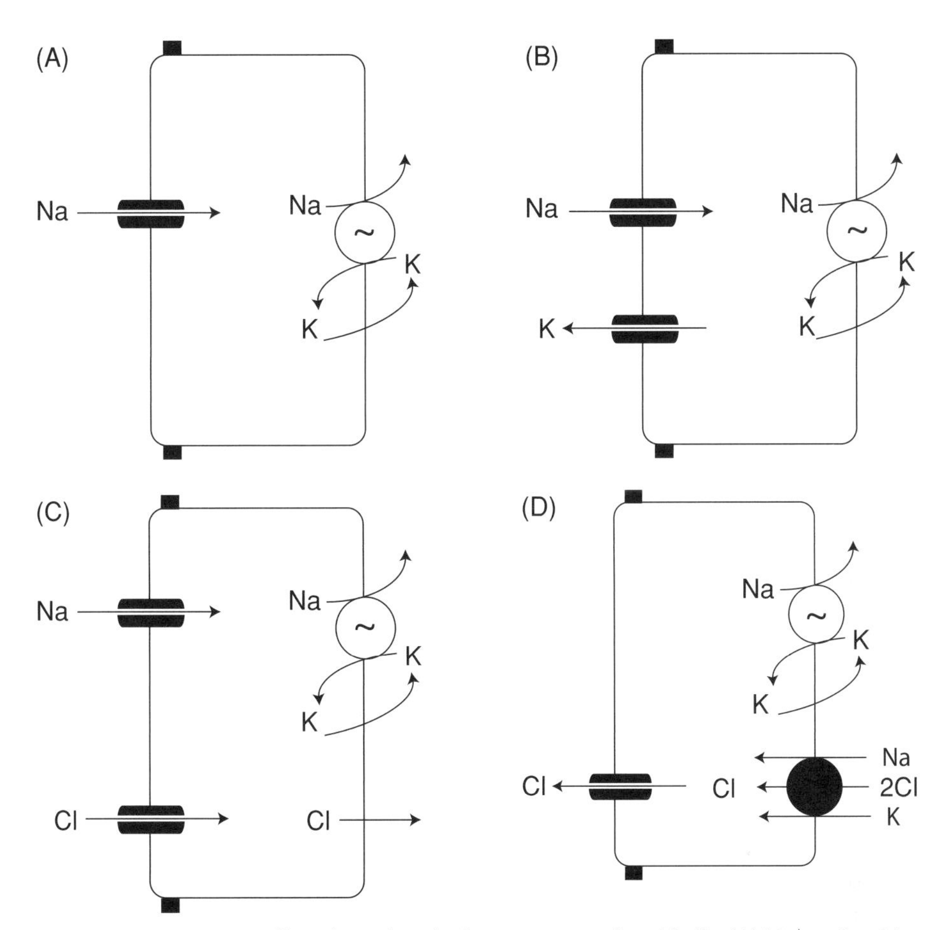

Fig. 12.1 Arrangement of ion channels and other transporters in epithelia. (A) Na^+-reabsorbing epithelium. Na^+ channels are expressed exclusively on the apical plasma membrane, which faces the urine, lumen of the gut etc., allowing Na^+ ions to enter the cell from the urine, the feces, or the ductal lumens. On the other side of the cell, called the basolateral membrane, a Na/K-ATPase or Na pump uses the energy of ATP to extrude Na^+ from the cell against unfavorable electrical and concentration differences. This completes the transport process and also keeps intracellular Na^+ concentrations low, maintaining the driving force for Na^+ to enter the cell through the channels. (B) K^+-secreting epithelia. The additional feature is a K^+-selective channel in the apical membrane, that allows the K^+ that enters the cell through the Na/K pump to exit the cell into the urine or the intestinal lumen. Na^+ and K^+ movements are electrically coupled since the movement of one creates a transmembrane voltage that increases the driving force on the other. (C) Cl^--reabsorbing epithelium. Here, Cl^- enters the cells in parallel with the movement of Na^+ through a separate channel. The Cl^-that enters the cells in this way leaves the cells across the basolateral membrane, either through another set of channels or in some cases through a cotransporter along with K^+. As in (B), the movements of Na^+ and Cl^- across the apical membrane are electrically coupled. (D) Cl^--secreting epithelium. The arrangement is very similar to the Cl^--absorbing cells in C, with two important differences. First, Na channels are absent. Second, a cotransporter that couples the inward movement of Na^+, K^+, and Cl^- is present on the basolateral membrane. These features combine to reverse the driving force on the Cl^- ion such that it moves out of the cells through the channels.

The generic case of K^+ transport is illustrated in Fig. 12.1B. Here, a single pathway involving an apically located K^+ channel has been added to the cartoon of the cell. Now some of the K^+ that enters the cells through the Na/K pump leaves across the apical membrane, resulting in the net *secretion* of the ion. This movement of K^+ can itself be desirable, since our diets are often rich in K^+ and the kidneys need to eliminate what is not required for replenishment or growth. In addition, as mentioned above, this transport helps to neutralize the charge that is moved when Na^+ is reabsorbed; transepithelial K^+ secretion stimulates the rate of Na^+ reabsorption and vice versa. This electrical coupling takes place to a large extent across the apical cell membrane. Efflux of K from the cell hyperpolarizes the membrane potential (makes it more negative), increasing the driving force for Na^+ entry into the cell.

As shown in Fig. 12.1C, addition of a channel for Cl^- movement across the apical cell membrane will also help to neutralize the charge movement associated with Na^+ transport. When combined with a mechanism for getting the Cl^- out of the cell across the basolateral membrane this will result in the reabsorption of NaCl. This efflux of Cl^- may be through channels or by cotransport with K^+. The net reabsorption of NaCl will decrease the osmolarity of the external fluid (e.g., the sweat). In epithelia which are permeable to water it results in the reabsorption of fluid along with the ions. As in the case for K^+ discussed above, the movements of Na^+ and of Cl^- across the epithelium as a whole, or across the apical membrane in particular, are electrically coupled. Increases in the flow of either ion will enhance the driving force for the other by changing the membrane potential.

A slightly different arrangement of these transporters can result in the secretion of Cl^- and as a result of Na^+ and fluid. This is illustrated in Fig. 12.1D. Here, the apical Cl^- channel remains but the Na^+ channel has disappeared. In addition, a new transporter appears on the basolateral membrane that carries Na^+, K^+, and Cl^- into the cell at the same time. The direct coupling of the movement of Na^+ and Cl^- in this system is key, since it uses the forces pulling Na^+ into the cell to drive the accumulation of Cl^-. Now, in contrast to the case of Fig. 12.1C, the electrochemical activity of Cl^- is higher in the cell than it is in the external fluid, and the anion moves out of the cells through the channels. The charge movement is balanced by that of Na^+, probably moving between the cells. Thus the same apical Cl^- channels (CFTR, as we shall see) can mediate either the reabsorption or the secretion of NaCl and water. In some sea-dwelling animals this mechanism serves to rid the body of excess salt. Two examples are the rectal gland of the shark and the nasal gland of duck. In humans and other terrestrial vertebrates, this same mechanism is used to provide fluid for many glandular secretions including those of the pancreas and the salivary, sweat and tear (lachrymal) glands. It also governs the secretion of fluid by the intestine, which can provide lubrication for the digestive process under normal conditions but which leads to excess loss of fluids in some forms of diarrhea, especially those caused by cholera and other infectious agents.

12.2 Structural and Evolution of Epithelial Channels

The major ion channels that participate in the transport schemes, shown in Fig. 12.1, have been identified at the molecular level. They all belong to larger families of genes and proteins but in some, cases have evolved into more specialized functions.

12.2.1 Epithelial Na^+ Channels (ENaC)

The proteins comprising the epithelial Na channel were identified using an expression-cloning approach (Canessa et al., 1993, 1994). There are three subunits, termed α, β, and γ ENaC (for *E*pithelial *Na* *C*hannel), all of which are necessary to reconstitute maximal activity in frog oocytes and other heterologous expression systems. Function is determined electrophysiologically as current or conductance that can be blocked by amiloride, a diuretic drug, which inhibits native channels with a K_d of around 100 nM. Only αENaC can produce channel activity by itself. However, all three subunits are homologous with each other and all three have the same predicted membrane topology, consisting of two predicted transmembrane helices, short cytoplasmic N- and C-termini and a large cysteine-rich extracellular domain (Fig. 12.2A). This suggests that the three subunits contribute in similar ways to the structure of the channel. The simplest models consist of a central pore surrounded by a pseudo-symmetrical arrangement of subunits. The nicotinic acetylcholine receptor provides a precedent for this structure, as it also consists of different but homologous membrane-spanning proteins constituting a single ion-conducting pore (Karlin, 2002). In addition, mutations in corresponding residues in the three ENaC subunits in some cases show some similar effects on ion conduction and block (Schild et al., 1997; Sheng et al., 2000). The stoichiometry of the channel is still a matter of debate. The most widely accepted structure, based on several experimental approaches, consists of 2α, 1β, and 1γ forming the holochannel (Firsov et al., 1998). However, a similar study suggested a larger complex with nine subunits and a 3:3:3 stoichiometry (Snyder et al., 1998). More recent evidence using fluorescence energy transfer is also consistent with a channel's having at least two of each subunit type (Staruschenko et al., 2005).

The ENaC proteins belong to a superfamily that includes channels which respond to mechanical and chemical stimuli. *Mec* and *deg* gene products were identified in *C. elegans* using screens of mutated worms having defects in their responses to touch (Chalfie, 1997). These proteins form channels when expressed in oocytes (O'Hagan et al., 2005). It is likely that in vivo they are sensory channels which open in response to a mechanical force, depolarizing neurons to initiate mechanosensory reflexes. The acid-sensing ion channels (ASICs) form another family of proteins with homology to ENaC (Waldmann et al., 1997) and are also considered to be sensory channels. They are expressed in neurons and are activated at low pH, presumably as a response to a noxious stimulus. FMRFamide-activated channels also belong to

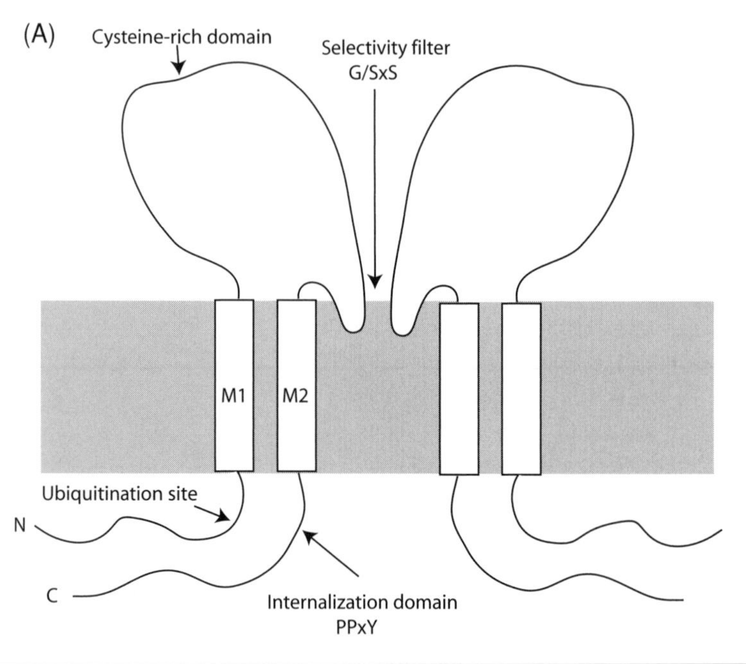

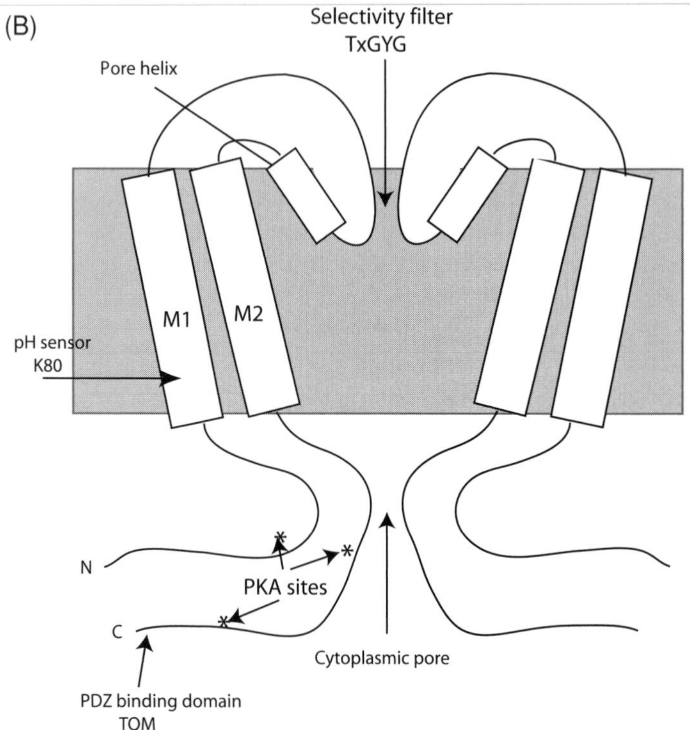

Fig. 12.2 Schematic drawings of the structures of ENaC (A), ROMK (B), and CFTR (C).

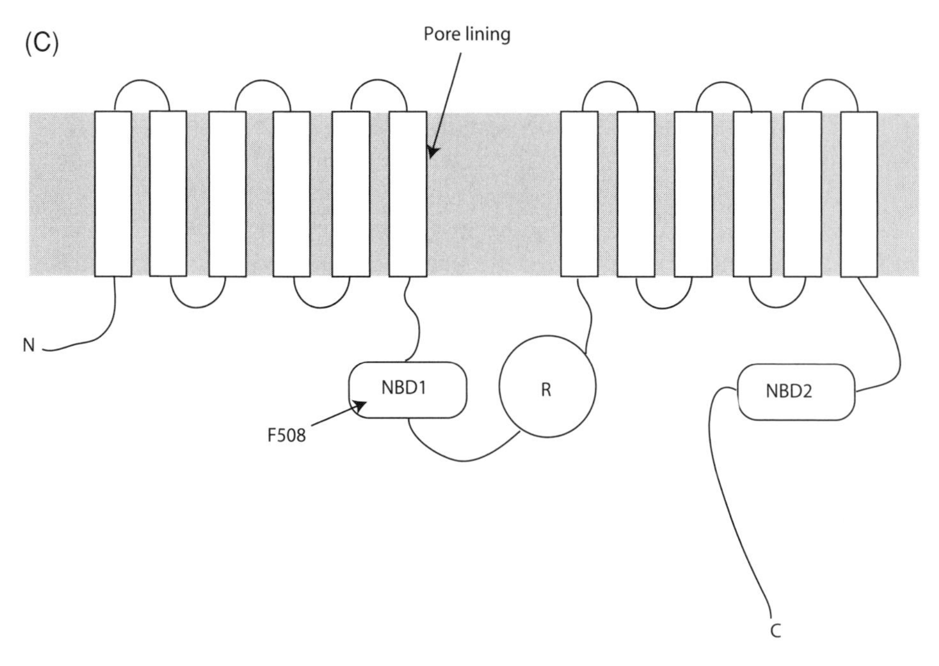

Fig. 12.2 *(continued)*

this superfamily. They are found in invertebrates such as snails and respond to low concentrations of peptides (Lingueglia et al., 1995). All of these channels appeared rather late in evolution; no bacterial orthologs have been identified. The sensory channels seem to be more widespread among the animal phyla than the ENaCs. It is possible that the latter were adapted from the former as the need for salt conservation and epithelial Na^+ transport arose. This might explain the large extracellular domains of the ENaC subunits, which have no clear role in epithelial transport but could be involved in the original function of sensing the external environment.

12.2.2 Epithelial K Channels (Kir1.1 = ROMK)

The K^+ channels residing in the apical membrane of renal epithelia (Fig. 12.1B) are part of a large superfamily that arose in bacteria. The function of these channels in bacteria remains obscure. However, in animals they have evolved into families that include voltage-gated, inward rectifier, and two-pore (background) channels. Each of these families contains a number of even more specialized channels which serve a huge range of physiological functions. All of the K^+ channels contain a signature sequence T-x-G-Y-G that forms the narrowest part of the conducting pore and that is responsible for establishing a high selectivity for K^+ over Na^+. The structure of several of these channels has been solved by crystallization and X-ray diffraction (Doyle et al., 1998; Jiang et al., 2002; Kuo et al., 2003). These crystals clearly demonstrate the expected quaternary structure of the proteins, with four usually identical subunits arranged around a central pore. They also indicate the selectivity filter near the interface of the membrane with the extracellular solution and a relatively large cavity in the center of the membrane. Both of these features had been predicted from hydrophobicity analysis and mutagenesis experiments.

Within this large group, the renal K^+ channels ROMK (for *R*enal *O*uter *M*edullary *K* channels) belong to the inward-rectifier (Kir) family of channels. In fact, they were the first of this family to be cloned (Ho et al., 1993) and are designated as Kir1.1. Other prominent members of this family include the anomalous rectifiers that set the resting membrane potential of skeletal muscle (Kir2), G-protein coupled channels that regulate the heart beat (Kir3) and ATP-sensitive channels that control insulin secretion by the pancreas in response to changes in plasma glucose concentrations (Kir6) (Bichet et al., 2003). Bacteria also have channels from this family, and a structure has been obtained for one of them (KirBac1.1) (Kuo et al., 2003). The basic selectivity filter and membrane cavity arrangement is similar to that of other K^+ channels. Each subunit contains two transmembrane helices, with a short extracellular loop containing the selectivity filter in between (Fig. 12.2B). The membrane cavity is formed mainly by the C-terminal helices. Two additional notable features of the channels are a cytoplasmic domain, formed by the N- and C-termini, that appears to form an extension of the transmembrane pore, and an additional helix on the N-terminus that lies at the interface between the membrane and the cytoplasm. The cytoplasmic portions of the channel proteins are involved in the regulation of channel function (see below).

Some Kir channels contain accessory subunits that may be required for proper activity. The best-studied case is that of Kir6. These form complexes with another set of transmembrane proteins, called sulfonylurea-binding proteins (SURs), named for their interactions with a class of antidiabetic agents (Inagaki et al., 1996). In this case the SURs are clearly necessary for full channel expression as well as for metabolic regulation by ADP. In the case of the renal K^+ channel, the Kir1.1 subunit appears to be sufficient to form a functional channel. However, there is evidence for interactions with SUR2b (Tanemoto et al., 2000) and with the SUR-related protein CFTR (McNicholas et al., 1996). It is not clear whether such interactions are important in the function of the channels in the kidney.

12.2.3 Epithelial Cl Channels (CFTR)

The most prominent of the apical Cl^- channels in epithelia is CFTR, the *c*ystic-*f*ibrosis *t*ransmembrane *c*onductance *r*egulator. This protein was identified as the site of mutations causing cystic fibrosis, a disease in which both the secretion and the absorption of fluids by epithelia can be defective (Riordan et al., 1989). CFTR is a larger and more complex protein than those formed by ENaC or Kir1.1 (Sheppard and Welsh, 1999). It is thought to have 12 membrane-spanning helices arranged in two domains. Each of these domains contains six helices and ends in a cytoplasmic region containing a nucleotide-binding domain (NBD) that interacts with ATP (Fig. 12.2C).

These features are shared by the family of ABC (ATP-binding cassette) transporters to which CFTR belongs. Curiously, CFTR is the only family member which clearly functions as an ion channel. The others generally carry out ATP-dependent active transport of nutrients and metabolites in bacteria, of a mating factor in yeast, and of xenobiotics, notably chemotherapeutic agents, in eukaryotic cells. CFTR has one structural feature unique to this family—a regulatory domain in the cytoplasm between the two sets of membrane-spanning helices. This is a site of phosphorylation by protein kinases that play an important role in the regulation of channel function.

The structure of the conducting pore in CFTR is unknown. The sixth transmembrane helix of the protein contains several positive charges which influence ion conduction, selectivity, and block. This segment very likely forms part of the lining of the pore. However, mutations in other transmembrane helices also have effects, and these may also contribute to the formation of the conduction pathway (Dawson et al., 1999; Sheppard and Welsh, 1999). The precise arrangement of the helices with respect to the pore is unknown. As with ENaC, the number of subunits required to form a functional channel is controversial. Evidence has been presented suggesting that a dimer is required (Zerhusen et al., 1999), but more recent data indicate that a single peptide can form a pore (Liu et al., 2004).

12.3 Functional Specializations of Epithelial Ion Channels

Epithelial ion channels are designed to transport solutes continuously over long periods of time. In contrast to many channels in excitable cells, they do not respond strongly to changes in membrane potential. Rather, they spend a significant fraction of time in the open state at all voltages. This makes a good deal of sense with respect to their physiological functions. One might also expect the conductances of individual channels to be relatively large, so as to minimize the number of proteins necessary to carry out these functions. This, however, turns out not to be the case, as the single-channel conductance of the epithelial ion channels discussed here tend to be at the lower end of the spectrum, which generally ranges from around 1 pS to several hundred pS for individual conducting units. In the section below I will review some of the basic permeation and selectivity properties of epithelial channels.

12.3.1 ENaC

ENaC channels are characterized by a single-channel conductance of about 5 pS when Na^+ is conducted at room temperature [see (Garty and Palmer, 1997) for a comprehensive review]. This is a small conductance even compared with voltage-gated Na^+ channels, which have around 15–20 pS unit conductance under similar conditions. At 37°C, the ENaC conductance increases to about 9 pS. One possible evolutionary advantage of the low conductance may be in conferring a high selectivity to the channels. They are almost perfectly selective for Na^+ over K^+; one estimate was around 1000:1 compared to about 10:1 for voltage-gated Na^+ channels. Indeed, the only ions that have a measurable conductance through the channels are Na^+, Li^+, and under extreme conditions H^+. Based on these results, it has been proposed that ions may be almost completely dehydrated as they go through the channels, which could then have a selectivity filter that discriminates on the basis of the nonhydrated radius of the ion (Palmer, 1987; Kellenberger et al., 1999). This selectivity is clearly an important aspect of epithelial function. As can be appreciated in Fig. 12.1, any permeability or conductance of K^+ through the Na^+ channels will result in K^+ secretion and loss of K^+ from the body. Although as discussed above this is often desirable, the kidney and other epithelia need to be able to keep track of body Na and K contents individually and regulate these levels separately. Having such a narrow constriction and high selectivity may limit the rate at which ions can pass through the channel. Although, as discussed below, there is no clear correspondence between conductance and selectivity in K^+ channels, this may be the most efficient way to create a high level of discrimination for Na^+ over K^+. Na^+ passes through the channels with nearly equal ease in both directions. Current–voltage relationships under physiological conditions exhibit "Goldman" type rectification that reflects differences in the concentrations of permeant ions on the two sides of the membrane. With symmetrical concentrations, I–V relationships are approximately linear.

Like most channels, ENaC channels make abrupt transitions between open and closed states. The lifetimes of these states are rather long—several seconds at room temperature—and to a first approximation insensitive to voltage. These kinetics, and indeed the open probability of the channels—are highly variable. In one study, P_o ranged from <0.05 to >0.95 under nominally the same experimental conditions (Palmer and Frindt, 1996). The reasons for this variability are not entirely clear but it suggests that the channels may be regulated through changes in P_o. As described in the next section, there is growing evidence that the major regulation of ENaC is through trafficking of the protein to and from the surface. Nevertheless several conditions modulate channel activity through alterations in P_o and gating kinetics. Conditions leading to increased P_o include strong hyperpolarization of the membrane, mechanical stress on the membrane, reduction of either extracellular or intracellular Na^+, and an increase in cytoplasmic pH (Garty and Palmer, 1997). The physical basis for the gating is unknown.

12.3.2 Kir1.1/ROMK

The unitary conductance of Kir1.1 channels is substantially higher than that of ENaC; with K^+ as the conducted ion the inward conductance is about 35 pS at room temperature and 45–50 pS at 37°C [see (Hebert et al., 2005) for a comprehensive review]. Furthermore, although again the channel's cycle between open and closed states, their open probability is quite high—around 0.9 independent of voltage. These two features seem well adapted to the physiological role of the channels. However, two aspects of the biophysical characteristics of the channels seem poorly suited to their function. First, the conductance is considerably smaller than that seen for other K^+ channels, which can reach values as high as 200 pS for the Ca-activated maxi-K or BK channels. These high conductances can be achieved without any sacrifice in ion selectivity; a very high preference for K^+ over Na^+ is observed in all channels with the TxGYG signature sequence, regardless of the unitary conductance. The lower conductance of the Kir1.1 channels means that at least four times as many channels must be manufactured to carry out the function of K^+ secretion than would be necessary if channels of maximum conductance were employed. Even stranger, the conductance in the physiological direction out of the cell is substantially smaller than that for inward flow of ions. This is a general characteristic of the inward rectifier channel family and is responsible for the family name. The rectification arises from a voltage-dependent block of the channels by intracellular multivalent cations including Mg^{2+} and polyamines such as spermine and spermidine (Lopatin et al., 1995). Inward rectification is much more pronounced in many of the family members, particularly in the Kir2 and Kir3 types. These channels are expressed in skeletal and cardiac muscle fibers and help make the membrane potential bi-stable. That is, in the absence of a strong depolarizing stimulus the channels are open and the resting membrane potentials are strongly negative. When a sufficient stimulus arrives and the membrane depolarizes past its threshold, the K^+ channels shut off, allowing the action potential to proceed. Kir1 channels do not need to function this way. Indeed their role is to allow a steady-state diffusion of K^+ out of the cells. In fact, their outward conductance is considerably higher than in its Kir2 and Kir3 cousins. However, it is still 2- to 3-fold lower than the inward conductance, furthering the requirement for a larger number of channels. In this sense the inward rectifier family seems like a curious choice for the selection of a channel designed for epithelial K^+ secretion.

As discussed above, inward rectifier K^+ channels share the same basic selectivity filter structure with the voltage-gated K^+ channels, and the ability of these channels to discriminate between K^+ and Na^+ is very large. Other ions which can pass through the pore are Rb^+, Tl^+, and NH_4^+ (Choe et al., 1998). Of these, only NH_4^+ is of physiological importance. The physical basis for this selectivity is the precise arrangement of carbonyl oxygen atoms within the selectivity filter (Zhou et al., 2001; Noskov et al., 2004). A K^+ ion can be surrounded by a cage of up to 8 of these oxygens, effectively replacing the hydration shell that stabilizes the ions in bulk solution. For Na^+ ions, which are smaller but which form larger hydration spheres in water, this arrangement of water-replacing contacts is not as effective and the ions are excluded from the filter on energetic grounds.

In addition to the narrow selectivity filter at the outer mouth of the pore and a relatively large cavity spanning the lipid bilayer, K_{ir} channels feature an extension of the pore deep into the cytoplasm. This "cytoplasmic pore" (Kuo et al., 2003) contains negatively charged amino acids that are important for binding impermeant cations such as polyamines and which help to confer the property of inward rectification. This pore is fairly narrow in places may also comprise a significant fraction of the total electrical resistance of the channel (Zhang et al., 2004).

Although Kir1 channels spend most of their time in the open state, they briefly visit a closed state that has a mean lifetime of around 1 ms (Choe et al., 1999). The physical basis of these transitions is unknown. However, they may represent fluctuations in the protein components surrounding the permeation pathway, as they are influenced both by the nature of the conducted ion and by the rate at which ions are passing through the pore. In addition to these transitions, the channels can undergo closures to a state that is relatively long-lived (lasting for many seconds or minutes). Such closures can be elicited, for example, by lowering the pH of the cytoplasm below 7.0. The physiological importance of this pH-dependent gating remains obscure, although it probably underlies the well-described phenomenon of K^+ retention by the kidneys during acidosis. However, it has parallels with other inward-rectifier K^+ channels. In Kir6.2 channels, for example, channels are shut down by high concentrations of ATP (or a high ATP/ADP ratio) (Enkvetchakul et al., 2000). This regulation ultimately couples the secretion of insulin by pancreatic beta cells in response to high levels of plasma glucose. Kir3 or GIRK channels are constitutively in a long-lived closed state but can be activated by the binding of the beta/gamma subunits of heterotrimeric G proteins (Sui et al., 1999). This regulation also has clear physiological significance in the process of slowing the heart rate by the parasympathetic nervous system.

12.3.3 CFTR

CFTR also forms pores with a low single-channel conductance—5 to 10 pS at room temperature with Cl^- as the conducted ion [see (Dawson et al., 1999; Sheppard and Welsh, 1999) for more comprehensive reviews]. Like the Na^+ channels, currents through open CFTR channels are approximately linear under symmetrical ion conditions, and in the presence of ion gradients across the membrane the I–V relationships show Goldman-type rectification. In addition, the open probability of the channels is not markedly dependent on voltage. Therefore, like the ENaC channels, the major effect of transmembrane voltage is to establish the driving force for ion movement. In the case of CFTR, this is an important property since, as discussed above, the direction of ion movement can be either inward or outward depending on the cell type and the physiological conditions.

CFTR is highly selective for negatively charged ions, but does permit the permeation of a wide variety of anions. Analysis of a wide range of anions showed that permeabilities followed the lyotropic series, suggesting that the basis for discrimination among them depends on the relative energies of dehydration and binding to sites within the conduction pathway (Smith et al., 1999). The size of the ion is much less important in these channels, at least up to a point; anions with diameters of 5.5 Å or even higher can permeate. Positively charged residues on the transmembrane domains of CFTR, particularly the sixth domain, are important in the formation of the conduction pathway and the establishment of ion selectivity and binding affinity. The detailed structure of the pore has not been determined.

As with the other channels, CFTR makes abrupt transitions between open and closed states. These events involve processes that operate on at least three different structural levels and time scales. First, the channels must be phosphorylated to have any activity at all (Sheppard and Welsh, 1999). This process will be discussed in the next section. Second, the opening of the phosphorylated channels is coupled to the binding and hydrolysis of ATP (Vergani et al., 2003). Opening of the channels requires binding of ATP to both nucleotide-binding domains. The nucleotide bound to the second (C terminal) NBD can be hydrolyzed to ADP and Pi, and this results in channel closure. This ATP-driven cycling between open and closed configuration may reflect the evolution of the channel protein from energy-dependent pumps that can move solutes against a concentration gradient. In this case, conformational changes determine whether the channels are open or not, with the direction of movement of ions determined by differences in their electrochemical activities. Finally, channels that have been opened by phosphorylation and ATP binding flicker back and forth between relatively short-lived open and closed states, similar to the Kir channels described above. The physical basis for these transitions is unknown.

12.4 Regulation of Epithelial Ion Channels

All of the epithelial ion channels discussed in this chapter are highly regulated. These regulatory processes are not nearly as fast as those that govern the activities of voltage-gated channels, which can be turned on and off in a matter of milliseconds through rapid changes in protein conformation. In contrast, the activities of epithelial channels are usually modulated over time scales of minutes or hours. The regulatory events include both covalent modification of the channel protein (e.g., phosphorylation) and translocation of channels to and from the membrane.

12.4.1 ENaC

Epithelial Na channels are strongly regulated by hormones. The most important of these is aldosterone, a steroid secreted by the adrenal gland in response to a deficit in the amount of Na or in the volume of the extracellular fluids in the body. Like that of other steroid hormones, the actions of aldosterone require alterations in gene expression. As a consequence the effects depend on the transcription of new mRNA and the synthesis of new proteins, processes which usually take an hour or more before the eventually physiological effects can be effected. In the continued presence of hormone, the effects persist indefinitely. The short-term effects (hours) and the long-term effects (days) may be different (Verrey et al., 2000).

Aldosterone appears to stimulate channel activity at least in part by increasing the number of channel proteins in the membrane. The mechanisms involved are different in different tissues. In the colon, αENaC subunits are expressed constitutively, but the amounts of β and γ subunits depend on the hormonal status. When aldosterone levels are low as the result of a high salt intake or of adrenalectomy, β and

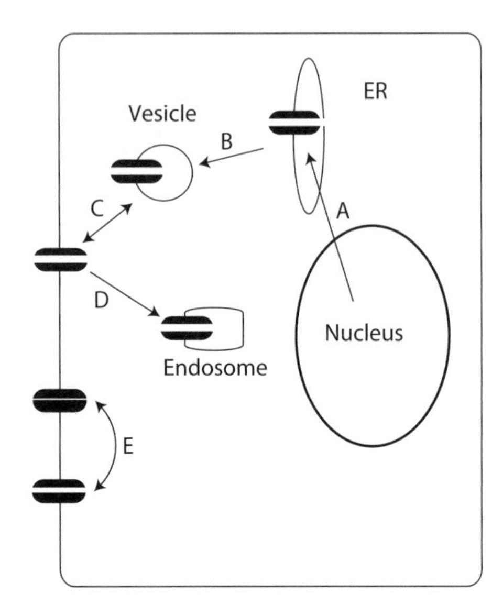

Fig. 12.3 Regulation of epithelial ion channels. (A) Regulation of the synthesis of the channels can occur by altering the rates of mRNA production (gene transcription) and/or protein translation. (B) Channel protein export from the endoplasmic reticulum may be regulated. (C) Channels may reside in submembrane vesicles and shuttled to and from the membrane by regulated exocytosis and endocytosis. (D) Channels in the membrane may be removed and degraded in endosomes or proteosomes through regulated processes. (E) Channels in the membrane may be activated by phosphorylation, membrane tension, or other chemical or mechanical processes.

γENaC are virtually undetectable. Aldosterone induces the transcription of mRNA encoding these subunits which in turn promotes the synthesis of new protein (Asher et al., 1996) (see Fig. 12.3, path A). These new subunits presumably combine with preexisting αENaC to form conducting channels in the apical plasma membrane. In the kidney, the same hormone (aldosterone) has the same ultimate effect (increase in ENaC protein in the apical membrane) but, remarkably, seems to act through different mechanisms. All three ENaC subunits are expressed constitutively in the kidney, but with low levels of hormone the protein appears to reside primarily in intracellular membranes (Masilamani et al., 1999; Loffing et al., 2000). These membrane sites have not been precisely identified, but they are distributed throughout the cytoplasm reminiscent of the distribution of the endoplasmic reticulum. A major effect of aldosterone is to induce the movement of the channels from these cytoplasmic stores to the apical membrane (Fig. 12.3, path B).

The turnover rate of ENaC in the plasma membrane is rapid; estimates of the half-life of channels at the surface of cultured epithelial cells range from around 20 minutes to about 2 hours (Weisz et al., 2000; Alvarez de la Rosa et al., 2002). Thus, an increase in the number of channels at the surface could result from either an increase in the rate of insertion into or a decrease in the rate of retrieval from the apical membrane. According to one hypothesis aldosterone induces the synthesis of a protein kinase (SGK) which phosphorylates a ubiquitin ligase (Nedd4-2), reducing its ability to bind to and eventually internalize ENaC (Debonneville et al., 2001) (Fig. 12.3, path D). Evidence exists for all of these events, but the model has not been fully tested in the in vivo situation. No such detailed schemes have been proposed for what pathway is involved in trafficking of channels to the apical membrane, or whether the hormone might increase the rates of this movement. In addition, there is evidence that in some epithelia the hormone may also act on channels already residing in the membrane (Garty and Edelman, 1983; Kemendy and Eaton, 1990). This would result in an increase in the open probability of the channels. Such a mechanism, which may occur during short-term stimulation, could be combined with that of channel translocation to increase the range of channel activities that can be achieved.

Liddle's syndrome is a rare form of hypertension that is transmitted as a dominant Mendelian genetic trait (Lifton et al., 2001). Its cause is a mutation in the C-terminal end of either the β or γENaC subunit which eliminates a putative internalization signal sequence (PPPxY). This sequence is thought to interact with Nedd4-2 (see above) which ubiquitinates the channel proteins and facilitates their removal from the membrane, possibly through the proteosome. Defects in this internalization process are expected to increase the residence time on the membrane and to promote the excess reabsorption of Na^+ (Fig. 12.3, path D). This accounts for the hypertension observed in these patients and further illustrates the importance of membrane trafficking events in the regulation of these channels.

In addition to aldosterone, Na^+ channels in some epithelia are activated by antidiuretic hormone, a peptide from the posterior pituitary gland that acts through increases in the intracellular concentration of cAMP. The physiological significance of this regulation is unclear, as the best-known function of ADH is to control water reabsorption by the kidney. ADH/cAMP likely also acts by stimulating the translocation of ENaC to the cell surface (Garty and Edelman, 1983; Morris and Schafer, 2002) (Fig. 12.3, path C) although again the mechanism is not understood. It is different from that of aldosterone as it does not require protein synthesis.

12.4.2 Regulation of Kir1.1

Epithelial K^+ channels are also influenced by dietary factors, particularly by the intake of K^+ (Palmer, 1999). In rats fed a diet very rich in K, the number of K^+ channels that can be detected by electrophysiological means on the apical membrane increases several-fold. This process can be observed after several hours but takes a few days to be complete. The precise signal involved in this upregulation has not been determined. It could be mediated by a hormone, although none has been identified (aldosterone appears to act in a permissive fashion). Alternatively, the kidney could respond to increases in plasma K^+ levels. Finally, it is possible that K sensors in the GI tract may detect increases in dietary K intake directly and signal the kidney through the nervous system (Rabinowitz, 1996).

The cellular mechanisms involved in regulation of these channels have been better studied in the opposite case of a decrease in dietary intake leading to K deprivation (Wang et al., 2000). The events here may or may not be the converse of those occurring with an increased dietary K load. Prolonged (1 week) K deprivation upregulates the activity of the protein tyrosine kinase *src* in the kidney and stimulates the phosphorylation of tyrosine residues on ROMK (Lin et al., 2002). This probably does not affect channel activity directly, but leads eventually to their internalization into endosomes (Lin et al., 2004).

The rate of insertion of channels into the membrane may also be regulated. In cultured cells, ROMK protein appears to be retained in the ER unless an N-terminal serine (S44) is phosphorylated, in which case the channels are translocated to the plasma membrane (O'Connell et al., 2005; Yoo et al., 2005). It is not clear whether or how this process might be linked to regulation by dietary K. As in the case of ENaC, apical K^+ channels are also regulated by ADH through cAMP-dependent mechanisms. Phosphorylation of the critical S44 PKA provides a possible connection to the actions of this hormone, although such a linkage remains to be established.

While the sections above have discussed the major events regulating Na^+ and K^+ channels as if they were separate and independent, this is clearly not the case. For example, acute increases in plasma K^+ increases aldosterone secretion by the adrenals, which will impact Na^+ channel activity. In addition, chronic increases in K intake stimulate Na^+ channels by a mechanism that appears to be independent of aldosterone levels. Superimposed on all of these interactions is the simple coupling of Na^+ and K^+ transport through changes in the apical membrane voltage described above. The kidney does not control Na^+ and K^+ excretion by keeping track of these ions separately. Rather, it seems to regulate the levels of these ions through complex and overlapping control systems.

12.4.3 Regulation of CFTR

Epithelial Cl^- secretion is in general regulated on a somewhat faster time scale, mediating fluid movements over a matter of minutes (Welsh, 1996). The secretion is controlled by hormones and neurotransmitters that act on membrane receptors which are coupled to adenylate cyclase. Increases in cAMP activate the protein kinase type A (PKA). In contrast to the case of ENaC and perhaps also to that of ROMK, the major regulation of CFTR involves the activation of channels residing in the membrane (Fig. 12.3, path E). This takes place primarily through the direct PKA-dependent phosphorylation of several serine and threonine residues on the regulatory domain of the channel protein. This process is faster than that which controls apical Na and K channels, presumably because it does not entail protein synthesis or protein translocation. It is analogous to these synthetic and trafficking events in that it is necessary but not sufficient for channel activity. In the case of CFTR the phosphorylated channel must still be opened by the binding of ATP to the NBD, as described above. The structural events entailed in phosphorylation-dependent channel activation are unknown. It is possible, for instance, that the R-domain acts as a pore-occluding plug in the nonphosphorylated state, and that phosphorylation permits the plug to move away from the inner mouth of the channel. There is no direct structural evidence for this or for any other mechanism.

While translocation of CFTR may be of secondary importance in the overall regulation of Cl conductance and transport, there are indications that some of the channels in epithelial cells are present in submembrane vesicles, and that these channels can recycle to the cell surface (Prince et al., 1993; Bradbury et al., 1994). In addition, the most common form of CFTR giving rise to cystic fibrosis is the ΔF508 deletion. The defect in Cl^- transport associated with this mutation results from a failure of the channel to translocate properly between the ER and the plasma membrane (Welsh, 1996). Thus, membrane trafficking issues are important at least in the pathophysiology of this channel.

12.5 Summary

Ion channels form important routes for the transepithelial movement of salt. Often, this movement of salt is accompanied by a parallel movement of water. These movements mediate a number of important physiological functions including maintenance of Na and K levels in the body fluids (homeostasis) and the formation of fluid for the delivery of digestive enzymes and mucous to the gut and the airways, as well as for sweat, saliva, and tears. These channels share basic principles of ion permeation with those of excitable cells. However, their regulation is quite different. In some cases, like CFTR, regulation depends on the generation of cAMP and the phosphorylation of the channel protein. In other cases, such as ENaC and possibly ROMK, translocation of channels between intracellular compartments and the plasma membrane underlies the modulation of channel activity. Defects in these processes lead to important diseases including cystic fibrosis and hypertension.

References

Alvarez de la Rosa, D., H. Li, and C.M. Canessa. 2002. Effects of aldosterone on biosynthesis, traffic, and functional expression of epithelial sodium channels in A6 cells. *J. Gen. Physiol.* 119:427–442.

Asher, C., H. Wald, B.C. Rossier, and H. Garty. 1996. Aldosterone-induced increase in the abundance of Na^+ channel subunits. *Am. J. Physiol.* 271:C605–C611.

Bichet, D., F.A. Haass, and L.Y. Jan. 2003. Merging functional studies with structures of inward-rectifier K(+) channels. *Nat. Rev. Neurosci.* 4:957–967.

Bradbury, N.A., J.A. Cohn, C.J. Venglarik, and R.J. Bridges. 1994. Biochemical and biophysical identification of cystic fibrosis transmembrane conductance regulator chloride channels as components of endocytic clathrin-coated vesicles. *J. Biol. Chem.* 269:8296–8302.

Campo, C., A. Mason, D. Maouyo, O. Olsen, D. Yoo, and P.A. Welling. 2005. Molecular mechanisms of membrane polarity in renal epithelial cells. *Rev. Physiol. Biochem. Pharmacol.* 153:47–99.

Canessa, C.M., J.-D. Horisberger, and B.C. Rossier. 1993. Epithelial sodium channel related to proteins involved in neurodegeration. *Nature* 361:467–470.

Canessa, C.M., L. Schild, G. Buell, B. Thorens, Y. Gautschi, J.-D. Horisberger, and B.C. Rossier. 1994. The amiloride-sensitive epithelial sodium channel is made of three homologous subunits. *Nature* 367:463–467.

Caplan, M.J. 1997. Membrane polarity in epithelial cells: Protein sorting and establishment of polarized domains. *Am. J. Physiol.* 272:F425–F429.

Chalfie, M. 1997. A molecular model for mechanosensation in *Caenorhabditis elegans*. *Biol. Bull.* 192:125.

Choe, H., L.G. Palmer, and H. Sackin. 1999. Structural determinants of gating in inward-rectifier K^+ channels. *Biophys. J.* 76:1988–2003.

Choe, H., H. Sackin, and L.G. Palmer. 1998. Permeation and gating of an inwardly rectifying potassium channel. Evidence for a variable energy well. *J. Gen. Physiol.* 112:433–446.

Dawson, D.C., S.S. Smith, and M.K. Mansoura. 1999. CFTR: Mechanism of anion conduction. *Physiol. Rev.* 79:S47–S75.

Debonneville, C., S. Flores, E. kamynina, P.J. Plant, C. Tauxe, M.A. Thomas, C. Münster, J.-D. Horisberger, D. Pearce, J. Loffing, and O. Staub. 2001. Phosphorylation of Nedd4-2 by Sgk1 regulates epithelial Na^+ channel cell surface expression. *EMBO J.* 20:7052–7059.

Doyle, D.A., J.M. Cabral, R.A. Pfuetzerl, A. Kuo, J.M. Gulbis, S.L. Cohen, B.T. Chait, and R. MacKinnon. 1998. The structure of the potassium channel: Molecular basis of K^+ conduction and selectivity. *Science* 280:69–77.

Enkvetchakul, D., G. Loussouarn, E. Makhina, S.L. Shyng, and C.G. Nichols. 2000. The kinetic and physical basis of K_{ATP} channel gating: Toward a unified molecular understanding. *Biophys. J.* 78:2334–2348.

Firsov, D., I. Gautschi, A.-M. Merillat, B.C. Rossier, and L. Schild. 1998. The heterotetrameric architecture of the epithelial sodium channel (ENaC). *EMBO J.* 17:344–352.

Garty, H., and I.S. Edelman. 1983. Amiloride-sensitive trypsinization of apical sodium channels. Analysis of hormonal regulation of sodium transport in toad bladder. *J. Gen. Physiol.* 81:785–803.

Garty, H., and L.G. Palmer. 1997. Epithelial Na^+ channels: Function, structure, and regulation. *Physiol. Rev.* 77:359–396.

Hebert, S.C., G. Desir, G. Giebisch, and W. Wang. 2005. Molecular diversity and regulation of renal potassium channels. *Physiol. Rev.* 85:319–371.

Ho, K.H., C.G. Nichols, W.J. Lederer, J. Lytton, P.M. Vassilev, M.V. Kanazirska, and S.C. Hebert. 1993. Cloning and expression of an inwardly rectifying ATP-regulated potassium channel. *Nature* 362:31–37.

Inagaki, N., T. Gonoi, J.P. Clement, C.Z. Wang, L. Aguilar-Bryan, J. Bryan, and S. Seino. 1996. A family of sulfonylurea receptors determines the pharmacological properties of ATP-sensitive K^+ channels. *Neuron* 16:1011–1017.

Jiang, Y., A. Lee, J. Chen, M. Cadene, B.T. Chait, and R. MacKinnon. 2002. Crystal structure and mechanism of a calcium-gated potassium channel. *Nature* 417:515–522.

Karlin, A. 2002. Emerging structure of the nicotinic acetylcholine receptors. *Nat. Rev. Neurosci.* 3:102–114.

Kellenberger, S., N. Hoffmann-Pochon, I. Gautschi, E. Schneeberger, and L. Schild. 1999. On the molecular basis of ion permeation in the epithelial Na^+ channel. *J. Gen. Physiol.* 114:13–30.

Kemendy, A.E., T.R. Kleyman, and D.C. Eaton. 1992. Aldosterone alters the open probability of amiloride-blockable sodium channels in A6 epithelia. *Am. J. Physiol.* 263:C825–C837.

Koefoed-Johnsen, V., and H.H. Ussing. 1958. On the nature of the frog skin potential. *Acta Physiol. Scand.* 42:298–308.

Kuo, A., J.M. Gulbis, J.F. Antcliff, T. Rahman, E.D. Lowe, J. Zimmer, J. Cuthbertson, F.M. Ashcroft, T. Ezaki, and D.A. Doyle. 2003. Crystal structure of the potassium channel KirBac1.1 in the closed state. *Science* 300:1922–1926.

Lifton, R.P., A.G. Gharavi, and D.S. Geller. 2001. Molecular mechanisms of human hypertension. *Cell* 104:545–556.

Lin, D.H., H. Sterling, K.M. Lerea, P. Welling, L. Jin, G. Giebisch, and W.H. Wang. 2002. K depletion increases protein tyrosine kinase-mediated phosphorylation of ROMK. *Am. J. Physiol. Renal Physiol.* 283:F671–F677.

Lin, D.H., H. Sterling, B. Yang, S.C. Hebert, G. Giebisch, and W.H. Wang. 2004. Protein tyrosine kinase is expressed and regulates ROMK1 location in the cortical collecting duct. *Am. J. Physiol.* 286:F881–F892.

Lingueglia, E., G. Champigny, M. Lazdunski, and P. Barbry. 1995. Cloning of the amiloride-sensitive FMRFamide peptide-gated sodium channel. *Nature* 378:730–733.

Liu, X., Z.R. Zhang, M.D. Fuller, J. Billingsley, N.A. McCarty, and D.C. Dawson. 2004. CFTR: A cysteine at position 338 in TM6 senses a positive electrostatic potential in the pore. *Biophys. J.* 87:3826–3841.

Loffing, J., L. Pietri, F. Aregger, M. Bloch-Faure, U. Ziegler, P. Meneton, B.C. Rossier, and B. Kaissling. 2000. Differential subcellular localization of ENaC subunits in mouse kidney in response to high- and low-Na diets. *Am. J. Physiol. Renal Physiol.* 279:F252–F258.

Lopatin, A.N., E.N. Makhina, and C.G. Nichols. 1995. The mechanism of inward rectification of potassium channels: "Long-pore plugging" by cytoplasmic polyamines. *J. Gen. Physiol.* 106:923–956.

Masilamani, S., G.H. Kim, C. Mitchell, J.B. Wade, and M.A. Knepper. 1999. Aldosterone-mediated regulation of ENaC alpha, beta, and gamma subunit proteins in rat kidney. *J. Clin. Invest.* 104:R19–R23.

McNicholas, C.M., W.B. Guggino, E.M. Schwiebert, S.C. Hebert, G. Giebisch, and M.E. Egan. 1996. Sensitivity of a renal K^+ channel (ROMK2) to the inhibitory sulfonylurea compound glibenclamide is enhanced by coexpression with the ATP-binding cassette transporter cystic fibrosis transmembrane regulator. *Proc. Natl. Acad. Sci. USA* 93:8083–8088.

Morris, R.G., and J.A. Schafer. 2002. cAMP increases density of ENaC subunits in the apical membrane of MDCK cells in direct proportion to amiloride-sensitive Na(+) transport. *J. Gen. Physiol.* 120:71–85.

Mostov, K.E., M. Verges, and Y. Altschuler. 2000. Membrane traffic in polarized epithelial cells. *Curr. Opin. Cell Biol.* 12:483–490.

Noskov, S.Y., S. Berneche, and B. Roux. 2004. Control of ion selectivity in potassium channels by electrostatic and dynamic properties of carbonyl ligands. *Nature* 431:830–834.

O'Connell, A.D., Q. Leng, K. Dong, G.G. MacGregor, G. Giebisch, and S.C. Hebert. 2005. Phosphorylation-regulated endoplasmic reticulum retention signal in the renal outer-medullary K^+ channel (ROMK). *Proc. Natl. Acad. Sci. USA* 102:9954–9959.

O'Hagan, R., M. Chalfie, and M.B. Goodman. 2005. The MEC-4 DEG/ENaC channel of *Caenorhabditis elegans* touch receptor neurons transduces mechanical signals. *Nat. Neurosci.* 8:43–50.

Palmer, L.G. 1987. Ion selectivity of epithelial Na channels. *J. Membr. Biol.* 96:97–106.

Palmer, L.G. 1999. Potassium secretion and the regulation of distal nephron K channels. *Am. J. Physiol.* 277:F821–F825.

Palmer, L.G., and G. Frindt. 1996. Gating of Na channels in the rat cortical collecting tubule: Effects of voltage and membrane stretch. *J. Gen. Physiol.* 107:35–45.

Prince, L.S., A. Tousson, and R.B. Marchase. 1993. Cell surface labeling of CFTR in T84 cells. *Am. J. Physiol.* 264:C491–498.

Rabinowitz, L. 1996. Aldosterone and potassium homeostasis. *Kidney Int.* 49:1738–1742.

Riordan, J.R., J.M. Rommens, B. Kerem, N. Alon, R. Rozmahel, Z. Grzelczak, J. Zelenski, S. Lok, N. Plavsik, J.L. Chao, M.L. Drumm, M.C. Iannuzzi, F.S. Collins, and L.-C. Tsui. 1989. Identification of the cystic fibrosis gene: Cloning and characterization of complementary DNA. *Science* 245:1006–1072.

Schild, L., E. Schneeberger, I. Gautschi, and D. Firsov. 1997. Identification of amino acid residues in the alpha, beta, and gamma subunits of the epithelial sodium channel (ENaC) involved in amiloride block and ion permeation. *J. Gen. Physiol.* 109:15–26.

Sheng, S., J. Li, K.A. McNulty, D. Avery, and T.R. Kleyman. 2000. Characterization of the selectivity filter of the epithelial sodium channel. *J. Biol. Chem.* 275:8572–8581.

Sheppard, D.N., and M.J. Welsh. 1999. Structure and function of the CFTR chloride channel. *Physiol. Rev.* 79:S23–S45.

Smith, S.S., E.D. Steinle, M.E. Meyerhoff, and D.C. Dawson. 1999. Cystic fibrosis transmembrane conductance regulator. Physical basis for lyotropic anion selectivity patterns. *J. Gen. Physiol.* 114:799–818.

Snyder, P.M., C. Cheng, L.S. Prince, J.C. Rogers, and M.J. Welsh. 1998. Electrophysiological and biochemical evidence that DEG/ENaC cation channels are composed of nine subunits. *J. Biol. Chem.* 273:681–684.

Staruschenko, A., E. Adams, R.E. Booth, and J.D. Stockand. 2005. Epithelial Na^+ channel subunit stoichiometry. *Biophys. J.* 88:3966–3975.

Sui, J.L., K. Chan, M.N. Langan, M. Vivaudou, and D.E. Logothetis. 1999. G protein gated potassium channels. *Adv. Second Messenger Phosphoprotein Res.* 33:179–201.

Tanemoto, M., C.G. Vanoye, K. Dong, R. Welch, T. Abe, S.C. Hebert, and J.Z. Xu. 2000. Rat homolog of sulfonylurea receptor 2B determines glibenclamide sensitivity of ROMK2 in Xenopus laevis oocyte. *Am. J. Physiol. Renal Physiol.* 278:F659–F666.

Vergani, P., A.C. Nairn, and D.C. Gadsby. 2003. On the mechanism of MgATP-dependent gating of CFTR Cl^- channels. *J. Gen. Physiol.* 121:17–36.

Verrey, F., E. Hummler, L. Schild, and B. Rossier. 2000. Control of Sodium Transport by Aldosterone. Lippincott Williams and Wilkins, Philadephia, pp. 1441–1471.

Waldmann, R., F. Bassilana, J. de-Weille, G. Champigny, C. Heurteaux, and M. Lazdunski. 1997. Molecular cloning of a non-inactivating proton-gated Na+ channel specific for sensory neurons. *J. Biol. Chem.* 272:20975–20978.

Wang, W.-H., K.M. Lerea, M. Chan, and G. Giebisch. 2000. Protein tyrosine kinase regulates the number of renal secretory K channels. *Am. J. Physiol.* 278:F165–F171.

Weisz, O.A., J.M. Wang, R.S. Edinger, and J.P. Johnson. 2000. Non-coordinate regulation of endogenous epithelial sodium channel (ENaC) subunit expression at the apical membrane of A6 cells in response to various transporting conditions. *J. Biol. Chem.* 275:39886–39893.

Welsh, M.J. 1996. Cystic fibrosis. *In*: Molecular Biology of Membrane Transport Disorders. S.G. Schultz, T.E. Andreoli, A.M. Brown, D.M. Fambrough, J.F. Hoffman, and M.J. Welsh, editors. Plenum, New York, pp. 605–623.

Yoo, D., L. Fang, A. Mason, B.Y. Kim, and P.A. Welling. 2005. A phosphorylation-dependent export structure in ROMK (KIR 1.1) channel overides an ER-localization signal. *J. Biol. Chem.* 280:35281–35289.

Zerhusen, B., J. Zhao, J. Xie, P.B. Davis, and J. Ma. 1999. A single conductance pore for chloride ions formed by two cystic fibrosis transmembrane conductance regulator molecules. *J. Biol. Chem.* 274:7627–7630.

Zhang, Y.Y., J.L. Robertson, D.A. Gray, and L.G. Palmer. 2004. Carboxy-terminal determinants of conductance in inward-rectifier K channels. *J. Gen. Physiol.* 124:729–739.

Zhou, Y., J.H. Morais-Cabral, A. Kaufman, and R. MacKinnon. 2001. Chemistry of ion coordination and hydration revealed by a K^+ channel-Fab complex at 2.0 A resolution. *Nature* 414:43–48.

Part III
Theoretical Approaches

13 Poisson–Nernst–Planck Theory of Ion Permeation Through Biological Channels

Rob D. Coalson and Maria G. Kurnikova

13.1 Introduction

The kinetics of an assembly of charged particles such as electrons, ions, or colloids, particularly when subjected to externally applied electric fields, has been of interest for many years and in many disciplines. In applied physics and electrical engineering, the motion of electrons and holes through semiconductor materials under the influence of an applied voltage plays an essential role in the function of modern electronic components such as transistors, diodes, and infrared lasers (Peyghambarian et al., 1993). Electrochemistry deals in large part with the motion of simple inorganic ions (e.g., Na^+, Cl^-) in electrolytic solutions and how this motion is influenced when electrodes are employed to generate an electric potential drop across the solution or a membrane interface (Bockris and Reddy, 1998). Larger macroions such as charged polystyrene spheres (radius 0.1–1 micron) can also be manipulated using applied electric fields (Ise and Yoshida, 1996). Many processes in molecular biology, from self-assembly of DNA strands into bundles (Wissenburg et al., 1995) to enzyme-ligand docking (Gilson et al., 1994), are steered by electrostatic forces between biological macroions which are mediated by the response of simple salt ions in the solution.

One particularly intriguing type of biological process that falls into this general category is the flow of ions (Na^+, Cl^-, K^+, Ca^{++}, etc.) through pores in lipid bilayer membranes. Lipid bilayers (Fig. 13.1) form the cell membrane as well as internal compartments, called organelles, in eukaryotic cells. The interior of a lipid bilayer, composed of alkane chains, is hydrophobic, and hence ions (being hydrophilic) cannot penetrate through it. Since many bioenergetic processes rely on separating charge across bilayer membranes and then transducing the energy thus stored, it is imperative that there be a mechanism for moving ions across these membranes in a controllable fashion. Nature has solved this problem by developing proteins which are, very roughly, cylindrical channels (pipes) possessing an aqueous pore. They insert themselves into the lipid bilayer, spanning it in a transverse fashion, so that ions can flow through the aqueous pore from one side of the membrane to the other (Fig. 13.1) when driven by an electrochemical gradient. Protein channels are extraordinary devices (Hille, 1992). In many cases, they can be opened and closed to the flow of ions reliably and reversibly by a specific stimulus (e.g., the binding

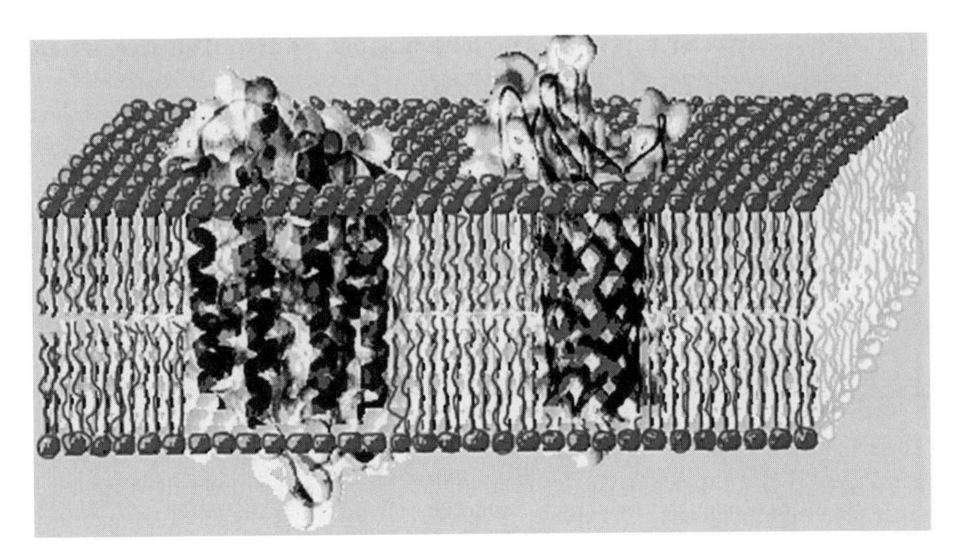

Fig. 13.1 Schematic picture of lipid bilayer with ion channel proteins in it.

of a specific molecule to the channel protein, or a change in the voltage across the membrane). Furthermore, in the open state, many ion channels are selective about the ions that they let through, sometimes passing only cations but not anions (or vice versa), or even allowing, for example, K^+ but not Na^+ to permeate.

In this review [which is an extension of a recent review on the same general topic (Coalson and Kurnikova, 2005)], we will consider the passage of ions through a protein channel in its open state. We seek to provide a practical solution to the following basic problem. Given the structure of the channel protein (and an estimation of the geometric features of the lipid bilayer), plus some details of the electrical properties of these objects (distribution of electrical charges in them and characteristic dielectric constants), we wish to compute the rate of ion flow through the channel as a function of experimentally controllable parameters, e.g., electric potential applied across the membrane and concentration of ions in the bathing solutions on either side of the membrane. That is, we want to calculate current–voltage (I–V) curves for the system at hand. There are several possible approaches that might be envisaged, including (if possible) all-atom molecular dynamics (MD) simulation (Crozier et al., 2001a,b; Aksimentiev and Schulten, 2005) and Brownian Dynamics (Chung et al., 1998, 1999, 2002; Graf et al., 2000; Mashl et al., 2001; Chung and Kuyucak, 2002; Burykin et al., 2002; Im and Roux, 2002; Graf et al., 2004; Noskov et al., 2004; Cheng et al., 2005) simulation of the ion motion. Here we concentrate on the simplest (most coarse-grained) level of treatment imaginable, namely a continuum electrodiffusion model in which the mobile ions are treated as a concentration profile whose distribution and motion are influenced by electrostatic forces (some of which are generated by the polarization of the mobile ion density). In the context of ion permeation through channel proteins, this approach is known as Poisson–Nernst–Planck (PNP) theory. In the next sections, we will present the basic ingredients of this theory as well as techniques for solving the relevant equations that comprise it. Fundamental limitations of the continuum electrodiffusion model (of which there are many!) will be discussed. Some illustrative applications will be presented. Attempts to improve the basic PNP theory will be reviewed. Finally, conclusions and prognosis for this type of analysis will be presented.

13.2 Basic (Primitive) PNP Theory (and Its Limitations)

13.2.1 The PNP Equations

In continuum theory electrolyte ions are treated as a continuous charge distribution characterized by the concentrations $\{c_i(\vec{r})\}$ of the ionic species (labeled by i) involved. It is the goal of electrodiffusion theories to determine the steady state concentrations of all mobile ion species, the electric fields they generate, and in the context of biological ion channels, the current of ions flowing through the channel as a function of experimentally controllable parameters such as the bulk concentrations of ions (electrolytes) and the electric potential applied across the embedding membrane, usually by means of a microelectrode apparatus. In the present subsection we will collect the working formulae that result from this approach, known generically as PNP theory. In Section 13.2.2, we will explore the conceptual underpinnings of this type of theory.

In electrodiffusion theory, the distribution of mobile ion concentrations is governed by a set of drift–diffusion equations, also called Nernst–Planck (NP) equations, one for each ionic species i present in solution. In particular, $\vec{j}_i$, the flux of species i at a given point in space is given by

$$\vec{j}_i(\vec{r}) = -D_i(\vec{r})\left[\frac{\partial c_i(\vec{r})}{\partial \vec{r}} + c_i(\vec{r})\frac{\partial}{\partial \vec{r}}(\beta\psi_i(\vec{r}))\right], \tag{13.1a}$$

and the concentration of species i evolves in accordance with the continuity equation $\frac{\partial c_i}{\partial t} = -\text{div}\,(\vec{j}_i)$. In Eq. 13.1a, D_i is the position-dependent diffusion coefficient of species i and $\beta = (kT)^{-1}$ is the inverse temperature, with k being Boltzmann's constant, and T the absolute temperature. Finally, $\psi_i(\vec{r})$ is the free energy of ions of species i in solution. At steady state,

$$\text{div}\,(\vec{j}_i) = 0, \tag{13.1b}$$

and thus all quantities in the NP equation (Eq. 13.1) are time-independent. The second term on the right-hand side of (Eq. 13.1a) is the drift term due to the forces on a charged particle of species i arising from both ion–ion interactions and other sources. The latter include charges on the protein system and the externally imposed electric field. Equation 13.1 is supplemented by concentration boundary conditions that account for the external bulk ionic concentrations of species i (which may be different on different boundary "faces," particularly if concentrations in the bathing solutions on the two sides of the membrane differ).

In a continuum model, $\psi_i(\vec{r})$ depends on the electrostatic charge distribution in the system and on the (generally position dependent) dielectric response function $\varepsilon(\vec{r})$. It is convenient to separate the ion free energy into two contributions:

$$\psi_i(\vec{r}) = q_i \phi_{\text{mobile}}(\vec{r}) + \Delta G^i_{\text{SIP}}(\vec{r}), \tag{13.2}$$

where q_i is the charge of an ion of species i, $\phi_{\text{mobile}}(\vec{r})$ is the electrostatic potential due to all mobile ions and the applied electric field associated with external electrodes, and $\Delta G^i_{\text{SIP}}(\vec{r})$ is the potential of mean force (PMF) (McQuarrie, 1976; Chandler, 1987) for a single test ion [hence "Single Ion Potential" (SIP)]. In an inhomogeneous dielectric medium $\phi_{\text{mobile}}(\vec{r})$ is determined by the Poisson equation[1] (PE):

$$\vec{\nabla} \cdot (\varepsilon(\vec{r}) \vec{\nabla} \phi_{\text{mobile}}(\vec{r})) = -4\pi \sum_i q_i c_i(\vec{r}), \tag{13.3}$$

subject to Dirichlet boundary conditions, i.e., values of the electrostatic potential (imposed by the electrodes) are fixed on the boundaries of the computational box (Kurnikova et al., 1999). In the simplest approximation that was introduced in the field of channel modeling by Eisenberg and coworkers (Barcilon, 1992; Barcilon et al., 1992) the term $\Delta G^i_{\text{SIP}}(\vec{r})$ is disregarded. In an obvious generalization $\Delta G^i_{\text{SIP}}(\vec{r})$ may include the electrostatic potential due to partial charges fixed on the protein and lipid atoms, i.e., $\Delta G^i_{\text{SIP}}(\vec{r}) = q_i \phi_{\text{protein}}(\vec{r})$ (Chen and Eisenberg, 1993a,b; Kurnikova et al., 1999; Cárdenas et al., 2000). Equations 13.1 and 13.3 are coupled nonlinearly via the c_i and ϕ_{mobile} variables. In the general case of a protein of arbitrary geometry and distribution of partial charges on protein atoms, these equations have no analytical solution and must be solved numerically to self-consistency (Kurnikova et al., 1999). Equations 13.1–13.3 with $\Delta G^i_{\text{SIP}}(\vec{r}) = q_i \phi_{\text{protein}}(\vec{r})$ comprise the standard PNP theory, which we shall refer to here as "primitive PNP" for reasons that will become obvious in the ensuing exposition.

13.2.2 Conceptual Framework of PNP Theory

Let us consider for the moment a somewhat simpler problem, namely motion of a structureless particle in an external force field $\vec{F}(\vec{r})$ derivable from a potential energy function $V(\vec{r})$, i.e., $\vec{F}(\vec{r}) = -\partial V(\vec{r})/\partial \vec{r}$. The particle is also subject to thermal agitation arising from incessant buffeting by other particles in the system. In the classic theory of Brownian motion (Chandrasekhar, 1943), the Brownian particle (the one whose motion we are explicitly tracking) is much larger than that of tiny particles which are colliding with it. For example, the Brownian particle might be

[1] In equation (Eq. 13.3) CGS Gaussian units are employed. To write the Poisson equation in SI units, substitute $4\pi \rightarrow 1/\varepsilon_0$ on the r.h.s., with ε_0 being the permittivity of free space (Kittel, 1996).

a colloid sphere (radius ca. 1 micron), with water molecules (molecular radius of a few Å) bouncing off of it in a rapid and stochastic fashion. When Brownian motion theory is applied to describe motion of one "tagged" atom, ion or small molecule in a condensed phase consisting of other such species, the separation of distance scales inherent in classic Brownian motion theory is not so clear. Nevertheless, this description seems to be quite successful for describing molecular level kinetics in many cases [e.g., the description of ion diffusion in bulk liquids (Koneshan et al., 1998; Mamonov et al., 2006)].

Assuming that Brownian motion theory can be applied to the case of interest, then in the high friction limit the probability distribution for the position of the Brownian particle (or, equivalently, the concentration profile achieved by a collection of such independently moving particles) is given by:

$$\frac{\partial c(\vec{r},t)}{\partial t} = -\mathrm{div}(\vec{j}) \tag{13.4a}$$

with particle flux

$$\vec{j}(\vec{r}) = -D(\vec{r})\left[\frac{\partial c(\vec{r})}{\partial \vec{r}} + c(\vec{r})\frac{\partial}{\partial \vec{r}}(\beta V(\vec{r}))\right] \tag{13.4b}$$

Here all symbols have the same meaning as in Section 13.2.1 above, except for the obvious substitution of the potential energy function $V(\vec{r})$ for the more complicated (and mysterious) free energy function invoked in Eq. 13.1a. (For now we will suppress the subscript that labels ion species and speak simply of the behavior of a generic species moving under a generic single-particle force-field). Equation 13.4 is known as the Smoluchowski equation (Chandrasekhar, 1943). The first term contributing to the flux (cf. Eq. 13.4b) is simply the concentration gradient that constitutes Fick's Law of diffusion. The second term, the so-called "drift" term, represents the influence of the systematic force $\vec{F}(\vec{r})$. Consistent with the high-friction assumption underlying the Smoluchowski equation, we imagine that at each point along its trajectory the Brownian particle is damped by the appropriate friction force characterized by friction constant γ and instantaneously reaches its terminal velocity $\vec{v}_{\mathrm{term}} = \vec{F}/\gamma = \beta D\vec{F}$. [Here we assume that the Stokes–Einstein relation $D = kT/\gamma$ connecting microscopic friction to macroscopic diffusion applies (McQuarrie, 1976)]. The flux associated with this drift process is thus $\vec{j}_{\mathrm{drift}} = c\vec{v}_{\mathrm{term}} = \beta Dc\vec{F}$. Since in general the diffusion constant and the systematic force are position-dependent, so is the drift flux, as indicated in the second term on the r.h.s. of Eq. 13.4b. Note one direct consequence of the structure of the drift–diffusion flux prescription given in Eq. 13.4b: the steady state solution to this equation corresponding to zero particle flux, i.e., thermal equilibrium, is simply the Boltzmann probability distribution $c(\vec{r}) \propto \exp[-\beta V(\vec{r})]$, as it should be. However, the most general boundary condition implies a *non*-equilibrium steady state, i.e., a steady-state solution of the Smoluchowski equation (Eq. 13.4) for which $\vec{j} \neq 0$ identically.

If we regard one permeating ion in the ion channel systems of direct interest in this article as the tagged Brownian particle, then the situation is clearly more complicated than that of Brownian motion of a collection of noninteracting particles all moving in the same static external force field. The tagged ion experiences a force due to each of the other moving ions, and indeed, due to each atom in the protein (and membrane), which are also fluctuating with time. To utilize the Brownian motion theory framework just sketched, we seek to identify an *optimal effective static single particle potential* that can play the role of $V(\vec{r})$ in Eq. 13.4. In the (temporary) absence of the other mobile ions and the externally applied electrical potential, the most reasonable candidate is the PMF (constrained free energy profile) $\Delta G^i_{\text{SIP}}(\vec{r})$ for a tagged particle of species i (McQuarrie, 1976). (We need to recall at this juncture that there are generally two or more ionic species in the electrolyte, each of which experiences different energetic interactions.) The PMF is obtained from the Boltzmann factor for the ion and all water, protein and membrane atoms particles in the system, given in terms of the full many-dimensional microscopic potential energy function U_i which describes the mutual interactions between these particles, by integrating over all degrees of freedom except those of the ion. Specifically:

$$\exp[-\beta \Delta G^i_{\text{SIP}}(\vec{r})] = \int d\vec{R} \exp[-\beta U_i(\vec{r}, \vec{R})] \Big/ \int d\vec{r} \int d\vec{R} \exp[-\beta U_i(\vec{r}, \vec{R})],$$

where $\vec{R}$ represents all "environmental" coordinates in the system, i.e., those of the water molecules and the atoms in the protein and the membrane. This choice ensures the correct reduced probability distribution (concentration profile), namely $\exp[-\beta \Delta G^i_{\text{SIP}}(\vec{r})]$, for the ion at thermal equilibrium taking into account the effects of water solvent and thermal fluctuation of the protein and membrane, but in the absence of ion–ion interactions and the external electric potential. To this effective potential we then add the influence of the applied external electric potential $\phi_{\text{ext}}(\vec{r})$, i.e., add the term $q_i \phi_{\text{ext}}(\vec{r})$ to $\Delta G^i_{\text{SIP}}(\vec{r})$. Finally, we need to account for the average force exerted by all other mobile ions in the system on the tagged ion. We assume that these other ions collectively generate an additional electric potential that acts on said ion. In reality, these other mobile ions are moving, so their distribution in space with respect to the tagged ion changes with time. We ignore these fluctuations and assume that the static average distribution of the ions other than the test ion can be used to calculate a meaningful time-averaged electric potential ϕ_{MI} at point $\vec{r}$. Further, we replace this conditional probability, i.e., fixing the test ion and averaging over all others, with the average ion density profile (including all ions without any constraints). These are the essential ingredients of a "mean field" approximation. [Similar strategies have long been utilized to understand the thermodynamic properties of ferromagnets (Chandler, 1987) and polymers (Doi, 1996)].

Within the context of this approximation, ϕ_{MI} can be calculated from the steady-state concentration profiles $c_i(\vec{r})$ characterizing each mobile ion species. In the case of a dielectrically uniform medium characterized by dielectric constant ε, then $\phi_{\mathrm{MI}}(\vec{r}) = \int d\vec{r}' \sum_j q_j c_j(\vec{r}')/\varepsilon|\vec{r} - \vec{r}'|$. In fact, biological ion channels are inherently characterized by several spatial regions with different dielectric constants. Hence, $\phi_{\mathrm{MI}}(\vec{r})$ must be obtained by solving the PE (cf. Eq. 13.3). (The relevant boundary condition for this computation zero electric potential on the boundaries of the computational box, since the polarization of $+/-$ mobile ion charge is confined to the channel region.) The complete effective single-particle potential thus becomes:

$$\psi_i(\vec{r}) = \Delta G^i_{\mathrm{SIP}}(\vec{r}) + q_i[\phi_{\mathrm{ext}}(\vec{r}) + \phi_{\mathrm{MI}}(\vec{r})]$$

Since the external electric potential typically supplied by microelectrodes can be computed by solving the Laplace equation [PE with zero free charge (Marion, 1965)] and fixed potential boundary conditions on the walls of the box (with a different potential on the two faces of the computational box which run parallel to the membrane surface), the sum $(\phi_{\mathrm{ext}} + \phi_{\mathrm{MI}}) \equiv \phi_{\mathrm{mobile}}$ can be obtained by solving a single PE with these "potential" boundary conditions, as specified in Section 13.2.1. Of course, the $c_i(\vec{r})$ must be obtained by self-consistent solution of both NP and Poisson equations. Note also that the size of the mobile ions is not taken into account in the mean field averaging procedure invoked here.

The calculation of $\Delta G^i_{\mathrm{SIP}}(\vec{r})$ is a prequel to the PNP-type calculation just outlined. In principle $\Delta G^i_{\mathrm{SIP}}(\vec{r})$ can be computed numerically using an all-atom model of the system (test ion, water, protein, and membrane). However, such calculations are quite time-consuming and require a high accuracy force field. In the absence of force fields, which include electronic polarizability effects properly, even recent high-level all atom single ion free energy profiles appear to give unrealistic results (Allen et al., 2004). An alternative strategy is to use semiempirical strategies based wholly or in part on a continuum theory description of the solvent, protein and membrane to obtain $\Delta G^i_{\mathrm{SIP}}(\vec{r})$ (Mamonov et al., 2003). A hierarchy of increasingly sophisticated strategies of this type will be presented in the course of this chapter. It is clear from the brief sketch of the "derivation" of PNP above that this type of theory is far from rigorous; the same sketch hopefully suggests ways to remove some of its deficiencies (Schuss et al., 2001; Mamonov et al., 2003; Graf et al., 2004; Gillespie et al., 2005; Wang et al., 2005).

13.2.3 Goldman–Hodgkin–Katz Theory of Ion Permeation Through Channel Proteins

A principal feature of the PNP equations is that the effective potential which enters into the NP equations, and hence determines the steady state concentration profile, is itself a function of the (unknown) concentration of mobile ions, as reflected in the structure of the PE (Eq. 13.3). Thus, as noted above, the NP and PEs must be solved self-consistently, which compounds the difficulty of solving them numerically and may cloud insight into the properties of the resultant solutions (steady state mobile ion distributions, ion currents, etc.). It would certainly be convenient if the effect of mobile ion–mobile ion interactions on the effective potential felt by a "tagged" mobile ion could be neglected, so that the NP equation for that ionic species would be governed by a simple externally prescribed potential energy function, namely: $V_i(\vec{r}) = \Delta G^i_{\mathrm{SIP}}(\vec{r}) + q_i \phi_{\mathrm{ext}}(\vec{r})$. One case where it is reasonable to expect such a situation to occur is when the ion channel is so narrow that only one ion is likely to be in it at any particular time. Numerical studies which include multi-ion kinetics at the Brownian dynamics (BD) level (i.e., not assuming a mean-field ion–ion interaction potential as is done in PNP theory) show that in narrow model channels ion–ion interactions have a relatively minor effect on net ion currents (Graf et al., 2000). Then, going one step further, since the ion channels under consideration in this subsection are presumed to be rather narrow, we may as well assume that they are cylindrical in nature with a cylinder radius R and a small ratio of cylinder radius to cylinder length L [see, for example, Fig. 1 of Ref. (Kurnikova et al., 1999)]. Because the cylinder is narrow, the potential energy profile inside it can be well-approximated as a function of the channel axis coordinate z only, i.e., $V_i(z)$.In this situation, the solution of the 3D NP equation yields a concentration profile that also depends only on the channel axis coordinate, and is a solution of the 1D NP equation (Barcilon, 1992; Barcilon et al., 1992). That is, suppressing the ionic species labels for notational convenience:

$$0 = \frac{\partial}{\partial z}\left\{D(z)\left[\frac{\partial c(z)}{\partial z} + c(z)\frac{\partial}{\partial z}(\beta V(z))\right]\right\}. \tag{13.5}$$

This differential equation can be solved explicitly for $c(z)$, given the values of the concentration at the two boundaries, $c(0) \equiv c_0$ and $c(L) \equiv c_L$. One finds in the case of a spatially homogeneous diffusion constant D that the current density (aligned with the channel axis) is prescribed by:

$$j = -D\frac{[c_L e^{\beta V_L} - c_0 e^{\beta V_0}]}{\int_0^L dz' e^{\beta V(z')}}, \tag{13.6}$$

where $V(0) \equiv V_0$ and $V(L) \equiv V_L$. Thus the problem of computing current flow through this class of simple model channels is reduced to a 1D quadrature for arbitrary $V(z)$.

Having simplified the complex phenomenon of ion permeation through a biological channel to this degree, it behooves us to ask what is the simplest meaningful model for $V_i(z)$ appropriate to ionic species i. In the absence of the channel (pore) in the membrane, the electric potential drop across the channel would be essentially linear, i.e., $\phi_{\mathrm{ext}}(z) = \phi_0 + z(\phi_L - \phi_0)/L$, where $\phi(0) = \phi_0$ and $\phi(L) = \phi_L$ are the values of the applied potential in the reservoirs abutting the channel on either side. (Only the applied potential difference $\phi_L - \phi_0 \equiv \phi_{\mathrm{ap}}$ is physically meaningful, as is apparent from Eq. 13.6). The channel pore represents only a tiny "pin prick" in the membrane, so we expect that the linear drop approximation remains reasonable inside the channel pore. By construction (of this model), we are ignoring ion–ion interactions. This still leaves the single-ion $\Delta G^i_{\mathrm{SIP}}(z)$ contribution to $V_i(z)$. In an ultra-simple model we can ignore this, assuming that it is small compared to the applied external potential term. We should emphasize that any $\Delta G^i_{\mathrm{SIP}}(z)$ can be added to the analysis in this section without changing its essential structure. The advantage to neglecting it is that the quadrature in Eq. 13.6 can be analytically computed in the case of the linear potential drop model, which is known historically as the Goldman–Hodgkin–Katz (GHK) model (Hille, 1992; Sten-Knudsen, 1978).

For concreteness, let us further specialize to the case that the bathing solutions consist of a monovalent electrolyte (e.g., NaCl). Then the GHK model predicts current densities for +/− ions of:

$$j_\pm = \frac{\mp D_\pm \Delta}{L}\left[\frac{c_L}{1-e^{\mp\Delta}} - \frac{c_0}{e^{\pm\Delta}-1}\right], \tag{13.7}$$

where $\Delta \equiv e_0\phi_{\mathrm{ap}}/kT$ and e_0 is the proton charge. Note also that the +/− ion species may have different effective diffusivities $D_\pm$, which describe their diffusion within the pore (and may differ in magnitude from their corresponding bulk solution values). To obtain absolute electric currents we simply multiply by the cross-sectional area of the cylinder and the ion charge. If $i_\pm$ is the (particle) current of ± ions (i.e., number of particles/sec), then

$$e_0 i_\pm \cong \mp 1000\frac{\pi R^2}{L} D_\pm \Delta\left[\frac{c_L}{1-e^{\mp\Delta}} - \frac{c_0}{e^{\pm\Delta}-1}\right]. \tag{13.8}$$

This formula gives as output current in pA, inputting D in cm^2/s, c in mM, and R and L in Å. Finally, the experimentally observed electric current is $I = e_0(i_+ - i_-)$.

The GHK equations can be used to predict ion current through a channel protein for a wide range of experimental conditions. One such plot is shown in Fig. 13.2 for parameters specified in the figure caption. This figure illustrates several generic characteristics of I–V curves obtained from GHK theory. In the general case, the current is much greater when the voltage has one sign than it is when the sign of the voltage is reversed, all other things being equal. This phenomenon, known as "rectification," is observed in many experimental measurements of ion channel I–V curves. Further, the current grows linearly with large applied voltage, again consistent with many experiments. Finally, the current vanishes at a particular applied voltage, known as the "reversal potential" ϕ_{rev}. In general ϕ_{rev} is not zero, but is given, according to GHK theory, by:

$$e^{\Delta_{\mathrm{rev}}} = \frac{D_+ c_0 + D_- c_L}{D_- c_0 + D_+ c_L},$$

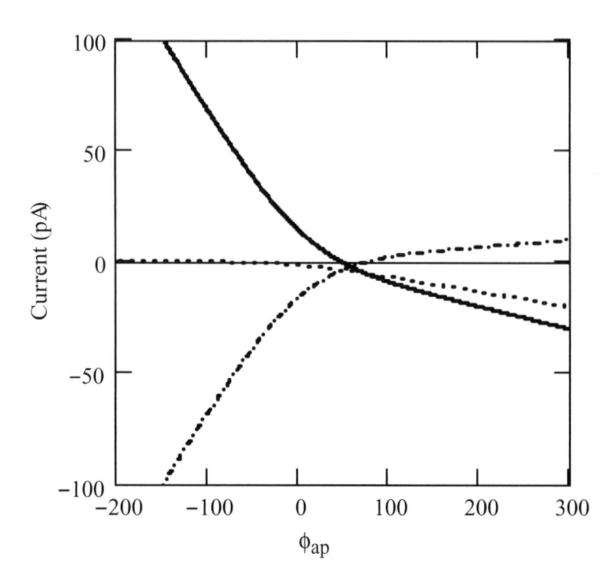

Fig. 13.2 GHK current–voltage characteristics (solid line) is shown for cylindrical model ion channel system, radius $R = 2$ Å, $L = 30$Å at room temperature ($T = 298$ K) characterized by: (internal) diffusion constants $D_+ = 10^{-5}$ cm^2/s, $D_- = 10^{-4}$ cm^2/s, and reservoir concentrations $c_0 = 20$ mM, $c_L = 400$ mM. Also shown are $e_0 i_+$ (dotted line) and $e_0 i_-$ (dot–dashed line), given by Eq. 13.8. The observed electrical current (solid line) I is given by the difference between these, i.e., $I = e_0 i_+ - e_0 i_-$. (Note: $\phi_{ap} \cong 25\Delta$ for monovalent ions at room temperature; cf. Eq. 13.8.)

with $\Delta_{rev} = e_0\phi_{rev}/kT$. This equation can be rearranged to read:

$$\frac{D_+}{D_-} = \frac{c_0 e^{\Delta_{rev}} - c_L}{c_0 - c_L e^{\Delta_{rev}}},$$

which provides a way to calculate the ratio of intrinsic diffusivities of $+/-$ ions (known as the "permeability ratio") by measuring the reversal potential experimentally under asymmetric bathing solution conditions.

One other important characteristic of GHK theory (or any 1D NP theory of the type under consideration here) is that the current scales are proportional to the concentrations in the bulk reservoirs. That is, if both c_0 and c_L double, all other factors being unchanged, then the electric current predicted by the GHK model doubles (cf. Eq. 13.8). This behavior is found experimentally at low bathing solution concentrations, but generally currents are observed to saturate as the concentration of the bathing solutions is increased. The origin of this discrepancy will be discussed in Section 13.6.

This brief survey of 1D NP theory and its application to ion channels in the case of a constant effective driving force (electric field) exposes advantages and disadvantages of GHK theory. The great advantage is the conceptual simplicity of the model and its amenability to analytical solution over a wide range of physical input parameters. But such simplicity comes at a price, namely some loss of realism:

GHK theory neglects numerous aspects of the 3D nature of the overall channel system (including the abutting bathing solutions), the atomic (chemical) details of the pore lining, and the interaction forces between all pairs of mobile ions. It is thus interesting to examine how well a fully 3D model of PNP theory, including modifications to primitive PNP theory that attempt to incorporate the effect of protein fluctuations on the permeation of ions through channel proteins, can correct some deficiencies of classical GHK theory. Such an examination is undertaken in the following sections.

13.3 Numerical Algorithms for Solving the 3D PNP Equations

The PNP equations consist of a set of coupled partial differential equations (PDEs), namely the NP equation (Eq. 13.1) (more precisely, one such NP equation for each ionic species), and the PE of electrostatics equation (Eq. 13.3). The 3D NP and the PE can only be solved analytically in a limited number of cases where the system possesses a high degree of geometric symmetry (thus resulting in an effective reduction in dimensionality). In the case of a general 3D system, these PDEs can be solved individually using a variety of numerical methods, two of the most popular being the method of Finite Differences (Kurnikova et al., 1999; Cardenas et al., 2000) and the method of Finite Elements (Hollerbach et al., 2000). For concreteness, we will focus here on the technically simpler Finite Difference approach. For example, the following strategy can be used to solve the 3D PE.

First we discretize onto a 3D cubic lattice an initial guess for the electric potential field (whose computation is the end result of the calculation). In the cases of interest to us the electric potential is specified on the boundaries of the computational box (lattice); it is unknown in the interior of the computational box (our goal is to determine it!). Hence, we set the known boundary values of the electric potential and make an arbitrary initial guess about the field values at the interior points of the lattice. In a similar fashion, we discretize the protein/membrane charge distribution and the spatially dependent dielectric constant profile, which are both assumed to be given. Then, we cycle around the interior lattice: the electric potential ϕ_k at each interior lattice point k is updated based on an appropriate average over its nearest neighbors (of which there are six for a 3D cubic lattice). For example, in the case where there is no free charge, i.e., the r.h.s. of Eq. 13.3 is identically zero [so that the PE reduces to Laplace equation (Marion, 1965)], and if the dielectric constant ε is the same everywhere in space, then the update value of the electric potential at lattice point k is simply the arithmetic average of the potential at the six nearest neighbor sites. If any of these neighbors is a boundary point, its value is known as a boundary condition. Otherwise, the neighboring field points depend on the current field configuration (as the relaxation procedure progresses). In the general case, when there is free charge in the system (the r.h.s. of Eq. 13.3 does not vanish) and the dielectric profile is spatially inhomogeneous, then the relevant average, which is still a linear function of the values of the "instantaneous" potential at the six nearest neighbor sites, is slightly more complicated [see Ref. (Kurnikova et al., 1999)for full details].

If we denote the appropriate average as $\bar{\phi}_k$, then the updated value of ϕ_k becomes $\phi_k \rightarrow w\phi_k + (1 - w)\bar{\phi}_k$, where w is a parameter (typically $0 < w < 2$) which is adjusted so as to accelerate convergence without sacrificing stability. (For w very small, only a small "portion" of $\bar{\phi}_k$ is mixed in with ϕ_k, thus preventing numerical instabilities from setting in). This process is repeated for several cycles until no further changes in the potential profile are obtained, i.e., the input potential field at the beginning of the cycle is the same as the output field at the end of the cycle. Strategies of this type are known as "relaxation techniques" (Coalson and Beck, 1998; Press. et al., 1986). They are well developed in applied PDE theory. Furthermore, for a linear PDE, such as the PE, relaxation techniques are guaranteed to converge for an arbitrary initial electric potential guess when w is chosen appropriately.

The NP equation is also a linear PDE. Thus the same relaxation methods can be used to solve it on a cubic grid. In the case of the concentration profile of an ionic species, the values at the external boundaries of the computational box are specified, analogous to the case of fixed electric potential in the PE. In addition, there may be interior walls (e.g., the pore of an ion channel) which do not allow ions to pass through them. The appropriate boundary condition at these interfaces is "zero flux": this boundary condition can easily be implemented by changing the averaging procedure slightly for interior grid points which abut such bounding surfaces. Analogous to charge and dielectric profile fields in the PE, the spatially-dependent diffusion constant profile as well as the potential energy function which enter into the NP equation must also be discretized, and these become ingredients in the nearest neighbor averaging prescription that is used to update the ion concentration field.

To illustrate the basic strategy, consider (for notational and pictorial simplicity) the 2D, one component analog shown in Fig. 13.3 [reproduced from (Cárdenas et al., 2000)]. Each of the flux contributions can be approximated by an appropriately symmetrized lattice discretization scheme. For example:

$$j^x_{i+1} = -[(D_{i+1,j} + D_{i,j})/2a][c_{i+1,j} - c_{i,j} + \beta(V_{i+1,j} - V_{i,j})(c_{i+1,j} + c_{i,j})/2],$$

where a is the lattice spacing. Then the lattice version of the NP equation $\mathrm{div}(\vec{j}) = 0$ is simply $j^x_{i+1} - j^x_{i-1} + j^y_{j+1} - j^y_{j-1} = 0$. The procedure for the 3D case is completely analogous: the lattice NP equation can then be rearranged to obtain the concentration of the central lattice point as a linear combination of the concentrations of its six nearest neighbors. Enforcement of zero-flux boundary conditions is done by setting the appropriate lattice flux to zero, thus altering the update formula for the central concentration point in a straightforward manner. For full details, see Refs. (Kurnikova et al., 1999; Cárdenas et al., 2000). The NP equation, like the PE, is a linear PDE, so that convergence of the relaxation technique is virtually guaranteed.

The PNP equations consist of coupled NP equations and a PE. These can only be solved analytically in special cases, namely, when the electric fields generated

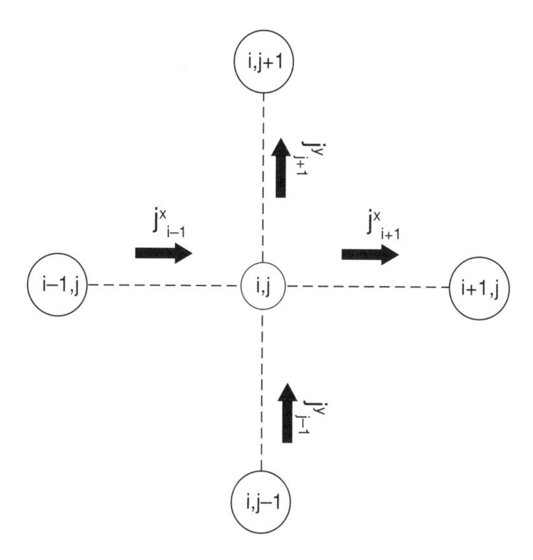

Fig. 13.3 Schematic description of current flow in/out of lattice point i, j in a 2D drift–diffusion process.

by mobile ions are small [Debye–Hückel theory being the most familiar example (McQuarrie, 1976)] and there is a degree of geometrical symmetry (as noted above in the case of NP and PEs individually). For the vast majority of interesting physical situations, the PNP equations must be solved numerically. This is done by solving the Poisson and NP equations (one NP equation for each ionic species) numerically and self-consistently. In practice, at each time step we use the current value of the electric potential as a (fixed) input into each NP equation, and vice versa. The coupled PNP equations are nonlinear, and therefore relaxation must be done delicately, using a small value of the weight parameter w in both NP and PEs. It is found numerically (Kurnikova et al., 1999; Cárdenas et al., 2000) that convergence can be obtained in this manner, although the process may be slow (depending on system conditions). Thus, there is room for improvement in the efficiency of numerical procedure, e.g., the use of variable mesh grids and multigrid methodologies.

13.4 Application of Primitive PNP to Gramicidin A in Charged/Dipolar Lipid Bilayers

13.4.1 The Model System

Gramicidin A (GA) is an antibiotic polypeptide widely used in single-channel experiments on passive ion–current permeation through a lipid membrane. GA is a small 15 amino acid β-helical peptide with an aqueous pore. Due to its unusual primary sequence of alternating L and D amino acids it forms a β-helix with all the amino acid side-groups extending away from the backbone helix, which forms the narrow (ca. 2 Å radius) channel. GA reconstructs into a lipid bilayer by forming head-to-head dimers. Therefore, the channel is lined with backbone carbonyl and amino groups, generating a hydrophilic environment inside the pore, and thus allows cations to flow through it. Its structure has been well-characterized by solution phase NMR (Arsen'ev et al., 1986), and it is readily available in large quantities. Consequently, it has been studied extensively, both experimentally (Andersen, 1984; O'Connell et al., 1990; CIFU et al., 1992) and theoretically (Roux and Karplus, 1993; Elber et al., 1995; Woolf and Roux, 1997; Allen et al., 2004; Andersen et al., 2005). In interesting experiments by Rostovtseva et al. (1998), single channel conductance was measured for GA in several types of lipid bilayers. The lipids used to form these bilayers were characterized by different molecular head groups, which have distinctly different electrostatic characteristics. Phosphatidylcholine (PC) and phosphatidylserine (PS) have dipolar head groups, while, in addition, PS can be charged (due to deprotonation of carboxyl groups on its surface). Busath et al. (Busath et al., 1998) performed similar studies on the uncharged dipolar diphytanoylphosphatidylcholine (DPhPC) membrane and on the uncharged nondipolar glycerilmonoolein (GMO) membrane. The data obtained from these experiments provide valuable information about the role of long-range electrostatic effects on ion permeation through functional protein channels.

13.4.2 Calculations

In an attempt to better understand these issues, an extensive set of 3D PNP ion permeation calculations was performed on the systems described above (Cárdenas et al., 2000). The geometric details of the lipid (thickness of the bilayer, perturbations of its structure at the regions of contact with the GA, etc.) were modeled based on known structural data (e.g., as obtained from NMR spectroscopy). Following standard arguments (Kurnikova et al., 1999), the dielectric constant of water (both in bulk solvent and in the aqueous pore) was taken to be 80, while that corresponding to protein and membrane regions was set to 2. Charges and dipoles were added as indicated in Fig. 13.4. The individual dipole magnitude and the surface density of dipoles for PC/PS are known. The surface density of titrating surfaces sites is also known. The degree to which these sites are protonated (and thus electrically neutral) or deprotonated (and thus, characterized by a charge of $-e_0$) is controlled by experimental conditions (solution pH and electrolyte concentrations): this is a "knob" which can be turned, experimentally, to control the degree of charging of the membrane surface from zero to a maximum (negative) surface charge density equal to the surface density of the titrating acidic (COOH) head groups.

Three-dimensional PNP results for I–V curves are presented in Figs. 13.5–13.7. The basic physical effects of membrane charge that explain the trends in these curves are as follows. For specified pH and bathing solution salt concentration, and for the same head group dipolar surface density, increasing the "bare" charge on the lipid (by deprotonation of COOH groups) increases the current through the channel. This is because Gramicidin is cation selective and negative surface charge helps attract positive ions in the bulk solution to the surface of the membrane, where they

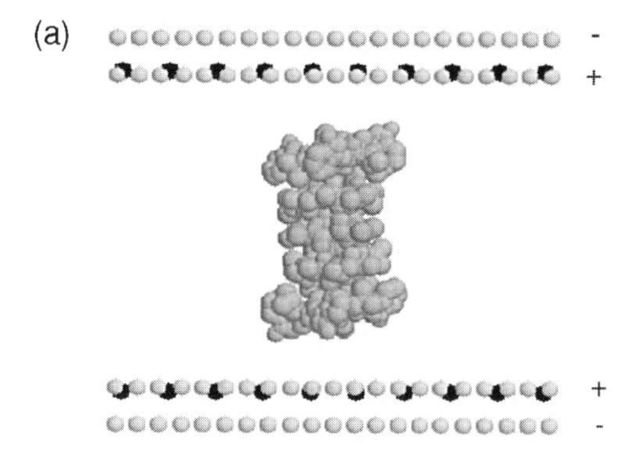

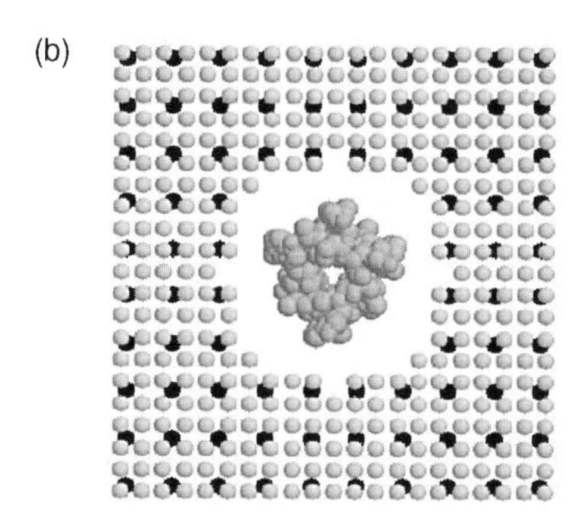

Fig. 13.4 Molecular representation of the GA dimer with negative charges (black) and dipoles (light gray) embedded in the membrane: (a) lateral view and (b) top view. The negative charges and the positive charges of the dipoles are placed inside the membrane [+ sign in (a)]. The negative charges of the dipoles are placed on the aqueous side of the membrane–liquid interface [−sign in (a)].

can be "sucked into" the mouth of the ion channel. In contrast, if all other factors are held constant, increasing the dipolar density decreases cation current because of the way the dipoles align with respect to the lipid surface: the positive part of the dipole is on the inside, and this positive surface charge presents a (mild) barrier to cation entry into the channel.

Figures 13.5–13.7 compare 3D PNP results to experiment for a variety of systems and conditions. In finalizing the computational output, there is one other critical parameter that needs to be set, namely the constant which governs diffusion within the channel. While the external (bulk) diffusion constants of ions like Na^+ and Cl^- are well known experimentally, the internal diffusion constant has not been measured. Because of the highly restricted motion of both ions and water in a narrow

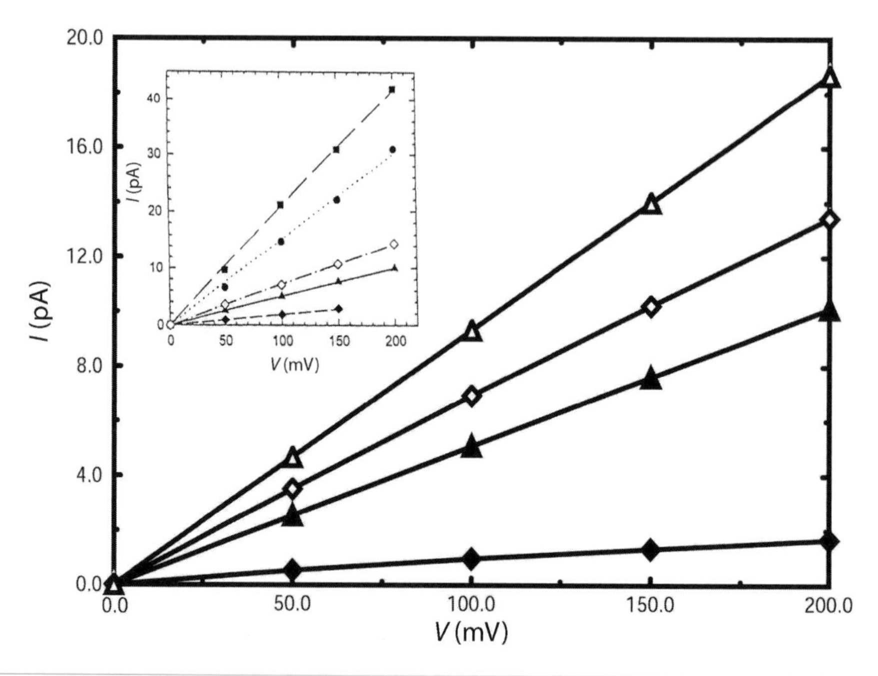

Fig. 13.5 Calculated current–voltage relationship for GA embedded in an uncharged PC membrane (solid symbols) and a charged PS membrane (open symbols) at neutral pH. The electrolyte concentrations are 0.1 M (diamonds) and 1.0 M (triangles). The inset shows experimental results (Rostovtseva et al., 1998) using the same symbol convention as in the main panel [it also shows experimental results at pH 1 (solid circles and squares)]. The value $D_{\text{int}} = 1.79 \times 10^{-6}$ cm^2/s was used in these PNP calculations.

channel like GA, it is plausible that the diffusion constant is significantly lower than its bulk value. There is some support for this conclusion from MD simulations in both artificial cavities (Lynden-Bell and Rasaiah, 1996) and in biological ion channels (Mamonov et al., 2003). In any case, it was found empirically that a value of $D_{\text{int}} \cong 1.5 \times 10^{-6}$cm^2/s for both anions and cations (about a factor of 10 less than the bulk value for K^+) leads to agreement between 3D PNP theory and experiment which is overall very good. Only in Fig. 13.7, at the high salt concentration of 2M (physiological concentrations of salt rarely exceed 1M), does PNP deviate from experiment significantly. Namely, the experimental current saturates (increasing the salt concentration does not increase the current), while the PNP current does not. This saturation effect is discussed in more detail below.

13.5 Incorporating Ion (De)Hydration Energy Effects into PNP: DSEPNP

It was recognized recently. that the change in solvation energy of a single ion when it moves in an inhomogeneous dielectric medium can provide an important contribution

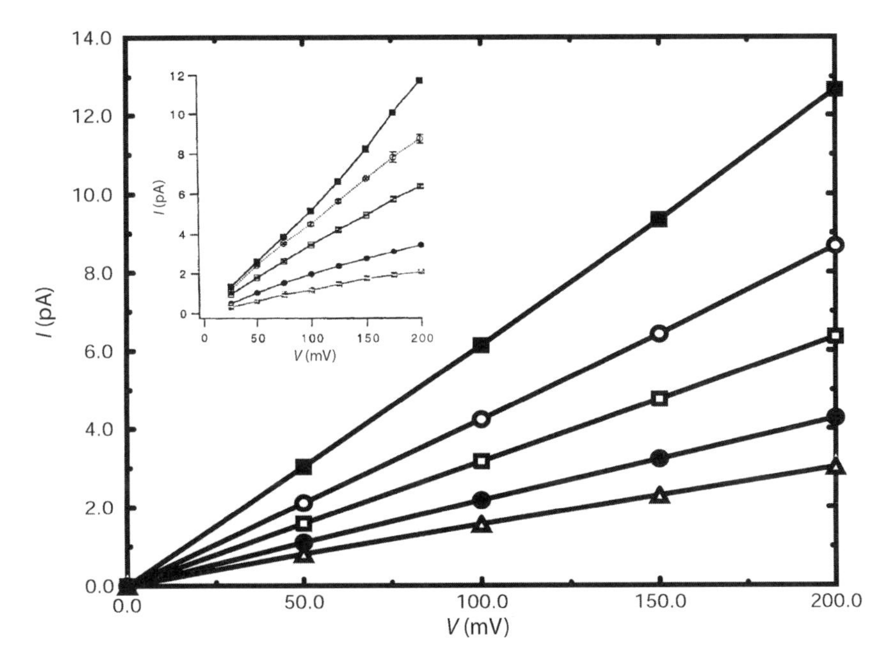

Fig. 13.6 Current–voltage relationship in GMO (nondipolar) membrane. The electrolyte concentrations are 0.1 (open triangle), 0.2 (closed circle), 0.5 (open square), 1.0 (open circle) and 2.0 M (closed square). The inset shows the experimental results (Busath et al., 1998) with the same symbol convention, except that these authors used dot-filled squares for 0.1 M. The value $D_{\text{int}} = 1.12 \times 10^{-6}$ cm^2/s was used in these PNP calculations.

to the drift flux term of Eq. 13.1 (Graf et al., 2000; Schuss et al., 2001) but is missing from the primitive PNP definition of $\Delta G^i_{\text{SIP}}(\vec{r})$. This change in the free energy of a single ion defined with respect to the free energy of that ion in a bulk solvent was termed the dielectric self-energy (or dielectric barrier) $\Delta G^i_{\text{DSE}}(\vec{r})$ (Graf et al., 2000, 2004). It can be calculated by solving the 3D PE for an ion with its center placed at one of the lattice points and, from the resultant electric potential field, evaluating the electrostatic energy of this point charge in the appropriate dielectric medium; the procedure is then repeated for all lattice points to map out the spatial dependence of the DSE [see (Dieckmann et al., 1999; Graf et al., 2000) for computational details]. When this contribution to the free energy is taken into account, $\Delta G^i_{\text{SIP}}(\vec{r})$ is modified to

$$\Delta G^i_{\text{SIP}}(\vec{r}) = q_i \phi_{\text{protein}}(\vec{r}) + \Delta G^i_{\text{DSE}}(\vec{r}). \tag{13.9}$$

Recent studies have shown that ΔG^i_{DSE} in a narrow channel strongly influences the resulting current (Graf et al., 2000, 2004). Therefore, a careful assessment of $\Delta G^i_{\text{SIP}}(\vec{r})$ is essential for modeling realistic channel behavior. PNP-like theory that implements $\Delta G^i_{\text{SIP}}(\vec{r})$ as defined in Eq. 13.4 will be termed Dielectric Self Energy–Poisson–Nernst–Planck (DSEPNP) theory (Graf et al., 2004). Comparison of 3D

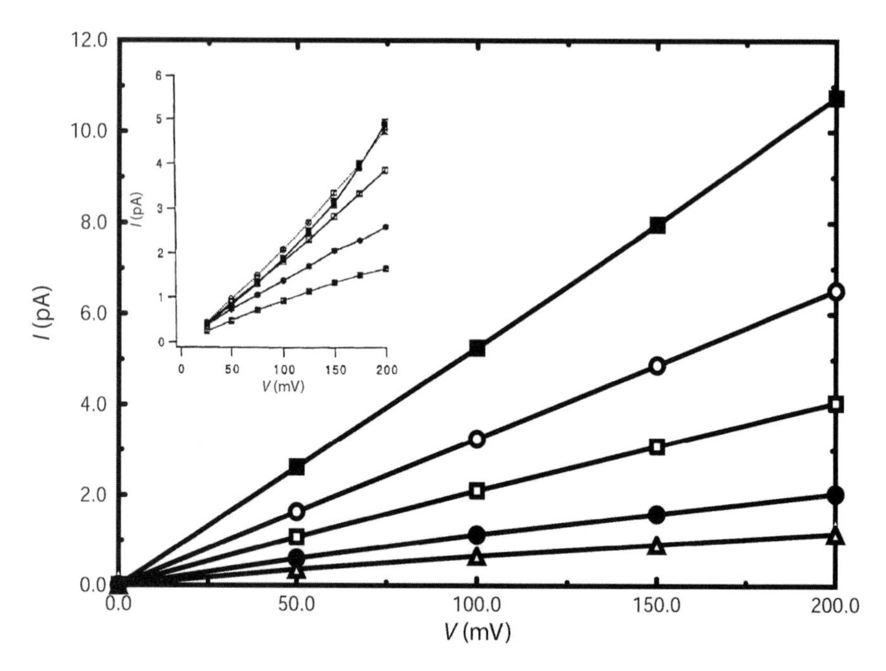

Fig. 13.7 Current–voltage relationship in DPhPC (dipolar) membrane. The symbol legend is the same as in the previous figure. The experimental results (Busath et al., 1998) are shown in the inset. The value $D_{\text{int}} = 1.12 \times 10^{-6}$ cm^2/s was used in these PNP calculations.

PNP and DSEPNP calculations to BD simulations in model cylindrical ion channel systems (cf. Fig. 13.8) shows that the simple procedure of adding the single-particle DSE term to the effective potential in the drift flux term of the PNP equations accounts, at least in these systems, for nearly the entire error inherent in the PNP. The BD model considers the same protein/membrane system, except that all the mobile ions are treated as spherical particles of finite size; water is again treated as a dielectric continuum, and the protein/membrane as an impenetrable dielectric medium. Each ion interacts with the electric field created by fixed ions in the protein/membrane slab, induced charge on dielectric boundaries, and with other ions in a pair-wise additive manner (this pair potential in general differs from a simple Coulomb potential: it is modulated by induced charge at dielectric interfaces), as well as with the external electric field generated by electrodes. Full details of the calculation of the instantaneous electrostatic force on each ion and the relevant kinetics algorithm used to produce the BD results shown in the present example may be found in (Graf et al., 2000).

Once the DSE potential is restored into PNP theory, the only "approximate" element in it, relative to a full many-ion BD simulation, is the mean-field approximation to ion–ion interactions. From the results shown in Figs. 13.9–13.10 it appears that this approximation is surprisingly accurate. Of course, further testing will be required to determine its full range of validity (Corry et al., 2003).

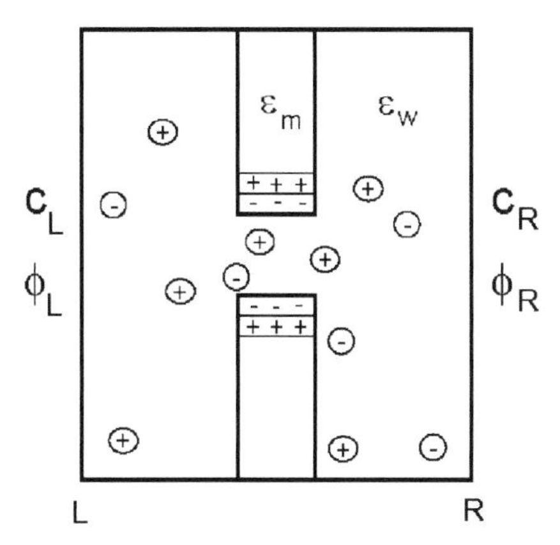

Fig. 13.8 2D cross section of the 3D BD simulation box depicting an assembly of free charges in a dielectrically inhomogeneous medium (ε_m = protein/membrane dielectric constant, ε_w = water dielectric constant). Note that some free charges (encircled) are mobile while others (in the dielectric region with ε_m) are fixed in space.

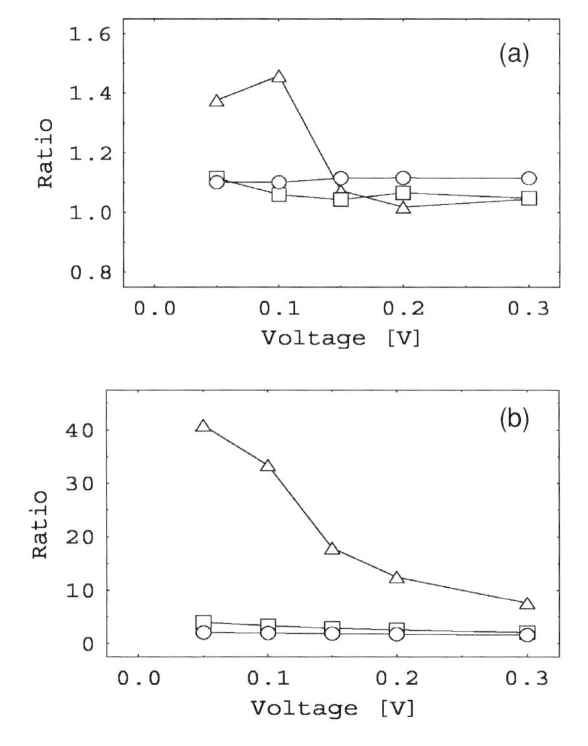

Fig. 13.9 (a) Ratio of DSEPNP/BD currents as a function of voltage for three channel radii: 0.4 nm (triangles); 0.75 nm (squares); 1.2 nm (circles); (b) Ratio of PNP/BD currents for the same channels, 0.4 nm (triangles); 0.75 nm (squares); 1.2 nm (circles).

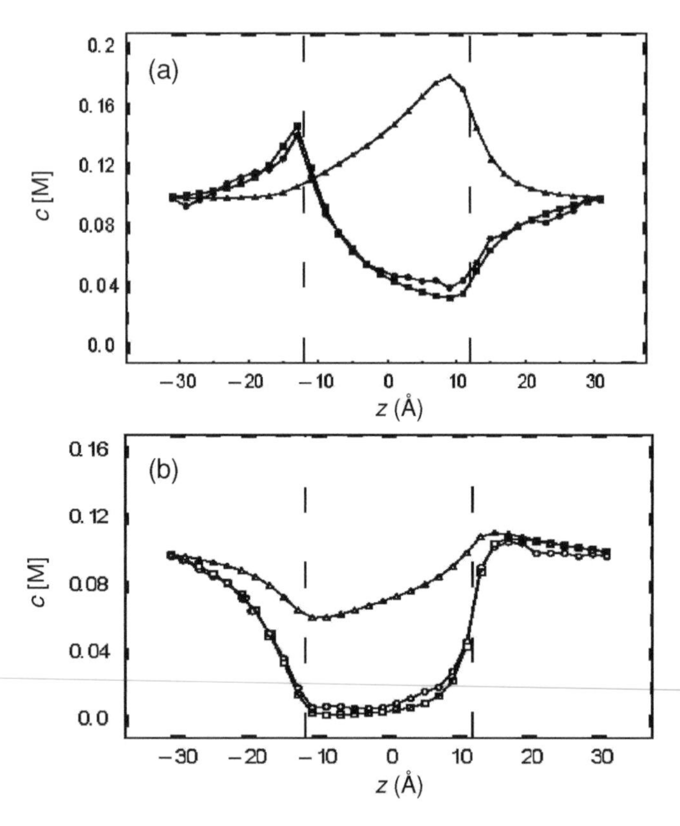

Fig. 13.10 Mobile ion concentrations calculated by BD, DSEPNP, and PNP plotted for the $R = 4$ Å channel at applied voltage of 0.3 V and 0.1 M reservoir concentration of salt. Circles show BD results, squares DSEPNP results, and triangles PNP results: (a) positive mobile ion concentration along the (channel) z-axis ($x = 0$, $y = 0$); (b) negative mobile ion concentration along z-axis. (Vertical dashed lines delineate membrane/channel boundaries.)

13.6 Incorporating Effects of Channel Protein Fluctuation in PNP: PMFPNP

13.6.1 Free Energy of Inserting an Ion into a Channel: General Considerations

In general, calculating free energy differences in biomolecular processes is a challenging task. Several approaches have been adopted for various problems in molecular modeling. These theoretical methodologies span a wide range of molecular resolution—from estimating electrostatic free energies on a continuum level by solving the PE (Dieckmann et al., 1999; Sharp and Honig, 1990; Luty et al., 1992) to full atomistic MD simulations (Roux and Karplus, 1993; Chung et al., 1998; Kollman et al., 2000). The electrostatic free energy of transferring an ion of species

i from the bulk solution into the channel is defined by

$$\Delta G^{i}_{\rm SIP}(\vec{r}) = G^{\rm complex}(\vec{r}) - G^{\rm protein} - G^{\rm ion}, \qquad (13.10)$$

where $G^{\rm complex}$ is the energy of an ion plus protein/membrane complex embedded in the solvent (water) with the ion located at a point $\vec{r}$ inside the channel, $G^{\rm protein}$ is the energy of the protein/membrane system (without the ion) embedded in the same solvent and $G^{\rm ion}$ is the energy of a single ion in the bulk solvent. The conventional continuum electrostatic approach for calculating $\Delta G^{i}_{\rm SIP}(\vec{r})$, based on Eq. 13.9, is reviewed in the next Section 13.6.2. A combined MD/continuum approach, which takes into account the channel flexibility, is presented in Section 13.6.3. Then, in Section 13.6.4, we present results of applying both methodologies and then simulate current through the GA channel.

13.6.2 A Continuum Approach to Calculate the Electrostatic Free Energy

In the absence of external fields, the electrostatic energy G of a collection of point charges can be found as $G = \frac{1}{2}\sum_i q_i \phi_i$, where the summation is over all electrostatic charges q_i in the system and ϕ_i is the value of the electrostatic potential at the position of charge i. The electrostatic potential $\phi(\vec{r})$ needed to calculate G can be obtained by solving the corresponding PE:

$$\vec{\nabla} \cdot (\varepsilon(\vec{r})\vec{\nabla}\phi(\vec{r})) = -4\pi \sum_j q_j \delta(\vec{r} - \vec{r}_j), \qquad (13.11)$$

supplemented by Dirichlet boundary conditions with the boundary potential set to zero. In Eq. 13.11, δ is the 3D Dirac delta-function and $\vec{r}_j$ is the position of charge q_j. As noted in the previous section, for channels as narrow as 4 Å in radius, a continuum description of ion permeation described by DSEPNP, i.e., Eqs. 13.1– 13.3, 13.9, compares well with results of BD simulations in which ions are treated as charged particles that diffuse in an inhomogeneous dielectric medium with a prescribed diffusion coefficient (Graf et al., 2000, 2004). Such particle-based simulation models of narrow rigid channels (Chung et al., 1999; Graf et al., 2000) typically exhibit very small superlinear currents for voltages up to 200 mV. The insignificance of these currents can be traced to the presence of a DSE barrier of several kT in such pores. In contrast, real biological channels of similar size and shape exhibit substantial ionic current at low voltages, with nearly linear or sublinear current–voltage characteristics. A detailed analysis of DSEPNP and BD particle simulations suggests that the effective polarizability of the channel environment (loosely defined as the ability of the local protein environment to adjust in order to stabilize an extra electric charge) must be higher than implied by the "standard" model utilized in both BD and DSEPNP studies. A major limitation of both approaches for simulating ion motions across channels is that the protein structure is taken to be rigid (usually at its average NMR or X-ray crystallographic configuration), while in reality the protein structure responds dynamically to an ion's presence. Below, we will investigate the consequences of the rigid protein assumption.

13.6.3 A Combined Molecular Dynamics/Continuum Electrostatics Approach to Calculate Free Energy

$\Delta G^{i}_{\text{SIP}}(\vec{r})$ can in principle be found from an atomistic simulation in which all atoms on the protein, the lipid membrane and the solvent are treated explicitly. Several attempts to calculate the free energy of an ion in a GA channel by MD simulation have been reported (Roux and Karplus, 1993; Woolf and Roux, 1997; Elber et al., 1995; Allen et al., 2004). Such calculations rely on a parameterized all-atom potential function (Elber et al., 1995; Roux and Berneche, 2002) and require complete sampling of the system configuration space. Improvements in the available parameterizations of potential functions have been slow in recent years (Roux and Berneche, 2002). Fortunately, an alternative method of dealing with this problem, namely limited sampling of the environment configurational space, has recently been introduced (Kollman et al., 2000). Since a large portion of the configuration space required for quantitative calculation of the free energy of an ion in a solvent is due to the solvent itself, it was recently proposed (Kollman et al., 2000) that the computationally expensive sampling of solvent configurations may be replaced by considering solvent effects via an appropriate approximate averaging procedure. In this approach a full-scale equilibrium MD trajectory of the protein in an atomistic solvent is generated to sample the protein conformational space (with and without an ion in the channel). The resulting sequence of N protein/water configurations is used to obtain a corresponding sequence of dielectric continuum models of these systems, in which the fixed protein charges are embedded at the appropriate atomic positions. These continuum dielectric configurations, obtained with the permeating ion fixed in a given position, are then used to compute the electrostatic free energy of inserting the ion at that position (Sharp and Honig, 1990). Adapting the procedure introduced in (Kollman et al., 2000), the free energy of ion–protein complex formation for ion species i is calculated as an average over all $n = 1, \ldots, N$ configurations:

$$\Delta G^{i}_{\text{SIP}} = \frac{1}{N} \sum_{n=1}^{N} \Delta G^{i(n)}_{\text{SIP}}, \tag{13.12}$$

where $\Delta G^{i(n)}_{\text{SIP}}$ has the same meaning as in Eq. 13.10, calculated for the nth configuration. The method thus combines an MD simulation to obtain atomistic configurations of the membrane–protein–ion complex with a continuum dielectric representation of each configuration in order to obtain a simple estimate of $\Delta G^{i(n)}_{\text{SIP}}$ for that configuration, followed by the average indicated in Eq. 13.12. This approach allows us to account for solvent effects on average, i.e., at a mean field level, and to reduce the noise in the free energy calculations due to insufficient sampling of solvent configurations. The procedure described above, in which the PMF ΔG_{SIP} is calculated via Eq. 13.12 and then used in the PNP formalism, will be termed PMF Poisson–Nernst–Planck (PMFPNP). We should note that this calculation still disregards contributions to the free energy due to changes in the protein internal energy and accounts only approximately (through the temperature dependence of the dielectric functions) for entropic contributions. These missing contributions are expected to be small because deformation of the protein is minimal during the ion permeation (see Section 13.6.4), and because the changes in configurational entropy in these processes are typically small. (A similar number of degrees of freedom are constrained independent of the ion position in the channel).

13.6.4 MD/Continuum Simulation of an Ion in the GA Channel

The approach outlined above was implemented in a series of calculations performed for a model GA channel. Figure 13.11 shows a 3D GA ion channel structure incorporated into a crude model of a lipid bilayer membrane, with the membrane/protein channel system solvated in water. This snapshot is taken from an MD simulation performed as described below. As has been noted above, .the dielectric self-energy is very large for channels less than 5 Å in radius, implying the conundrum discussed above in modeling their permeability. Working with GA, the narrowest known ion channel, emphasizes the goal of understanding the permeability of such narrow channels (Dieckmann and DeGrado, 1997; Roux and MacKinnon, 1999; Graf et al., 2000; Mamonov et al., 2003).

A set of MD simulations of i) a single potassium ion and ii) a single chloride ion fixed at various positions in a GA channel was performed. GA was incorporated into a slab of heavy (mass = 100 au) spheres with Lennard–Jones parameters ε = 0.05 kcal/mol and $R_M = 2.5$ Å, and no partial charge. The slab of these dummy spheres represents a lipid bilayer by providing a nonpolar environment for the channel molecule. This channel-membrane model system was then immersed in a box of 738 SPC/E water molecules (Leach, 2001). Eight water molecules in random configurations were placed inside the GA pore. This system was subjected to energy minimization followed by a 200 ps constant pressure MD equilibration run at 300 K. Positions of the dummy atoms and GA atoms were constrained in space with 200 kcal/mol/Å^2 harmonic spring forces. After the GA-water equilibration was completed, an ion (K^+or Cl^-) was introduced into the channel. A force constant of 200 kcal/mol/ Å^2 was again applied to the positions of the dummy atoms and a 10 kcal/mol/Å^2 force constant was applied to the backbone atoms of the GA. The energy of each system thus prepared was minimized, followed by a 30 ps equilibration period when the harmonic constraints on the GA backbone atoms were gradually reduced from 10 kcal/mol/Å^2 to 0.5 kcal/mol/Å^2. Subsequently, 300 ps production runs were performed with constant volume dynamics at 300 K. 0.5 kcal/mol/ Å^2 harmonic constraints were maintained on each of the backbone C and N atoms of GA. The coordinate of the ion along the channel axis (z-axis) was held fixed, while its x,y coordinates were allowed to fluctuate. The coordinates of the protein atoms

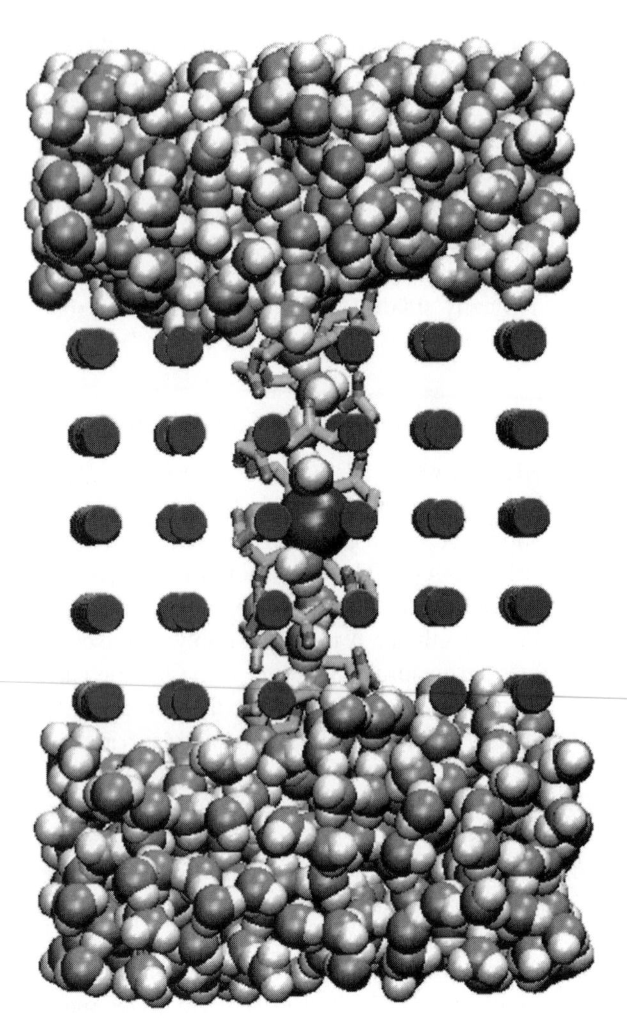

Fig. 13.11 Snapshot of the GA channel with a K^+ ion embedded in a model membrane and solvated with water after a 300 ps MD simulation as described in text. The model lipid bilayer is represented by an array of flat grey spheres (the radius of such sphere in a picture does not reflect its Lennard–Jones parameters). The K^+ ion is shown as the large grey sphere in the center of the channel. Water atoms are also shown as spheres with oxygens and hydrogens colored in grey and white respectively. Only backbone atoms of the peptide chains are shown in stick representation. [For a color version of this figure, see Ref. (Mamonov et al., 2003).]

were collected every 2 ps. For every such time point along the MD trajectory the coordinates of the protein molecule and the ion were used to calculate the appropriate electrostatic free energy by solving the PE as described above.

An MD trajectory of GA without K^+ or Cl^- was also generated as described above. All MD simulations were performed using the AMBER 6 software package and Cornell et al. force field (Cornell et al., 1995). The Lennard–Jones parameters for the potassium ion were taken from work of (Åquist, 1990). Bonds involving hydrogen atoms were constrained via the SHAKE algorithm. A 12 Å cut-off distance was used for all nonbonded interactions. The MD time step was set to 2 fs.

For the continuum electrostatics calculations, partial charges on the GA atoms were taken from the Cornell et al. force field (Cornell et al., 1995). The dielectric response profile $\varepsilon(\vec{r})$ and the positions of the partial charges represent the molecular system in a continuum representation. In this study, the dielectric constant of the membrane and the protein was set to 4, while the value characterizing both the bulk solvent and the aqueous pore was taken as 80 [for an extensive discussion of how these parameters were chosen, consult (Mamonov et al., 2003)]. In the numerical solution of the PE (Eq. 13.3), these functions are discretized on a uniform 3D grid as described in (Kurnikova et al., 1999). The radii of potassium and chlorine ions, estimated by fitting experimental enthalpies of hydration, were chosen to be $R_{K^+} = 2.17$ Å (Dieckmann et al., 1999) and $R_{K^-} = 1.81$ Å (Dasent, 1982). For all results reported below, the grid dimensions of the simulation box were 151^3 with a linear scale of 3 grid points per Å. The width of the membrane was set to 33 Å to mimic a GMO bilayer. The set of calculations described above was repeated with the potassium ion fixed at 18 different positions along one GA monomer at spatial increments of 1 Å, and the chloride ion fixed at seven different positions at spatial increments of 3 Å.

The results of the MD/free energy calculations outlined above show the following basic features. Since GA is a tightly-bound β-helical structure, it is not surprising that the overall structure of peptide, embedded in an artificial lipid bilayer, does not change significantly over the course of the MD trajectory. Consequently, the DSE contribution to the overall single-particle free energy of insertion (PMF, or in the present notation, SIP) does not vary much over the course of the MD simulation. It remains close to the value associated with the static average protein configuration and provides, as expected, a large energy barrier (of about $20kT$) to the passage of ions of either sign. A more interesting finding is that small local distortions of pore-lining parts of the peptide (especially carbonyl groups) significantly stabilizes cations as they move through it; cf. Fig. 13.12. This large energetic stabilization is possible because electrostatic forces are strong, and the permeating ion is very close to partially charged groups of the protein that face the aqueous pore—hence, changes in the positions of these protein groups by only fractions of an Ångstrom can change the direct Coulomb interaction with the ion by many kT. The situation here is reminiscent of polaron formation in polar crystalline solids (Kittel, 1996). In that phenomenon, an electron migrating through the crystal distorts the lattice of ions that define the crystal locally and instantaneously (on the slow time scale of the electron's migration) in such a way as to lower the overall energy of the system (electron + lattice) and thus stabilize it. The resultant species, an electron surrounded by displaced positive ions (with the negative ions distorting away from the electron), is termed a polaron. Its conduction and optical properties are altered by the "phonon cloud" that surrounds it. The case of an ion moving through a narrow protein channel is somewhat similar; again, cf. Fig. 13.12. In this system, polaronic stabilization of cations approximately cancels out the DSE barrier and allows them to permeate

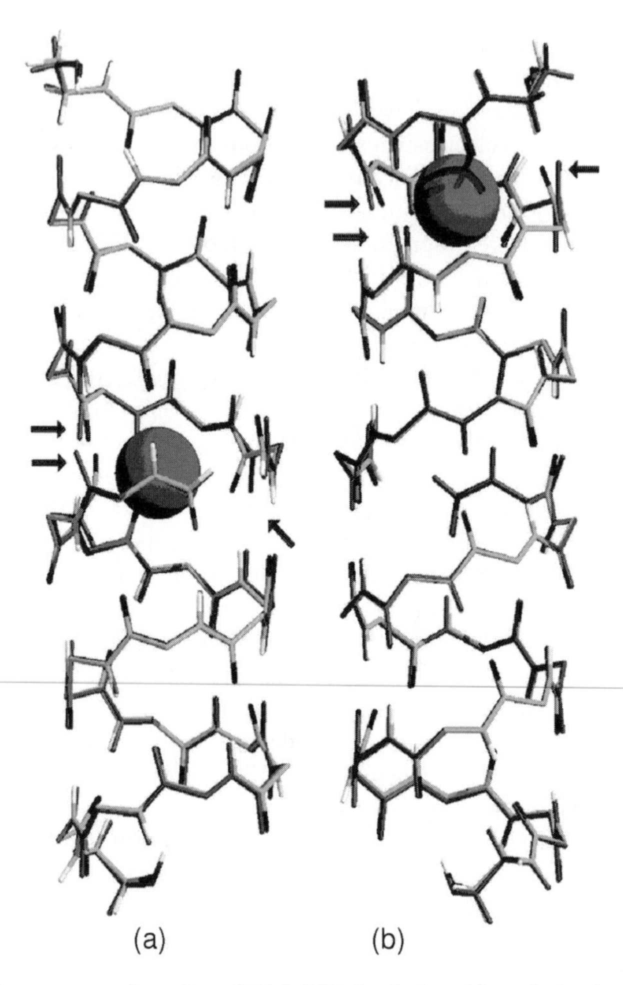

Fig. 13.12 The average configuration of GA in MD simulation without the ion (mono color grey) is superimposed with the average configuration of GA with the K^+ ion present in the simulation (mainly grey with backbone oxygens colored in black and nitrogens in white). K^+ is shown as a large sphere. Arrows indicate the carbonyl oxygens that bend toward the K^+ due to favorable electrostatic interactions. During the MD simulation a K^+ ion was (a) in the center of the channel, (b) at 9 Å from the center of the channel, the predicted position of the cation binding site. [For a color version of this figure, see Ref. (Mamonov et al., 2003).]

with relative ease (as was found in primitive PNP calculations which omitted both effects!).

13.6.5 Application of PMFPNP to Calculate Ion Currents Through the GA Channel

Some computed PMFPNP I–V curves are shown in Fig. 13.13. The currents obtained in these calculations are typically within a factor of two of currents measured

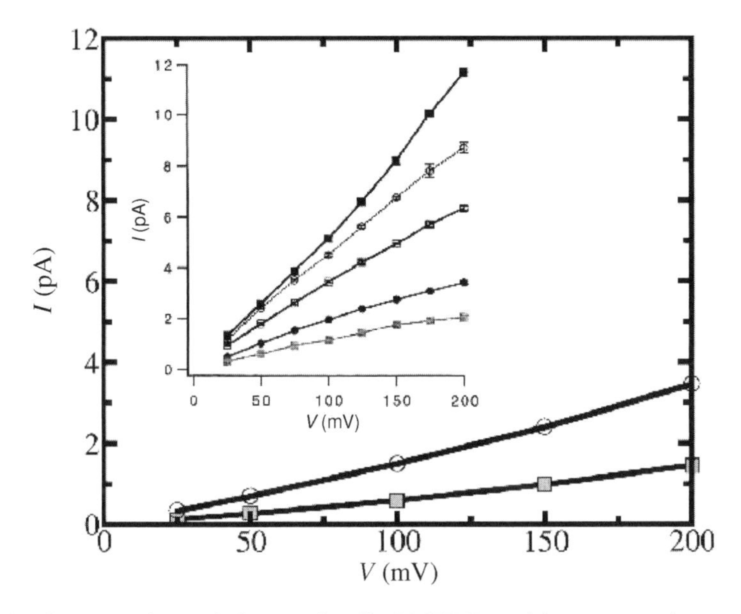

Fig. 13.13 Current–voltage relations predicted by PMFPNP model are compared to experimental results (Busath et al., 1998) (upper left inset). Bulk KCl concentrations of 0.1 M (shaded square) and 1.0 M (open circle) were used in the simulations. The experimental curves in the inset correspond to the following concentrations of bulk KCl solutions: shaded square—0.1 M, filled circle—0.2 M, open square—0.5 M, open circle—1.0 M, and filled square—2.0 M. The analogous experimental and calculated curves are labeled with the same symbols.

experimentally (Busath et al., 1998). By adjusting the internal diffusion constant an even better (nearly perfect) fit could have been obtained. However, in these calculations, the internal ion diffusion constants were actually calculated by processing MD simulation data (based on all-atom simulations of the protein and water, plus one ion fixed inside the channel) in a standard fashion. The calculated reduction of the internal diffusion from its bulk value is comparable to that obtained by fitting ("reverse-engineering") the internal diffusion constant (Edwards et al., 2002) to obtain agreement with experimental GA data, as discussed above.

A final important result of this study is the demonstration that PMFPNP theory is able to account for effects that are beyond the reach of primitive PNP theory, namely, saturation of ion current through the channel as the concentration of bathing solutions increases to a sufficiently high value (see Fig. 13.14). Physically, once the channel becomes filled up with ions, the dwell time of these ions before they escape from the exit side of the pore becomes the rate determining factor, rather than the rate of attempted entry into the channel (the latter being proportional to the concentration of ions in the external reservoir on the entry side, while the former is independent of this concentration). In PNP theory, a build-up of positive charge density in the channel would be expected to generate an electric field that prevents other positive ions from entering the channel. However, in primitive PNP this positive charge has the unphysical effect of attracting negative mobile

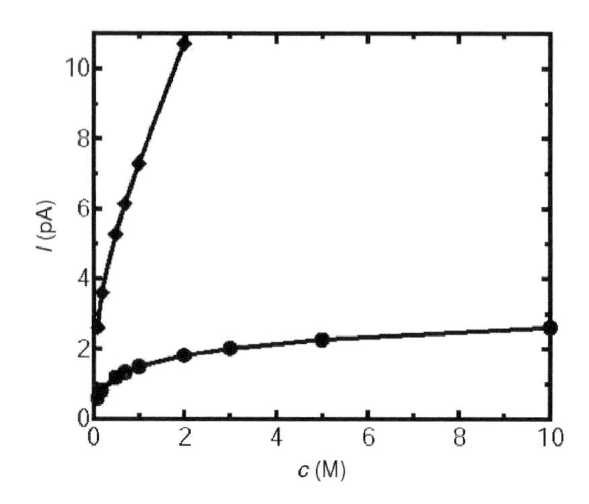

Fig. 13.14 Current–concentration relations as predicted by PNP (diamonds) and PMFPNP (circles) models. The external potential difference was set to 100 mV.

charge into the channel. Note the behavior of the anions shown in Fig. 13.15c: at low electrolyte concentration anions do not enter the pore, while at high bulk electrolyte concentration (10 M), the apparent concentration of Cl^- in parts of the channel exceeds 5 M. As positive and negative charge build up in the same region (again, an unphysical process), the aqueous region of the pore is effectively rendered charge neutral, thus allowing more positive charge to enter and flow through it—the higher the bathing concentration of ions, the higher the rate of ion permeation (Fig. 13.14). In particular, Fig. 13.16a,c shows that at high concentrations the potential drop across the channel becomes roughly linear, implying that positive and negative mobile ions pile into the channel in such a way as to cancel out all electrostatic driving forces except for the applied voltage. Now we are back to the 1D NP model (essentially the GHK model discussed in Section 13.2.3), which predicts ion current proportional to bathing solution concentration. This is exactly what we see in Fig. 13.14.

In PMFPNP, by contrast, the DSE added to the unfavorable protein–anion interaction potential forms a very high barrier to anion entry into the channel—even the build-up of positive ion density cannot compensate for this (see Fig 13.16d). Anions never enter into the channel (Fig. 13.15d), while cation charge continues to build up as the bathing solution salt concentration is increased (notice in Fig. 13.15b that unlike primitive PNP result [Fig. 13.15a] cation concentration build-up happens only at particular locations in the channel, which can be loosely regarded as cation binding sites), until the tendency toward greater cation flow rate into the channel with increasing electrolyte concentration is counterbalanced by the electrostatic repulsion generated by cations already in the channel—the density profile of cations in the channel saturates, ultimately causing the saturation behavior in current flow illustrated in Fig. 13.14. This saturation mechanism can be appreciated by examining

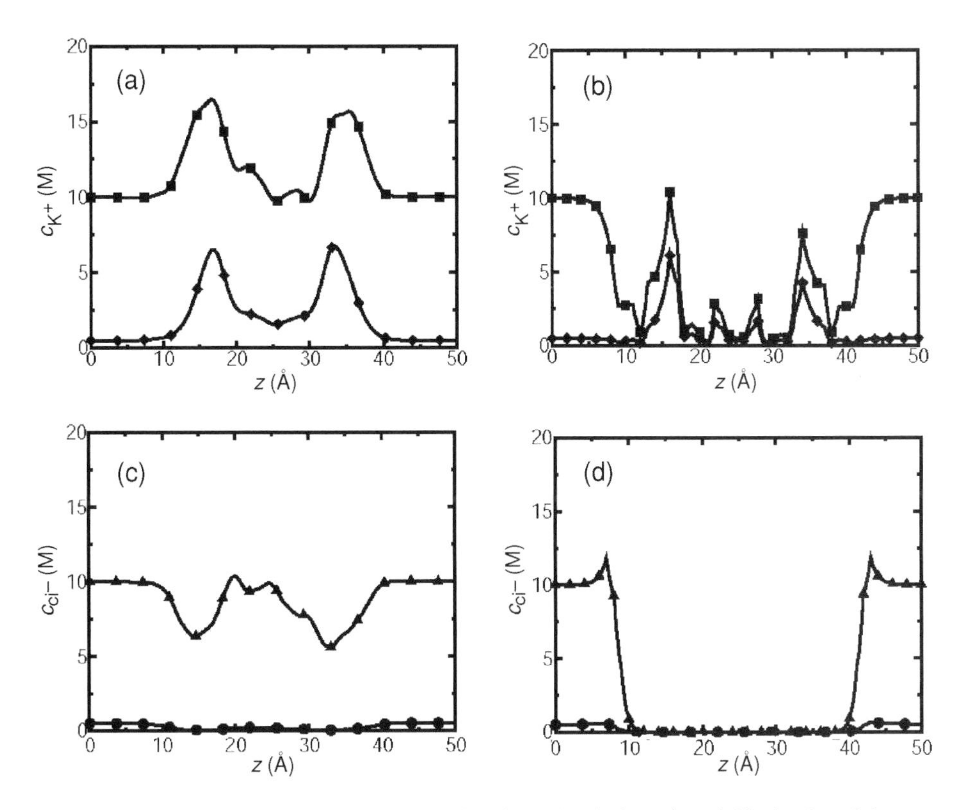

Fig. 13.15 Ion concentration profile along the channel axis for K^+ and Cl^- is plotted for two (high and low) bulk electrolyte concentrations: (a), (c) calculated using PNP; (b), (d) calculated using PMFPNP. The curves with diamonds and circles are for 0.5 M, the curves with squares and triangles are for 10 M electrolyte concentrations.

the concentration dependence of the effective driving potentials seen by anions vs. cations, as illustrated in Fig. 13.16a,c for primitive PNP and in Fig. 13.16b,d for PMFPNP.

13.7 Conclusions and Outlook

Three-dimensional PNP Theory has an intuitive appeal due to its conceptual simplicity. It relies on a caricature of the microscopic world in which background media are treated as dielelectric slabs and the primary particles of interest, mobile ions like Na^+ and Cl^-, are "smeared out" into a continuous charge distribution. The polarization of this mobile charge distribution in response to concentration gradients in boundary reservoirs and electrostatic forces arising from both internal (ion–ion, ion–protein, etc.) and external (electrode-generated) sources is described in terms of drift–diffusion equations. These are coupled naturally to the PE of electrostatics which must be utilized to calculate the relevant electric fields (self-consistently

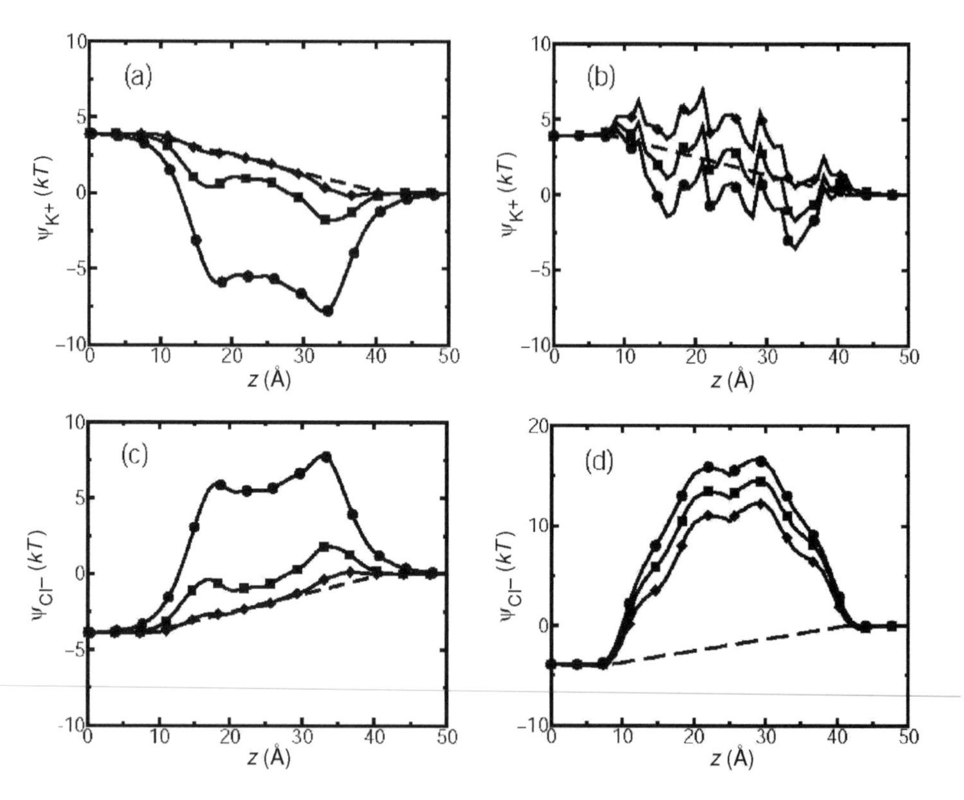

Fig. 13.16 $\psi_i(\vec{r})$ profile along the channel axes for K^+ and Cl^- is plotted for several bulk electrolyte concentrations and 100 mV applied voltage: (a), (c) calculated using PNP; (b), (d) calculated using PMFPNP. The curve with circles is for 0 M, the curve with squares is for 0.5 M and the curve with diamonds is for 10 M electrolyte concentrations. The dashed line is the result of a calculation in which the protein molecule has no partial charges on the atoms. It corresponds to the linear ramp potential caused by the high resistivity of the membrane.

with the solution of the drift–diffusion equations). This "engineering flavor" is transferred to numerical solution techniques, e.g., PDE solvers relying on well-developed finite difference or finite element methods. As advances in these techniques (such as variable meshes and multigridding (Beck, 1997; Tsonchev et al., 2004)) become available, 3D PNP solvers will surely become fast enough that they can be distributed as software, analogous, say, to the FEMLAB program (FEMLAB, 2004) that solves standard PDEs of many types. In the context of understanding structure–function relations in biological ion channels, 3D PNP solvers may soon serve as a Computer Assisted Design (CAD) tool, which allows the user to vary inputs and get accurate output solutions (within the range of validity of PNP theory) quickly. One can imagine high throughput scanning, for example, of the effects on current flow through an ion channel due to changes in critical amino acids. Such calculations could then guide experimentalists who wish to generate the mutated channels (using site-directed mutagenesis techniques) to alter channel function.

Before getting carried away with the potential uses of 3D PNP solvers, we should bear in mind the inherent restrictions of the theory. The caricature outlined in the preceding paragraph may simply be unrealistic for treating certain properties of certain ion channels. Many ion channels have a narrow segment through which ions can only flow in single file fashion (this is a common feature of ion channel selectivity filters, for example). These regions are characterized by electrostatic traps (binding sites), where ions reside, one ion per binding site, temporarily. A site-bound ion is then knocked out of its binding site (and through the channel) by another ion entering the channel. These kinds of mechanisms are not included in primitive PNP. We have seen above that the consequences of many such effects (e.g., saturation of ion current with increasing bathing solution concentration) can in fact be attained from modified versions of PNP, but this may be due to judicious cancellation of errors in a description which is fundamentally inadequate to accurately describe the underlying dynamics.

Despite these concerns, we expect that in years to come PNP type theories will continue to play a useful role in computing and understanding the kinetics of ion permeation through biological channels, especially in wider channels (Im and Roux, 2002; Noskov et al., 2004), synthetic channels [e.g., based on carbon nanotubes (Hummer et al., 2001)], etc. Consider the fate of other continuum electrostatic theories which describe dynamical processes in solution, such as the Born theory of solvation (Dill and Bromberg, 2003) and the Marcus theory of electron transfer (Marcus, 1956, 1965). Although they are much criticized for not possessing sufficient microscopic detail, they have proven remarkably robust in estimating, semi-quantitatively, the complicated phenomena that they were developed to model, with only back of the envelope calculations. Thus they remain invaluable to the present day. We suspect that the same will prove true of electrodiffusion theories, and PNP in particular, for understanding ion permeation through nano-pores.

Acknowledgments

We thank our collaborators A. Nitzan, P. Graf, A. Cárdenas, A. Mamonov, and M. Cheng for sharing their insights on this subject with us over the years. Our work on 3D PNP theory and its application to ion permeation through biological channels has been supported by grants from PRF, NSF, NIH, and ARO.

References

Aksimentiev, A. and K. Schulten. 2005. Imaging alpha-hemolysin with molecular dynamics: Ionic conductance, osmotic permeability, and the electrostatic potential map. *Biophys. J.* 88:3745–3761.

Allen, T.W., O.S. Andersen, and B. Roux. 2004. Energetics of ion conduction through the gramicidin channel. *Proc. Natl. Acad. Sci. USA* 101:117–122.

Åquist, J. 1990. Ion water interaction potentials derived from free-energy perturbation simulations. *J. Phys. Chem.* 94:8021–8024.

Andersen, O.S. 1984. Gramicidin channels. *Ann. Rev. Physiol.* 46:531–548.

Andersen, O.S., R.E. Koeppe, and B. Roux. 2005. Gramicidin channels. *IEEE Trans. Nanobiosci.* 4:10–20.

Arsen'ev, A.S., A.L. Lomize, I.L. Barsukov, and V.F. Bystrov. 1986. Gramicidin A transmembrane ion-channel. Three-dimensional structure reconstruction based on NMR spectroscopy and energy refinement. *Biol. Membr.* 3:1077–1104.

Barcilon, V. 1992. Ion flow through narrow membrane channels. 1. *SIAM J. Appl. Math.* 52:1391–1404.

Barcilon, V., D.P. Chen, and R.S. Eisenberg. 1992. Ion flow through narrow membrane channels. 2. *SIAM J. Appl. Math.* 52:1405–1425.

Beck, T.L. 1997. Real-space multigrid solution of electrostatics problems and the Kohn-Sham equations. *Int. J. Quantum Chem.* 65:477–486.

Bockris, J.O., and A.K.N. Reddy. 1998. Modern Electrochemistry 1: Ionics. Premium Press, New York.

Burykin, A., C.N. Schutz, J. Vill·, and A. Warshel. 2002. Simulations of ion current in realistic models of ion channels: The KcsA potassium channel. *Proteins Struct. Funct. Genet.* 43:265–280.

Busath, D.D., C.D. Thulin, R.W. Hendershot, L.R. Phillips, P. Maughan, C.D. Cole, N.C. Bingham, S. Morrison, L.C. Baird, R.J. Hendershot, M. Cotten, and T.A. Cross. 1998. Noncontact dipole effects on channel permeation. I. Experiments with (5F-indole)trp(13) gramicidin A channels. *Biophys. J.* 75:2830–2844.

Cárdenas, A.E., R.D. Coalson, and M.G. Kurnikova. 2000. Three-dimensional Poisson–Nernst–Planck theory studies: Influence of membrane electrostatics on gramicidin A channel conductance. *Biophys. J.* 79:80–93.

Chandler, D. 1987. Introduction to Modern Statistical Mechanics. Oxford University Press, New York, 274pp.

Chandrasekhar, S. 1943. Stochastic problems in physics and astronomy. *Rev. Mod. Phys.* 15:1–89.

Chen, D.P., and R. Eisenberg. 1993a. Charges, currents, and potentials in ionic channels of one conformation. *Biophys. J.* 64:1405–1421.

Chen, D.P., and R.S. Eisenberg. 1993b. Flux, coupling, and selectivity in ionic channels of one conformation. *Biophys. J.* 65:727–746.

Cheng, M.H., M. Cascio, and R.D. Coalson. 2005. Theoretical studies of the M2 transmembrane segment of the glycine receptor. Models of the open pore structure and current–voltage characteristics. *Biophys. J.* 88:110a.

Chung, S.H., T.W. Allen, M. Hoyles, and S. Kuyucak. 1999. Permeation of ions across the potassium channel: Brownian dynamics studies. *Biophys. J.* 77:2517–2533.

Chung, S.H., T.W. Allen, and S. Kuyucak. 2002. Conducting-state properties of the KcsA potassium channel from molecular and Brownian dynamics simulations. *Biophys. J.* 82:628–645.

Chung, S.H., M. Hoyles, T. Allen, and S. Kuyucak. 1998. Study of ionic currents across a model membrane channel using Brownian dynamics. *Biophys. J.* 75:793–809.

Chung, S.H., and S. Kuyucak. 2002. Recent advances in ion channel research. *BBA Biomembr.* 1565:267–286.

Cifu, A.S., R.E. Koeppe, and O.S. Andersen. 1992. On the superamolecular organization of gramicidin channels—the elementary conducting unit is a dimer. *Biophys. J.* 61:189–203.

Coalson, R.D., and T.L. Beck. 1998. Numerical Methods for Solving Poisson and Poisson–Boltzman type Equations. John Wiley & Sons, New York, pp. 2086–2100.

Coalson, R.D., and M.G. Kurnikova. 2005. Poisson–Nernst–Planck theory approach to the calculation of current through biological ion channels. *IEEE Trans. Nanobiosci.* 4:81–93.

Cornell, W.D., P. Cieplak, C.I. Bayly, I.R. Gould, K.M. Merz, D.M. Ferguson, D.C. Spellmeyer, T. Fox, J.W. Caldwell, and P.A. Kollman. 1995. A 2nd generation force-field for the simulation of protein, nucleic-acid, and organic-molecules. *J. Am. Chem. Soc.* 117:5179–5197.

Corry, B., S. Kuyucak, and S.H. Chung. 2003. Dielectric self-energy in Poisson–Boltzmann and Poisson–Nernst–Planck models of ion channels. *Biophys. J.* 84:3594–3606.

Crozier, P.S., D. Henclerson, R.L. Rowley, and D.D. Busath. 2001a. Model channel ion currents in NaCl-extended simple point charge water solution with applied-field molecular dynamics. *Biophys. J.* 81:3077–3089.

Crozier, P.S., R.L. Rowley, N.B. Holladay, D. Henderson, and D.D. Busath. 2001b. Molecular dynamics simulation of continuous current flow through a model biological membrane channel. *Phys. Rev. Lett.* 86:2467–2470.

Dasent, W.E. 1982. Inorganic Energetics, 2nd Ed. Cambridge University Press, New York.

Dieckmann, G.R., and W.F. DeGrado. 1997. Modeling transmembrane helical oligomers. *Curr. Opin. Struct. Biol.* 7:486–494.

Dieckmann, G.R., J.D. Lear, Q.F. Zhong, M.L. Klein, W.F. DeGrado, and K.A. Sharp. 1999. Exploration of the structural features defining the conduction properties of a synthetic ion channel. *Biophys. J.* 76:618–630.

Dill, K.A., and S. Bromberg. 2003. Molecular Driving Forces: Statistical Thermodynamics in Chemistry and Biology. Garland Science, New York. 666pp.

Doi, M. 1996. Introduction to Polymer Physics. Clarendon Press, Oxford. 120pp.

Edwards, S., B. Corry, S. Kuyucak, and S.H. Chung. 2002. Continuum electrostatics fails to describe ion permeation in the gramicidin channel. *Biophys. J.* 83:1348–1360.

Elber, R., D. Rojewska, D.P. Chen, and R.S. Eisenberg. 1995. Sodium in gramicidin – an example of a permion. *Biophys. J.* 68:906–924.

FEMLAB: Multiphyics modeling. 2004. http://www.comsol.com/products/femlab.

Gillespie, D., L. Xu, Y. Wang, and G. Meissner. 2005. (De)constructing the ryanodine receptor: Modeling ion permeation and selectivity of the calcium release channel. *J. Phys. Chem. B* 109:15598–15610.
Gilson, M.K., T.P. Straatsma, J.A. McCammon, D.R. Ripoll, C.H. Faerman, P.H. Axelsen, I. Silman, and J.L. Sussman. 1994. Open "back door" in a molecular dynamics simulation of acetylcholinesterase. *Science* 263:1276–1278.
Graf, P., M. Kurnikova, R. Coalson, and A. Nitzan. 2004. Comparison of dynamic lattice Monte Carlo simulations and the dielectric self-energy Poisson–Nernst–Planck continuum theory for model ion channels. *J. Phys. Chem. B* 108:2006–2015.
Graf, P., A. Nitzan, M.G. Kurnikova, and R.D. Coalson. 2000. A dynamic lattice Monte Carlo model of ion transport in inhomogeneous dielectric environments: Method and implementation. *J. Phys. Chem. B* 104:12324–12338.
Hille, B. 1992. Ionic Channels of Excitable Membranes. Sinauer Associates, Cambridge, NY. 814pp.
Hollerbach, U., D.P. Chen, D.D. Busath, and B. Eisenberg. 2000. Predicting function from structure using the Poisson–Nernst–Planck equations: Sodium current in the gramicidin A channel. *Langmuir* 16:5509–5514.
Hummer, G., J.C. Rasaiah, and J.P. Noworyta. 2001. Water conduction through the hydrophobic channel of a carbon nanotube. *Nature* 414:188–190.
Im, W., and B. Roux. 2002. Ions and counterions in a biological channel: A molecular dynamics simulation of OmpF porin from *Escherichia coli* in an explicit membrane with 1 M KCl aqueous salt solution. *J. Mol. Biol.* 319:1177–1197.
Ise, N., and H. Yoshida. 1996. Paradoxes of the repulsion-only assumption. *Acc. Chem. Res.* 29:3–5.
Kittel, C. 1996. Introduction to Solid State Physics, 7th Ed. John Wiley & Sons, New York, 665pp.
Kollman, P.A., I. Massova, C. Reyes, B. Kuhn, S.H. Huo, L. Chong, M. Lee, T. Lee, Y. Duan, W. Wang, O. Donini, P. Cieplak, J. Srinivasan, D.A. Case, and T.E. Cheatham. 2000. Calculating structures and free energies of complex molecules: Combining molecular mechanics and continuum models. *Acc. Chem. Res.* 33:889–897.
Koneshan, S., R.M. Lynden-Bell, and J.C. Rasaiah. 1998. Friction coefficients of ions in aqueous solution at 25 degrees C. *J. Amer. Chem. Soc.* 120:12041–12050.
Kurnikova, M.G., R.D. Coalson, P. Graf, and A. Nitzan. 1999. A lattice relaxation algorithm for three-dimensional Poisson–Nernst–Planck theory with application to ion transport through the gramicidin A channel. *Biophys. J.* 76:642–656.
Leach, A.R. 2001. Molecular Modelling: Principles and Applications, 2nd Ed. Prentice Hall, Harlow, England. 744pp.
Luty, B.A., M.E. Davis, and J.A. McCammon. 1992. Solving the finite-difference non-linear Poisson–Boltzmann equation. *J. Comp. Chem.* 13:1114–1118.
Lynden-Bell, R.M., and J.C. Rasaiah. 1996. Mobility and solvation of ions in channels. *J. Chem. Phys.* 105:9266–9280.

Mamonov, A.B., R.D. Coalson, A. Nitzan, and M.G. Kurnikova. 2003. The role of the dielectric barrier in narrow biological channels: A novel composite approach to modeling single-channel currents. *Biophys. J.* 84:3646–3661.
Mamonov, A.B., M.G. Kurnikova, and R.D. Coalson. 2006. Diffusion constant of K^+ inside Gramicidin A: A comparative study of four computational methods. *Biophys. Chem.* doi:10.1061/j.bpc.2006.03.019.
Marcus, R.A. 1956. On the theory of oxidation–reduction reactions involving electron transfer. I. *J. Chem. Phys.* 24:966–978.
Marcus, R.A. 1965. On the theory of electron-transfer reactions. VI. Unified treatment for homogeneous and electrode reactions. *J. Chem. Phys.* 43:679–701.
Marion, J.B. 1965. Classical Electromagnetic Radiation. Academic Press, New York, 479pp.
Mashl, R.J., Y.Z. Tang, J. Schnitzer, and E. Jakobsson. 2001. Hierarchical approach to predicting permeation in ion channels. *Biophys. J.* 81:2473–2483.
McQuarrie, D.A. 1976. Statistical Mechanics, Harper Collins Publishers, New York, 641pp.
Noskov, S.Y., W. Im, and B. Roux. 2004. Ion permeation through the alpha-hemolysin channel: Theoretical studies based on Brownian dynamics and Poisson–Nernst–Plank electrodiffusion theory. *Biophys. J.* 87:2299–2309.
O'Connell, A.M., R.E. Koeppe, and O.S. Andersen. 1990. Kinetics of gramicidin channel formation in lipid bilayers—transmembrane monomer association. *Science* 250:4985.
Peyghambarian, N., S.W. Koch, and A. Mysyrowicz. 1993. Introduction to Semiconductor Optics. Prentice Hall, Englewood Cliffs, NJ.
Press, W.H., B.P. Flannery, S.A. Teukolsky, and W.T. Vetterling. 1986. Numerical Recipes. Cambridge University Press, New York, 848pp.
Rostovtseva, T.K., V.M. Aguilella, I. Vodyanoy, S.M. Bezrukov, and V.A. Parsegian. 1998. Membrane surface-charge titration probed by gramicidin A channel conductance. *Biophys. J.* 75:1783–1792.
Roux, B., and S. Berneche. 2002. On the potential functions used in molecular dynamics simulations of ion channels. *Biophys. J.* 82:1681–1684.
Roux, B., and R. MacKinnon. 1999. The cavity and pore helices the KcsA K^+ channel: Electrostatic stabilization of monovalent cations. *Science* 285:100–102.
Roux, B., and M. Karplus. 1993. Ion transport in the Gramicidin channel: Free energy of the solvated ion in a model membrane. *J. Am. Chem. Soc.* 115:3250–3260.
Schuss, Z., B. Nadler, and R.S. Eisenberg. 2001. Derivation of Poisson and Nernst–Planck equations in a bath and channel from a molecular model. *Phys. Rev. E* 64:036116:1–14.
Sharp, K.A., and B.H. Honig. 1990. Electrostatic interactions in macromolecules—theory and applications. *Ann. Rev. Biophys. Biophys. Chem.* 19:301–332.
Sten-Knudsen, O. 1978. Passive transport processes. *In:* Membrane Transport in Biology. G. Giebish, D.C. Tosteson, and H.H. Ussing, editors. Springer-Verlag, New York, pp. 5–114.

Tsonchev, S., R.D. Coalson, A. Liu, and T.L. Beck. 2004. Flexible polyelectrolyte simulations at the Poisson–Boltzmann level: A comparison of the kink-jump and multigrid configurational-bias Monte Carlo methods. *J. Chem. Phys.* 120:9817–9821.

Wang, Y., L. Xu, D.A. Pasek, D. Gillespie, and G. Meissner. 2005. Probing the role of negatively charged amino acid residues in ion permeation of skeletal muscle ryanodine receptor. *Biophys. J.* 89:256–265.

Wissenburg, P., T. Odjik, P. Cirkel, and M. Mandel. 1995. Multimolecular aggregation of mononucleosomal DNA in concentrated isotropic solutions. *Macromol* 28:2315–2328.

Woolf, T.B., and B. Roux. 1997. The binding site of sodium in the gramicidin A channel: Comparison of molecular dynamics with solid-state NMR data. *Biophys. J.* 72:1930–1945.

14 A Mesoscopic–Microscopic Perspective on Ion Channel Permeation Energetics: The Semi-Microscopic Approach[1]

Peter C. Jordan

14.1 Introduction

Understanding how physiological ion channels simultaneously exhibit the apparently contradictory properties of high throughput and great discrimination is a long-standing theoretical problem. These nanodevices all operate on the same basic principle: ions, solvated by bulk water, lose a significant part of their hydration shell as they pass through a constriction where a chemical selection process occurs (Hille, 2001). High throughput requires that the chosen ion faces no significant energy barrier, which would forbid its entry. On first blush, it seems that falling into a deep well is also forbidden, since that would apparently trap it in the channel and block further passage. While generally true, some channels function in multi-ion mode, so that they are permanently ion-occupied; permeation then occurs with the entry of a second (or third) ion, repelling the prior occupant and leading to conduction. In all instances, high selectivity requires that there is a mechanism by which all other physiologically prevalent ions face significant energetic discrimination.

The theoretical tools for studying permeation are of two basic types. Continuum approaches describe ions by Poisson–Boltzmann theory, as charge distributions with associated ion atmospheres, in essence extending the standard description of electrodiffusion to account for the influence of bulk electrolyte. This Poisson–Nernst–Planck (PNP) theory is computationally efficient (Chen et al., 1992; Eisenberg, 1999). It directly assesses how changes in a protein's electrical structure and physical geometry affect channel conductance. However, it is parameter rich, requiring input assumptions regarding channel geometry, ionic diffusion coefficients and the local channel dielectric constant, $\varepsilon(\mathbf{r})$, all of which hard to substantiate. A local ε is an exceptionally nebulous concept; it cannot be rigorous, since the dielectric constant is a generalized susceptibility (Partenskii and Jordan, 1992). In nonuniform surroundings, it is not surprising that different measures of electrical response lead to different ε estimates (Jordan et al., 1997). The PNP approach is a reliable tool for wider channels (Im and Roux, 2002). For narrow, selective channels, in addition to

[1] Portions reprinted, with permission, from P.C. Jordan, "Semimicroscopic Modeling of Permeation Energetics in Ion Channels," *IEEE Transactions on Nanobioscience*, 4:94–101 [2005], IEEE.

properly choosing ε, there is another serious concern. A permeant ion cannot carry an ion atmosphere with it into the interior of a narrow channel (Corry et al., 2000; Moy et al., 2000).

Brownian dynamics (BD) is an alternate way to directly compute I–V profiles (Chung et al., 2002a). Here ions are discrete, and a channel's charge distribution effectively forbids counterion entry into selective confines. Ions move in a field determined by protein charges and are impeded by viscous drag due to the surrounding (continuum) water. The influence of bulk electrolyte is relatively easily taken into account. As with PNP, BD requires assumptions with respect to a channel's physical and dielectric geometry. Carefully parameterized it is a valuable correlational tool for analyzing electrical behavior in narrow channels (Chung et al., 1999, 2002b; Corry et al., 2004).

Given the enormous advances in computational power, and the undoubted success of computational chemistry in the analysis of the properties of (water soluble) proteins, molecular dynamics (MD) would appear ideal for relating structure and function. However, realizing this goal has been difficult, in part reflecting technical problems, in part for more fundamental reasons. Even with massively parallel computing, direct monitoring of ion transit through a highly selective channel is not yet possible. To simulate an ion channel embedded in a phospholipid membrane and sandwiched between electrolyte layers typically involves some 10^5 atoms; heroic computations follow dynamics for $\leq$100 ns, easily an order of magnitude less than the fastest ion transit times. Efforts to circumvent these difficulties in the prototype potassium channel, KcsA, have been made using a microscopic–mesoscopic approach (Burykin et al., 2002). Current has however been directly simulated in the large bore system, α-hemolysin, where simulations of 1 ns duration are adequate for reliable statistics (Aksimentiev and Schulten, 2005). Most often, rather than attempting to directly observe ion translocation, MD is used to compute the free energy for transferring an ion from electrolyte into the channel, the so-called potential of mean force (PMF) (Bernèche and Roux, 2003). Combined with an ionic diffusion coefficient (also obtained from MD) a PMF can then indirectly determine the ion channel conductance (Allen et al., 2004) .

However, this protocol is only as reliable as the underlying PMF. Gramicidin, a structurally superbly well-characterized, extremely narrow ion channel (Ketchem et al., 1997), is a case where MD misleads. In this system, which accommodates ~9 water molecules in a linear array, either of two conventional force fields yields a translocational free energy barrier for potassium passage of ~50 kJ mol^{-1} at 300 K; in 0.1 M electrolyte this implies a single-channel conductance of ~10^{-7} pS, about seven orders of magnitude less than that observed (Allen et al., 2003). What may have gone wrong? Simulations are never totally faithful representations of macroscopic reality. Only a thin layer of electrolyte is simulated, but a technical trick, the use of periodic boundary conditions, to a large extent compensates for this limitation. More problematical are the force fields themselves. They are classical substitutes for quantum mechanics. Atoms are assigned fixed effective partial charges, determined by a variety of parameterization methods. But during permeation ions pass close to channel moieties, polarizing charge distributions. Water models most clearly illustrate the possible difficulties in ion channel simulation. They are designed to reproduce behavior of bulk liquid water, where four or five near neighbors typically surround each molecule. Their charge distributions, with dipole moments of ~2.2–2.4 Debye, reflect mean liquid properties. As water's gas phase dipole moment is 1.85 Debye, water–water interaction has substantially altered the molecules' charge distributions. In a narrow ion channel like gramicidin, where water molecules line up single file, the environment is vastly different from the network structure of bulk water. Here polarization may be expected to have affected water's charge distribution differently, with uncertain consequences. In other words, the more the simulational surroundings differ from ambient water, the less reliable a conventional, nonpolarizable water model is likely to be.

The semi-microscopic (SMC) approach is designed to circumvent many of these problems (Dorman et al., 1996; Dorman and Jordan, 2003, 2004; Garofoli and Jordan, 2003). It is a conceptually transparent simplification of biophysical reality, one that is readily extendable. It exploits the time scale separation between electronic and structural contributions to dielectric stabilization, dealing with electronic polarization by embedding the channel in surroundings that account for polarization in a mean sense. The reorganizational contribution to dielectric stabilization is ascribed to a limited set of mobile entities: ions, water, and selected peptide moieties. This approach forms a bridge between continuum electrostatic modeling and MD, avoiding two of MD's major limitations: force fields that don't account for electronic polarization; potential electrostatic artifacts introduced by periodic boundary conditions. It permits ready deconvolution of individual moieties' influence on permeation free energies and relatively easily contrasts the behavior of different permeant species, thus directly addressing selectivity. Advantages and limitations of the technique are described. Methodological details are outlined and applied to three convenient dielectric geometries. Practical aspects of the SMC procedure are explained, highlighting some areas ripe for further development. Finally, some specific applications are considered.

14.2 Electronic Polarization and the Dielectric Background

The physical observation central to the SMC method is that dielectric reorganization, which is responsible for the solvating ability of polar media, cleanly separates into structural and electronic components. This is always true, regardless of whether a system is uniform (like bulk water) or massively nonuniform (like the surroundings of ion channel selectivity filters). Ligand response to permeant species movement is described by structural terms. Electronic terms account for permeation-induced reorganization of the solvent (here water, peptide, and lipid) charge distribution. The two time scales differ vastly. Electrons rearrange in femtoseconds, while structural relaxation occurs over a wide time span. But even the fastest such contributions, due to translation and vibration, are at least 100 times slower. Water is typical. At low frequencies (<100 GHz) it exhibits bulk dielectric properties. At higher frequencies water rotates or translates too slowly to respond to electrical stress; in effect these degrees of freedom are frozen and ε drops to its optical value, ~ 1.8 (Hasted, 1973). In an essentially Born–Oppenheimer-like approximation structural change takes place with the electron distribution relaxed. A first approximation treats the electronic background as uniform, with a roughly universal polarizability representative of lipids (Fettiplace et al., 1971) as well as water; in essence molecular realignment occurs under conditions where $\varepsilon \sim 2$.

The SMC treatment exploits this time scale separation, extending semi-microscopic approaches discussed in the study of Helmholtz layers at electrode–electrolyte interfaces, where explicit reorientable electrical sources are placed in interfacial background dielectrics with an assigned ε, $\sim$2–3 (Kornyshev, 1988). Our SMC work accounts for electronic polarization by locating the mobile moieties in a dielectric background with $\varepsilon = 2$. It treats structural influences to dielectric shielding by explicitly including those charged features of the peptide–lipid–water ensemble that interact strongly with the permeant. As these contributions to charge stabilization arise from translation and rotation of charged and polar elements in the surroundings (the generalized solvent), a sufficient number of these must be mobile for this simplification to be effective. Thus, for the SMC approach to be more generally applicable requires extensive parameterizations, similar to those familiar in constructing the force fields of computational chemistry (van Gunsteren and Berendsen, 1987; Cornell et al., 1995; MacKerell et al., 1998).

14.3 Long-Range Electrostatics and Model Geometries

Standard MD simulations impose severe computational loads. In order to model both cytoplasmic and extracellular water, the channel–lipid assembly is surrounded by extended regions of aqueous solvent; simulating these is generally the most costly part of the computation. Due to boundary effects, simulations with as many as $\sim 10^5$ atomic centers may still not be representative of bulk. Periodic boundary conditions are used to reduce the associated errors; however as electrostatic forces are long-range, great care is required to correct for electrical cutoffs in treating the reaction field.

The SMC approach avoids this problem by embedding the membrane–peptide–ion channel ensemble in electrical geometries where generalized image methods rigorously determine the reaction field. This is the crux of the technique, and is described in detail. Figure 14.1 illustrates a specific SMC simplification, one designed to approximate the potassium channel selectivity filter. It is based on the structure of the KcsA potassium channel (Doyle et al., 1998), and generates a tractable computational model. Explicitly modeled features are in the interior, low ε region. The

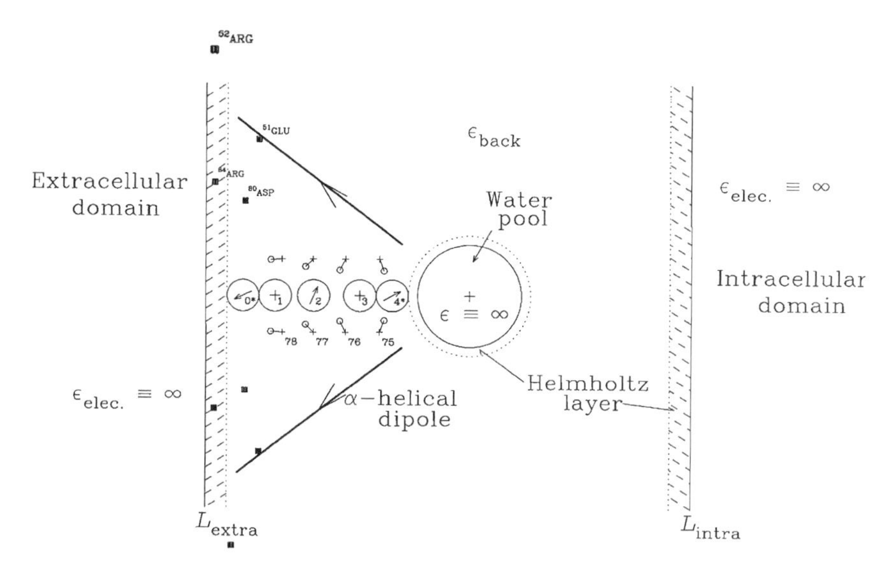

Fig. 14.1 SMC model of an ion, water, KcsA filter, lipid, and electrolyte system (Garofoli and Jordan, 2003). The planar projection shows single file ions and waters, the coordinating carbonyl groups, peptide dipoles, selected charged residues, the aqueous cavity, and an included ion. Ions, waters, and carbonyl oxygens (numbered 75–78) are mobile. Helmholtz layers account for water orientationally immobilized by interaction with polar and/or ordered species at dielectric boundaries; explicit sources are forbidden from penetrating these regions. The pool accommodates ~20 water molecules. Numbering follows the convention of Morais-Cabral et al. (2001). Reprinted from Garofoli and Jordan (2003) with permission.

water pool approximates a major feature of potassium channels, their mid-membrane water-filled cavities. Sharp electrical boundaries demarcate the different regions.

Three tractable computational geometries are illustrated in Fig. 14.2. As for the case of Fig. 14.1, the low permittivity, ε_1 domain contains all explicitly modeled charges; the ε_2 surroundings dielectrically mimic (or closely approximate) bulk water. Figure 14.2a, with its parallel boundaries, is simplest and exactly treatable. The reaction field for a charge q, located at $\mathbf{R} = z\mathbf{k}$ in a slab of width W, is rigorously described by a set of oscillating image charges, their locations determined by successive reflection at alternating planar boundaries,

$$q_n = (-)^n[(\varepsilon_2 - \varepsilon_1)/(\varepsilon_2 + \varepsilon_1)]q, \quad \mathbf{R}_n^{\pm} = [\pm nW + (-)^n z]\mathbf{k}, n > 0; \quad (14.1)$$

from these, the charge's polarization energy and the reaction field contribution to interaction with other explicit charges is found by summing the coulombic interaction energy of every explicit charge with the whole image set.

Reaction field energetics for the Fig. 14.2b and c geometries only has approximate image equivalents. However, in all applications of interest, the two dielectric domains are vastly dissimilar, $\varepsilon_2 \gg \varepsilon_1$, and simplification, by assuming that $\varepsilon_2 \equiv \infty$, introduces insignificant error (Dorman and Jordan, 2003). In this limit,

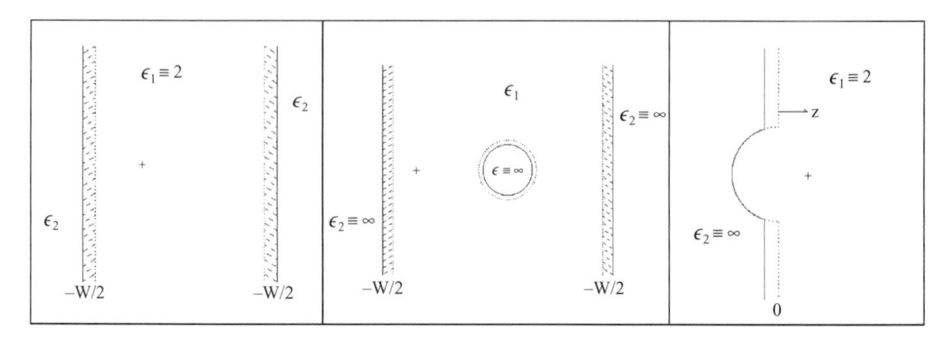

Fig. 14.2 Three electrical geometries used in SMC studies. Explicit electrical sources are in the low permittivity region; for specificity a single positive source is included. All geometries yield analytical expressions for the reaction field. (a) Slab of width W. (b) Slab of width W with a water-like cavity. (c) Semi-infinite bubble geometry. Hatched regions in (a) and (b), and region between solid and dotted lines in (c) are Helmholtz layers (see text and Fig. 14.1 caption). Reprinted from Jordan (2005) with permission.

image methods are again reliable. The geometry of Fig. 14.2b includes the potassium channel's mid-channel water pool (Doyle et al., 1998), illustrated in Fig. 14.1. Instead of explicitly describing the ~20–25 water molecules that occupy this region, this inner domain is assigned a high permittivity, $\varepsilon \rightarrow \infty$. Images for a charge at $\mathbf{R}$ in the ε_1 region are generated iteratively, and must include the influence of the two bulk electrolyte regions as well as that of the water pool. For the geometry illustrated, a sphere of radius a_o, centered in the middle of a slab of width W, the first images arising from interaction with bulk water are the solutions of Eq. 14.1 for $n = 1$ and $\varepsilon_2 \equiv \infty$. Interaction with the spherical cavity creates an image charge q' at $\mathbf{S}$,

$$q' = -q(a_o/R), \qquad \mathbf{S} = (a_o/R)^2\mathbf{R}. \tag{14.2}$$

As image interaction makes the cavity effectively dipolar, not charged, a compensating charge is required at the cavity center.

An explicit charge generates four primary images, one in each bulk region and a dipolar pair in the cavity. Repeating the procedure, each of these generates further images with oscillating polarity. Bulk phase images spawn three new images, one in the other bulk phase and two in the cavity, one of which is at the origin. Each cavity image spawns two new ones, one in each bulk phase. The total number grows very rapidly and doesn't sum to closed form. However, the cavity dipolar images attenuate rapidly; only the first two must be considered. The series of purely planar images oscillates rapidly; beyond tenth order they can be ignored (Garofoli and Jordan, 2003).

The Fig. 14.2c geometry was designed for a much different purpose. SMC model geometries have sharp electrical discontinuities between continuum, bulk-like dielectric regions and the low permittivity, ε_1 domain, which harbors the explicit, mobile charges. This is problematical for two reasons: charges near a boundary interact directly with molecules in the bulk solvent domain in ways that are sensitive to structural details; in reality electrical boundaries are never sharp. Neither is accounted for by the model geometries of Fig. 14.2a and b. The closer a charge is to a boundary, the more its reaction field is subject to discontinuity-related artifacts and the less faithfully it reflects ensemble averaged details of interaction between a charged or polar moiety and nearby molecular bulk water. Consequently, we introduced Helmholtz layers as buffer domains (Kornyshev, 1988), but these do not fully eliminate errors arising when sources approach an electrical discontinuity.

An obvious approach is to ensure that charged species exiting the pore or near bulk water domains are fully hydrated. This can be done in two ways. Either insert a large explicit water region between the membrane–protein assembly and the dielectric boundary, which increases the computational load substantially (Miloshevsky and Jordan, 2004), or graft a small water-filled bubble, large enough to fully hydrate an ion at or near the pore mouth, onto the channel mouth. This is illustrated in Fig. 14.2c (Dorman and Jordan, 2004). It is particularly simple; an explicit charge has a single electrical image because there is only one electrical dividing surface. However, the geometry is semi-infinite and only useful if, as is often the case, selectivity and permeation are controlled by structures on one side of the assembly. Fig. 14.3 maps it into an equivalent electrical problem, a right-angled wedge, where the high ε domain subtends one-quarter of the space, for which the electrical potential can be computed (Dorman and Jordan, 2004). The SMC approach, by introducing buffer domains and limiting the computational system to those few water molecules and peptide groups explicitly solvating the ion(s), shares similarities with constraint methods introduced by Warshel (Warshel, 1979; King and Warshel, 1989). Our simplifying decomposition, treating a few solvent molecules explicitly and modeling the surroundings as dielectric continua, also strongly resembles an approach developed by Roux (Beglov and Roux, 1994); however, they don't exploit time scale separation, assigning the region containing explicit moieties an ε of 1.

This reaction field problem is difficult and doesn't have a simple solution like Eqs. 14.1 or 14.2. The confocal transformation of Fig. 14.3 turns the hemispherical bubble separating low and high ε regions into a right-angled wedge, which is an exactly soluble electrostatic problem. R_{elec} is the bubble's electrical radius and the transformation is based on the invariant point where the dashed circle is tangent to the bubble. The electrical equivalents of points on or below the dividing surface are found by inversion through a surface defined by the dashed circle of radius $2R_{\mathrm{elec}}$, centered at point I. The upper panel translates a generic point $\mathbf{r} = (x, y, z)$ along x by $-R_{\mathrm{elec}}$; the translated vector, $\mathbf{r}_{\mathrm{T}} = (x + R_{\mathrm{elec}}, y, z)$ is then inverted, a procedure which requires charge and potential rescaling (Dorman and Jordan, 2003; Jackson, 1962). The inverted vector is

$$\mathbf{r}_{\mathrm{I}} = (2R_{\mathrm{elec}})^2 \mathbf{r}_{\mathrm{T}} / |r_{\mathrm{T}}|^2; \tag{14.3}$$

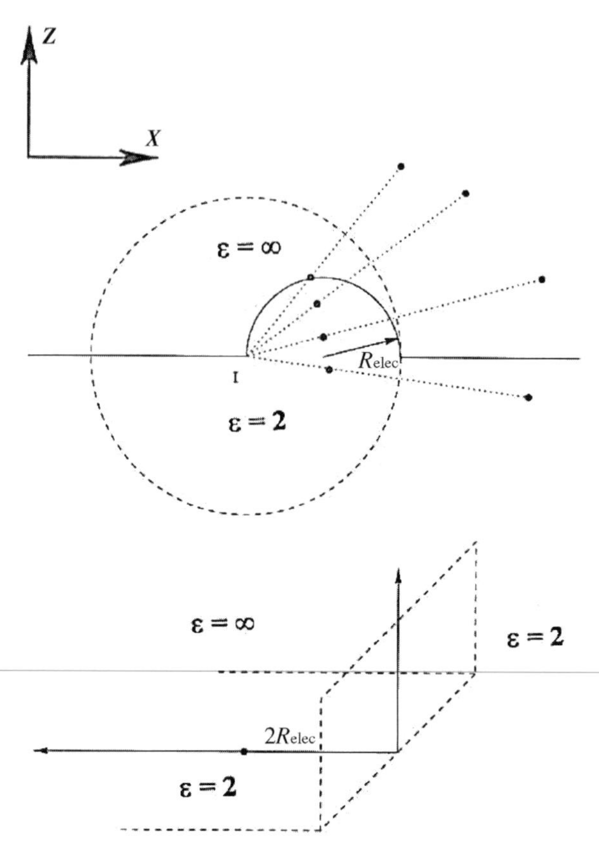

Fig. 14.3 Mapping of bubble geometry of Fig. 14.2c into a wedge. The point I is the inversion center for the transformation. After transformation points on the bubble surface lie on the plane at $x = 2R_{\text{elec}}$. The invariant point is where the bubble intersects the inversion sphere. Reprinted from Dorman and Jordan (2004) by permission.

charge and potential rescaling yields

$$q_{\text{I}} = -2R_{\text{elec}}q/r_{\text{T}}, \quad \varphi_{\text{I}} = (r_{\text{T}}/2R_{\text{elec}})\varphi. \tag{14.4}$$

The x–z projection of the new geometry is illustrated in the lower panel. The hemisphere and the original plane become semi-infinite planes, the new surfaces are $(x_{\text{I}} = 2R_{\text{elec}}, -\infty < y_{\text{I}} < +\infty, z_{\text{I}} > 0)$ and $(x_{\text{I}} < 2R_{\text{elec}}, -\infty < y_{\text{I}} < +\infty, z_{\text{I}} = 0)$, respectively. Representative transformed points within the region containing all explicit electrical sources, the hemisphere ($z > 0$) and cylinder ($-R_{\text{elec}} < \rho < R_{\text{elec}}$, $z < 0$), are shown in the upper panel. Points in the low ε region map into exterior points. Points on the bubble map to points on the surface $x = 2R_{\text{elec}}$. In this planar projection the high ε region is the second quadrant. The original bubble geometry is equivalent to a wedge, a solved electrical problem (Smythe, 1968).

If the angle subtended by the low ε region is $\beta = \pi/\mathrm{p}$, the potential at (ρ, α, z) from an electrical source at (ρ_0, α_0, z_0) is

$$\varphi_{\mathrm{p}} = \pi^{-1}[rr_0]^{-1/2} \int_{\eta}^{\infty} dx\, f(p, x, \alpha, \alpha_0) K(x, \eta), \tag{14.5}$$

where η, the kernel $K(x, \eta)$ and the function $f(p, x, \alpha, \alpha_0)$ are respectively

$$\cosh \eta = \left[r^2 + r_0^2 + (z - z_0)^2\right]/2rr_0, \tag{14.6}$$

$$K(x, \eta) = (\cosh x - \cosh \eta)^{-1/2}, \tag{14.7}$$

$$\begin{aligned} f(p, x, \alpha, \alpha_0) = {} & p \sin h \ \sinh px[(\cosh px - \cos p(\alpha - \alpha_0))^{-1} \\ & - (\cosh px - \cos p(\alpha + \alpha_0))^{-1}]. \end{aligned} \tag{14.8}$$

Here, $\beta = 3\pi/2$ and $p = 2/3$. While Eq. 14.5 is well defined, its integrand has singularities requiring careful handling. In determining the source's "self-energy," $[r, z, \alpha] \to [r_0, z_0, \alpha_0]$ where the denominator of the first term in Eq. 14.8 approaches zero. This requires integration over the putative singularity before effecting the limit $[r, z, \alpha] \to [r_0, z_0, \alpha_0]$. The denominator of the second term approaches zero as both $\eta \to 0$ and $\alpha \to \alpha_0 \to 3\pi/2$ and in Eq. 14.7 the kernel, $K(x, \eta)$, is singular as both $\eta \to 0$ and $x \to 0$. These singularities are integrable, and are handled by a reformulation that treats them analytically, all remaining contributions to the reaction field being computed numerically (Dorman and Jordan, 2004). A "dimple" geometry (a wedge angle $\beta = \pi/2$) is much simpler; it can be treated by extensions of standard image approaches since, for all $\beta < \pi$, a point within the low ε region has a unique image. This isn't the case for the "bubble" treated here.

Reflecting the left-hand boundary of Fig. 14.2c, by grafting another bubble onto the right-hand dividing surface of Fig. 14.2a, would generate the "ideal" SMC electrical geometry. Both channel mouths would then be in direct contact with solvent containing bubbles that mimic explicit bulk water. However, this would create two electrical dividing surfaces and each explicit charge, just as in Fig. 14.2a, would generate an infinite set of oscillating images. Unfortunately, this geometry can't be mapped simply; there is no easy way to catalogue images or determine the reaction field potential.

14.4 The Born Energy

SMC modeling of permeation energetics requires transferring ions between dielectrics. A Born charging energy is associated with such processes. Removing an ion (or a polar species) from bulk ambient water and placing it in a dielectric with

$\varepsilon = 2$ is substantially endoergic. For an ion the energy penalty is

$$\Delta G^{\text{ion}}_{\text{aqueous}\rightarrow\varepsilon=2} = -\Delta G_{\text{hydration}} - \frac{1}{2\varepsilon_1 R^{\text{ion}}_{\text{Cav}}}, \tag{14.9a}$$

where the overall free energy change is $\Delta G^{\text{ion}}_{\text{aqueous}\rightarrow\varepsilon=2}$, the ionic hydration free energy is $\Delta G_{\text{hydration}}$ and the radius of the ionic solvation cavity in the ε_1 domain is $R^{\text{ion}}_{\text{Cav}}$. The corresponding expression for water molecule transfer is

$$\Delta G^{\text{water}}_{\varepsilon=2\rightarrow\text{aqueous}} = -\Delta G_{\text{vaporization}} + G^{\text{water}}_{\text{Born}}\left(\varepsilon_1, R^{\text{water}}_{\text{Cav}}\right) \tag{14.9b}$$

(Beveridge and Schnuelle, 1975; Dorman and Jordan, 2003; Dorman and Jordan, 2004); the second term in Eq. 14.9b is analogous to the ionic Born energy of Eq. 14.9a. This transfer of polar moieties between water and a solvation cavity differs fundamentally from that in the Born model of ionic solvation (Born, 1920; Grunwald, 1996). There solvent is a structureless continuum and its microscopic reorientational influence on dielectric stabilization is incorporated into a Born radius, a parameter in many ways decoupled from the ion's physical size (Papazyan and Warshel, 1998). SMC modeling explicitly treats solvent reorientation, rearranging these surroundings as water is exchanged for an ion. The Born-like term only includes the influence of the dielectric background, i.e., the electronic polarizability. Consequently, and in stark contrast to the Born process that simultaneously accounts for solvent reorganization and electronic polarization, both $R^{\text{ion}}_{\text{Cav}}$ and $R^{\text{water}}_{\text{Cav}}$ can be interpreted simply. They closely approximate an ion's (or a water's) physical radius (Dorman and Jordan, 2003, 2004) and, in general, are only slightly altered by changes in their immediate surroundings, a point to be discussed later.

14.5 Force Field Parameterization

SMC modeling sites explicit electrical sources in an unconventional dielectric background, one with $\varepsilon = 2$. Consequently, standard force fields cannot be used. Ions are described as charged hard spheres, with crystallographic radii (Pauling, 1960). The water charge distribution is a simple point charge (SPC)-like (Berendsen et al., 1981), with charges chosen to reproduce water's permanent dipole moment, a choice required since electronic polarization has been implicitly treated by introducing the background dielectric. The water charges are placed at the atomic centers, embedded in a sphere with the radius of ambient bulk water. This primitive model reproduces the hydration free energies of the smaller halides and all physiologically important alkali cations (Dorman and Jordan, 2003). Carbonyls and amide groups are viewed as rigid dipoles, with standard bond lengths of 1.24 and 0.96 Å, respectively. In all applications to date ions and waters have been in direct contact only with one another and with peptide backbone atoms; the carbonyl oxygens, amide nitrogens, and α-carbons are treated as hard spheres (Dorman and Jordan, 2004). Just as with water, the other dipolar moieties must be assigned charges representative of their permanent dipoles. Peptide backbones are rigid and CO and NH libration is described by harmonic restoring forces (Garofoli and Jordan, 2003; Dorman and Jordan, 2004).

This procedure has an ad hoc component, and limits the predictive capacity of the SMC approach. Model improvement is possible, but requires a major research investment. To go further entails proceeding as is done in parameterizing more familiar force fields (van Gunsteren and Berendsen, 1987; Cornell et al., 1995; MacKerell et al., 1998). Bonded terms in a force field as well as effective partial charges reflect quantum considerations. These should not change in a dielectric background with $\varepsilon \equiv 2$, rather than vacuum. Instead of assigning hard-core radii, short-range, nonbonded interactions between the various moieties should be described by standard two parameter functions; this could be done by developing a 6–12 (or other) formalism suitable for SMC studies.

14.6 The Thermodynamic Cycle

Permeation is a concerted process. As an ion enters from one side of the channel, another ion or a water molecule departs on the other side. Expressing ion–water exchange in chemical terms yields

$$\text{Ion}_{\text{aqeous}} + \text{Water}_{\text{channel}} \rightarrow \text{Ion}_{\text{channel}} + \text{Water}_{\text{aqeous}}, \tag{14.10}$$

where the subscripts refer to aqueous and channel solvation environments. A channel water molecule gradually mutates into an ion; simultaneously a hydrated ion turns into a water molecule. Each polar moiety is surrounded by a solvation cavity, with which is associated a Born energy. The overall free energy change can be decomposed into an equivalent multistep thermodynamic cycle. Two somewhat different procedures have been devised, depending on the electrical geometry used.

For the KcsA water pool geometry (Figs. 14.1 and 14.2b), there are three intermediate stages in the mutation, illustrated in Fig. 14.4. An ion in bulk is exchanged for water in vacuum, then the gas phase ion is exchanged for a water in the ε_1 dielectric background and finally the ion in this background is exchanged for a water in the channel (Garofoli and Jordan, 2003). This can be expressed as sequential reactions:

$$\text{Ion}_{\text{aqeous}} + \text{Water}_{\text{vacuum}} \rightarrow \text{Ion}_{\text{vacuum}} + \text{Water}_{\text{aqeous}}, \tag{14.11a}$$
$$\text{Ion}_{\text{vacuum}} + \text{Water}_{\text{background}} \rightarrow \text{Ion}_{\text{background}} + \text{Water}_{\text{vacuum}}, \tag{14.11b}$$
$$\text{Ion}_{\text{background}} + \text{Water}_{\text{channel}} \rightarrow \text{Ion}_{\text{channel}} + \text{Water}_{\text{background}}, \tag{14.11c}$$

and the new subscripts refer to intermediate residence in the dielectric background and vacuum. In the first two steps the permittivity of the solvent changes; the total free energy change is given by summing Eq. 14.9a and Eq. 14.9b. The last step, which involves interaction with the channel peptide, takes place in the uniform, ε_1,

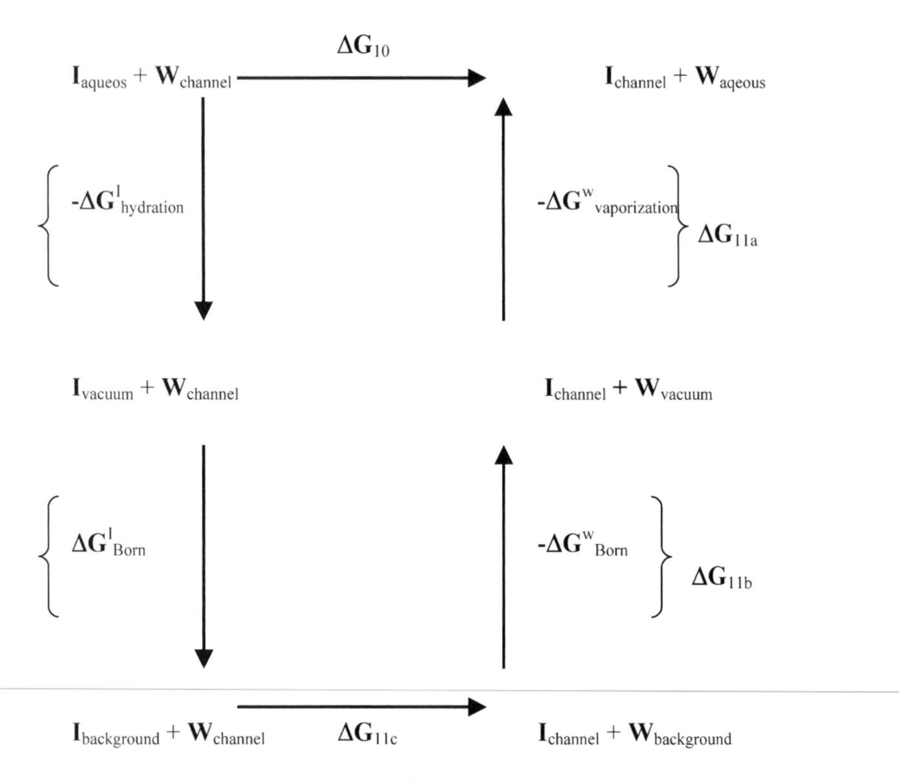

Fig. 14.4 Equivalent thermodynamic cycle of Eqs. 14.10 and 14.11. The steps corresponding to Eqs. 14.11a, 14.11b, and 14.11c are illustrated separately. Adapted from Garofoli and Jordan (2003) with permission.

dielectric. A water molecule, at a specified axial distance from the channel–water interface, but otherwise free to move, mutates into an ion; simultaneously an ion in the dielectric, but far from the channel, turns into a water molecule. Standard thermodynamic perturbation determines the associated free energy, thus accounting for how differences in a peptide's interaction with water and ions leads to structural reorganization (Dorman and Jordan, 2003; Garofoli and Jordan, 2003; Dorman and Jordan, 2004)

$$\Delta G_{\text{structural}} = \int_0^1 d\lambda \left\langle \frac{dH}{d\lambda} \right\rangle, \tag{14.12a}$$

where

$$\langle(\cdots)\rangle = \frac{1}{Z}\int_{(\Omega)} d\Omega(\cdots)\exp[-\beta H(\Omega,\lambda)] \tag{14.12b}$$

is the canonical statistical average. $H(\Omega, \lambda)$ is the system Hamiltonian, Ω the complete set of phase space variables, λ the mutational variable and Z the corresponding partition function with $\beta \equiv 1/k_B T$ (k_B is Boltzmann's constant and T is absolute temperature). In the perturbation process, one channel water mutates to an ion, gradually changing its charge distribution according to the prescription

$$\lambda q_{\text{ion}} + (1 - \lambda)\rho_{\text{water}}, \tag{14.13}$$

where ρ_{water} is the charge distribution in the model water molecule, q_{ion} is the ionic charge and λ varies from 0 to 1. The Hamiltonian for channel reorganization is

$$\begin{aligned} H_{\text{channel}}(\lambda) = {} & H_{\text{V}}(\lambda) + H_{\text{W}} + H_{\text{pep}} + H_{\text{W,pep}} \\ & + H_{\text{V,W}}(\lambda) + H_{\text{V,pep}}(\lambda) + H_{\text{hard-core}}, \end{aligned} \tag{14.14}$$

where $H_{\text{V}}(\lambda)$ describes interaction of the variable species with its reaction field. H_{W}, H_{pep}, and $H_{\text{W,pep}}$ account for water–peptide interactions, $H_{\text{V,W}}(\lambda)$ and $H_{\text{V,pep}}(\lambda)$ describe interaction between the variable species, water and peptide, in all cases including reaction field influences. The hard-core term keeps species apart.

In the bubble geometry of Fig. 14.2c, the perturbation procedure is different. The parameterization procedure relies on solvating an ion in a small, but sufficiently large "aqueous sphere" containing a limited number of model water molecules (Dorman and Jordan, 2003). Instead of mutation passing through a vacuum intermediate, channel water changes to an ion and an ion in the computational sphere simultaneously turns into a water molecule; the corresponding Hamiltonian is analogous to Eq. 14.14, with peptide contributions deleted. All changes take place in an electrical background with $\varepsilon \equiv 2$. The thermodynamic exchange process is

$$\text{Ion}_{\text{sphere}} + \text{Water}_{\text{channel}} \rightarrow \text{Ion}_{\text{channel}} + \text{Water}_{\text{sphere}}. \tag{14.15}$$

There is still a Born-like contribution, since ion and water solvation cavities differ slightly in size in the "aqueous sphere" and channel environments. The total free energy change in transferring an ion from bulk water to the channel is given by Eq. 14.12a plus an elaboration of Eqs. 14.9a and 14.9b. Instead of the hydration and vaporization terms of Fig. 14.4 there are cavity terms describing the ionic and water environments in the "aqueous sphere," i.e., the changes in the size of the solvation cavities. The associated Born energy is

$$\Delta G_{\text{Born}} = \frac{e_o^2}{2\varepsilon}\left(\frac{1}{R_{\text{channel}}} - \frac{1}{R_{\text{sphere}}}\right) + \Delta G_{\text{water cavity}}(\text{channel} \rightarrow \text{sphere}), \tag{14.16}$$

where R_{sphere} and R_{channel} are cavity radii for the ion in the two regions; the second term in Eq. 14.16 accounts for the energy needed to change the size of water's solvation cavity (Beveridge and Schnuelle, 1975). These are determined by evaluating the mean distance between the variable species and its surroundings at the start and conclusion of the complementary thermodynamic perturbations (Dorman and Jordan, 2003, 2004).

14.7 Applications

14.7.1 Deconvoluting Structural Influences

The SMC method is ideal for "*gedanken*" experiments based on reengineered channel architectures, allowing an assessment of how individual structural features influence ion permeation. While protein parameters are constrained, hypothetical scenarios can provide insight into structural evolution.

One example, focusing on the role of the water cavity in potassium channels, is illustrative. The five single file sites of Fig. 14.1 model possible ion binding locations (Morais-Cabral et al., 2001; Zhou et al., 2001), of which sites 1–3 are most important (Doyle et al., 1998). In treating potassium channel permeation energetics with the geometry of Fig. 14.1, these three positions are better modeled because explicit water molecules act as buffers separating them from the nearby Helmholtz layer. To stress the conditional nature of predictions for the outer sites, where energetics is sensitive to the nearby dielectric discontinuity, they are designated 0* and 4*. Ion(s), channel waters, and the selectivity filter carbonyls are mobile; more distant electrical sources are fixed (Garofoli and Jordan, 2003).

The spherical water pool is only a rough approximation to the inner pore of potassium channels. The KcsA crystal captured the channel in a nonconductive form, its inner mouth too narrow to permit ion entry (Doyle et al., 1998). Gating opens the channel, altering its structure drastically; the small water pool must deform into a vestibule into which bulk water can diffuse freely. The crystal structure of such a channel illustrates this clearly (Jiang et al., 2002). On its intracellular side, the channel is greatly deformed, but the selectivity filter region outside the water pool remains essentially unchanged (Jiang et al., 2002). Consequently, using the structure of the "closed" state filter to analyze permeation energetics in the "open" state is not unreasonable. The SMC approach does this in two ways: (1) by reducing membrane thickness, bringing the cytoplasm ever closer to the cavity, in essence shortening the inner pore; (2) by deforming the cavity into a cylindrical tube extending to the cytoplasm and filled with 72 explicit waters. Regardless of the size of the buffer domain or the amount of water it contains, ion energetics in the carbonyl binding pockets is essentially unaltered (Garofoli and Jordan, 2003). The cavity insulates ions in the filter from changes in inner pore geometry, guaranteeing that filter operation is essentially unperturbed by structural changes in the channel's inner pore. Only at site 4*, the filter-cavity boundary, does altering inner pore structure have any effect, which is totally consistent with observed behavior in the potassium channel family. Unitary conductances, which depend on the size of the inner pore (Chung et al., 2002b), vary by two orders of magnitude but are essentially uncorrelated with channel selectivity (Hille, 2001).

14.7.2 Selectivity in the Potassium Channel Filter

SMC modeling permits ready analysis of filter selectivity, by mutational transformation among the four larger alkali cations. The focus remains on the interior ionic sites numbered 1–3, which are most reliably modeled. Such studies provide useful insights at modest computational cost.

While there is compelling evidence that the filter is always multiply occupied in KcsA, SMC analysis permits another thought experiment: direct comparison of alkali cation energetics in the interior of the singly occupied filter, a condition that is possibly experimentally realized in the delayed rectifier channel, $K_V2.1$ (Consiglio and Korn, 2004; Immke et al., 1999). At the three interior sites sodium occupancy is unfavorable relative to potassium by $\sim$10–15kT, for $T = 300$ K, values in reasonable agreement with standard simulational analysis (Allen et al., 2000; Bernèche and Roux, 2001; Luzhkov and Åqvist, 2001). In the SMC approach the energetic influence of solvent structural reorganization is separable from that due to ion transfer between surroundings of different permittivity; the two terms are given by Eqs. 14.12 and 14.9, respectively. SMC comparison of sodium and potassium energetics indicates that the ion transfer term exerts the larger effect on the filter's ability to discriminate (Garofoli and Jordan, 2003). As seen in Table 14.1, sodium is less well cradled in its solvation cavity than potassium. This is consonant with MacKinnon's original idea (Doyle et al., 1998) and the earlier "close fit" hypothesis (Bezanilla and Armstrong, 1972), but differs from a recent simulational interpretation (Noskov et al., 2004) that is conceptually similar to Eisenman's early interpretation of selectivity (Eisenman, 1962).

SMC analysis of a multiply occupied filter yields no surprises (Garofoli and Jordan, 2003). Consonant with experiment (Hille, 2001) rubidium occupancy is roughly as likely as that of potassium, but cesium is significantly less stable in the filter, mainly reflecting a relatively unfavorable reorganizational free energy. Ionic energetics at site 4*, the filter-cavity boundary, is illuminating. Here the filter no longer imposes a severe structural constraint and the scaffolding can deform more

Table 14.1 Mean ionic cavity radii (in Å) for various cations at the KcsA filter sites of Fig. 14.1.

	Site 1	Site 2	Site 3	Site 4*
Na^+	1.23	1.24	1.23	1.03
K^+	1.38	1.43	1.40	1.32
Rb^+	1.57	1.48	1.50	1.44
Cs^+	1.65	1.61	1.66	1.56
Ba^{2+}	1.43	1.44	1.43	1.38

readily. Both sodium and possibly barium can compete effectively with potassium, consistent with experiment (Bezanilla and Armstrong, 1972; French and Wells, 1977; Yellen, 1984; Neyton and Miller, 1988a,b; Heginbotham et al., 1999; Jiang and MacKinnon, 2000).

14.7.3 Gramicidin Permeability

The inability of standard simulational methods to account for permeation energetics in gramicidin A has been perplexing to theorists. As discussed in the introduction, PMF determinations with conventional force fields yield free energy barriers of $\sim 20kT\,(T = 300\,\text{K})$, implying conductances $\sim 10^7$ times too small. Since gramicidin is structurally far better characterized than any other channel (Arseniev et al., 1985; Ketchem et al., 1997), this discrepancy is dismaying. SMC analysis provides an alternative perspective to more familiar methods. Unlike the case of conventional MD simulations, electronic polarization is taken into account and the reaction field problem can be solved without approximation.

Gramicidin's permeation pathway is a water-filled pore bounded by a cylindrical scaffolding formed by the peptide's hydrogen bonded backbone, with its amino acid residues embedded in the surrounding lipid (Arseniev et al., 1985; Ketchem et al., 1997). SMC studies using the electrical geometry of Fig. 14.2c, treat the ion, channel waters, and carbonyls and amides of the backbone as mobile; the peptide's aryl residues and the membrane dipoles (Thompson et al., 2001) impose a fixed background potential. A correction is used to compensate for Fig. 14.2c's semi-infinite geometry. SMC calculations of permeation free energies for the larger alkali cations and chloride yield results, which, while imperfect, are in fair agreement with experiment. Cation binding occurs at the proper location, ~ 1.0 nm from the channel midpoint. The correct selectivity sequence is observed with rubidium the most permeable cation and chloride passage forbidden. Instead of cation conductances seven orders of magnitude low, discrepancies are greatly reduced, to no more than ~ 100, as shown in Table 14.2 (Dorman and Jordan, 2004).

Table 14.2 Ionic conductances (pS) of gramicidin A in 0.1 M electrolyte, based on total permeation free energy profiles determined using GROMOS and CHARMM19 partial charge parameters. Conductances were computed using Levitt's (Levitt, 1986) single-file algorithm. Simulational values are estimated from the free energy profiles of Edwards et al. (2002).

	GROMOS	CHARMM19	Experiment[a]
K^+	0.24	0.025	11
Rb^+	12.	0.40	20
Cs^+	1.0	0.074	18
Cl^-	2.4×10^{-8}	6.0×10^{-6}	0
Simulation, K^+	1.8×10^{-7}	4.2×10^{-7}	11

[a] Data extrapolated from Fig. 5 of Andersen (1983).

The electrical decomposition procedure illustrates how the aryl residues and membrane dipoles influence cation conductance. Ionic interaction with the aryl residues creates the binding site. The membrane dipoles substantially reduce cation conductance. A thought experiment based on variation of the SMC parameters helps explain why standard simulational approaches are plagued with difficulties. Reducing ε_1 to 1 eliminates the mean contribution of electronic polarization; this increases the internal permeation free energy barrier by $\sim 8kT$, nearly half the error in computations based on standard force fields (Allen et al., 2003).

14.8 Final Observations

SMC modeling is a blunt tool. It is an alternative to standard simulational procedures, affording complementary insights. It accommodates electronic polarization in a mean field sense and, by introducing specially designed dielectric geometries, treats long-range electrostatics exactly. However, it must be used cautiously, as electronic environments in nonuniform surroundings like peptide–lipid–water ensembles are far from uniform. To further exploit this idea requires development of a better SMC force field. The procedure used here is too simple. The peptide force field is rudimentary and the water model is primitive. As a result only the larger cations and halides have been reliably parameterized and conclusions must be viewed as qualitative. Extensions lifting these limitations so that peptides are more accurately described and so that sodium and the alkaline earth cations could be more reliably modeled would be valuable.

Acknowledgments

I thank Michael Partensky and Gennady Miloshevsky for helpful commentary. This work was supported by a grant from the National Institutes of Health, GM-28643.

References

Aksimentiev, A., and K. Schulten. 2005. Imaging α-hemolysin with molecular dynamics: Ionic conductance, osmotic permeability, and the electrostatic potential map. *Biophys. J.* 88:3745–3761.

Allen, T.W., O.S. Andersen, and B. Roux. 2004. Energetics of ion conduction through the gramicidin channel. *Proc. Natl. Acad. Sci. USA* 101:117–122.

Allen, T.W., T. Bastug, S. Kuyucak, and S.H. Chung. 2003. Gramicidin A channel as a test ground for molecular dynamics force fields. *Biophys. J.* 84:2159–2168.

Allen, T.W., A. Bilznyuk, A.P. Rendell, S. Kuyucak, and S.H. Chung. 2000. The potassium channel: Structure, selectivity and diffusion. *J. Chem. Phys.* 112:8191–8204.

Andersen, O.S. 1983. Ion movement through gramicidin A channels. Single-channel measurements at very high potentials. *Biophys. J.* 41:119–133.

Arseniev, A.S., I.L. Barsukov, V.F. Bystrov, A.L. Lomize, and Y.A. Ovchinnikov. 1985. 1H-NMR study of gramicidin A transmembrane ion channel. Head-to-head right-handed, single-stranded helices. *FEBS Lett.* 186:168–174.

Beglov, D., and B. Roux. 1994. Finite representation of an infinite bulk system—solvent boundary potential for computer-simulations. *J. Chem. Phys.* 100:9050–9063.

Berendsen, H.J.C., J.P.M. Postma, W.F. van Gunsteren, and J. Hermans. 1981. Interaction models for water in relation to protein hydration. *In*: Intermolecular Forces. B. Pullman editor. Reidel, Dordrecht, pp. 331–342.

Bernèche, S., and B. Roux. 2001. Energetics of ion conduction through the K^+ channel. *Nature* 414:73–77.

Bernèche, S., and B. Roux. 2003. A microscopic view of ion conduction through the K^+ channel. *PNAS* 100:8644–8648.

Beveridge, D.L., and G.W. Schnuelle. 1975. Free energy of a charge distribution in concentric dielectric continua. *J. Phys. Chem.* 79:2562–2566.

Bezanilla, F., and C.M. Armstrong. 1972. Negative conductance caused by entry of sodium and cesium ions into the potassium channels of squid axon. *J. Gen. Physiol.* 60:588–608.

Born, M. 1920. Volumen und hydrationswarme der Ionen. *Zeit. für Physik.* 1:45–48.

Burykin, A., C.N. Schutz, J. Villa, and A. Warshel. 2002. Simulations of ion current in realistic models of ion channels: The KcsA potassium channel. *Proteins* 47:265–280.

Chen, D.P., V. Barcilon, and R.S. Eisenberg. 1992. Constant fields and constant gradients in open ionic channels. *Biophys. J.* 61:1372–1393.

Chung, S.H., T.W. Allen, M. Hoyles, and S. Kuyucak. 1999. Permeation of ions across the potassium channel: Brownian dynamics studies. *Biophys. J.* 77:2517–2533.

Chung, S.H., T.W. Allen, and S. Kuyucak. 2002a. Conducting-state properties of the KcsA potassium channel from molecular and Brownian dynamics simulations. *Biophys. J.* 82:628–645.

Chung, S.H., T.W. Allen, and S. Kuyucak. 2002b. Modeling diverse range of potassium channels with Brownian dynamics. *Biophys. J.* 83:263–277.

Consiglio, J.F., and S.J. Korn. 2004. Influence of permeant ions on voltage sensor function in the Kv2.1 potassium channel. *J. Gen. Physiol.* 123:387–400.

Cornell, W.D., P. Cieplak, C.I. Bayly, I.R. Gould, K.M. Merz, D.M. Ferguson, D.C. Spellmeyer, T. Fox, J.W. Caldwell, and P.A. Kollman. 1995. A 2nd generation force-field for the simulation of proteins, nucleic-acids, and organic-molecules. *J. Am. Chem. Soc.* 117:5179–5197.

Corry, B., S. Kuyucak, and S.H. Chung. 2000. Tests of continuum theories as models of ion channels. II. Poisson–Nernst–Planck theory versus Brownian dynamics. *Biophys. J.* 78:2364–2381.

Corry, B., M. O'Mara, and S.-H. Chung. 2004. Conduction mechanisms of chloride ions in ClC-type channels. *Biophys. J.* 86:846–860.

Dorman, V., M.B. Partenskii, and P.C. Jordan. 1996. A semi-microscopic Monte Carlo study of permeation energetics in a gramicidin-like channel: The origin of cation selectivity. *Biophys. J.* 70:121–134.

Dorman, V.L., and P.C. Jordan. 2003. Ion–water interaction potentials in the semi-microscopic model. *J. Chem. Phys.* 118:1333–1340.

Dorman, V.L., and P.C. Jordan. 2004. Ionic permeation free energy in gramicidin: A semimicroscopic perspective. *Biophys. J.* 86:3529–3541.

Doyle, D.A., J. Morais-Cabral, R.A. Pfuetzner, A. Kuo, J.M. Gulbis, S.L. Cohen, B.T. Chait, and R. MacKinnon. 1998. The structure of the potassium channel: Molecular basis of K^+ conduction and selectivity. *Science* 280:69–77.

Edwards, S., B. Corry, S. Kuyucak, and S.H. Chung. 2002. Continuum electrostatics fails to describe ion permeation in the gramicidin channel. *Biophys. J.* 83:1348–1360.

Eisenberg, R.S. 1999. From structure to function in open ionic channels. *J. Membr. Biol.* 171:1–24.

Eisenman, G. 1962. Cation selective glass electrodes and their mode of operation. *Biophys. J.* 2(2, Pt 2):259–323.

Fettiplace, R., D.M. Andrews, and D.A. Haydon. 1971. Thickness, composition and structure of some lipid bilayers and natural membranes. *J. Membr. Biol.* 5:277–296.

French, R.J., and J.B. Wells. 1977. Sodium ions as blocking agents and charge carriers in the potassium channel of squid giant axon. *J. Gen. Physiol.* 70:707–724.

Garofoli, S., and P.C. Jordan. 2003. Modeling permeation energetics in the KcsA potassium channel. *Biophys. J.* 84:2814–2830.

Grunwald, E. 1996. Thermodynamics of Molecular Species. Wiley-Interscience, New York.

Hasted, J.B. 1973. In Dielectric Properties. F. Franks, editor. Water, a Comprehensive Treatise, Vol. 1. Plenum, New York, pp. 405–458.

Heginbotham, L., M. LeMasurier, L. Kolmakova-Partensky, and C. Miller. 1999. Single *Streptomyces lividans* K^+ channels: Functional asymmetries and sidedness of proton activation. *J. Gen. Physiol.* 114:551–560.

Hille, B. 2001. Ionic Channels of Excitable Membranes, 3rd Ed. Sinauer Associates, Sunderland, MA.

Im, W., and B. Roux. 2002. Ion permeation and selectivity of OmpF porin: A theoretical study based on molecular dynamics, Brownian dynamics, and continuum electrodiffusion theory. *J. Mol. Biol.* 322:851–869.

Immke, D., M. Wood, L. Kiss, and S.J. Korn. 1999. Potassium-dependent changes in the conformation of the Kv2.1 potassium channel pore. *J. Gen. Physiol.* 113:819–836.

Jackson, J.D. 1962. Classical Electrodynamics. John Wiley, New York.

Jiang, Y., A. Lee, J. Chen, M. Cadene, B.T. Chait, and R. MacKinnon. 2002. The open pore conformation of potassium channels. *Nature* 417:523–526.

Jiang, Y., and R. MacKinnon. 2000. The barium site in a potassium channel by X-ray crystallography. *J. Gen. Physiol.* 115:269–272.

Jordan, P.C. 2005. Semimicroscopic modeling of permeation energetics in ion channels. *IEEE Trans. Nanobiosci.* 4:94–101.

Jordan, P.C., M.B. Partenskii, and V. Dorman. 1997. Electrostatic influences on ion–water correlation in ion channels. *Prog. Cell Res.* 6:279–293.

Ketchem, R., B. Roux, and T. Cross. 1997. High-resolution polypeptide structure in a lamellar phase lipid environment from solid state NMR derived orientational constraints. *Structure* 5:1655–1669.

King, G., and A. Warshel. 1989. A surface constrained all-atom solvent model for effective simulations of polar solutions. *J. Chem. Phys.* 91:3647–3661.

Kornyshev, A.A. 1988. Solvation of a metal surface. *In*: The Chemical Physics of Solvation. R.R. Dogonadze, E. Kálmán, A.A. Kornyshev, and J. Ulstrup, editors. Elsevier, Amsterdam, pp. 355–400.

Levitt, D.G. 1986. Interpretation of biological ion channel flux data—Reaction-rate versus continuum theory. *Annu. Rev. Biophys. Biophys. Chem.* 15:29–57.

Luzhkov, V.B., and J. Åqvist. 2001. K^+/Na^+ selectivity of the KcsA potassium channel from microscopic free energy perturbation calculations. *Biochim. Biophys. Acta* 1548:194–202.

MacKerell, A.D., D. Bashford, M. Bellott, R.L. Dunbrack, J.D. Evanseck, M.J. Field, S. Fischer, J. Gao, H. Guo, S. Ha, D. Joseph-McCarthy, L. Kuchnir, K. Kuczera, F.T.K. Lau, C. Mattos, S. Michnick, T. Ngo, D.T. Nguyen, B. Prodhom, W. E. Reiher, B. Roux, M. Schlenkrich, J.C. Smith, R. Stote, J. Straub, M. Watanabe, J. Wiorkiewicz-Kuczera, D. Yin, and M. Karplus. 1998. All-atom empirical potential for molecular modeling and dynamics studies of proteins. *J. Phys. Chem. B* 102:3586–3616.

Miloshevsky, G.V., and P.C. Jordan. 2004. Anion pathway and potential energy profiles along curvilinear bacterial ClC Cl^- pores: Electrostatic effects of charged residues. *Biophys. J.* 86:825–835.

Morais-Cabral, J.H., Y. Zhou, and R. MacKinnon. 2001. Energetic optimization of ion conduction rate by the K^+ selectivity filter. *Nature* 414:37–42.

Moy, G., B. Corry, S. Kuyucak, and S.H. Chung. 2000. Tests of continuum theories as models of ion channels. I. Poisson–Boltzmann theory versus Brownian dynamics. *Biophys. J.* 78:2349–2363.

Neyton, J., and C. Miller. 1988a. Discrete Ba^{2+} block as a probe of ion occupancy and pore structure in the high-conductance Ca^{2+} -activated K^+ channel. *J. Gen. Physiol.* 92:569–586.

Neyton, J., and C. Miller. 1988b. Potassium blocks barium permeation through a calcium-activated potassium channel. *J. Gen. Physiol.* 92:549–567.

Noskov, S.Y., S. Berneche, and B. Roux. 2004. Control of ion selectivity in potassium channels by electrostatic and dynamic properties of carbonyl ligands. *Nature* 431:830–834.

Papazyan, A., and A. Warshel. 1998. Effect of solvent discreteness on solvation. *J. Phys. Chem. B* 102:5348–5357.

Partenskii, M.B., and P.C. Jordan. 1992. Theoretical perspectives on ion-channel electrostatics: Continuum and microscopic approaches. *Q. Rev. Biophys.* 25:477–510.

Pauling, L. 1960. The Nature of the Chemical Bond, 3rd Ed. Cornell University Press, Ithaca, NY.

Smythe, W.R. 1968. Static and Dynamic Electricity, 3rd Ed. McGraw-Hill, New York.

Thompson, N., G. Thompson, C.D. Cole, M. Cotten, T.A. Cross, and D.D. Busath. 2001. Noncontact dipole effects on channel permeation. IV. Kinetic model of 5F-Trp_{13} gramicidin A currents. *Biophys. J.* 81:1245–1254.

van Gunsteren, W.F., and H.J.C. Berendsen. 1987. Groningen Molecular Simulations (GROMOS) Library Manual. Biomos, Groningen, NL.

Warshel, A. 1979. Calculations of chemical processes in solutions. *J. Phys. Chem.* 83:1640–1652.

Yellen, G. 1984. Relief of Na^+ block of Ca^{2+}-activated K^+ channels by external cations. *J. Gen. Physiol.* 84:187–199.

Zhou, Y., J. H. Morais-Cabral, A. Kaufman, and R. MacKinnon. 2001. Chemistry of ion coordination and hydration revealed by a K^+ channel-Fab complex at 2.0 Å resolution. *Nature* 414:43–48.

15 Brownian Dynamics: Simulation for Ion Channel Permeation[1]

Shin-Ho Chung and Vikram Krishnamurthy

15.1 Introduction

All living cells are surrounded by a thin membrane, composed of two layers of phospholipid molecules, called the lipid bilayer. This thin membrane effectively confines some ions and molecules inside and exchanges others with outside and acts as a hydrophobic, low dielectric barrier to hydrophilic molecules. Because of a large difference between the dielectric constants of the membrane and electrolyte solutions, no charged particles, such as Na^+, K^+, and Cl^- ions, can jump across the membrane. The amount of energy needed to transport one monovalent ion, in either direction across the membrane, known as the *Born energy*, is enormously high. For a living cell to function, however, the proper ionic gradient has to be maintained, and ions at times must move across the membrane to maintain the potential difference across the membrane and to generate synaptic and action potentials. The delicate tasks of regulating the transport of ions across the membrane are carried out by biological nanotubes called "ion channels," water-filled conduits inserted across the cell membrane through which ions can freely move in and out when the gates are open. These ion channels can be viewed as biological sub-nanotubes, the typical pore diameters of which are $\sim 10^{-9}$ m or 10 Å.

Ionic channels in lipid membranes play a crucial role in the existence of living organisms. All electrical activities in the nervous system, including communication between cells and the influence of hormones and drugs on cell function, are regulated by the opening and closing of these membrane proteins. Because these channels are elementary building blocks of brain function, understanding their mechanisms at a molecular level is a fundamental problem in biophysics. Moreover, the elucidation of how single channels work will ultimately help us find the causes of, and potentially cures for, a number of neurological and muscular disorders.

In the past few years, there have been enormous strides in our understanding of the structure–function relationships in biological ion channels. This sudden

[1] This chapter is an extended version of Krishnamurthy and Chung (2005) that appeared in IEEE Transactions Nanobioscience, March 2005.

advance has been brought about by the combined efforts of experimental and computational biophysicists, who together are beginning to unravel the working principles of these exquisitely designed biological nanotubes that regulate the flow of charged particles across living membranes. In recent breakthroughs, the crystal structures of the potassium channels, mechanosensitive channel, chloride channel, and nicotinic acetylcholine receptor have been determined from crystallographic analysis (Doyle et al., 1998; Bass et al., 2002; Chang et al., 1998; Dutzler et al., 2002, 2003; Long et al., 2004a,b; Unwin, 2005). It is expected that crystal structures of other ion channels will follow these discoveries, ushering in a new era in ion channel studies, where predicting function of channels from their atomic structures will become the main quest. Parallel to these landmark experimental findings, there have also been important advances in computational biophysics. As new analytical methods have been developed and the available computational power increased, theoretical models of ion permeation have become increasingly sophisticated. Now it has become possible to relate the atomic structure of an ion channel to its function through the fundamental laws of physics operating in electrolyte solutions. Many aspects of macroscopic observable properties of ion channels are being addressed by molecular and stochastic dynamics simulations. Quantitative statements based on rigorous physical laws are replacing qualitative explanations of how ions permeate across narrow pores formed by the protein wall and how ion channels allow one ionic species to pass while blocking others. The computational methods of solving complex biological problems, such as permeation, selectivity and gating mechanisms of ion channels, will increasingly play prominent roles as the speed of computers increases and theoretical approaches that are currently underdevelopment become further refined.

Here we give a brief account of Brownian dynamics (BD), one of the several theoretical computational methods that are being used for treating time-dependent, nonequilibrium processes that underlie the flow of currents across biological ion channels. We first give a simple, intuitive explanation of how one traces the trajectories of ions in electrolyte solutions interacting with a low dielectric boundary. We briefly illustrate how the BD simulation algorithm has been employed in elucidating the mechanisms of ion permeation in the KcsA potassium channel and ClC chloride channels. We then outline the principles underlying BD, its statistical consistency and algorithms for practical implementation. We also describe one novel extension of BD, called adaptive controlled BD simulations, that circumvents some of the caveats to computing current flow across ion channels using the conventional method. Three other computational approaches—the Poisson–Nernst–Planck theory, semi-microscopic Monte Carlo method, and molecular dynamics—are summarized in the preceding and following chapters (Coalson and Kurnikova, 2006; Grottesi et al., 2006; Jordan, 2006). The reader is also referred to recent review articles (Eisenberg, 1999; Partenskii and Jordan, 1992; Roux et al., 2000; Tieleman et al., 2001; Chung and Kuyucak, 2002) for further details of recent advances in ion channel research.

15.2 Stochastic Dynamics Simulations

15.2.1 Overview

One of the ultimate aims of theoretical biophysicists is to provide a comprehensive physical description of biological ion channels. Such a theoretical model, once successfully formulated, will link channel structure to channel function through the fundamental processes operating in electrolyte solutions. It will also concisely summarize the data, by interlacing all those seemingly unrelated and disparate observations into a connected whole. The theory will elucidate the detailed mechanisms of ion permeation—where the binding sites are in the channel, how fast an ion moves from one biding site to another, and where the rate-limiting steps are in conduction. Finally, it will make predictions that can be confirmed or refuted experimentally.

The tools of physics employed in this endeavor, from fundamental to phenomenological, are ab initio and classical molecular dynamics, BD, and continuum theories. These approaches make various levels of abstractions in replacing the complex reality with a model, the system composed of channel macromolecules, lipid bilayer, ions, and water molecules. One of the important criteria of successful modeling is that macroscopic observables remain invariant when the real system is replaced by the model. Each of these approaches has its strengths and limitations, and involves a degree of approximation.

At the lowest level of abstraction we have the ab initio quantum mechanical approach, in which the interactions between the atoms are determined from first-principle electronic structure calculations. As there are no free parameters in this approach, it represents the ultimate approach to the modeling of biomolecular systems. But because of the extremely demanding nature of computations, its applications are limited to very small systems at present. A higher level of modeling abstraction is to use classical molecular dynamics (Grottesi et al., 2005, 2006). Here, simulations are carried out using empirically-determined pairwise interaction potentials between the atoms, and their trajectories are followed using Newton's equation of motion. Although it is possible to model an entire ion channel in this way, it is not feasible to simulate the system long enough to see permeation of ions across a model channel and to determine its conductance, which is the most important channel property.

For that purpose, one has to go up one further step in abstraction to stochastic dynamics, of which BD is the simplest form, where water molecules that form the bulk of the system in ion channels are integrated out and only the ions themselves are explicitly simulated. Thus, instead of considering the dynamics of individual water molecules, one considers their average effect as a random force or Brownian motion on the ions. This treatment of water molecules as implicit water can be viewed as a functional central limit theorem approximation. In BD, it is further assumed that the protein is rigid and its dynamics are not considered. Thus, in this approach, the motion of each individual ion is modeled as the evolution of a stochastic differential equation, known as the Langevin equation.

A still higher level of abstraction is the Poisson–Nernst–Planck theory (Coalson and Kurnikova, 2005, 2006), which is based on the continuum hypothesis of electrostatics. In this and other electrodiffusion theories, one makes a further simplification, known as the mean-field approximation. Here, ions are treated not as discrete entities but as continuous charge densities that represent the space–time average of the microscopic motion of ions. In the Poisson–Nernst–Planck theory, the flux of an ionic species is described by the Nernst–Planck equation that combines Ohm's law with Fick's law of diffusion, and the potential at each position is determined from the solution of Poisson's equation using the total charge density (ions plus fixed charges). The Poisson–Nernst–Planck theory thus incorporates the channel structure, and its solution yields the potential, concentration, and flux of ions in the system in a self-consistent manner.

There is one other approach that has been fruitfully employed to model biological ion channels, namely, the reaction rate theory (Jordan, 1999; McCleskey, 1999; Hille, 2001). In this approach, an ion channel is represented by a series of ion binding sites separated by barriers, and ions are assumed to hop from one biding site to another, with the probability of each hop determined by the height of the energy barrier. Although the model parameters have no direct physical relation to the channel structure, many useful insights have been gleaned in the past about the mechanisms of ion permeation using this approach.

15.2.2 General Description of Brownian Dynamics

Brownian dynamics offers one of the simplest methods for following the trajectories of interacting ions in a fluid. Figure 15.1 shows a schematic illustration of a BD simulation assembly. An ion channel representing the potassium channel is placed at the center of the assembly. The positions in three-dimensional space of all the atoms forming the channel are given by its X-ray structure, and the charge on each atom is assigned. Then, a large cylindrical reservoir with a fixed number of K^+ (or Na^+) and Cl^- ions is attached at each end of the channel (Fig. 15.1A) to mimic the extracellular or intracellular space. The membrane potential is imposed by applying a uniform electric field across the channel (Fig. 15.1B). This is equivalent to placing a pair of large plates far away from the channel and applying a potential difference between them. Since the space between the voltage plates is filled with electrolyte solution, each reservoir is in iso-potential. That is, the average potential anywhere in the reservoir is identical to the applied potential at the voltage plate on that side, and the potential drop occurs almost entirely across the channel.

The algorithm for performing BD simulations is conceptually simple. The velocity of the ion with mass m and charge q located at a given position is determined by the force acting on it at time t. This velocity is computed by integrating the equation of motion, known as the Langevin equation. Once its velocity is determined at time t, the position this ion will occupy in three-dimensional space at time $t + \Delta t$ can be specified. The calculation is repeated for each ion in the assembly, and the new distribution of the positions of all ions at time $t + \Delta t$ are assigned. At each

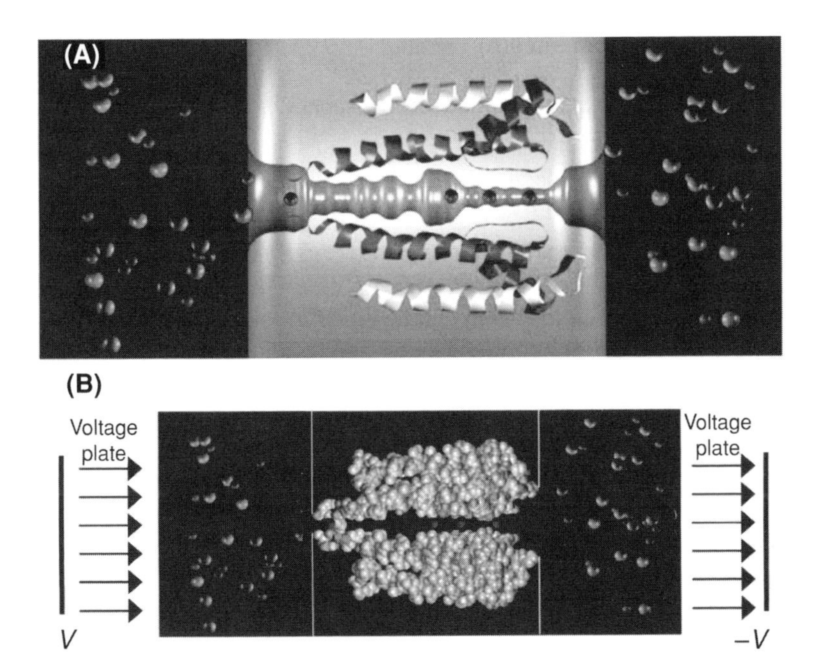

Fig. 15.1 Brownian dynamics setup. (A) The KcsA K^+ channel is placed at the center of the simulation and a large reservoir containing K^+ and Cl^- ions is attached at each end of the protein. The intracellular aspect of the channel is on the left. (B) A uniform electric field is applied across the channel to mimic the membrane potential. This arrangement is equivalent to having two voltage plates far away from the channel.

time-step, usually 2 fs, the forces acting on each ion are calculated and the Langevin equation is used to determine where it will move in the next time-step. By repeating this process many billions of times, usually for a simulation period lasting between 10 and 100 μs, we can trace the movement of each ion in space during a simulation period, and count how many ions have crossed from one side of the channel to the other.

What ultimately determines the motion of ions, and hence the current flowing across the channel, is the total force acting on charged particles. It is important, therefore, to specify all the components of the forces accurately. Two main sources of the forces influencing the motion of ions in or in the vicinity of an ion channel are the "stochastic" force and electric force. The former arises from the effects of collisions between ions and water molecules. Ions in electrolyte solutions are tightly bound by shells of water molecules and these hydrated ions collide incessantly with surrounding water molecules. As a result of such bombardments, the motion of an ion is retarded, and it undergoes random fluctuations from an equilibrium position.

The electric field inside or outside of the channel originates from four different sources. First, there is a field resulting from the membrane potential, which is generated by diffuse, unpaired, ionic clouds on each side of the membrane. Second, there are fixed charges in the channel protein and the electric field emanating from them adds to the field generated by the membrane potential. Third, charges carried by all the ions in electrolyte solutions contribute to the total electric field. Whenever any of these ions come near the protein wall, it induces surface charges of the same polarity at the water–protein interface. These induced surface charges, the fourth source contributing to the electric field, stem from the fact that polar or carbonyl groups on the protein wall cannot rotate as freely as water molecules. Each of these four components has to be computed and added together to obtain the total electric force experienced by an ion at any given position at any given time. The stochastic force and electrical force acting on a charged particle together determine to which position it will have moved to in a short time interval.

To carry out BD simulations of ion channels, one needs to specify the boundaries of the system. This is a relatively simple problem for one-dimensional BD simulations (Cooper et al., 1985; Jakobsson and Chiu, 1987; Bek and Jakobsson, 1994), but requires the addition of reservoirs to the channel system in the more realistic case of three-dimensional BD simulations. In several recent studies, a simple stochastic boundary has been used successfully in applications of BD simulations to a number of ion channels (Chung et al., 1998, 1999; Corry et al., 2001). When an ion strikes the reservoir boundary during simulations, it is elastically scattered back into the reservoir, equivalent to letting an ion enter the reservoir whenever one leaves the simulation system. Thus the concentrations of ions in the reservoirs are maintained at the desired values at all times. During simulations of current measurements, the chosen concentration values in the reservoirs are maintained by recycling ions from one side to the other whenever there is an imbalance due to a conduction event, mimicking the current flow through a closed circuit.

15.2.3 Two Simplifying Assumption of Brownian Dynamics

The ability to compute current flow across ion channels confers a distinct advantage to BD compared to other simulation techniques. To trace the trajectories of about 100 ions interacting with a dielectric boundary for many microseconds, a period long enough to deduce the conductance of an ion channel, BD makes two simplifying assumptions. First, water is not treated explicitly but as a continuum. In reality, ions collide with neighboring water molecules incessantly and the net effects of these collisions are lumped together and treated as the frictional and random forces. Second, the atoms forming the channel are considered to be rigid whereas in reality they will undergo rapid thermal fluctuations. Several independent lines of evidence suggest that root-mean-square fluctuations of typical proteins are of the order of 0.75 Å, suggesting that the transmembrane passage through which ions traverse may be quite flexible (Allen et al., 2004; Noskov et al., 2004). By making these simplifications it is possible to measure channel conductances under various conditions and compare these measurements with experimental findings with only a modest amount of computational power.

Since the water and protein in BD are already represented as continuous media, the forces acting on charged particles are most often calculated by solving Poisson's equation. A crucial issue is whether such a continuum approximation can be justified in a narrow, biological nanotube. In bulk water, molecules polarize so as to shield electrostatic interactions by a factor of approximately 1/80. However, given the likely preferential alignment of water in narrow pores and regions of high charge, this shielding is likely to be far less effective in an ion channel. Thus, one should use a lower value of the dielectric constant for the water in the channel when solving Poisson's equation. But exactly what value of the dielectric constant should be used is unknown. Determining the appropriate values using molecular dynamics simulations or otherwise would be a useful project.

Assigning the appropriate value of the dielectric constant of protein is also non-trivial. Unlike water and lipids, which form homogeneous media, proteins are quite heterogeneous, exhibiting large variations in polarizability depending on whether we are dealing with the interior or exterior of a protein (Schutz and Warshel, 2001). There are several molecular dynamics studies of the dielectric constant of protein (Smith et al., 1993; Simonson and Brooks, 1996; Pitera et al., 2001). The dielectric constant for the whole protein varies between 10 and 40, but when only the interior region of the protein consisting of the backbone and uncharged residues is considered, the value drops to 2 or 4. The effects of changing the dielectric constant of protein from 3.5 to 5 were examined by Chung et al. (2002a), using the KcsA potassium channel. They showed that the precise value adopted in solving Poisson's equation has negligible effects on the macroscopic properties derived from BD simulations.

The validity of treating the channel protein as a static structure in BD also deserves further investigation. It should be noted that thermal fluctuations of proteins occur in the time-scale of femtoseconds, whereas a conduction event across a typical ionic channel takes place once in 100 ns—approximately 6 to 7 orders of magnitude slower time-scale. Thus, it is likely that rapid thermal fluctuations of the atoms forming the channel are not important for channel selectivity and conduction. This can be formally proved using stochastic averaging methods in nonlinear dynamical systems (e.g., Sanders and Verhulst, 1985). Alterations in the average positions of the protein atoms caused by the presence of permeating ions may play a role, and their effects should be examined both experimentally and by using molecular dynamics simulations. If found to be important, some of the motions of the protein, such as the bending of carbonyl groups, can readily be incorporated in BD modeling of ion channels. Finally, size-dependent selectivity among ions with the same valence cannot be easily understood within the BD framework, and one has to appeal to molecular dynamics or semi-microscopic Monte Carlo simulations (Garofoli and Jordan, 2003; Jordan, 2005) for that purpose.

15.3 Application of Brownian Dynamics in Ion Channels

Despite the caveats to the use of BD, as outlined in the previous section, the technique has been fruitfully utilized in studying the dynamics of ion permeation in a number of ion channels. An obvious application of BD is the calculation of current–voltage and conductance–concentration curves, which can be directly compared to the physiological measurements to assess the reliability and predictive power of the method. In addition to simple counting of ions crossing the channel, one can carry out a trajectory analysis of ions in the system to determine the steps involved in conduction. It is useful to find out the binding sites and the average number of ions in the channel, both of which are experimentally observable quantities. It is also possible to study the mechanisms of blocking of channels by larger molecules or other ion species. We summarize here some of the computational studies carried out on two important classes of biological ion channels—the KcsA potassium channel and ClC Cl^- channel.

15.3.1 Potassium Channels

KcsA K^+ Channel: To determine currents flowing across the channel, Chung et al. (1999, 2002a) and others (Mashl et al., 2001; Burykin et al., 2002) have performed BD simulations on the KcsA channel using the experimentally-determined channel structure. The shape of the ion-conducting pathway across the KcsA protein is illustrated in Fig. 15.1. The KcsA structure determined from X-ray diffraction consists of 396 amino acid residues, or 3504 atoms excluding polar hydrogens. The channel is constructed from four subunits of a tetramer of peptide chains, each subunit consisting of an outer helix, inner helix, pore helix, and a TVGYG (threonine–valine–glycine–tyrosine–glycine) amino acid sequence that forms the selectivity filter. The protein atoms form a central pore between these subunits. An outline of the pore reveals that the channel is composed of three segments—a long intracellular region of length 20 Å lined with hydrophobic amino acids extending toward the intracellular space (left-hand side in the Inset), a wide water-filled chamber of length 10 Å, and a narrow selectivity filter of length 12 Å, extending toward the extracellular space. The selectivity filter is the most important element in this structure as it can distinguish K^+ ions from those of Na^+ on the basis of their sizes (the crystal radius of K^+ is 1.33 Å and that of Na^+ is 0.95 Å). BD simulations show that there are three regions in the selectivity filter and cavity where K^+ ions dwell preferentially (see Fig. 15.1). There is also another prominent binding site near the intracellular entrance of the channel. The preferred positions where ions dwell preferentially are in close agreement with the positions observed in Rb^+ X-ray diffraction maps (Doyle et al., 1998).

To illustrate the permeation mechanism across the potassium channel, the channel is bisected such that ions in the chamber and filter are consigned to the right side, and the rest to the left side. The most common situation in the conducting state of the channel has one ion in the left half, and two ions in the right half. This configuration is referred to as the [1, 2] state. A typical conduction event consists of the following transitions: [1, 2]→[0, 3]→[0, 2]→[1, 2]. In other words, the ion waiting near the intracellular mouth overcomes a small energy barrier in the intracellular pore to enter the chamber region. Because this system is unstable in the presence of an applied potential, the right-most ion is ejected from the channel. Another ion enters the intracellular mouth, leaving the system in its original configuration. The precise sequence of events taking place for conduction of ions depends on their concentration, applied potential and the ionization state of charged residues at the intracellular gate, and many other states can be involved in the conduction process depending on the values of these variables. Simulations also reveal that permeation across the filter is much faster than in other parts of the channel. That is, once a third ion reaches the oval cavity, the outermost ion in the selectivity filter is expelled almost instantaneously. Thus, although the filter plays a crucial role in selecting the K^+ ions, its role in influencing their conductance properties is minimal.

In Fig. 15.2A and B, we show the current–voltage and current–concentration curves obtained from BD simulations (Chung et al., 2002a). The results of BD simulations are in broad agreement with those determined experimentally (Coronado et al., 1980; Schrempf et al., 1995; Cuello et al., 1998; Heginbotham et al., 1999; Meuser et al., 1999; LeMasurier et al., 2001). When the radius of the intracellular gate of the crystal structure is expanded to 4 Å, the conductances at +150 mV and −150 mV are, respectively, 147 ± 7 and 96 ± 4 pS. The relationship is linear when the applied potential is in the physiological range but deviates from Ohm's law at a higher applied potential, especially at high positive potentials. The current saturates with increasing ionic concentrations, as shown in Fig. 15.2B. This arises because ion permeation across the channel is governed by two independent processes: the time it takes for an ion to enter the channel mouth depends on the concentration, while the time it takes for the ion to reach the oval chamber is independent of the concentration but depends solely on the applied potential.

Modeling other potassium channels: There are many different types of potassium channels, which differ widely in their conductances and gating characteristics while having a similar primary structure. Conductance levels of various types of potassium channels range from 4 to 270 pS (1 pS equals 0.1 pA of current across the channel with the driving force of 100 mV). Despite this diversity, they all share the common feature of being highly selective to potassium ions and display broadly similar selectivity sequences for monovalent cations.

To understand this feature, Chung et al. (2002b) investigated the possible structural differences that could give rise to different potassium channels. Using the experimentally determined potassium channel structure as a template, as shown in Fig. 15.3A, they systematically changed the radius of the intracellular pore entrance, leaving the dimensions of the selectivity filter and cavity unaltered. As the intrapore radius is increased from 2 to 5 Å, the channel conductance changes from 0.7 to 197 pS (0.17 to 48 pA). In Fig. 15.3B, the simulated current across the model ion channel determined from BD is plotted against the radius of the intrapore gate. By examining the energy profiles and the probabilities of ion occupancies in various segments of the channel, they deduce the rate-limiting step for conduction in the potassium channels. Ion distributions revealed that the selectivity filter is occupied by two K^+ ions most of the time. Potential energy profiles encountered by a third

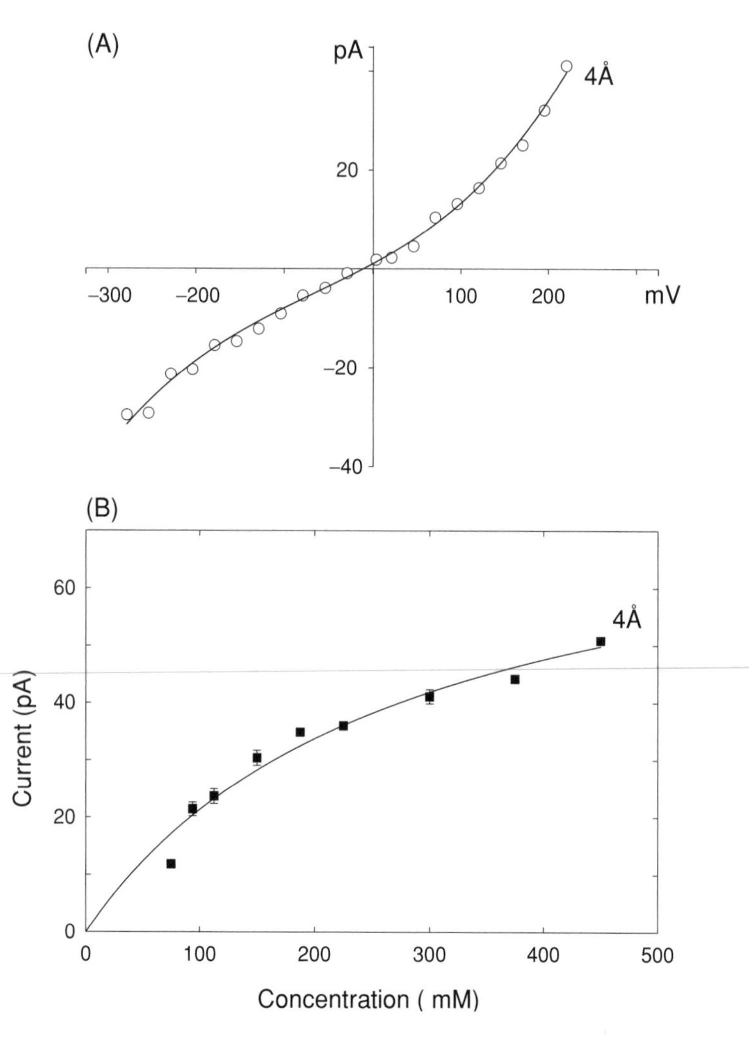

Fig. 15.2 The current–voltage–concentration profile of the K^+ channel with the intrapore radius of 4 Å. (A) The magnitude of current passing through the channel with symmetrical solution of 300 mM KCl in both reservoirs is plotted against the applied potential. (B) The outward currents are obtained with symmetrical solutions of varying concentrations of KCl in the reservoirs.

ion traversing along the central axis of the channel when there are two ions in or near the selectivity filter are shown for the channels with radii 2 Å (solid line in Fig. 15.3C), 3 Å (long-dashed line) and 4 Å (dashed line). Ions need to climb over the energy barrier, whose height is denoted as ΔU, to move across the channel. This barrier is the rate-limiting step in the permeation process: as its height increases with a decreasing intrapore radius, the channel conductance drops exponentially. Thus, the diversity of potassium channels seen in nature is achieved by slightly altering the geometry of the intracellular aspect of the channel macromolecule.

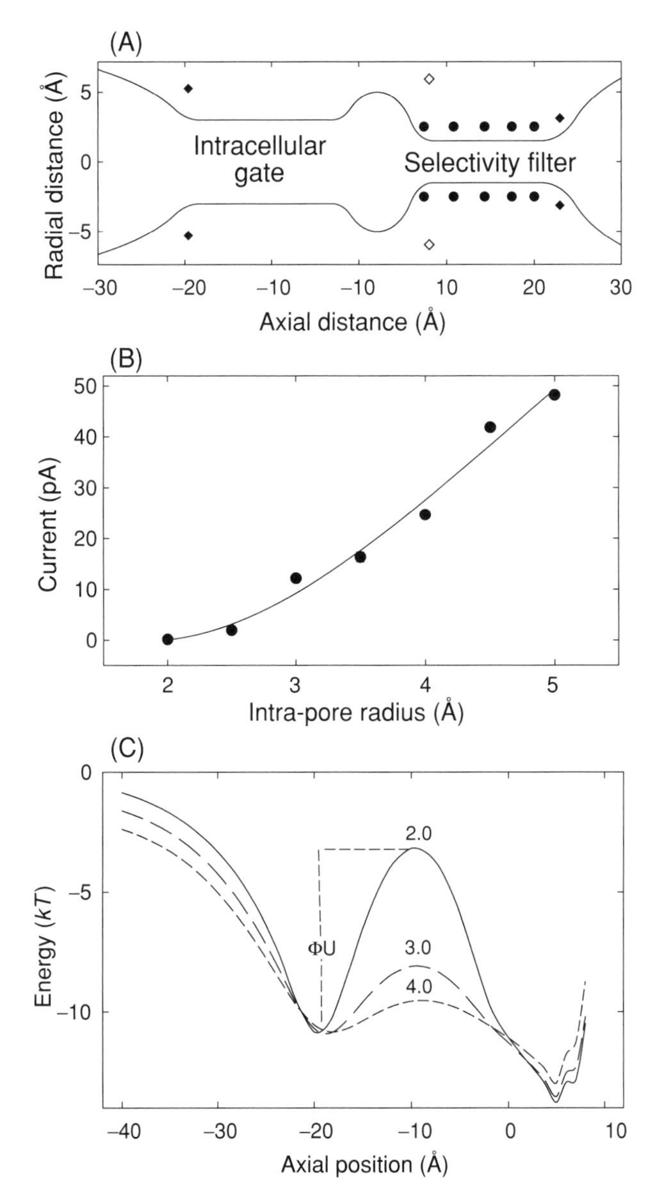

Fig. 15.3 Diversity of the potassium channels. (A) The shape of the KscA potassium is modified such that the minimal radius of the intracellular gate is 3 Å. The solid line shows the outline of a simplified model channel. The positions of dipoles on the channel wall are indicated. Filled circles are 10 of the 20 carbonyl oxygen atoms, open diamonds are N-termini of the helix dipole, and filled diamonds are mouth dipoles. (B) The dependence of outward channel currents on the intrapore radius of the channel is illustrated. The applied field to obtain the current is 2×10^7 V/m. (C) Potential energy profiles encountered by an ion traversing along the central axis of the channel when there are two other ions in or near the selectivity filter are shown for the channels with radii of 2 Å (solid line), 3 Å (long-dashed line) and 4 Å (dashed line). Ions need to climb over the energy barrier, whose height is denoted as ΔU, to move across the channel.

15.3.2 ClC Chloride Channels

BD simulations were similarly applied to elucidate the dynamics of ion permeation across ClC-type channels (Corry et al., 2004a,b). The prototype channel, known as ClC-0, first discovered and characterized by Miller (1982), is found in *Torpedo electroplax*. Since then, nine different human ClC genes and four plant and bacterial ClC genes have been identified. The ClC family of Cl^- channels is present in virtually all tissues—in muscle, heart, brain, kidney, and liver—and is widely expressed in most mammalian cells. By allowing Cl^- ions to cross the membrane, ClC channels perform diverse physiological roles, such as control of cellular excitability, cell volume regulation, and regulation of intracellular pH (Jenstsch et al., 1999; Maduke et al., 2000; Fahlke, 2001). Dutzler et al. (2002, 2003) determined the X-ray structure of a transmembrane ClC protein in bacteria, that has subsequently been shown to be a transporter, not an ion channel (Accardi and Miller, 2004). Nevertheless, many amino acid sequences of the bacterial ClC protein are conserved in their eukaryotic ClC relatives, which are selectively permeable to Cl^- ions.

Because the bacterial ClC protein shares many signature sequence identities with the eukaryotic ClC channels, it is possible to build homology models of these channels based on the structural information provided by Dutzler et al. (2002, 2003). With this aim in mind, Corry et al. (2004b) first altered the X-ray structure of the bacterial ClC protein using molecular dynamics to create an open-state configuration. They then converted to an open-state homology model of a eukaryotic ClC channel, ClC-0, using the crystal structure of the prokaryotic protein as a basis. As illustrated in Fig. 15.4A, the ionic pathway of ClC-0 takes a tortuous course through the protein, unlike that of the potassium channel, which is straight and perpendicular to the membrane surface. The channel is quite narrow, having a minimum radius of 2.5 Å near the center, but opens up quite rapidly at each end. The distance from one end of the pore to the other is 55 Å and it is lined with many charged and polar amino acid residues. Incorporating this homology model into BD, they determined the current–voltage–concentration profile of ClC-0. A current–voltage relationship obtained with symmetrical solutions of 150 mM in both reservoirs is shown in Fig. 15.4B. The relationship is linear, with a conductance of 11.3 ± 0.5 pS that agrees well with experimental measurements reported by Miller (1982) (superimposed open circles). The slope conductance determined from the experimental data is 9.4 ± 0.1 pS. The current–concentration relationship obtained from the homology model using BD (filled circles) is also accord with the experimental observations (obtained by Tsung-Yu Chen, personal communication) as shown in open circles in Fig. 15.4C. The lines fitted through the data points are calculated from the Michaelis–Menten equation. There is a reasonable agreement between the simulated data and experimental measurements for ClC-0.

BD simulations also reveal the steps involved in permeation of Cl^- ions across the ClC channel. The pore is normally occupied by two Cl^- ions. When a third ion enters the pore from the intracellular space (left-hand side in the Inset of Fig. 15.4A),

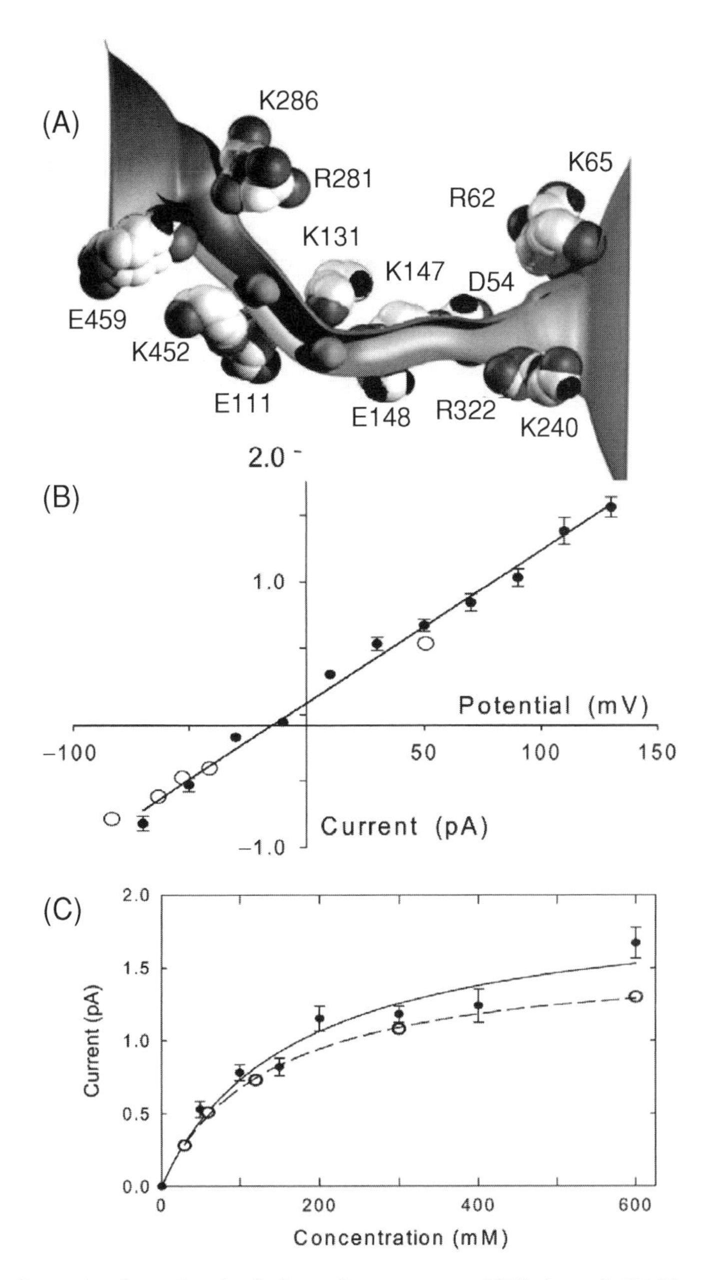

Fig. 15.4 Brownian dynamics simulations of an open-state ClC0 channel. (A) The water-filled pore of the channel through which Cl^- ions move is lined with both acidic and basic residues. The channel is normally occupied by two Cl^- ions, shown here in green. (B) The current–voltage relationship obtained from Brownian dynamics simulations (filled circles) is compared with the experimental data (open circles). (C) The current–concentration curve obtained with symmetrical solutions of varying concentrations of NaCl in the reservoirs under an applied potential of −80 mV (filled circles) is fitted with the Michaelis–Menten equation. The experimental measurements are shown in open circles. The half-saturation points determined from the fitted curves are 163 ± 51 mM for the simulated data and 136 ± 8 mM for the experimental data.

the stable equilibrium is disrupted, and the outermost Cl^- ion is expelled to the extracellular space.

15.4 Mathematical Formulation of Brownian Dynamics Algorithm

15.4.1 Overview

Here we provide a rigorous and mathematically complete formulation of the BD system for determining currents across a membrane ion channel. We prove that the continuous-time stochastic dynamical system, in which ions propagate via the Langevin equation, has a well-defined unique stationary distribution. We then show that the current across an ionic channel can be formulated in terms of mean passage rates of the ionic diffusion process, satisfying a boundary-valued partial differential equation, similar to the Fokker–Planck equation. We show that BD simulations can be viewed as a randomized algorithm for solving this partial differential equation to yield statistically consistent estimates of the currents flowing across an ionic channel.

Figure 15.5 shows the block diagram of BD simulation for permeation of ions through an ion channel. An iterative approach is used as follows: First, an initial estimate of the structural information of the channel, namely, the channel geometry and charges on the ionizable and polar residues in the protein are used to determine the parameters of Poisson's equation. Numerically solving Poisson's equation yields the potential of mean force (PMF) or energy landscape an ion traveling through the ion channel will experience. This in turn feeds into the BD simulation that

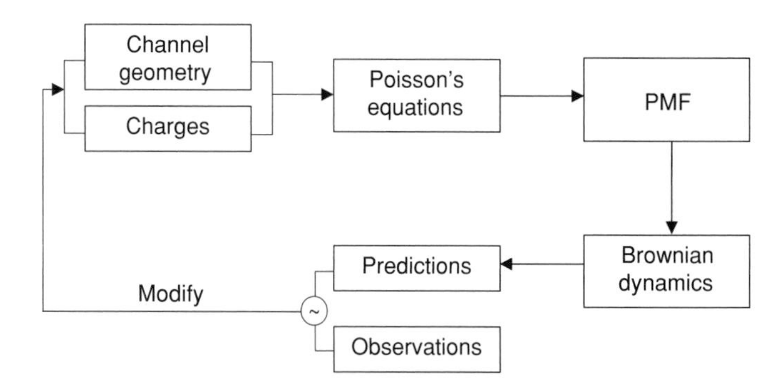

Fig. 15.5 A block diagram of the Brownian dynamics simulations. Using the channel shape of the channel and the charges on the atoms forming the protein, the profile of the potential of mean force along the central axis of the ion-conducting pathway is constructed by solving Poisson's equation. The currents derived from Brownian dynamics simulations are compared with those obtained experimentally. If there is a discrepancy between the simulated and experimental results, the channel geometry or the charges on atoms are modified and the procedure is repeated.

governs the stochastic evolution of all the ions. As a result of ions modeled by BD permeating through the ion channel, a simulated ion channel current is obtained. This simulated ion channel current is compared with the experimentally observed ion channel current. The difference between the two currents is used to refine our model of the channel geometry and charges and the process is repeated until the error between the simulated (predicted) ion channel current and experimentally determined ion channel current is minimized.

15.4.2 Mesoscopic Brownian Dynamics Formulation

The permeation model for the ion channel comprises two cylindrical reservoirs $\mathcal{R}_1$ and $\mathcal{R}_2$ connected by the ion channel C as depicted in Fig. 15.6, in which $2N$ ions are inserted (N denotes a positive integer). As an example we have chosen the gramicidin pore, although the results below hold for any ion channel. Throughout, we index the $2N$ ions by $i = 1, 2, \ldots, 2N$. These $2N$ ions comprise

- N positively charged ions indexed by $i = 1, 2, \ldots, N$. Of these, $N/2$ ions indexed by $i = 1, 2, \ldots N/2$ are in $\mathcal{R}_1$ and $N/2$ ions indexed by $i = N/2 + 1, \ldots, 2N$ are in $\mathcal{R}_2$. Each Na^+ ion has charge q^+, mass $m^{(i)} = m^+ = 3.8 \times 10^{-26}$ kg and frictional coefficient $m^+\gamma^+$, and radius r^+.
- N negatively charged ions. We index these by $i = N + 1, N + 2, \ldots, 2N$. Of these, $N/2$ ions indexed by $i = N + 1, \ldots, 3N/2$ are placed in $\mathcal{R}_1$ and the remaining $N/2$ ions indexed by $i = (3N/2) + 1, \ldots, 2N$ are placed in $\mathcal{R}_2$. Each negative ion has charge $q^{(i)} = q^-$, mass $m^{(i)} = m^-$, frictional coefficient $m^-\gamma^-$, and radius r^-.

Specifying the height of each reservoir to be N Å guarantees that the concentration of ions in them is at the physiological concentration of 150 mM.

Let $t \geq 0$ denote continuous time. Each ion i moves in three-dimenensional space over time. Let $\mathbf{x}_t^{(i)} = (x_t^{(i)}, y_t^{(i)}, z_t^{(i)})' \in \mathcal{R}$ and $\mathbf{v}_t^{(i)} \in \mathbb{R}^3$ denote the position and velocity of ion i at time t. Here and throughout this chapter all vectors are column vectors and denoted by the boldface font. Also we use $'$ to denote the transpose of a vector or matrix. The three components $x_t^{(i)}, y_t^{(i)}, z_t^{(i)}$ of $\mathbf{x}_t^{(i)} \in \mathbf{R}$ are, respectively, the x, y, and z position coordinates. Similarly, the three components of $\mathbf{v}_t^{(i)} \in \mathbb{R}^3$ are the x, y, z velocity components.

At time $t = 0$, the position $\mathbf{x}_0^{(i)}$ and velocity $\mathbf{v}_0^{(i)}$ of each of the $2N$ ions in the two reservoirs are randomly initialized as follows: The upper reservoir is divided into N cells of equal volume. In each cell is placed either one K^+ (or Na^+) or one Cl^- ion, each with probability half. The initial position $\mathbf{x}_0^{(i)}$ of ion i is chosen according to the uniform distribution within its cell. Similarly, the remaining $N/2$ K^+ ions $\{(N/2) + 1, \ldots, N\}$ and remaining $N/2$ Cl^- ions $\{(3N/2) + 1, \ldots, 2N\}$ are placed uniformly in the lower reservoir. This initialization of $\mathbf{x}_0^{(i)}$ emulates ensures that two particles are not placed too close to each other. The initial velocity vectors $\mathbf{v}_0^{(i)}$

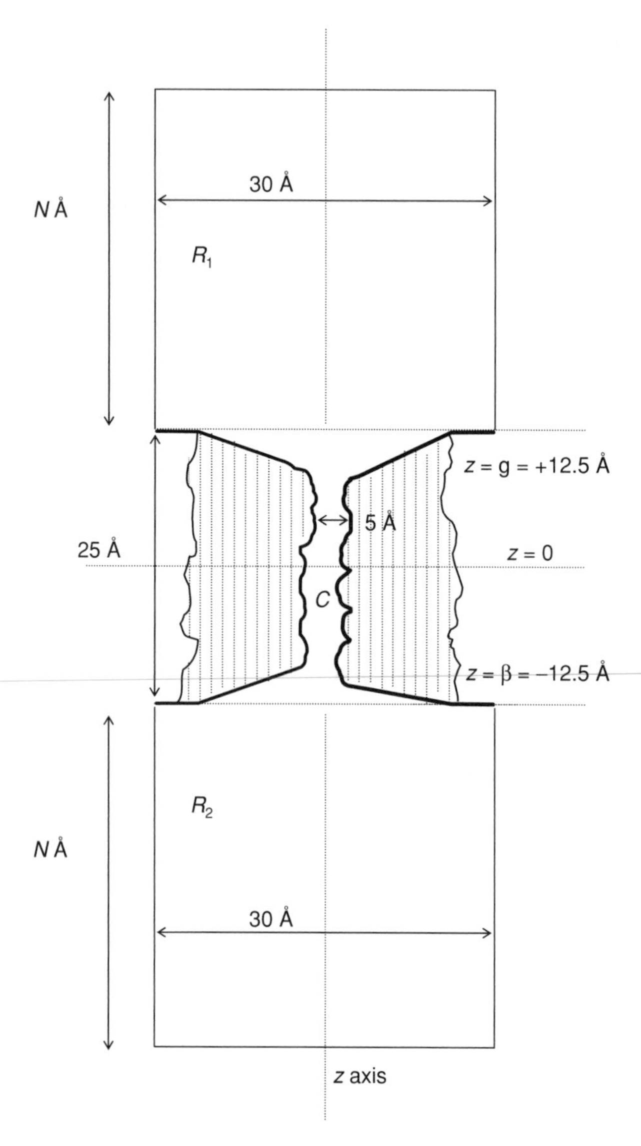

Fig. 15.6 A schematic illustration of a simulation assembly. The protein forming an ion channel, indicated with vertical dotted lines, is placed at the center of the assembly. For illustration, we use the gramicidin pore, whose length is approximately 25 Å. A reservoir, $\mathcal{R}_1$R and $\mathcal{R}_2$, containing ions is attached at each end of the channel. The dimensions of each reservoir are indicated. The two reservoirs are connected via a conduit formed by the channel protein $\mathcal{C}$.

of the $2N$ ions are typically initialized according to a three-dimensional Gaussian distribution with zero mean, and 3×3 diagonal positive definite covariance matrix. Thus the distribution of the magnitude of the initial velocity $|\mathbf{v}_0^{(i)}|$ has a Maxwell density.

An external potential $\Phi(\mathbf{x})$ is applied along the z axis of Fig. 15.6, i.e., with $\mathbf{x} = (x, y, z)$,

$$\Phi_\lambda^{\text{ext}}(\mathbf{x}) = -\text{E}^{\text{ext}} z \tag{15.1}$$

where $-\mathbf{E}^{\text{ext}}$ is the external field in V/m in z direction. Applied potential λ is related to $\mathbf{E}^{\text{ext}}$ by $\mathbf{E}^{\text{ext}} = -\lambda/l$, where l is the length of the channel, and $\lambda \in \Lambda$. Here, Λ denotes a finite set of applied potentials. Typically, $\Lambda = \{-200, -180, \ldots, 0, \ldots, 180, 200\}$ mV/m. Due to this applied external potential, K^+ ions drift from reservoir $\mathcal{R}_1$ to $\mathcal{R}_2$ via the ion channel $\mathcal{C}$ in Fig. 15.6.

Let $\mathbf{X}_t = \left(\mathbf{x}_t^{(1)\prime}, \mathbf{x}_t^{(2)\prime}, \mathbf{x}_t^{(3)\prime}, \ldots, \mathbf{x}_t^{(2N)\prime}\right)' \in \mathcal{R}^{2N}$ denote the positions and $\mathbf{V}_t = \left(\mathbf{v}_t^{(1)\prime}, \mathbf{v}_t^{(2)\prime}, \mathbf{v}_t^{(3)\prime}, \ldots, \mathbf{v}_t^{(2N)\prime}\right)' \in \mathbb{R}^{6N}$, denote the velocities of all the $2N$ ions at time $t \geq 0$. The position and velocity of each individual ion evolves according to the following continuous time stochastic dynamical system (recall $i = 1, 2, \ldots, N$ denote positive ions and $i = N + 1, \ldots, 2N$ denote negative ions):

$$\mathbf{x}_t^{(i)} = \mathbf{x}_0^{(i)} + \int_0^t \mathbf{v}_s^{(i)}\, ds, \tag{15.2}$$

$$m^+ \mathbf{v}_t^{(i)} = m^+ \mathbf{v}_0^{(i)} - \int_0^t m^+ \gamma^+(\mathbf{x}_s^{(i)}) \mathbf{v}_s^{(i)}\, ds + \int_0^t F_{\theta,\lambda}^{(i)}(\mathbf{X}_s)\, ds + \int_0^t b^+(\mathbf{x}_s^{(i)}) \mathbf{w}_s^{(i)}, \tag{15.3}$$

$$m^- \mathbf{v}_t^{(i)} = m^- \mathbf{v}_0^{(i)} - \int_0^t m^- \gamma^+(\mathbf{x}_s^{(i)}) \mathbf{v}_s^{(i)}\, ds + \int_0^t F_{\theta,\lambda}^{(i)}(\mathbf{X}_s)\, ds + \int_0^t b^-(\mathbf{x}_s^{(i)}) \mathbf{w}_s^{(i)}, \tag{15.4}$$

where $\gamma^\pm(\mathbf{x}_s^{(i)}) = \gamma^\pm$ (defined in the beginning of this section) if the ion is in the reservoir, and $\gamma(\mathbf{x}_s^{(i)})$ is determined by molecular dynamics simulation when the ion is in the ion channel (Allen et al., 2000). Eq. 15.2 says that velocity is the time derivative of the position. Eqs. 15.3 and 15.4 constitute the well-known *Langevin* equations. We now describe the various quantities in the above equations.

In Eqs. 15.3 and 15.4, the process $\{\mathbf{w}_t^{(i)}\}$ denotes a three-dimensional zero mean Brownian motion, which is component-wise independent. The constants b^+ and b^- are, respectively,

$$b^{+2}(\mathbf{x}_s^{(i)}) = 2m^+ \gamma^+(\mathbf{x}_s^{(i)}) kT, \qquad b^{-2}(\mathbf{x}_s^{(i)}) = 2m^- \gamma^-(\mathbf{x}_s^{(i)}) kT. \tag{15.5}$$

The noise processes $\{\mathbf{w}_t^{(i)}\}$ and $\{\mathbf{w}_t^{(j)}\}$, that drive any two different ions, $j \neq i$, are assumed to be statistically independent.

In Eqs. 15.3 and 15.4, $F_{\theta,\lambda)}^{(i)}(\mathbf{X}_t) = -q^{(i)} \nabla_{\mathbf{x}_t^{(i)}} \Phi_{\theta,\lambda}^{(i)}(\mathbf{X}_t)$ represents the *systematic force* acting on ion i, where the scalar-valued process $\Phi_{\theta,\lambda}^{(i)}(\mathbf{X}_t)$ is the total electric potential experienced by ion i given the position $\mathbf{X}_t$ of the $2N$ ions. The subscript λ is the applied external potential in Eq. 15.1. The subscript θ is a parameter vector that characterizes the PMF, which is an important component of $\Phi_{\theta,\lambda}^{(i)}(\mathbf{X}_t)$. As described below, $F^{(i)}_{\theta,\lambda}(\mathbf{X}_t)$ includes an ion–wall interaction force that ensures that position $\mathbf{x})^{(i)}_t$ of each ion lies in $\mathcal{R}$—see Eq. 15.9 below.

It is notationally convenient to represent the above system, Eqs. 15.2, 15.3 and 15.4 as a vector stochastic differential equation. Define the following vector-valued variables:

$$\mathbf{V}_t = \begin{bmatrix} \mathbf{V}^+_t \\ \mathbf{V}^-_t \end{bmatrix}, \text{ where } \mathbf{V}^+_t = \begin{bmatrix} \mathbf{v}^{(1)}_t \\ \vdots \\ \mathbf{v}^{(N)}_t \end{bmatrix}, \mathbf{V}^-_t = \begin{bmatrix} \mathbf{v}^{(N+1)}_t \\ \vdots \\ \mathbf{v}^{(2N)}_t \end{bmatrix}, \mathbf{w}_t = \begin{bmatrix} \mathbf{0}_{2N\times 1} \\ \mathbf{w}^{(1)}_t \\ \vdots \\ \mathbf{w}^{(2N)}_t \end{bmatrix},$$

$$\zeta_t = \begin{bmatrix} \mathbf{X}_t \\ \mathbf{V}^+_t \\ \mathbf{V}^-_t \end{bmatrix}, \mathbf{F}^+_{\theta,\lambda}(\mathbf{X}_t) = \begin{bmatrix} F^{(1)}_{\theta,\lambda}(\mathbf{X}_t) \\ \vdots \\ F^{(N)}_{\theta,\lambda}(\mathbf{X}_t) \end{bmatrix}, \quad \mathbf{F}^-_{\theta,\lambda}(\mathbf{X}_t) = \begin{bmatrix} F^{(N+1)}_{\theta,\lambda}(\mathbf{X}_t) \\ \vdots \\ F^{(2N)}_{\theta,\lambda}(\mathbf{X}_t) \end{bmatrix}. \quad (15.6)$$

The above system, namely, Eqs. 15.2, 15.3, 15.4 can be written as

$$\zeta_t = \zeta_0 + \int_0^t \mathbf{A}(\mathbf{X}_\tau)\zeta_\tau \, d\tau + \int_0^t \mathbf{f}(\zeta_\tau)\, d\tau + \int_0^t \Sigma^{1/2}(\mathbf{X}_\tau)\mathbf{w}_{\tau s}\, d\tau \quad (15.7)$$

where $\Sigma^{1/2}(\mathbf{X}_\tau) = \operatorname{diag}(\mathbf{0}_{6N\times 6N}, b^+(\mathbf{X}_\tau)/m^+, b^-(\mathbf{X}_\tau)/m^-)$, I_{6N} denotes the $6N \times 6N$ identity matrix,

$$\mathbf{A} = \left[\begin{array}{c|cc} \mathbf{0}_{6N\times 6N} & \multicolumn{2}{c}{\mathbf{I}_{6N}} \\ \hline \mathbf{0}_{6N\times 6N} & -\gamma^+(\mathbf{X}_\tau) & \mathbf{0}_{3N\times 3N} \\ & \mathbf{0}_{3N\times 3N} & -\gamma^-(\mathbf{X}_\tau) \end{array}\right], \quad \mathbf{f}(\zeta_t) = \begin{bmatrix} \mathbf{0}_{6N\times 1} \\ \frac{1}{m^+}\mathbf{F}^+_{\theta,\lambda}(\mathbf{X}_t) \\ \frac{1}{m^-}\mathbf{F}^-_{\theta,\lambda}(\mathbf{X}_t) \end{bmatrix}. \quad (15.8)$$

We will subsequently refer to Eqs. 15.7 and 15.8 as the BD equations for a biological ion channel.

Remark: Another way of ensuring that the positions $\mathbf{x}^{(i)}_t$ of all ions are in $\mathcal{R}^o$ is to introduce a reflection term $\mathbf{Z}_t$ that models elastic collisions at the boundary of $\mathcal{R}$. However, as described below, the ion wall systematic interaction force ensures that all ions remain in $\mathcal{R}^o$. Hence we do not consider a reflected diffusion formulation in this chapters.

15.4.3 Systematic Force Acting on Ions

As mentioned after Eq. 15.4, the systematic force experienced by each ion i is

$$\mathbf{F}^{(i)}_{\theta,\lambda}(\mathbf{X}_t) = -q^{(i)}\nabla_{\mathbf{x}^{(i)}_t}\Phi^{(i)}_{\theta,\lambda}(\mathbf{X}_t),$$

where the scalar-valued process $\Phi^{(i)}_{\theta,\lambda}(\mathbf{X}_t)$ denotes the total electric potential experienced by ion i given the position $\mathbf{X}_t$ of all the $2N$ ions. We now give a detailed formulation of these systematic forces.

The potential $\Phi^{(i)}_{\theta,\lambda}(\mathbf{X}_t)$ experienced by each ion i comprises the following five components:

$$\Phi^{(i)}_{\theta,\lambda}(\mathbf{X}_t) = U_\theta(\mathbf{x}^{(i)}_t) + \Phi^{\text{ext}}_\lambda(\mathbf{x}^{(i)}_t) + \Phi^{IW}(\mathbf{x}^{(i)}_t) + \Phi^{C,i}(\mathbf{X}_t) + \Phi^{SR,i}(\mathbf{X}_t). \qquad (15.9)$$

Just as $\Phi^{(i)}_{\theta,\lambda}(\mathbf{X}_t)$ is decomposed into five terms, we can similarly decompose the force $F^{(i)}_{\theta,\lambda}(\mathbf{X}_t) = -q\nabla_{\mathbf{x}^{(i)}_t}\Phi^{(i)}_{\theta,\lambda}(\mathbf{X}_t)$ experienced by ion i as the superposition (vector sum) of five force terms, where each force term is due to the corresponding potential in Eq. 15.9—however, for notational simplicity we describe the scalar-valued potentials rather than the vector-valued forces.

Note that the first three terms in Eq. 15.9, namely $U_z\theta(\mathbf{x}^{(i)}_t)$, $\Phi^{\text{ext}}_\lambda(\mathbf{x}^{(i)}_t)$, $\Phi^{IW}(\mathbf{x}^{(i)}_t)$ depend only on the position $\mathbf{x}^{(i)}_t$ of ion i, whereas the last two terms in Eq. 15.9 $\Phi^{C,i}(\mathbf{X}_t)$, $\Phi^{SR,i}(\mathbf{X}_t)$ depend on the distance of ion i to all the other ions, namely, the position $\mathbf{X}_t$ of all the ions. The five components in Eq. 15.9 are now defined.

(i) *Potential of mean force* (PMF), denoted $U_\theta(\mathbf{x}^{(i)}_t)$ in Eq. 15.9, comprises electric forces acting on ion i when it is in or near the ion channel (nanotube) $\mathcal{C}$ in Fig. 15.6. The PMF U_θ is a smooth function of the ion position $\mathbf{x}^{(i)}_t$ and depends on the structure of the ion channel. Therefore, estimating $U_\theta(\cdot)$ yields structural information about the ion channel. The PMF U_θ originates from fixed charges in the channel protein and surface charges induced by mobile ions.

(ii) *External applied potential*: In the vicinity of living cells, there is a strong electric field resulting from the membrane potential, which is generated by diffuse, unpaired, ionic clouds on each side of the membrane. Typically, this resting potential across a cell membrane, whose thickness is about 50 Å, is 70 mV, the cell interior being negative with respect to the extracellular space.

For ion i at position $\mathbf{x}^{(i)}_t = (x, y, z)$, $\Phi^{\text{ext}}_\lambda(\mathbf{x}) = \lambda z$ (see Eq. 15.1) denotes the potential on ion i due to the applied external field. The electrical field acting on each ion due to the applied potential is therefore $-\nabla_{\mathbf{x}^{(i)}_t}\Phi^{\text{ext}}_\lambda = (0, 0, \lambda)$ V/m at all $\mathbf{x} \in \mathcal{R}$. It is this applied external field that causes a drift of ions from the reservoir $\mathcal{R}_1$ to $\mathcal{R}_2$ via the ion channel $\mathcal{C}$. As a result of this drift of ions within the electrolyte in the two reservoirs, eventually the measured potential drop across the reservoirs is zero and all the potential drop occurs across the ion channel.

(iii) *Inter–ion Coulomb potential*: In Eq. 15.9, $\Phi^{C,i}(\mathbf{X}_t)$ denotes the Coulomb interaction between ion i and all the other ions.

$$\Phi^{C,i}(\mathbf{X}_t) = \frac{1}{4\pi\epsilon_0}\sum_{j=1,j\neq i}^{2N}\frac{q^{(j)}}{\epsilon_w\|\mathbf{x}^{(i)}_t - \mathbf{x}^{(j)}_t\|} \qquad (15.10)$$

(iv) *Ion–wall interaction potential*: The ion–wall potential Φ^{IW}, also called the $(\sigma/r)^9$ potential, ensures that the position of all ions $i = 1, \ldots, 2N$ lie in $\mathcal{R}^o$. With $\mathbf{x}_t^{(i)} = (x_t^{(i)}, y_t^{(i)}, z_t^{(i)})'$, it is modeled as

$$\Phi^{IW}(\mathbf{x}_t^{(i)}) = \frac{F_0}{9} \frac{(r^{(i)} + r_w)^9}{\left[r_c + r_w - \left(\sqrt{(x_t^{(i)2} + y_t^{(i)2}}\right)\right]^9}, \tag{15.11}$$

where for positive ions $r^{(i)} = r^+$ (radius of K^+ atom) and for negative ions $r^{(i)} = r^-$ (radius of Cl^- atom), $r_w = 1.4$ Å is the radius of atoms making up the wall, r_c denotes the radius of the ion channel, and $F_0 = 2 \times 10^{-10}$ N which is estimated from the ST2 water model used in molecular dynamics (Stillinger and Rahman, 1974). This ion–wall potential results in short-range forces that are only significant when the ion is close to the wall of the reservoirs $\mathcal{R}_1$ and $\mathcal{R}_2$ or anywhere in the ion channel $\mathcal{C}$ (since the narrow segment of an ion channel can be comparable in radius to the ions).

(v) *Short-range potential*: Finally, at short ranges, the Coulomb interaction between two ions is modified by adding a potential $\Phi^{SR,i}(\mathbf{X}_t)$, which replicates the effects of the overlap of electron clouds. Thus,

$$\Phi^{SR,i}(\mathbf{X}_t) = \frac{F_0}{9} \sum_{j=1, j \neq i}^{2N} \frac{(r^{(i)} + r^{(j)})}{\|\mathbf{x}_t^{(i)} - \mathbf{x}_t^{(j)}\|^9}. \tag{15.12}$$

Similar to the ion–wall potential, $\Phi^{SR,i}$ is significant only when ion i gets very close to another ion. It ensures that two opposite charge ions attracted by inter-ion Coulomb forces (Eq. 15.10) cannot collide and annihilate each other. Molecular dynamics simulations show that the hydration forces between two ions add further structure to the $1/|\mathbf{x}_t^{(i)} - \mathbf{x}_t^{(j)}\|^9$ repulsive potential due to the overlap of electron clouds in the form of damped oscillations (Guàrdia et al., 1991a, b). Corry et al. (2001) incorporated the effect of the hydration forces in Eq. 15.12 in such a way that the maxima of the radial distribution functions for Na^+–Na^+, Na^+–Cl^-, and Cl^-–Cl^- would correspond to the values obtained experimentally.

15.5 Probabilistic Characterization of Channel Conductance

Thus far, Eqs. 15.7–15.9 give a complete description of the stochastic dynamics of the ions. We now demonstrate here that the mean ion channel current satisfies a boundary-valued partial differential equation related to the Fokker–Planck equation. There are two main results in this section that logically progress toward deriving this partial differential equation. Theorem 1 shows that the BD system, given in Eq. 15.7, converges exponentially fast to a unique stationary distribution. Theorem 3 gives a characterization for the ion channel current in terms of the mean first passage time of the diffusion process (Eq. 15.7).

To motivate these results, we first formalize mathematically the construction of the BD simulation. There are two key requirements that a mathematical construction of the BD simulation should take into account: First, the concentration of ions in each reservoir $\mathcal{R}_1$ and $\mathcal{R}_2$ should remain approximately constant and equal to the physiological concentration. Note that if the system was allowed to evolve for an infinite time with the channel open, then eventually due to the external applied potential, more ions will be in $\mathcal{R}_2$ than $\mathcal{R}_1$. This would violate the condition that the concentration of particles in $\mathcal{R}_1$ and $\mathcal{R}_2$ remain constant.

Second, the dynamics of the BD simulation has an inherent two-time scale property. Typically, the time for an ion to enter and propagate through the ion channel is at least an order of magnitude larger compared to the time it takes for an ion to move within a reservoir. That is, the time constant for the particles in the reservoirs to attain steady state is much smaller than the time it takes for a particle to enter and propagate through the channel.

The following two step probabilistic construction formalizes the above two requirements and ensures that they are satisfied.

Procedure 1: Probabilistic construction of Brownian dynamics ion permeation in ion channels

- *Step 1*: The $2N$ ions in the system are initialized as described in Eq. 15.1 and the ion channel $\mathcal{C}$ is closed. The system evolves and attains stationarity. Theorem 1 below shows that the probability density function of the $2N$ particles converges exponentially fast to a unique stationary distribution. Theorem 3 shows that in the stationary regime, all positive ions in reservoir $\mathcal{R}_1$ have the same stationary distribution and so are statistically indistinguishable (similarly for $\mathcal{R}_2$).
- *Step 2*: After stationarity is achieved, the ion channel is opened. The ions evolve according to Eq. 15.7. As soon as an ion from $\mathcal{R}_1$ crosses the ion channel $\mathcal{C}$ and enters $\mathcal{R}_2$, the experiment is stopped. Similarly, if an ion from $\mathcal{R}_2$ crosses $\mathcal{C}$ and enters $\mathcal{R}_1$, the experiment is stopped. Theorem 3 gives partial differential equations for the mean time an ion in $\mathcal{R}_1$ takes to cross the ion channel and reach $\mathcal{R}_2$ (and for the time it takes an ion to cross from $\mathcal{R}_2$ to $\mathcal{R}_1$). From this a theoretical expression for the mean ion channel current is constructed (Eq. 15.24).

These two steps constitute one iteration of the BD simulation Algorithm 1. The construction of restarting the simulation each time an ion crosses the channel ensures that the random amount of time for an ion to cross the ion channel in any BD simulation iteration is statistically independent of the time for any other iteration. This statistical independence will be exploited in Theorem 3 to show that the BD algorithm yields statistically consistent estimates of the ion channel current.

Remarks. The above construction is a mathematical idealization. In actual BD algorithms, the ion channel is kept open and ions that cross the channel are simply removed and replaced in their original reservoir. However, as described later (following Algorithm 1), the above mathematical construction is an excellent approximation due to the fact that by virtue of Step 1, the system of particles with the newly replaced ion converges exponentially fast to its stationary distribution, and by virtue of the two time scale property, the time taken to attain this stationary distribution is much less than the time it takes for a single ion to cross the ion channel.

With the above mathematical construction of the BD simulation, we now proceed to stating and proving the main results. Let

$$\pi_t^{(\theta,\lambda)}(\mathbf{X},(\mathbf{V}) = p^{(\theta,\lambda)}\big(\mathbf{x}_t^{(1)},\mathbf{x}_t^{(2)},\ldots,\mathbf{x}_t^{(2N)},\mathbf{v}_t^{(1)},\mathbf{v}_t^{(2)},\ldots,\mathbf{v}_t^{(2N)}\big) \tag{15.13}$$

denote the joint probability density function of the position and velocity of all the $2N$ ions at time t. We explicitly denote the θ, λ dependence of the probability density functions since they depend on the PMF U_θ and applied external potential λ. Note that the marginal probability density function $\pi_t^{(\theta,\lambda)}(\mathbf{X}) = p_t^{(\theta,\lambda)}(\mathbf{x}_t^{(1)},\mathbf{x}_t^{(2)},\ldots,\mathbf{x}_t^{(2N)})$ of the positions of all $2N$ ions at time t is obtained as

$$\pi_t^{(\theta,\lambda)}(\mathbf{X}) = \int_{\mathbb{R}^{6N}} \pi_t^{(\theta,\lambda)}(\mathbf{X},(\mathbf{V})\,d\mathbf{V}.$$

The following result, the proof of which is not given here, states that for the above stochastic dynamical system, $\pi_t^{(\theta,\lambda)}(\mathbf{X},\mathbf{V})d\mathbf{V}$ converges exponentially fast to its stationary (invariant) distribution $\pi_\infty^{\theta,\lambda)}(\mathbf{X},V)$. That is, the ions in the two reservoirs attain steady state exponentially fast.

Theorem 1. *Consider Step 1 of the BD probabilistic construction in Procedure 1. For the BD system, represented in Eqs. 15.7 and 15.8, comprising* $2N$ *ions, with* $\zeta = (\mathbf{X},\mathbf{V})$*, there exists a unique stationary distribution* $\pi_\infty^{(\theta,\lambda)}(\zeta)$*, and constants* $K > 0$ *and* $0 < \rho < 1$*, such that*

$$\sup_{\zeta\in\mathcal{R}^{2N}\times\mathbb{R}^{6N}} |\pi_t^{(\theta,\lambda)}(\zeta) - \pi_\infty^{(\theta,\lambda)}(\zeta)| \le K\mathcal{V}(\zeta)\rho^t. \tag{15.14}$$

Here $\mathcal{V}(\zeta) > 1$ *is an arbitrary measurable function on* $\mathcal{R}^{2N}\times\mathbb{R}^{6N}$.

The next result to establish is, under the conditions of Step 1, the ions in the two reservoirs are statistically indistinguishable. Let us first introduce the following notation and the *Fokker–Planck* equation.

Notation. For $\zeta = (\zeta^{(1)},\ldots,\zeta^{(4N)})'$, define the gradient operator

$$\nabla_\zeta = \left(\frac{\partial}{\partial\zeta^{(1)}},\frac{\partial}{\partial\zeta^{(2)}},\ldots,\frac{\partial}{\partial\zeta^{(4N)}}\right)'$$

For a vector field $\mathbf{f}(\zeta) = \left[f^{(1)}(\zeta)\ f^{(2)}(\zeta)\ \cdots\ f^{(4N)}(\zeta)\right]'$ defined on $\mathbb{R}^{4N}$, define the divergence operator

$$\operatorname{div}(\mathbf{f}_{\theta,\lambda}) = \frac{\partial f^{(1)}}{\partial \zeta^{(1)}} + \frac{\partial f^{(2)}}{\partial \zeta^{(2)}} + \cdots + \frac{\partial f^{(4N)}}{\partial \zeta^{(4N)}}. \tag{15.15}$$

For the stochastic dynamical system (Eq. 15.7) comprising of $2N$ ions, define the backward elliptic operator (infinitesimal generator) $\mathcal{L}$ and its adjoint $\mathcal{L}^*$ for any test function $\phi(\zeta)$ as

$$\mathcal{L}(\phi) = \frac{1}{2}\operatorname{Tr}[\Sigma \nabla_\zeta^2 \phi(\zeta)] + (\mathbf{f}_{\theta,\lambda}(\zeta) + \mathbf{A}\zeta)' \, \nabla_\zeta \phi(\zeta) \tag{15.16}$$

$$\mathcal{L}^*(\phi) = \frac{1}{2}\operatorname{Tr}\left[\nabla_\zeta^2(\Sigma\phi(\zeta))\right] - \operatorname{div}[(\mathbf{A}\zeta + \mathbf{f}_{\theta,\lambda}(\zeta))\phi(\zeta)].$$

Here, $\mathbf{f}_{\theta,\lambda}$ and Σ are defined in Eq. 15.8. We refer the reader to Karatzas and Shreve (1991) for an exposition of stochastic differential equations driven by Brownian motion.

It is well known that the probability density function $\pi_t^{(\theta,\lambda)}(\cdot)$ of the $2N$ ions where $\zeta_t = (\mathbf{X}_t', \mathbf{V}_t')'$ (defined in Eq. 15.13) satisfies the Fokker–Planck equation (Wong and Hajek, 1985):

$$\frac{\partial \pi_t^{\theta,\lambda}}{\partial t} = \mathcal{L}^* \pi_t^{(\theta,\lambda)}, \tag{15.17}$$

where $\pi_0^{(\theta,\lambda)}$ is initialized as described in Eq. 15.1. We refer the reader to Wong and Hajek (1985) for an excellent treatment of the Fokker–Planck equation. Briefly, the Fokker–Planck equation may be merely viewed as a partial differential equation (involving derivatives with respect to the state ζ and time t) that determines the time evolution of the probability density function $\pi_t^{(\theta,\lambda)}$.

Also, the stationary probability density function $\pi_\infty^{(\theta,\lambda)}(\cdot)$ satisfies

$$\mathcal{L}^*(\pi_\infty^{(\theta,\lambda)}) = 0, \quad \int_{\mathbb{R}^{6N}} \int_{\mathcal{R}^{2N}} \pi_\infty^{(\theta,\lambda)}(\mathbf{X}, \mathbf{V})\, d\mathbf{X}\, d\mathbf{V} = 1 \tag{15.18}$$

The intuition behind this is that if $\pi_t^{(\theta,\lambda)}$ attains "steady state" (stationarity), it no longer evolves with time, i.e., its derivative with respect to time is zero. Hence, setting the left-hand side of Eq. (15.17) to zero yields the above equation.

We next show that once stationarity has been achieved in Step 1, the N positive ions behave statistically identically, i.e., each ion has the same stationary marginal distribution. Define the stationary marginal density $\pi_{\infty}^{(\theta,\lambda)}(\mathbf{x}^{(i)}, \mathbf{v}^{(i)})$ of ion i as

$$\pi_{\infty}^{(\theta,\lambda)}(\mathbf{x}^{(i)}, \mathbf{v}^{(i)}) = \int_{\mathbb{R}^{6N-3}} \int_{\mathcal{R}^{2N-1}} \pi_{\infty}^{(\theta,\lambda)}(\mathbf{X}, \mathbf{V}) \prod_{j=1, j \neq i}^{2n} d\mathbf{x}^{(j)} d\mathbf{v}^{(j)} \tag{15.19}$$

We state the following result without the proof.

Theorem 2. *Consider Step 1 of the BD probabilistic construction in Procedure 1. Then the stationary marginal densities for the positive ions in* $\mathcal{R}_1$ *are identical:*

$$\pi_{\infty}^{(\theta,\lambda),\mathcal{R}_1} \equiv \pi_{\infty}^{(\theta,\lambda)}(\mathbf{x}^{(1)}, \mathbf{v}^{(1)}) = \pi_{\infty}^{(\theta,\lambda)}(\mathbf{x}^{(2)}, \mathbf{v}^{(2)}) = \cdots = \pi_{\infty}^{(\theta,\lambda)}(\mathbf{x}^{(N)}, \mathbf{v}^{(N/2)}). \tag{15.20}$$

Similarly, the stationary marginal densities for the positive ions in $\mathcal{R}_2$ are identical:

$$\begin{aligned}\pi_{\infty}^{(\theta,\lambda),\mathcal{R}_2} \equiv \pi_{\infty}^{(\theta,\lambda)}(\mathbf{x}^{(N/2+1)}, \mathbf{v}^{(N/2+1)}) &= \pi_{\infty}^{(\theta,\lambda)}(\mathbf{x}^{(N/2+2)}, \mathbf{v}^{(N/2+2)}) \\ &= \cdots = \pi_{\infty}^{(\theta,\lambda)}(\mathbf{x}^{(N)}, \mathbf{v}^{(N)}).\end{aligned} \tag{15.21}$$

Theorem 2 is not surprising—as Eqs. 15.2, 15.3 and 15.4 are symmetric in i, one would intuitively expect that once steady state has been attained, all the positive ions behave identically—similarly with the negative ions. Due to above result, once the system has attained steady state, any positive ion is representative of all the N positive ions, and similarly for the negative ions.

Having discussed Step 1, we now proceed to Step 2 of the BD probabilistic construction of Procedure 1. Assume that the system (Eq. 15.7) comprising $2N$ ions has attained stationarity with the ion channel $\mathcal{C}$ closed according to Step 1. Now in Step 2 of Procedure 1, the ion channel is opened so that ions can diffuse into it. Our key result below is to give a boundary-valued partial differential equation for the mean first passage time for an ion to cross the ion channel—this immediately yields an equation for the ion channel current.

Let $\tau_{\mathcal{R}_1,\mathcal{R}_2}^{(\theta,\lambda)}$ denote the mean first passage time for any of the $N/2$ K^+ ions in $\mathcal{R}_1$ to travel to $\mathcal{R}_2$ via the channel $\mathcal{C}$, and $\tau_{\mathcal{R}_2,\mathcal{R}_1}^{(\theta,\lambda)}$ denote the mean first passage time for any of the $N/2$ K+ ions in $\mathcal{R}_2$ to travel to $\mathcal{R}_1$:

$$\begin{aligned}\tau_{\mathcal{R}_1,\mathcal{R}_2}^{(\theta,\lambda)} &= \mathbf{E}\{t_\beta\} \text{ where } t_\beta \equiv \inf\left\{t : \max\left(z_t^{(1)}, z_t^{(2)}, \ldots, z_t^{(N/2)}\right) \geq \beta\right\}, \\ \tau_{\mathcal{R}_2,\mathcal{R}_1}^{(\theta,\lambda)} &= \mathbf{E}\{t_\alpha\} \text{ where } t_\alpha \equiv \inf\left\{t : \min\left(z_t^{(N/2+1)}, z_t^{(N/2+2)}, \ldots, z_t^{(2N)}\right) \leq \alpha\right\}.\end{aligned} \tag{15.22}$$

In cationic channels, for example, only K^+ or Na^+ ions flow through to cause the channel current—so we do not need to consider the mean first passage time of the

Cl^- ions. To give a partial differential equation for $\tau^{(\theta,\lambda)}_{\mathcal{R}_1,\mathcal{R}_2}$ and $\tau^{(\theta,\lambda)}_{\mathcal{R}_2,\mathcal{R}_1}$, it is convenient to define the closed sets

$$\mathcal{P}_2 = \left\{\zeta : \{z^{(1)} \geq \beta\} \cup \{z^{(2)} \geq \beta\} \cup \cdots \cup \{z^{(N/2)} \geq \beta\}\right\}$$
$$\mathcal{P}_1 = \left\{\zeta : \{z^{(N/2+1)} \leq \alpha\} \cup \{z^{(N/2+2)} \leq \alpha\} \cup \cdots \cup \{z^{(2N)} \leq \alpha\}\right\}. \quad (15.23)$$

Then it is clear that $\zeta_t \in \mathcal{P}_2$ is equivalent to $\max\left(z_t^{(1)}, z_t^{(2)}, \ldots, z_t^{(N/2)}\right) \geq \beta$ since either expression implies that at least one ion has crossed from $\mathcal{R}_1$ to $\mathcal{R}_2$. Similarly, $\zeta_t \in \mathcal{P}_1$ is equivalent to $\min\left(z_t^{(N/2+1)}, z_t^{(N/2+2)}, \ldots, z_t^{(2N)}\right) \leq \alpha$. Thus, t_β and t_α defined in Eq. 15.22 can be expressed as $t_\beta = \inf\{t : \zeta_t \in \mathcal{P}_2\}$, $t_\alpha = \inf\{t : \zeta_t \in \mathcal{P}_1\}$.

In a typical ionic channel, $\tau^{(\theta,\lambda)}_{\mathcal{R}_2,\mathcal{R}_1}$ is much larger compared to $\tau^{(\theta,\lambda)}_{\mathcal{R}_1,\mathcal{R}_2}$. In terms of the mean passage rate $\tau^{(\theta,\lambda)}_{\mathcal{R}_2,\mathcal{R}_1}$, $\tau^{(\theta,\lambda)}_{\mathcal{R}_1,\mathcal{R}_2}$, the mean current flowing from $\mathcal{R}_1$ via the ion channel $\mathcal{C}$ into $\mathcal{R}_2$ is defined as

$$I^{(\theta,\lambda)} = q^+ \left(\frac{1}{\tau^{(\theta,\lambda)}_{\mathcal{R}_1,\mathcal{R}_2}} - \frac{1}{\tau^{(\theta,\lambda)}_{\mathcal{R}_2,\mathcal{R}_1}}\right). \quad (15.24)$$

The following result, adapted from Gihman and Skorohod (1972, pp. 306) shows that the mean passage times $\tau^{(\theta,\lambda)}_{\mathcal{R}_1,\mathcal{R}_2}$, $\tau^{(\theta,\lambda)}_{\mathcal{R}_2,\mathcal{R}_1}$ satisfy a boundary-valued partial differential equation. In particular, the expressions for the mean first passage time below, together with Eq. 15.24, give a complete characterization of the ion channel current. Of course, the partial differential equation cannot be solved in closed form—so later on in this chapter we use BD simulation as a randomized numerical method for solving this partial differential equation.

Theorem 3. *Consider the two step BD probabilistic construction in Procedure 1. Then the mean passage times $\tau^{(\theta,\lambda)}_{\mathcal{R}_1,\mathcal{R}_2}$ and $\tau^{(\theta,\lambda)}_{\mathcal{R}_2,\mathcal{R}_1}$ (defined in Eq. 15.24) for ions to diffuse through the ion channel are obtained as*

$$\tau^{(\theta,\lambda)}_{\mathcal{R}_1,\mathcal{R}_2} = \int \tau^{(\theta,\lambda)}_{\mathcal{R}_1,\mathcal{R}_2}(\zeta)\pi^{(\theta,\lambda)}_\infty(\zeta)\, d\zeta \quad (15.25)$$

$$\tau^{(\theta,\lambda)}_{\mathcal{R}_2,\mathcal{R}_1} = \int \tau^{(\theta,\lambda)}_{\mathcal{R}_2,\mathcal{R}_1}(\zeta)\pi^{(\theta,\lambda)}_\infty(\zeta)\, d\zeta \quad (15.26)$$

where

$$\tau^{(\theta,\lambda)}_{\mathcal{R}_1,\mathcal{R}_2}(\zeta) = \mathbf{E}\{\inf\{t : \zeta_t \in \mathcal{P}_2 | \zeta_0 = \zeta\}\},$$
$$\tau^{(\theta,\lambda)}_{\mathcal{R}_2,\mathcal{R}_1}(\zeta) = \mathbf{E}\{\inf\{t : \zeta_t \in \mathcal{P}_1 | \zeta_0 = \zeta\}\}.$$

Here $\tau^{(\theta,\lambda)}_{\mathcal{R}_1,\mathcal{R}_2}(\zeta)$ and $\tau^{(\theta,\lambda)}_{\mathcal{R}_2,\mathcal{R}_1}(\zeta)$ satisfy the following boundary value partial differential equations:

$$\begin{aligned} \mathcal{L}\tau^{(\theta,\lambda)}_{\mathcal{R}_1,\mathcal{R}_2}(\zeta) &= -1 \quad \zeta \notin \mathcal{P}_2, \quad \tau^{(\theta,\lambda)}_{\mathcal{R}_1,\mathcal{R}_2}(\zeta) = 0 \quad \zeta \in \mathcal{P}_2 \\ \mathcal{L}\tau^{(\theta,\lambda)}_{\mathcal{R}_2,\mathcal{R}_1}(\zeta) &= -1 \quad \zeta \notin \mathcal{P}_1, \quad \tau^{(\theta,\lambda)}_{\mathcal{R}_2,\mathcal{R}_1}(\zeta) = 0 \quad \zeta \in \mathcal{P}_1 \end{aligned} \tag{15.27}$$

where $\mathcal{L}$ denotes the backward operator defined in Eq. 15.16.

15.6 Brownian Dynamics Simulation

It is not possible to solve the boundary-valued partial differential equations, given in Eq. 15.27, to obtain explicit closed form expressions. The aim of BD simulation is to obtain estimates of these quantities by directly simulating the stochastic dynamical system Eq. 15.7. Thus, BD simulation can be viewed as a randomized numerical method for solving this partial differential equation.

To implement the BD simulation algorithm described below on a digital computer, it is necessary to discretize the continuous-time dynamical equation of the $2N$ ions Eq. 15.7. A two-time scale time discretization is used in the BD simulation algorithm. For dynamics of ions within the ion channel, The BD simulation algorithm uses a sampling interval of $\Delta = 2 \times 10^{-15}$ s. For dynamics of ions within the reservoirs a sampling interval of $\Delta = 2 \times 10^{-12}$ s is used in the reservoirs. There are several possible methods for time discretization of the stochastic differential equation Eq. 15.7, as described in detail by Kloeden and Platen (1992). Our BD simulation algorithm uses the second-order discretization approximation of van Gunsteren et al. (1981).

In the BD simulation algorithm below, we use the following notation:

The algorithm runs for L iterations where L is user specified. Each iteration l, $l = 1, 2, \ldots, L$, runs for a random number of discrete-time steps until an ion crosses the channel. We denote these random times as $\hat{\tau}^{(l)}_{\mathcal{R}_1,\mathcal{R}_2}$ if the ion has crossed from $\mathcal{R}_1$ to $\mathcal{R}_2$ and $\hat{\tau}^{(l)}_{\mathcal{R}_2,\mathcal{R}_1}$ if the ion has crossed from $\mathcal{R}_2$ to $\mathcal{R}_1$. Thus

$$\hat{\tau}^{(l)}_{\mathcal{R}_1,\mathcal{R}_2} = \min\{k : \zeta^{(d)}_k \in \mathcal{P}_2\}, \qquad \hat{\tau}^{(l)}_{\mathcal{R}_2,\mathcal{R}_1} = \min\{k : \zeta^{(d)}_k \in \mathcal{P}_1\}.$$

The positive ions $\{1, 2, \ldots, N/2\}$ are in $\mathcal{R}_1$ at steady state $\pi^{\theta,\lambda}_{\infty}$, and the positive ions $\{N/2 + 1, \ldots, 2N\}$ are in $\mathcal{R}_2$ at steady state.
$L_{\mathcal{R}_1,\mathcal{R}_2}$ is a counter that counts how many K^+ ions have crossed from $\mathcal{R}_1$ to $\mathcal{R}_2$ and $L_{\mathcal{R}_2,\mathcal{R}_1}$ counts how many K^+ ions have crossed from $\mathcal{R}_2$ to $\mathcal{R}_1$. Note

$$L_{\mathcal{R}_1,\mathcal{R}_2} + L_{\mathcal{R}_2,\mathcal{R}_1} = L.$$

In the algorithm below, to simply notation, we only consider passage of K^+ ions $i = 1, \ldots, N$ across the ion channel.

Algorithm 1. Brownian dynamics simulation algorithm for ion permeation (for fixed θ *and* λ*)*

- Input parameters θ for PMF and λ for applied external potential.
- For $l = 1$ to L iterations:
  - *Step 1.* Initialize all $2N$ ions according to the stationary distribution $\pi_\infty^{(\theta,\lambda)}$ defined in Eq. 15.18.
    Open ion channel at discrete time $k = 0$ and set $k = 1$.
  - *Step 2.* Propagate all $2N$ ions according to the time discretized BD system until time k^* at which an ion crosses the channel.
    * If ion crossed ion channel from $\mathcal{R}_1$ to $\mathcal{R}_2$, i.e., for any ion $i^* \in \{1, 2, \ldots, N/2\}$, $z_{k^*}^{(i^*)} \geq \beta$ then set $\hat{\tau}_{\mathcal{R}_1,\mathcal{R}_2}^{(l)} = k^*$.
      Update number of crossings from $\mathcal{R}_1$ to $\mathcal{R}_2$: $L_{\mathcal{R}_1,\mathcal{R}_2} = L_{\mathcal{R}_1,\mathcal{R}_2} + 1$.
    * If ion crossed ion channel from $\mathcal{R}_2$ to $\mathcal{R}_1$, i.e., for any ion $i^* \in \{N/2 + 1, \ldots, N\}$, $z_{k^*}^{(i)} \leq \alpha$ then set $\hat{\tau}_{\mathcal{R}_2,\mathcal{R}_1}^{(l)} = k^*$.
      Update number of crossings from $\mathcal{R}_2$ to $\mathcal{R}_1$: $L_{\mathcal{R}_2,\mathcal{R}_1} = L_{\mathcal{R}_2,\mathcal{R}_1} + 1$.
  - End for loop.
- Compute the mean passage time and mean current estimate after L iterations as

$$\hat{\tau}_{\mathcal{R}_1,\mathcal{R}_2}^{(\theta,\lambda)}(L) = \frac{1}{L_{\mathcal{R}_1,\mathcal{R}_2}} \sum_{l=1}^{L_{\mathcal{R}_1,\mathcal{R}_2}} \hat{\tau}_{\mathcal{R}_1,\mathcal{R}_2}^{(l)}, \quad \hat{\tau}_{\mathcal{R}_2,\mathcal{R}_1}^{(\theta,\lambda)}(L) = \frac{1}{L_{\mathcal{R}_2,\mathcal{R}_1}} \sum_{l=1}^{L_{\mathcal{R}_2,\mathcal{R}_1}} \hat{\tau}_{\mathcal{R}_1,\mathcal{R}_2}^{(l)}. \tag{15.28}$$

$$\hat{I}^{\theta,\lambda}(L) = q^+ \left(\frac{1}{\hat{\tau}_{\mathcal{R}_1,\mathcal{R}_2}^{(\theta,\lambda)}(L)} - \frac{1}{\hat{\tau}_{\mathcal{R}_2,\mathcal{R}_2}^{(\theta,\lambda)}(L)} \right) \tag{15.29}$$

The following result shows that the estimated current $\hat{I}^{(\theta,\lambda)}(L)$ obtained from a BD simulation run over L iterations is strongly consistent. This means that if the BD simulation is run for a large number of iterations, i.e., as $L \to \infty$, the estimate of the current obtained from the BD simulation converges with probability one (w.p.1) to the actual ion channel current that was theoretically obtained in Eq. 15.24 in terms of mean passage rates. Thus, this theorem shows that BD simulation is a statistically valid algorithm for estimating the ion channel current.

Theorem 4. *For fixed* PMF $\theta \in \Theta$ *and applied external potential* $\lambda \in \Lambda$*, the channel current estimate* $\hat{I}^{\theta,\lambda}(L)$ *obtained from the BD simulation Algorithm 1 over L iterations is strongly consistent, that is,*

$$\lim_{L \to \infty} \hat{I}^{\theta,\lambda}(L) = I^{(\theta,\lambda)} \quad \text{w.p.1} \tag{15.30}$$

where $I^{(\theta,\lambda)}$ is the mean current defined in Eq. 15.24.

Proof. Since by construction in Algorithm 1, each of the L iterations are statistically independent, and $\mathbf{E}\{\hat{\tau}^{(l)}_{\mathcal{R}_1,\mathcal{R}_2}\}$, $\mathbf{E}\{\hat{\tau}^{(l)}_{\mathcal{R}_2,\mathcal{R}_1}\}$ are finite, it then follows by Kolmogorov's strong law of large numbers (Billingsley, 1986)

$$\lim_{L\to\infty} \hat{\tau}^{(\theta,\lambda)}_{\mathcal{R}_1,\mathcal{R}_2}(L) = \tau^{(\theta,\lambda)}_{\mathcal{R}_1,\mathcal{R}_2}, \qquad \lim_{L\to\infty} \hat{\tau}^{(\theta,\lambda)}_{\mathcal{R}_2,\mathcal{R}_1}(L) = \tau^{(\theta,\lambda)}_{\mathcal{R}_2,\mathcal{R}_1} \quad \text{w.p.1.}$$

Thus, $q^+\left(\frac{1}{\tau^{(\theta,\lambda)}_{\mathcal{R}_1,\mathcal{R}_2}(L)} - \frac{1}{\tau^{(\theta,\lambda)}_{\mathcal{R}_2,\mathcal{R}_1}(L)}\right) \to I^{\theta,\lambda)}$ w.p.1 as $L \to \infty$.

15.7 Adaptive Controlled Brownian Dynamics Simulation

In this section, we briefly describe a new extension of BD simulation for estimating the PMF of an ion channel. This extension involves a novel simulation-based learning control algorithm that dynamically adapts the evolution of the BD simulation. It is based on our current and on-going research. The complete formalism, convergence proofs, and numerical results will be presented elsewhere (Krishnamurthy and Chung, 2005).

We estimate the PMF U_θ parameterized by some finite-dimensional parameter θ (e.g., θ are the means, variances, and mixture weights of a Gaussian basis function approximation), by computing the parameter θ that optimizes the fit between the mean current $I(\theta.\lambda)$ (defined above in Eq. 15.24) and the experimentally observed current $y(\lambda)$ defined below. There are two reasons why estimating the PMF U_θ is useful. First, by directly estimating PMF, the need for solving Poisson's equation is obviated. Thus, the problem of assigning the effective dielectric constants of the pore and of the protein is avoided. Second, no assumption about the ionization state of some of the residues lining the pore has to be made.

Unfortunately, it is impossible to explicitly compute $I(\theta.\lambda)$ from Eq. 15.24. For this reason we resort to a *stochastic optimization problem formulation* below, where consistent estimates of $I(\theta.\lambda)$ are obtained via the BD simulation algorithm.

From experimental data, an accurate estimate of the current–voltage–concentration profiles of an ion channel can be obtained. These curves depict the actual current $y(\lambda)$ flowing through an ion channel for various external applied potentials $\lambda \in \Lambda$ and ionic concentrations. For a fixed applied field $\lambda \in \Lambda$ at a given concentration, define the square error loss function as

$$\mathcal{Q}(\theta,\lambda) = \mathbf{E}\left\{\hat{I}_n(\theta,\lambda) - y(\lambda)\right\}^2, \quad \mathcal{Q}(\theta) = \sum_{\lambda\in\Lambda} \mathcal{Q}(\theta,\lambda). \tag{15.31}$$

Note that the total loss function $\mathcal{Q}(\theta)$ is obtained by adding the square error over all the applied fields $\lambda \in \Lambda$ on the current–voltage or current–concentration

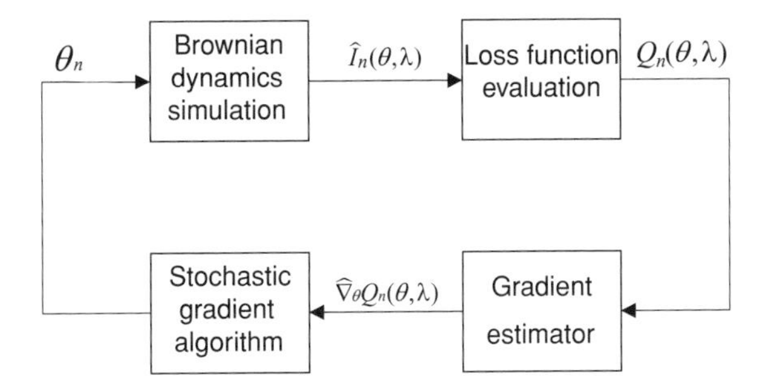

Fig. 15.7 Block diagram of controlled Brownian dynamics simulations for estimating PMF. The currents obtained from Brownian dynamics simulations using the parameters of the initial PMF are compared with the experimental measurements. The parameters for the next iteration is modified such that it would reduce the difference between the simulated and experimental results. The procedure is iterated many times.

curve. The optimal PMF U_{θ^*} is determined by the parameter θ^* that best fits the mean current $I(\theta, \lambda)$ to the experimentally determined curves of an ion channel, i.e.,

$$\theta^* = \arg\min_{\theta \in \Theta} Q(\theta) \tag{15.32}$$

Suppose that the BD simulation Algorithm is run in batches indexed by batch number $n = 1, 2, \ldots$. In each batch n, the PMF parameter θ_n is selected (as described below), the BD Algorithm is run over L iterations, and the estimated current $\hat{I}_n(\theta, \lambda)$ is computed using (Eq. 15.29). In summary, Eqs. 15.32 and 15.31 define the stochastic optimization problem we will solve in this section.

To solve the stochastic optimization problem by a simulation-based optimization approach, we need to evaluate unbiased estimates $Q_n(\theta, \lambda)$ of the loss function and derivative estimates $\widehat{\nabla}_\theta Q_n(\theta, \lambda)$. The estimation of the derivative $\widehat{\nabla}_\theta Q_n(\theta, \lambda)$ involves using recent sophisticated techniques in Monte-Carlo gradient estimation (see, Vazquez and Krishnamurthy, 2003). Krishnamurthy and Chung (2005) present several such algorithms including the Kiefer–Wolfowitz algorithm which evaluates derivate estimates as finite differences, simultaneous perturbation stochastic approximation (SPSA) which evaluates the derivatives in random directions (and thus saves computational cost), and pathwise infinitesimal perturbation analysis (IPA) gradient estimators.

The controlled BD simulation algorithm for estimating the PMF is schematically depicted in Fig. 15.7. Using an initial parameter set θ, several BD simulations are carried out under various applied potentials and concentrations. From the results of these simulations, the loss function for each BD simulation is computed, using Eq. 15.31. From the total loss function, a new parameter set θ is estimated, based on a stochastic gradient algorithm, and the entire process is repeated. The iterative algorithm for carrying out controlled BD simulation is summarized below.

- *Step 0*: Set batch index $n = 0$, and initialize $\theta_0 \in \Theta$.
- *Step 1 (Evaluation of loss function)*: At batch n, evaluate loss function $\mathcal{Q}_n(\theta_n, \lambda)$ for each external potential $\lambda \in \Lambda$.
- *Step 2 (Gradient estimation)*: Compute gradient estimate $\widehat{\nabla}_\theta \mathcal{Q}_n(\theta, \lambda)[\theta_n]$.
- *Step 3 (Stochastic approximation Algorithm)*: Update PMF estimate:

$$\theta_{n+1} = \theta_n - \epsilon_{n+1} \sum_{\lambda \in \Lambda} \widehat{\nabla}_\theta \mathcal{Q}_n(\theta_n, \lambda), \tag{15.33}$$

  where $\epsilon_n = 1/n$ denotes a decreasing step size.
- Set n to $n + 1$ and go to Step 1.

A crucial aspect of the above algorithm is the gradient estimation Step 2. In this step, an estimate $\widehat{\nabla}_\theta \mathcal{Q}_n(\theta, \lambda)$ of the gradient $\nabla_\theta \mathcal{Q}_n(\theta, \lambda)$ is computed. This gradient estimate is then fed to the stochastic gradient algorithm (Step 3) which updates the PMF. It can be proved via standard arguments in stochastic approximations that the above algorithm converges to the optimal PMF θ^* w.p.1. For a more detailed exposition and numerical studies of adaptive Brownian dynamics for estimating the potential of mean force of the gramicidin channel, see Krishnamurthy and Chung (2006) and Krishnamurthy et al. (2006).

15.8 Concluding Remarks

Three computational tools discussed in this volume—Poisson–Nernst–Planck theory (Coalson and Kurnikova, 2006), molecular dynamics (Grottesi et al., 2006), and Brownian dynamics—will play increasingly prominent roles in understanding how biological ion channel work. Each of these approaches has its strengths and limitations, and involves a degree of approximation. The main defects of Poisson–Nernst–Planck theory are errors stemming from the mean-field assumption. In particular, it ignores the effects of induced surface charges created as a charged particle in electrolyte solutions approaches the protein boundary. The magnitude of the errors introduced by the mean-field approximation become large when the theory is applied to a narrow ionic channels. By incorporating a term in the PNP equations to account for the barrier created by induced surface charges, the magnitude of the errors can be reduced somewhat (Corry et al., 2003). However, doing this removes much of the simplicity of the theory, one of its main advantages over the other approaches, and also it is still hard to know the accuracy of the results without comparison to a more detailed model.

The greatest limitations of molecular dynamics is the computational power required that limits the possible simulation times. While the calculation of free energy profiles provides useful information on ion permeation, it is not a substitute for a direct estimation of conductance from simulations. Thus, virtually no predictions derived from molecular dynamics simulations can be directly compared with experimental data. If no such comparisons can be made, there can only be a limited interaction between experimenters and theoreticians. With the current doubling of computer speeds every 2 years, this computational limitation will eventually be overcome. Then, the force fields employed in molecular dynamics simulations may need to be improved to include polarization effects, perhaps using ab initio molecular dynamics as a guide.

One of the main caveats to the application of Brownian dynamics to biological ion channels is the use of Poisson's equation to estimate the forces encountered by permeant ions. The issue here is whether one can legitimately employ macroscopic electrostatics in regions that are not much larger than the diameters of the water molecules and ions. In the narrow constricted region of the channel, such as in the selectivity filter of the potassium channel, the representation of the channel contents as a continuous medium is a poor approximation.

All three theoretical approaches are useful in elucidating the mechanisms underlying selectivity and permeation of ions across biological nanotubes. For ion channels with large pore radii, such as mechanosensitive channels, Poisson–Nernst–Planck theory can be fruitfully utilized. Also, if one is interested in simply obtaining order-of-magnitude estimates of conductances of various model channels, this simple theory will provide the answers with little computational cost. To study the mechanisms underlying the selectivity sequences of monovalent ions or to determine the precise conformational changes of the protein when a channel undergoes the transition from the closed to the open state, one has to rely on molecular dynamics simulations.

The ability to compute current flow across ion channels confers a distinct advantage to Brownian dynamics compared to molecular dynamics. Because ions are treated as discrete entities, induced surface charges are correctly accounted for. Thus, an obvious application of Brownian dynamics is the calculation of current–voltage and conductance–concentration curves, which can be directly compared to the physiological measurements to assess the reliability and predictive power of the method. In addition to simple counting of ions crossing the channel, one can carry out a trajectory analysis of ions in the system to determine their average concentrations and the steps involved in conduction. This is useful in finding the binding sites and the average number of ions in the channel, both of which are experimentally observable quantities. It is also possible to study the mechanisms of blocking of channels by larger molecules or other ion species.

Brownian dynamics has been extensively used in the past to simulate the current flowing across a variety of model ion channels (Allen et al., 1999, 2000; Allen and Chung, 2001; Chung et al., 1999, 2002a, b; Corry et al., 2001, 2004a, b; Im and Roux, 2002a, b; Noskov et al., 2004; O'Mara et al., 2003, 2005; Vora et al., 2004). Here we show that BD simulation is a statistically valid algorithm for estimating the ion channel current, placing this nonequilibrium method used by the previous authors in studying model ion channels on a firm mathematical foundation. In BD, the propagation of ions in the ion channel is modeled as a large-scale, multi-particle continuous-time, stochastic dynamical system satisfying the Langevin equation. The key idea here is that instead of considering the dynamics of individual water molecules, which is computationally intractable, the BD system considers the average effect of water molecules as a random force acting on individual ions. This treatment of water molecules can be viewed as an approximation of the central-limit theorem by using stochastic averaging of water molecules. We then provide the proof that ions drifting executing Brownian motion in a simulation assembly in which a channel protein is imbedded achieve a stationary distribution (steady state) exponentially fast (Theorem 1). We also demonstrate that the current across the model pore is related to the mean first passage time of ions, which satisfies a boundary-valued partial differential equation related to the Fokker–Planck equation. Thus, BD can be construed as a randomized algorithm for numerically solving this partial differential equation. The simulated current converges to the explicit solution of the partial differential equation w.p.1 (Theorem 4).

The mathematical formulation and statistical analysis of BD we provide in this chapter are essential for further extension and refinement of the method. With the BD method placed on a firm theoretical ground, we are now in the position to further refine and extend it by applying the state-of-the-art, novel stochastic estimation algorithms, thus making this approach far more versatile than in its current form. One of the major caveats to the use of BD in studying the permeation dynamics in biological ion channels is the use of Poisson's equation to calculate the forces encountered by permeant ions. The issue here is whether one can legitimately employ macroscopic electrostatics in regions that are not much larger than the diameters of the water molecules and ions. In the narrow, constricted region of the channel, such as in the selectivity filter of the potassium channel, the representation of the channel contents as a continuous medium is a poor approximation. The method of adaptive controlled BD, which we discussed briefly, is designed to circumvent the limitations posed in the conventional simulation approach. Using the learning-based dynamic control algorithm, we are able to solve the inverse problem. That is, given the three-dimensional shape of a channel, we can deduce the potential of mean force encountered by an ion traversing the channel that correctly replicates experimental findings, thus obviating the need to solve Poisson's equation. The BD algorithm thus can now be used to study the propagation of individual ions through a mesoscopic system where continuum electrostatics breaks down (Edwards et al., 2002) and molecular dynamics fails to yield a sensible profile of the PMF (Allen et al., 2003, 2004). Alternatively, if continuum electrostatic is to be applied, we can pinpoint, using a stationary stochastic optimization algorithm, the effective dielectric constants of the pore and of protein that need to be used to replicate experimental measurements.

The combined techniques of statistical signal processing and stochastic control of large-scale dynamic systems of interacting particles will help us unravel the structure–function relationships in ion channels. Also, by combining the state-of-the-art dynamic control algorithms with BD, it should be possible to predict the open-state structure of an ion channel with a fair degree of certainty and also design new nanotubes that can be utilized as antifungal or antibacterial agents. Now and in the near future, as we attempt to understand membrane channels in terms of rigorous molecular physics, there will be an increasing interplay between experiment and theory, the former providing hints and clues for building and refining models and the later making testable predictions.

References

Accardi, A., and C. Miller. 2004. Secondary active transport mediated by a prokaryotic homologue of ClC Cl^- channels. *Nature* 427:803–807.

Allen, T.W., O.S. Andersen, and B. Roux. 2004. On the importance of atomic fluctuations, protein flexibility, and solvent in ion permeation. *J. Gen. Physiol.* 124:679–690.

Allen, T.W., T. Bastug, S. Kuyucak, and S.H. Chung. 2003. Gramicidin A channel as a test ground for molecular dynamics force fields. *Biophys. J.* 84:2159–2168.

Allen, T.W., and S.H. Chung. 2001. Brownian dynamics study of an open-state KcsA potassium channel. *Biochim. Biophys. Acta Biomembr.* 1515:83–91.

Allen, T.W., M. Hoyles, S. Kuyucak, and S.H. Chung. 1999. Molecular and Brownian dynamics study of ion permeation across the potassium channel. *Chem. Phys. Letts.* 313:358–365.

Allen, T.W., S. Kuyucak, and S.H. Chung. 2000. Molecular dynamics estimates of ion diffusion in model hydrophobic and KcsA potassium channels. *Biophys. Chem.* 86:1–14.

Bass, R.B., P. Stropo, M. Baraclay, and D.C. Reece. 2002. Crystal structure of *Escherichia coli* MscS, a voltage-modulated and mechanosensitive channel. *Science* 298:1582–1587.

Bek, S., and E. Jakobsson. 1994. Brownian dynamics study of a multiply occupied cation channels: Application to understanding permeation in potassium channel. *Biophys. J.* 66:1028–1038.

Billingsley, P. 1986. Probability and Measure. Wiley, New York.

Burykin, A., C.N. Schutz, J. Villa, and A. Warshel. 2002. Simulations of ion current realistic models of ion channels: KcsA potassium channel. *Proteins Struct. Funct. Genet.* 47:265–280.

Chang, G., R.H. Spencer, A.T. Lee, M.T. Barclay, and D.C. Rees. 1998. Structure of the MscL homolog from mycobacterium tuberculosis: A gated mechanosensitive channel. *Science* 282:2220–2226.

Chung, S.H., T.W. Allen, M. Hoyles, and S. Kuyucak. 1999. Permeation of ions across the potassium channel: Brownian dynamics studies. *Biophys. J.* 77:2517–2533.

Chung, S.H., T.W. Allen, and S. Kuyucak. 2002a. Conducting-state properties of the KcsA potassium channel from molecular and Brownian dynamics simulations. *Biophys. J.* 82:628–645.

Chung, S.H., T.W. Allen, and S. Kuyucak. 2002b. Modeling diverse range of potassium channels with Brownian dynamics. *Biophys. J.* 83:263–277.

Chung, S.H., and S. Kuyucak. 2002. Recent advances in ion channel research. *Biochim. Biophys. Acta Biomembr.* 1565:267–286.

Chung, S.H., M. Hoyles, T.W. Allen, and S. Kuyucak. 1998. Study of ionic currents across a model membrane channel using Brownian dynamics. *Biophys. J.* 75:793–809.

Coalson, R., and M.G. Kurnikova. 2005. Poisson–Nernst–Planck theory approach to the calculation of current through biological ion channels. *IEEE Trans. Nanobiosci.* 4:81–93.

Coalson, R., and M.G. Kurnikova. 2006. Poisson–Nernst–Planck theory of ion permeation through biological channels. In: Handbook of Ion Channels: Dynamics, Structure and Application. S.H. Chung, O.S. Andersen, and V. Krishnamurthy, editors. Springer-Verlag, New York.

Cooper, K.E., E. Jakobsson, and P. Wolynes. 1985. The theory of ion transport through membrane channels. *Prog. Biophys. Mol. Biol.* 46:51–96.

Coronado, R., R.L. Rosenberg, and C. Miller. 1980. Ionic selectivity, saturation, and block in a K^+-selective channel from sarcoplasmic reticulum. *J. Gen. Physiol.* 76:425–446.

Corry, B., T.W. Allen, S. Kuyucak, and S.H. Chung. 2001. Mechanisms of permeation and selectivity in calcium channels. *Biophys. J.* 80:195–214.

Corry, B., S. Kuyucak, and S.H. Chung. 2003. Dielectric self-energy in Poisson–Boltzmann and Poisson–Nernst–Planck models of ion channels. *Biophys. J.* 84:3594–3606.

Corry, B., M. O'Mara, and S.H. Chung. 2004a. Permeation dynamics of chloride ions in the ClC-0 and ClC-1 channels. *Chem. Phys. Lett.* 386:233–238.

Corry, B., M. O'Mara, and S.H. Chung. 2004b. conduction mechanisms of chloride ions in ClC-type channels. *Biophys. J.* 86:846–860.

Cuello, L.G., J.G. Romero, D.M. Cortes, and E. Perozo. 1998. pH dependent gating in the *Streptomyces lividans* K^+ channel. *Biochemistry* 37:3229–3236.

Doyle, D.A., J.M. Cabral, R.A. Pfuetzner, A. Kuo, J.M. Gulbis, S.L. Cohne, B.T. Chait, and R. MacKinnon. 1998. The structure of the potassium channel: Molecular basis of K^+ conduction and selectivity. *Science* 280:69–77.

Dutzler, R., E.B. Campbell, M. Cadene, B.T. Chait, and R. MacKinnon. 2002. X-ray structure of a ClC chloride channel at 3.0 Å reveals the molecular basis of anion selectivity. *Nature* 415:287–294.

Dutzler, R., E.B. Campbell, and R. MacKinnon. 2003. Gating the selectivity in ClC chloride channels. *Science* 300:108–112.

Edwards, S., B. Corry, S. Kuyucak, and S.H. Chung. 2002. Continuum electrostatics fails to describe ion permeation in the gramicidin channel. *Biophys. J.* 83:1348–1360.

Eisenberg, R. S. 1999. From structure to function in open ionic channels. *J. Membr. Biol.* 171:1–24.

Fahlke, C. 2001. Ion permeation and selectivity in ClC-type chloride channels. *Am. J. Renal Physiol.* 280:F748–F758.

Garofoli, S., and P.C. Jordan. 2003. Modeling permeation energetics in the KcsA potassium channel. *Biophys. J.* 84:2814–2830.

Gihman, I., and A. Skorohod. 1972. Stochastic Differential Equations. Springer-Verlag, Berlin.

Grottesi, A., C. Domene, S. Haider, and M.S.P. Sansom. 2005. Molecular dynamics simulation approaches to K channels: Conformational flexibility and physiological function. *IEEE Trans. Nanobiosci.* 4:112–120.

Grottesi, A., S. Haider, and M.S.P. Sansom. 2006. Molecular dynamics simulation approaches to K channels. In: Handbook of Ion Channels: Dynamics, Structure and Application. S.H. Chung, O.S. Andersen, and V. Krishnamurthy, editors. Springer-Verlag, New York.

Guàrdia, E., R. Rey, and J. Padró. 1991a. Na^+–Na^+ and Cl^-–Cl^- ion pairs in water: Mean force potentials by constrained molecular dynamics. *J. Chem. Phys.* 95:2823–2831.

Guàrdia, E., R. Rey, and J. Padró. 1991b. Potential of mean force by constrained molecular dynamics: A sodium chloride ion-pair in water. *J. Chem. Phys.* 155:187–195.

Heginbotham, L., M. LeMasurier, L. Kolmakova-Partensky, and C. Miller. 1999. Single *Streptomyces lividans* K^+ channels: Functional asymmetries and sidedness of proton activation. *J. Gen. Physiol.* 114:551–559.

Hille, B. 2001. Ionic Channels of Excitable Membranes, 3rd Ed. Sinauer Associates, Sunderland, MA.

Im, W., and B. Roux. 2002a. Ion permeation and selectivity of ompf porin: A theoretical study based on molecular dynamics, Brownian dynamics, and continuum electrodiffusion theory. *J. Mole. Biol.* 322:851–869.

Im, W., and B. Roux. 2002b. Ions and counterions in a biological channel: A molecular dynamics simulation of ompf porin from *Escherichia coli* in an explicit membrane with 1 M KCl aqueous salt solution. *J. Mol. Biol.* 319:1177–1197.

Jakobsson, E., and W.W. Chiu. 1987. Stochastic theory of singly occupied ion channels. *Biophys. J.* 52:33–45.

Jentsch, T.J., T. Friedrich, A. Schriever, and H. Yamada. 1999. The ClC chloride channel family. *Pflügers Arch.* 437:783–795.

Jordan, P.C. 1999. Ion permeation and chemical kinetics. *J. Gen. Physiol.* 114:601–604.

Jordan, P.C. 2005. Semimicroscopic modeling of permeation energetics in ion channels. *IEEE Trans. Nanobiosci.* 4:94–101.

Jordan, P.C. 2006. A mesoscopic-microscopic perspective on ion channel permeation energetics: The semi-microscopic approach. In: Handbook of Ion Channels: Dynamics, Structure and Application. S.H. Chung, O.S. Andersen, and V. Krishnamurthy, editors. Springer-Verlag, New York.

Karatzas, I., and S.E. Shreve. 1991. Brownian Motion and Stochastic Calculus. Springer-Verlag, New York.

Kloeden, P.E., and E. Platen. 1992. Numerical Solution of Stochastic Differential Calculus. Springer-Verlag, Berlin.

Krishnamurthy, V., and S.H. Chung. 2005. Brownian dynamics simulation for modeling ion permeation across bio-nanotubes. *IEEE Trans. Nanobiosci.* 4:102–111.

Krishnamurthy, V., and S.H. Chung. 2006. Adaptive Brownian dynamics simulation for estimating potential of mean force in ion channel permeation. *IEEE Trans. Nanobiosci.* 5:126–138.

Krishnamurthy, V., M. Hoyles, R. Saab, and S.H. Chung. 2006. Permeation in gramicidin ion channels by directly estimating the potential of mean force using Brownian dynamics simulation. *J. Comput. Theoret. Nanosci.* (in press).

LeMasurier, M., L. Heginbotham, and C. Miller. 2001. KcsA: It's a potassium channel. *J. Gen. Physiol.* 118:303–313.

Long, S.B., E.B. Campbell, and R. MacKinnon. 2004a. Crystal structure of a mammalian voltage-dependent *Shaker* family K^+ channel. *Science* 309:897–903.

Long, S.B., E.B. Campbell, and R. MacKinnon. 2004b. Voltage sensor of Kv1.2: Structural basis of electromechanical coupling. *Science* 309:903–908.

Maduke, M., C. Miller, and J.A. Mindell. 2000. A decade of ClC chloride channels: Structure, mechanism, and many unsettled questions. *Annu. Rev. Biophys. Biomol. Struct.* 29:411–438.

Mashl, R.J., Y. Tang, J. Schnitzer, and E. Jakobsson. 2001. Hierarchical approach to predicting permeation in ion channels. *Biophys. J.* 81:2473–2483.

McCleskey, E.M. 1999. Calcium channel permeation: A field in flux. *J. Gen. Physiol.* 113:765–772.

Meuser, D., H. Splitt, R. Wagner, and H. Schrempf. 1999. Explorign the open pore of the potassium channel from *Steptomyces lividans*. *FEBS Letts.* 462:447–452.

Miller, C. 1982. Open-state substructure of single chloride channels from *torpedo electroplax*. *Phil. Trans. R. Soc. Lond.* B 299:401–411.

Noskov, S.Yu., S. Bernèche, and B. Roux. 2004. Control of ion selectivity in potassium channels by electrostatic and dynamic properties of carbonyl ligands. *Nature* 431:830–834.

O'Mara, M., P.H. Barry, and S.H. Chung. 2003. A model of the glycine receptor deduced from Brownian dynamics studies. *Proc. Natl. Acad. Sci. USA* 100:4310–4315.

O'Mara, M., B. Cromer, M. Parker, and S.H. Chung. 2005. Homology model of $GABA_A$ channel examined with Brownian dynamics. *Biophys. J.* 88:3286–3299.

Partenskii, M.B., and P.C. Jordan. 1992. Theoretical perspectives on ion-channel electrostatics: Continuum and microscopic approaches. *Q. Rev. Biophys.* 25:477– 510.

Pitera, J.W., M. Falta, and W.F. van Gunsteren. 2001. Dielectric properties of proteins from simulation: The effects of solvent, ligands, pH, and temperature. *Biophys. J.* 80:2546–2555.

Roux, R., S. Bernèche, and W. Im. 2000. Ion channels, permeation, and electrostatics: Insight into the function of KcsA. *Biochemistry* 39:13295–13306.

Sanders, J.A., and F. Verhulst. 1985. Averaging Methods in Nonlinear Dynamical Systems. Springer-Verlag, Berlin.

Schrempf, H., O. Schmidt, R. Kümerlen, S. Hinnah, D. Müller, M. Betzler, T. Steinkamp, and R. Wagner. 1995. A prokaryotic potassium ion channel with two predicted transmembrane segment from *Streptomyces lividans*. *EMBO J.* 14:5170–5178.

Schutz, C.N., and A. Warshel. 2001. What are the dielectric "constants" of proteins and how to validate electrostatic model. *Proteins* 44:400–417.

Simonson, T., and C.L. Brooks III. 1996. Charge screening and the dielectric constant of proteins: Insights from molecular dynamics. *J. Am. Chem. Soc.* 118:8452–8458.

Smith, P.E., R.M. Brunne, A.E. Mark, and W.F. van Gunsteren. 1993. Dielectric properties of trypsin inhibitor and lysozyme calculated from molecular dynamics simulations. *J. Phys. Chem.* 97:2009–2014.

Stillinger, F.H., and A. Rahman. 1974. Improved simulation of liquid water by molecular dynamics. *J. Chem. Phys.* 60:1545–1557.

Tieleman, D.P., P.C. Biggin, G.R. Smith, and M.S.P. Sansom. 2001. Simulation approaches to ion channel structure-function relationships. *Q. Rev. Biophys.* 34:473–561.

Unwin, N. 2005. Refined structure of the nicotinic acetylcholine receptor at 4 Å resolution. *J. Mol. Biol.* 346:967–989.

van Gunsteren, W., H. Berendsen, and J. Rullman. 1981. Stochastic dynamics for molecules with constraints Brownian dynamics of n-alkalines. *Mol. Phys.* 44:69–95.

Vazquez, F.J., and V. Krishnamurthy. 2003. Implementation of gradient estimation to a constrained Markov decision problem. *In:* Proceedings of 42nd IEEE Conference on Decision and Control, pp. 4841–4846.

Vora, T., B. Corry, and S.H. Chung. 2004. A model of sodium channels. *Biochim. Biophys. Acta Biomembr.* 1668:106–116.

Wong, E., and B. Hajek. 1985. Stochastic Processes in Engineering System, 2nd Ed. Springer-Verlag, Berlin.

16 Molecular Dynamics Simulation Approaches to K Channels

Alessandro Grottesi, Shozeb Haider, and Mark S. P. Sansom

16.1 Introduction: Potassium Channels

Ion channels are proteins that form pores of nanoscopic dimensions in cell membranes. As a consequence of advance in protein crystallography we now know the three-dimensional structures of a number of ion channels. However, X-ray diffraction techniques yield an essentially static (time- and space-averaged) structure of an ion channel, in an environment often somewhat distantly related to that which the protein experiences when in a cell membrane. Thus, additional techniques are required to fully understand the relationship between channel structure and function. Potassium (K) channels (Yellen, 2002) provide an opportunity to explore the relationship between membrane protein structure, *dynamics*, and function. Furthermore, K channels are of considerable physiological and biomedical interest. They regulate K^+ ion flux across cell membranes. K channel regulation is accomplished by a conformational change that allows the protein to switch between two alternative (closed vs. open) conformations, a process known as *gating*. Gating is an inherently dynamic process that cannot be fully characterized by static structures alone.

The elucidation of the structures of several K^+ channels (Mackinnon, 2003; Gulbis and Doyle, 2004) has shed light on the structural basis of the mechanisms of ion selectivity and permeation (Doyle et al., 1998; Morais-Cabral et al., 2001; Jiang et al., 2002a,b, 2003; Kuo et al., 2003; Zhou and MacKinnon, 2003; Lee et al., 2005; Long et al., 2005). All K channels share a common fold in their pore-forming domain, while exhibiting differences in their structures corresponding to their various gating mechanisms: KcsA is gated by low pH; MthK is gated by Ca^{2+} ions; KvAP and Kv1.2 are gated by transmembrane voltage; and KirBac is presumed to be gated by binding of ligands to its intracellular domain. The core pore-forming domain of K channels is tetrameric, with the monomers surrounding a central pore. The domain is formed of four M1-P-M2 motifs, where M1 and M2 are transmembrane (TM) helices (corresponding to S5 and S6 in Kv channels), with the short P-helix and extended filter (F) region forming a re-entrant loop between the two TM helices (see Fig. 16.1). In voltage-gated potassium (Kv) channels each subunit also contains a voltage sensor domain composed of four TM helices (S1 to S4). In inward rectifier channels, exemplified by the KirBac1.1 structure, there is an additional helix in the pore domain, called "the slide helix," that runs parallel to the cytoplasmic face of

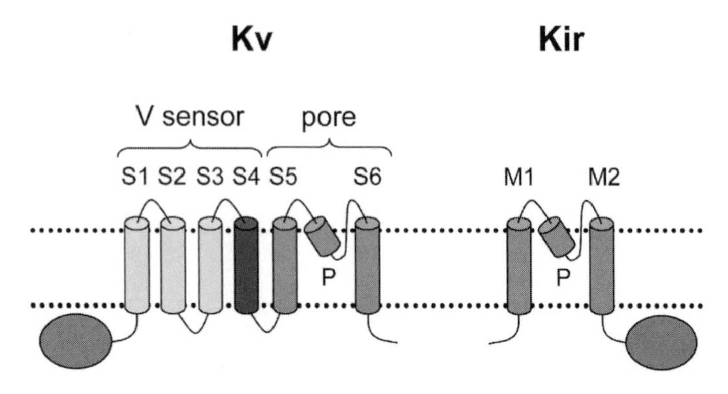

Fig. 16.1 K channel topologies: schematic transmembrane topologies are shown for Kv and Kir channels. The voltage-sensor (S1 to S4 in Kv channels) and pore-forming domains (i.e., S5-P-S6 in Kv channels, M1-P- M2 in Kir channels) are indicated. The intact channel is formed by four subunits.

the membrane. Along with an extensive C-terminal intracellular domain, the slide helix is thought to play a role in the gating mechanism of the channel (Kuo et al., 2003).

16.2 Homology Modeling

The X-ray structures of K channels are all, except for that of Kv1.2 (Long et al., 2005), of bacterial homologues of mammalian channels. It is therefore of some interest to what extent homology modeling (Marti-Renom et al., 2000) and related techniques may be used to extrapolate from bacterial K channel structures to the structures of their mammalian and human homologues (Capener et al., 2002). There have been a number of studies that have employed homology modeling of K channels to help to explain structure/function relationships (Capener et al., 2000; Ranatunga et al., 2001; Eriksson and Roux, 2002; Durell et al., 2004; Laine et al., 2004; Antcliffe et al., 2005). Here we will focus on one particular application of this technique, to the pore-forming domain of the much-studied *Shaker* Kv channel from *Drosophlia* (Tempel et al., 1987). Models of two states of the *Shaker* Kv pore domain have been generated: one, based on the X-ray structure of KcsA, represents a closed state of the channel; the other, based on KvAP, represents an open state pore domain (Fig. 16.2).

The main steps followed in building a homology model of channel are as follows:

1. Sequence alignment of target sequence against the sequence of a suitable template structure.
2. Identification of TM helices within the target sequence using a range of prediction methods (Chen and Rost, 2002).

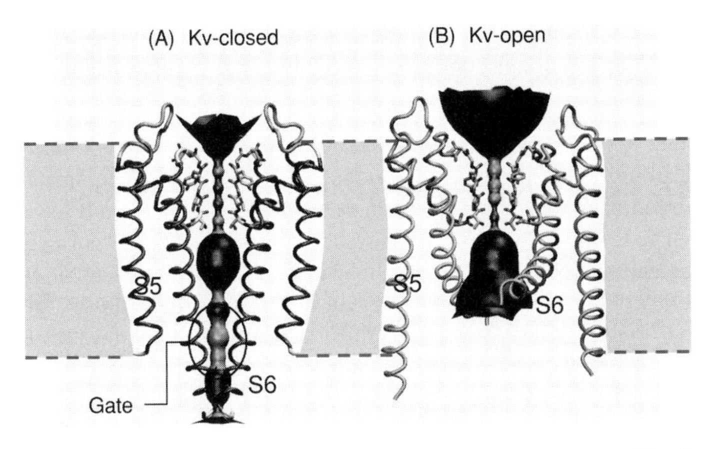

Fig. 16.2 Homology models of the *Shaker* Kv pore domain, based on KcsA [for the Kv-closed model in (A)] and on KvAP [for the Kv-open model in (B)]. In both cases the model is restricted to the core pore-forming domain (i.e., S5-P-S6) and only two of the four subunits are shown for clarity. The approximate location of the lipid bilayer is shown by the horizontal band. The surface of the pore [calculated using HOLE (Smart et al., 1996)] is shown, and the region of the hydrophobic gate is indicated for the Kv-closed model.

3. Generation of a model via one of a number of homology modeling programs [e.g., Modeler (Sali and Blundell, 1993; Fiser et al., 2000)], using the sequence alignment, template structure, and predicted TM helices as inputs.
4. Validation of the resultant model by, e.g., molecular dynamics (MD—see below) simulations in a bilayer-like environment (Capener et al., 2000; Capener and Sansom, 2002).

The first step is crucial: a poor alignment will inevitably yield a poor model. It has been shown that there is a strong correlation between the percentage sequence identity between two proteins and the similarity of their three-dimensional structures (Chothia, 1984). In general, it can be difficult to obtain a good homology model when the percentage sequence identity is <30%. However, the presence of conserved sequence motifs provides a good indicator of structural and functional similarity among apparently distant-related proteins. In the case of K channels, the sequence motif TVGYG in the selectivity filter region of the pore domain is highly conserved, and thus aids alignment. Along with prediction of TM helices, this means it is in general possible to obtain a reasonable alignment of the pore domain (i.e., M1-P-M2 or S5-P-S6) of a target K channel sequence to a template, even though the overall percentage residue identity may be quite low (e.g., <20%). As several structures of bacterial K channels are available, it is also possible to model both closed state (e.g., using KcsA as a template) and open state (e.g., using MthK or KvAP as a template) conformations of a K channel pore domain (Holyoake et al., 2003). This is illustrated in Fig. 16.2, where models of a closed and open state of the pore domain of the *Shaker* Kv are shown. It is evident that in the Kv-closed model (based on a KcsA template) the intracellular mouth of the channel, which is believed to be the location of the channel gate, is much narrower than in the Kv-open model. The functional consequence of this will be discussed in more detail below.

16.3 MD Simulations of Ion Channels

Molecular dynamics simulations have been used to over 25 years to simulate the conformational dynamics of biomolecules (Karplus and McCammon, 2002). In MD simulations a trajectory (i.e., a series of conformations evolving in time) of a molecular system is generated by simultaneous integration of Newton's equations of motion for all the atoms in the system. To achieve this, MD simulations require a potential energy function, which describes the interactions of all the atoms in the system. This energy function includes terms representing interactions between atoms that are covalently bonded to one another, and also terms for van der Waals and electrostatics interactions between nonbonded atoms. The potential energy is evaluated via an empirical force field, in which atoms are treated as van der Waals particles carrying a point charge. Bonds between atoms are modeled by simple harmonic functions for bond lengths and bond angles, and simple sinusoidal functions for torsion angles (Leach, 2001). Integration of the equations of motion is performed in small time steps, typically 1 or 2 fs. Equilibrium quantities can be calculated by averaging over a trajectory, which therefore should be of sufficient length to sample a representative ensemble of the state of the system.

A simulation starts from initial atomic coordinates and velocities. The initial coordinates are generally obtained by embedding in membrane protein structure in a pre-equilibrated model of a lipid bilayer (Tieleman et al., 1997), followed by addition of water molecules (>30 per lipid molecule) and ions. The initial velocities are taken from a Maxwellian distribution, corresponding to the desired temperature of the simulation (usually 300 K).

Several issues determine the "quality" of a simulation. One of these is the nature of the empirical force field used to describe interactions between atoms of the simulation system. There are a number of force fields that are routinely used in macromolecular simulations, which continue to evolve and improve. Without addressing the details of comparison between different force fields, it is important to be aware of their relative strengths and limitations, especially in the context of membrane simulations. A further issues, especially if simulations are being employed to explore possible conformational changes of a membrane protein, is the extent to which the duration of the simulation allows available conformations of the protein to be sampled (Faraldo-Gómez et al., 2004).

The treatment of long-range electrostatic interactions, and the method used to control the system temperature are also of importance in governing the accuracy of a simulation. There are two principal approaches to approximating long-range electrostatic interactions in membrane simulations. The simplest approach is to neglect the small interactions between any pair of atoms further apart than a given cut-off distance (e.g., 1.0 nm). This method is efficient from a computational point of view, but introduces inaccuracies that may detract from the quality of a membrane simulation (Tobias, 2001). These inaccuracies may be avoided by using Ewald summation-based methods to treat long-range electrostatic interactions (Darden et al., 1993; Essmann et al., 1995; Sagui and Darden, 1999). However, Ewald methods may introduce artifacts in some simulations (Hunenberger and J.A., 1999; Hunenberger and McCammon, 1999; Weber et al., 2000; Bostick and Berkowitz, 2003), and so should not be employed uncritically. With respect to control of simulation temperature, in most protein simulations the system is coupled to an external thermostat with which it can exchange energy. This may be achieved simply via use of, e.g., a Berendsen thermotstat (Berendsen et al., 1984), or via the Nosé-Hoover method (Nose and Klein, 1983).

For simulations of ion channels (Roux, 2005) and related membrane proteins (Ash et al., 2004), there are two main choices of environment for the protein. The first is to embed the protein within a model of a lipid bilayer. This provides an accurate representation of the environment of the protein, at least in vitro, and is of particular use if, e.g., protein–lipid interactions are to be analyzed (Domene et al., 2003). However, the effective viscosity of a lipid bilayer is quite high, and this may restrict the ability to observe conformational changes in channel proteins on the time scales ($\sim$20 ns) currently available to MD simulations. A lower viscosity approximation to a lipid bilayer is provided by a "slab" of octane molecules. Combined with water molecules on either side, this provides a membrane-mimetic environment that has been successfully employed in a number of K channel simulations (Capener et al., 2000; Capener and Sansom, 2002; Arinaminpathy et al., 2003). An example of such a system, showing an octane slab within which the TM domain of KirBac1.1 is embedded, is provided in Fig. 16.3.

16.4 Essential Dynamics of Ion Channels

One of the challenges of analysis of MD simulations is to extract functionally relevant motions of the protein from the simulation "noise." Recent studies have suggested that such motions occur along directions of a few collective coordinates, which may dominate simulated atomic fluctuations (Garcia, 1992; Amadei et al., 1993; Kitao and Go, 1999). It can be difficult to extract functionally relevant motions from simulation results, mainly because simulation times are generally too short to yield proper sampling of the corresponding conformational changes. A possible solution of this problem is to make use of collective coordinates. A collective coordinate can be calculated by determining a set of eigenvectors via diagonalization of a second moment matrix. This approach forms the basis of principal component analysis (PCA). In the case of a simulation of a protein, a covariance matrix is derived from a set of Cartesian coordinates as a function of time:

$$a_{ij} \equiv \langle (x_i - \langle x_i \rangle)\left(x_j - \langle x_j \rangle\right) \rangle ,$$

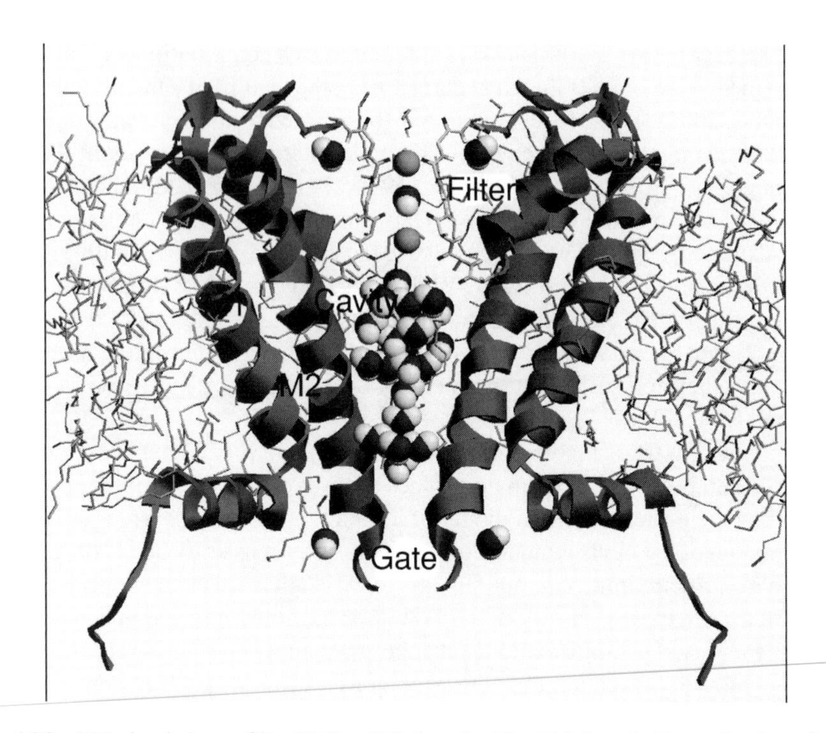

Fig. 16.3 MD simulations of the KirBac TM domain. The TM domain (two subunits only are shown, for clarity) is shown embedded with a membrane-mimetic octane slab. Water molecules and ions within the filter, and water molecules within the cavity of the channel are shown.

where $<>$ denotes a time average over the instantaneous structures sampled during the simulation, and x_i is a mass-weighted atomic coordinate. A set of eigenvalues and eigenvectors is calculated by solving the standard eigenvalue equation:

$$\mathbf{A\Gamma} = \mathbf{\Gamma}\Lambda,$$

where $\Lambda = \text{diag}(\lambda_m)$ is a diagonal matrix whose mth diagonal element is the eigenvalue λ_m, Γ is a matrix whose mth column vector is the eigenvector of λ_m. It can be shown that the eigenvalues correspond to mean square fluctuation of the collective coordinates along selected components (Amadei et al., 1993). Collective coordinates are a powerful tool to investigate conformational changes in proteins. They are calculated by projecting the MD trajectories onto the mth eigenvector.

This technique has been useful in analyzing the dynamics of proteins, especially in probing, e.g., large conformational changes related to protein folding/unfolding (de Groot et al., 1996; Roccatano et al., 2001, 2003; Daidone et al., 2003, 2004). As shown by Amadei et al., the overall conformational space sampled by a protein can be divided into a low-dimensional essential subspace of $\sim$10 eigenvectors, within which the principal collective motions of proteins are usually confined, and

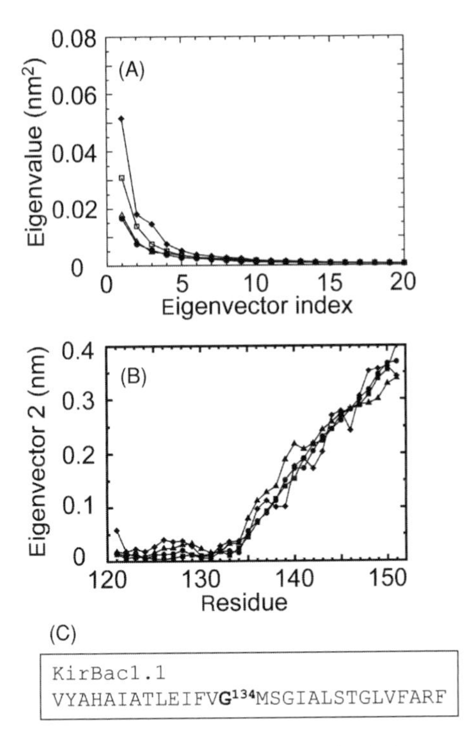

Fig. 16.4 (A) Eigenvalues vs. eigenvector index from principal components analysis (PCA) of the motions of the M2 helices in a simulation of KirBac1.1 in a phospholipid bilayer. Each line represents the M2 helix of a different subunit, according to the following symbols: circles = subunit 1; squares = subunit 2; diamonds = subunit 3; and triangles = subunit 4. (B) PCA analysis of M2 helix motions: mean square displacements of the Cα atoms according to eigenvector 2 are shown as a function of residue number. The fitting of the helices uses only the Cα atoms of the N-terminal half (residues 121–134) of M2. Thus, the mean square displacements increase after the molecular hinge at residue 134. (C) Sequence of the M2 helix from KirBac1.1, highlighting the hinge residue at Gly134.

a near-constraint space within which simple small fluctuations occur. Indeed, the first 10 eigenvectors usually account for 70–80% of the total fluctuations spanned by the principal components. This implies that it may be possible to capture the nature of large amplitude biological relevant motions (as might occur in potassium channels).

As an example of PCA, in Fig. 16.4, we illustrate its application to analysis of the motion of the pore-lining M2 helices in a simulation of KirBac1.1 in a lipid bilayer. A comparable analysis has been performed for the related channel KirBac3.1 (Grottesi et al., 2005). As discussed in more detail below, a major element of the motion of the M2 helices in this and related simulations appears to be a bending motion about a central glycine residue (Gly134) which forms a molecular hinge. In Fig. 16.4A, the first 10 eigenvalues are shown, as derived from PCA of the motions of the M2 helices in the simulation. From this it can be estimated that the first 10 eigenvectors account for ∼80% of the motion of the M2 helices. For each eigenvector, the resultant motions may be analyzed by plotting the mean square displacement along the eigenvector as a function of residue number. This is shown in Fig. 16.4B for the second eigenvector, after fitting the Cα atoms of M2 residues N-terminal the location of the molecular hinge. The results of this analysis clearly support the existence of a molecular hinge at residue Gly134 (Fig. 16.4C), midway down the M2 helix. The functional consequences of this are discussed in more detail below.

The application of the PCA technique to protein dynamics analysis is limited by the time scale accessible to current MD techniques and computers power. However, by means of ED sampling techniques it is possible to try overcoming this limitation and to investigate conformations that are normally not accessible by classic MD (de Groot et al., 1996).

16.5 What Can Simulations Tell Us?

In considering MD simulations of ion channels, and of related membrane proteins, it is useful to consider the time scales involved. For a simulation of an ion channel embedded in a lipid bilayer plus waters, which yields a system size of ∼50,000 atoms, simulation times of ∼20 ns are readily achievable. This time is of the same order of magnitude as the mean time for the passage of a single ion through a channel. Thus, MD simulations may be used directly to examine the nature of ion permeation, but multiple simulations are required to lend statistical significance to the observations made. In contrast, channel gating and associated conformational changes occur on time scales of microseconds to milliseconds. Thus, MD simulations cannot be used to address such events directly, and so alternative approaches, both to analysis and to simulation, are required.

MD simulations can also provide details of the interaction of channel proteins with their membrane environment. Given the importance of interactions with lipids on the stability, function (Valiyaveetil et al., 2002; Demmers et al., 2003), and regulation (Du et al., 2004) of K channels, this aspect of analysis via MD simulations is increasingly important. However, considerations of space preclude a more detailed discussion in the current paper, and the interested reader if referred to e.g. (Domene et al., 2003; Deol et al., 2004) for further details.

16.5.1 Ion Permeation

Simulation studies of K channels from a number of laboratories have focused on the events during ion permeation through the selectivity filter, both in KcsA (Allen et al., 1999; Bernèche and Roux, 2000; Guidoni et al., 2000; Shrivastava and Sansom, 2000, 2002; Domene and Sansom, 2003) and in related K channels (Capener et al., 2000; Capener and Sansom, 2002; Domene et al., 2004). The results of these simulations

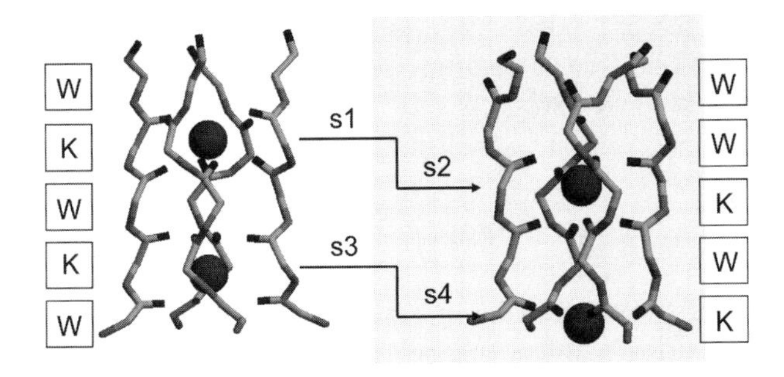

Fig. 16.5 Two snapshots from a simulation of KirBac1.1 in a POPC bilayer (Domene et al., 2004). The snapshots show just the selectivity filter backbone, plus two K^+ ions, and correspond to structures 0.1 ns apart, just before and after the concerted translocation of the two ions from sites S1 and S3, to sites S2 and S4. Note the small changes in backbone conformation coupled to translocation of the K^+ ions. Schematics of the two patterns of occupancy of the five sites (S0 at the top to S4 at the bottom) are given on either side; W = water; K = potassium ion.

have confirmed that ion permeation occurs via concerted single file motion of K^+ ions and water molecules through the filter. The oxygen atoms that line the filter region form five distinct binding sites for K^+ ions, from S0 at the extracellular mouth of the filter to S4 at the opposite end of the filter next to the central cavity. Ion translocation through the filter occurs via a switch between two patterns of occupancy of the filter: between ions at sites S0, S2, and S4, and ions at sites S1 and S3. This result from simulations is in good agreement with structural data (Morais-Cabral et al., 2001; Zhou et al., 2001).

As an example of the results from such simulations, Fig. 16.5 shows two snapshots (separated in time by 0.1 ns) of the filter from a simulation of the bacterial inward rectifier homologue Kirbac1.1 embedded in a lipid bilayer (Domene et al., 2004). It can be seen that there is a spontaneous, concerted switch between K^+ ions at sites S1 and S3, to ions at sites S2 and S4. Close examination of the structures reveals small changes in the conformation of the filter, with the oxygen atoms lining the pore moving by <0.1 nm as the ion switches from site to site. Such limited flexibility of the pore is needed to facilitate rapid translocation of the ions between adjacent binding sites.

It is also possible to use MD simulations to examine the energetics of ion translocation along the filter. Early studies by Åqvist and colleagues (Åqvist and Luzhkov, 2000) established that configurations with ions at S1, S3 and at S2, S4 were of similar energy. More extensive exploration of the free energy surface of ion translocation along the filter has been performed by Roux et al. (Bernèche and Roux, 2001), who have also examined the role of filter flexibility in ion translocation and ion selectivity (Allen et al., 2004; Noskov et al., 2004). These results indicated that inclusion vs. omission of thermal atomic fluctuations resulted in significant differences in permeation energy profiles.

In addition to considerations of the effects of flexibility, it must be recalled that such energetic analyses are an approximation, as the underlying force fields do not allow for any electronic repolarization as the cation moves relative to the oxygen atoms lining the filter. Treatment of the latter effect by density functional theory calculations (Guidoni and Carloni, 2002) suggests it may play an important role in the mechanism of permeation and selectivity.

16.5.2 Filter Flexibility

The conformational dynamics of the selectivity filter of K channels are also of interest in the context of possible gating mechanisms. For example, MD simulations of KcsA and of KirBac1.1 provide evidence both for limited (<0.1 nm) filter flexibility during the concerted motion of ions and water molecules within the filter (see above), but also for more substantial distortions (Fig. 16.6). In particular, in simulations of both KcsA and KirBac, occasional peptide bond "flips" are seen, especially for the valine carbonyl of the TVGYG motif, such that one of the peptide backbone oxygen atoms point away from the filter instead of toward it. Clearly, such events will result in a significant change in the permeation energy landscape for the duration of the "flip." If simulations are performed in the *absence* of K^+ ions, then a more profound distortion of the filter occurs, resulting in, e.g., three of the four valine carbonyls

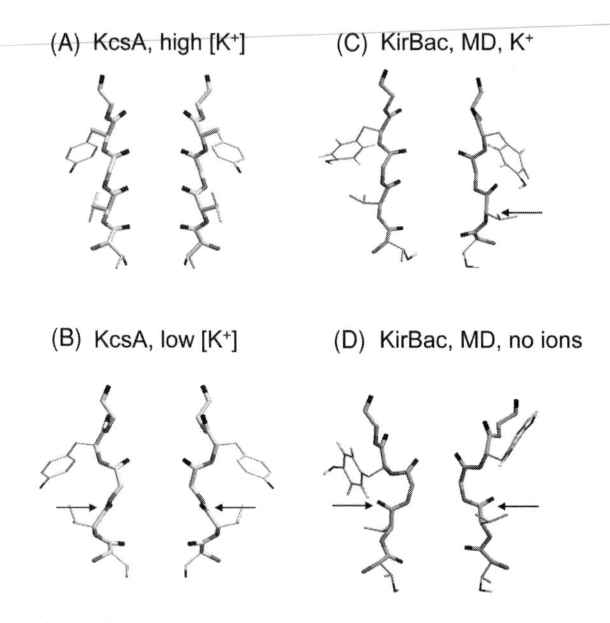

Fig. 16.6 Comparison of the structures of K channel selectivity filter in X-ray crystallographic structures and MD simulations. In each case the backbone (thick bonds) and side chain (thin bonds) atoms of just two subunits of the filter are shown. (A) KcsA, crystallized in the presence of a high concentration of K^+ ions (PDB code 1K4C); (B) KcsA, crystallized in the presence of a low concentration of K^+ ions (PDB code 1K4D); (C) KirBac, midway through a 10 ns MD simulation in the *presence* of K^+ ions (at sites S2 and S4 of the filter—see Fig. 16.4); and (D) KirBac, at the end of a 10 ns MD simulation in the *absence* of K^+ ions. The flipped carbonyls of the valine residue of TVGYG are indicated by arrows.

pointing away from the pore. Such changes in filter conformation are expected to lead to functional closure of the channel. These simulations suggest that filter distortions may provide a mechanism of K channel gating, in addition to larger scale changes in the conformation of the hydrophobic gate formed at the intracellular crossing point of the M2 helices. Indeed, certain mutations of Kir6.2 channels in the filter region have been shown to result in changes in "fast gating" of the channel (Proks et al., 2001), and simulations based on models of such mutants suggest they may promote a change in the conformational dynamics of the filter region (Capener et al., 2003).

It is of particular interest that the simulation results on filter flexibility correlate well with structural studies. Thus, comparison of the filter conformations generated during simulations in the absence of K^+ ions (Fig. 16.6D) with the structural change in KcsA crystals resulting from a low concentration of K^+ ions (Fig. 16.6B) shows that in both cases the change in conformation relative to the X-ray structure in the presence of a high concentration of K^+ involves movement of the carbonyl oxygen of the valine away from the pore. The conformational change is less pronounced in the X-ray structure/reflecting the presence of K^+ ions, albeit at reduced occupancy. The relationship of such local changes in conformation, which will result in changes in the energy landscape of permeation, to changes in channel conductance and fast gating continues to be an active area of research. For example, simulation-based estimates of ion permeation free energies based on the low K^+ conformation of KcsA suggest it is nonconducting (Bernèche and Roux, 2005). It is clear from all of these studies that relatively minor changes in channel conformation can have a profound effect on channel properties.

16.5.3 Channel Gating

As noted above, the time scale of channel gating is too long to enable this process to be studied directly by MD simulation. However, a combination of simulations of simple models, homology modeling of various channel states, and PCA analysis of K channel simulations has provided information on the nature of the conformational transitions that may underlie K channel gating.

Examination of the structures of K channels suggests that the main gate is at the intracellular mouth of the channel, where the pore-lining M2 (or S6) helices come together in the closed state of the channel to form a narrow hydrophobic region. Comparison of the structures of KcsA (closed state) and of MthK (crystallized in an open state) indicated a change in conformation and orientation of the M2 helices resulted in opening of the gate, seen as a loss of the hydrophobic constriction at the intracellular mouth.

The fundamental properties of hydrophobic gating in ion channels have been studied using simulations of simple nanopore models (Beckstein et al., 2001, 2003, 2004; Beckstein and Sansom, 2003, 2004), in particular to establish the relationship between pore radius and hydrophobicity of residues lining the ion pathway. The results of these studies suggest that a purely hydrophobic pore of radius $<\sim 0.8$ nm may be functionally closed, even though it is not sterically occluded. This is because even at a radius of ∼0.6 nm there is a considerable energetic barrier to water entering a hydrophobic pore, which therefore prevents a complete solvation shell entering alongside an ion (Beckstein et al., 2004).

As mentioned above, comparison of X-ray structures suggests a role for the conformational dynamics of the pore-lining M2 (or S6) helices in the gating mechanism of K channels. This is supported by spectroscopic data (Perozo et al., 1998, 1999; Liu et al., 2001), by early simulations (Biggin et al., 2001), and by mutation studies on a number of K channels. In particular, it is suggested (Jiang et al., 2002b) that a glycine residue in the M2 helix of, e.g., KcsA and MthK channels forms a molecular hinge that enables hinge-bending of the M2 helices so as to switch the channels from a closed to an open conformation. The proposed molecular hinge corresponds to a conserved glycine residue (Jiang et al., 2002b) present in the sequence of most K channels. In Kv channels the situation may be somewhat more complex, with a possible second hinge lower down the S6 helix associated with a conserved PVP sequence motif (Camino and Yellen, 2001; Bright et al., 2002; Beckstein et al., 2003; Bright and Sansom, 2004; Webster et al., 2004). In the case of Kv channel mutations in the vicinity of both the conserved glycine hinge (Magidovich and Yifrach, 2004) and the second (PVP) hinge result in perturbations of channel gating (Hackos et al., 2002; Labro et al., 2003; Sukhareva et al., 2003). For Kir channels, proline scanning mutations combined with molecular modeling (Jin et al., 2002) support the proposed M2 helix glycine molecular hinge. Mutational studies of the more distantly related bacterial sodium channel NaChBac suggest a glycine hinge may also be present in the pore-lining helices of sodium channels (Zhao et al., 2004).

A number of modeling and simulation studies, ranging from early in vacuo simulations (Kerr et al., 1996) to more recent simulations in either a membrane mimetic octane slab and/or in a lipid bilayer (Shrivastava et al., 2000; Tieleman et al., 2001; Bright et al., 2002), have been performed on models of the isolated S6 helix from *Shaker* Kv channels. All of these indicated the presence of a dynamic hinge (Sansom and Weinstein, 2000) in the S6 helix induced by the PVP motif, demonstrating that this is a region of intrinsic flexibility in the pore-lining helix. More recent studies have focused on simulations of a model of the pore domain [i.e. $(S5\text{-}P\text{-}S6)_4$] of *Shaker* Kv channels, again in a membrane mimetic environment (Bright and Sansom, 2004). These also yield S6 structures kinked in the vicinity of the PVP motif. Comparison with the corresponding S6 helix from the X-ray structure of Kv1.2 (see Fig. 16.7) reveals that the conformations of S6 from the simulated model are very similar to those captured in the experimental crystal structure. It is also informative to compare the structure of the S6 tetrameric bundle from the modeling and simulation study with that present in the X-ray structure (Fig. 16.8). From this it is evident that not only are the S6 helices kinked in both structures, but the way in which the helices are packed around the central pore is very similar. This suggests that a combined modeling/simulation approach can be genuinely predictive in terms of understanding how pore-lining helices behave during channel gating.

Based on these studies of modeling the S6 helix of Kv channels, more extended simulation studies have been used to explore the S6 (or M2) hinge-bending

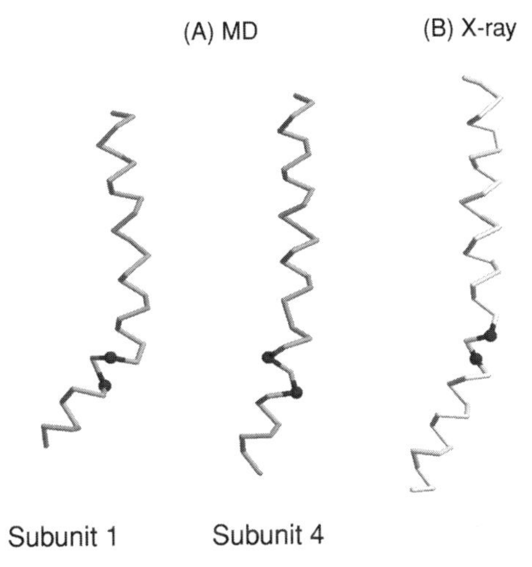

Fig. 16.7 Kv S6 helices from simulations and X-ray structures compared. In (A) two S6 helices from a simulation of the *Shaker* Kv pore domain [i.e. $(S5\text{-}P\text{-}S6)_4$] are shown (Bright and Sansom, 2004); in (B) the corresponding S6 helix from the X-ray structure of Kv1.2 (Long et al., 2005) is shown. The Cα atoms of the prolines of the PVP motif are shown as dark grey spheres.

hypothesis of gating of Kv (or Kir) channels. By simulating just the transmembrane pore-forming domain of Kv or KirBac channels in a membrane-mimetic octane slab, it is possible (within the 10–20 ns time scale accessible to such simulations) to explore the intrinsic flexibility of the S6 (or M2) helices. Combined with PCA of the simulation data, such studies enable us to characterize possible conformational changes underlying the channel-gating mechanism, uncoupled from the "gate-keeping" role of the of the voltage-sensor S1–S4 domain (in Kv channels) or the ligand-binding intracellular domain (in Kir channels).

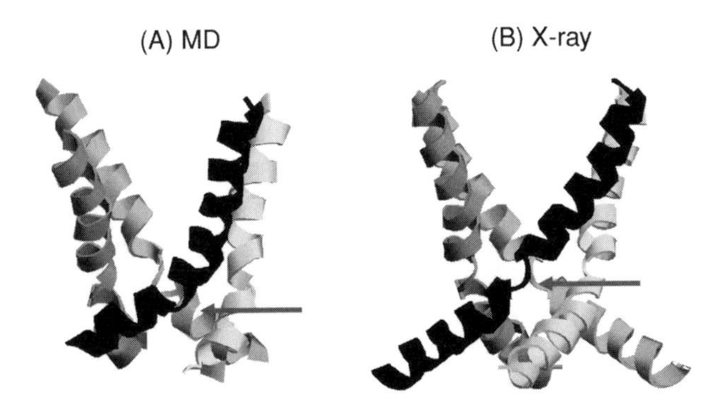

Fig. 16.8 Comparison of the S6 helix bundle from (A) a simulation of the *Shaker* Kv pore domain (Bright and Sansom, 2004); and (B) the X-ray structure of Kv1.2 (Long et al., 2005). The location of the PVP motif-induced hinge is indicated by the grey arrows.

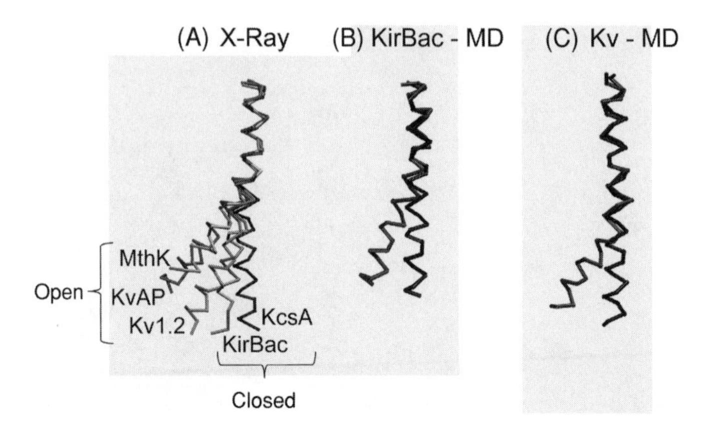

Fig. 16.9 Comparison of X-ray crystallographic and MD simulation-generated structures for the pore-lining M2 or S6 helices of K channels. In (A) the M2 helices from the crystal structures of KcsA and of KirBac1.1 (both closed pores), along with the M2 helix from MthK, and the S6 helices from KvAP and Kv1.2 (all three with open pores) are shown. In (B) selected structures are shown of M2 helices from the start (0 ns) and end (20 ns) of a simulation of KirBac1.1 in an octane slab (Grottesi et al., 2005) are shown. In (C) selected structures are shown of S6 helices from the start (0 ns) and end (9 ns) of a simulation of *Shaker* Kv channel in an octane slab. In all three diagrams, the N-terminal halves (i.e., before the molecular hinge) of the helices are used for fitting.

The results of such studies are summarized in Fig. 16.9, where snapshots from simulations are presented alongside superimposed X-ray structures of pore-lining helices from K channels. For KirBac, it can be seen that the M2 helix has a molecular hinge associated with the conserved Gly134 residue (Fig. 16.9B). PCA of the motions of M2 in KirBac simulations reveals that the first two eigenvectors correspond to helix kinking and swivelling about the glycine hinge (Grottesi et al., 2005). A comparable analysis of Kv channel simulations (Grottesi and Sansom, unpublished results) suggests that in this case the molecular hinge is associated with the PVP motif lower down the S6 helix from the conserved glycine (Fig. 16.9C). This is in agreement with the modeling and simulation studies of S6 helices in membrane and membrane-like environments discussed above.

One may attempt a simple calculation of the energetic barrier presented by the hydrophobic gate region in the Kv-closed model and how this is altered in the Kv-open model. One may employ a Born energy calculation to obtain a first approximation to the barrier height presented by a hydrophobic gate (Beckstein et al., 2004). Compared to calculations of the barrier height based on atomistic (MD) simulations and umbrella sampling, such as Born energy calculation underestimates the barrier height for hydrophobic pores of radius $>\sim 3.5$ Å, and overestimates it for pores of radius $<\sim 3.5$ Å. The pore radius profile of the Kv-closed model (discussed above) has an average hydrophobic gate radius of ~ 1.4 Å, extending over a length of ~ 20 Å. The minimum radius is ~ 1 Å, at the cytoplasmic end of

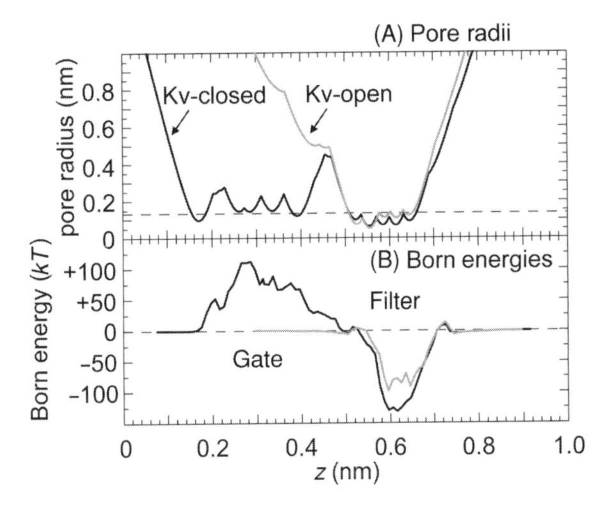

Fig. 16.10 Comparison of Kv-closed and Kv-open models. (A) Pore radius profiles and (B) Born energy profiles for the two models shown in Fig. 16.2. In each graph the black line corresponds to Kv-closed, and the grey line to K-open. [The Born energy calculation was performed essentially as described in (Beckstein et al., 2004).]

the gate (Fig. 16.10A). A Born energy calculation for the Kv-closed model yields a barrier of height $\sim 70kT$ in the center of the gate, rising to $\sim 100kT$ at the narrowest region of the gate (Fig. 16.10B). Even if, based on the comparison of Born energies and atomistic simulation-based free energies for simple nanopore models, we allow for an approximately twofold overestimation of the Kv-closed barrier height by the Born energy calculation, this still yields an estimated barrier height of $\sim 50kT$ for the Kv-closed channel. This model clearly corresponds to a fully closed conformation of the channel. For comparison, there is virtually no barrier present in the Kv-open model. Thus, the S6 helix bending motions seen in the Kv simulations are sufficient to switch fully the pore from a functionally closed to a functionally open state.

16.5.4 Regulation of Channel Gating

It is also possible to probe the motions of the "gate-keeping" domains of K channels by MD simulation. Of course, caveats regarding simulation time scales vs. gating time scales similar to those expressed for gating per se apply, but the simulations do reveal aspects of the intrinsic flexibility of domains, and hence provide valuable clues as to the overall channel-gating mechanism. For example, the X-ray structure of the intracellular domain of the Kir channel Kir3.1 has been determined (Nishida and MacKinnon, 2002), enabling simulations of the conformational dynamics of the intracellular domains of Kir3.1 and of related Kir channels. Like the TM domain, the intracellular domain forms a tetramer. Multiple 10 ns duration simulations of two

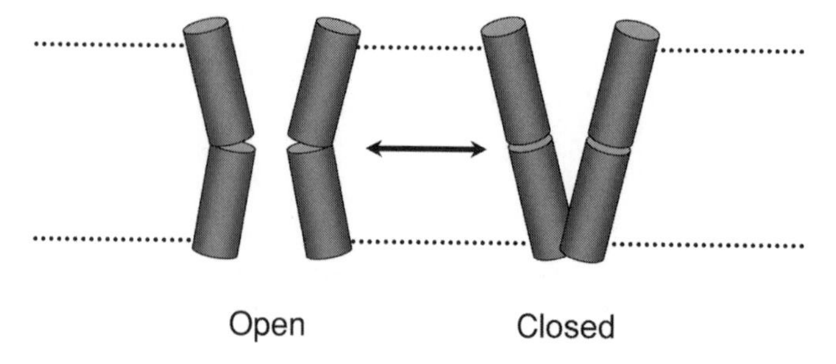

Fig. 16.11 Proposed gating mechanism of K channels. The pore-lining S6 or M2 helices of two opposite subunits are shown. In the open state these helices are kinked; in the closed state the helices are undistorted and cross at their C-termini to form a hydrophobic constriction.

different Kir intacellular domain tetramers (of Kir3.1 and Kir6.2) combined with PCA reveal loss of exact fourfold symmetry (Haider et al., 2005). This is consistent with "dimer-of-dimers" motion of the subunits in the intracellular domains. Combining this analysis with the results of simulations of the TM domain (see above), enables us to propose a Kir-gating model in which a transition between exact tetrameric symmetry (channel open) and dimer-of-dimers symmetry (channel closed) is associated with a change in M2 helix packing and kinking coupled to gating of the channel (Grottesi et al., 2005). This provides a simple example of how combining simulation and modeling studies may provide clues as to the gating mechanisms of complex ion channels. Overall, structural and simulation studies suggest a general model of K channel gating (Fig. 16.11) in which the "gate-keeper" domain (e.g., the voltage sensor in Kv channels or the intracellular domain in Kir channels) regulates the switching of the pore-lining helices between an open (kinked) conformation and a closed (distorted) conformation.

16.6 Future Perspectives

The results reviewed above indicate how MD simulations help to contribute to our current understanding of K channels. In particular, they have provided information on the nature of ion permeation through the selectivity filter, and on the flexibility of the filter in relationship to ion permeation and to "fast gating." More recently, combining MD simulations of K channel components with PCA has enabled us to isolate key motions that provide clues as to the underlying gating mechanisms of Kv and Kir channels.

Future studies will have to address at least two aspects of K channel simulations. One is that of time and length scales. There is a pressing need for simulation approaches that allow us to explore larger time and length scale motions. One possibility lies in the use of Gaussian network simulations (Erkip and Erman, 2003). However, this and other coarse-grained simulation approaches need to be explored further, and tested on a number of membrane protein systems. A second area is that of extending models and simulations to a wider range of K channels, and to extending the realism of the biological environment present in the simulations. To date, most simulation studies have been of bacterial K channels in a simple model membrane (e.g., phosphatidyl choline). However, from a physiological perspective, one would like to extend simulations to realistic models of mammalian K channels in a complex (i.e., mixed lipid) membrane environment. This is of particular importance given the role of some lipids (e.g., phosophatidyl inositol phosphate) in regulation of channel gating. Recent simulation studies [e.g., of KcsA (Deol et al., 2005; Sansom et al., 2005)] provide some encouragement that extended simulations in more complex lipid bilayers will be able to capture at least some aspects of specific channel–lipid interactions.

Acknowledgments

We wish to thank all of our colleagues for their interest in this work, which is supported by grants from the Wellcome Trust (to MSPS).

References

Allen, T.W., O.S. Andersen, and B. Roux. 2004. On the importance of atomic fluctuations, protein flexibility, and solvent in ion permeation. *J. Gen. Physiol.* 124:679–690.

Allen, T.W., S. Kuyucak, and S.H. Chung. 1999. Molecular dynamics study of the KcsA potassium channel. *Biophys. J.* 77:2502–2516.

Amadei, A., A.B.M. Linssen, and H.J.C. Berendsen. 1993. Essential dynamics of proteins. *Proteins Struc. Funct. Genet.* 17:412–425.

Antcliffe, J.F., S. Haider, P. Proks, M.S.P. Sansom, and F.M. Ashcroft. 2005. Functional analysis of a structural model of the ATP-binding site of the K_{ATP} channel Kir6.2 subunit. *EMBO J.* 24:229–239.

Åqvist, J., and V. Luzhkov. 2000. Ion permeation mechanism of the potassium channel. *Nature* 404:881–884.

Arinaminpathy, Y., P.C. Biggin, I.H. Shrivastava, and M.S.P. Sansom. 2003. A prokaryotic glutamate receptor: Homology modelling and molecular dynamics simulations of GluR0. *FEBS Lett.* 553:321–327.

Ash, W.L., M.R. Zlomislic, E.O. Oloo, and D.P. Tieleman. 2004. Computer simulations of membrane proteins. *Biochim. Biophys. Acta* 1666:158–189.

Beckstein, O., P.C. Biggin, P.J. Bond, J.N. Bright, C. Domene, A. Grottesi, J. Holyoake, and M.S.P. Sansom. 2003. Ion channel gating: Insights via molecular simulations. *FEBS Lett.* 555:85–90.

Beckstein, O., P.C. Biggin, and M.S.P. Sansom. 2001. A hydrophobic gating mechanism for nanopores. *J. Phys. Chem. B* 105:12902–12905.

Beckstein, O., and M.S.P. Sansom. 2003. Liquid–vapor oscillations of water in hydrophobic nanopores. *Proc. Natl. Acad. Sci. USA* 100:7063–7068.

Beckstein, O., and M.S.P. Sansom. 2004. The influence of geometry, surface character and flexibility on the permeation of ions and water through biological pores. *Phys. Biol.* 1:42–52.

Beckstein, O., K. Tai, and M.S.P. Sansom. 2004. Not ions alone: Barriers to ion permeation in nanopores and channels. *J. Am. Chem. Soc.* 126:14694–14695.

Berendsen, H.J.C., J.P.M. Postma, W.F. van Gunsteren, A. DiNola, and J.R. Haak. 1984. Molecular dynamics with coupling to an external bath. *J. Chem. Phys.* 81:3684–3690.

Bernèche, S., and B. Roux. 2000. Molecular dynamics of the KcsA K^+ channel in a bilayer membrane. *Biophys. J.* 78:2900–2917.

Bernèche, S., and B. Roux. 2001. Energetics of ion conduction through the K^+ channel. *Nature* 414:73–77.

Bernèche, S., and B. Roux. 2005. A gate in the selectivity filter of potassium channels. *Structure* 13:591–600.

Biggin, P.C., I.H. Shrivastava, G.R. Smith, and M.S.P. Sansom. 2001. Non-equilibrium molecular dynamics study of KcsA gating. *Biophys. J.* 80:514.

Bostick, D.L., and M.L. Berkowitz. 2003. The implementation of slab geometry for membrane-channel molecular dynamics simulations. *Biophys. J.* 85:97–107.

Bright, J.N., and M.S.P. Sansom. 2004. The Kv channel S6 helix as a molecular switch: Simulation studies. *IEE Proc. Nanobiotechnol.* 151:17–27.

Bright, J.N., I.H. Shrivastava, F.S. Cordes, and M.S.P. Sansom. 2002. Conformational dynamics of helix S6 from Shaker potassium channel: Simulation studies. *Biopolymers* 64:303–313.

Camino, D.D., and G. Yellen. 2001. Tight steric closure at the intracellular activation gate of a voltage-gated K^+ channel. *Neuron* 32:649–656.

Capener, C.E., H.J. Kim, Y. Arinaminpathy, and M.S.P. Sansom. 2002. Ion channels: Structural bioinformatics and modelling. *Human Mol. Genet.* 11:2425–2433.

Capener, C.E., P. Proks, F.M. Ashcroft, and M.S.P. Sansom. 2003. Filter flexibility in a mammalian K channel: Models and simulations of Kir6.2 mutants. *Biophys. J.* 84:2345–2356.

Capener, C.E., and M.S.P. Sansom. 2002. MD Simulations of a K channel model—sensitivity to changes in ions, waters and membrane environment. *J. Phys. Chem. B* 106:4543–4551.

Capener, C.E., I.H. Shrivastava, K.M. Ranatunga, L.R. Forrest, G.R. Smith, and M.S.P. Sansom. 2000. Homology modelling and molecular dynamics simulation studies of an inward rectifier potassium channel. *Biophys. J.* 78:2929–2942.

Chen, C.T., and B. Rost. 2002. State-of-the-art in membrane protein prediction. *Appl. Bioinformatics* 1:21–35.

Chothia, C. 1984. Principles that determine the structure of proteins. *Ann. Rev. Biochem.* 53:537–572.

Daidone, I., A. Amadei, D. Roccatano, and A. Di Nola. 2003. Molecular dynamics simulation of protein folding by essential dynamics sampling: Folding landscape of horse heart cytochrome *c*. *Biophys. J.* 85:2865–2871.

Daidone, I., D. Roccatano, and S. Hayward. 2004. Investigating the accessibility of the closed domain conformation of citrate synthase using essential dynamics sampling. *J. Mol. Biol.* 339:515–525.

Darden, T., D. York, and L. Pedersen. 1993. Particle mesh Ewald—an N.log(N) method for Ewald sums in large systems. *J. Chem. Phys.* 98:10089–10092.

de Groot, B.L., A. Amadei, R.M. Scheek, N.A. van Nuland, and H.J.C. Berendsen. 1996. An extended sampling of the configurational space of HPr from *E. coli*. *Proteins Struct. Funct. Genet.* 26:314–322.

Demmers, J.A.A., A. van Dalen, B. de Kruijiff, A.J.R. Heck, and J.A. Killian. 2003. Interaction of K channel KcsA with membrane phospholipids as studied by ESI mass spectrometry. *FEBS Lett.* 541:28–32.

Deol, S.S., P.J. Bond, C. Domene, and M.S.P. Sansom. 2004. Lipid–protein interactions of integral membrane proteins: A comparative simulation study. *Biophys. J.* 87:3737–3749.

Deol, S.S., C. Domene, P.J. Bond, and M.S.P. Sansom. 2006. Anionic phospholipids interactions with the potassium channel KcsA: Simulation studies. *Biophys. J.* 90:822–830.

Domene, C., P.J. Bond, S.S. Deol, and M.S.P. Sansom. 2003. Lipid–protein interactions and the membrane/water interfacial region. *J. Am. Chem. Soc.* 125:14966–14967.

Domene, C., A. Grottesi, and M.S.P. Sansom. 2004. Filter flexibility and distortion in a bacterial inward rectifier K^+ channel: Simulation studies of KirBac1.1. *Biophys. J.* 87:256–267.

Domene, C., and M.S.P. Sansom. 2003. A potassium channel, ions and water: Simulation studies based on the high resolution X-ray structure of KcsA. *Biophys. J.* 85:2787–2800.

Doyle, D.A., J.M. Cabral, R.A. Pfuetzner, A. Kuo, J.M. Gulbis, S.L. Cohen, B.T. Cahit, and R. MacKinnon. 1998. The structure of the potassium channel: Molecular basis of K^+ conduction and selectivity. *Science* 280:69–77.

Du, X.O., H.L. Zhang, C. Lopes, T. Mirshahi, T. Rohacs, and D.E. Logothetis. 2004. Characteristic interactions with phosphatidylinositol 4,5-bisphosphate determine regulation of Kir channels by diverse modulators. *J. Biol. Chem.* 279:37271–37281.

Durell, S.R., I.H. Shrivastava, and H.R. Guy. 2004. Models of the structure and voltage-gating mechanism of the *Shaker* K^+ channel. *Biophys. J.* 87:2116–2130.

Eriksson, M.A.L., and B. Roux. 2002. Modeling the structure of agitoxin in complex with the Shaker K^+ channel. *Biophys. J.* 83:2595–2609.

Erkip, A., and B. Erman. 2003. Dynamics of large-scale fluctuations in native proteins. Analysis based on harmonic inter-residue potentials and random external noise. *Polymer* 45:641–648.

Essmann, U., L. Perera, M.L. Berkowitz, T. Darden, H. Lee, and L.G. Pedersen. 1995. A smooth particle mesh Ewald method. *J. Chem. Phys.* 103:8577–8593.

Faraldo-Gómez, J.D., L.R. Forrest, M. Baaden, P.J. Bond, C. Domene, G. Patargias, J. Cuthbertson, and M.S.P. Sansom. 2004. Conformational sampling and dynamics of membrane proteins from 10-nanosecond computer simulations. *Proteins Struct. Funct. Bioinformatics* 57:783–791.

Fiser, A., R. Kinh Gian Do, and A. Sali. 2000. Modeling of loops in protein structures. *Prot. Sci.* 9:1753–1773.

Garcia, A.E. 1992. Large-amplitude nonlinear motions in proteins. *Phys. Rev. Lett.* 68:2696–2699.

Grottesi, A., C. Domene, and M.S.P. Sansom. 2005. Conformational dynamics of M2 helices in KirBac channels: Helix flexibility in relation to gating via molecular dynamics simulations. *Biochem.* 44:14586–14594.

Guidoni, L., and P. Carloni. 2002. Potassium permeation through the KcsA channel: A density functional study. *Biochim. Biophys. Acta* 1563:1–6.

Guidoni, L., V. Torre, and P. Carloni. 2000. Water and potassium dynamics in the KcsA K^+ channel. *FEBS Lett.* 477:37–42.

Gulbis, J.M., and D.A. Doyle. 2004. Potassium channel structures: Do they conform? *Curr. Opin. Struct. Biol.* 14:440–446.

Hackos, D.H., T.H. Chang, and K.J. Swartz. 2002. Scanning the intracellular S6 activation gate in the shaker K^+ channel. *J. Gen. Physiol.* 119:521–531.

Haider, S., A. Grottesi, F.M. Ashcroft, and M.S.P. Sansom. 2005. Conformational dynamics of the ligand-binding domain of inward rectifier K channels as revealed by MD simulations: Towards an understanding of Kir channel gating. *Biophys. J.* 88:3310–3320.

Holyoake, J., C. Domene, J.N. Bright, and M.S.P. Sansom. 2003. KcsA closed and open: Modelling and simulation studies. *Eur. Biophys. J.* 33:238–246.

Hunenberger, P.H., and J.A. McCammon. 1999. Effect of artificial periodicity in simulations of biomolecules under Ewald boundary conditions: A continuum electrostatics study. *Biophys. Chem.* 78:69–88.

Hunenberger, P.H., and J.A. McCammon. 1999. Ewald artifacts in computer simulations of ionic solvation and ion–ion interaction. *J. Chem. Phys.* 110:1856–1872.

Jiang, Y., A. Lee, J. Chen, M. Cadene, B.T. Chait, and R. MacKinnon. 2002a. Crystal structure and mechanism of a calcium-gated potassium channel. *Nature* 417:515–522.

Jiang, Y., A. Lee, J. Chen, M. Cadene, B.T. Chait, and R. MacKinnon. 2002b. The open pore conformation of potassium channels. *Nature* 417:523–526.

Jiang, Y., A. Lee, J. Chen, V. Ruta, M. Cadene, B.T. Chait, and R. Mackinnon. 2003. X-ray structure of a voltage-dependent K^+ channel. *Nature* 423:33–41.

Jin, T., L. Peng, T. Mirshahi, T. Rohacs, K.W. Chan, R. Sanchez, and D.E. Logothetis. 2002. The $\beta\gamma$ subunits of G proteins gate a K^+ channel by pivoted bending of a transmembrane segment. *Mol. Cell.* 10:469–481.

Karplus, M.J., and J.A. McCammon. 2002. Molecular dynamics simulations of biomolecules. *Nat. Struct. Biol.* 9:646–652.

Kerr, I.D., H.S. Son, R. Sankararamakrishnan, and M.S.P. Sansom. 1996. Molecular dynamics simulations of isolated transmembrane helices of potassium channels. *Biopolymers* 39:503–515.

Kitao, A., and N. Go. 1999. Investigating protein dynamics in collective coordinate space. *Curr. Opin. Struct. Biol.* 9:164–169.

Kuo, A., J.M. Gulbis, J.F. Antcliff, T. Rahman, E.D. Lowe, J. Zimmer, J. Cuthbertson, F.M. Ashcroft, T. Ezaki, and D.A. Doyle. 2003. Crystal structure of the potassium channel KirBac1.1 in the closed state. *Science* 300:1922–1926.

Labro, A.J., A.L. Raes, I. Bellens, N. Ottschytsch, and D.J. Snyders. 2003. Gating of Shaker-type channels requires the flexibility of S6 caused by prolines. *J. Biol. Chem.* 278:50724–50731.

Laine, M., D.M. Papazian, and B. Roux. 2004. Critical assessment of a proposed model of Shaker. *FEBS Lett.* 564:257–263.

Leach, A.R. 2001. Molecular Modelling. Principles and Applications, 2nd Ed. Prentice Hall, Harlow, England.

Lee, S.Y., A. Lee, J. Chen, and R. MacKinnon. 2005. Structure of the KvAP voltage-dependent K^+ channel and its dependence on the lipid membrane. *Proc. Natl. Acad. Sci. USA* 102:15441–15446.

Liu, Y., P. Sompornpisut, and E. Perozo. 2001. Structure of the KcsA channel intracellular gate in the open state. *Nat. Struct. Biol.* 8:883–887.

Long, S.B., E.B. Campbell, and R. MacKinnon. 2005. Crystal structure of a mammalian voltage-dependent *Shaker* family K^+ channel. *Science* 309:897–902.

Mackinnon, R. 2003. Potassium channels. *FEBS Lett.* 555:62–65.

Magidovich, E., and O. Yifrach. 2004. Conserved gating hinge in ligand- and voltage-dependent K^+ channels. *Biochemistry* 43:13242–13247.

Marti-Renom, M.A., A. Stuart, A. Fiser, R. Sanchez, F. Melo, and A. Sali. 2000. Comparative protein structure modelling of genes and genomes. *Ann. Rev. Biophys. Biomol. Struct.* 29:291–325.

Morais-Cabral, J.H., Y. Zhou, and R. MacKinnon. 2001. Energetic optimization of ion conduction by the K^+ selectivity filter. *Nature* 414:37–42.

Nishida, M., and R. MacKinnon. 2002. Structural basis of inward rectification: Cytoplasmic pore of the G protein-gated inward rectifier GIRK1 at 1.8 Å resolution. *Cell* 111:957–965.

Nose, S., and M.L. Klein. 1983. Constant pressure molecular-dynamics for molecular-systems. *Mol. Phys.* 50:1055–1076.

Noskov, S.Y., S. Bernèche, and B. Roux. 2004. Control of ion selectivity in potassium channels by electrostatic and dynamic properties of carbonyl ligands. *Nature* 431:830–834.

Perozo, E., D.M. Cortes, and L.G. Cuello. 1998. Three-dimensional architecture and gating mechanism of a K^+ channel studied by EPR spectroscopy. *Nat. Struct. Biol.* 5:459–469.

Perozo, E., D.M. Cortes, and L.G. Cuello. 1999. Structural rearrangements underlying K^+-channel activation gating. *Science*. 285:73–78.

Proks, P., C.E. Capener, P. Jones, and F. Ashcroft. 2001. Mutations within the P-loop of Kir6.2 modulate the intraburst kinetics of the ATP-sensitive potassium channel. *J. Gen. Physiol.* 118:341–353.

Ranatunga, K.M., R.D. Law, G.R. Smith, and M.S.P. Sansom. 2001. Electrostatics studies and molecular dynamics simulations of a homology model of the Shaker K^+ channel pore. *Eur. Biophys. J.* 30:295–303.

Roccatano, D., I. Daidone, M.A. Ceruso, C. Bossa, and A. Di Nola. 2003. Selective excitation of native fluctuations during thermal unfolding simulations: Horse heart cytochrome c as a case study. *Biophys. J.* 84:1876–1883.

Roccatano, D., A.E. Mark, and S. Hayward. 2001. Investigation of the mechanism of domain closure in citrate synthase by molecular dynamics simulation. *J. Mol. Biol.* 310:1039–1053.

Roux, B. 2005. Ion conduction and selectivity in K^+ channels. *Ann. Rev. Biophys. Biomol. Struct.* 34:153–171.

Sagui, C., and T.A. Darden. 1999. Molecular dynamics simulations of biomolecules: Long-range electrostatic effects. *Ann. Rev. Biophys. Biomol. Struct.* 28:155–179.

Sali, A., and T.L. Blundell. 1993. Comparative protein modeling by satisfaction of spatial restraints. *J. Mol. Biol.* 234:779–815.

Sansom, M.S.P., P.J. Bond, S.D. Deol, A. Grottesi, S. Haider, and Z.A. Sands. 2005. Molecular simulations and lipid/protein interactions: Potassium channels and other membrane proteins. *Biochem. Soc. Transac.* 33:916–920.

Sansom, M.S.P., and H. Weinstein. 2000. Hinges, swivels and switches: The role of prolines in signalling via transmembrane α-helices. *Trends Pharm. Sci.* 21:445–451.

Shrivastava, I.H., C. Capener, L.R. Forrest, and M.S.P. Sansom. 2000. Structure and dynamics of K^+ channel pore-lining helices: A comparative simulation study. *Biophys. J.* 78:79–92.

Shrivastava, I.H., and M.S.P. Sansom. 2000. Simulations of ion permeation through a potassium channel: Molecular dynamics of KcsA in a phospholipid bilayer. *Biophys. J.* 78:557–570.

Shrivastava, I.H., and M.S.P. Sansom. 2002. Molecular dynamics simulations and KcsA channel gating. *Eur. Biophys. J.* 31:207–216.

Smart, O.S., J.G. Neduvelil, X. Wang, B.A. Wallace, and M.S.P. Sansom. 1996. Hole: A program for the analysis of the pore dimensions of ion channel structural models. *J. Mol. Graph.* 14:354–360.

Sukhareva, M., D.H. Hackos, and K. Swartz. 2003. Constitutive activation of the Shaker Kv channel. *J. Gen. Physiol.* 122:541–556.

Tempel, B.L., D.M. Papazian, T.L. Schwarz, Y.N. Jan, and L.Y. Jan. 1987. Sequence of a probable potassium channel component encoded at *Shaker* locus of *Drosophila*. *Science* 237:770–775.

Tieleman, D.P., S.J. Marrink, and H.J.C. Berendsen. 1997. A computer perspective of membranes: Molecular dynamics studies of lipid bilayer systems. *Biochim. Biophys. Acta* 1331:235–270.

Tieleman, D.P., I.H. Shrivastava, M.B. Ulmschneider, and M.S.P. Sansom. 2001. Proline-induced hinges in transmembrane helices: Possible roles in ion channel gating. *Proteins Struct. Funct. Genet.* 44:63–72.

Tobias, D.J. 2001. Electrostatics calculations: Recent methodological advances and applications to membranes. *Curr. Opin. Struct. Biol.* 11:253–261.

Valiyaveetil, F.I., Y. Zhou, and R. MacKinnon. 2002. Lipids in the structure, folding and function of the KcsA channel. *Biochem.* 41:10771–10777.

Weber, W., P.H. Hunenberger, and J.A. McCammon. 2000. Molecular dynamics simulations of a polyalanine octapeptide under Ewald boundary conditions: Influence of artificial periodicity on peptide conformation. *J. Phys. Chem. B* 104:3668–3675.

Webster, S.M., D. del Camino, J.P. Dekker, and G. Yellen. 2004. Intracellular gate opening in Shaker K^+ channels defined by high-affinity metal bridges. *Nature* 428:864–868.

Yellen, G. 2002. The voltage-gated potassium channels and their relatives. *Nature* 419:35–42.

Zhao, Z., V. Yarov-Yarovoy, T. Scheuer, and W.A. Catterall. 2004. A gating hinge in Na^+ channels: A molecular switch for electrical signalling. *Neuron* 41:859–865.

Zhou, Y., and R. MacKinnon. 2003. The occupancy of ions in the K^+ selectivity filter: Charge balance and coupling of ion binding to a protein conformational change underlie high conduction rates. *J. Mol. Biol.* 333:965–975.

Zhou, Y., J.H. Morais-Cabral, A. Kaufman, and R. MacKinnon. 2001. Chemistry of ion coordination and hydration revealed by a K^+ channel-Fab complex at 2.0 Å resolution. *Nature* 414:43–48.

Part IV

Emerging Technologies

17 Patch-Clamp Technologies for Ion Channel Research

Fred J. Sigworth and Kathryn G. Klemic

The electrical activity of living cells can be monitored in various ways, but for the study of ion channels and the drugs that affect them, the patch-clamp techniques are the most sensitive. In this chapter the principles of patch-clamp recording are reviewed, and recent developments in microfabricated patch-clamp electrodes are described.Technical challenges and prospects for the future are discussed.

17.1 Introduction

The human genome contains more than 400 genes that code for ion channels. How are all of these channel types to be characterized? The standard tests of a new channel type—its ion selectivity, conductance, voltage dependence, ligand sensitivity—traditionally require weeks of effort by a PhD-level scientist. Then, besides the basic characterization, are questions about accessory proteins or posttranslational protein modifications that may alter an ion channel's behavior in significant ways. The functional understanding of each ion channel type presents a difficult but important problem.

Ion channels are also contributors to various diseases (Kass, 2005), and they are important targets for drugs. The large number of distinct ion channel types is an advantage for therapeutics, because it means that a drug targeted to a particular ion channel type that is expressed in one tissue is less likely to have undesired side effects in another tissue. Thus for example an immunosuppressive drug targeted to the potassium channels of lymphocytes will not affect the different potassium channels found in the nervous system (Vennekamp et al., 2004). The screening of compounds to find specific ion-channel blockers or modulators is therefore of great interest.

The understanding of biophysical transduction mechanisms of ion channels is also important. How do changes in membrane potential, or mechanical stretch, or the concentration of transmitter molecules, result in the activation of ion channels? Studies of these mechanisms, and ways to modulate them are discussed in other chapters of this volume. Progress in these areas would be aided by more sensitive and higher-throughput methods for recording ion channel currents.

In mechanism and, in some cases, in structure, ion transporters are related to ion channels. These "pumps" and "carriers" are studied with some of the same tools as ion channels, including voltage-clamp and patch-clamp measurements of electrical current. However, the current produced by the operation of an ion transporter is orders of magnitude smaller than that of an ion channel. There are substantial technical challenges in being able to record these small currents.

17.1.1 The Need for High Throughput

The functional analysis of ion-channel genes and the screening of pharmaceutical compounds that affect particular ion channels require much higher-throughput assays of ion channel activity than the traditional patch-clamp technique. Emerging chip-based technologies, especially planar patch-clamp technology, are beginning to make large-scale screens of genes and compounds possible.

An illustrative example of the need for higher-throughput evaluation of channel behavior is the problem of drug-induced cardiac arrhythmias. Some people are susceptible to the acquired long-QT syndrome (Roden, 2004) arising from the drug-induced blockade of the hERG potassium channel (Fig. 17.1). The Q-T interval of the electrocardiogram is a measure of the duration of the action potentials in the ventricles of the heart. If the Q-T interval of one action potential is too long, the ventricular muscle will not be able to respond uniformly to the subsequent beat, and the desynchronization of the ventricular cells results in chaotic electrical activity and sudden death.

Several common drugs have been taken off the market, or their use curtailed, because they have caused sudden death due to cardiac arrhythmias (Brown, 2004;

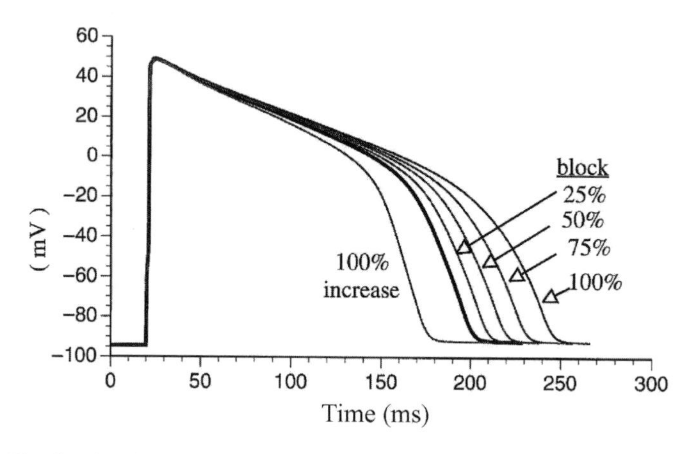

Fig. 17.1 Simulated action potentials in a ventricular muscle cell of the heart, from Zeng et al. (1995). Blocking the I_{Kr} current carried by hERG channels produces a lengthening of the action potential. The duration of the ventricular action potential is reflected in the Q-T interval of the electrocardiogram.

Roden, 2004). The U.S. Federal Drug Administration now requires that every drug be screened for an effect on the Q-T interval, and recommends a test for drug effects on hERG channel function.

17.1.2 The Need for Higher Sensitivity

The best-studied ion channels have conductances of 100 pS or more. With driving forces of 100 mV, these channels carry currents of more than 10 pA, which is easily recorded with the conventional patch-clamp technique, which yields measurements with an rms noise level of about 0.1 pA with a 1 kHz recording bandwidth.

There are however very important channels that carry smaller currents—the voltage-gated calcium channels of neurons and the chloride channels whose dysfunction causes Cystic Fibrosis, to name two. The currents carried by these channels are on the order of 0.1 pA, fewer than 10^6 elementary charges per second. Much smaller currents, about $100e_0$/s, are carried by ATP-driven ion pumps and other ion transporters. The grand challenge would be to resolve the displacement of a single elementary charge across the membrane. With such resolution, the individual movements of the voltage sensors in a potassium or sodium channel could be monitored, and the sequence of displacements in the transport cycle of an ion pump or transporter protein would be open to inspection.

As we shall see in Section 17.5, there are two major determinants of the noise level of a patch-clamp system. One is the thermal noise in the seal resistance and the electrode dielectric. The second is a noisy current that arises from the voltage noise in the amplifier as it is imposed on the capacitance of the input circuit, including the electrode itself. The controlled geometry of chip electrodes can be exploited to especially reduce the second noise source, so that their low capacitance, combined with high-performance amplifier devices, may allow single-elementary-charge resolution to be approached.

17.2 Recordings of Ion Channel Activity

17.2.1 Measurement of Membrane Potential

The activity of ion channels is reflected in a cell's membrane potential. The membrane potential is defined to be the voltage difference from the inside to the outside of the cell's bounding membrane. The membrane potential is conventionally measured directly using a microelectrode, which is a saline-filled glass micropipette that impales the cell. Alternatively, the membrane potential can be monitored (with somewhat less precision) using voltage-sensitive optical probes.

An electrical model for a cell membrane (Fig. 17.2) is a capacitance in parallel with various current sources that represent the ion channel currents. The membrane

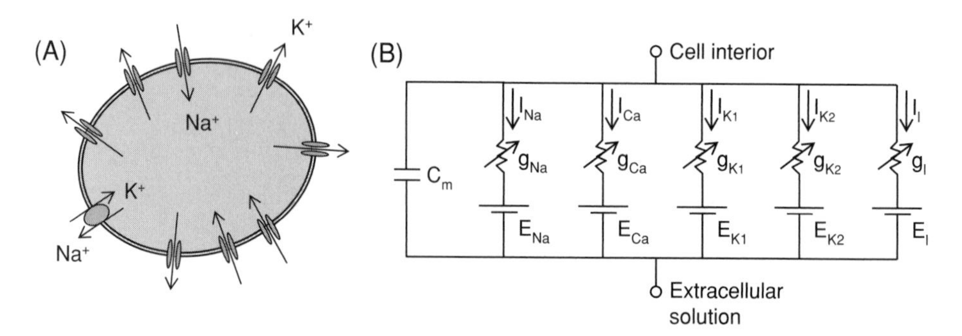

Fig. 17.2 Models for electrical activity in a cell. (A) About 15 μm in diameter, a "typical" mammalian cell contains hundreds of ion channels of several different types in its plasma membrane, as well as many copies of transport proteins including the ATP-driven sodium–potassium pump diagrammed here. (B) Electrical model of the cell. The membrane presents a capacitance of about 10 fF/$\mu m^2$of surface area, roughly 20 pF for a mammalian cell. The total current from a population of ion channels is represented by a conductance, that varies according to the fraction of channels open, in series with a battery representing the reversal potential of the channel type. At rest the total conductance may be 1 nS, but at the peak of an action potential the conductance can be 1000 times larger. The reversal potentials are generally nonzero because of differences in ion concentration across the membrane.

potential E depends on the ion channel currents I_j according to

$$C_{\mathrm{m}}\frac{dE}{dt} + \sum_j I_j = 0. \tag{17.1}$$

The current carried by the jth ion channel type is then modeled with a conductance g_j,

$$I_j = g_j(E - E_j), \tag{17.2}$$

where g_j itself is typically a function of both E and time. The reversal potential E_j represents the electrochemical driving force on the ions that pass through the channels. For example, for potassium channels $E_{\mathrm{K}} \approx -90$ mV while the reversal potentials E_{Na} for sodium, and E_{Ca} for calcium channels typically are about +60 mV. At rest, nerve and muscle cells have a membrane potential in the range of –70 to –90 mV that reflects a preponderance of potassium conductance. Action potentials occur when large sodium or calcium conductances drive the membrane potential transiently to values of +50 mV or so.

The fact that multiple channel types contribute to the membrane potential makes it a poor reporter of ion channel activity in some cases. A cell typically has several potassium channel types in its membrane, all of which contribute to the resting potential. The change in conductance due to the activation or block of one of these channel types can have a very small effect on the membrane potential due to the parallel conductance of other channel types.

Voltage-gated sodium channels, on the other hand, are responsible for the fast action potentials in neurons and muscle cells, including cardiac muscle. The presence of action potentials in these cells, and their "fatigue" with repetitive stimulation, is strongly dependent on the properties of the sodium channels. Methods for recording membrane potentials with high time resolution can therefore provide useful information, for example, about use-dependent block of sodium channels. New, highly sensitive optical probes of membrane potential are approaching the required time resolution (Gonzalez and Tsien, 1997).

17.2.2 Extracellular Measurements

In some cases action potentials can be detected with electrodes placed outside a cell. The signals are very small, typically less than 1 mV in amplitude, and result from voltage drops in the bathing solution due to extracellular current flow. The amplitude of these signals is highly variable, and depends on the exact geometry of the cell and the extracellular current pathways. Thus as an assay of channel activity extracellular measurements contain less information than membrane potential measurements. Extracellular microelectrode arrays are nevertheless useful in monitoring the action potential activity from a population of cells, for example, in small networks of neurons, in brain slices, and in heart tissue.

Might it be possible to measure directly the membrane potential with an extracellular device? The cell membrane is a self-assembled bilayer of lipid molecules, containing a hydrophobic core about 5 nm thick. The membrane potential produces a large electric field within the membrane core, which might be coupled directly to a field-effect semiconductor device. This approach has been pursued by two groups (Hutzler and Fromherz, 2004; Ingebrandt et al., 2005) (Fig. 17.3). Unfortunately, a layer of ionic solution is interposed between the cell and the device, whose thickness is large compared to the Debye length (about 1 nm) of the solution; thus the electric field at the device surface greatly attenuated. Indeed, the small signals amplified by

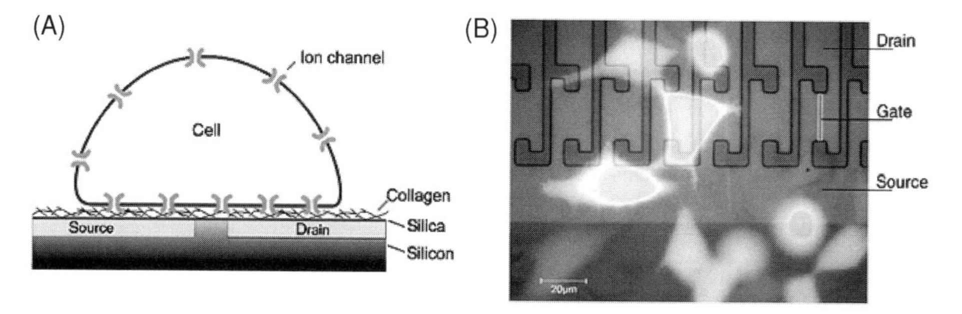

Fig. 17.3 Cell–FET interface. An open-gate field-effect transistor (A) senses the potential in the vicinity of the cell membrane. (B) A photograph of cells on an FET array. From Fromherz (2002).

these field-effect transistors arise mainly from extracellular current flow rather than the direct field effect (Braun and Fromherz, 2004).

17.2.3 Measurements of Ion Flux

Another way to monitor ion channel activity is by measuring the transport of ions across the cell membrane. The activity of ion channels generally produce only small and slow changes in ion concentration in a cell, making this measurement difficult for most ion channel types. There are however two notable exceptions. In the case of most potassium channels, rubidium ions can serve as an excellent tracer, and fluorescent dyes sensitive to rubidium report the total permeability of the channels (Terstappen, 1999).

The situation is even better for calcium channels, because local Ca^{2+} fluxes can be measured with very high sensitivity. The cytoplasmic free calcium concentration is very low, roughly 100 nM, maintained by buffering and active pumping. Meanwhile high-affinity Ca^{2+}-sensitive fluorescent dyes can readily detect micromolar Ca^{2+} concentration changes. The result is the ability to detect the local calcium "sparks" accompanying the opening of single Ca^{2+}-permeable channels. This has been extremely useful in studying the activity of the channels that release calcium from intracellular stores (Cheng et al., 1993; Baylor, 2005), but has also been used to observe the opening and closing of single neurotransmitter-receptor channels (Demuro and Parker, 2005). The disadvantage is that sophisticated optical system (confocal or total internal reflection excitation) must be used to keep the background fluorescence low; however a very interesting feature of Ca^{2+}-imaging is that the activity of hundreds of individual channels can be recorded simultaneously, even with millisecond time resolution.

17.2.4 The Patch-Clamp Techniques

A direct electrical measurement of ion channel activity is provided by voltage-clamp techniques, in which an injected current I_{inj} is supplied to balance the ion channel currents (Fig. 17.4),

$$C_{\mathrm{m}}\frac{dE}{dt} + \sum_j I_j = I_{\mathrm{inj}}. \tag{17.3}$$

If E is held constant, the injected current becomes equal to the sum of the channel currents. I_{inj} is then recorded as a direct measure of the ion channel current. A voltage-clamp system (Fig. 17.4) uses an operational amplifier to set the potential while measuring the current, and requires a low value of the series resistance R_{s} to make a high-quality recording.

The patch-clamp methods were originally developed by Neher, Sakmann and colleagues (Neher et al., 1978) to observe the activity of single ion channels. The

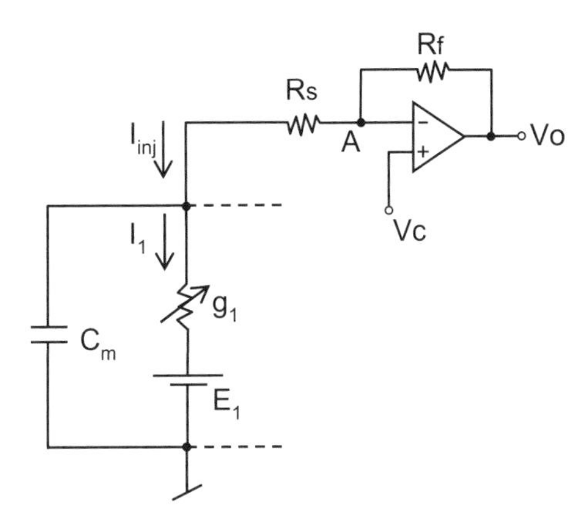

Fig. 17.4 Model of a cell and voltage-clamp system. The current I_{inj} is equal to the total ionic current through channels I_{i} provided the voltage across the membrane capacitance C_{m} remains constant. An operational amplifier forces node A to be at the potential V_{C} while also providing the current monitor output voltage $V_{\mathrm{O}} = R_f I_{\mathrm{inj}} + V_{\mathrm{C}}$. Thus the injected current is measured while the membrane potential is held approximately at V_{C}. The series resistance R_{s} of the electrode however causes the actual membrane potential to be in error by the amount $R_{\mathrm{s}} I_{\mathrm{inj}}$, so it is important that the series resistance be kept relatively small.

idea was to electrically isolate a small area of cell membrane (containing only a few ion channels) and to minimize the thermal noise in the recording system. To establish a patch-clamp recording, the tip of the glass pipette is gently placed against the cell membrane (Fig. 17.5A). The pipette tip is visualized in a microscope and is moved by a micromanipulator controlled by a skilled operator. When sealed against the cell membrane, the pipette collects most of the current flowing through the patch of membrane, delivering it to a current-measuring amplifier (Fig. 17.5B). A high resistance pipette-membrane seal allows the background noise to be low. For example, a seal leakage resistance $R_{\mathrm{L}} = 10\ \mathrm{G}\Omega$ yields a noise standard deviation (rms value) of 0.13 pA at 10 kHz bandwidth. A poor seal with resistance $R_{\mathrm{L}} = 100\ \mathrm{M}\Omega$ produces ten times the noise, 1.3 pA rms.

The key feature of the modern patch-clamp technique is the tight glass-to-membrane seal (Hamill et al., 1981) whose mechanical stability means that the seal remains intact even if the membrane patch is ruptured (by suction or voltage pulses) or excised, in the latter case resulting in a cell-free membrane patch. Cell-free patches are ideal for studying the effects of solution changes on channel activity, as the solution bathing the pipette can be changed on a millisecond timescale without disturbing the patch membrane. Changing the solution inside the pipette is however more difficult, and is a challenge that is being met in microfabricated devices.

Rupturing the membrane patch produces the most-used variant of the patch-clamp technique, whole-cell recording (Fig. 17.5C). This provides access to the cell interior with a series resistance typically in the range of 3–10 MΩ. The result is the ability to measure membrane potentials and to perform voltage-clamp measurements. The series resistance is not always negligible, because an error voltage equal

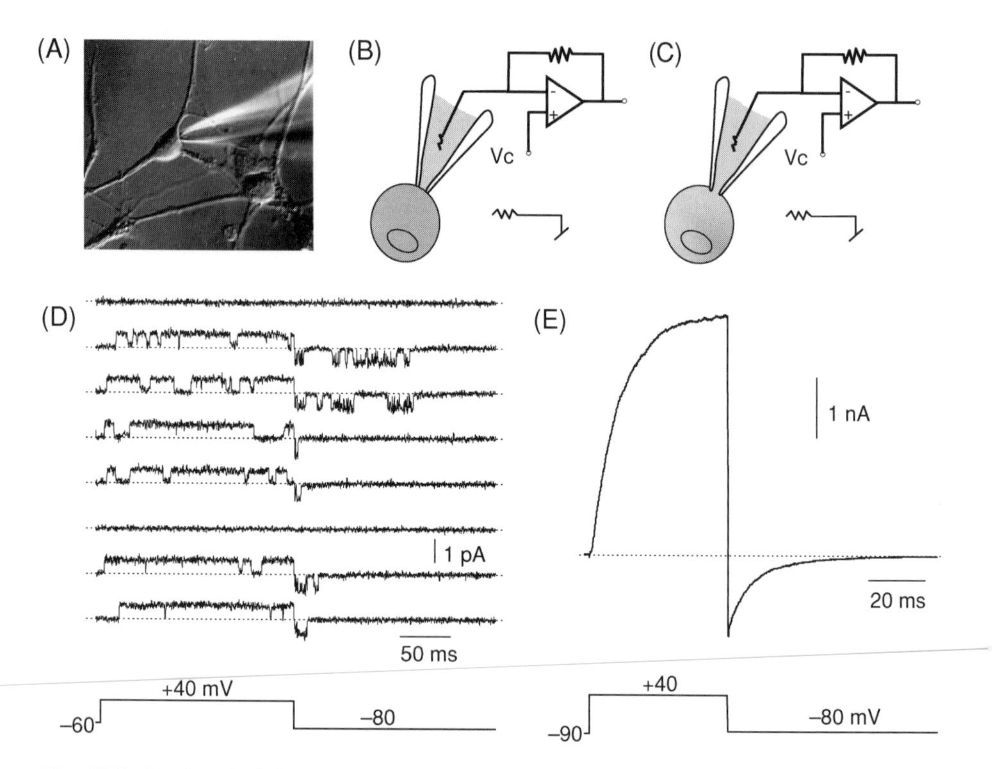

Fig. 17.5 Patch and whole-cell variants of the patch-clamp technique. (A) Photograph of a patch pipette sealed to a cultured neuron. (B) Schematic of an on-cell patch recording, where current is collected by the pipette from a small area of membrane. (C) Whole-cell recording, in which the patch membrane is ruptured, giving the pipette access to the cell interior. (D) Example recordings from a patch containing a single Kv2.1 voltage-gated channel, stimulated by the applied voltage waveform shown at the bottom. The reversal potential of this channel was about—30 mV, such that open channel currents are positive (outward) at +40 mV and negative (inward) at—80 mV. The successive sweeps were all obtained with the same voltage stimulus, demonstrating the stochastic behavior of single channels. (E) Whole-cell recording showing the current through a population of about 10^4 Kv2.1 channels. The depolarization to +40 mV elicits a gradual opening of channels; repolarization to—80 mV causes the channels to close on a time scale of about 20 ms. Leakage and capacitive currents have been subtracted from the recordings in (D) and (E), leaving only the ion channel currents.

to $R_s I_{inj}$ develops across it. For example, with currents of several nanoamperes the $R_s I_{inj}$ error is on the order of 10 mV, which can be a serious error in studying voltage-dependent channel activity. As Hodgkin et al. (1952) showed, these errors can be mitigated by using appropriate feedback, called series-resistance compensation. In the case of a patch-clamp recording, a voltage equal to $R_s I_{inj}$ is estimated and added to the pipette voltage, effectively canceling the series-resistance error (Sigworth, 1995).

The origin of the all-important tight glass-to-membrane seal is not entirely understood. What is known is that a very clean glass surface is required, and the presence of divalent cations, particularly Ca^{2+}, helps in its formation. Under good conditions the measured seal resistance rises spontaneously, soon after the pipette touches the cell, as if the membrane is "zippering" up to the pipette surface. Otherwise, gentle suction applied to the back of the pipette can encourage sealing to occur. The seal, typically with a resistance greater than 1 gigaohm, is commonly called a "gigaseal."

17.3 Planar Patch-Clamp Technologies

The economic driving force for developing planar patch-clamp technologies has been the need for highly parallel, automated voltage-clamp recordings from cultured mammalian cells. A wide variety of approaches have been pursued, but in all of these devices the glass pipette is replaced with a micromachined, insulating partition that separates two chambers filled with saline. The partition contains an aperture, 1–2 μm in diameter that is the topological equivalent of the opening at the tip of a glass pipette. A cell seals against the partition such that a patch of its membrane covers the aperture. A pulse of pressure or voltage breaks the patch membrane, establishing the whole-cell recording configuration.

17.3.1 The Cell-Guidance Problem

The biggest technical challenge for automated patch-clamp recording has been the "precision guidance" problem, of providing a way for cells to be guided directly into contact with a recording site. Experience has shown that once a cell (or cellular debris) seals to the tip of a glass pipette or to a planar recording site, a residue remains that is very difficult to remove, and which prevents the subsequent formation of a gigaseal. Thus all patch-clamp devices use disposable chips, and systems are designed to make the first contact of a cell with the active site to be as successful as possible in forming a gigaseal.

As a microchip replacement for the glass pipette, a device made by Stett et al. (Stett et al., 2003; van Stiphout et al., 2005) is perhaps the most elegant (Fig. 17.6). Suction of fluid into the 10 μm outer channel guides a cell into contact with the inner (about 1 μm diameter) contact channel. The contact channel functions like the interior of a patch pipette; positive pressure is initially applied to prevent debris from approaching its surface, and then suction is applied to encourage seal formation when a cell is in place. This is the same procedure that is used with conventional pipettes.

A cross-section of this device (Fig. 17.6C and D) shows a complex profile that is created on a silicon wafer through many processing steps including the deposition and etching of SiO_2. The SiO_2 surface forms excellent gigaseals with cells, with a success rate for establishing seal resistances above 1 GΩ of about 90%. Cytocentrics CCS GmbH has announced a parallel, automated patch-clamp system

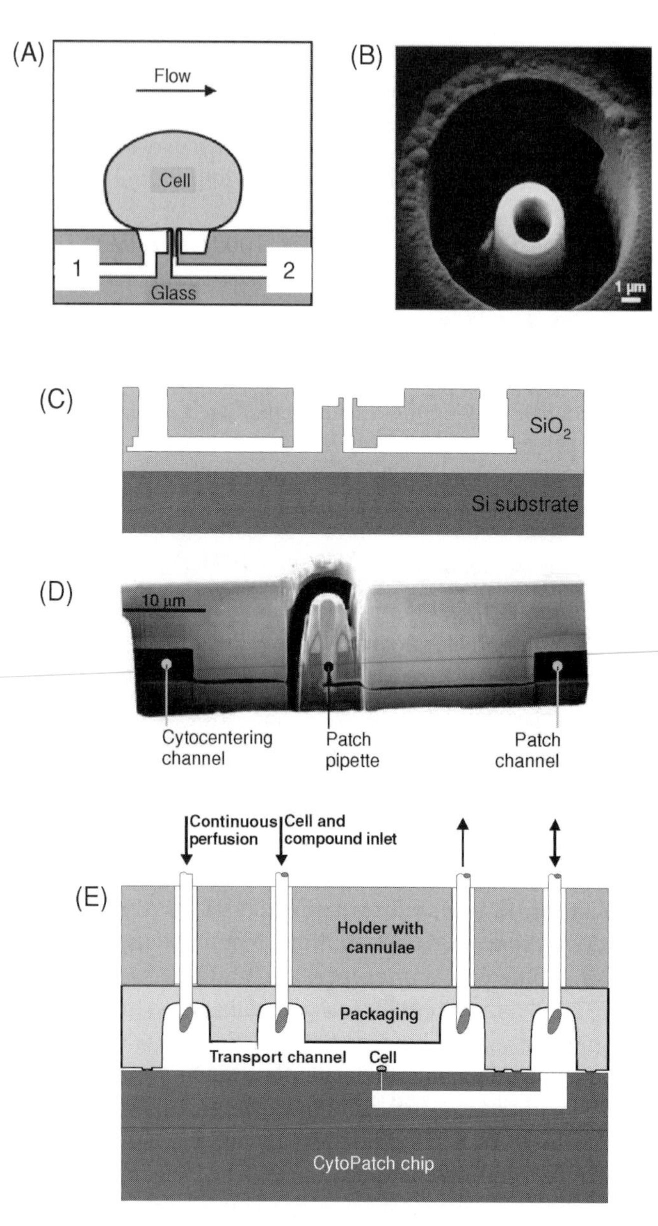

Fig. 17.6 The "cytocentering" chip. (A) Cross section. A cell is trapped by suction applied to the large port 1 of the device. Subsequent suction on the central port 2 forms a seal, and currents are recorded through port 2. (B) Scanning electron microscope (SEM) image of the device from above. (C) Cross section diagram of the device, showing the fluidic compartments. (D) Corresponding SEM cross-sectional view. (E) Cross section diagram of the packaged chip with fluidic ports. From van Stiphout et al. (2005).

using disposable chips based on this design. Another device that replaces the pipette tip with a micronozzle has been described by Lehnert et al. (2002).

Another technology for guiding cells toward apertures is dielectric focusing (Schmidt et al., 2000; Guia et al., 2002). A potential difference of several hundred millivolts is imposed across a silicon nitride membrane about 100 nm thick. The membrane has a micron-sized aperture, and the resulting electric field guides small nonconducting particles into the aperture. This system works well for liposomes, which are small artificial lipid-membrane structures. With an additional 20 nm SiO_2 layer on the membrane that is modified by an aminosilane or by adsorption of polylysine, gigaohm seals to liposomes were demonstrated. Unfortunately, for larger structures such as mammalian cells no success has been reported. We speculate that the larger electric fields required to manipulate cells would cause cell damage.

17.3.2 Loose Patch Clamp

A very different solution to the "precision guidance" problem is not to attempt the formation of gigaseals at all. The earliest patch-clamp recordings (Neher and Sakmann, 1976; Neher et al., 1978) were made under conditions where the seal resistance was only a few tens of megaohms; in this situation the pipette surface is separated from the cell membrane by an aqueous film on the order of 1 nm in thickness, and glass and membrane are held together by continuous suction. Similarly, in the planar patch-clamp system by Molecular Devices (Kiss et al., 2003) suction draws cells into contact with micron-sized holes in a polymer sheet. The resulting seal resistances are on the order of 100 MΩ. This yields substantial leakage currents which are however tolerable if the ionic currents to be measured are sufficiently large. Breaking through the patch membrane to allow whole-cell recordings is done with the pore-forming antibiotic Amphotericin B (Rae et al., 1991). The IonWorks HT instrument uses disposable, 384-well "PatchPlates" from which the instrument makes 48 recordings simultaneously.

Even with this simple system, cell-to-cell variability results in a success rate of only about 70% in practical screening for ion-channel effects of drugs. An astonishingly effective improvement on the original PatchPlate is the "Population Patch Clamp" system, also by Molecular Devices, Fig. 17.7. There the single aperture at the bottom of each well is replaced by an array of 64 apertures. If each aperture seals onto a cell, then the resulting parallel recording yields a sum of the current in the entire population of cells. But what if one or more apertures are left open? The low resistance of an open aperture produces a large leakage current. Surprisingly, in practice it is found that the leakage current is not excessively large (nearly all the apertures are plugged by something!) and the quality of the recordings is as good as a single-well recording, and much more reproducible. A success rate better than 95% is reported for wells yielding useful recordings (Finkel et al., 2006).

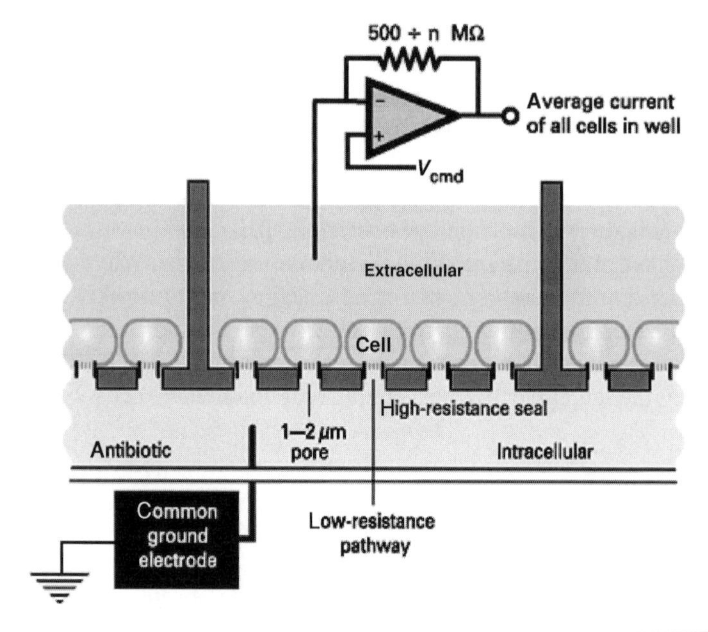

Fig. 17.7 Population patch-clamp device from Molecular Devices (Finkel et al., 2006). At the bottom of a well are multiple apertures, against which cells are pressed by suction, and permeabilized by Amphotericin treatment. The current from the entire population of cells is monitored.

17.3.3 Planar Patch-Clamp Chips

Several groups have developed chips in which the cell seals against a planar surface having a simple round aperture. Lacking special fluidics, these systems rely on the use of a very dense and debris-free suspension of cells to provide a high likelihood of a cell coming to rest over the partition aperture (Fig. 17.8). Suction through the aperture helps in the final docking and sealing of the cell to the partition.

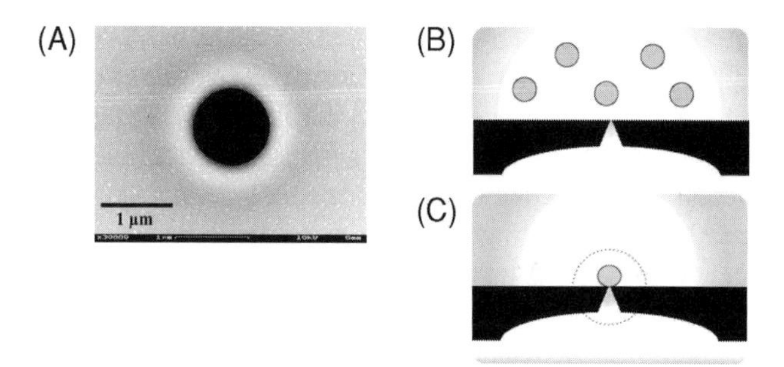

Fig. 17.8 A basic planar patch-clamp device made from glass or fused quartz. (A) Aperture in the planar "top" surface of a glass chip. (B) A cell suspension is applied to the top surface and suction is applied. (C) A cell seals over the aperture, allowing electrical recording. Unattached cells are washed away. Illustrations from Fertig et al. (2000, 2003).

The fabrication of a representative planar patch-clamp chip is diagrammed in Fig. 17.9. Here the partition is a self-supporting insulating membrane containing a micron-sized hole; it spans a window in a larger etched cavity in the bulk silicon substrate. Fig. 17.9A demonstrates how such a structure is made. The process starts with a silicon wafer having a thin layer of SiO_2 or Si_3N_4 grown or deposited on one surface. A micron-sized hole is etched in this layer to form the final aperture. Then an anisotropic etch of the back side of the wafer, forms the pyramidal cavity and leaves the suspended membrane. The first reported devices of this sort had a relatively thin, 120 nm Si_3N_4 membrane (Fertig et al., 2000). The formation of gigaseals in these devices was never observed, perhaps because the thin membranes provided insufficient sidewall area for forming the membrane seal. Such devices also show a large capacitance (hundreds of picofarads) across the partition, which would impair the noise performance even if gigaseals were formed.

An improved device of this kind has been described by Sophion Bioscience (Asmild et al., 2003). After forming a suspended membrane, a thick layer of SiO_2 is deposited, forming a gently-rounded opening (Fig. 17.9C). The chips are employed in a 16-channel patch-clamp system, the Sophion QPatch. This system shows outstanding solution-exchange because of integrated microfluidics. After the silicon wafer is fabricated, a glass microfluidics wafer is bonded to it. The glass layer contains etched microfluidic channels which provide a pathway for the application of cells and also the rapid exchange of solutions. The bath solution chamber has a volume of 0.5 μl and can be exchanged completely in about 150 ms. This rapid exchange is very useful for the study of ligand-activated channels (Asmild et al., 2003)

In some devices the partition is made in silicon instead. The strategy is to form a micron-sized aperture in bulk silicon, and then grow silicon nitride and oxide layers to provide insulation. After a surface treatment these chips form gigaseals readily (Schmidt et al., 2000). Unfortunately, the quality of the recordings is limited by the large capacitance between the aqueous solutions and the bulk silicon, due to the relatively thin (<1 μm) insulating layer. In another silicon chip described by Pantoja et al. (2004) deep reactive-ion etching is used to define a micron-sized pore, onto which was deposited a SiO_2 layer. PDMS microfluidic layers are pressure-bonded to each side of the device, which allowed cell delivery and solution exchange while also reducing the electrical capacitance. Formation of gigaseals was rare, perhaps due to the roughness of the chip surface.

The most successful planar patch-clamp chips have been made of quartz or glass. The fabrication processes of commercial chips—in particular the way in which apertures are made—are proprietary. The only device of this kind whose exotic fabrication process has been described in detail (Fertig et al., 2002) is shown in Fig. 17.8A. A 200 μm fused-quartz wafer is thinned locally by wet etching to 20 μm thickness. The wafer is then penetrated by a single high-energy gold ion, which leaves a latent track that is selectively etched from one side. The etch yields a conical cavity that terminates in a micron-sized hole. These chips form gigaseals with cells and can be used for whole-cell and patch recordings (Fertig et al., 2002). Glass planar

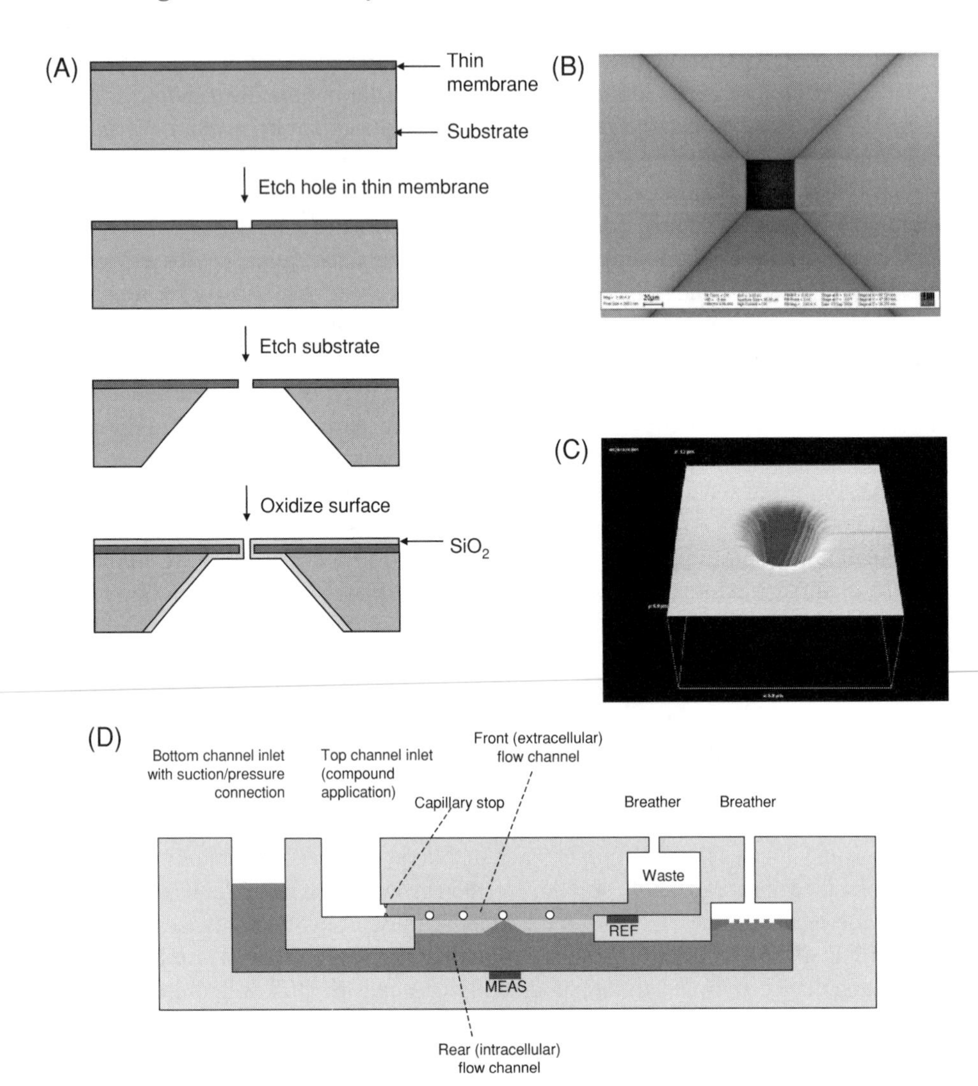

Fig. 17.9 An example of planar patch-clamp chip fabrication. (A) Process flow for fabricating a patch-clamp partition in a substrate with a thin membrane. The substrate is silicon with a thin silicon nitride or silicon oxide layer on top, only a few microns thick. The difference in composition permits preferential etching of either the substrate or the thin layer which will be used to form a membrane. (A) Micron-sized hole is etched in the membrane layer to form the final aperture. Then an anisotropic etch of the back side of the wafer, forms the pyramidal cavity and leaves the suspended membrane. (B) SEM view of the thin membrane of a finished device (Sophion QPatch chip), looking up through the pyramidal cavity in the substrate. A tiny dark spot marks the aperture. (C) AFM image of the top surface near the aperture, after deposition of an SiO_2 layer several microns thick. (D) The QPatch chip (center) is mounted in a carrier with microfluidic channels. Illustrations courtesy of Sophion Biosciences, www.sophion.com.

chips based on this general design are made by Nanion Technologies GmbH are used in single-well and 16-well recording systems (Fertig et al., 2003).

As a different sort of solution to the precision-targeting problem, Aviva Biosciences makes glass planar chips having a proprietary coating that greatly enhances the ability to form gigaseals. With this coating, simple suction through the aperture is sufficient to dock and seal a cell with a probability of about 75% (Xu et al., 2003). Chips from Aviva are used in the Molecular Devices PatchXpress instrument, which is a highly automated device that makes 16 simultaneous whole-cell recordings.

17.3.4 PDMS Patch Partitions

Polydimethylsiloxane (PDMS) is an elastomer that has many applications in microfluidics and biotechnology (Sia and Whitesides, 2003). Our laboratory has pursued the fabrication of simple patch-clamp partitions using micromolding of PDMS. The present fabrication method (Fig. 17.10) uses a stream of air to define a 2 μm hole in a PDMS sheet. Subsequent plasma oxidation of the cured PDMS forms a thin silica surface layer that is suitable for forming gigaseals with cells. Patch and whole-cell recordings can be made with the devices (Klemic et al., 2002, 2005). The probability of successful gigaseal recordings relatively low, presently about 25%; however, the fabrication method is easily scaled to form arrays of partitions, and is simple enough that it can be carried out in an electrophysiology laboratory.

Ionescu-Zanetti et al. (2005) have described a PDMS patch-clamp device in which the aperture is formed in the lateral wall of a microfluidic channel. This geometry allows very flexible transport of cells and solutions to the recording site. The formation of the lateral-wall aperture is challenging, and the success rate for forming gigaohm seals is low, about 5%, but in view of the easy integration with

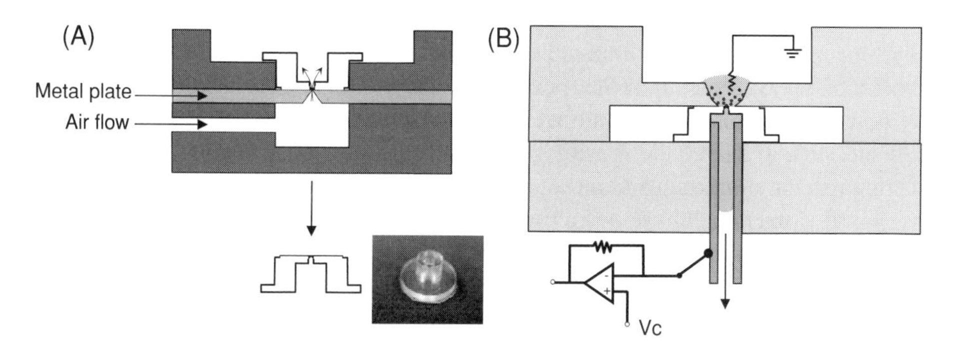

Fig. 17.10 PDMS planar patch-clamp device. (A) Fabrication. Air is forced through a 2 μm hole in a micromachined metal plate. A layer ~20 μm thick of PDMS resin is applied to the top of the plate and is surrounded by a pre-formed PDMS button. The plate is heated to cure the PDMS layer, the button is removed and treated in an oxygen plasma. (B) Cells are applied to the planar button surface. A chlorided silver tube inserted into the back of the button serves both as a suction port and electrode.

microfluidics this system is promising for situations in which a low-resistance seal is acceptable.

17.4 Simultaneous Electrical and Fluorescence Measurements

Fluorescent probes are an important tool in monitoring the molecular movements of proteins. The simultaneous recording of fluorescence changes and ion channel currents has been particularly useful in studying the movements of the voltage sensor of voltage-gated channels (Bezanilla, 2002). These measurements have been carried out on populations of channels, but particularly informative would be the ability to correlate spectroscopic changes of single reporter fluorophores with single channel currents. This has been done in one case (Borisenko et al., 2003) where fluorescently-labeled gramicidin A channels were detected in an artificial lipid membrane.

For channels in cell membranes, single-molecule fluorescence studies have been performed with cells under whole-cell voltage clamp (Sonnleitner et al., 2002). To avoid background fluorescence, the excitation light is provided by the evanescent wave, which illuminates a very thin ($\sim$100 nm) region above a glass surface. If a cell adheres to the surface, its membrane can be probed, but there is no room for a local patch electrode; hence whole-cell recording has been used up to now. It may be possible that the controlled geometry of planar patch-clamp devices might allow simultaneous electrical and optical recording from a small membrane patch with low fluorescence background.

17.5 The Grand Challenge: Single-Charge Detection

Starting in the 1970s patch-clamp recording provided the first biological "single molecule" measurements, as the open-closed conformational changes of single channels switched on and off picoampere-sized currents that could be recorded. Might it be possible to extend the sensitivity of membrane current recordings to be able to distinguish the movements of a few electronic charges?

With current patch-pipette technology the smallest detectable current pulse is on the order of 100 elementary charges. The sensitivity is limited by noise sources that have been analyzed in detail (Levis and Rae, 1998) and are summarized in Fig. 17.11. Current noise is most usefully described by the current spectral density $S_I(f)$ which has units of A^2/Hz. It is the variance in the measurement of current, making use of a narrow filter centered on the frequency f, and normalized by the bandwidth of the filter. The current spectral densities from independent sources are additive, so that the noise sources in a complete system consisting of patch membrane, electrode, and amplifier, can be separated into individual contributions. In an electrical model of the noise sources in a patch-clamp system (Fig. 17.11), the thermal noise in each

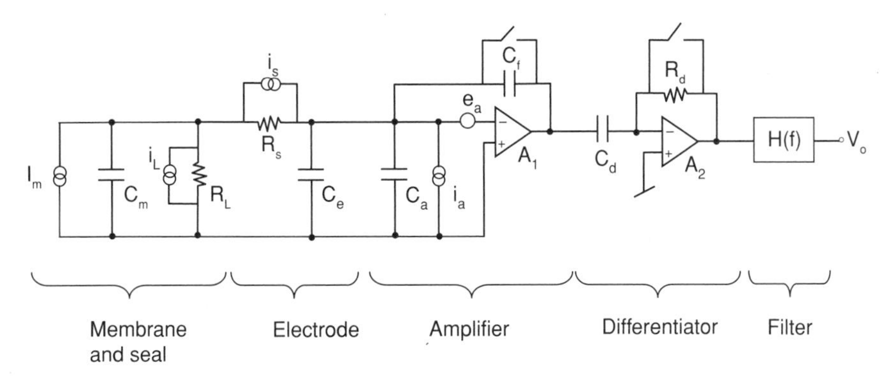

Fig. 17.11 Simplified equivalent circuit for a high-performance patch-clamp recording system using a capacitive-feedback amplifier for low noise. The output voltage is proportional to the patch membrane current I_m according to $V_o = R_d C_d I_m / C_f$ with filtering by a filter response $H(f)$. Switches allow periodic discharging of the integrating capacitor to allow DC currents to be recorded. Thermal noise sources in resistors (i_L, i_s) and amplifier input noise sources (i_a, e_a) are shown. Approximate component values for an optimized glass-pipette recording setup, along with values (in parentheses) for a hypothetical low noise planar electrode system are: patch membrane capacitance $C_m = 20(1)$ fF, seal resistance $R_L = 0.2(1)$ TΩ, electrode series resistance $R_s = 3(2)$ MΩ and electrode capacitance $C_e = 1(0.1)$ pF, amplifier input capacitance $C_a = 6(0.1)$ pF and amplifier noise $e_a = 2(0.5)$ nV/$\sqrt{\text{Hz}}$ and $I_a < 0.1$ fA/$\sqrt{\text{Hz}}$, $C_f = 1(0.1)$ pF. Another noise source is dielectric loss in the electrode capacitance, not modeled here.

actual resistance is modeled as a small random current generator in parallel with an idealized, noiseless resistance. The current generator for a resistance R has spectral density $4kT/R$, where k is Boltzmann's constant and T the absolute temperature. An amplifier is similarly modeled by an idealized amplifier with a noise voltage generator e_n and current generator i_n attached to its input.

The sum of all the noise sources in Fig. 17.11 yields an expression for the current spectral density having four terms,

$$S_I(f) = 4kT/R_L + i_a^2 + 16\pi^2 kTR_s C_m^2 f^2 + 4\pi^2(C_m + C_e + C_a + C_f)e_n^2 f^2. \tag{17.4}$$

The first two terms give a frequency-independent spectral density (white noise). The first is the thermal noise in the seal leakage resistance R_L; with glass pipettes this resistance can be well over 10 GΩ and is comparable to the expected resistance of a few square microns of plasma membrane. The thermal noise current is $4kT/R_L$, which works out to about 1.6×10^{-30} A^2/Hz for $R_L = 10$ GΩ. Integrated over a bandwidth of 10 kHz this yields a variance of 1.6×10^{-26} A^2, or an rms noise of 0.13 pA. Smaller electrode apertures generally yield higher R_L values. The second term is the amplifier current noise spectral density i_a^2, which is expected to arise from shot noise in the gate leakage current (typically 1 pA or less) of the input transistor.

The third and fourth terms represent current spectral densities that increase steeply with frequency. The third comes from the thermal noise in the electrode series resistance R_s as it is reflected in the patch membrane capacitance C_m. The fourth term is the reflection of the amplifier voltage noise in the total input capacitance. When all four terms are added together, the result is a spectral density that is constant at low frequencies but increases as f^2 at high frequencies. Fig. 17.12A shows the spectral density of the presently lowest-noise patch-clamp amplifier, the Molecular Devices AxoPatch 200B, measured with and without a low-noise fused-quartz pipette (Levis and Rae, 1998).

For practical purposes we would like to know the rms noise (i.e., the standard deviation of the noise) when the recording is filtered to a known extent, for example, by a Gaussian filter (Colquhoun and Sigworth, 1995) of a particular bandwidth. If the Gaussian filter's response is $H(f)$, then the rms current noise σ_I is obtained from the integral

$$\sigma_i^2 = \int_0^\infty S_i(f)\,|H(f)|^2\,df.$$

For the detection of very small charge movements, let us compare σ_I with the peak amplitude of the response to an impulse of current that represents a given number of elementary charges. Figure 17.12B shows the sensitivity of the recording system

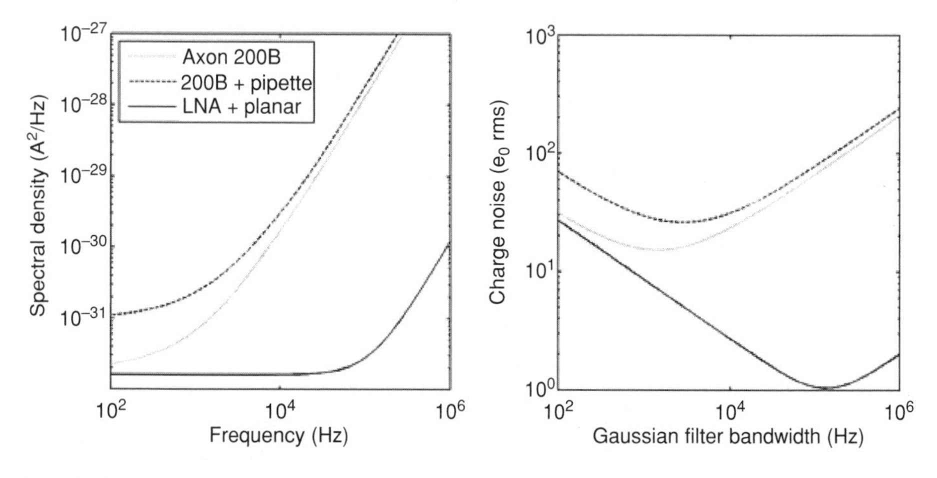

Fig. 17.12 Noise in patch-clamp recordings. A state-of-the-art amplifier (Axopatch 200B) alone and with an optimized fused-quartz pipette, is compared with a hypothetical low-noise amplifier and planar electrode. The left panel shows the input-referenced current spectral density as a function of frequency. The right panel is the equivalent rms charge noise as a function of filter bandwidth, computed as the ratio of rms current noise to the peak output of a Gaussian filter given an input of $1e_0 \times \delta(t)$. The pipette-based system has a charge noise of $25e_0$ rms at the optimum Gaussian filter bandwidth of 5 kHz. The hypothetical system (parameters given in the Fig. 17.9 legend) has a charge noise of $1e_0$ at a bandwidth of 200 kHz and would be able to detect individual charge pulses about $5e_0$ in magnitude.

evaluated in this way. What is plotted is the magnitude of a charge impulse whose expected peak value is equal to the standard deviation of the noise. This is plotted as a function of the bandwidth of a Gaussian low-pass filter. It can be seen that the present state-of-the-art amplifier and pipette combination yields a minimum charge noise of about 25 elementary charges rms.

Chip electrode technology makes improvements in electrode geometry that in principle can be exploited to make a very great decrease in noise levels. The legend of Fig. 17.11 gives parameters, and Fig. 17.12 gives noise performance, of a hypothetical electrode chip and amplifier that could reach a charge noise of $1e_0$ rms. How could this low noise level be reached? First, an electrode partition fabricated in quartz or PDMS could have a very small electrode solution volume and therefore decrease the electrode capacitance. A small aperture but a large solid-angle of convergence could allow a large seal resistance R_L and small membrane capacitance C_m to be obtained while keeping R_s low. Thus the flexibility of geometry afforded by planar electrode designs should allow the first and third terms of (Eq. 17.4) to be kept small. The second term, the amplifier current noise, is basically the shot noise in the input transistor's leakage current. Small MOS transistors can have sufficiently small leakage currents, in the femtoampere range.

The big challenge is the last term, involving the capacitance and the amplifier voltage noise. Given an electrode with greatly reduced capacitance C_e, it should be matched with smaller amplifier capacitances. An order of magnitude reduction in all the capacitances seems possible. A decrease in amplifier voltage noise is also required, perhaps obtainable with very short channel field-effect transistors. Then if other noise sources not modeled here are also kept under control—dielectric noise being one of the most serious (Levis and Rae, 1998)—it may be possible to meet the grand challenge of detecting the movements of a few elementary charges.

The ability to detect such small charge movements would be of great interest in working out the mechanisms of ion pumps and co-transporters. These are proteins which transport ions individually or in small groups, where a conformational change, and sometimes the expenditure of metabolic energy, accompanies each ion movement. Small charge movements are also involved in the "gating charges" of voltage-gated ion channels. To observe these charge movements would illuminate the detailed kinetics of the voltage-transduction mechanism.

17.6 Future Prospects

Chip-based, planar patch-clamp instruments are already in use in the pharmaceutical industry for screening of compounds targeted to ion channels. There is still much to be done to improve this technology: the throughput of measurements is still one or two orders of magnitude below what is needed to do primary "high-throughput" screening of drug compounds. Increased levels of integration, involving multiple electrodes per chip and integrated microfluidics, will allow an increase in the number of measurements made in parallel. Integration of the electronics will also help: monolithic patch-clamp amplifiers and arrays, at least for whole-cell recording, should be possible to make.

Small chip-based instruments are becoming available also for research laboratories, where many large and expensive patch-clamp rigs will eventually be replaced by these benchtop instruments. In addition to the more routine measurements in ion-channel research, there are great benefits to be gained through the design of special microfabricated devices. One benefit is the possibility of rapid solution exchange on both sides of the partition, thanks to the integration of microfluidics. Another will be the possibility of single-molecule experiments in which both ionic current and fluorescence measurements are made from the same ion channel protein. New microfabricated devices will also make possible measurements of smaller currents thanks to reductions in capacitance that come from miniaturization.

Acknowledgment

This work was supported by National Institutes of Health Grant number EB004020.

References

Asmild, M., N. Oswald, K.M. Krzywkowski, S. Friis, R.B. Jacobsen, D. Reuter, R. Taboryski, J. Kutchinsky, R.K. Vestergaard, R.L. Schroder, C.B. Sorensen, M. Bech, M.P.G. Korsgaard, and N.J. Willumsen. 2003. Upscaling and automation of electrophysiology: Toward high throughput screening in ion channel drug discovery. *Receptors Channels* 9(1):49–58.

Baylor, S.M. 2005. Calcium sparks in skeletal muscle fibers. *Cell Calcium* 37(6):513–530.

Bezanilla, F. 2002. Voltage sensor movements. *J. Gen. Physiol.* 120(4):465–473.

Borisenko, V., T. Lougheed, J. Hesse, E. Fureder-Kitzmuller, N. Fertig, J.C. Behrends, G.A. Woolley, and G.J. Schutz. 2003. Simultaneous optical and electrical recording of single gramicidin channels. *Biophys. J.* 84(1):612–622.

Braun, D., and P. Fromherz. 2004. Imaging neuronal seal resistance on silicon chip using fluorescent voltage-sensitive dye. *Biophys. J.* 87(2):1351–1359.

Brown, A.M. 2004. Drugs, hERG and sudden death. *Cell Calcium* 35(6):543–547.

Cheng, H., W.J. Lederer, and M.B. Cannell. 1993. Calcium sparks—Elementary events underlying excitation–contraction coupling in heart-muscle. *Science* 262(5134):740–744.

Colquhoun, D., and F.J. Sigworth. 1995. Fitting and statistical analysis of single-channel records. *In*: Single-Channel Recording, 2nd Ed. B. Sakmann and E. Neher, editors. Plenum, New York. 483–587.

Demuro, A., and I. Parker. 2005. "Optical patch-clamping": Single-channel recording by imaging Ca^{2+} flux through individual muscle acetylcholine receptor channels. *J. Gen. Physiol.* 126(3):179–192.

Fertig, N., R.H. Blick, and J.C. Behrends. 2002. Microstructered glass chip for ion-channel electrophysiology. *Biophys. J.* 82(1):161A.

Fertig, N., M. George, M. Klau, C. Meyer, A. Tilke, C. Sobotta, R.H. Blick, and J.C. Behrends. 2003. Microstructured apertures in planar glass substrates for ion channel research. *Receptors Channels* 9(1):29–40.

Fertig, N., A. Tilke, R. Blick, and J. Behrends. 2000. Nanostructured suspended aperture for patch clamp recording and scanning probe application on native membranes. *Biophys. J.* 78(1):266A.

Finkel, A., Wittel, A., Yang, N., Handran, S., Hughes, J., and Costantin, J. June, 2006. Population patch clamp improves data consistency and success rates in the measurement of ionic currents. *J. Biomol. Screen. Epub.*

Fromherz, P. 2002. Electrical interfacing of nerve cells and semiconductor chips. *Chemphyschem* 3(3):276–284.

Gonzalez, J.E., and R.Y. Tsien. 1997. Improved indicators of cell membrane potential that use fluorescence resonance energy transfer. *Chem. Biol.* 4(4):269–277.

Guia, A., Y.B. Wang, J.Q. Xu, K. Sithiphong, Z.H. Yang, C.L. Cui, L. Wu, E. Han, and J. Xu. 2002. Micro-positioning enabled patch clamp recordings on a chip. *Biophys. J.* 82(1):161A.

Hamill, O.P., A. Marty, E. Neher, B. Sakmann, and F.J. Sigworth. 1981. Improved patch-clamp techniques for high-resolution current recording from cells and cell-free membrane patches. *Pflugers Arch.* 391(2):85–100.

Hodgkin, A.L., A.F. Huxley, and B. Katz. 1952. Measurement of current–voltage relations in the membrane of the giant axon of loligo. *J. Physiol. Lond.* 116(4):424–448.

Hutzler, M., and P. Fromherz. 2004. Silicon chip with capacitors and transistors for interfacing organotypic brain slice of rat hippocampus. *Eur. J. Neurosci.* 19(8):2231–2238.

Ingebrandt, S., C.K. Yeung, M. Krause, and A. Offenhausser. 2005. Neuron-transistor coupling: Interpretation of individual extracellular recorded signals. *Eur. Biophys. J. Biophys. Lett.* 34(2):144–154.

Ionescu-Zanetti, C., R.M. Shaw, J.G. Seo, Y.N. Jan, L.Y. Jan, and L.P. Lee. 2005. Mammalian electrophysiology on a microfluidic platform. *Proc. Natl. Acad. Sci. USA* 102(26):9112–9117.

Kass, R.S. 2005. The channelopathies: Novel insights into molecular and genet mechanisms of human disease. *J. Clin. Invest.* 115(8):1986–1989.

Kiss, L., P.B. Bennett, V.N. Uebele, K.S. Koblan, S.A. Kane, B. Neagle, and K. Schroeder. 2003. High throughput ion-channel pharmacology: Planar-array-based voltage clamp. *Assay Drug Dev. Technol.* 1(1):127–135.

Klemic, K.G., J.F. Klemic, M.A. Reed, and F.J. Sigworth. 2002. Micromolded PDMS planar electrode allows patch clamp electrical recordings from cells. *Biosens. Bioelectron.* 17(6/7):597–604.

Klemic, K.G., J.F. Klemic, and F.J. Sigworth. 2005. An air-molding technique for fabricating PDMS planar patch-clamp electrodes. *Pflugers Arch.* 449(6):564–572.

Lehnert, T., M.A.M. Gijs, R. Netzer, and U. Bischoff. 2002. Realization of hollow SiO_2 micronozzles for electrical measurements on living cells. *Appl. Phys. Lett.* 81(26):5063–5065.

Levis, R.A., and J.L. Rae. 1998. Low-noise patch-clamp techniques. *Methods Enzymol.* 293:218–266.

Neher, E., and B. Sakmann. 1976. Single-channel currents recorded from membrane of denervated frog muscle fibres. *Nature* 260(5554):799–802.

Neher, E., B. Sakmann, and J.H. Steinbach. 1978. The extracellular patch clamp: A method for resolving currents through individual open channels in biological membranes. *Pflugers Arch.* 375(2):219–228.

Pantoja, R., J.M. Nagarah, D.M. Starace, N.A. Melosh, R. Blunck, F. Bezanilla, and J.R. Heath. 2004. Silicon chip-based patch-clamp electrodes integrated with PDMS microfluidics. *Biosens. Bioelectron.* 20(3):509–517.

Rae, J., K. Cooper, P. Gates, and M. Watsky. 1991. Low access resistance perforated patch recordings using amphotericin-B. *J. Neurosci. Methods* 37(1):15–26.

Roden, D.M. 2004. Drug-induced prolongation of the QT interval. *N. Engl. J. Med.* 350(10):1013–1022.

Schmidt, C., M. Mayer, and H. Vogel. 2000. A chip-based biosensor for the functional analysis of single ion channels. *Angew. Chem. Int. Ed.* 39(17):3137–3140.

Sia, S.K., and G.M. Whitesides. 2003. Microfluidic devices fabricated in poly(dimethylsiloxane) for biological studies. *Electrophoresis* 24(21):3563–3576.

Sigworth, F.J. 1995. Electronic design of the patch clamp. *In*: Single-Channel Recording, 2nd Ed. B. Sakmann and E. Neher, editors. Plenum, New York, pp. 95–127.

Sonnleitner, A., L.M. Mannuzzu, S. Terakawa, and E.Y. Isacoff. 2002. Structural rearrangements in single ion channels detected optically in living cells. *Proc. Natl. Acad. Sci. USA* 99(20):12759–12764.

Stett, A., C. Burkhardt, U. Weber, P. van Stiphout, and T. Knott. 2003. Cytocentering: A novel technique enabling automated cell-by-cell patch clamping with the CytoPatch (TM) chip. *Receptors Channels* 9(1):59–66.

Terstappen, G. 1999. Functional analysis of native and recombinant ion channels using a high-capacity nonradioactive rubidium efflux assay. *Anal. Biochem.* 272:149–155.

van Stiphout, P., T. Knott, T. Danker, and A. Stett. 2005. 3D Microfluidic Chip for Automated Patch-Clamping. *In*: Mikrosystemtechnik Kongress 2005 (pp. 435–438). Edited by: VDE Verband der Elektrotechnik Elektronik Informationstechnik e. V. and VDI/VDE Innovation + Technik GmbH (VDI/VDE-IT). Berlin.

Vennekamp, J., H. Wulff, C. Beeton, P.A. Calabresi, S. Grissmer, W. Hansel, and K.G. Chandy. 2004. Kv1.3-blocking 5-phenylalkoxypsoralens: A new class of immunomodulators. *Mol. Pharmacol.* 65(6):1364–1374.

Xu, J., A. Guia, D. Rothwarf, M.X. Huang, K. Sithiphong, J. Ouang, G.L. Tao, X.B. Wang, and L. Wu. 2003. A benchmark study with SealChip (TM) planar patch-clamp technology. *Assay Drug Dev. Technol.* 1(5):675–684.
Zeng, J.L., K.R. Laurita, D.S. Rosenbaum, and Y. Rudy. 1995. 2 Components of the delayed rectifier K^+ current in ventricular myocytes of the guinea-pig type—theoretical formulation and their role in repolarization. *Circ. Res.* 77(1):140–152.

18 Gated Ion Channel-Based Biosensor Device

Frances Separovic and Bruce A. Cornell

18.1 Introduction

A biosensor device based on the ion channel gramicidin A (gA) incorporated into a bilayer membrane is described. This generic immunosensing device utilizes gA coupled to an antibody and assembled in a lipid membrane. The membrane is chemically tethered to a gold electrode, which reports on changes in the ionic conduction of the lipid bilayer. Binding of a target molecule in the bathing solution to the antibody causes the gramicidin channels to switch from predominantly conducting dimers to predominantly nonconducting monomers. Conventional a.c. impedance spectroscopy between the gold and a counter electrode in the bathing solution is used to measure changes in the ionic conductivity of the membrane. This approach permits the quantitative detection of a range of target species, including bacteria, proteins, toxins, DNA sequences, and drug molecules.

18.2 Membrane-Based Biosensors

The use of a synthetic molecular membrane as the basis for a chemical sensor was proposed 40 years ago (Toro-Goyco et al., 1966). The effect of protein–protein interactions between a bilayer lipid membrane (BLM) and species in the bathing solution was reported via the admittance of the BLM. However, a major practical issue was the stability of the membrane against mechanical damage. Membranes made from polymerized lipids were considered to be more stable (Ligler et al., 1988) and bacteriorhodopsin, the light sensitive bacterial proton pump, was incorporated into polymerized liposomes and observed to function (Yager, 1988; Zaitsev et al., 1988). A membrane-based biosensor device for the detection of biological toxins (Ligler et al., 1998) also used polymerized lipids, with alamethicin and a calcium channel complex in a BLM supported on a porous silicon substrate. The porous silicon gave mechanical support for the lipid membrane as well as acting as a reservoir for the transmembrane flow of ions. The range of applications of the device was limited by the stability of the receptor–membrane complex: from the inherently labile calcium channel complex and the mechanical instability of the supporting membrane. We have addressed this essential instability by using gA as the ion channel and a tethered membrane on a gold surface as a biosensor device (Cornell et al.,

1997). Once the first lipid monolayer is physisorbed and/or chemically attached to the surface, a second layer of mobile lipids is fused onto the tethered monolayer to form a tethered BLM (see reviews by Sackman, 1996; Plant, 1999). Receptors are attached to gA in order to gate the ionophore.

18.3 The Ion Channel Switch ICS^{TM} Biosensor

A biosensor based on gA, a low molecular weight bacterial ion channel, has been used for the detection of low molecular weight drugs, proteins, and microorganisms (Cornell et al., 1997; Anastasiadis and Separovic, 2003). Gramicidin A is a 15 residue peptide of alternating L- and D- amino acids and two gA molecules form a monovalent cation channel across the lipid bilayer (Wallace, 1998). The Ion Channel Switch (ICS^{TM}) biosensor (Fig. 18.1) employs a disulfide lipid monolayer tethered to a gold surface by a polar spacer, which provides a reservoir for ions permeating through the membrane. The transduction mechanism depends on the gA properties within a BLM. Gramicidin A monomers diffuse within the individual monolayers of the BLM. The flow of ions through gA occurs when two nonconducting monomers align to form a conducting dimer. In the biosensor, however, the gA molecules

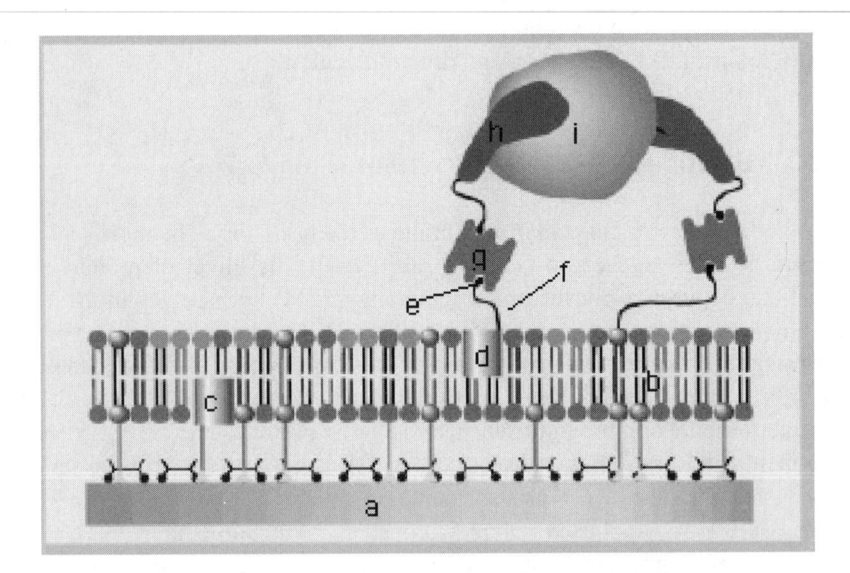

Fig. 18.1 The Ion Channel Switch^{TM} biosensor consisting of a gold electrode (a) with tethered lipid bilayer (b). Embedded in the lipid bilayer are two gA analogues, one fixed (c) and a mobile gA (d) with biotin (e), attached through an aminocaproyl linker (f). Biotin binds strongly to streptavidin (g). When the biotinylated receptor (h) binds to a desired target molecule (i), the mobile gA (d) in the upper monolayer is immobilized and, therefore, is not able to align with the fixed gA (c) in the lower monolayer to form an ion channel. Hence ion current flow to the gold electrode below is decreased when an analyte is detected.

within the tethered inner leaflet of the lipid bilayer are also tethered whilst those on the outer monolayer are free to diffuse and are attached to antibodies. The binding of an analyte cross-link antibodies attached to the membrane spanning lipid tethers to those attached to the mobile outer layer channels. Due to the low density of tethered gA molecules within the inner membrane leaflet, this anchors the cross-linked gA distant, on average, from their immobilized inner layer channel partners. Gramicidin dimer conduction is thus prevented and the admittance of the membrane is decreased. Application of a small alternating potential between the gold substrate and a reference electrode in the test solution generates a charge at the gold surface and causes electrons to flow in an external circuit.

Biological lipid membranes are liquid crystalline and the outer leaflet of the biosensor membrane is effectively a 2D liquid crystal. The antibodies attached to the mobile ion channels are able to scan a significant membrane area in a few minutes and thus have access to multiple antibody sites attached to membrane spanning lipid tethers. Hence, the ICS^{TM} biosensor responds more rapidly than if simply binding to antibodies attached to the mobile gA channels triggered the transduction mechanism. The response speed of the biosensor is improved in direct proportion to the number of binding sites accessible to the mobile channels, analogous to an electronic multiplexer.

Tethering the inner monolayer to the gold surface enhances the stability of the membrane but additional stability is achieved by using "archaebacterial lipids." These are lipids modeled on constituents found in bacteria capable of surviving extremes of temperature and hostile chemical environments. The hydrocarbon chains of these lipids span the entire membrane, (Kushawaha et al., 1981) and ether linkages replace the esters (de Rosa et al., 1983). BLM films formed from archaebacterial lipids have resulted in membranes that are stable to temperatures in excess of 90°C (Gliozzi et al., 1983).

Although most studies of the ICS^{TM} biosensor have used antibody F_{ab} fragments as the receptor, the approach is generic and has been demonstrated to operate using oligonucleotide sequences, heavy metal chelates, and cell surface receptors. The specificity of the response is dependent on the receptor. A commercial device is under development and is primarily aimed at rapid, quantitative determination of time-critical diagnostic measures in whole blood. Operation of the device, however, is not constrained to just blood and could be used with any electrolyte containing biological fluid including serum, saliva, and urine.

18.4 Fabrication of a Membrane-Based Biosensor

Self-assembly of a stable membrane incorporating ion channels on a clean, smooth gold surface is possible using a combination of sulfur–gold chemistry and physisorption (Philp and Stoddart, 1996). The stages of biosensor fabrication are considered in the following sections.

18.4.1 Preparation of Gold Surface

An initial step in the fabrication of a reliable biosensing device is the production of a molecularly smooth gold surface free of contaminants and oxides. An extensive literature exists on how to produce good quality gold surfaces and is briefly discussed here.

The quality of vapor-deposited and sputter-coated thin gold films on mica, heated in ultrahigh vacuum up to 500°C, has been reported (Guo et al., 1994). The gold surface, as shown by AFM, became atomically flat at 450–500°C. Surface contamination of the gold by organics and gold oxide was more significant in obtaining good high impedance films than simply the molecular smoothness of the gold. Grain boundaries in the "pebbled" gold surfaces observed in the nonannealed gold were suggested to capture impurities, which may cause leakiness in alkanethiol films. Immersion in hot piranha solution (H_2SO_4/H_2O_2) for 10–15 min, followed by electrochemical stripping of any resulting oxide resulted in gold that supports a high impedance self-assembled membrane (SAM).

Exposure to oxidants such as piranha solution, oxygen plasma or UV/ozone removes organic contaminants but causes the formation of gold oxide. Ron and Rubenstein (1994) discuss the risk of forming alkane thiol monolayers over preoxidized gold. Gold oxide may be removed by prolonged immersion in ethanol or by using electrochemical stripping. Failure to remove all traces of oxide prior to depositing an alkanethiol monolayer causes the oxide to be trapped beneath the alkanethiol film. As seen below, the ICS^{TM} sensor technology employs disulfides which requires rigor in eliminating gold oxide

Gold surfaces exposed to a laboratory atmosphere for only minutes were found by XPS to develop a layer of contamination approximately 0.6 nm thick composed of oxygen and carbon (Ron and Rubenstein, 1998). This problem was addressed by using an electrochemical technique, which oxidizes surface contaminants, cleans the surface of oxide and then accelerates the deposition of the alkane monolayer (Ron et al., 1998). Initially, the gold is electrochemically oxidized in water. At −0.3 V, the gold oxide is reduced in ethanol. Switching the potential at the gold to an oxidizing +1.45 V and by using a fast liquid exchange flow cell to synchronously introduce an alkanethiol into the solution a high quality, electrically sealing monolayer is formed in 1–2 s.

Alkanethiolates on gold, oxidize in air in the dark to form sulfinates and sulfonates. The kinetics of oxidation depends on the morphology of the underlying gold, increasing dramatically with a decrease in grain size and the amount of Au(111) on the surface and may explain the variation in the quality of alkanethiol SAM as reported by different groups (Lee et al., 1998). Grain boundaries of evaporated polycrystalline gold films were identified as an important catalytic site for oxidation. To prepare a smooth surface, the gold was annealed with a small butane flame. This cleaned the gold and facilitated its recrystallization into large regions of Au(111). "Epitaxially" grown gold surfaces were produced by slow (1 Å/s) thermal evaporation on "scratch free" mica at 2.5×10^{-7} mbar at 300°C until 100 nm of gold had been deposited. The surface was transformed into defect free Au(111) over 150 × 150 nm areas. These regions were substantially more robust against oxidation and its effect on destabilizing electrical leakage of alkanethiol SAM.

The state of the gold surface and, in particular, the surface roughness is important for the reproducible formation of high quality self-assembled monolayers on gold. A pulsed potential pretreatment, which results in a twofold reduction in roughness of mechanically polished surfaces, has been reported (Hoogvliet et al., 2000). A flow cell and a 100 ms triple pulse sequence of +1.6, 0.0, and −0.8 V relative to a counter electrode are used. Pulsing for 2000–5000 s under flow conditions is required to achieve a smooth gold surface.

18.4.2 Formation of Tethered Bilayer Lipid Membrane

In order to increase the stability of the membrane, which forms the basis of the biosensor, a tethered monolayer of an alkane disulfide tethers the BLM to the gold electrode surface. In addition, a significant fraction of ether linked, membrane-spanning lipids are also tethered to the gold. These membrane-spanning lipids further stabilize the lamellar phase of the membrane lipids and possibly prevent insertion of additional material once the membrane is assembled. In our hands, these tethered membranes possess excellent stability over many months and can be stored dry and retain function when rehydrated for use.

18.4.2.1 Assembly of the Monolayer

The convenience and extensive experience available on sulfur–gold chemistry and tethering mechanism (Lingler et al., 1997) provide excellent groundwork for making a tethered BLM. The early literature on the formation of SAM of gold-tethered alkanethiols has been reviewed widely (Ulman, 1996).

The adsorption induced by Au_2O_3 interactions with polar species such as –OH, –COOH and $–PO_3H_2$ on the terminating groups of the SAM can cause serious packing disorder in the alkanethiol films (Tsai and Lin, 2001). This effect may be minimized by plasma, oxidizing the entire surface and then reducing it to metallic gold by ultrasonic agitation in oxygen free ethanol. Thus, the assembly process needs to be well controlled in order to achieve a reproducible SAM. Although disulfides have been detected, steric effects may play a significant role in the reduction of disulfides to thiols at gold surfaces. The major difference between disulfide and thiol SAM is in the kinetics of competitive film formation, which favors the thiols by 75:1 (Bain et al., 1989).

It is important to have control over the relative composition of the adsorbed film when fabricating devices based on SAM. The relationship between the composition of two-component SAM on gold and that of the solutions from which they were formed has been examined (Folker et al., 1994). A two-component monolayer of $HS(CH_2)_{21}CH_3$ and $HS(CH_2)_{11}OH$ was formed from solutions in ethanol. Phase separated monolayers were not observed: a single phase is preferred at equilibrium for a two-component system of alkanethiols on gold, well equilibrated with alkane thiols in solution. Thus, under the ideal conditions of strongly cooperative film formation at the gold surface and equilibrium film growth, little or no control would exist over the relative ratios of component species in a mixed film on the gold. To counteract this effect, the film could be assembled under kinetic conditions and/or the cooperativity of the assembly process could be reduced. The ICS^{TM} biosensor membrane comprises phytanyl chain lipids and a substantial ethyleneglycol spacer group, and comprises four tethered species, which reduces the cooperativity of the film formation. The bulk and dynamic disorder of the phytanyl side chains result in a better seal against electrochemical stripping. The electrochemical sealing ability of the phytanyl groups further reduces the cooperativity of the film formation (Braach-Maksvytis and Raguse, 2000).

Potentially driven assembly of SAM favors the kinetic, metastable regime (Ma and Lennox, 2000). Self-assembled membrane formed under an applied potential produced an excellent seal in less than 10 min compared to no seal existing in the absence of a potential. Similarly, the composition of the binary film was uncontrolled without an applied potential but with 0.6 V applied to the gold during assembly, the composition was controlled in proportion to the relative concentration of the two components. Potential driven film formation can be used to create patterned SAM. An electrochemically directed adsorption process was used to selectively form SAM on one gold electrode in the presence of another nearby electrode (Hsueh et al., 2000). The monolayers formed were very similar to analogous SAM formed by chemisorption with added advantages of spatial control over coverage, short time required for coating (<1 min) and an ability to form SAM on gold that is not freshly evaporated. Difference in film properties across the gold surface are associated with lattice defects and oxide sites, at which thiol desorption is 2–3 times easier (Walczak et al., 1995).

During the self-assembly process, the sulfur atoms adopt an hexagonal lattice commensurate with the Au(111) structure but rotated 30° relative to the gold lattice (Dishner et al., 1996). The chains of the thiolates extend into space with the same all-*trans* conformation as observed in crystalline paraffins. Scanning Tunneling Microscopy studies have revealed that the chemisorption (presumably desorption) of the thiol is accompanied by the formation of 3–10 nm pits (Dishner et al., 1996). These pits can be eliminated by UV photolysis, which oxidizes the thiol to sulfonate and which can be rinsed from the surface with ethanol or water, leaving a smooth, pit-free surface. The pits, typically 3.3 nm diameter, involve nearly 100 gold atoms and represent approximately 10–15% of the surface area and are filled in by lateral diffusion of the gold. We have observed the introduction of gold in ethanol solutions resulting from soaking gold-coated slides in alkane-thiol and -disulfide containing solutions. Defect-free films have also been produced by heating SAM of *n*-octadecylthiol to almost the boiling point of ethanol at 77°C (Bucher et al., 1994) although no impedance measures were reported to indicate whether damage had occurred to the electrical seal of the film at these temperatures.

18.4.2.2 Assembly of the Lipid Bilayer

A number of reviews on the development of supported membranes are available (Sackmann, 1996; Meuse et al., 1998b). Stabilized BLMs have been prepared comprising an alkanethiol inner monolayer and a dimyristoylphosphatidylcholine, mobile outer layer (Meuse et al., 1998a). These structures have been termed hybrid bilayer membranes and may be formed from lipid vesicles or by transfer from air–water interfaces. These bilayers are noninterdigitated and possess a more ordered structure, equivalent to the effect of lowering temperature. Tethered and supported bilayer membranes have been generated from both thiolated and nonthiolated phospholipids (Steinem et al., 1996). Three techniques were employed: (1) the gold surface was initially covered with a chemisorbed monolayer of an alkanethiol such as octadecane thiol (ODT) or a thiolated phospholipid such as dimyristoylphosphatidyl-thiolethanolamine (DMPTE). A second lipid layer was deposited: (i) by Langmuir–Schaefer technique, (ii) from lipid solution in *n*-decane/isobutanol, (iii) by lipid–detergent dilution, or (iv) by fusion of vesicles; (2) charged molecules with thiol-anchors for attachment to the gold surface by chemisorption; or (3) direct deposition of lipid bilayer vesicles containing a thiolated phospholipid. The ion channel gA was codeposited using the second technique described above. Membrane capacitances of $0.55\ \mu F/cm^2 \pm 10\%$ were obtained independent of the technique employed. Typically membrane seals were obtained to frequencies of $\sim$30 Hz and when doped with gA this was raised to $\sim$100 Hz.

Hybrid bilayers have been formed by the interaction of phospholipids with the hydrophobic surface of a self-assembled alkanethiol monolayer on gold (Meuse et al., 1998a). These membranes were characterized in air using atomic force microscopy, spectroscopic ellipsometry and reflection–absorption infrared spectroscopy. The added phospholipid was one monolayer thick, continuous, and exhibited molecular order similar to that seen in phospholipid bilayers. When characterized in water using neutron reflectivity and impedance spectroscopy, these hybrid bilayers possessed essentially the same properties as normal phospholipid bilayers, although the bilayer leakage level was not reported below 10 Hz. The capacitance of phospholipid/alkanethiol bilayers was found to closely mimic that obtained from solvent free phospholipid bilayers (Plant, 1993). The introduction of 10^{-6}M melittin, a pore-forming peptide, changed the bilayer from being impermeable to being highly conductive.

A biosensor based on a supported bilayer membrane was proposed (Stelzle et al., 1993). Positively charged vesicles of dioctadecyldimethylammonium bromide were fused to form a supported bilayer membrane on a monolayer of a carboxy mercaptan, which was deposited onto a 100 nm thick evaporated gold film on a silanized glass substrate. Impedance spectroscopy (2000–1 Hz) revealed sealing membranes down to $\sim$100 Hz, at which point defects in the membranes required a more complex network to explain the data. A tethered BLM on a silicon oxide surface or membrane on a chip has been described recently and included functional gA channels (Atanasov et al., 2005).

18.4.3 Membrane Components

The chemistries of the species making up the lipid membrane of the ICS™ biosensor have been described (Raguse et al., 1998; Burns et al., 1999a,b,c; Raguse et al., 2000; Anastasiadis et al., 2001). Representative examples of the membrane components are shown in Fig. 18.2. These include two classes of compounds: one that is tethered to the gold surface and another that is physically absorbed to the surface but free to diffuse in the plane of the membrane.

Typically, the concentration used is 350 μM of **1** and 2 mM for **5**; while the mole ratios of **1**:**2**(R_1):**2**(R_2):**3** are 40,000:400:1:1 and for **5**:**6**:**4** are 28,000:12,000:1. Attachment of the membrane to the gold substrate is via a disulfide moiety in **1**–**3**. The disulfide compound acts as a tether and includes a benzyl spacer group that reduces the 2D packing density of the assembled membrane. The lower packing density acts to facilitate the entry of ions into the space between the gold surface and the membrane, or the reservoir, which is discussed further below. As discussed above, the phytanyl chain lipids used here are chemically stable at high temperatures (Stetter, 1996), and the bulky methyl substituents reduce the temperature dependence of the membrane disorder around the normal operating temperatures for the biosensor of 20–30°C. The phytanyl chains also provide an excellent electrical seal based on impedance measurements. Biphenyl linkers at the mid-plane of the bilayer, **2**, adds rigidity and prevents the membrane spanning lipids entering and emerging on the same side of the membrane (Kang et al., 1999).

Another series of tether compounds are shown in Fig. 18.3. The C11 series (compound **7**) have both the sulfur attachment chemistry and the van der Waals attraction of the C11 sequence for each other, giving additional binding energy and protection to the gold surface from the electrolyte solution. The reduction in surface charge also significantly alters the reservoir performance. The all-ether reservoir linkers in compounds **7** and **8** replace the succinate groups of **1**–**4** and eliminate instabilities that can arise from the hydrolysis of ester groups. The all-ether reservoir significantly extends the storage lifetime of the membrane as well as improving the reservoir properties.

The area per lipid molecule is determined by a number of contributions, including the gold–sulfur interface, the spacer molecules, and the interaction of the hydrocarbon chains within the membrane. Membrane spanning lipids such as **2** and half membrane spanning lipids such as **1**, **5** and **6** are mixed in different ratios to adjust the membrane packing. The packing within the membrane may also be adjusted using compounds such as **9** and other spacer molecules such as dodecanethiol.

18.4.4 The Gramicidin A Ion Channel

Gramicidin A is a pentadecapeptide (1882 Da) produced by the soil bacteria *Bacillus brevis* (Katz and Demain, 1977). An extensive literature exists on the ion transport properties of the linear gramicidins (Koeppe and Andersen, 1996; Wallace, 1998). The conformation of gramicidin ranges from random coil to a family of intertwined

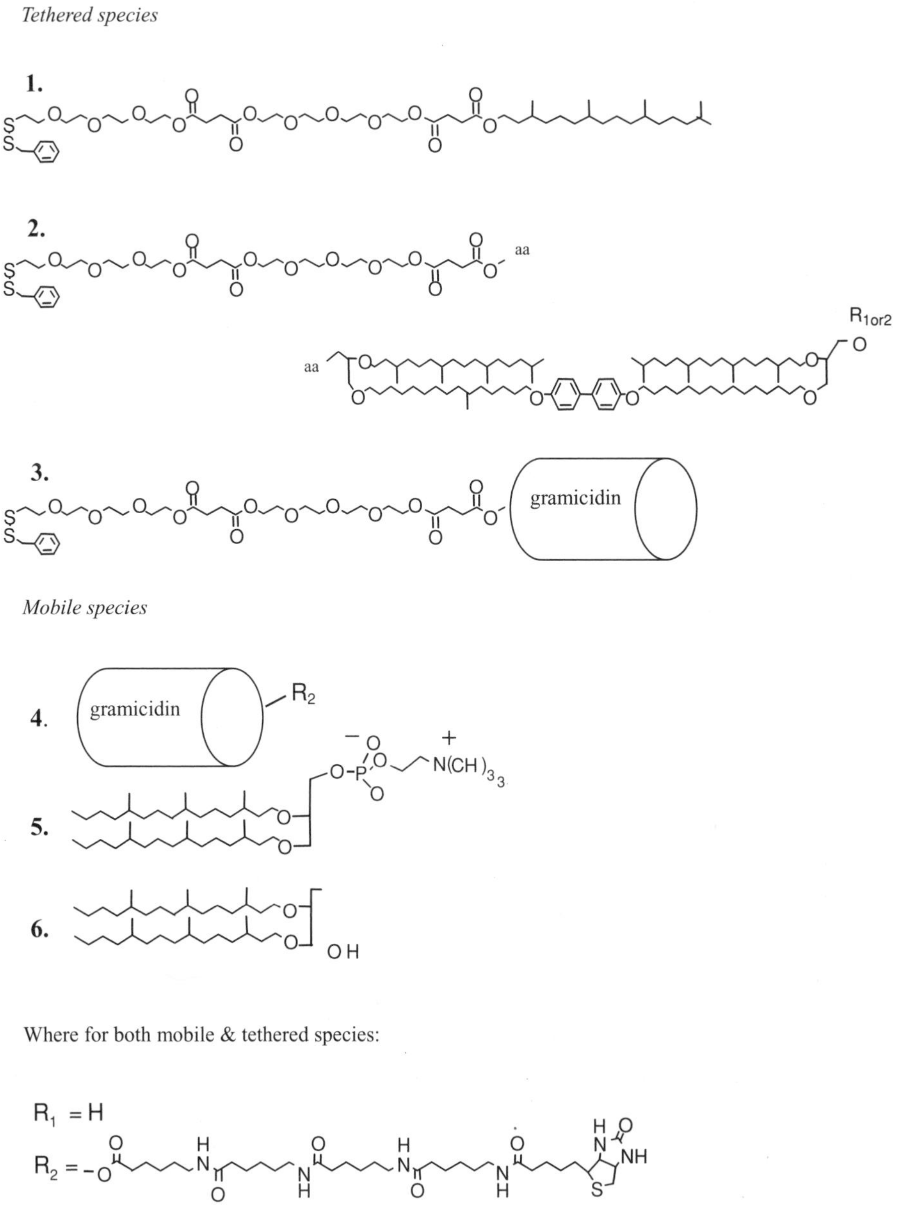
Tethered species
1.
2.
aa
aa
R1or2
3.
gramicidin
Mobile species
4.
gramicidin
R2
5.
N(CH)33
6.
OH
Where for both mobile & tethered species:
R1 = H
R2 =

Fig. 18.2 Chemical structures of components that form the tethered BLM. Compounds **1–3** are tethered species while **4–6** are mobile. A minor portion of **2** and all of **4** are linked to R_2, an aminocaproyl-linked biotin.

7.

8.

9.

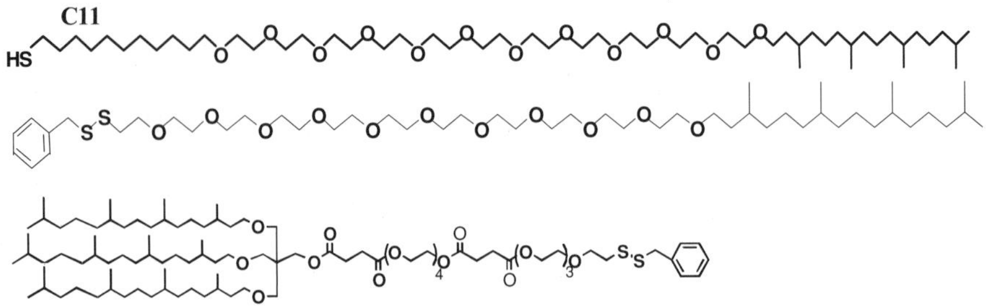

Fig. 18.3 Additional tethered compounds used in assembly of the BLM.

helices, depending on the solvent. However, in a lipid bilayer lipid gA forms a single coil, right-handed $\beta^{6.3}$ helix, anchored at the lipid–water interface by four tryptophans in positions 9, 11, 13, and 15 at the C terminus. The structure is further stabilized by nine internal hydrogen bonds between the adjacent turns of the helix, made possible by the alternating L-D amino-acid sequence. The conducting form of gA is an *N–N* dimer of $\beta^{6.3}$ helices, which forms six hydrogen bonds stabilizing the dimer across the hydrocarbon core of the BLM. Gramicidin selectively facilitates the transport of monovalent cations across the bilayer (Myers and Haydon, 1972).

Unlike large mammalian channels, gA is very stable both chemically and structurally. Gramicidin can be modified using synthetic organic chemistry techniques and can spontaneously fold into the ion conducting form when incorporated into a BLM. Below mM concentrations in ethanol gA has a random coil structure and converts to the $\beta^{6.3}$ helix in a BLM. It is important to ensure that gA has not been transferred from a solvent such as dioxane, which may require many hours to convert from the conformation adopted in dioxane to the channel form.

Many analogues of gA have been synthesized. In order to utilize gA as an ion channel switch it is necessary to attach linkers for both the tethered and the mobile species. The linkers are attached to the ethanolamine hydroxyl group and have been mainly variants on aminocaproyl and tetraethyleneglycol groups for the mobile and tethered gA, respectively. Provided the linker attachment was to the C terminus ethanolamine hydroxyl, the conductance fell within a range of approximately $\pm$ 10% of the native gA (Fig. 18.4). Modifications too near the channel entrance can eliminate conduction. Similar "C" terminus biotin labeled gA has been reported (Suarez et al., 1998; Separovic et al., 1999; Anastasiadis et al., 2001) and we have shown that biotinylated gA analogues preserve essentially the same 3D structure of native gA (Separovic et al., 1999; Cornell et al., 1988).

18.4.5 Receptor Component: Antibody Fragments

Rather than using whole antibodies, which can cause species cross reactivity, F_{ab} fragments are used in the ICS™ biosensor. Antibodies are enzymatically cleaved at the hinge region and biotinylated. The F_{ab} fragments are attached to the membrane using a streptavidin–biotin complex. Although other linker chemistries have proven

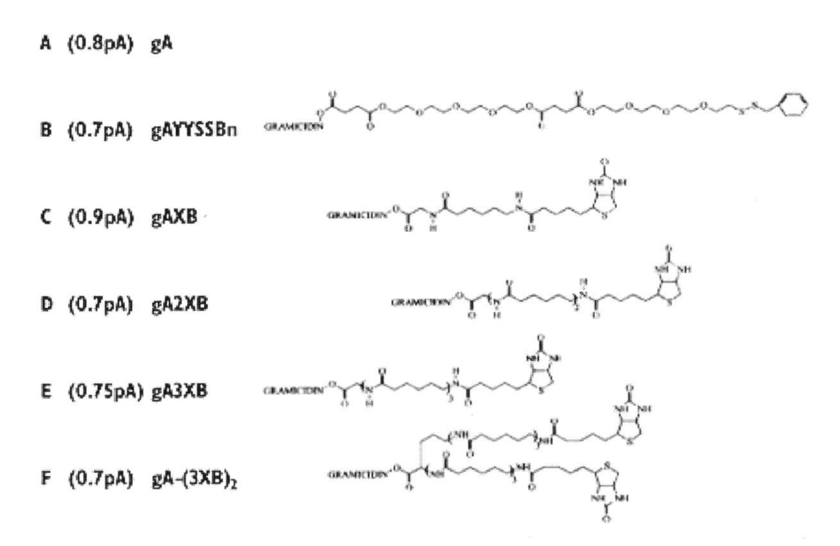

Fig. 18.4 Examples of native (A) and modified gramicidins containing: (B) tetra-ethylene glycol, or (C–F) aminocaproyl linker groups. The single channel current is shown in parenthesis adjacent to each structure.

satisfactory for certain applications, the streptavidin–biotin linkage is convenient and readily available.

18.4.6 Assembly of the Biosensor

In the ICS^{TM} biosensor the bilayer membrane is assembled during an ethanol/water rinse as shown schematically in Fig. 18.5. Assembly of the biosensor depends on forming a stable BLM using a tethered hydrocarbon layer on a gold substrate to achieve a tight bonding.

18.5 Mechanism of Operation

Although there have been many attempts to engineer receptor-based gated ion channels, the proposed mechanisms have had a limited range of applications and require reengineering for each new analyte. Gating mechanisms include molecules that block the channel entrance (Lopatin et al., 1995) and antichannel antibodies that disrupt ion transport (Bufler et al., 1996). Whilst the ICS^{TM} biosensor uses ion channel transduction, however, it is able to be adapted to detect many different classes of target.

18.5.1 Detection of Large Analytes

The large analyte class includes microorganisms, proteins, polypeptides, hormones, oligonucleotides, and DNA segments. If a suitable antibody pair is available, the

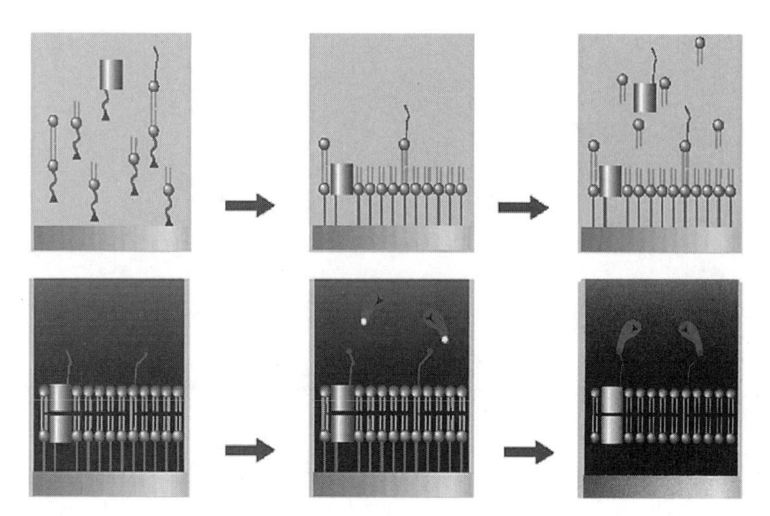

Fig. 18.5 A fresh gold surface is exposed to an ethanol solution of the tethering species **1–3**, for 10 min, which produces the inner and part of the outer monolayer of the BLM. After an alcohol rinse, a second ethanol solution of the mobile membrane species, **4** and **5**, is added. A lipid bilayer structure forms spontaneously when rinsed with water. Some lipids span the membrane, whilst others are mobile within the membrane plane. Antibody fragments are then added in the aqueous solution and attached using a streptavidin–biotin linkage (as shown in Fig. 18.1). For large analyte detection, which involves a "sandwich assay" configuration, an equimolar mixture of the two antibody fragments is added.

biosensor may be adapted to the detection of any antigenic target. The ion channel gA is assembled into a tethered lipid membrane and coupled to an antibody targeting a particular compound of diagnostic interest. Analyte binding causes the gA channels to switch from predominantly conducting dimers to predominantly nonconducting monomers (Fig. 18.6). A competitive assay has also been devised in which the analyte causes the population of channel dimers to increase.

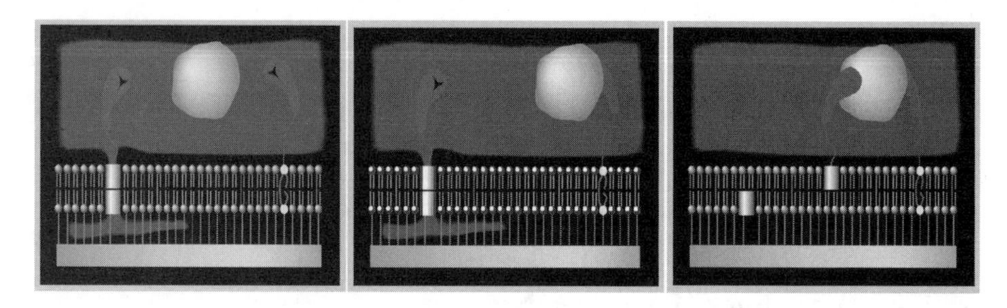

Fig. 18.6 Analyte binding to the antibody fragments causes the conformation of gA to shift from conducting dimers to nonconducting monomers, which causes a loss of ion conduction across the membrane. Scale = 5 nm.

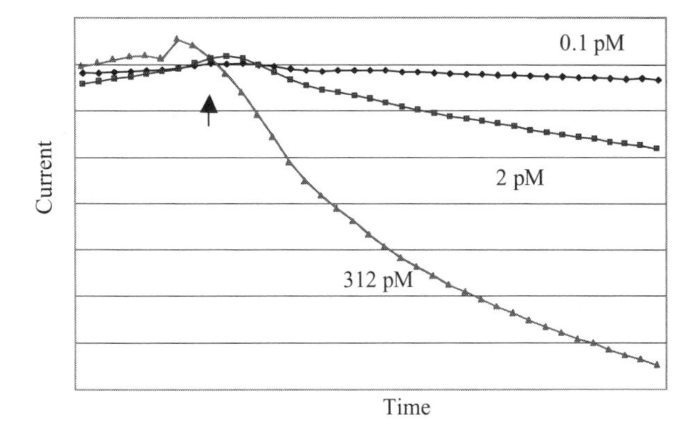

Fig. 18.7 Biosensor response to Thyroid Stimulating Hormone (TSH) over the initial 300 s following analyte addition (arrow).

18.5.2 Examples of Operation

(i) *Thyroid Stimulating Hormone* (TSH) is a 28 kDa protein with α and β subunits to which a matched pair of antibodies is used, each targeting different nonoverlapping epitopes. The biosensor response to TSH is shown in Fig. 18.7.
(ii) *Ferritin* is the principal iron transporting protein in human serum and, having a molecular weight of ∼450 kDa, is one of the largest soluble proteins regularly measured clinically. With 24 equivalent subunits each accessible to cross-linking, only a single F_{ab} type is required to elicit a response. Figure 18.8 shows a titration curve of the biosensor response to ferritin in patient serum.

A comparison of the results of the ICS™ biosensor with a Diagnostics Products Corporation (Los Angeles, CA) analyzer for measuring ferritin concentration in the serum taken from 100 patients in given in Fig. 18.9. The biosensor performs well over a wide range of analyte concentrations under clinical conditions.

18.5.2.1 Performance Considerations

Assuming an adequate mass transport of analyte to the membrane and that the surface density of tethered antibodies significantly exceeds that of antibodies linked to the mobile channels, the idealized behavior of the biosensor may be modeled by a family of coupled equations. The 2D reactions on the membrane surface are generally faster than the 3D reaction rates between the analyte in solution and the surface (Hardt, 1979). Thus for low and medium analyte concentrations, the 3D processes will be rate limiting while at high concentrations, 2D processes will become important. A further limiting condition is the lifetime of the dimeric channel. It is thus straightforward to numerically simulate the device behavior for large analyte detection (Woodhouse et al., 1999; Cornell, 2002). The quantitative measure of analyte concentration may be taken as the initial rate of current increase. At low analyte concentrations 2–3 decades of linear response to analyte are available while at higher concentrations,

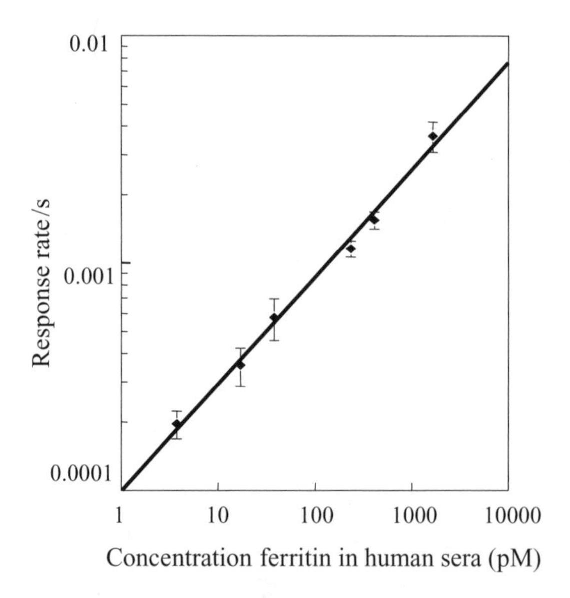

Fig. 18.8 Biosensor response rate for a range of concentrations of ferritin in human serum. The rate is obtained by normalizing the initial slope of the response curve to the initial admittance. The initial slope of the response is obtained within the first 180–300 s depending on the analyte concentration.

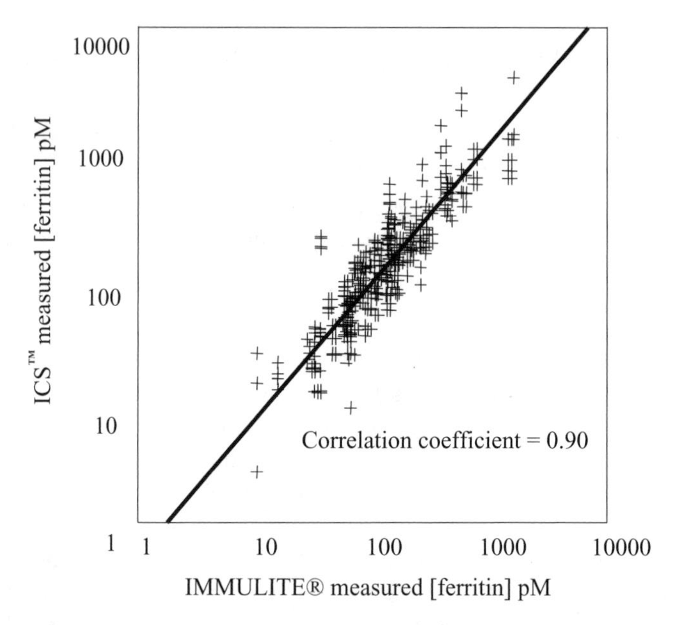

Fig. 18.9 A comparison of estimated ferritin concentrations from 100 patients measured by the ICS™ biosensor and a Diagnostics Products IMMULITE® analyzer. The measurements were made in serum at 30°C. The patients were chosen to provide a wide spread of clinical values.

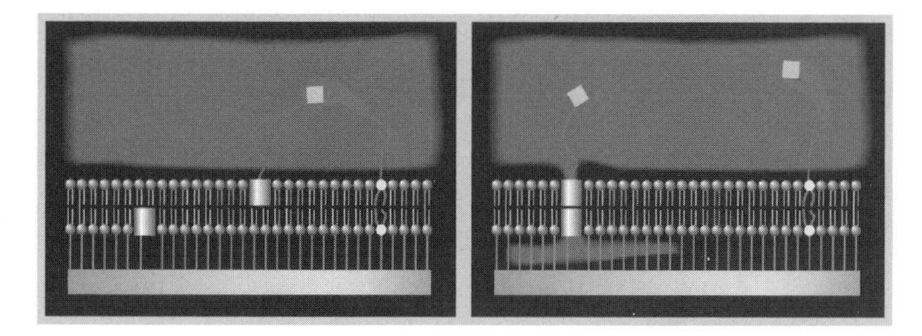

Fig. 18.10 Small analyte detection: the mobile channels are cross-linked to anti-digoxin F_{ab} anchored at the tether sites in the absence of analyte (e.g. digoxin). Dimer formation is prevented and the conductance of the membrane falls. The introduction of analyte competes off the hapten (digoxin analogue) and increases the membrane conductance.

approaching nanomolar, the gA dimer lifetime or the cross-linking rate of the mobile ion channels to the tethered antibodies will be limiting.

The variables available in establishing the biosensor operating conditions are: the surface density of tethered antibody sites $[T]$, the surface density of antibodies attached to mobile channels $[G_M]$, and the "on rate" of the chosen antibody k_{3D}. The "on rate" of the antibodies attached to the mobile channels in general need not be as large as those on the primary tethered capture sites $[T]$ since 2D processes are more effective. The ratio $[T]/[G_M]$ "amplifies" the apparent capture rate of analyte $[A]$ from what would otherwise be the simple first order rate constant, $k_{3D}[A]$ to $k_{3D}[A][T]/[G_M]$. The maximum amplification in this configuration approaches 10^2–10^3 and is dependent on the length of the linkers and whether streptavidin has been used as a coupling protein. At higher analyte concentrations, in the range nM–μM, the amplification may be adjusted downwards by lowering the $[T]/[G_M]$ ratio. With the introduction of flow, 1 μL/min, to the analyte stream the mass transfer limitation can be overcome and the capture density $[T]$ increased by an order of magnitude with a proportionate increase in the response.

18.5.3 Detection of Small Analytes

For low molecular weights analytes such as therapeutic drugs where the target is too small to use a two-site sandwich or cross-linking assay, a competitive adaptation of the ICS™ biosensor is available (Fig. 18.10). In this case the biosensor conduction increases when the analyte binds.

18.5.3.1 Small Analyte Response

The ICS™ system has been configured for the cardiac stimulant digoxin (781 Da) linked at the 3′ position to gramicidin through a flexible tetra-aminocaproyl group. During fabrication when the biotinylated F_{ab} fragment is added, the conduction is reduced due to cross-linking of mobile gA-digoxin with tethered antibody fragments.

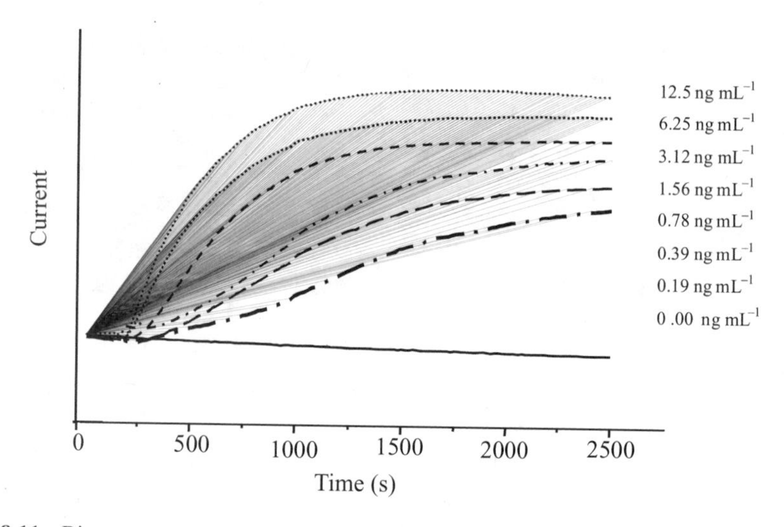

Fig. 18.11 Biosensor reponses to a range of digoxin concentrations that span the clinical range of ~1 ng/mL.

The addition of digoxin displaces the gA-digoxin and the conduction increases (Fig. 18.11).

18.5.3.2 Performance Considerations

Competition for the tethered antibodies $[T]$ between analyte and hapten on the mobile gA $[G^{\mathrm{h}}_{\mathrm{M}}]$ establishes an equilibrium, which determines the surface density of the complex $[T^{*}\,G^{\mathrm{h}}_{\mathrm{M}}]$. As with the large analyte example, the current output from the sensor is determined by the surface density of gramicidin dimers $[G_{\mathrm{D}}]$. The quantitative measure of analyte concentration may be taken as the initial rate of current increase or the endpoint gating ratio (Fig. 18.12). With the competitive, small analyte system, the 'amplification' effect described above is not available (Woodhouse et al., 1999). However, it is possible to adjust the biosensor sensitivity over a considerable range by manipulation of component surface densities.

18.5.4 Biosensor Applicability

The ICS^{TM} biosensor has a wide range of applications based on a common transduction mechanism. Over 50 different antibodies as well as extracellular cell surface receptors for growth factor detection, oligonucleotide probes for DNA strand detection, lectins for glucose detection, and metal chelates for heavy metal detection have been utilized in the ICS^{TM} biosensor. The major requirement is an attachment chemistry that will not inactivate the receptor for a particular analyte. Figure 18.13 shows the biosensor response to a range of analytes.

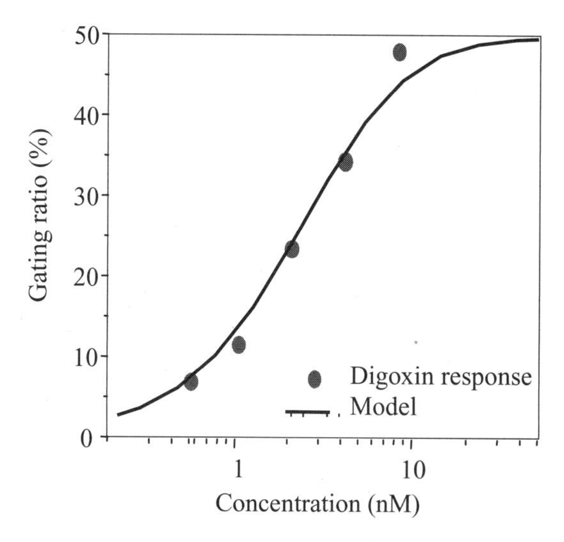

Fig. 18.12 Endpoint gating ratios, i.e., % change in conduction upon the addition of digoxin at various concentrations shown superimposed on a numerical model (Woodhouse et al., 1999). The response time is essentially fixed by the off rate $k^{-1}A^{h}$ of the hapten from the F'_{ab}, which for the present example of digoxin is ≈50 s.

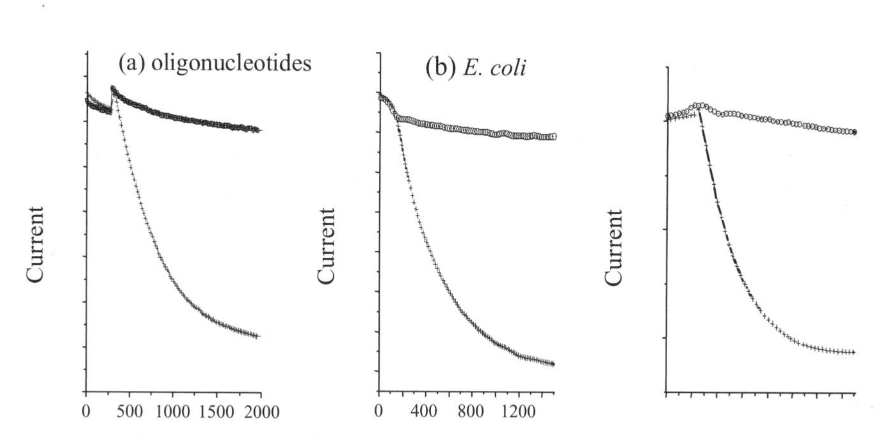

Fig. 18.13 Biosensor responses to different classes of target analyte: (a) 19-base oligonucleotide probe biotinylated at the 5′ end via a 23 atom phosphoamidase linker. The response to a 5 nM challenge of a 52-mer target sequence is shown. The double stranded target was heated to 95°C and cooled to ~60°C. The biotinylated probe on the gA hybridized faster than its complementary sequence resulting in the gating shown. (b) Biotinylated F_{ab} directed to *E. coli* following a challenge of 10^5 cells/mL. (c) Biotinylated complementary pair of F_{ab} directed to two epitopes on the α subunit of TSH after addition of 100 pM TSH. The upper curve is a control using an anti-theophylline antibody and the lower is the active measure.

18.6 Biosensor Characterization

18.6.1 Ellipsometry

Ellipsometry was used to determine the biosensor membrane thickness. Direct calibration using alkane thiols from C_8 to C_{18} yielded a thickness for the BLM of 4 nm. The thickness of the mobile outer layer was determined to be 2 nm by rinsing off the layer with ethanol. This is close to the value expected for a fluid monolayer. The overall thickness of the tethered BLM was measured to be ~12–15 nm, although in estimating this value assumptions are made of the dielectric constant for a tethered membrane separated from the gold electrode by a polar linker.

18.6.2 Impedance Spectroscopy

The membrane thickness can also be determined from the membrane capacitance, C_m, based on modeling the impedance spectrum over a frequency range swept over typically 0.1–1000 Hz. Both the phase and modulus are measured and fitted. The electrical equivalent circuit of a tethered BLM is shown in Fig. 18.14. Simple RC networks provide a fit with <2% residual. A measured membrane capacitance of 0.5 $\mu F/cm^2$ was obtained and, assuming a relative dielectric constant of 2.2 for the membrane chains, gives an overall membrane thickness of ~4 nm.

The validity of this equivalent circuit as a good description of the conductive tethered BLM is shown in Fig. 18.15, which shows the impedance profile and phase relationship.

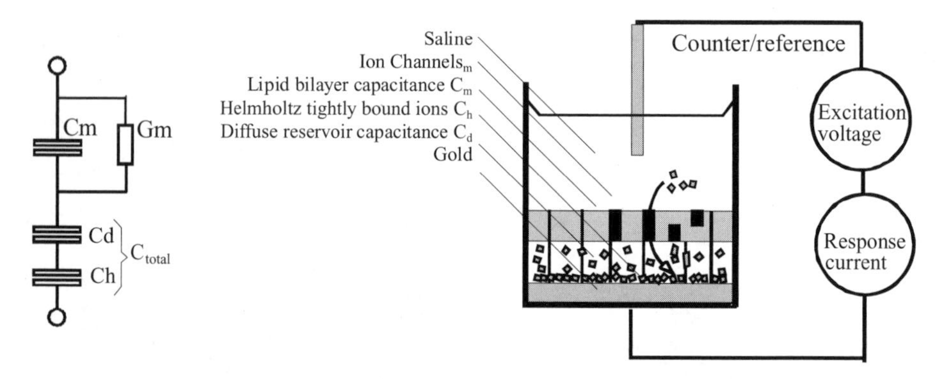

Fig. 18.14 Equivalent electrical circuit used to characterize the tethered BLM (Krishna et al., 2001). The main elements are the conduction G_m, which describes the ion flow across the membrane dominated by the gramicidin channels, the membrane capacitance C_m, the Helmholtz capacitance C_h of ions crowding against the gold electrode, and the diffuse capacitance C_d of ions within a concentration cloud decaying back into the reservoir space. For an excitation of typically 50 mV the equivalent circuit may be approximated as an effective Helmholtz capacitance, C_h of 3.5 $\mu F/cm^2$, in series with a membrane capacitance of 0.6 $\mu F/cm^2$.

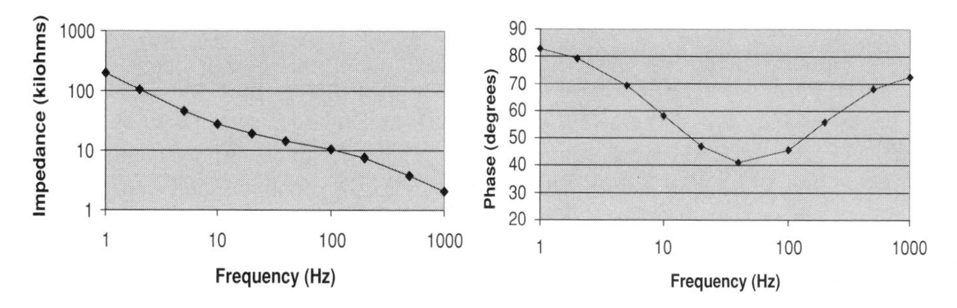

Fig. 18.15 The impedance of a capacitor varies according to $Z = 1/2\pi fC$. A plot of $\log_{10}Z$ against $\log_{10}f$ yields a linear relationship with a slope of -1 and an intercept at $1/2\pi$Hz of $\log_{10}C$. (Left): At high frequencies $C \approx C_{\mathrm{m}}$ and at low frequencies $C \approx C_{\mathrm{h}}$ where C_{m} and C_{h} refer to the capacitance of the membrane and Helmholtz layers, respectively. The frequency at which the impedance profile crosses from C_{m}-dominated to C_{h}-dominated behavior follows the change in G_{m} arising from the ion channels. (Right): The crossing point is conveniently identified by the frequency f_{ϕ} at which the phase angle ϕ between the excitation and the resultant current is minimized.

18.6.3 Membrane Composition

When fabricating self-assembling systems an important issue is to determine and control the ratio of components in the film. Many factors influence the membrane configuration including the composition of the coating solutions, the condition of the surface, and the conditions under which the deposition occurs. A useful measure available in the gA-based system is the ability to measure directly the ratio of gramicidin to nongramicidin species within the film, based on changes in the electrical conductivity, e.g., G_{m} can be titrated against the gA concentration in the depositing solution. The conduction is found to be dependent on the insertion of gA into the inner, outer, or both layers. The derived K_{2D} for the gramicidin monomer–dimer interaction was $7 \times 10^{9}\mathrm{cm}^{-2}$ if the inner layer gramicidins are taken as reference and $1 \times 10^{9}\mathrm{cm}^{-2}$ if the outer layer gramicidins are taken as reference (Cornell, 2002). This suggests that the tethered gA surface density is lower than nominal, although absolute estimates of the membrane composition require other techniques such as radiolabeling.

18.6.4 Characterization of the Reservoir

The conductance of the tethered BLM assembled from the compounds shown in Fig. 18.2 is insensitive to ion type, indicating that it is not limited by the conductivity of the gA channel. The reservoir properties, however, dominate the ion channel conduction. The linker composition of the reservoir significantly influences the magnitude and properties of the diffuse layer capacitance (Krishna et al., 2001) and the magnitude and voltage dependence of the apparent ion channel conductance. The ion channel conductance appears to be a factor of 8–10 greater for the all-ether reservoirs (compound **8** in Fig. 18.3). Spacer molecules improve the apparent channel conduction but also add complexity to the coating solutions and introduce variation in the composition of the tethered BLM.

The application of a 100–300 mV potential, negative on the gold electrode, improves the ability of the all-ether and succinate-linked reservoirs to store ions, and increases the apparent conductivity of the ion channels. The dependence on potential of the reservoir species with the C11 linkers is much reduced, in proportion to the lowered surface charge at the gold arising from the C11 coating. As described earlier, additional advantages that result from using the C11 family of compounds is the protection of the sulfur–gold surface against chemical degradation from the test solution and a greater stability of the attachment chemistry. The C11–C11 groups interact through van der Waals forces adding to the sulfur–gold interaction. As a result, the C11 family of compounds is the preferred attachment chemistry for long-term storage of the ICS^{TM} biosensor.

18.6.5 Nonspecific Binding

Ellipsometry was used to determine the level of nonspecific binding to the tethered BLM, using streptavidin–biotin as a model for binding on the surface. The effect of surface type on nonspecific binding of 42 nM streptavidin was examined (Cornell, 2002). A tethered bilayer of predominantly a 70:30 mix of compounds **5** and **6** in Fig. 18.2 resulted in binding below the level of detection. A substantial fraction of phosphatidylcholine groups within the tethered bilayer surface significantly reduces nonspecific binding of streptavidin. When biotinylated lipids are included in the BLM, the apparent thickness of the film progressively increases reflecting the binding of streptavidin to the membrane surface, i.e., specific binding increases linearly with the inclusion of biotinylated lipids. The low nonspecific binding characteristics of the phosphatidylcholine headgroup tethered BLM results in an ability to detect target analytes in whole blood without prepreparation of the sample and minimal matrix effects (Fig. 18.16).

18.6.6 BLM Stability

The stability of the tethered BLM depends on several factors including the lifetime of the sulfur–gold bond, the chemical stability of the various linkers, and the ability of the membrane to resist mechanical damage and loss of the mobile outer layer. The use of a C11 segment with a thiol or disulfide species (compound **7** in Fig. 18.3) provides substantial additional retention of the monolayer components. Assemblies of thiols employing the C11 sequence are able to withstand extensive storage in ethanol at 50°C with no loss of material to solution compared to films without C11, which lost a substantial amount (Cornell, 2002). Replacement of succinate by all ether linkers (compound **8** in Fig. 18.3) has a further substantial effect on the stability of the membrane. The sealed membranes fabricated from the ether linkers are stable beyond 3 months when fully hydrated.

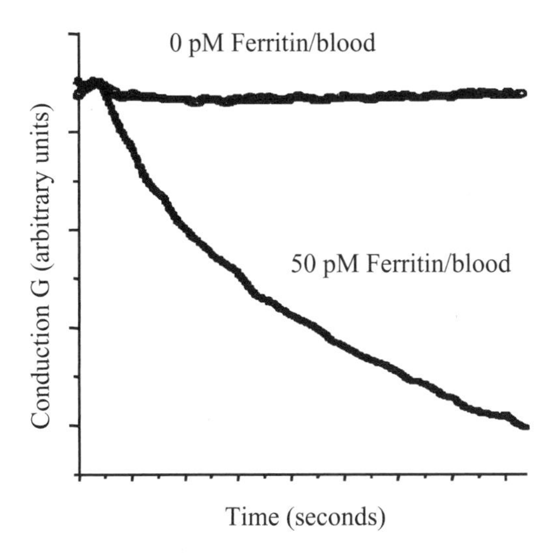

Fig. 18.16 Biosensor response to whole blood containing 50 pM ferritin compared to a control without ferritin.

A further strategy for stabilizing the biosensor is to return function on rehydrating a fully assembled, freeze-dried sensor. Following rehydration sensor function returns. This requires assembling a hydrated sensor up to the stage where analyte is to be added. The biosensor is rinsed with a storage solution and is then freeze dried. Freezing a fully hydrated sensor also appears to cause no loss of function upon being returned to operating temperature (DARPA report, 2001). Under these circumstances no cryoprotectants are needed and excellent recovery has been obtained with phosphate-buffered saline. These approaches can be used for the long-term storage of BLM devices.

18.7 Further Developments

Major opportunities in technology developments using BLM include miniaturization and patterning. Nanometer-sized gold electrodes have been made by using nm size latex spheres as a lift-off mask (Padeste et al., 1996). Cell adhesion and growth can be controlled with micro-patterned supported BLM by using phosphatidylserine to specify regions for growth of cells (Groves et al., 1997, 2000). Patterned SAM have been formed by patterning the topography of their metallic supports (Aizenberg et al., 1998), while microcontact printing of lipophilic SAM has been used for the attachment of biomimetic lipid layers to surfaces (Jenkins et al., 1999). Microstructures of solid-supported lipid layers have been made using SAM pattern by scanning electrochemical microscopy (Ufheil et al., 2000) and techniques for rapid prototyping and production of nanometer-sized arrays have been described (Fan et al., 2000).

Once a microarray has been produced, the arrays can be patterned with selected functionalities. The ICS™ biosensor is able to employ a technique in which the streptavidin–biotin linker is blocked unless irradiated by UV. The irradiation can be applied through a patterned mask to permit streptavidin attachment to selected sites so that UV irradiation is used to selectively deprotect biotin sites and permit their functionalization using a streptavidin–biotin attachment.

A range of functionalities has been incorporated within tethered BLM, including peptide nanotubes (Motesharei and Ghadiri, 1997), crown ethers (Terrettaz et al., 1998), antigenic peptides (Scheibler et al., 1999), and the Ompf porin channel (Stora et al., 1999). Cytochrome C has been incorporated into a tethered membrane as one of the first examples of a functionally active biomimetic surface (Naumann et al., 1999). A chip-based sensor using etched silicon has been fabricated to measure single state currents of alamethicin (Schmidt et al., 2000), while the pore-forming toxin α-hemolysin has been reconstituted in supported bilayers (Glazier et al., 2000). Although the detection of single ion channel currents has not yet been reported, with the reduction of electrode dimensions to the micrometer scale, however, single channel noise from gA can be detected in tethered BLM. The combination of miniaturization and selective patterning can lead to higher sensitivities and detection of multiple analytes.

Acknowledgments

This work was in part supported by the Cooperative Research Centre for Molecular Engineering. Ambri Ltd. has licensed certain fields of application of the Ion Channel Switch technology to Biosensors Enterprise Ltd. (BEL). BEL has undertaken further development of the technology with Ambri. BEL is a joint venture between Dow Corning and Genencor International.

References

Aizenberg, J., A.J. Black, and G.M. Whitesides. 1998. Controlling local disorder in self-assembled monolayers by patterning the topography of their metal supports. *Nature* 394:868–871.

Anastasiadis, A., and F. Separovic. 2003. Solid-state NMR structural determination of components in an ion channel switch biosensor. *Aust. J. Chem.* 56:163–166.

Anastasiadis, A., F. Separovic, and J. White. 2001. Synthesis of deuterated aminocaproyl linkers. *Aust. J. Chem.* 54:747–750.

Atanasov, V., N. Knorr, R.S. Duran, S. Ingebrandt, A. Offenhausser, W. Knoll, and I. Koper. 2005. Membrane on a chip: A functional tethered lipid bilayer membrane on silicon oxide surfaces. *Biophys. J.* 89:1780–1788.

Bain, C.D., J. Evall, and G.M. Whitesides. 1989. Formation of monolayers by the coadsorption of thiols on gold: Variation in the head group, tail group, and solvent. *J. Am. Chem. Soc.* 111:7155–7164.

Braach-Maksvytis, V., and B. Raguse. 2000. Highly impermeable 'soft' self-assembled monolayers. *J. Am. Chem. Soc.* 122:9544–9545.

Bucher, J.-P., L. Santesson, and K. Kern. 1994. Thermal healing of self-assembled organic monolayers: Hexane- and octadecanethiol on Au(111) and Ag(111). *Langmuir Lett.* 10:979–983.

Bufler, J., S. Kahlert, S. Tzartos, K.V. Toyka, A. Maelicke, and C. Franke. 1996. Activation and blockade of mouse muscle nicotinic channels by antibodies directed against the binding site of the acetylcholine receptor. *J. Physiol. (Lond.)* 492:107–114.

Burns, C.J., L.D. Field, K. Hashimoto, B.J. Petteys, D.D Ridley, and M. Rose. 1999a. Synthesis of steroisomerically pure mono-ether lipids. *Aust. J. Chem.* 52:387–396.

Burns, C.J., L.D. Field, K. Hashimoto, B.J. Petteys, D.D. Ridley, and S. Sandanayake. 1999b. A convenient synthetic route to differentially functionalized long chain polyethylene glycols. *Synthetic Comm.* 29:2337–2347.

Burns, C.J., L.D. Field, J. Morgan, D.D. Ridley, and V. Vignevich. 1999c. Preparation of cyclic disulfides from bisthiocyanates. *Tetrahedron Lett.* 40:6489–6492.

Cornell, B.A. 2002. Membrane-based biosensors. *In*: Optical Biosensors. F.S. Ligler and C.A. Rowe Taitt, editors. Elsevier Science B.V., Amsterdam, pp. 457–495.

Cornell, B.A., V.L.B. Braach-Maksvytis, L.G. King, P.D. Osman, B. Raguse, L. Wieczorek, and R.J. Pace. 1997. A biosensor that uses ion-channel switches. *Nature* 387:580–583.

Cornell, B.A., F. Separovic, R. Smith, and A.J. Baldassi. 1988. Conformation and orientation of gA in oriented phospholipid bilayers measured by solid state carbon13 NMR. *Biophys. J.* 53:67–76.

De Rosa, M., A. Gambacorta, B. Nicolaus, B. Chappeand, and P. Albrecht. 1983. Isoprenoidethers: Backbone of complex lipids of the archaebacterium *Sulfolobus solfataricus*. *Biochim. Biophys. Acta* 753:249–256.

DARPA Report number N65236-98-1-5412. 2001.

Dishner, M.H., F.J. Feher, and J.C. Hemminger. 1996. Formation and photooxidation of n-dodecanethiol self-assembled monolayers on Au(111): 'Pits' formed during chemisorption disappear upon oxidation *J. Chem. Soc. Chem. Comm.* 1971–1972.

Fan, H., Y. Lu, A. Stump, S.T. Reed, T. Baer, R. Schunk, V. Perez-Luna, G.P. López, and C.J Brinker. 2000. Rapid prototyping of patterned functional nanostructures. *Nature* 405:56–60.

Folker, J.P., P.E. Laibinis, G.M. Whitesides, and J. Deutch. 1994. Phase behavior of two-component self-assembled monolayers of alkanethiolates on gold. *J. Phys. Chem.* 98:563–571.

Glazier, S.A., D.J. Vanderah, A.L. Plant, H. Bayley, G. Valincius, and J.J. Kasianowicz. 2000. Reconstitution of the pore-forming toxin, α-hemolysin, in phospholipid/1-thiahexa(ethylene-oxide) alkane-supported bilayers. *Langmuir* 16:10428–10435.

Gliozzi, A., R. Rolandi, M. De Rosa, and A. Gambacort. 1983. Monolayer black membranes from bipolar lipids of archaebacteria and their temperature-induced structural changes. *J. Membr. Biol.* 75:45–56.

Groves, J.T., L.K. Mahal, and C.R Bertozzi. 2001. Control of cell adhesion and growth with micropatterned supported lipid membranes. *Langmuir* 17:5129–5133.

Groves, J.T., N. Ulman, and S.G. Boxer. 1997. Micropatterning fluid lipid bilayers on solid supports. *Science* 275:651–653.

Guo, L.H., J. Facci, R. Moser, and G. McLendon. 1994. Reactivity of alkanethiols toward gold films deposited on silica and mica.*Langmuir* 10:4588–4593.

Hardt, S.L. 1979. Rates of diffusion controlled reactions in one, two, and three dimensions. *Biophys. Chem.* 10:239–243.

Hoogvliet, J.C., M. Dijksma, B. Kamp, and W.P. van Bennekom. 2000. Electrochemical pretreatment of polycrystalline gold electrodes to produce a reproducible surface roughness for self-assembly: A study in phosphate buffer pH 7.4. *Anal. Chem.* 72:2016–2021.

Hsueh, C.C., M.T. Lee, M.S. Freund, and G.S. Ferguson. 2000. Electrochemically directed self-assembly on gold. *Angew. Chem. Int. Ed. Engl.* 39:1227–1230.

Jenkins, A.T.N., N. Boden, R.J. Bushby, S.D. Evans, P.F. Knowles, R.E. Miles, S.D. Ogier, H. Schoenherr, G.J. Vancso. 1999. Microcontact printing of lipophilic self-assembled monolayers for the attachment of biomimetic lipid bilayers to surfaces. *J. Amer. Chem. Soc.* 121:5274–5280.

Kang, J.F., A. Ulman, and R. Jordan. 1999. Mixed self-assembled monolayers of highly polar rigid biphenyl thiols. *Langmuir* 15:2095–2098.

Katz, E., and A.L. Demain. 1977. The peptide antibiotics of Bacillus: Chemistry, biogenesis and possible functions. *Bacteriol. Rev.* 41:449–474.

Koeppe, R.E., and O.S. Andersen. 1996. Engineering the gramicidin channel. *Ann. Rev. Biophys. Biomol. Struct.* 25:231–258.

Krishna, G., J. Schulte, B.A. Cornell, R. Pace, L. Wieczorek, and P.D. Osman. 2001. Tethered bilayer membranes containing ionic reservoirs: The interfacial capacitance. *Langmuir* 17:4858–4866.

Kushwaha, S.C., M. Kates, G.D. Sprott, and I.C. Smith. 1981. Novel complex polar lipids from the methanogenic archaebacterium *Methanospirillum hungatei. Science* 211:1163–1164.

Lee, M.T., C.C. Hsueh, M.S. Freund, and G.S. Ferguson. 1998. Air oxidation of self-assembled monolayers on polycrystalline gold: The role of the gold surface. *Langmuir* 14:6419–6423.

Ligler, F.S., G.P. Anderson, P.T. Davidson, R.J. Foch, J.T. Ives, K.D. King, G. Page, D.A. Stenger, and J.P. Whelan. 1998. Remote sensing using an airborne biosensor. *Environ. Sci. Technol.* 32:2461–2466.

Ligler, F.S., T.L. Fare, E.E. Seib, J.W. Smuda, A. Singh, P. Ahl, M.E. Ayers, A.W. Dalziel, and P. Yager. 1988. Fabrication of key components of a receptor-based biosensor. *Med. Instrum.* 22:247–256.

Lingler, S., I. Rubenstein, W. Knoll, and A. Offenhausser. 1997. Fusion of small unilamellar lipid vesicles to alkanethiol and thiolipid self-assembled monolayers on gold. *Langmuir* 13:7085–7091.

Lopatin, A.N., E.N. Makhina, and C.G Nichols. 1995. The mechanism of inward rectification of potassium channels: "long-pore plugging" by cytoplasmic polyamines. *J. Gen. Physiol.* 106:923–955.

Ma, F., and R.B. Lennox. 2000. Potential-assisted deposition of alkanethiols on Au: Controlled preparation of single and mixed-component SAMs. *Langmuir* 16:6188–6190.

Meuse, C.W., S. Krueger, C.F. Majkrzak, J.A. Dura, J. Fu, J.T. Connor, and A.L. Plant. 1998a. Hybrid bilayer membranes in air and water: Infrared spectroscopy and neutron reflectivity studies. *Biophys. J.* 74:1388–1398.

Meuse, C.W., G. Niaura, M.L. Lewis, and A.L. Plant. 1998b. Assessing the molecular structure of alkanethiol monolayers in hybrid bilayer membranes with vibrational spectroscopies. *Langmuir* 14:1604–1611.

Motesharei, K., and M.R. Ghadiri. 1997. Diffusion-limited size-selective ion sensing based on SAM-supported peptide nanotubes. *J. Am. Chem. Soc.* 119:11306–11312.

Myers, V.B., and D.A. Haydon. 1972. Ion transfer across lipid membranes in the presence of gA. II. The ion selectivity. *Biochim. Biophys. Acta* 274:313–322.

Naumann, R., E.K. Schmidt, A. Jonczyk, K. Fendler, B. Kadenbach, T. Liebermann, A. Offenhausser, and W. Knoll. 1999. The peptide-tethered lipid membrane as a biomimetic system to incorporate cytochrome c oxidase in a functionally active form. *Biosens. Bioelectron.* 14:651–662.

Padeste, C., S. Kossek, H.W. Lehmann, C.R. Musil, J. Gobrecht, and L.J. Tiefenauer. 1996. Fabrication and characterization of nanostructured gold electrodes for electrochemical biosensors. *J. Electrochem. Soc.* 143:3890–3895.

Philp, D., and J.F. Stoddart. 1996. Self-assembly in natural and unnatural systems. *Angew. Chem. Int. Ed. Engl.* 35:1154–1196.

Plant, A.L. 1993. Self-assembled phospholipid/alkanethiol biomimetic bilayers on gold. *Langmuir* 9:2764–2767.

Plant, A.L. 1999. Supported hybrid bilayer membranes as rugged cell membrane mimics. *Langmuir* 15:5128–5135.

Raguse, B., V.L.B. Braach-Maksvytis, B.A. Cornell, L.G. King, P.D. Osman, R.J. Pace, and L. Wieczorek. 1998. Tethered lipid bilayer membranes: Formation and ionic reservoir characterisation. *Langmuir* 14:648–659.

Raguse, B., P.N. Culshaw, J.K. Prashar, and K. Raval. 2000. The synthesis of archaebacterial lipid analogs. *Tetrahedron Lett.* 41:2971–2974.

Ron, H., S. Matlis, and I. Rubenstein. 1998. Self-assembled monolayers on oxidized metals. 2. Gold surface oxidative pretreatment, monolayer properties, and depression formation. *Langmuir* 14:1116–1121.

Ron, H., and I. Rubenstein. 1994. Alkanethiol monolayers on preoxidized gold. Encapsulation of gold oxide under an organic monolaye. *Langmuir* 10:4566–4573.

Ron, H., and I. Rubenstein. 1998. Self-assembled monolayers on oxidized metals. 3. Alkylthiol and dialkyl disulfide assembly on gold under electrochemical conditions. *J. Am. Chem. Soc.* 120:13444–13452.

Sackmann, E. 1996. Supported membranes. Scientific and practical applications. *Science* 271:43–48.

Scheibler, L., P. Dumy, M. Boncheva, K. Leufgen, H.J. Mathieu, M. Mutter, and H. Vogel. 1999. Functional molecular thin films: Topological templates for the chemoselective ligation of antigenic peptides to self-assembled monolayers. *Angew. Chem. Int. Ed. Engl.* 38:696–699.

Schmidt, C., M. Mayer, and H. Vogel. 2000. A chip-based biosensor for the functional analysis of single ion channels. *Angew. Chem. Int. Ed. Engl.* 39:3137–3140.

Separovic, F., S. Barker, M. Delahunty, and R. Smith, R. 1999. NMR structure of C-terminally tagged gramicidin channels. *Biochim. Biophys. Acta.* 1416: 48–56.

Steinem, C., A. Janshoff, W.P. Ulrich, M. Sieber, and H.J. Galla. 1996. Impedance analysis of supported lipid bilayer membranes: A scrutiny of different preparation techniques. *Biochim. Biophys. Acta* 1279:169–180.

Stelzle, M., G. Weissmuller, and E. Sackmann. 1993. On the application of supported bilayers as receptive layers for biosensors with electrical detection. *J. Phys. Chem.* 97:2974–2981.

Stetter, K.O. 1996. Hyperthermophilic prokaryotes. *FEMS Microbiol. Revs.* 18:149–158.

Stora, T., J. Lakey, and H. Vogel. 1999. Ion-channel gating in transmembrane receptor proteins: Functional activity in tethered lipid membranes. *Angew. Chem. Int. Ed. Engl.* 38:389–392.

Suarez, E., E.D. Emmanuelle, G. Molle, R. Lazaro, and P. Viallefont. 1998. Synthesis and characterization of a new biotinylated gramicidin. *J. Peptide Sci.* 4:371–377.

Terrettaz, S., H. Vogel, and M. Grätzel. 1998. Determination of the surface concentration of crown ethers in supported lipid membranes by capacitance measurements. *Langmuir* 14:2573–2576.

Toro-Goyco, E., A. Rodriguez, and J. del Castillo. 1966. Detection of antiinsulin antibodies with a new electrical technique: Lipid membrane conductometry. *Biochem. Biophys. Res. Comm.* 23:341–346.

Tsai, M.Y., and J.C. Lin. 2001. Preconditioning gold substrates influences organothiol self-assembled monolayer (SAM) formation. *J. Colloid Interface Sci.* 238:259–266.

Ufheil, J., F.M. Boldt, M. Börsch, K. Borgwarth, and J. Heinze. 2000. Microstructuring of solid-supported lipid layers using sam pattern generation by scanning electrochemical microscopy (SECM) and the chemical lens. *Bioelectrochemistry* 52:103–107.

Ulman, A. 1996. Formation and structure of self-assembled monolayers. *Chem. Rev.* 96:1533–1554.

Walczak, M.M., C.A. Alves, B.D. Lamp, and M.D. Porter. 1995. Electrochemical and X-ray photoelectron spectroscopic evidence for differences in the binding sites of alkanethiolate monolayers chemisorbed at gold. *J. Electroanal. Chem.* 396:103–114.

Wallace, B.A. 1998. Recent advances in the high resolution structures of bacterial channels: Gramicidin A. *J. Struct. Biol.* 121:123–141.

Woodhouse, G., L. King, L.Wieczorek, P. Osman, and B. Cornell. 1999. The ion channel switch biosensor. *J. Mol. Recogn.* 12:328–335.

Yager, P. 1988. Development of membrane-based biosensors: Measurement of current from photocycling bacteriorhodopsin on patch clamp electrodes. *Adv. Exp. Med. Biol.* 238:257–267.

Zaitsev, S.Yu., S.V. Dzekhitser, and V.P. Zubov. 1988. Polymer monolayers with immobilized bacteriorhodopsin. *Bioorg. Khim.* 14:850–852.

19 Signal Processing Based on Hidden Markov Models for Extracting Small Channel Currents

Vikram Krishnamurthy and Shin-Ho Chung

19.1 Introduction

The measurement of ionic currents flowing through single channels in cell membranes has been made possible by the giga-seal patch-clamp technique (Neher and Sakmann, 1976; Hamill et al., 1981). A tight seal between the rim of the electrode tip and the cell membrane drastically reduces the leakage current and extraneous background noise, enabling the resolution of the discrete changes in conductance that occur when single channels open or close. Although the noise from a small patch is much less than that from a whole-cell membrane, signals of interest are often obscured by the noise. Even if the signal frequently emerges from the noise, low-amplitude events such as small subconductance states can remain below the noise level and there may be little evidence of their presence. It is desirable, therefore, to have a method to measure and characterize not only relatively large ionic currents but also much smaller current fluctuations that are obscured by noise.

Extracting the real signal from a limited set of imperfect measurements is a problem that commonly occurs in scientific experiments and techniques have been developed to overcome this difficulty. Following digitization, a single-channel record consists of a sequence of data points. Each data point contains a mixture of the signal and extraneous noise. The challenge is to remove the noise leaving the biological signal untouched. Some of the methods that have been used to do this are linear and nonlinear filtering (Chung and Kennedy, 1991) and transition detectors (Patlak, 1988, 1993; Tyerman et al., 1992; Queyroy and Verdetti, 1992). Although both linear and nonlinear filtering suppress noise and nonlinear filtering produces little distortion of rapid transitions in underlying signals, neither method utilizes all of the knowledge available about the nature of the signal and interfering noise. Using such information improves the probability of recovering the underlying signal accurately. Broadly speaking, this is the strategy used in the hidden Markov models (HMM) processing technique. The HMM processing technique has been fruitfully utilized in electrical engineering in the disciplines of artificial speech recognition and target tracking in defense systems. The technique was then applied for the analysis of single-channel recordings and to extract small channel currents contaminated by random and deterministic noise (Chung et al., 1990, 1991; Krishnamurthy et al.,

1991, 1993; Venkataramanan et al., 1998a,b, 2000). With this signal processing method, the underlying parameters of the HMM could be obtained to a remarkable precision despite the extremely poor signal to noise ratio.

The aim of this chapter is to review the construction and use of HMM for estimating the dynamics of ion channel gating. We first provide a brief intuitive explanation and then a rigorous account of the underlying principles of the processing method. We also outline state-of-the-art results in HMM that are the subject of recent research in mathematical statistics and signal processing in electrical engineering. Some of these techniques are relatively new and not yet known in the biophysics community. These include ideas such as estimating the model order of a HMM, jump Markov linear systems (which is a generalization of HMM to deal with digitally filtered Markov chains and correlated noise), and recursive (online) HMM parameter estimation. We refer the reader to Ephraim and Merhav (2002) for a comprehensive review of HMM with a stronger mathematical flavor compared to this chapter.

An HMM is an example of a partially observed stochastic dynamical system. Because opening and closing of an ion channel is random, recordings of single-channel current may be modeled probabilistically as a finite-state, random realization of a Markov chain. Since the underlying ion channel current is corrupted by large amounts of thermal, capacitance and other deterministic noise, the underlying state of the dynamical system is only partially observed. HMM and their generalizations are extremely versatile in capturing the response of complex dynamical systems such as ion channels.

19.2 General Description of the HMM Method

19.2.1 Signal Model and Assumptions

To apply a digital signal processing technique based on HMM to records of single-channel currents contaminated by noise, we first make a plausible guess about the origin of the observation sequence and then construct a signal model. It is assumed that the pure single-channel signal, not contaminated by noise, can be represented as a Markov process with the following characteristics.

In this chapter we deal exclusively with *discrete time* HMM. By discrete time, we mean that the noisy ion channel current is observed at discrete time instants $k = 0, 1, 2, \ldots$ after suitable anti-aliasing filtering and sampling. The advantage of using discrete time HMM is that powerful algorithms can be derived with a fairly elementary background in probability (Papoulis and Pillai, 2002) involving manipulation of conditional probability density functions and Bayes' rule. An analogous theory can be developed for continuous time HMM although the mathematical tools are more difficult since they require the use of stochastic differential equations for a unifying treatment of both discrete and continuous time HMM (see, James et al., 1996).

For each discrete time k, the signal s_k is assumed to be at one of the finite number of states, $q_1, q_2, \ldots, q_N$. Each q_i, where $i = 1, 2, \cdots, N$, is called a state of the process and such a process is called an N-state Markov chain. In the context of channel currents, the Markov state, s_k, represents the true conductance level (or current amplitude) uncontaminated by noise at time k. The observed value at time k, y_k, contains the signal, s_k, random noise w_k, and possibly deterministic interferences d_k, such as sinusoidal interferences from electricity mains and baseline drift. We note here that the meaning of the term *state* differs from that adopted in the Colquhoun–Hawkes model of channel dynamics (Colquhoun and Hawkes, 1981), in which *state* refers to a hypothetical, not directly observable, conformation of the channel macromolecule.

We also assume that the probability of the current being at a particular level (state) at time $k+1$ depends solely on the state at time k and that the transition probabilities are invariant of time k. In other words, the process is construed as a first-order, homogenous Markov chain. The transition probabilities of passing from state level q_i at time k to state level q_j at time $k+1$ are expressed as

$$a_{ij} = P(s_{k+1} = q_j | s_k = q_i) \tag{19.1}$$

and form a state transition probability matrix $\mathbf{A} = \{a_{ij}\}$. Thus, $\mathbf{A}$ is an $N \times N$ stochastic matrix, with its diagonal elements denoting the probabilities of remaining in the same state at time $k+1$ as at time k.

Finally, we define the noise, know also as *the probabilistic function* of the Markov chain or the *symbol probability* as $\mathbf{B} = b_i(y_k)$. It is convenient to assume that the noise is Gaussian. In reality, noise superimposed on single-channel currents is not white but tends to be colored. Its spectral power, instead of being flat, rises steeply at high frequencies. This issue was addressed by use of an autoregressive noise model to represent temporal correlation in the background noise contained in patch-clamp recordings and then the algorithm for handling such correlated and state-dependent excess noise was formulated (Venkataramanan et al., 1998a,b; Venkataramanan and Sigworth, 2002). It was demonstrated that the performance of the algorithm was markedly improved when the background noise was modeled realistically.

19.2.2 Example of a Signal Model

Suppose we know that a record contains four current levels but we are not sure of the exact levels nor the exact signal sequence. We can set up an initial model with the following assumed characteristics. First, we make a reasonable guess, and say that the baseline level is 0 pA and that there are three open states at -1, -2, and -3 pA. Second, we assume that the noise is zero-mean Gaussian, with a standard deviation of, say, 0.25 pA. Third, we provide our initial guesses of transition probabilities from one state level at time k to another state level at time $k+1$. For a four-state Markov chain, these transition probabilities form a 4×4 transition probability matrix. In the example given here, the first entry of the first row of the matrix represents the

Table 19.1 An example of the signal model.

State levels	$q_1 = 0$ pA $q_2 = -1$ pA $q_3 = -2$ pA $q_4 = -3$ pA
Noise characteristics	Gaussian with $\sigma = 0.25$ pA
Transition matrix	a_{11} a_{12} a_{13} a_{14} a_{21} a_{22} a_{23} a_{24} a_{31} a_{32} a_{33} a_{34} a_{41} a_{42} a_{43} a_{44}
Initial probability	$\pi_1 = \pi_2 = \pi_3 = \pi_4 = 0.25$

probability that the process remained in the closed state at time $k+1$ given that it was closed at time k, whereas the second entry of the first row represents the probability of transiting to the first open level at time $k+1$ given that it was closed at time k. Similarly, the last entry of the last row represents the probability that the process remained in the fourth level, or the -3 pA level, at time $k+1$ given that it was at this level at time k. Finally, we stipulate that the probability of the signal being at each one of the four levels at time $k=1$ is 0.25.

These assumptions can be represented as shown in Table 19.1.

We can also stipulate, if needed, that there is AC hum (50 or 60 Hz and its odd harmonics) embedded in the data but, for simplicity, this is not included in this example. We compactly write all these initial guesses as:

$$\lambda^{(0)} = (\mathbf{q}, \mathbf{A}, \mathbf{B}, \pi), \tag{19.2}$$

our first signal model.

19.2.3 Iterative Algorithm for Estimating HMM Parameters

Here we explain in simple terms how the expectation maximization (EM) algorithm can be used to estimate the HMM parameters. A more rigorous formulation together with other numerical algorithms is given in Section 19.5. The signal model is compared with the data. Essentially, the probabilities of all possible pathways between adjacent data points are calculated, both forward and backward, and the true current levels derived from the highest probabilities. Because the initial parameters we have supplied (e.g., the transition probability matrix and conductance levels) are only guesses, there is going to be a mismatch between the model and data. The model is revised so that it will be more consistent with the data. Using the revised model, the observation sequence is compared again with the new model, and the model is again revised. This iterative process continues, as shown in Fig. 19.1.

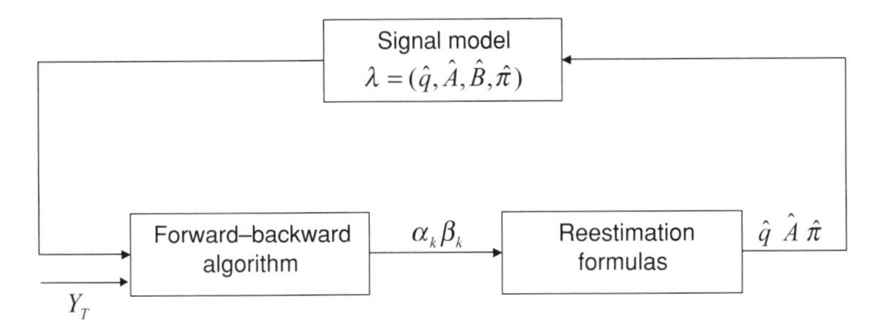

Fig. 19.1 A block diagram of processing method. On the basis of the initial signal model λ, the observation sequence Y_T is processed, and the forward and backward variables, α_k and β_k are computed for each discrete time k and each Markov state q_i. By using these variables, the parameters of the signal model are revised according to the reestimation formulas. The entire process is repeated many times.

In this iterative process, the HMM processing technique utilizes two mathematical principles. The first is the forward–backward procedure, known also as the *E Step* of the EM algorithm, to be discussed in detail in Section 19.5.2 For each data point k, the algorithm evaluates, using Bayes' rule, the forward and backward variables, α and β. In words, the forward variable $\alpha_k(i)$ is the joint probability of the past and present observations with the present signal in state q_i, given the model λ, and $\beta_k(i)$ is the probability of the future observation given that the present state is q_i and given the model λ. The forward variable is calculated in a forward recursion and the backward variable in a backward recursion.

Then, using the forward and backward variables, the initial model $\lambda^{(0)}$ is reestimated, using the Baum–Welsh reestimation formulas, the *M Step* of the EM algorithm. Estimation formulas stipulate how the model parameters should be revised, given the forward and backward variables. Loosely stated, the E step of the EM algorithm maximizes the expectation of the likelihood function if the parameters of the initial model $\lambda^{(0)}$ are replaced by a new model $\lambda^{(1)}$. The same segment of data is now processed with the second signal model, and then the parameters of this model are replaced, again according to the reestimation formulas, in the third signal model $\lambda^{(2)}$. This process is iterated again and again until the difference in the estimates $\lambda^{(n)}$ and $\lambda^{(n+1)}$ in two successive iterations is sufficiently small.

After each iteration, we can compute from the forward variables a numerical value that we call the *likelihood function*—that is, how likely are the model parameters given the data sequence. The closer the model parameters are to the true parameters, the higher the likelihood function. If the likelihood functions were to increase or decrease erratically with successive iterations, this procedure would have been a waste of time. The rationale behind the iterative procedure rests on the elegant reestimation theorem formulated by Baum and colleagues (Baum and Petrie, 1966; Baum et al., 1970; Baum, 1972), which states:

$$P(Y_T|\lambda^{n+1} \geq P(Y_T|\lambda^n). \tag{19.3}$$

In words, the probability of the observation sequence Y_T, given the reestimated signal model, is greater than or equal to the probability of Y_T, given the previous signal model. Thus, the signal sequence estimated using the revised model is more consistent with the data than that estimated using the previous signal model. When the iterative procedure converges, then $P(Y_T|\lambda^{n+1}) = P(Y_T|\lambda^n)$, and λ^n is termed the *maximum likelihood estimate* of the HMM. This important theorem, the proof of which is based on Jensen's inequality (Baum and Petrie, 1966), is the core of the HMM processing scheme.

There is a choice of numerical methods of calculating the maximum likelihood estimates, as discussed in Section 19.5.1. One approach is the Newton–Ralphson (NR) algorithm, which, when it converges, does so quadratically and thus rapidly. The EM algorithm, on the other hand, converges linearly, and so convergence can be slow. However, successive iterations with the NR algorithm do not necessarily improve the likelihood function. In contrast, the EM algorithm is simple to implement and has the appealing property that the likelihood function is always improved after each iteration.

19.2.4 Estimating the Number of States

Perhaps the most subjective part of the HMM processing method, like any other parameter estimation scheme, is finding the state dimension—or the number of conductance states in our example—in hidden Markov chain processes. The error in fitting a model to a given set of data decreases with the number of free parameters in the model. Thus, it makes sense, in selecting a model from a set of models with different numbers of parameters, to penalize models having too many parameters. The question of how to penalize HMM for having an excessive number of free parameters is an area of current research (see, for example, Poskitt and Chung, 1996) and one proposed criterion for model-order selection is the compensated likelihood approach (Finesso, 1990; Liu and Narayan, 1994; Rydén, 1995). See Section 19.6.2 for further discussion of this issue.

In practice, however, it is relatively easy to identify the number of states present in the underlying Markov chains. One of the several ways of doing this is by constructing the most likely amplitude histogram. Here we assume that the signal can be represented as a Markov chain with a large number of equally spaced states, say 100 states, and then estimate the most likely signal sequence. We construct an amplitude histogram from the estimated signal sequence under this assumption. After a number of iterations, the maximum likelihood histogram clearly shows prominent peaks, even when no meaningful information can be gleaned from the amplitude histogram of the original record.

Alternatively, we can appeal to the principle of parsimony in deciding the number of conductance states. We measure the goodness of fit by evaluating the likelihood of the model, and weigh this against what is to be gained by increasing the number of parameters, which generally increases the likelihood. Thus, we process the same data segment under the assumption that the underlying signal has a different number of conductance states. If a plot of log likelihood against model order (the number of states) shows a "knee" for a certain model order, we would prefer this model to one of higher order. This approach has been used to determine the number of conductance substates in channel currents activated γ-aminobutyric acd (Gage and Chung, 1994).

19.3 HMM Formulation and Estimation Problems

What follows is a rigorous formulation of the HMM processing techniques. We begin by formalizing the definition of a HMM process.

19.3.1 Definitions

A discrete time HMM process is a stochastic process comprising two ingredients:

1. A *stochastic dynamical system* modeled as an S state discrete time Markov chain s with state space

$$\mathcal{S} = \{1, 2, \ldots, S\}. \tag{19.4}$$

   This Markov chain evolves probabilistically according to the $S \times S$ *transition probability matrix* A. The elements of A are the transition probabilities

$$a_{ij} = \mathbf{P}(s_{k+1} = j | s_k = i), \quad 0 \leq a_{ij} \leq 1, \sum_{j=1}^{S} a_{ij} = 1, \quad i, j \in \{1, \ldots, s\}. \tag{19.5}$$

   The Markov chain s is initialized at time $k = 0$ with

$$\pi_0 = (\pi_0(i),\ i \in \mathcal{S}) = \mathbf{P}(s_0 = i). \tag{19.6}$$

   The Markov chain s models the actual pure, unobserved ion channel current.
2. *Partially observed state*: In an HMM, the Markov chain state (i.e., channel current not contaminated by noise) s is not directly observed. Instead, the observation process y is a noisy corrupted version of s. The observation y is modeled as a random process generated from the conditional probability density (or the probability mass function if y_k is discrete valued)

$$b_i(y_k) = \mathbf{p}(y_k | s_k = i). \tag{19.7}$$

   This conditional probability density is called the *observation likelihood function* in the statistical inference literature. Throughout this chapter we assume that the observation likelihood function $\mathbf{p}$ is parameterized. More precisely, θ denotes the sufficient statistic for the probability density $\mathbf{p}$ by some finite vector θ. For example, as described below, if the observation likelihood b is Gaussian, then θ

comprises the mean and variance since the mean and variance completely specify a Gaussian probability density function.

The above HMM is thus completely modeled by initial probability distribution π_0, the transition probability matrix A, and observation likelihoods b (or equivalently θ). Since we are primarily interested in the evolution and estimation of the HMM over long time scales, the initial distribution π_0 is unimportant. Indeed it can be shown that most HMM forget their initial condition geometrically fast. Most HMM estimation algorithms also forget their initial condition geometrically fast—this is a consequence of "geometric ergodicity" and requires that the transition probability be aperiodic and irreducible (LeGland and Mevel, 2000). To summarize, an HMM is completely parameterized by the model parameter

$$\lambda = (A, \theta). \tag{19.8}$$

In this chapter we are interested in estimating λ given a sequence of N observations of the HMM, where $N > 0$ is a large positive integer denoting the data size (typically several thousand or larger). Denote this N-length HMM observation sequence as

$$Y_N = (y_1, y_2, \ldots, y_N). \tag{19.9}$$

19.3.2 Modeling Ion Channel Current as an HMM

Here we illustrate how to model the noisily observed ion channel current from a patch clamp experiment as an HMM.

A typical trace of the ion channel current measurement from a patch-clamp experiment (after suitable anti-aliasing filtering and sampling) shows that the channel current is a piecewise constant discrete time signal that randomly jumps between two values—zero amperes, which denotes the *closed state* of the channel, and q amperes which denotes the *open* state. Figure 19.2 shows a computer-generated example of a patch-clamp record. To the pure channel current (Fig. 19.2A), noise from various sources is added to mimic the observation sequence, shown in Fig. 19.2B. The *open-state* current level is denoted as q. Sometimes the current recorded from single ion channel dwells on one or more intermediate levels, known as conductance substates. The *pure* ion channel current, uncontaminated by noise, is modeled as the Markov chain s with state space $\mathcal{S} = \{1, 2, 3\}$. These states correspond to the physical ion channel current of

$$\mathbf{q} = (q(1), q(2), q(3))' = \{\mathrm{C}, \mathrm{O}_1, \mathrm{O}_2\} \tag{19.10}$$

corresponding to the physical states of *closed state*, *partially-open state* and *fully-open state*. Subsequently, we will refer to $\mathbf{q}$ as the physical state levels of the Markov chain. When the channel is in the closed state, no currents flow across it. In the open state, the ion channel current has a value of q pA. Figure 19.2 also shows a computer simulated clean ion channel current s.

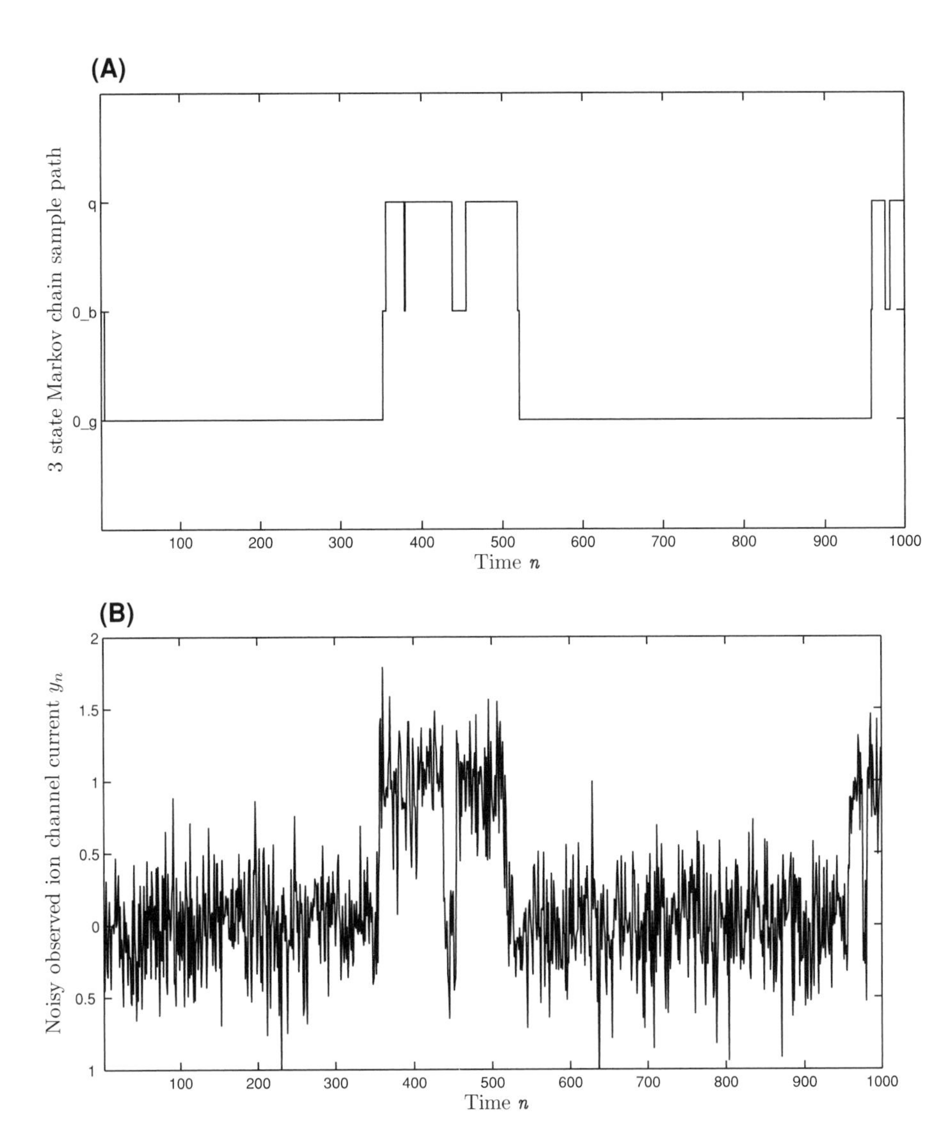

Fig. 19.2 A computer-generated patch-clamp record. To a three-state Markov chain (A), a Gaussian noise was added to mimic a channel current contaminated by amplifier and other noises (B).

The (3×3) transition probability matrix A of the Markov chain s, which governs the probabilistic behavior of the channel current, is given by

$$A = \begin{array}{c|c|c|c|} & \text{C} & \text{O}_1 & \text{O}_2 \\ \hline \text{C} & a_{11} & a_{12} & 0 \\ \hline \text{O}_1 & a_{21} & a_{22} & a_{23} \\ \hline \text{O}_2 & 0 & a_{32} & a_{33} \\ \hline \end{array} \tag{19.11}$$

The elements of A are the transition probabilities $a_{ij} = \mathbf{P}(s_{n+1} = j | s_n = i)$ where $i, j \in \mathcal{S}$ or equivalently the physical state of the ion channel in $\{C, O_1, O_2\}$. The zero probabilities in the matrix A, given as an example, state that an ion channel current cannot directly jump from the close state to the fully-open state, or vice versa.

The *observed* noisy ion channel current y from a patch-clamp experiment can be modeled as the Markov chain s corrupted by additive thermal noise with noise variance depending on the state of the ion channel:

$$y_k = q(s_k) + w_k, \quad k = 0, 1, \ldots. \tag{19.12}$$

Here $q(s_k) \in \{C, O_1, O_2\}$, given in (Eq. 19.10), are the physical state levels of the ion channel current, and w_k is a zero mean independent and identically distributed (*iid*) Gaussian scalar noise process with variance σ_w^2. That is, the probability density function of w_k is

$$\mathbf{p}_W(w) = \frac{1}{\sqrt{2\pi}\sigma_w} \exp\left(-\frac{1}{2}\frac{w^2}{\sigma_w^2}\right).$$

In terms of the HMM observation likelihood (Eq. 19.7),

$$b_i(y_k) = \frac{1}{\sqrt{2\pi}\sigma_w} \exp\left(-\frac{1}{2}\frac{(y_k - q(i))^2}{\sigma_w^2}\right). \tag{19.13}$$

In summary, the noisily observed ion channel current y is modeled as a hidden Markov model sequence parameterized by the model

$$\lambda = \{A, q, \sigma_w^2\}. \tag{19.14}$$

Our aim is to devise algorithms for estimating λ given an N length sequence of noisy ion channel observations Y_N [defined in (Eq. 19.9)].

Remarks. It is possible to extend the above HMM in several ways.

1. *Higher-order Markov chains*: The model assumption (Eq. 19.5) that s_k depends probabilistically only on its state at the previous time instant, i.e., s_{k-1} means that s is a first-order Markov chain. This assumption can be straightforwardly generalized to higher-order Markov chains with s_k depending probabilistically on the previous Δ time points $s_{k-1}, s_{k-2}, \ldots, s_{k-\Delta}$ where $\Delta \geq 1$. Then define a new first order Markov chain $\bar{s}_k = (s_k, s_{k-1}, \ldots, s_{k-\Delta+1})$ on the state space $\mathcal{S} \times \mathcal{S} \times \mathcal{S}$. The HMM processing algorithms presented in this chapter straightforwardly apply to this Markov chain.
2. *Correlated observations*: A critical assumption in constructing an HMM is that the observation process y is "conditionally independent." This means that the

conditional probability density

$$\mathbf{p}(y_k|s_k, \text{past values of } s, y) = \mathbf{p}(y_k|s_k), \tag{19.15}$$

i.e., given s_k, the HMM y_k is independent of the past. This conditional independence holds for the HMM ion channel observations (Eq. 19.12) providing that the corrupting noise w_k is an *iid* process. It is possible to generalize this conditional independence to allow y_k to depend on Δ past values of the observation and state. Such a generalization facilitates dealing with correlated noise. In such a case the observation likelihood of (Eq. 19.7) generalizes to

$$\mathbf{p}(y_k|s_{k-1}, s_{k-2}, \ldots, s_{k-\Delta}, y_{k-1}, y_{k-2}, \ldots, y_{k-\Delta}).$$

Such generalized HMM are widely used in, for example, econometric modeling and fault tolerant systems and are called Markov modulated autoregressive processes or jump Markov autoregressive processes (Krishnamurthy and Rydén, 1998).

3. *State-dependent noise variance*: A generalization of the observation equation (Eq. 19.12) is to model the thermal noise as $w_k(s_k)$, where the noise variance σ_w^2 is a function of the state s_k of the Markov chain. This generalization can easily be incorporated in the HMM.
4. *Additional deterministic interference*: Often the recorded ion channel currents have a deterministic drift and sinusoidal alternating current hum, which corrupts the Markov chain, in addition to the thermal noise. The observation equation (Eq. 19.12) can be formulated as

$$y_k = q(s_k) + w_k + d_k(\theta), \tag{19.16}$$

where $d_k(\theta)$ is the deterministic interference parameterized by some parameter vector θ. For example, in the case of deterministic sinusoidal interference, θ would include the amplitudes, phases and frequencies of the odd harmonics comprising the sinusoidal signal. The observation likelihood (Eq. 19.13) now becomes

$$b_i(y_k) = \frac{1}{\sqrt{2\pi}\sigma_w} \exp\left(-\frac{1}{2}\frac{(y_k - q(i) - d_k(\theta))^2}{\sigma_w^2}\right). \tag{19.17}$$

The HMM parameter is then $\lambda = \{A, q, \theta, \sigma_w^2\}$.
5. *Continuous state space*: The implicit assumption in the above HMM is that the underlying ion channel current s is a finite state process that randomly jumps between a finite number of values according to a Markov chain. It is worthwhile mentioning that there is an equivalently well defined theory for continuous valued states s. For example, if s is represented as a Gaussian continuous-state Markov process, then the Kalman filter and associated parameter estimation algorithms can be used.

19.3.3 Estimation Problems for HMM

Given an N point noisily observed ion channel current sequence Y_N defined in (19.9), there are two HMM estimation problems that are of interest:

Problem 1. *Bayesian state estimation problem.* Compute the optimal state estimate s_k at each time $k = 1, 2, \ldots, T$. The term Bayesian reflects the fact that the optimal estimator (defined below) is based on the a posteriori density function of the state and this a posteriori density function is computed via Bayes' rule.

Problem 2. *Maximum likelihood parameter estimation problem.* Compute the model parameter λ that best fits the HMM data Y_N with respect to the maximum likelihood criterion.

In the application to patch-clamp recordings, we are primarily interested in Problem 2 since our ultimate goal is to estimate the model λ that best fits this data. In particular, the transition probability matrix A and state level q yield important information about the kinetics of the ion channel gating. However, Problems 1 and 2 are intimately linked in HMM. Solving Problem 2 involves solving Problem 1 as an intermediate step.

In solving Problem 2, we are looking for the best model λ within the class of models Λ where

$$\Lambda = \{A\},\ q(i) \in [-M, M],\ \sigma_w^2 \in [\sigma_{\min}^2, \sigma_{\max}^2],$$

where A is the transition probability matrices satisfying Eq. 19.5 and M, $\sigma_{\min}^2$ and $\sigma_{\max}^2$ are finite positive constants. Mathematically speaking, for the maximum of a continuous function to exist, the function needs to be defined over a compact set. The above restriction merely restricts the likelihood function to a compact set. Naturally, there are several cost functions that can be used to define the "best" model. The most widely used criterion is the model log likelihood criterion. The log likelihood is more convenient to work with compared to the likelihood. Naturally, since log is a monotone function, maximizing the likelihood is equivalent to maximizing the log likelihood. The log likelihood of a model λ given Y_N is

$$L_N(\lambda) = \log \mathbf{p}(Y_N|\lambda). \tag{19.18}$$

The maximum likelihood estimate (MLE) is defined as the model λ^* that maximizes $L_N(\lambda)$, i.e.,

$$\lambda^* = \arg\max_{\lambda \in \Lambda} L_N(\lambda). \tag{19.19}$$

The log likelihood is the most widely used criterion for estimating HMM because under quite general conditions it has two asymptotic properties that are attractive from a statistician's point of view. First, the MLE is asymptotically consistent, i.e., as $N \to \infty$, the MLE λ_N^* converges to the true model with probability one (Leroux, 1992). Second, the asymptotic error (i.e., as $N \to \infty$) between the true model and the estimate λ_N^* has a Gaussian distribution—this property is called asymptotic normality (Bickel et al., 1998). This further implies that the MLE of an HMM is an asymptotically efficient estimator.

The formulation of the MLE problem is essentially an off-line estimation problem. We collect a batch of observations Y_N and then aim to compute the MLE λ_N^*.

Given that the MLE is a useful parameter estimate, how does one compute the MLE λ_N^* given a block of data Y_N of a HMM? For HMM it is not possible to explicitly solve the maximization problem (Eq. 19.18) and one must resort to a numerical optimization algorithm to compute the MLE. There are two widely used classes of numerical optimization algorithms for computing the MLE, namely the EM algorithm and the NR algorithm. An essential requirement for carrying out any numerical optimization algorithm for optimizing a function (log likelihood in our case) is to be able to evaluate the function at any value. That is, we first need to figure out a way of evaluating the log likelihood $L_N(\lambda)$ for any valid model $\lambda \in \Lambda$. It turns out that evaluating $L_N(\lambda)$ involves solving the state estimation Problem 1.

19.4 Problem 1: Bayesian State Estimation of HMM

At any time $k = 1, 2, \ldots, N$, define the *observation history* of the HMM as

$$Y_k = (y_1, \ldots, y_k). \tag{19.20}$$

The aim is to compute an estimate of the Markov chain s_l at any time $l = 1, 2, \ldots$ given the observation sequence Y_k. More precisely, the aim is to construct a state estimator (function) $\sigma_l(Y_k)$ where $\sigma_l \in \Sigma$ denotes the estimation algorithm and Σ denotes the space of all possible estimation algorithms. By an *optimal state estimator* or *Bayesian state estimator* for an HMM, we shall mean an estimator $\sigma_l^* \in \Sigma$ that minimizes the mean square state estimation error, i.e.,

$$\mathbf{E}\{s_l - \sigma_l^*(Y_k)\}^2 \leq \mathbf{E}\{s_l - \sigma_l(Y_k)\}^2, \quad \sigma_l \in \Sigma.$$

Here $\mathbf{E}\{\cdot\}$ denotes mathematical expectation. Since the metric $\mathbf{E}\{s_l - \sigma_l(Y_k)\}^2$ is simply the variance of the state estimation error, the optimal state estimator is also called the *minimum variance state estimator* or *minimum mean square error (MMSE) state estimator.*

We denote the optimal filtered state estimate as

$$\hat{s}_{l|k} = \sigma_l^*(Y_k). \tag{19.21}$$

The subscript $l|k$ is a reminder that the estimate at time l involves observations up to time k. At first sight it may appear that computing the optimal σ^* to minimize

$\mathbf{E}\{s_k - \sigma(Y_k)\}^2$ is a formidable task. However, by the mean square optimality property of conditional expectations (Jazwinski, 1970) it turns out that the optimal state estimate is

$$\hat{s}_{l|k} = \sigma_l^*(Y_k) = \mathbf{E}\{s_l | Y_k\}. \tag{19.22}$$

In words: *the optimal estimate $\hat{s}_l$ of the state s_l of the HMM at any time l, given the observation history Y_k (Eq. 19.20), is the conditional mean (conditional expectation) of the state s_l given Y_k*. For such a simple result, Eq. 19.22 is quite profound. All of recursive Bayesian estimation, optimal filtering theory merely deals with computing this conditional mean recursively. Indeed the Kalman filter, HMM filter, and particle filter are simply numerical algorithms for computing this conditional mean for different types of partially observed stochastic dynamical systems. The term "Bayesian" reflects the fact that in recursively computing $\hat{s}_{l|k} = \mathbf{E}\{s_l | Y_k\}$, Bayes' rule is used.

Depending on the choice of k and l, there are three types of optimal Bayesian state estimators:

- *Filtering*: If $k = l$, then the estimate $\hat{s}_{k|k}$ is the Bayesian estimate of the state at time k given observations up to time k. Such an estimate is called the filtered state estimate.
- *Prediction*: If $k < l$, then $\hat{s}_{l|k}$ is the Bayesian state estimate at some future time l, given observations up to time k. Such an estimate is called a predicted state estimate.
- *Smoothing*: If $k > l$, then $\hat{s}_{l|k}$ is the Bayesian state estimate of the problem and involves computing the state estimate at some past time l, given the past, present, and future observations up to time k.

In fact to solve Problem 2 (HMM parameter estimation problem), we will require solving the smoothing problem for estimating s_l, $l = 1, 2, \ldots, N$, given the observation sequence Y_N. However, as we show below, the smoothing problem is easily solved once we can solve the filtering problem.

19.4.1 HMM Filtering

The aim here is to derive a real time algorithm for estimating the filtered state estimate $s_{k|k} = \mathbf{E}\{s_K | Y_k\}$. The resulting HMM filter evolves recursively over time k. For notational convenience, we denote $\hat{s}_{k|k} = \hat{s}_k$ for the filtering problem.

Computing the filtered estimate $\hat{s}_k$ can be naturally broken into two steps:

- *Step 1*: Recursively compute the joint probability density $\alpha_k(i)$ defined as

$$\alpha_k(i) = \mathbf{p}(s_k = i, Y_k), \quad k = 1, 2, \ldots, N \tag{19.23}$$

for the HMM Eqs. 19.5 and 19.7. This can be implemented recursively according to the following algorithm:

$$\alpha_{k+1}(j) = b_j(y_{k+1}) \sum_{i=1}^{S} a_{ij}\alpha_k(i), \tag{19.24}$$

initialized by $\alpha_0(i) = \pi_0(i)$, $i = 1, 2, \ldots, S$. The derivation of Eq. 19.24 uses elementary algebra of marginal, conditional probabilities and Bayes' rule as follows:

$$\begin{aligned}\alpha_{k+1}(j) &= \mathbf{p}(s_{k+1} = j, Y_{k+1}) = \mathbf{p}(s_{k+1} = j, y_{k+1}, Y_k)\\ &= \mathbf{p}(y_{k+1}|s_{k+1} = j, Y_k)\mathbf{p}(s_{k+1} = j, Y_k)\\ &= \mathbf{p}(y_{k+1}|s_{k+1} = j, Y_k)\sum_{i=1}^{S}\mathbf{P}(s_{k+1} = j|s_k = i, Y_k)\mathbf{p}(s_k = i, Y_k)\end{aligned} \tag{19.25}$$

Then, using the conditional independence of the observations Eq. 19.15, the Markovian property Eq. 19.5 and noting that $\alpha_k(i) = \mathbf{p}(s_k = i, Y_k)$, directly yields Eq. 19.24.
Recall that $b_j(y_{k+1})$ are the observation likelihoods defined in Eq. 19.7 and are explicitly evaluated in the Gaussian noise case as Eq. 19.13 and deterministic interference case as Eq. 19.17. The above recursion is popularly termed the "Forward algorithm" or the "hidden Markov model state filter."

- *Step 2*: Compute the conditional mean estimate from $\alpha_k(i)$ by summation over all the Markov chain states:

$$\hat{s}_{k+1|k+1} = \mathbf{E}\{s_{k+1}|Y_{k+1}\} = \sum_{i=1}^{S} q(i)\mathbf{p}(s_{k+1}|Y_{k+1}) = \frac{\sum_{i=1}^{S} q(i)\alpha_{k+1}(j)}{\sum_{j=1}^{S}\alpha_{k+1}(j)}. \tag{19.26}$$

It is convenient to express the HMM filter Eq. 19.24 in matrix vector notation. Let $B(y_k)$ denote the $S \times S$ diagonal matrix with (i, i) elements $\mathbf{p}(y_k|s_k = i)$, $i = 1, \ldots, S$. Then the above HMM filter can be conveniently expressed as

$$\alpha_{k+1} = B(y_{k+1})A'\alpha_k, \quad \alpha_0 = \pi_0. \tag{19.27}$$

Here at each time k $\alpha_k = (\alpha_k(1), \ldots, \alpha_k(S))'$ is an S-dimensional column vector with nonnegative elements. Also, applying Step 2 in Eq. 19.26 yields the conditional mean estimate of the HMM filter as

$$\hat{s}_{k+1} = \mathbf{E}\{s_{k+1}|Y_{k+1}\} = \frac{\sum_{i=1}^{S} q(i)\alpha_{k+1}(i)}{\sum_{i=1}^{S} \alpha_{k+1}(i)} = \frac{\alpha_{k+1}}{\mathbf{1}'\alpha_{k+1}} = \frac{B(y_{k+1})A'\alpha_k}{\mathbf{1}'B(y_{k+1})A'\alpha_k}, \tag{19.28}$$

where $\mathbf{1}$ denotes the S-dimensional vector of ones.

Before one can implement the above equations on a computer, one slight modification is required. It is necessary to scale α_k to prevent numerical underflow. The numerical underflow occurs because α_{k+1} is the product of the transition probabilities and observation likelihood, which are smaller than one in magnitude. Performing the above recursion, thus, leads to all the components of α_k decaying to zero exponentially fast—eventually leading to an underflow error on a computer. Since we are ultimately interested in the normalized filtered density $\mathbf{P}(s_k = i|Y_k)$ and the state estimate $\hat{s}_k = \mathbf{E}\{s_k|Y_k\}$, the underflow problem is straightforwardly remedied by scaling all the elements of α_k by any arbitrary positive number. Since $\hat{s}_k$ involves the ratio of α_k with $\mathbf{1}'\alpha_k$, this scaling factor cancels out in the computation of $\hat{s}_k$ and hence can be chosen arbitrarily. One particularly convenient scaling factor is obtained by normalizing α_k at each iteration. This results in the HMM filter

$$\bar{\alpha}_{k+1} = \frac{B(y_{k+1})A'\bar{\alpha}_k}{\mathbf{1}'B(y_{k+1})A'\bar{\alpha}_k}, \quad \bar{\alpha}_0 = \pi_0. \tag{19.29}$$

$$\hat{s}_{k+1} = \mathbf{q}'\bar{\alpha}_{k+1}, \tag{19.30}$$

where $\mathbf{q}$ are the physical state levels of the Markov chain as defined in Eq. 19.10. The HMM filter, defined in Eqs. 19.29 and 19.30, is straightforwardly implementable on a computer. The main computational cost is in evaluating $\bar{\alpha}_k$ at each iteration. This requires $O(S^2)$ multiplications at each time k.

19.4.2 HMM Smoothing

So far we have shown how to compute the HMM filtered estimate $\hat{s}_k = \mathbf{E}\{s_k|Y_k\}$. Here we show how to compute the HMM smoothed estimate $\hat{s}_{k|N} = \mathbf{E}\{s_k|Y_N\}$ given a batch of data Y_N.

The smoothing algorithm involves the forward (filtering) recursion given in Eq. 19.23, or equivalently Eq. 19.29, and a backward recursion. Define the smoothed state density as

$$\gamma_{k|N}(i) = \mathbf{P}(s_k = i|Y_N), \quad i = 1, 2, \ldots, S. \tag{19.31}$$

Then similar to Eq. 19.26, the smoothed state estimate is computed as

$$\hat{s}_{k|N} = \mathbf{E}\{s_k|Y_N\} = \sum_{i=1}^{S} q(i)\gamma_{k|N}(i) \tag{19.32}$$

Our task is now to present the forward–backward algorithm for computing $\gamma_{k|N}$. By elementary application of Bayes' rule, this density is computed as

$$\gamma_{k|N}(i) = \frac{\alpha_k(i)\beta_k(i)}{\sum_{i=1}^{S} \alpha_k(i)\beta_k(i)}, \tag{19.33}$$

where α_k is computed via the forward algorithm given in Eq. 19.24, and the backward density $\beta_k(i)$ for $i = 1, 2, \ldots, S$ is defined as

$$\beta_k(i) = \mathbf{p}(Y_{k+1,N}|s_k = i) \tag{19.34}$$

By using a similar argument to Eq. 19.25 it can be shown that β_k can be computed via the backward recursion

$$\beta_k(i) = \sum_{j=1}^{S} \beta_{k+1}(j) a_{ij} b_j(y_{k+1}), \quad k = N, N-1, \ldots 1 \tag{19.35}$$

initialized with $\beta_N(i) = 1, i = 1, 2, \ldots, S$. The above recursion is termed the "Backward algorithm."

In summary, the forward algorithm (Eq. 19.23) together with the backward algorithm (Eq. 19.35) substituted into Eq. 19.33 yields the smoothed density $\gamma_{k|N}(i)$. For N data points, the forward–backward algorithm requires $O(S^2N)$ computations and $O(SN)$ memory.

19.5 Problem 2: HMM Maximum Likelihood Parameter Estimation

Here we present two classes of algorithms for solving Problem 2, i.e., computing the MLE λ^* defined in Eq. 19.19. As mentioned previously, the algorithms use the estimates generated by Problem 1 as an intermediate step.

19.5.1 Newton–Raphson and Related Algorithms

The NR algorithm is a general purpose numerical optimization algorithm that can be used to optimize the likelihood function and thus compute the MLE λ^*. It proceeds iteratively as follows:

- Initialize $\lambda^{(0)} \in \Lambda$.
- For iterations $n = 1, 2, \ldots,$
  - Update parameter estimate as:

$$\lambda^{(n+1)} = \lambda^{(n)} + \left[\nabla_\lambda^2 L(\lambda)\right]^{-1} \nabla_\lambda L(\lambda)\Big|_{\lambda=\lambda^{(n)}}. \tag{19.36}$$

Here $\nabla_\lambda L(\lambda)$ and $\nabla_\lambda^2 L(\lambda)$ denote the first and second derivatives of the likelihood function with respect to the parameter vector λ. The matrix $\nabla_\lambda^2 L(\lambda)$ is called the Hessian matrix.

The main advantage of the NR algorithm is that it has a quadratic convergence rate. One of its main disadvantages is that the Hessian $\nabla_\lambda^2 L(\lambda^{(n)})$ needs to be evaluated and inverted. Moreover, additional constraints need to be introduced to ensure that the transition probabilities are nonnegative and add up to one, i.e., Eq. 19.5 holds for the estimates obtained from the NR algorithm.

Two variations of the above NR algorithm that avoid this inversion are:

(i) First-order gradient descent: The first-order gradient algorithm is a special case of Eq. 19.36 with the Hessian matrix step size $\left[\nabla_\lambda^2 L(\lambda^{(n)})\right]^{-1}$ replaced by a scalar step size of the form $1/n$. Naturally, the convergence rate using a scalar step size is much slower.

(ii) Quasi–Newton–Raphson: The inverse of the Hessian is replaced by a matrix that is easier to compute and invert.

The NR algorithm, Eq. 19.36, requires evaluation of the likelihood function $L(\lambda)$ and its first and second derivatives at $\lambda = \lambda^{(i)}$, $i = 1, \ldots, I$. These can be evaluated in terms of the optimal HMM filter as follows: Consider the HMM filtered density $\alpha_k(i) = \mathbf{p}(s_k = i, Y_k)$ defined in Eq. 19.23 and computed recursively according to Eq. 19.24. At time N (and showing the explicit dependence of α on λ):

$$\alpha_N^\lambda(i) = \mathbf{p}^\lambda(s_N = i, Y_N).$$

The likelihood can then be computed by summing the unnormalized filtered density and time N:

$$L(\lambda) = \mathbf{p}^\lambda(Y_N) = \sum_{i=1}^{S} \alpha_N^\lambda(i). \tag{19.37}$$

Indeed, this is precisely the normalization term in Eq. 19.26.

Consider now evaluating the derivative in Eq. 19.36. Define the sensitivity of the HMM filter as

$$R_k^\lambda(i) = \frac{d}{d\lambda}\alpha_k^\lambda(i), \quad k = 1, \ldots, T.$$

From Eq. 19.37, assuming sufficient regularity to bring the derivative inside the integral

$$\nabla_\lambda L(\lambda) = \sum_{i=1}^{S} R_k^\lambda(i). \tag{19.38}$$

This can be evaluated recursively by differentiating the optimal filter:

$$R_{k+1}^{\lambda}(i) = \left(\nabla_{\lambda}(b_j^{\lambda}(y_{k+1})\right) \sum_{j=1}^{S} a_{ij}^{\lambda}\alpha_k^{\lambda}(i) + b_j^{\lambda}(y_{k+1}) \sum_{i=1}^{S} (\nabla_{\lambda} a_{ij}^{\lambda})\alpha_k^{\lambda}(i)$$
$$+ b_j^{\lambda}(y_{k+1}) \sum_{i=1}^{S} a_{ij}^{\lambda}) R_k^{\lambda}(i).$$

The second-order derivative (Hessian) can be evaluated similarly.

19.5.2 Expectation Maximization Algorithm

The EM algorithm is one of the most widely used numerical methods for computing the ML parameter estimate of a partially observed stochastic dynamical system. The seminal paper by Dempster et al. (1977) formalizes the concept of EM algorithms. Actually, before EM algorithms were formalized in 1977, it was applied in the 1960s by Baum and colleagues (Baum and Petrie, 1966; Baum et al., 1970) to compute the ML parameter estimate of HMM—thus when applied to HMM, the EM algorithm is also called the *Baum–Welch algorithm*.

Similar to the NR algorithm, the EM algorithm is an iterative algorithm. However, instead of directly working on the log likelihood function, the EM algorithm works on an alternative function called the auxiliary or complete likelihood at each iteration. The nice property of the EM algorithm is that by optimizing this auxiliary likelihood at each iteration, the EM algorithm climbs up the surface of the log likelihood, i.e., each iteration yields a model with a better or equal likelihood compared to the previous iteration.

Starting from an initial parameter estimate $\lambda^{(0)}$, the EM algorithm iteratively generates a sequence of estimates $\lambda^{(n)}$, $n = 1, 2, \ldots$ as follows.
Each iteration n consists of two steps:

- *Expectation step*: Evaluate auxiliary (complete) likelihood

$$Q(\lambda^{(n)}, \lambda) = E\{\log \mathbf{p}(X_N, Y_N; \lambda) | Y_N, \lambda^{(n)}\}.$$

The auxiliary likelihood $Q(\lambda^{(n)}, \theta)$ for a HMM can be computed as

$$Q(\lambda^{(n)}, \lambda) = -\frac{N}{2} \ln \sigma_w - \frac{1}{2\sigma_w} \sum_{t=1}^{N} \sum_{i=1}^{S} \mathbf{E}\{(y_k - q(i))^2\} \gamma_k^{\lambda^{(n)}}(i)$$
$$+ \sum_{t=1}^{N} \sum_{i=1}^{S} \sum_{j=1}^{S} \gamma_k^{\lambda^{(n)}}(i, j) \log a_{ij},$$

where $\gamma_k^{\lambda^{(n)}}(i) = \mathbf{P}(x_k = q(i) | Y_N; \lambda^{(n)})$ denotes the smoothed state estimate (see, Eq. 19.31) computed using model $\lambda^{(n)}$ via the forward–backward recursions (Eq. 19.33). $\gamma_k^{\lambda^{(n)}}(i, j) = \mathbf{P}(s_k = i, s_{k+1} = j | Y_N; \lambda^{(n)})$ is computed using the forward and backward variables according to the following equation:

$$\gamma_k^{\lambda^{(n)}}(i, j) = \frac{\alpha_k(i) a_{ij} \beta_{k+1}(j) b_j(y_{k+1})}{\sum_i \sum_j \alpha_k(i) a_{ij} \beta_{k+1}(j) b_j(y_{k+1})}. \tag{19.39}$$

- *Maximization step*: Maximize auxiliary (complete) likelihood, i.e, compute

$$\lambda^{(n+1)} = \max_{\lambda} Q(\lambda^{(n)}, \lambda).$$

This maximization is performed by setting $\partial Q / \partial \lambda = 0$, which yields

$$a_{ij} = \frac{\sum_{k=1}^N \gamma_k^{\lambda^{(n)}}(i, j)}{\sum_{k=1}^N \gamma_k^{\lambda^{(n)}}(i)} = \frac{\mathbf{E}\{\#\text{jumps from } i \text{ to } j | Y_N, \lambda^{(n)}\}}{\mathbf{E}\{\#\text{of visits in } i | Y_N, \lambda^{(n)}\}} \tag{19.40}$$

$$q(i) = \frac{\sum_{k=1}^N \gamma_k^{\lambda^{(n)}}(i) y_k}{\sum_{t=1}^N \gamma_k^{\lambda^{(n)}}(i)} \tag{19.41}$$

$$\sigma_w^2 = \frac{1}{N} \sum_{k=1}^N \sum_{i=1}^S \gamma_k^{\lambda^{(n)}}(i)(y_k - q(i))^2. \tag{19.42}$$

19.5.3 Advantages and Disadvantages of EM

The EM Algorithm described above has several advantages compared to the NR algorithm.

- *Monotone property*: The estimate generated in any iteration n is always better or equal to the model in the previous iteration, i.e., $L(\lambda^{(n+1)}) \geq L(\lambda^{(n)})$ with equality holds at a local maximum (see Eq. 19.3). NR does not have monotone property. We refer the reader to Wu (1983) for a rigorous convergence proof of the EM algorithm.
- In many cases, EM is conceptually simpler to apply than NR. For example, the transition probability estimates generated by Eq. 19.40 are automatically nonnegative and add to one. In other words, Eq. 19.5 holds. Similarly, the variance estimate Eq. 19.42 is automatically nonnegative by construction.
- EM is often numerically more robust than NR; inverse of Hessian is not required in EM.
- There are recent variants of the EM that speed up convergence, such as SAGE, AECM (Meng and van Dyk, 1997).

The following are some of the disadvantages of EM Algorithm.

- Typically the convergence of EM can be excruciatingly slow. In comparison, NR often has a quadratic convergence rate, which is much faster than EM. However,

with increasing computing power, the slow convergence is usually not a problem for moderately sized HMM.
- NR automatically yields estimates of parameter estimate variance, i.e., the Hessian, whereas EM does not.

19.6 Discussion

In this chapter we have shown how HMMs can be used to model the noisily observed ion channel current. We then described HMM signal processing algorithms for estimating the state and parameters (such as transition probabilities) of the HMM given the noisy observations.

A key advantage of the HMM approach is that it has rigorously provable performance bounds rooted in deep results in mathematical statistics. From a practical point of view, the HMM approach uses all the macroscopic information about the underlying dynamics of the ion channel current to compute the state and parameter estimates: that is, it uses the fact that the underlying ion channel current is piecewise constant, that the gating is approximately Markovian, and that the Markov chain is corrupted by noise with a known distribution. The HMM approach is in contrast to the more ad hoc approach of plotting dwell time histograms, which does not systematically use the above information. In addition, the ML parameter estimate of a HMM is known to be statistically efficient, i.e., it achieves the Cramer–Rao bound (Bickel et al., 1998)—or equivalently for large data lengths N, the resulting ML parameter estimate has the smallest error covariance among the class of asymptotically unbiased parameter estimators.

The conventional approach for processing patch-clamp ion channel current data comprises first eyeballing the noisy ion channel current and rounding off the noisy current to a finite number of values. Then dwell-time histograms are constructed of how long the rounded off process spends in the various states. Such a histogram, plotted on a logarithmic scale, reveals how many exponential functions are needed to fit the observed distributions, or the number of hidden states in the open or closed conformation. The same information can be derived more reliably by adopting the HMM approach. By representing the observed currents as an aggregated Markov chain, the number of hidden states and their transition probability can be directly estimated.

Given the power and elegance of the HMM processing technique, it is not surprising that it has been an active area of research in statistics, electrical engineering, and other areas during the last 15 years. Below we summarize recent developments and extensions.

19.6.1 Recent Developments

The EM algorithm has been the subject of intense research during the last 20 years. We briefly summarize some of the recent developments.

- It has been shown that the EM algorithm can be implemented using a forward step only, i.e., without computation of the backward variable β (Elliott et al., 1995; James et al., 1996). This saves memory requirements, but the computational cost becomes $O(S^4 N)$ compared to the forward–backward EM computational cost of $O(S^2 N)$ per iteration.
- EM algorithm, like all hill climbing algorithms, converges to a local maximum of the likelihood surface. Thus, one needs to initialize and run EM from several starting points in order to determine the global optimizer of the MLE. During the last 10 years, Markov Chain Monte Carlo methods have been developed that can be combined with the EM algorithm to yield algorithms that converge to the global optimum. For further details, we refer the reader to Liu (2001).

19.6.2 Extensions

Model-order estimation: Throughout this chapter, we have assumed that the model order, i.e., the number of states S of the Markov chain is known. However, in reality there could be several substates when the ion channel is open—and the number of states of the Markov chain may not be known a priori. One way of estimating the number of states of the Markov chain is to introduce a penalized likelihood function

$$L_N(\lambda, S) = L_N(\lambda) + p(S), \tag{19.43}$$

where the penalty function $p(S)$ is a decreasing function of the number of states S of the Markov chain. This function penalizes by choosing a large-dimensional Markov chain. The penalized MLE is then

$$(\lambda^*, S^*) = \arg\max_{\lambda, S} L_N(\lambda, S). \tag{19.44}$$

Conventionally, for model-order estimation, different penalty functions such as the Akaike information criterion (AIC), Bayesian information criterion (BIC), and Minimum description length (MDL) are widely used. Rydén (1995) and Liu and Narayan (1994) present different choices of the penalty function that result in asymptotically consistent model-order estimates.

On-line (recursive) HMM parameter estimation: The algorithms we have proposed so far are off-line. They operate on a batch of data Y_N and assume that there is a fixed underlying model λ that does not change with time. However, in some cases the transition probabilities of the ion channel gating evolve slowly with time. In such cases, it is necessary to devise on-line (recursive) HMM parameter estimation algorithms that operate in real time and adaptively track the slowly time varying parameters of the HMM.

Several such recursive HMM estimators have been proposed (Krishnamurthy and Moore, 1993; Collings et al., 1994; Dey et al., 1994; Krishnamurthy and Yin,

2002). The algorithms are based on applying a stochastic gradient algorithm to either maximize the expected likelihood or the expected prediction error. They are of the form

$$\lambda_{k+1} = \lambda_k + \epsilon \nabla_\lambda e_k(\lambda_k). \tag{19.45}$$

Here, λ_k denotes the HMM parameter estimate at time k, e_k is either the instantaneous log likelihood or prediction error (computed in terms of the forward variable α), ϵ is a step size, and ∇_λ denotes the derivative with respect to the model parameter λ. Choosing ϵ as a small positive constant results in the algorithm tracking slowly time varying parameters. A rigorous weak convergence proof is given by Krishnamurthy and Yin (2002). Such recursive algorithms fall under the general class of "stochastic approximation" algorithms and have been the subject of much research during the last 20 years. We refer to Kushner and Yin (1997) for a mathematically rigorous treatment of stochastic approximation algorithms and their convergence.

Jump Markov linear systems: Jump Markov linear systems are a significant generalization of HMM. They permit modeling correlated noise with linear dynamics and also filtered Markov chains. For example, miniature end-plate potentials in a muscle fiber or neuronal cell body recorded with an intracellular electrode comprised of exponentially decaying signals (modeled as a digitally filtered Markov chain) corrupted by noise as follows:

$$\begin{aligned} z_k &= a z_{k-1} + \delta(s_k - s_{k-1}) q(s_k) \\ y_k &= z_k + w_k, \end{aligned}$$

where $\delta(s_k - s_{k-1}) = 1$ if $s_k = s_{k-1}$ and 0 otherwise. Fig. 19.3 shows an example of an exponentially-decaying Markov process embedded in noise, y_k.

For the above model, the conditional independence assumption (Eq. 19.15) does not hold as y_k given s_k depends on the entire history of previous states. The above model is a special case of a jump Markov linear system of the form

$$x_{k+1} = a(s_k)x_k + b(s_k)v_k \tag{19.46}$$

$$y_k = c(s_k)x_k + d(s_k)w_k, \tag{19.47}$$

where x_k is a continuous valued state, s_k is a finite state Markov chain (Eq. 19.5), and v_k and w_k are *iid* noise processes typically assumed to be Gaussian. In such models given the observation sequence $\{y_k\}$, the aim is to construct estimates of the finite state Markov chain s_k and continuous state process x_k. Note that the above dynamical system is a linear system whose parameters $a(s)$, $b(s)$, $c(s)$, $d(s)$ evolve in time according to the realization of the jump Markov chain s—hence the name jump Markov linear system. It is clear that in the special case $a(s) = 1$, $b(s) = 0$, then y_k is a HMM. Also, in the special case s is a 1 state Markov chain

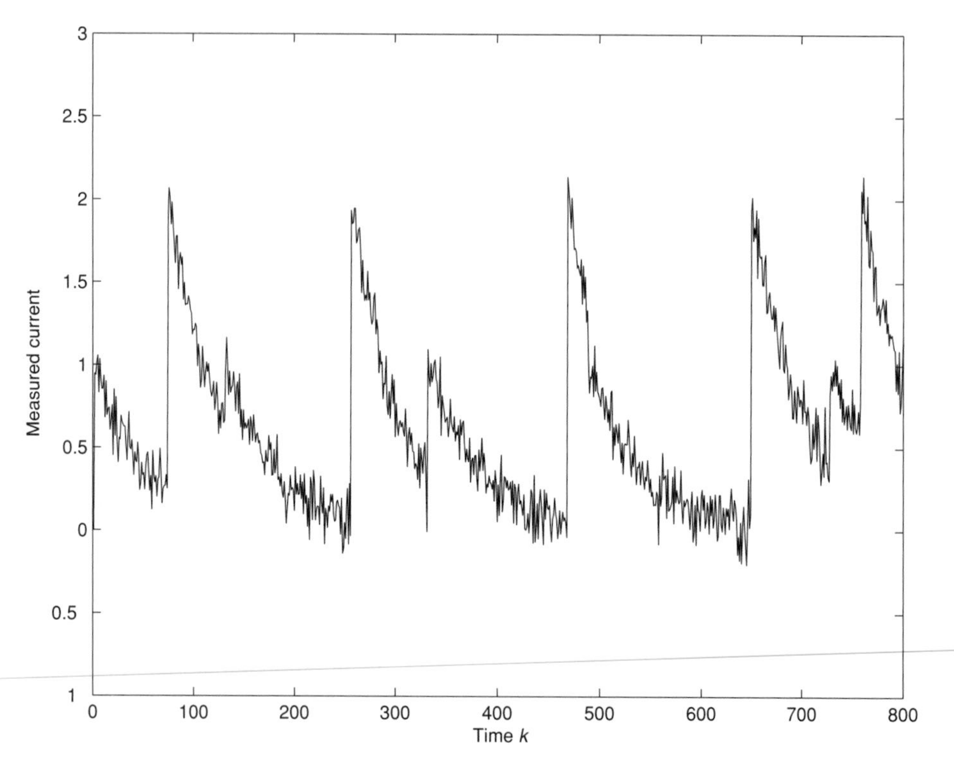

Fig. 19.3 A computer-generated intracellular record. An exponentially-decaying Markov chain, mimicking intracellularly recorded miniature end-plate potentials, is embedded in noise.

(i.e., a constant), the above model becomes a linear state space model. In this special case, if v_k and w_k are Gaussian noise, the conditional mean state estimator of x_k given the observation history is given by the Kalman filter.

Unlike the special cases of the HMM and linear Gaussian state space model, for general jump Markov linear models the optimal Bayesian state estimation problem requires exponential computational complexity (exponential in the data length N). However, there are several high performance non-Bayesian schemes such as maximum a posteriori state estimators which can be used to compute the maximum a posteriori state estimate, rather than the Bayesian conditional mean state estimate (see, Logothetis and Krishnamurthy, 1998). Alternatively, Markov chain Monte Carlo methods can be used to compute approximations of the Bayesian state estimate (see, Ducet et al., 2000, 2001). In particular, Ducet et al. (2001) proposes the so-called "particle filter" which is a sequential Markov chain Monte Carlo algorithm for computing the approximate Bayesian state estimate. Particle filters are widely used in complex Bayesian state estimation problems (Arulampalam et al. 2001).

Automatic control of patch-clamp experiments: Another extension of the basic HMM problem in this chapter is to dynamically control the HMM. For example, it is of interest to dynamically control the patch-clamp experiment to estimate the Nernst potential of the current–voltage curve of an ion channel. The Nernst potential is the applied external potential at which the ion channel current is zero—i.e., it is the applied external potential difference required to maintain electrochemical equilibrium across the ion channel. We refer the reader to Krishnamurthy and Chung (2003) for discrete stochastic optimization-based control algorithms for efficiently estimating the Nernst potential of an ion channel. More generally, the control of HMM with discrete valued observations falls under the class of problems called partially observed Markov decision processes (see for further details, Lovejoy, 1991).

Reference

Arulampalam, S., S. Maskell, N. Gordon, and T. Clapp. 2001. A tutorial on particle filters for on-line non-linear/non-gaussian Bayesian tracking. *IEEE Trans. Signal Proc.* 50:174–188.

Baum, L.E. 1972. An inequality and associated maximization technique in statistical estimation for probabilistic functions of Markov processes. *Inequalities* 3:1–8.

Baum, L.E., and T. Petrie. 1966. Statistical inference for probabilistic functions of finite Markov chain. *Ann. Math. Stat.* 37:1554–1563.

Baum, L.E., T. Petrie, G. Soules, and N. Weiss. 1970. A maximization technique occurring in the statistical analysis of probabilistic functions of Markov chains. *Ann. Math. Stat.* 41:164–171.

Bickel, P.J., Y. Ritov, and T. Rydén. 1998. Asymptotic normality of the maximum-likelihood estimator for general hidden Markov models. *Ann. Stat.* 26:1614–1635.

Chung, S.H., and R.A. Kennedy. 1991. Nonlinear forward-backward filtering technique for extracting small biological signals from noise. *J. Neurosci. Methods* 40:71–86.

Chung, S.H., V. Krishnamurthy, and J.B. Moore. 1991. Adaptive processing techniques based on hidden Markov models for characterizing very small channel currents in noise and deterministic interferences. *Phil. Trans. R. Soc. Lond.* B 334:243–284.

Chung, S.H., J.B. Moore, L.G. Xia, L.S. Premkumar, and P. W. Gage. 1990. Characterization of single channel currents using digital signal processing techniques based on hidden Markov models. *Phil. Trans. R. Soc. Lond.* B 329:265–285.

Collings, I.B., V. Krishnamurthy, and J.B. Moore. 1994. On-line identification of hidden Markov models via recursive prediction error techniques. *IEEE Trans. Signal Proc.* 42:3535–3539.

Colquhoun, D., and A.G. Hawkes. 1981. On stochastic properties of single ion channels. *Proc. R. Soc. Lond.* B 211:205–235.

Dempster, A.P., N.M. Laird, and D.B. Rubin. 1977. Maximum likelihood from incomplete data via the EM algorithm. *J. R. Stat. Soc.* B 39:1–38.

Dey, S., V. Krishnamurthy, and T. Salmon-Legagneur. 1994. Estimation of Markov modulated time-series via the EM algorithm. *IEEE Signal Proc. Lett.* 1:153–155.

Ducet, A., N. Gordon, and V. Krishnamurthy. 2001. Particle filters for state estimation of jump Markov linear systems. *IEEE Trans. Signal Proc.* 49:613–624.

Ducet, A., A. Logothetis, and V. Krishnamurthy. 2000. Stochastic sampling algorithms for state estimation of jump Markov linear systems. *IEEE Trans. Auto. Control* 45:188–200.

Elliott, R.J., L. Aggoun, and J.B. Moore. 1995. Hidden Markov Models—Estimation and Control. Springer-Verlag, New York.

Ephraim, Y., and N. Merhav. 2002. Hidden Markov processes. *IEEE Trans. Inform. Theory* 48:1518–1569.

Finesso, L. 1990. Consistent estimation of the order of Markov and hidden Markov chains. Ph.D. Dissertation, University of Maryland, MD.

Gage, P.W., and S.H. Chung. 1994. Influence of membrane potential on conductance sublevels of chloride channels activated by GABA. *Proc. R. Soc. Lond.* B 255:167–172.

Hamill, O.P., A. Marty, E. Neher, B. Sakmann, and F.J. Sigworth. 1981. Improved patch-clamp techniques for high-resolution current recording from cells and cell-free membrane patches. *Pflügers Arch.* 391:85–100.

James, M.R., V. Krishnamurthy, and F. LeGland. 1996. Time discretization of continuous-time filters and smoothers for HMM parameter estimation. *IEEE Trans. Inform. Theory* 42:593–605.

Jazwinski, A.H. 1970. Stochastic Processes and Filtering Theory. Academic Press, New Jersey.

Krishnamurthy, V., and S.H. Chung. 2003. Adaptive learning algorithms for Nernst potential and current–voltage curves in nerve cell membrane ion channels. *IEEE Trans. Nanobiosci.* 2:266–278.

Krishnamurthy, V., and J.B. Moore. 1993. On-line estimation of hidden Markov model parameters based on the Kullback-Leibler information measure. *IEEE Trans. Signal Proc.* 41:2557–2573.

Krishnamurthy, V., J.B. Moore, and S.H. Chung. 1991. On hidden fractal model signal processing. *IEEE Trans. Signal Proc.* 24:177–192.

Krishnamurthy, V., J.B. Moore, and S.H. Chung. 1993. Hidden Markov model signal processing in the presence of unknown deterministic interferences. *IEEE Trans. Automatic Control* 38:146–152.

Krishnamurthy, V., and T. Rydén. 1998. Consistent estimation of linear and non-linear autoregressive models with Markov regime. *J. Time Series Anal.* 19:291–308.

Krishnamurthy, V., and G. Yin. 2002. Recursive algorithms for estimation of hidden Markov models and autoregressive models with Markov regime. *IEEE Trans. Inform. Theory* 48:458–476.

Kushner, H.J., and G. Yin. 1997. Stochastic Approximation Algorithms and Applications. Springer-Verlag, New york.

LeGland, F., and L. Mevel. 2000. Exponential forgetting and geometric ergodicity in hidden Markov models. *Math. Controls Signals Syst.* 13:63–93.

Leroux, B.G. 1992. Maximum-likelihood estimation for hidden Markov models. *Stochastic Proc. Appl.* 40:127–143.

Liu, J.S. 2001. Monte Carlo Strategies in Scientific Computing. Springer-Verlag, New York.

Liu, C., and P. Narayan. 1994. Order estimation and sequential universal data compression of a hidden Markov source by the method of mixtures. *IEEE Trans. Inform. Theory* 40:1167–1180.

Logothetis, A., and V. Krishnamurthy. 1998. De-interleaving of quantized AR processes with amplitude information. *IEEE Trans. Signal Proc.* 46:1344–1350.

Lovejoy, W.S. 1991. A survey of algorithmic methods for partially observed Markov Decision processes. *Ann. Operations Res.* 28:47–66.

Meng, X.L., and D. van Dyk. 1997. The EM algorithm—an old folk-song sung to a fast new tune. *J. R. Statist. Soc.* B 59:511–567.

Neher, E., and B. Sakmann. 1976. Single-channel currents recorded from membrane of denervated frog muscle fibers. *Nature* 260:799–802.

Papoulis, A., and S. Pillai. 2002. Probability, Random Variables and Stochastic Processes, 4th Ed. McGraw Hill, New York.

Patlak, J.B. 1988. Sodium channel subconductance levels measured with a new variance-mean analysis. *J. Gen. Physiol.* 92:413–430.

Patlak, J.B. 1993. Measuring kinetics of complex single ion channel data using mean-variance histograms. *Biophys. J.* 65:29–42.

Poskitt, D.S., and S.H. Chung. 1996. Markov chain models, time series analysis and extreme value theory. *Adv. Appl. Probab.* 28:405–425.

Queyroy, A., and J. Verdetti. 1992. Cooperative gating of chloride channels subunits in endothelial cells. *Biochim. Biophys. Acta* 1108:159–168.

Rydén, T. 1995. Estimating the order of hidden Markov models. *Statistics* 26:345–354.

Tyerman, D., B.R. Terry, and G.P. Findlay. 1992. Multiple conductances in the large K^+ channel from *Chara corallina* shown by a transient analysis method. *Biophys. J.* 61:736–749.

Venkataramanan, L., R. Kuc, and F.J. Sigworth. 1998a. Identification of hidden Markov models for ion channel currents—Part II: State-dependent excess noise. *IEEE Trans. Signal Proc.* 46:1916–1929.

Venkataramanan, L., R. Kuc, and F.J. Sigworth. 2000. Identification of hidden Markov models for ion channel currents—Part III: Bandlimited, sampled data. *IEEE Trans. Signal Proc.* 48:376–385.

Venkataramanan, L., and F.J. Sigworth. 2002. Applying hidden Markov models to the analysis of single channel activity. *Biophys. J.* 82:1930–1942.

Venkataramanan, L., L.J. Walsh, R. Kuc, and F.J. Sigworth. 1998a. Identification of hidden Markov models for ion channel currents—Part I: Colored background noise. *IEEE Trans. Signal Proc.* 46:1901–1915.
Wu, C. F.J. 1983. On the convergence properties of the EM algorithm. *Ann. Stat.* 11:95–103.

Index

Index

Index

Index

Index

Index

Index

Volumes Published in This Series:

The Physics of Cerebrovascular Diseases, Hademenos, G.J., and Massoud, T.F., 1997

Lipid Bilayers: Structure and Interactions, Katsaras, J., 1999

Physics with Illustrative Examples from Medicine and Biology: Mechanics, Second Edition, Benedek, G.B., and Villars, F.M.H., 2000

Physics with illustrative Examples from Medicine and Biology: Statistical Physics, Second Edition, Benedek, G.B., and Villars, F.M.H., 2000

Physics with Illustrative Examples from Medicine and Biology: Electricity and Magnetism, Second Edition, Benedek, G.B., and Villars, F.M.H., 2000

The Physics of Pulsatile Flow, Zamir, M.,2000

Molecular Engineering of Nanosystems, Rietman, E.A., 2001

Biological Systems Under Extreme Conditions: Structure and Function, Taniguchi, Y., et al., 2001

Intermediate Physics for Medicine and Biology, Third Edition, Hobbie, R.K., 2001

Epilepsy as a Dynamic Disease, Milton, J.,and Jung, P. (Eds), 2002

Photonics of Biopolymers, Vekshin, N.L., 2002

Photocatalysis: Science and Technology, Kaneko, M.,and Okura, I., 2002

E. coli in Motion, Berg, H.C., 2004

Biochips: Technology and Applications, Xing, W.-L., and Cheng, J. (Eds.), 2003

Laser-Tissue Interactions: Fundamentals and Applications, Niemz, M., 2003

Medical Applications of Nuclear Physics, Bethge, K., 2004

Biological Imaging and Sensing, Furukawa, T. (Ed.), 2004

Biomaterials and Tissue Engineering, Shi, *D.*,2004

Biomedical Devices and Their Applications, Shi, D., 2004

Microarray Technology and Its Applications, Müller, U.R., and Nicolau, D.V. (Eds), 2004

Emergent Computation: Emphasizing Bioinfomatics, Simon, M., 2005

Molecular and Cellular Signaling, Beckerman, M., 2005

The Physics of Coronary Blood Flow, Zamir, M., 2005

The Physics of Birdsong, Mindlin, G.B., and Laje, R., 2005

Radiation Physics for Radiation Physicists, Podgorsak, E.B., 2005

Neutron Scattering in Biology: Techniques and Applications, Fitter, J., Gutberlet, T., and Katsaras, J. (Eds.), 2006

Structural Approaches to Sequence Evolution: Molecules, Networks, Populations, Bastolla, U., Porto, M., Roman, H.E., and Vendruscolo, M. (Eds.), 2006

Physics of the Human Body: A Physical View of Physiology, Herman, I.P., 2007

Printed in the United States of America